Springer Series in Optical Sciences Volume 69
Editor: Theodor Tamir

Springer Series in Optical Sciences

42 **Principles of Phase Conjugation**
By B. Ya. Zel'dovich, N. F. Pilipetsky, and V. V. Shkunov

43 **X-Ray Microscopy**
Editors: G. Schmahl and D. Rudolph

44 **Introduction to Laser Physics**
By K. Shimoda 2nd Edition

45 **Scanning Electron Microscopy**
Physics of Image Formation and Microanalysis
By L. Reimer

46 **Holography and Deformation Analysis**
By W. Schumann, J.-P. Zürcher, and D. Cuche

47 **Tunable Solid State Lasers**
Editors: P. Hammerling, A. B. Budgor, and A. Pinto

48 **Integrated Optics**
Editors: H. P. Nolting and R. Ulrich

49 **Laser Spectroscopy VII**
Editors: T. W. Hänsch and Y. R. Shen

50 **Laser-Induced Dynamic Gratings**
By H. J. Eichler, P. Günter, and D. W. Pohl

51 **Tunable Solid State Lasers for Remote Sensing**
Editors: R. L. Byer, E. K. Gustafson, and R. Trebino

52 **Tunable Solid-State Lasers II**
Editors: A. B. Budgor, L. Esterowitz, and L. G. DeShazer

53 **The CO_2 Laser** By W. J. Witteman

54 **Lasers, Spectroscopy and New Ideas**
A Tribute to Arthur L. Schawlow
Editors: W. M. Yen and M. D. Levenson

55 **Laser Spectroscopy VIII**
Editors: W. Persson and S. Svanberg

56 **X-Ray Microscopy II**
Editors: D. Sayre, M. Howells, J. Kirz, and H. Rarback

57 **Single-Mode Fibers** Fundamentals
By E.-G. Neumann

58 **Photoacoustic and Photothermal Phenomena**
Editors: P. Hess and J. Pelzl

59 **Photorefractive Crystals in Coherent Optical Systems**
By M. P. Petrov, S. I. Stepanov, and A. V. Khomenko

60 **Holographic Interferometry in Experimental Mechanics**
By Yu. I. Ostrovsky, V. P. Shchepinov, and V. V. Yakovlev

61 **Millimetre and Submillimetre Wavelength Lasers** A Handbook of cw Measurements
By N. G. Douglas

62 **Photoacoustic and Photothermal Phenomena II**
Editors: J. C. Murphy, J. W. Maclachlan Spicer, L. C. Aamodt, and B. S. H. Royce

63 **Electron Energy Loss Spectrometers**
The Technology of High Performance
By H. Ibach

64 **Handbook of Nonlinear Optical Crystals**
By V. G . Dmitriev, G. G. Gurzadyan, and D. N. Nikogosyan

65 **High-Power Dye Lasers**
Editor: F. J. Duarte

66 **Silver Halide Recording Materials for Holography** and Their Processing
By H. I. Bjelkhagen

67 **X-Ray Microscopy III**
Editors: A. G. Michette, G. R. Morrison, and C. J. Buckley

68 **Optical Phase Conjugation**
Editors: M. C. Gower and D. Proch

69 **Photoacoustic and Photothermal Phenomena III**
Editor: D. Bićanić

Volumes 1-41 are listed at the end of the book

D. Bićanić (Ed.)

Photoacoustic and Photothermal Phenomena III

Proceedings of the 7th International Topical Meeting, Doorwerth, The Netherlands, August 26–30, 1991

With 501 Figures

Springer-Verlag
Berlin Heidelberg GmbH

DANE BIĆANIĆ
Laser Photoacoustic Laboratory
Department of Agricultural Engineering and Physics
Wageningen Agricultural University
Bomenweg 4, 6703 HD Wageningen, The Netherlands

DOI 10.1007/978-3-540-47269-8

Originally published by Springer-Verlag Berlin Heidelberg New York in 1992
MyCopy version of the original edition 1992

Typesetting: Camera ready by author
54/3140-5 4 3 2 1 0 - Printed on acid-free paper
www.springer.com/mycopy

United by sympathy and understanding, STIMPAP attendees gathered frequently for discussions around the old willow tree on the conference site, which was emancipated from the bondage of "de Branding's" garden. The tree's leaves and branches represent the new ideas that were proposed and assimilated during the conference.

Foreword

The field of photoacoustic and photothermal studies has exploded since the first international meeting was organized nearly thirteen years ago. Nowadays this multidisciplinary field brings together physicists, engineers, chemists, and biologists, as well as medical, environmental, and agricultural scientists.

It is worth analyzing why scientists belonging to such a broad community share a common interest in understanding contributions that can be far from their own field: all participants are concerned with learning and improving experimental techniques, understanding the physical phenomena, and developing better theories leading to quantitative results.

The present book provides a very deep insight into important aspects of photoacoustic and photothermal sciences: new trends leading to the exploration of new subjects (terra incognita) and the consolidation of more classical approaches by the improvement of methodologies, instrumentation, and theories.

We must then point out how close the topics of these proceedings are to the needs of high-tech industries, environmental studies, and the fundamental sciences: Dane Bićanić and his team have succeeded in demonstrating their efficiency by gathering a large number of new topics, which will contribute to the maintenance of the fast growth of the subject.

This book will certainly play an important role, both for specialists, through the detailed content of the contributions, and for newcomers, providing them with a picture of international activities in this non-destructive approach to the problems of today and tomorrow.

Paris
March 1992

A.C. Boccara
Laboratoire d'Optique Physique,
ESPCI, Ecole Superieure de Physique et
de Chimie Industrielle de la ville de Paris

Preface

This volume contains the proceedings of the Seventh Topical International Meeting on Photoacoustic and Photothermal Phenomena (STIMPAP), which was held from August 26 to 30, 1991, in the Parkhotel "de Branding", in Doorwerth, the Netherlands. This seventh meeting in a series of biannual conferences (Ames 1979, Berkeley 1981, Paris 1983, Montreal 1985, Heidelberg 1987, and Baltimore 1989), which coincided with a period of major changes in the world, was the largest and most international thus far. The collapsing borders certainly helped to add a new dimension to this event. As a result, some 250 scientists – including 53 students from 40 countries – attended STIMPAP, where they not only heard 4 tutorials, 10 invited lectures, and 98 contributed papers but also had the opportunity to view more than 100 displayed posters and to discuss at ease their experiences and their work. The participants included physicists, biologists, and chemists, agricultural, environmental, and material scientists, engineers and others, all sharing one common interest: to develop new methodologies, techniques, and instrumentation, and also to improve existing ones, in order to better understand photoacoustic and photothermal phenomena. The flurry of activities and growing interest in the multidisciplinary field of photoacoustic and photothermal sciences observed throughout the last two decades continues, as was demonstrated by many new scientific results presented at STIMPAP.

The material of these proceedings is organized in eight parts, with the texts of lengthy plenary lectures and very useful tutorials, wherever presented, at the beginning of each part. These are followed by the contributions from other authors arranged according to topic. Part I, entitled "Environmental and Trace Analysis, Agricultural, Biological, and Medical Applications", is followed by "Spectroscopy" (Part II) and "Non-Destructive Evaluation and Material Characterization" (Part III). A survey of activities dealing with "Ultrasound and Kinetics, Including Ultrafast Phenomena" is given in Part IV, while the contributions of those engaged in work on "Electronic Materials and Semiconductors" make up Part V. "New Topics, Experimental Techniques, and Instrumentation" (Part VI) emerged as the sector attracting the largest number of papers. The investigations concerned with "Phase Transitions; Heat and Mass Transfer" are discussed in Part VII. Finally, Part VIII deals with the latest progress in "Surfaces, Interfaces, Thin Films, and Heterogeneous Media".

The extra features of STIMPAP were very well attended evening tutorials, which were essentially open to anyone interested, and an interesting exhibition of instrumentation. It was certainly encouraging to realize that most of the manufacturers and exhibitors were the scientists themselves, who demonstrated the latest version of often very competitive setups to their colleagues. It seems that conditions for transfer of instrumentation from the laboratory to industry are now more favorable than ever before.

The organization of an international event such as STIMPAP is impossible without the tremendous, selfless input of many tireless individuals. In particular, as STIMPAP Chairman I wish to express my deepest gratitude to the domestic (M. Lubbers, P. Torfs, H. Jalink, J. Reuss, H. Gorter, A. Verhaegen, H. Sauren, I. Bićanić, J. Quist, J.P. Favier, J. Buys, C. Sikkens, J. Oomens, W. Wolters, S. Stolte, A. Dymanus, J. v.d. Waals) and foreign (J. Angster, D. Fournier, S. Braslavsky, C. Boccara, J. Thoen, A. Miklós, D. Cahen, P. Hess, and J. Pelzl) members of the committees concerned with the local organization. I further thank members of the International Advisory Board (A. Tam, D. Almond, H. Vargas, F. Scudieri, P. Helander, R. Palmer, V. Gusev, S. Thakur, T. Toyoda, H. Coufal, M. Sigrist, J. Murphy, L. Bertrand, and J. Harris), who provided very firm and perpetual assistance throughout all STIMPAP phases, maintaining high quality standards. All the colleagues who presented invited lectures, tutorials, contributed papers, and posters deserve credit. H. Jalink helped to set up an interesting exhibition of instrumentation. My thanks go to K. Timmer and WNO Printers, Nijmegen, for demonstrating skilled work in producing the Conference Digest and Technical Abstracts, J. van Trijp, who provided the illustrations (one of which is included in this book), and J. Quist, for his assistance in designing the Programme's cover page and the STIMPAP shirt. I am also most indebted to W. Snelarts and his enthusiastic team from "de Branding", L. Siesling and the Veldboer family at "de Bosbeek", B. Hanekam and A. van Kampen from the Dutch Red Cross, W. Oele, Queen's Governor for the province of Gelderland, and H. v.d. Plas, Rector Magnificius of Wageningen Agricultural University, for the fantastic work they all did to make STIMPAP a success. I am thankful to my wife Vesna, daughter Iva, son Timi and my parents, who despite wondering what I was doing all the days and nights, showed much understanding and gave me the necessary spiritual and physical support.

Finally, I wish to acknowledge the indispensable financial support provided by quite a sizeable number of domestic and international institutions. STIMPAP was sponsored by: Royal Academy of Science, Amsterdam; Commission of the European Communities, Brussels; Netherlands Convention Bureau, Amsterdam; Ministry of Agriculture and Fisheries, The Hague; Wageningen Agricultural University; Shell Nederland B.V., Amsterdam; Therma-Wave, Fremont, CA, USA; Foundation Physica, Utrecht; Philips Research Laboratories, Eindhoven; Foundation for Fundamental Research on Matter, FOM, Utrecht; Lucht, Society for Clean Air in The Netherlands, Delft; IBM Nederland B.V., Amsterdam; and AKZO, Arnhem; and supported by: Intechmij B.V., Diemen; Phototherm Dr. Petry GmbH, Saarbrücken, FRG; Unilever Research, Vlaardingen; Nashua Nederland B.V., s'Hertogenbosch; Aktibolaget Varilab, Huddinge, Sweden; MTEC Photoacoustics Inc., Ames, IA, USA; Lambdaphysik, Göttingen, FRG; Applied Laser Technology, Maarheze; Ithaco, Ithaca, NY, USA; Royal Dutch Airlines, KLM, Amstelveen; Oog-Loep, Eindhoven; and Hans Strassner GmbH, Leverkusen, FRG.

The international photoacoustic community was saddened by the sudden deaths of Professor E. Strauss and Professor S. Akhmanov, who stood at the forefront of the development of photoacoustics. And here in the Netherlands, we shall miss the newcomers Dr. N. Beganović and Dr. H. Boon, with whom the scheduled initial photoacoustic experiments in the field of neonatology will never be performed. I am sure that in the emptiness that is left, their unfinished work and ideas will serve as a stimulus for many others.

My colleague Marcel Lubbers deserves much credit for his perpetual and very efficient support provided during the preparation of these proceedings. Valuable help received from Ms D. Hollis, Mrs. C. Pendl, and Dr. H. Lotsch from Springer-Verlag, Heidelberg, is acknowledged with great pleasure.

Nijmegen/Wageningen *D. Bićanić*
April 1992

Contents

Part I Environmental and Trace Analysis, Agricultural, Biological, and Medical Applications

Exploration of FTIR-Based PAS
for On-Site Analysis of Volatile Contaminants in Air
By D.P. Baldwin, R.W. Jones, and J.F. McClelland (With 4 Figures) . . . 3

Photothermal Electrostatics
of the Pd-PVDF Photopyroelectric Hydrogen Gas Sensor
By A. Mandelis and C. Christofides (With 1 Figure) 6

Photoacoustic Spectra of Chlorinated Ethylenes
at CO_2 Laser Frequencies
By A. Olafsson, M. Hammerich, and J. Henningsen (With 2 Figures) . . . 9

The Photothermal Deflection Technique (PDT):
Fast Trace Gas Detection in the Atmosphere
By H.S.M. de Vries, F.J.M. Harren, and J. Reuss (With 4 Figures) 12

Photoacoustic Measurement of Vertical Ammonia Gradients
in the Atmosphere
By A.J.L. Verhage, R.A. Rooth, and L.W. Wouters (With 2 Figures) . . . 16

Interfacing Photoacoustic and Photothermal Techniques
for New Hyphenated Methodologies and Instrumentation
Suitable for Agricultural, Environmental and Medical Applications
By D. Bićanić, H. Jalink, M. Chirtoc, H. Sauren, M. Lubbers, J. Quist,
E. Gerkema, K. van Asselt, A. Miklós, A. Sólyom, Gy.Z. Angeli,
P. Helander, and H. Vargas (With 6 Figures) 20

Photoacoustic In Situ Monitoring of Trace Gases in a Rural Environment
By A. Thöny, B. Brand, M. Ferro-Luzzi, and M.W. Sigrist
(With 2 Figures) 28

Photoacoustic Field Measurements of Methane
By M. Fiedler, Ch. Gölz, and U. Platt (With 2 Figures) 31

A Liquid Nitrogen Cooled CO Laser
in a Photoacoustic Set-Up Monitors Low Gas Concentrations
By F.G.C. Bijnen, T. Brugman, F.J.M. Harren, and J. Reuss
(With 3 Figures) 34

Photothermal Detection of Trace Chemicals by Fiber-Optic Interferometric Probe
By G. Vegetti, M. Martinelli, L. Balconi, and F. Sigon (With 2 Figures) 38

Fiber Optic Modified Laser Induced Photoacoustic Spectroscopy for the Detection of Organic Pollutants in Solutions
By K. Adelhelm, W. Faubel, and H.J. Ache (With 4 Figures) 41

Remote Photothermal Deflection Spectroscopy for Trace Analysis of Pesticides in Aqueous Solutions
By B. Bohnert, W. Faubel, and H.J. Ache (With 3 Figures) 46

Laser Optoacoustic Probing of Biopolymer Solutions
By S.V. Egerev, O.M. Zozulya, A.E. Pashin, and O.V. Puchenkov (With 2 Figures) 50

Photoacoustic Spectroscopy of Proteins in the Visible: Human Serum Albumin
By A. Figari, L. Sabbioneda, and A. Zanella (With 3 Figures) 53

Photoacoustic Measurements of C_2H_4 Production and Entrapment in Plants: A Comparison with Gas-Chromatographic Results
By F.J.M. Harren, J. Reuss, L.A.C.J. Voesenek, R.H. Thier, and C.W.P.M. Blom (With 2 Figures) 56

Interpretation of the Photoacoustic Effect on Leaves by Evolution and Transport of Heat and Oxygen
By P. Korpiun and S. Malkin (With 2 Figures) 59

Photoacoustically Monitored Energy Dissipation and Xanthophyll Cycle Compared in Two Durum Wheat Varieties Exposed to a High-Light Treatment
By M. Eyletters and R. Lannoye 62

Deconvolution of Photobaric and Photothermal Signals from Spinach Leaves
By J.F.H. Snel, M.W. Polm, W.F. Buurmeijer, and W.J. Vredenberg (With 3 Figures) 65

In Vivo State Transitions as Followed by Photoacoustically Measured Energy Storage
By M. Charland, K. Veeranjaneyulu, D. Charlebois, and R.M. Leblanc (With 2 Figures) 69

Laser-Induced Optoacoustic Spectroscopy (LIOAS) of Bacteriorhodopsin (BR): A Mathematical Model and Temperature Studies
By M. Rohr, W. Gärtner, and S.E. Braslavsky (With 4 Figures) 73

Sub-ps and ns Laser-Induced Optoacoustic Spectroscopy (LIOAS) of Bacteriorhodopsin (BR)
By M. Rohr, W. Gärtner, G. Schweitzer, A.R. Holzwarth, and S.E. Braslavsky (With 3 Figures) 77

Fourier Transform Infrared Photoacoustic Spectra of Bacteria
By M. Ardeleanu, R. Morisset, and L. Bertrand (With 4 Figures) 81

New and Versatile Method for Thermophysical Characterization of Materials: The Inverse Photopyroelectric Technique
By M. Chirtoc, D. Bićanić, D. Dadarlat, and I. Chirtoc (With 2 Figures) 85

Use of an Open Photoacoustic Cell for some Applications in Agriculture
By A.M. Sólyom, D. Bićanić, Gy.Z. Angeli, A. Miklós, and M. Lubbers (With 3 Figures) 88

In Vivo Evaluation of the Human Skin Permeation of Topically Applied Light Absorbing Agents by a Mirage Effect Spectrometer
By K. Plamann, K. Giese, B. Sennhenn, N. Harendt, and K. Kölmel (With 3 Figures) 92

Optothermal Characterisation of Human Skin
By R.M.S. Bindra, G.M. Eccleston, R.E. Imhof, and D.J.S. Birch (With 2 Figures) 95

Part II **Spectroscopy**

Standard Versus New Photopyroelectric (PPE) Techniques: Reflection (RPPE) and Inverse (IPPE) Configurations
By M. Chirtoc, D. Dadarlat, and D. Bićanić (With 7 Figures) 103

A Tutorial on the State-of-the Art of FTIR Photoacoustic Spectroscopy
By J.F. McClelland, S. Luo, R.W. Jones, and L.M. Seaverson (With 12 Figures) 113

Transient Infrared Spectroscopy
By R.W. Jones and J.F. McClelland (With 4 Figures) 125

Photoacoustic Spectra of Polymer Fibres and Textile Materials
By M. Kaplanová (With 1 Figure) 128

Study of Absorption Properties of Dark Coloured Textile Fabrics
By A. Eickmeier and E. Schollmeyer (With 2 Figures) 131

Photoacoustic Response to X-Ray Absorption in CuO and Cu_2O
By T. Toyoda, T. Masujima, H. Shiwaku, A. Iida, and M. Ando (With 2 Figures) 135

Photoacoustic Response to X-Ray Absorption in Copper, Phosphor Bronze, and Brass
By T. Toyoda, T. Masujima, H. Shiwaku, A. Iida, and M. Ando (With 2 Figures) 138

Theory of the Photoacoustic Effect in Photoactive Samples
By P. Poulet, M. Ouzafe, and J. Chambron (With 1 Figure) 141

Photoacoustic Detection of Triplet State and Singlet Oxygen
By M. Ouzafe, P. Poulet, and J. Chambron (With 1 Figure) 144

Origin of Photoacoustic Signals in the Visible Spectrum of I_2 Vapour
By K. Narayanan and S.N. Thakur (With 3 Figures) 147

Fourier Transform Infrared Photoacoustic Spectroscopy of Amorphous Carbon Films
By A. Raveh, L. Martinu, A. Domingue, M.R. Wertheimer, and L. Bertrand (With 3 Figures) 151

An Excitation Energy Transfer Between Monomer and Dimer of Rhodamine 6G in a Viscous Solution
By J. Szurkowski (With 3 Figures) 155

Vibrational Spectroscopy of Monolayers by Pulsed Photothermal Beam Deflection (PBD)
By C.D. Bain, P.B. Davies, and T.H. Ong (With 2 Figures) 158

Reverse Mirage Photothermal Beam Deflection: Theory and Experiment
By C.J. Manning, F. Charbonnier, J.L. Chao, and R.A. Palmer (With 4 Figures) 161

Piezophotoacoustic Spectroscopy of Gyrotropic Layered Structures
By G.S. Mityurich, V.P. Zelyony, and A.N. Serdyukov (With 1 Figure) . 165

Photoacoustic Signal Generation and Spectra of Powders
By J. Buijs, J.P. Favier, O. Dóka, and A. Miklós (With 2 Figures) 168

Part III **Non-Destructive Evaluation and Material Characterization**

Recent Developments in the Thermal Diffusivity Measurements of Diamond Materials
By P.K. Kuo, L. Wei, and R.L. Thomas 175

Photoacoustic Study of the Thermal Effusivity of Cellulose and Paper
By M. Kaplanová and G. Katuščáková (With 2 Figures) 180

Photoacoustic Characterization of Calcinated and Non-Calcinated Zeolitic Chromosilicate
By O. Nakamura, A.M. Mansanares, H. Vargas, H.O. Pastore, E.J.S. Vichi, N.F. Leite, and L.C.M. Miranda (With 1 Figure) 183

Photoacoustic Depth Profiling
of Continuously Inhomogeneous Condensed Phases
and Application to the Liquid Crystal Octylcyanobiphenyl (8 CB)
By A. Mandelis, E. Schoubs, S.B. Peralta, and J. Thoen (With 1 Figure) 186

Computer Simulation of the Thermal Diffusivity Measurements
of Oriented Polymer Samples Using the Mirage Effect
By J. Rantala, J. Jaarinen, and P.K. Kuo (With 3 Figures) 189

The Determination of Thermal Diffusivity of Oriented Polymer Films
by Means of Optical Beam Deflection
By B. Skagerberg, J. Jaarinen, and J. Rantala (With 1 Figure) 193

Photothermal Measurement of Temperature Dependent Absorbance
of Infrared Components
By P. Eckardt, H.G. Walther, and W. Richter (With 3 Figures) 196

Numerical Analysis of Simultaneous Surface and Subsurface Features
in Homogeneous Solids and in Coatings by the Mirage Technique
By A. Salazar, A. Sánchez-Lavega, and J. Fernández (With 3 Figures) .. 199

Dichroic Ratio and Thermal Diffusivity Measurements on Polydiacetylene
Single Crystal Films by Photothermal Probe Beam Deflection
By X. Quelin, C. Lapersonne-Meyer, J. Berrehar, G. Louis, and P. Peretti
(With 1 Figure) 202

Photothermal Analysis of Layered Materials Under Random Excitation
By P. Egée, E. Mérienne, K. Hakem, M. Heuret, and M. Egée
(With 3 Figures) 205

Opto-Thermal Study of Glass Properties
By R.E. Imhof, R.M.S. Bindra, D.J.S. Birch, P.H. Willson, J. Locke,
D.G. Sanger, and N.D. Cowan (With 2 Figures) 209

Non-Invasive Opto-Thermal Measurement
of Clear Lacquer Coating Thickness
By M.E. Pita de Jesus, R.E. Imhof, D.J.S. Birch, J.F. Webb, P.H. Willson,
and T.A. Strivens (With 3 Figures) 213

Thermal Characterization of Low Thermal Diffusivity Glasses
By M. Bertolotti, R. Li Voti, G. Liakhou, C. Sibilia, A. Montenero, and
G. Gnappi (With 1 Figure) 217

Temperature Dependence of Photothermally Modulated FMR Signals:
Investigation of Magnetostatic Modes in Yttrium Iron Garnet Films
By M. Hoffmann, Th. Orth, O. von Geisau, and J. Pelzl (With 2 Figures) 221

Analytical Relations Between the Phase of the Photothermal Signal
and the Thermal Wavelength
By A. Figari (With 2 Figures) 224

A Highly Sensitive Photothermal Method for Measuring the Optical Losses in the Mirrors of the Gravitational Antenna Virgo
By D. Ivanov, N. Man, J.P. Roger, and A.C. Boccara (With 1 Figure) .. 228

Effect of Photoelasticity on Photothermal Beam Deflection in Transparent Solids
By H.J. Seo and S.I. Yun (With 2 Figures) 231

Contrast Analysis in Thermal Wave Materials Inspection. Part I: Pulsed Excitation
By P.M. Patel, S.K. Lau, and D.P. Almond (With 3 Figures) 235

Contrast Analysis in Thermal Wave Materials Inspection. Part II: Periodic Excitation
By P.M. Patel, S.K. Lau, and D.P. Almond (With 4 Figures) 238

Transient Thermal Wave Scattering from Disc Air Gap Defects in Homogeneous Media
By P.M. Patel, S.K. Lau, and D.P. Almond (With 5 Figures) 241

Dimensionality Viewed from a Thermal Perspective
By L.C. Aamodt (With 3 Figures) 245

Thickness Determination for Two-Layer Systems of Paint on Polymers with Photothermal Radiometry
By W. Karpen, A. Bohnacker, H.G. Walther, K. Friedrich, U. Seidel, and G. Busse (With 3 Figures) 248

Thermal Diffusivity Measurement of Diamond Materials
By L. Wei, P.K. Kuo, and R.L. Thomas (With 2 Figures) 251

Part IV **Ultrasound and Kinetics, Including Ultrafast Phenomena**

Femtosecond Thermoreflectivity
By A. Lörincz, R.A. Bakker, A. Miklós, and J. Somlói (With 1 Figure) 257

Photoacoustic Waveforms Generated by Fluid Bodies
By G.J. Diebold, T. Sun, and M.I. Khan (With 3 Figures) 263

Study of Chemical Relaxation of Dimeric Formic Acid by Photoacoustic Resonance Spectroscopy
By A. Winkler, H. Jung, M. Fiedler, and P. Hess (With 2 Figures) 270

Characterization of Chemical Reactions by Thermal Lens Spectrometry
By M. Franko and C.D. Tran (With 1 Figure) 273

Methane Vibrational Rates from Deconvoluting a Large Data Set with Varying Pressure, Buffer Gases, Temperature, Signal-to-Noise Ratio, and Instrumental Response
By J.A. Burt, K. Yang, J.J. Caldwell, R.N. Halthore, and J.E. Allen (With 2 Figures) 276

New Aspects of Sound Generation by Light Absorption in Gases
By A. Miklós 280

Photoacoustic Studies of Non-Radiative Energy Transfer
in the Flexible Styrylcyanine-Rhodamine B Dye System
By M.S.A. Abdel-Mottaleb, S. Negm, and H. Talaat (With 2 Figures) . . 283

Material Characterization by Interferometric Detection
of Acoustically Induced Surface Displacements
By P. Weiss and M.W. Sigrist (With 2 Figures) 287

An Optical Calorimeter Based on an Ultrasonic Resonator
By O. Oehler, J. Wieland, D. Raillard, and M. Schumacher
(With 1 Figure) 290

Optoacoustic Source as a Tool
for the Investigation of Nonlinear Acoustic Effects
By T.A. Dunina, S.V. Egerev, and A.T. Skvortsov (With 2 Figures) 294

Nonlinear Photoeffect
at the Boundary of a Semiconductor and Liquid Helium
By V.E. Gusev, S.A. Telenkov, and B.V. Zhdanov (With 1 Figure) 297

Probe Beam Detection of IR and UV Laser Polymer Ablation
on the Nanosecond Time Scale
By L.M. Kukreja and P. Hess (With 2 Figures) 300

Picosecond Photoacoustics in Thin Films of SiO_2 and Si_3N_4
By O.B. Wright, T. Hyoguchi, and K. Kawashima (With 1 Figure) 303

Nonlinear Regimes of Longitudinal Acoustic Pulse Excitation
in Piezocrystals by Ultrashort Laser Action
By V.E. Gusev and L.N. Makarova (With 1 Figure) 306

Pulsed Ultrasonic Spectroscopy
Based on Laser Thermooptical Generation
By A.A. Karabutov, M.P. Matrosov, and N.B. Podymova
(With 3 Figures) 309

Optical Observation of Laser Induced Lamb Wave Dispersion
By R. Hrovatin and J. Možina (With 5 Figures) 313

Precision Measurement of Laser Induced Surface Acoustic Wave
Velocities on Silicon Single Crystals
By A. Neubrand, P. Hess, H. Coufal, and R.K. Grygier
(With 2 Figures) 317

Photoelastic Law Explains Optically Induced Changes
in the Index of Refraction in $Ti^{3+}:Al_2O_3$
By H. Eilers, F. Kieseling, E. Strauss, and W.M. Yen (With 1 Figure) . . 320

On the Possibility of Rayleigh-Type Surface Acoustic Waves Excitation at Frequencies up to 10–100 GHz by Laser Action on a Normally Cut Superlattice
By V.E. Gusev (With 1 Figure) 323

Some Applications of SAW Pulses Photogenerated in Silicon
By A.A. Kolomenskii and A.A. Maznev (With 3 Figures) 326

Broad-Band Detection of Laser Excited Surface Acoustic Wave Pulses by a Novel Transducer Employing Ferroelectric Polymers
By A. Neubrand, P. Hess, H. Coufal, and R.K. Grygier
(With 2 Figures) 331

Optoacoustic Investigation of IR Absorption by Diffraction Gratings
By A.A. Karabutov, I.A. Kudinov, V.T. Platonenko, and M.A. Sogoyan
(With 2 Figures) 334

Part V **Electronic Materials and Semiconductors**

Competition Between Thermal and Plasma Waves for the Determination of Electronic Parameters in Semiconductor Samples
By D. Fournier (With 7 Figures) 339

Photoacoustic Calibration of Silicon Photodiodes
By E. Halaška, M. Dienstbier, and P. Sladky (With 5 Figures) 350

The Comparative Study of Semiconductor Structures Using Simultaneous Detection of Photoacoustic Signal and Photovoltage
By R.M. Burbelo, Yu.B. Utrobin, G.I. Bulakh, A.L. Gulyaev, and I.Ya. Kucherov (With 3 Figures) 354

Photoacoustic Characterization of Ion-Implantation Damage in Silicon
By T. Matsumori, M. Uchida, H. Yoshinaga, J. Kawai, T. Izumi, and F. Uehara (With 2 Figures) 357

Photoacoustic Effect in CdS Under Combined Optical Exposure
By G.I. Bulakh, R.M. Burbelo, O.V. Volchanski, and I.Ya. Kucherov
(With 2 Figures) 360

Electron-Induced Ionization and Thermalization in CdS Using Optical Beam Deflection Imaging
By G.Y. Chang, R.B. Givens, J.W.M. Spicer, and J.C. Murphy
(With 3 Figures) 363

Topography of Integrated Circuits Detected by Phase Selection in Modulated Optical Reflectance Microscopy
By H.S. Li, S.Y. Zhang, J.C. Cheng, Q. Shen, and Y.S. Lu
(With 2 Figures) 366

Nonlinear Response of Modulated Photoreflectance in Silicon
By S.Y. Zhang, Y.S. Lu, and H.S. Li (With 1 Figure) 369

Photomodulated Optical-Reflectance Studies of Germanium: Diagnostic Separation of Temperature and Free-Carrier Effects
By R.E. Wagner and A. Mandelis (With 2 Figures) 372

Surface Characterization by Modulated Photoreflectance
By B.C. Forget and D. Fournier (With 3 Figures) 375

Solar Cells Studied with Modulated Reflectance
By B. Büchner, R. Osiander, G. Meyer-Berg, and P. Korpiun (With 3 Figures) 378

Application of the Photothermal Beam Deflection Technique to the Measurement of Si Thermal and Electronic Parameters
By A.O. Salnick, A.N. Petrovsky, D.O. Mukhin, and V.V. Zuev 381

Spatially Resolved Determination of Photogenerated Temperature and Excess Carrier Distribution in GaAsP
By G. Li, A. Leyk, and L.J. Balk (With 4 Figures) 384

Subgap Photothermal Deflection Spectroscopy of Silicon Wafers
By G. Amato, G. Benedetto, L. Boarino, and R. Spagnolo (With 3 Figures) 387

Theory of Pulsed Photothermal Beam Deflection Signal Formation in Semiconductors
By A.N. Petrovsky, A.O. Salnick, V.V. Zuev, V.V. Grigoryev, and M.M. Mekhtiev 390

Photothermal Deflection Spectroscopy Study of the Effect of Ion Dose on the Subgap Absorption of Defects in Ion Implanted Layers of GaAs and Si
By U. Zammit, F. Gasparrini, M. Marinelli, R. Pizzoferrato, A. Agostini, F. Mercuri, F. Scudieri, and S. Martellucci (With 2 Figures) 393

Influence of Carrier Recombination on the Pulsed Photothermal Beam Deflection Signal in Semiconductors
By A.N. Petrovsky, A.O. Salnick, V.V. Zuev, D.O. Mukhin, M.M. Mekhtiev, J. Pelzl, A.C. Boccara, and D. Fournier (With 3 Figures) 396

Thermal Diffusivity Measurements in Semiconductors – Influence of the Carrier Diffusion and Recombination
By S. Surnev and D. Ivanov (With 2 Figures) 399

Low Frequency, Photothermal Measurement of Transport Properties of Crystalline Solar Cells
By B. Büchner, N. Cella, and D. Cahen (With 3 Figures) 403

Photothermal Characterization of Materials Suitable for Thin Film Solar Cells
By B. Büchner and H.W. Schock (With 2 Figures) 406

Photothermal Deflection Spectroscopy and Electron Spin Resonance in Amorphous Silicon Based Alloys
By G. Amato, G. Benedetto, L. Boarino, and R. Spagnolo (With 3 Figures) . . . 408

Single Frequency Thermal Diffusivity Measurements in Semiconductors by Photothermal Deflection Spectroscopy
By F. Gasparrini, U. Zammit, M. Marinelli, R. Pizzoferrato, and A. Agostini (With 2 Figures) . . . 411

Reflection and Transmission of Thermal Waves and Plasma Waves in Silicon Wafers
By B.C. Forget and D. Fournier (With 4 Figures) . . . 414

Out-of-Phase Photopyroelectric (PPE) Spectroscopy on Amorphous Semiconductors
By C. Christofides and A. Mandelis (With 1 Figure) . . . 417

Electron Acoustic Microscopy of Defects in Plastically Deformed Si Crystals
By V.G. Eremenko (With 2 Figures) . . . 420

Nonlinear Photoacoustic Effect in Semiconductors: Second Harmonic Wave Behavior
By J.C. Cheng, S.Y. Zhang, Y.S. Lu, and Z.Q. Wang (With 2 Figures) . . . 423

Part VI **New Topics, Experimental Techniques and Instrumentation**

Some Theoretical and Experimental Aspects of Photothermal Imaging
By H.G. Walther, K. Friedrich, R. Göring, K. Haupt, and U. Seidel (With 7 Figures) . . . 429

Photoacoustic and Photothermal Investigations of Thin Films
By E. Matthias, H. Grönbeck, E. Hunger, J. Jauregui, H. Pietsch, M. Reichling, S. Petzoldt, E. Welsch, and Z.L. Wu (With 12 Figures) . . . 436

Signal Enhancement and Noise Suppression Considerations in Photothermal Spectroscopy
By A.C. Tam (With 4 Figures) . . . 447

Analytical Thermal Lens Spectrometry: Past, Present and Future Prospects
By C.D. Tran (With 5 Figures) . . . 463

Stimulated Spectroscopic Ellipsometry and Imaging in the Space, Time and Frequency Domain: A New Photothermal Probe
By J. Gang, J.P. Roger, A.C. Boccara, and J.L. Stehlé (With 3 Figures) . . . 474

Photothermal Induced Frequency Conversion
of a Double Modulated Laser Beam in Layer Structures
By H.D. Geiler, M. Wagner, P. Kowalski, and R. Bleiß
(With 3 Figures) 477

Cavity-Enhanced Photothermal Spectroscopy:
A Tool for High Sensitivity Experiments
By H.A. Schuessler, J. Tang, and E.C. Benck (With 2 Figures) 482

A New Differential Probe System
for the Detection of Thermal and Plasma Waves
By M. Liu, M.B. Suddendorf, and M.G. Somekh (With 3 Figures) 485

Pulsed Photoacoustic Technique
for the Measurement of Laser Damage Threshold in Bulk Polymers
By A.V.R. Kumar, G. Padmaja, P. Radhakrishnan, V.P.N. Nampoori, and
C.P.G. Vallabhan (With 3 Figures) 489

Depth Profiling of Polymer Laminates
by Step-Scan Fourier Transform Infrared Photoacoustic Spectroscopy
By R.M. Dittmar, J.L. Chao, and A. Palmer (With 4 Figures) 492

Methodical Studies of the Alignment Problem
and Data Filtering in Photothermal Beam Deflection Experiments
By B.K. Bein, M. Woelker, J. Pelzl, and H.G. Walther
(With 3 Figures) 497

Nondestructive Testing of Electric Contacts
by Time-Resolved Infrared Radiometry
By U. Netzelmann (With 3 Figures) 501

Radiometric Detection of Small-Amplitude Thermal Waves
Between 500 K and 1000 K
By R. Hüttner, B.K. Bein, J.H. Gu, J. Pelzl, D.L. Balageas, and
A.A. Déom (With 4 Figures) 504

Photothermal Imaging of Electric Current Distributions
in Thin Metallic Films
By M. Reichling, E. Welsch, C. Göbel, Z.L. Wu, D. Schäfer, and
E. Matthias (With 3 Figures) 509

Induction Heating as a Selective Heat Source
in Fast Thermal Non-Destructive Evaluation
By R. Lehtiniemi, J. Hartikainen, J. Varis, and M. Luukkala
(With 3 Figures) 512

Competition Between Normal and Transverse Heat Flow
in Time-Resolved Infrared Radiometry (TRIR)
By J.W.M. Spicer, L.C. Aamodt, W.D. Kerns, and J.C. Murphy
(With 3 Figures) 516

Real Time Thermal Wave Tomography
By L.D. Favro, H.J. Jin, P.K. Kuo, R.L. Thomas, and Y.X. Wang
(With 6 Figures) 519

A Quantitative Analysis of Pulsed Video Thermographic Imaging
of Subsurface Defects
By S.K. Lau, D.P. Almond, and P.M. Patel (With 3 Figures) 522

Photodeflection and Interferometric Methods
in Thermal Wave Spectroscopy and Microscopy of Solids
By A.L. Glazov and K.L. Muratikov (With 1 Figure) 525

Photodeflection Signal Formation from Solids
Within the Framework of Wave Optics
By A.L. Glazov and K.L. Muratikov (With 2 Figures) 528

Pulsed Laser-Induced Temperature Changes
Measured by the Photothermal Deformation Technique
By J. Jauregui, Z.L. Wu, and E. Matthias (With 4 Figures) 531

Measurement of Thermal Diffusivities Using Transient Thermal Gratings
By J. Jauregui and E. Matthias (With 4 Figures) 534

Measurement of the Thermal Radial Diffusivity of Anisotropic Materials
by the Converging Thermal Wave Technique – Application to NDE:
First Results in the Detection of Normal Cracks
By F. Enguehard, A. Déom, and D. Balageas (With 3 Figures) 537

Optimization of a Mirage Setup for Thermal Diffusivity Measurements
By Ph. Forge, F. Lepoutre, and M.Z. Silva (With 5 Figures) 541

Inversion of Pulsed Thermal Wave Images
for Recovery of the Shape of the Object
By D.J. Crowther, L.D. Favro, P.K. Kuo, and R.L. Thomas
(With 6 Figures) 544

Energy Dependent Characteristics of Optoacoustic Pulses
Measured by Optical Probe
By J. Diaci and J. Možina (With 2 Figures) 547

Photopyroelectric Thermal Wave Spatial and Depth Resolved Imaging
with Ray Optic Tomographic Reconstruction
By M. Munidasa and A. Mandelis (With 2 Figures) 550

Evaluation of Magnetic Disk Heads by Photoacoustic Microscope
By T. Hoshimiya, M. Ohta, and M. Kojima (With 2 Figures) 553

Rate-Window Transient Thermomodulation Spectrometry:
Technique and Measurements in Semiconductors
By Z. Chen and A. Mandelis (With 3 Figures) 556

Open Photoacoustic Cell Spectroscopy of Transparent Samples
By N.F. Leite and L.C.M. Miranda 559

A Reliable Optothermal Sensor
By P. Helander (With 1 Figure) 562

Transferable Measurement System for Fast Non-Destructive Evaluation
By J. Varis, J. Hartikainen, R. Lehtiniemi, J. Rantala, and M. Luukkala
(With 2 Figures) 565

High-Sensitivity Cell for Pulsed Photoacoustic Spectroscopy
in Gases and Liquids
By I. Carrer, L. Fiorina, and E. Zanzottera (With 3 Figures) 568

Modification of a Conventional Fourier Transform Infrared Spectrometer
for Enhanced Photoacoustic Detection
By S.H. Gordon, R.V. Greene, and C. Pareja (With 4 Figures) 572

Why Phase in Phase Fourier Transform Infrared Photoacoustic
Spectroscopy?
By L. Bertrand, A. Cournoyer, A. Domingue, and H. Marchand
(With 5 Figures) 578

Photoacoustic Investigation of the Evaporation Resistance
of Monomolecular Lipid Layers
By W. Micheler, R. Osiander, P. Korpiun, J. Schmidt, and W. Knoll
(With 4 Figures) 583

Photoacoustic Detection of Temperature Phenomena in Beeswax
By H. Jalink, D. Bićanić, W.J. van der Burg, R.A. van Zwol, and
J.W. Aartse (With 2 Figures) 586

Three Dimensional Acoustical Modelling
of Resonant Photoacoustic Cells
By Z. Bozóki, I. Gáspár, and A. Miklós (With 2 Figures) 590

Photoacoustic Signal Generation in a Long Open Tube
By A. Miklós, Z. Bozóki, I. Gáspár, and Gy.Z. Angeli
(With 1 Figure) 593

Dependence of the Photoacoustic Cell Constant on the Sample Gas
By Gy.Z. Angeli, A.M. Sólyom, A. Miklós, and D. Bićanić
(With 2 Figures) 596

Characterisation of a Photochromic Triarylmethane Dye Sulphite
and Its Application to the Visualisation of Water Flows
By P. Douglas, R.D. Enos, B. Azzopardi, and C.B. Hope
(With 2 Figures) 600

Photorefractive Lock-In Amplifier and Thermal Wave Microscope
By A. Lörincz and O. Dóka (With 3 Figures) 603

Neural Networks in the Framework of Non-Destructive Evaluation
from Photothermal Experiments
By S. Knerr, D. Fournier, G. Dreyfus, and A.C. Boccara
(With 3 Figures) 607

Pulsed Laser Machining by Training Artificial Neural Networks with the Help of the Thermal Diffusion Equation
By T. Szakács and A. Lörincz (With 1 Figure) 610

Raster Thermowave Technique of Laser Spectroscopy
By S.A. Vinokurov and A.M. Skaleris (With 1 Figure) 614

Part VII Phase Transitions, Heat and Mass Transfer

Photothermal Study of Heat Conduction, Diffusion, and Sorption in Thin Films and Porous Materials
By P. Korpiun and R. Osiander (With 8 Figures) 619

Analysis of Photothermal Sensitivity in the Detection of Phase Transitions
By A. Hadj-Sahraoui, G. Louis, P. Peretti, and J. Billard (With 2 Figures) 628

Laser Induced Photoacoustic Effect During Ferroelectric Phase Transitions
By J.O. Tocho, F. Cussó, R. Ramirez, and J.A. Gonzalo (With 2 Figures) 631

A Nonlinearity in the Photoacoustic Effect at a First Order Phase Transition
By P. Korpiun and J. Weiser (With 3 Figures) 634

The Anomalous Behaviour of the PPE Signal in the Critical Region of a Phase Transition – New Support for the General Theoretical Model
By D. Dadarlat, M. Chirtoc, R.M. Candea, A. Frandas, and D. Bićanić (With 3 Figures) 637

Photopyroelectric Study of a Phase Transition in Liquid Crystals
By F. Mercuri, M. Marinelli, U. Zammit, R. Pizzoferrato, and F. Scudieri (With 6 Figures) 641

Photopyroelectric (PPE) Detection of the Antiferromagnetic 3D–2D Phase Transition in Single Crystal La_2CuO_4
By D. Dadarlat, R.M. Candea, A. Frandas, M. Chirtoc, and D. Bićanić (With 1 Figure) 644

Acousto-Thermal Detection of the Magnetocaloric Effect
By W. Otowski, C. Glorieux, R. Hofman, and J. Thoen (With 3 Figures) 647

Physical Adsorption Detected by the Mirage Effect
By M.Z. Silva, Ph. Forge, F. Lepoutre, and P. Korpiun (With 2 Figures) 651

Part VIII Surfaces, Interfaces, Thin Films, and Heterogeneous Media

Photothermal Detection at Surfaces and Interfaces: Developments in Non-Conventional Diagnostics
By A. Mandelis 657

Heat Diffusion at a Micronic Scale
By F. Lepoutre, S. Bouchoule, G. Bäckström, and D. Balageas (With 5 Figures) 664

Photothermal Investigations of Realistic Interfaces
By H.G. Walther, K. Friedrich, U. Seidel, G. Busse, and W. Karpen (With 4 Figures) 669

Pulsed Photothermal Radiometry to Measure the Properties of Thin Layered Films
By A.C. Tam, W.P. Leung, and H. Sontag (With 8 Figures) 672

Thermal "Superdiffusion" in the Vicinity of a Rough Surface
By A.C. Boccara and D. Fournier (With 2 Figures) 679

Thermal Diffusivities of Thin Gold Films
By J. Jauregui, Z.L. Wu, D. Schäfer, and E. Matthias (With 3 Figures) . . 682

Photoacoustic Investigation of Thin Solid Films in Contact with Liquids
By C. Glorieux, E. Mens, and J. Thoen (With 2 Figures) 685

In Situ Measurement of Thin Film Growth by Photothermal Induced Frequency Conversion
By H.D. Geiler, N. Winkler, and D. Schiller (With 4 Figures) 688

Thermal Characterization of Thin Layers
By M. Beyfuss, H. Reichl, and J. Baumann (With 2 Figures) 692

Photodeflection Method Applied to Loss Characterization of Channel Waveguides
By M. Bertolotti, L. Fabbri, R. Li Voti, and C. Sibilia (With 2 Figures) 695

High Frequency Photothermal Reflectivity and Displacement Measurements on Thin Film Samples
By M. Reichling, Z.L. Wu, E. Welsch, D. Schäfer, and E. Matthias (With 3 Figures) 698

Photoacoustic Measurement of the Energy Absorption of a Pulsed Nd:YAG Laser at Cu and Al Surfaces
By Y. Ito, S. Nakamura, and F. Hirakawa (With 1 Figure) 702

Directivity Pattern of OA Signals Accompanying Laser Induced Damage of a Thin Cr Film
By D. Maksimović, M. Terzić, and J. Možina (With 4 Figures) 705

Investigation of Surface Laser-Ablation Plumes
Using High-Sensitivity Cavity-Enhanced Detection
By E.C. Benck and H.A. Schuessler (With 3 Figures) 708

The Thermal Resistance of Grain Boundaries
Determined by Modulated Optical Reflectance
By G. Meyer-Berg, R. Osiander, P. Korpiun, P. Kakoschke, and
H. Joswig (With 3 Figures) 711

Determination of the Thickness, Density and Elastic Properties
of Thin Films with Laser Generated Surface Acoustic Waves
By A. Neubrand, A. Mayer, and P. Hess (With 1 Figure) 714

Macro-, Meso- and Microscopic Heat Diffusion in Ceramics
By F. Lepoutre, D. Fournier, and A.C. Boccara (With 3 Figures) 717

Non-Destructive Evaluation of Microdefects in Ceramic Materials
by Means of Photoacoustic Spectroscopy
By K. Iwasaki (With 3 Figures) 721

Anomalous Thermal Waves in Powdered Samples
By R. Osiander, J. Lobermeier, and P. Korpiun (With 1 Figure) 725

Index of Contributors 729

Part I

Environmental and Trace Analysis, Agricultural, Biological, and Medical Applications

Exploration of FTIR-Based PAS for On-Site Analysis of Volatile Contaminants in Air

D.P. Baldwin, R.W. Jones, and J.F. McClelland

Ames Laboratory-USDOE, Spedding Hall,
Iowa State University, Ames, IA 50011, USA

Abstract. The use of FTIR-based PAS to quantify low concentrations of volatile organics in air has been examined. Band intensities are accurately proportional to concentration and sensitivities are good. Typical limits of detection are a few parts per million.

1. Introduction

The growing efforts in pollution monitoring and remediation have increased the need for instruments that can identify and quantify low contaminant concentrations in small-volume environmental samples, while being small and robust enough to be used in the field. PAS of gases requires much smaller sample volumes than traditional long-pathlength-cell transmission spectroscopy. Laser-based PAS systems have the high sensitivity and broad applicability required [1], but their size and fragility make them impractical for on-site use. We have been examining the abilities of FTIR-based PAS to detect and quantify low-concentration contaminants in air. Unlike laser- and filter-based systems, an FTIR-based instrument provides a full-spectrum analysis, which is particularly useful when the identities of contaminants are not known or there are spectroscopic interferences.

2. Experimental Section

The dual-microphone photoacoustic cell from a Brüel & Kjær Type 1302 Multi-gas Monitor was mounted in the sample compartment of a Perkin-Elmer Model 1800 FTIR Spectrometer. The microphone outputs were preamplified and summed using circuits from the multi-gas monitor, and then the signal was supplied to the spectrometer through its standard photoacoustic-detector input. Samples were prepared by diluting measured flows of certified gas mixtures (Alphagaz, ±2% accuracy) in vehicle-emission zero air. All of the spectra shown here are single beam and were taken at 0.05 cm/s optical-path-difference velocity with 32 scans coadded. The data were taken at 8 cm^{-1} resolution except where specified.

3. Results and Discussion

Figure 1 shows the spectrum we observe for 50 ppmV carbon tetrachloride. CCl_4 has essentially one strong feature at 795 cm^{-1}; the rest of the structure arises from water

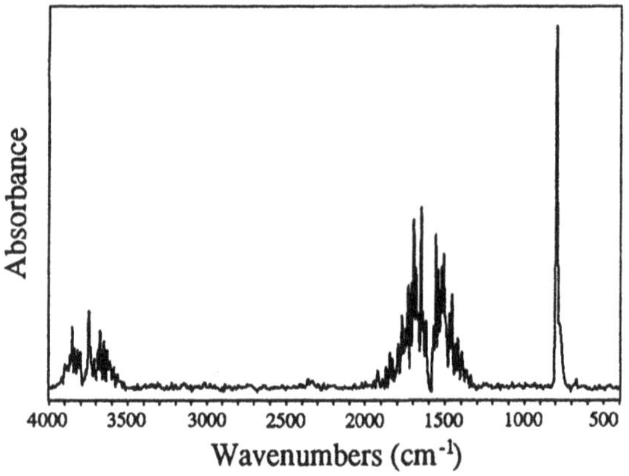

Figure 1. 50 ppmV carbon tetrachloride in air.

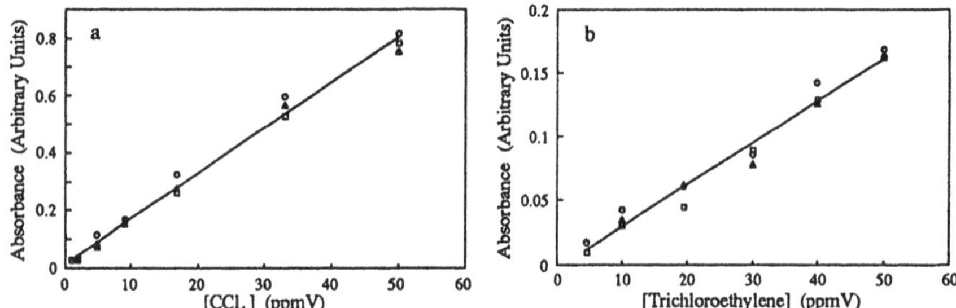

Figure 2. Correlations between peak height and concentration for (a) the 795 cm^{-1} band of carbon tetrachloride, and (b) the 942 cm^{-1} band of trichloroethylene. The three different symbols correspond to data taken on three different occasions.

and a small amount of carbon dioxide. The presence of a single strong, broad feature means an excellent signal-to-noise ratio is attained in only a few scans. Figure 2a shows the correlation between the peak height (arbitrary absorbance units) and the CCl_4 concentration. It is apparent that the correlation is perfectly linear and highly reproducible. We estimate a 2 ppmV limit of detection (LOD) under these conditions.

The fact that CCl_4 has only a single strong band results in a high sensitivity, but makes it especially susceptible to interferences in mixtures. Fortunately, most molecules have more than one band from which to select. Trichloroethylene, whose spectrum is shown in Figure 3, is more typical. It has two main bands at 850 and 942 cm^{-1}, along with weaker bands at 782, 1255, and 3100 cm^{-1}. This choice of bands allows the avoidance or deconvolution of interferences. Figure 2b shows the correlation of the 942 cm^{-1} band intensity with concentration. This band is intrinsically weaker than the CCl_4 band, resulting in somewhat more scatter among the data points. Using this peak gives a LOD of 5 ppmV. FTIR with a long-pathlength cell typically gives LOD's between 1 and 0.01 ppm [2].

The main bands of CCl_4 and $Cl_2C{=}CHCl$ are sufficiently broad that improving the resolution beyond 8 cm^{-1} does not significantly alter their heights. Sharper bands, though, can be improved by better resolution. Benzene has bands of both types, as

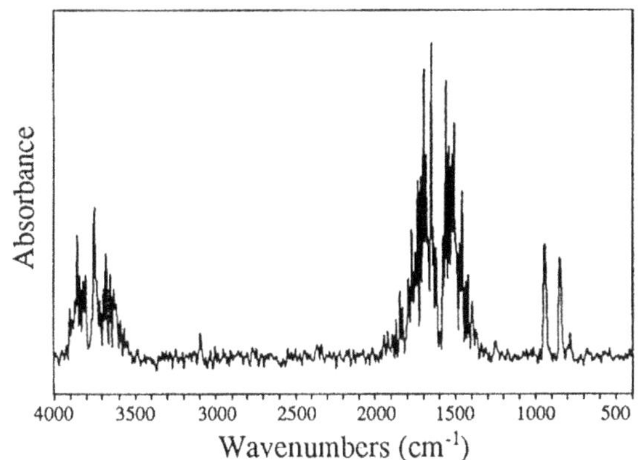

Figure 3. 50 ppmV trichloroethylene in air.

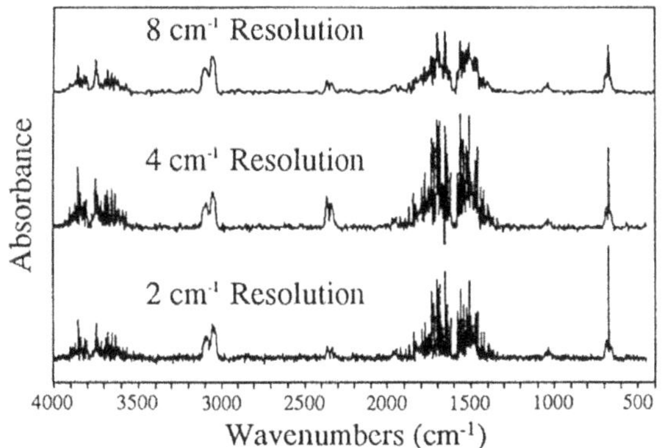

Figure 4. 102 ppmV benzene observed at three different instrumental resolutions.

shown in Figure 4. The heights of the bands at 3050 and 3100 cm^{-1} do not change with improving resolution, but the height of the sharp 674 cm^{-1} peak increases significantly.

4. References

1. F. Harren, J. Reuss, E. Woltering, D. Bicanic: Appl. Spectrosc. **44**, 1360 (1990)
2. Y. Li-Shi, S. Levine, C. Strang, W. Herget: Am. Ind. Hyg. Assoc. J. **50**, 354 (1989)

Photothermal Electrostatics of the Pd-PVDF Photopyroelectric Hydrogen Gas Sensor

A. Mandelis and C. Christofides

Photoacoustic and Photothermal Sciences Laboratory,
Department of Mechanical Engineering, and
Center for Hydrogen and Electrochemical Studies,
University of Toronto, Toronto, Ontario, Canada M5S 1A4

Abstract: A detailed photothermal electrostatic consideration of the Pd-pyroelectric junction H_2 sensor is presented. Experimental evidence is in agreement with the fundamental features of the theory, which supports two possible mechanisms of ac-mode device operation: pyroelectric coefficient dependence on the hydrogenic dipole-induced charge density at the Pd-insulator (polyvinylidene fluoride, PVDF) interface; upon H_2 adsorption, absorption and diffusion into the Pd matrix, and Pd work-function shift due to hydrogen absorption. The dominant operating mechanism is found to depend on the experimental conditions. The concept of image dipole thermostatistical vibration and libration in the pyroelectric matrix is further successfully used to explain the temperature dependence of the photopyroelectric signal in support of the former above-mentioned mechanism.

The geometry of the recently reported Pd-PVDF hydrogen sensor [1] is shown in Fig. 1(a). This amounts to a capacitor consisting of Pd-electrode (ground; anode)/PVDF dielectric/Al-Ni electrode (cathode). **D**, **P** and **E** stand for the electric displacement, polarization, and field vectors, respectively; ω denotes their modulation frequency dependence due to the pyroelectric effect. It can be shown that the average ac potential difference generated photopyroelectrically across the capacitor is

$$\Delta V(\omega,t) \approx -I_0 \frac{i\eta\, p_z\, [1 - e^{-(1+i)\,a_p(\omega)\,d}]}{4\chi_e\, \varepsilon_0\, k_p\, a_p^2(\omega)}\, e^{i\omega t} \tag{1}$$

η is the nonradiative (optical-to-thermal) energy conversion efficiency; k_p is the thermal conductivity of the pyroelectric; p_z is the z-component of the pyroelectric vector; χ_e is the electric susceptibility of PVDF; and $a_p(\omega) \equiv (\omega/2\alpha_p)^{1/2}$ is the thermal diffusion coefficient, with α_p being the diffusivity of the pyroelectric. In the presence of hydrogen in the catalytic metal Pd, the energy band diagram of the structure in Fig. 1(a) is shown in Fig. 1(b). The various contributions to the electrostatic potential field are a) the anode-cathode contact potential; b) the PPE potential; and c) the hydrogen-induced potential. Taking all these sources into account one finds electrostatically:

$$\Delta V(\omega,[H];t) = \left[\frac{d}{\varepsilon}\right] \sigma_f^{(PPE)}(\omega,t) + \left[\frac{d}{2\varepsilon}\right] \sigma_f^{(+)} + \Delta\phi_0[H] \tag{2}$$

Springer Series in Optical Sciences, Vol. 69 Photoacoustic and Photothermal Phenomena III
Editor: D. Bićanić

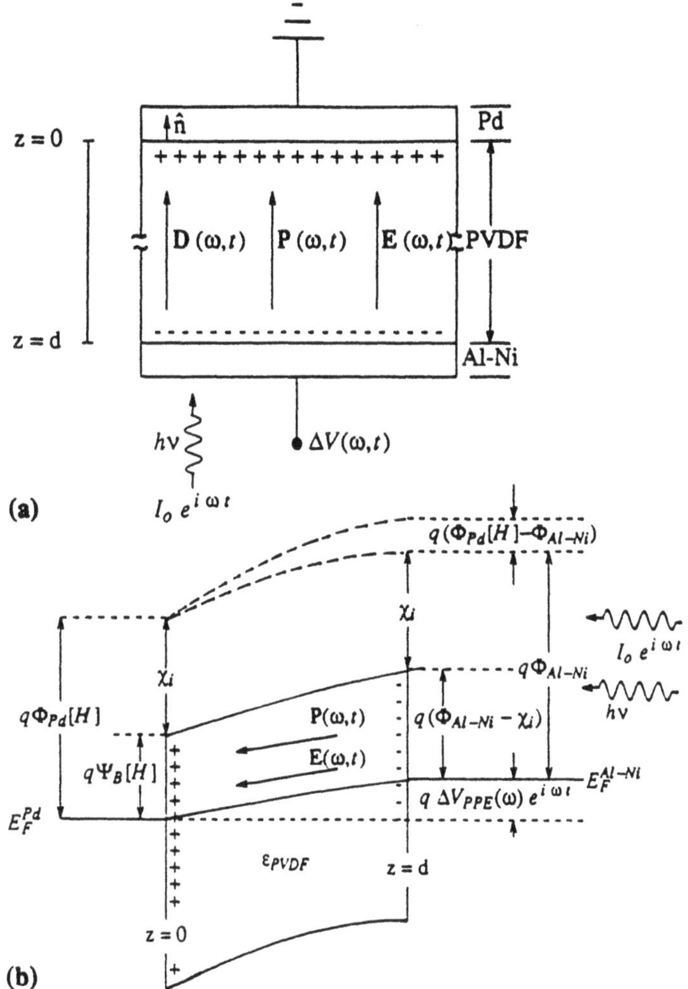

Fig. 1: (a) PPE H_2 capacitive sensor. The Pd anode (ground electrode) is in contact with the Inficon^{TM} housing of the sensor. The cathode (back electrode) is the standard Pennwalt $Al-Ni$ layered electrode. Intermittent photothermal excitation at frequency $f = \omega/2\pi$ is assumed from the rear-side of the capacitor. (b) Energy band structure of the PPE H_2 sensor device. E_F^j: Fermi level of metal j ($j = Pd,\ Al-Ni$); χ_i: electron affinity of the insulator (PVDF film); $\phi_{Pd}\ [H]$: Pd work function, a function of hydrogen concentration in the catalytic metal; $\Psi_B\ [H]$: potential barrier height between the Pd Fermi level and the conduction band of the insulator; ϕ_{Al-Ni}: work function of the layered electrode of the anode.

where $\Delta\phi_0 = \phi_{Al-Ni} - \phi_{Pd}[H]$, Fig. 1(b); σ_f is the free charge density at the Pd-PVDF interface due to either the PPE effect or the hydrogen dipoles; and ε is the PVDF dielectric constant. Electrostatic field considerations show [1] that the pyroelectric vector component in the z-direction, p_z, depends on N_H, the total number density of hydrogenic dipoles thermally averaged over all thermodynamic microstates:

$$p_z[H] \approx p_z[0] + N_H \left[\frac{\partial}{\partial T} < m_z >_T \right] \qquad (3)$$

where $< m_z >_T$ is the statistically-averaged component of the hydrogenic dipole moment along the z-axis at temperature T. Given that $\sigma_f^{(PPE)}(\omega, t) \propto p_z [H]$, synchronous detection of PPE signals responds to $[H]$ concentrations in the Pd matrix and the first term in Eq. (2), $\Delta V_I(\omega, t)$, renders the device sensitive to ambient H_2. It can be further shown [1] that the *ac* temperature oscillation of the thermal wave can modulate the Pd work function and, consequently, the contact potential difference $\Delta\phi_0([H],\omega)$. Central to these considerations is the frequency dependence of ΔV_I and $\Delta\phi_0$: the former potential is due to the PPE effect, with ω^{-1} amplitude dependence in the thermally thick region; the latter is due to the exponential tail distribution of the thermal wave introduced from the back surface of the capacitor, Fig. 1(a), with $\omega^{-1} \exp[-(\omega/2\alpha_p)^{1/2} d]$ amplitude dependence. Experimentally we found a power frequency dependence of the saturation PPE signal [2,3] at 100% H_2 flows, provided the hydrogen remains in the Pd matrix throughout the experiment, in support of the former (ΔV_I-dominated) mechanism; and a definite exponential decay, provided the flow cell is emptied after each frequency measurement, in support of the latter ($\Delta\phi_0$-dominated) mechanism. The temperature dependence of the PPE saturation voltage at 100% H_2 exposure was

$$\Delta V_{sat}(T) = const. \times T^{-1.4 \pm 0.15}; \qquad -63°C \leq T \leq 51°C \qquad (4)$$

Statistical mechanical considerations of Eq. (3), under the assumption of independent vibrational and librational motions of the hydrogenic dipoles in the Pd matrix and at the Pd-PVDF interface, yield $\partial < m_z >_T / \partial T \propto T^{-1.5}$, in good agreement with experiment and in support of the mechanism of Eq. (3), implying the dependence of the pyroelectric coefficient on the number density of hydrogenic dipoles.

The $PPE-H_2$ sensor has been found to respond to H_2 concentrations in N_2 down to 40 ppm, the highest room-temperature sensitivity reported for any solid-state hydrogen sensor [4]. This sensor was further shown to respond to ambient H_2 at temperatures down to 77K; no other H_2 sensor has been known to exhibit a response to H_2 (under room pressure) below 17°C [4].

[1] A. Mandelis and C. Christofides, J. Appl. Phys. **70**, 4496 (1991).

[2] C. Christofides and A. Mandelis, J. Appl. Phys. **66**, 3975 (1989).

[3] A. Mandelis and C. Christofides, J. Vac. Sci. Technol. **A8**, 3980 (1990).

[1] C. Christofides and A. Mandelis, J. Appl. Phys. **68**, R1, (1990).

Photoacoustic Spectra of Chlorinated Ethylenes at CO_2 Laser Frequencies

A. Olafsson[1], *M. Hammerich*[2], *and J. Henningsen*[3]

[1]Science Institute, University of Iceland, Dunhaga 3, IS-107 Reykjavik, Iceland

[2]Physics Laboratory, Ørsted Institute, Universitetsparken 5, DK-2100 Copenhagen, Denmark

[3]Danish Institute of Fundamental Metrology, Bldg. 307, Lundtoftevej 100, DK-2800 Lyngby, Denmark

Abstract. Photoacoustic Doppler limited spectra of mono–, tri– and per–chloroethylene in windows around the CO_2 laser frequencies are presented. Of these three molecules C_2H_3Cl alone exhibits well resolved line spectra. Measurements of pressure broadening for one C_2H_3Cl transition are reported.

1. Introduction

Chlorinated compounds are widely used in modern industry. Because of the environmental impact of emission of vapors of these compounds to the atmosphere, there is a growing need for analytical methods to monitor their presence.

The aim of this work is to explore the feasibility of using photoacoustic methods to identify trace amounts of chlorinated ethylenes, by recording their spectral signatures in windows around the CO_2 laser lines. This method has been successfully applied to detection of ammonia against a high background of CO_2 and water vapor in smoke stack emission [1]. Kornilov et al. [2] have applied a similar method to record signatures of some medium sized molecules using photothermal detection.

The spectra are measured at a reduced pressure in an effort to resolve the structure of the ro–vibrational lines. Line resolved spectra offer an increased molecular selectivity in comparison to higher pressure spectra with interfering lines. The molecular identification is thus based on the line structure as opposed to the band structure utilized in higher pressure samples. The advantage of the reduced pressure fades away with increasing line density, when neighbouring Doppler profiles start to overlap. In the high line density limit the absorption coefficient scales roughly proportional to the pressure as we are effectively probing the number of lines covered within the collisional width.

2. Results and discussion

Due to the high natural isotope ratio of $[^{37}Cl]/[^{35}Cl] \simeq 1/3.1$, the spectra of chlorinated compounds are more complex than the molecular structure alone would indicate. The isotope shift is much less than the rotational bandwidth, leading to interwoven spectra for the different isotopomers.

The spectra are presented in figure 1 as "window" spectra, covering ± 250MHz around each CO_2 laser line, and in a compact discrete representation

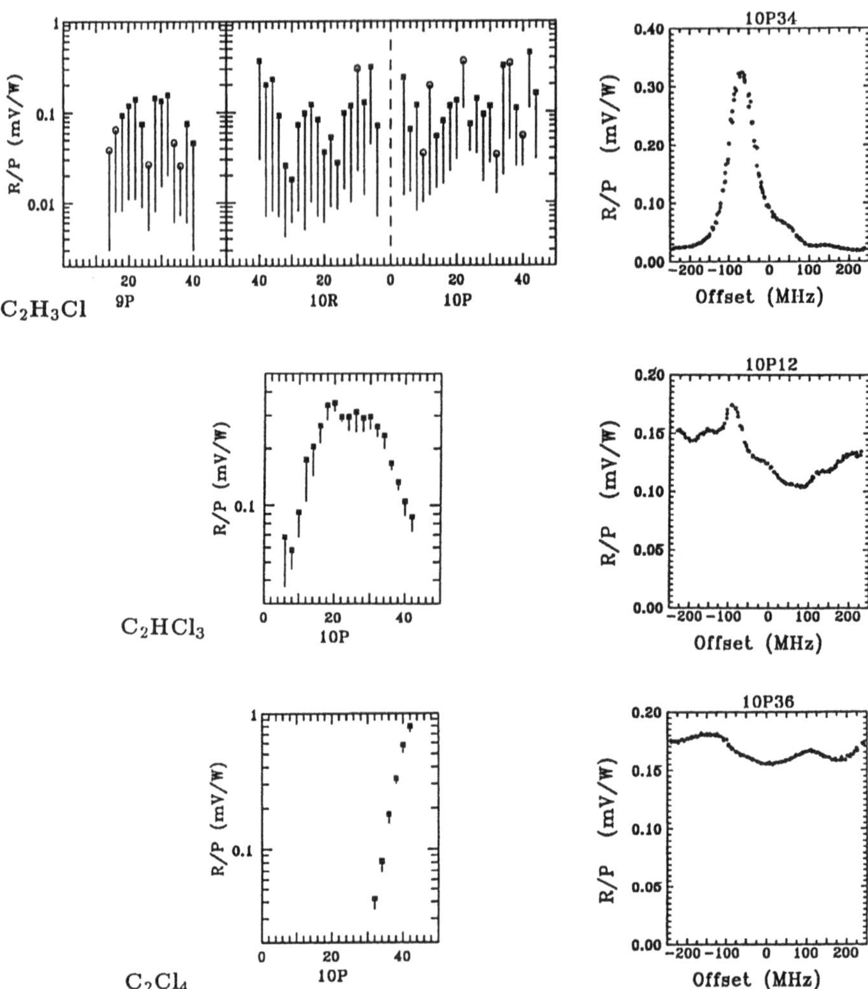

Fig. 1. (Left) Discrete representation of the window spectra showing the signal span. Where the dominant linecenter is outside the window, the maximum signal is marked with an unfilled circle, otherwise it is marked with a filled square. (Right) Examples of window spectra.

showing the span of the signal strength within each window. The span indicates the contrast in each spectrum.

Monochloro-ethylene (vinylchloride) has two vibrational bands overlapping the CO_2 laser frequencies. The PA spectra, which were recorded at 5mBar for a lean mixture of vinylchloride in N_2, show a mostly resolved line structure even though they are superpositions of contributions from the two chlorine isotopomers. Signatures in many single windows are well suited for identification of vinylchloride. The maximum response is found within the 10P42 window, corresponding to an absorption coefficient $\alpha = 0.10cm^{-1}$ when scaled

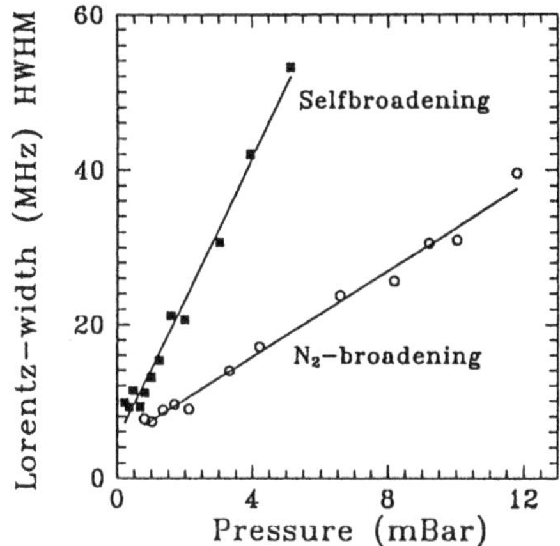

Fig. 2. Lorentz-width as function of pressure for the dominant C_2H_3Cl transition in the 10R12 window.

to 100% vinylchloride. None of the measured lines has as yet been assigned to a particular transition. The signatures are pressure dependent due to collision broadening. Measurements of the broadening rate of a dominating line in the 10R12 window are shown in figure 2. The rates were found to be 9.1MHz/mBar and 2.8MHz/mBar for selfbroadening and N_2–broadening respectively.

Trichloro–ethylene, C_2HCl_3, shows a much higher line density than vinylchloride, partly due to smaller rotational constants and partly due to the greater variety of chlorine isotopomers in the sample. The spectra were recorded at 2.5mBar. The strongest absorption coefficient is found in the 10P20 window, but the contrast in the signal is very low. The signature in the 10P12 window is better suited for trace detection and identification, showing higher contrast but lower absorption. The discrete representation in figure 1 reveals the band structure of the large scale spectrum as one would expect for high line density molecules.

Perchloro–ethylene, C_2Cl_4, shows even less structure than trichloroethylene, with smaller rotational constants but similar number of isotopomers. The spectra were recorded at 5mBar and the highest absorption is found in the 10P42 window. The contrast within each window is too low to form a basis for identification of the molecule in complex mixtures, but the band structure can still be used. The absorption coefficients of tri–, and perchloro–ethylene scale roughly linearly with pressure. At 5mBar they are of the order of $\alpha = 0.1cm^{-1}$.

References

1. A. Olafsson, M. Hammerich, J. Bülow, and J. Henningsen: Applied Physics B **49**, 91 (1989).
2. S.T. Kornilov, I.V. Ostrejkovskij and E.D. Protsenko: in Photoacoustic and Photothermal Phenomena II, Murphy, Spicer, Aamodt and Royce eds., Springer Series in Opt. Sci. **62**, 517, (1990).

The Photothermal Deflection Technique (PDT): Fast Trace Gas Detection in the Atmosphere

H.S.M. de Vries, F.J.M. Harren, and J. Reuss

Department of Physics, University of Nijmegen,
Toernooiveld, 6525 ED Nijmegen, The Netherlands

1 Introduction

Photothermal deflection is a sensitive technique for detecting very low concentrations of trace gases, such as NH_3, C_2H_4 and CO_2. These gases can have an important influence on the ecosystem of plants. To get a better understanding of this influence it is desirable to measure local concentrations in the surroundings of plants. These local measurements make it possible to determine the activity of the different parts of the plants. For example, C_2H_4 exhaust in leafs and stems [1], CO_2 and H_2O emission through stomata, and NH_3 emission in barleyplants [2] can be measured.

In comparison with the photoacoustic technique PDT is less sensitive. However, with this method measurements can be carried out very close to the plants where the concentrations of trace gases are higher, i.e. undiluted by the sampling air flow, and with better spatial resolution.

2 The experiment

The gases, C_2H_4, CO_2, NH_3 and H_2O, absorb CO_2 laser radiation. By collisional relaxation the absorption region is heated and the refractive index changes. Deflection of a probe beam, passing through this region, is proportional to the concentration. The experimental set-up to measure this deflection is given in figure 1. For excitation

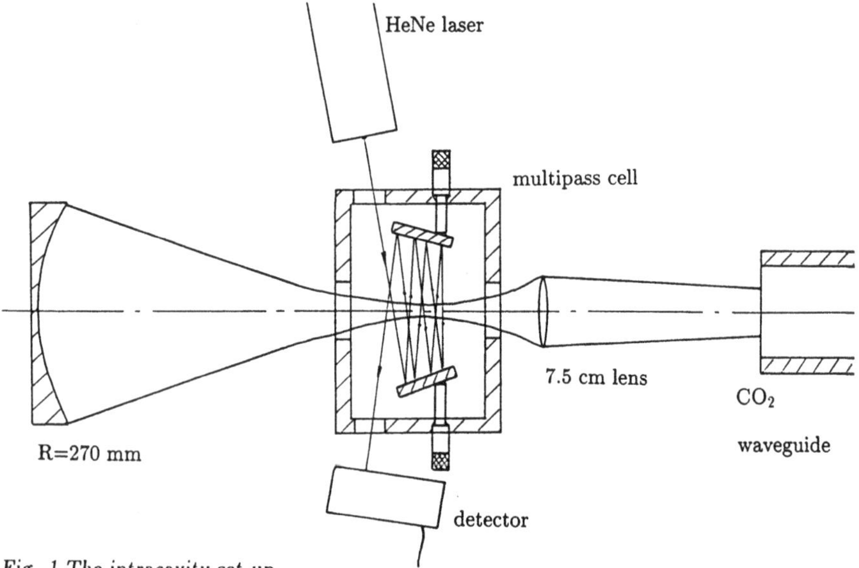

Fig. 1 The intracavity set-up.

Springer Series in Optical Sciences, Vol. 69 Photoacoustic and Photothermal Phenomena III
Editor: D. Bićanić

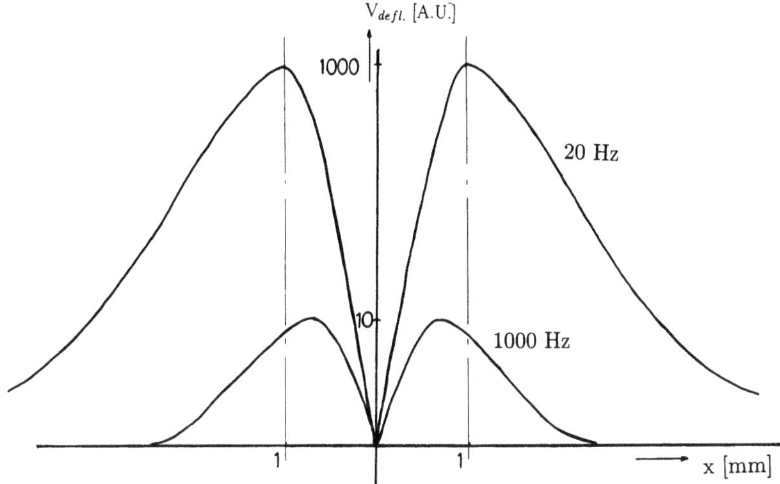

Fig. 2 The deflection signal as a function of the distance from the HeNe laser to the CO_2 laser focus.

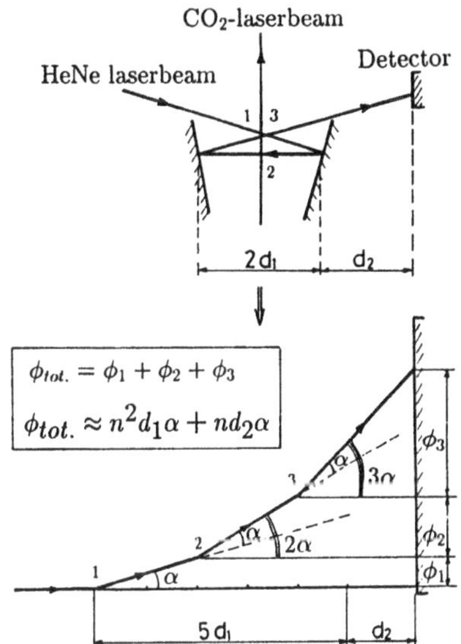

Fig. 3 The total deflection for three passes.

of specific gases a CO_2 waveguide laser is used in two configurations, an intracavity and an extracavity set-up. In both cases the beam is sharply focused (waist ω_0 = 0.221 mm; $\omega \leq$ 0.292 mm over the width of the sampling zone) by a 7.5 cm lens. For the intracavity set-up a hollow output mirror (98.5%, R=270 mm) is used. The distances between discharge tube, lens and output mirror are fixed [3].

The CO_2 laser beam is modulated by a rotating blade chopper that is mechanically decoupled from the set-up to avoid vibrations. Around the CO_2 laser focus a temperature profile and thus a refractive index gradiënt profile is generated [4]; see figure 2. A HeNe laser is used as probe laser to measure this profile.

The influence of the chopping frequency is clearly seen; using a high chopping frequency gives a relatively short heating time and a short time for the heat to spread out. This results in a narrow S-curve. Using a low frequency gives a relatively long heating time and the heat is also widely spread out, yielding a wider S-curve.

At a chopping frequency of 20 Hz the HeNe laser is maximally deflected at points 1 mm from the focus. The HeNe laser passes through these maxima about 40 times, by means of a multipass system consisting of two silver coated mirrors (reflectivity of 99.4% at 632.8 nm), tilted from being parallel by a small angle. The beam divergence of the HeNe laser is 0.78 mrad. After 1 pass the waist is 0.89 mm and after 39 passes the waist is 2.66 mm. For n passes, the total deflection is given by $\phi_{tot.} \approx n^2 d_1 \alpha + n d_2 \alpha$, as can be seen in figure 3 for n = 3. The maximum deflection signal as a function of the number of passes is given in figure 4. The total deflection is monitored by a linear position-sensing detector (length 1 cm, width 4 mm). To enhance the sensitivity the angle (α) between the normal of the detector plane and the incoming probe beam is increased to typically 70°, so that the detector surface fully interupts the laser beam.

3 Results and discussion

The theoretical and experimental results in figure 4 do not agree with each other for two reasons: First, the large beam divergence of the HeNe laser; after n passes the HeNe laser is smeared out over the top of the S-curve. Second, the theoretical model is based on the fact that the HeNe laser crosses each time the maximum of the S-curve. Thus the mirrors should be adjusted in three directions, a difficult procedure not yet mastered by us satisfactorily.

At a chopping frequency of 20 Hz the heat is widely spread out, typically 2 mm (see fig. 2). Thus it is clear that the discrepancy is not caused by the change in waist of the CO_2 laser beam over the width of the sampling zone. This is experimentally verified.

We have measured C_2H_4 concentrations in a cell in the intracavity set-up down to 250 ppt at atmospheric pressure, using 39 passes. The mechanical instability,

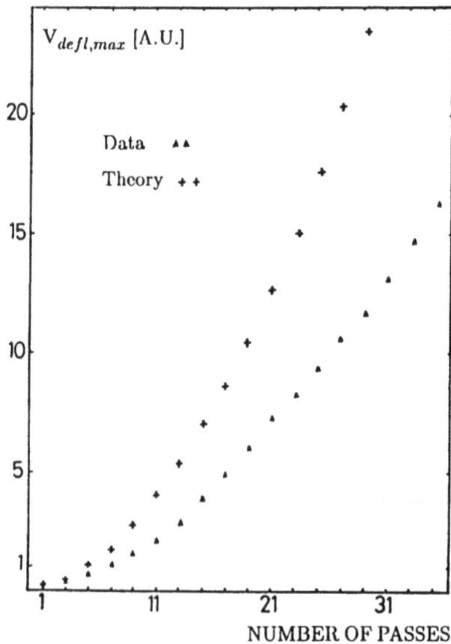

Fig. 4 The maximum deflection signal as a function of the number of passes. The theoretical curve is normalized to the maximum deflection signal at 1 pass.

the limiting factor at this moment, can be decreased by about a factor of 5 in a rigid body frame. Preliminary measurements in the open air detected NH_3 in concentrations of $\geq$ 18 ppb using the 9R30, 10R6 and 10R8 CO_2 laser lines; H_2O and CO_2 were measured on the 10R20 and the 10R18, respectively. Note that these open air measurements are our real aim because they do not require the use of a sample cell with its wall adsorption problems. Response times were below 1 second and the spatial resolution amounted to 8 mm^3.

In the near future a more rigid body frame, combining both lasers, will be used, in which it is also possible to align the mirrors in three directions. Interference effects of different gases will be examined. CO_2 isotopes will be used too in the CO_2 waveguide laser. To measure other trace gases, e.g. NO_x, a CO laser will be utilized for PDT measurements.

References:

[1] F.B. Abeles, "Ethylene in Plant Biology", Academic Press, New York and London (1973)

[2] T.D. Sharkey, H.A. Mooney and E.A. Holland, eds., "Trace gas emissions by plants", Academic Press, New York (in press)

[3] F.J.M. Harren, F.G.C. Bijnen, J. Reuss, L.A.C.J. Voesenek, and C.W.P.M. Blom, Appl. Phys. B 50 (1990), 137-144

[4] W.B. Jackson, N.M. Amer, A.C. Boccara, and D. Fournier, Appl. Opt. 20 (1981), 1333-1344

Photoacoustic Measurement of Vertical Ammonia Gradients in the Atmosphere

A.J.L. Verhage, R.A. Rooth, and L.W. Wouters

N.V. KEMA, Utrechtseweg 310, 6812 AR Arnhem, The Netherlands

A mobile photo-acoustic spectrometer is used to measure concentration differences of ammonia above heathland. Absorption by carbon dioxide and water vapour is discussed and a measuring scheme for concentration gradients is given.

1 Introduction

Recent figures indicate that ammonia is the main source of acidification in the Netherlands, due to its conversion into nitrates. The residence time of ammonia in the atmosphere being short, the ambient concentration is on the average only about 8 ppbv. This low value, and the property of ammonia to adhere to almost any surface, make it difficult to measure this gas accurately. To determine gradients a multiple inlet is required to be connected to one and the same measuring system. Gas transport to this system must be swift in order to avoid adhesion, and any form of pretreatment is prohibited. A photo-acoustic cell made of fluor polymer, together with a CO_2-laser emitting in the fingerprint region of the NH_3-spectrum, is a suitable instrument to measure NH_3-concentrations and -gradients [1]. Simultaneous absorption of atmospheric water vapour and carbon dioxide inside the photo-acoustic cell require a reference measurement. A suitable reference is provided by the 9R28 CO_2-laser line adjacent to the line where NH_3-absorption is high, the 9R30-line. In this way absolute NH_3-concentrations are obtained. Concentration differences do not demand the knowledge of absolute concentrations at every measured height, so an accelerated measuring scheme is possible. Deposition rates of ammonia are evaluated from the gradients in the concentrations using an Eddy diffusion model and meteorological data [2].

2 Absorption by background gases

The photo-acoustic signal at the CO_2-laser line, for which ammonia has the strongest absorption, the 9R30-line, is expressed as:

$$p(30) = C[\alpha_{H_2O}(30)x_{H_2O} + \alpha_{NH_3}(30)x_{NH_3} + \alpha_{CO_2}(30)x_{CO_2}F] \tag{1}$$

Here α is the absorption coefficient, x the concentration, C is a calibration factor and F a correction function that accounts for various relaxation processes [1]. The measured signal is characterized by both amplitude and phase.

Springer Series in Optical Sciences, Vol. 69 Photoacoustic and Photothermal Phenomena III
Editor: D. Bićanić

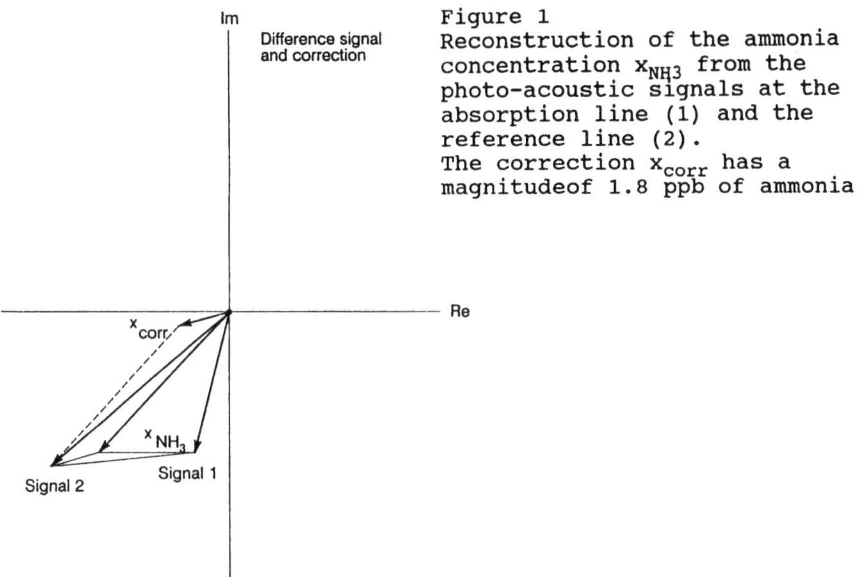

Figure 1
Reconstruction of the ammonia concentration x_{NH3} from the photo-acoustic signals at the absorption line (1) and the reference line (2).
The correction x_{corr} has a magnitudeof 1.8 ppb of ammonia.

Similarly for the reference line 9R28, where the absorption due to water is equal to that at the 9R30, and no ammonia absorption occurs, the signal reads:

$$p(28) = C[\alpha_{H_2O}(28)x_{H_2O} + \alpha_{CO_2}(28)x_{CO_2}F] \tag{2}$$

The difference between the two signals yields the ammonia concentration:

$$p(30) - p(28) = C[\alpha_{NH_3}(30)x_{NH_3} + (\alpha_{CO_2}(30) - \alpha_{CO_2}(28))x_{CO_2}F] \tag{3}$$

$$x_{NH_3} = \frac{(p(30) - p(28)}{C\alpha_{NH_3}} - x_{corr} \tag{4}$$

where

$$x_{corr} = \frac{(\alpha_{CO_2}(28) - \alpha_{CO_2}(30))}{\alpha_{NH_3}} x_{CO_2}F \tag{5}$$

The ammonia concentration is the difference of the two measured signals after correction, as shown in figure 1.

The correction term x_{corr} is a result of the 15% difference in CO_2-absorption for the two lines. Assuming a constant CO_2-concentration of 360 ppm the magnitude of the correction term is 1.8 ppb of NH_3. Due to the dependence of F on the water vapour concentration this term varies slightly under different outdoor conditions. Its magnitude changes by not more than 5%, its phase is such that the final ammonia concentration is a real number, i.e. zero phase.

An independent check of this phase is provided by a water vapour measurement using a polymer transducer. The phase of F

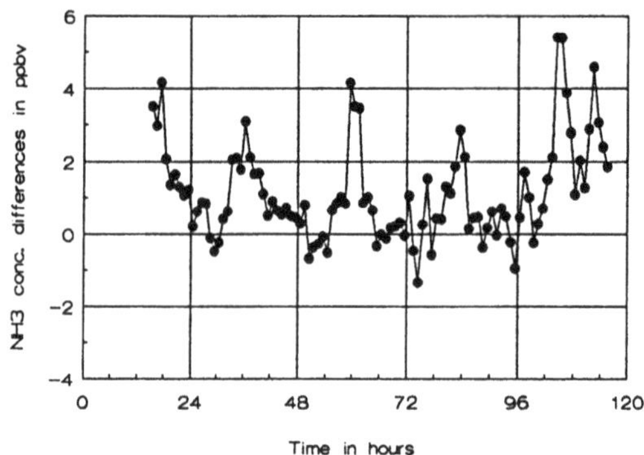

Figure 2
The difference between the ammonia concentrations at the heights of 4 m and 0.5 m above heathland during the course of 5 days. Time 0 is at midnight.

derived from the measurement is compared with the computed phase.

If the CO_2- and H_2O-concentrations do not vary as a function of height, a reasonable assumption for the vertical range (4 m, 2 m, 1 m, 0.5 m) above low vegetation, the concentration difference between two heights i and j can be obtained directly from equation (3):

$$x_{NH_3}(i) - x_{NH_3}(j) = \frac{p_i(30) - p_j(30)}{C\alpha_{NH_3}} \qquad \textbf{(6)}$$

In a forest this is not generally true. The measuring strategy above flat vegetation will be: tune the laser to the 9R30-line and scan along the vertical measuring points, occasionally checking the background variations on the 9R28-line. If water- or CO_2-concentrations vary strongly with height the dual line scheme, equation (4), is recommended. Rapid time variations in the concentrations necessitate interpolation procedures.

Turbulence during a photo-acoustic measurement is avoided when the rate of flow of air through the cell is set at 1 l/min. In the conditioning stage the flow is increased to 5 l/min, which reduces the response time for a step concentration of 25 ppb NH_3 + 1% water vapour to 2 minutes. For a measuring period of 1 minute and a conditioning time of 2 minutes for each position, the cycle of 4 gradient measurements takes 12 minutes. In the dual-line scheme some extra time is required for stabilization, demanding 15 minutes for a cycle.

In the latter set-up a series of measurements has been carried out in Elspeet (Netherlands) above heathland yielding realistic deposition rates and an error in the concentration difference of less than 0.2 ppbv (see figure 2).

Thanks are due to C.M.G. Heffels and F. Bijnen for help and discussions.

References

[1] R.A. Rooth, A.J.L. Verhage and L.W. Wouters, 1990. Applied Optics, 29, No. 25, pp. 3643-3653.

[2] J.W. Erisman et al, 1988. Amos. Environ., 22, pp. 1153-1160.

Interfacing Photoacoustic and Photothermal Techniques for New Hyphenated Methodologies and Instrumentation Suitable for Agricultural, Environmental and Medical Applications

D. Bićanić[1], *H. Jalink*[2], *M. Chirtoc*[3], *H. Sauren*[4], *M. Lubbers*[5], *J. Quist*[1], *E. Gerkema*[1], *K. van Asselt*[1], *A. Miklós*[6], *A. Sólyom*[7], *Gy.Z. Angeli*[7], *P. Helander*[8], and *H. Vargas*[9]

[1]Laser Photoacoustic Laboratory, Department of Agricultural Engineering and Physics, Wageningen Agricultural University, Agrotechnion, Bomenweg 4, 6703 HD Wageningen, The Netherlands
[2]Centre for Plant Breeding and Reproduction Research CPRO-DLO, P.O. Box 16, 6700 AA Wageningen, The Netherlands
[3]Institute of Isotopic and Molecular Technology, P.O. Box 700, R-3400 Cluj-Napoca 5, Romania
[4]Department of Chemistry, Queens University, Kingston, Ontario, Canada
[5]Department of Soil Science and Geology, Wageningen Agricultural University, P.O. Box 37, 6700 AA Wageningen, The Netherlands
[6]Institute of Isotopes, Hungarian Academy of Sciences, P.O. Box 77, H-1525 Budapest, Hungary
[7]Analtron Applied Research Company, P.O. Box 63, H-1311 Budapest, Hungary
[8]AB Varilab, Fullerstavägen 32, S-14144 Huddinge, Sweden
[9] Instituto de Fysica, Universidada Estadual de Campinas, 13100 Campinas, SP, Brazil

1. Introduction

During the past few years existing (laser) photoacoustic (PA) and photothermal (PT) techniques have gradually diffused into variety of fields including these of agricultural, medical and environmental sciences. In addition, there has been a quite substantial development of new, versatile and competitive instrumentation and of experimental methodologies suitable for use in a daily practice. This encouraging progress is mainly due to perpetual improvements made in a development of sources, transducers and equipment for data collection and processing.

Current applications in agricultural, medical and environmental sectors include the optical as well as the thermal characterization of gaseous, liquid and solid specimens. This paper is concerned with few PA and PT methodologies that when combined with other techniques lead to hyphenated methods the performances of which surpass those of intended for a stand-alone use.

Springer Series in Optical Sciences, Vol. 69 Photoacoustic and Photothermal Phenomena III
Editor: D. Bićanić

2. Experimental

The potential that laser PA and PT methods have for high sensitivity environmental trace detection has been recognized for quite some time. However, when dealing with PA and PT analysis of realistic and complex mixtures (such as for example the atmospheric air), one often faces the problem of spectral interferences especially when using only poorly tunable lasers. Very high sensitivity, compound specificity and fast response are very rarely provided by one technique or available in a single piece of the instrumentation. An elegant approach (IMPASS) to the enhancement of spectral discrimination of ammonia (and other molecules with appreciable electric dipole moment) has recently been demonstrated by Sauren et al. [1-3]. The IMPASS is a hyphenated technique that combines well known concepts of intermodulation (IM) and Stark spectroscopy (SS) at reduced pressure values (200 mbar) with that of CO_2 laser PAS in order to exploit their complimentary advantages in a best possible way. While modulating at the same time the radiation of the CO_2 laser (mechanically at $f_1=40$ Hz) and the external electric field (3 kV/cm electronically at $f_2=1640$ Hz) the lock-in detection is attempted at the frequency f_1-f_2 that equals the first longitudinal acoustic resonance frequency of the PA cell shown in Fig. 1. In this way the effects of a background gas and of window noise are eliminated. Detection sensitivity of 0.5 ppbv of ammonia (with a gaseous mixture flowing at 60 lit/h rate) has already been reached in an extracavity configuration while fully supressing the effect of major interfering constituents water vapour and carbon dioxide.

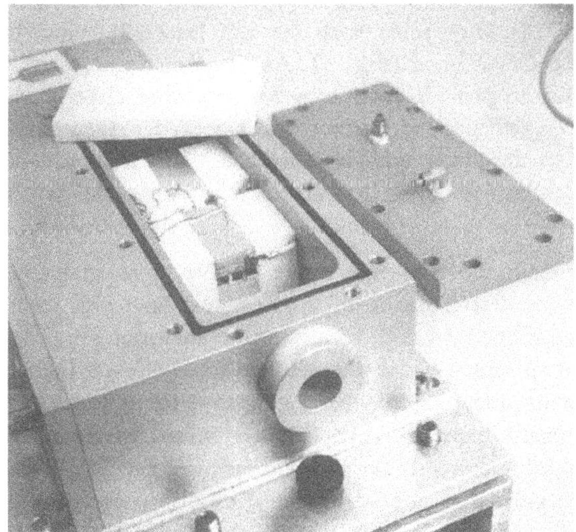

Fig. 1. The photograph showing the resonant part of the PA cell, two Stark plates and one of the two "open" sections that serve for suppression of the window noise [1]. The laser beam enters the cell through the front window.

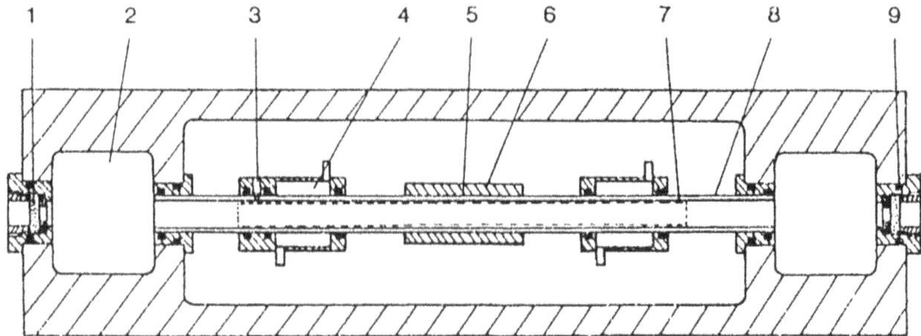

Fig. 2. The photoacoustic heat-pipe (PAHP) with windows and "O" ring seals (1 and 9), one of the two acoustic buffer volumes (2), miniature electret microphone with the battery power supply (3), cooling jacket (4), thermocouple (5), heating coil (6), metal wick (7) and quartz tube (8).

Another example of a hyphenated instrument is the resonant PA cell based on the principle of a heat-pipe (PAHP). A device, originally proposed by Jalink and Bicanic [4] enables one to obtain photoacoustically valuable structural informations from liquid and powdered specimens having low volatility. The cell, actually a quartz tube (Fig. 2) with the length selected to support the third longitudinal of the generated acoustic wave, operates due to a perpetual evaporation and condensation of the fluid of interest along the heated zone irradiated by the laser light. Selection of microphone's location within the cell and the way how it is being cooled permit inexpensive detectors to be used despite high temperatures and in addition efficiently prevents the contamination of optical components. The PAHP has been successfully used to record the IR vapour phase spectra of geraniol [4] fragrance (with the CO_2 laser) and of longer (up to C18) fatty acids with the CO laser [5] tuned to wavelength coinciding with the C=O bond (Fig. 3). An important feature of a PAHP cell is its intrinsic ability to provide the means for recording directly the vapour pressure-temperature dependence of various substances across a large temperature range [6]. As far as the upper limit is concerned, there is in fact no restriction as long as the temperature at which the substance will dissociate is not exceeded.

In another independent experiment the heat-pipe was used [7] to estimate the potential of the PA technique for determination of total organic (and inorganic) content of carbon in aqueous solutions by a high temperature combustion (Fig. 4). Upon injecting small amount of specimen under investigation into hot oven (above 1200 K) all the carbonaceous matter is rapidly oxydized and converted to a CO_2 gas and the concentration is traditionally measured (in a flowing regime) by means of the absorption spectroscopy at 4.27 microns. In the above experiment the PAHP was used in a series with the conventional absorption detector. The CO_2 gas leaving the oven and passing the absorption cell was directed through the heated PAHP and irradiated by an arbitrary transition of the CO_2 laser. Although the maximum absorption cross section of the CO_2 gas (600 x 10^{-5} atm^{-1} cm^{-1}) in a 10 micron region is typically much less favourable than that at 4.2 micron (600 atm^{-1} cm^{-1}) detection

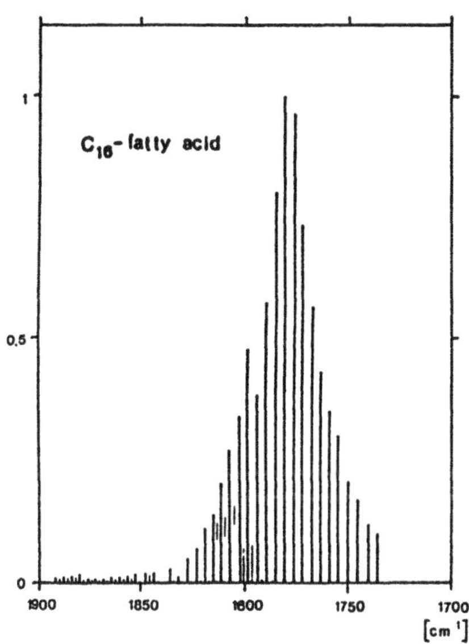

Fig. 3. The normalized photoacoustic signal (arbitrary units) recorded (at 1 atmosphere and 446 K) with the PAHP loaded with palmitic acid $C_{16}H_{32}O_2$ plotted versus the wavenumber of the CO laser.

sensitivity obtained with PAHP was 4 mmol/l which is about 400 times inferior to that achievable by the conventional technique. The attempts to integrate the oven and detector in one and the same compact size practical PAHP instrument are currently in progress.

The PAHP offers an attractive opportunity for observing the temperature dependence of the absorption cross section for a given gas (as demonstrated with a CO_2 gas [8] at 10 microns wavelength in 300-900 K range). A steady increase in the magnitude of the PA signal (and hence the cross section) was observed as the temperature was raised from 300 K to about 820 K due to a correspondingly larger population of the lower level of the transition. Above that temperature the signal maximizes and higher energy levels became more populated [9]. At temperatures near 900 K the absorption cross section exhibits a decreasing tendency.

In recent experiments the PAHP was used with the CO_2 laser to record IR vapour spectra of volcanic gases, eugenol pheromone [10] and malathion a well known organophosphorus pesticide of very low volatility [11]. Likewise, PAHP in conjunction with the CO laser has been employed to estimate the potential of PA method in detecting isoprene [12], a reactive biogenic forest emittant and constituent in rural and remote areas. This olefin alkadiene is considered as a dominant non-methane hydrocarbon released by foliage emission during the volatilization of the essential oils and serves as a precursor for photochemical oxidation. The absorption cross section of 5.8×10^{-20} cm^2 has been found at 1780 cm^{-1} transition emitted by CO laser. Detection sensitivity on the order of few ppbv (adequate for present

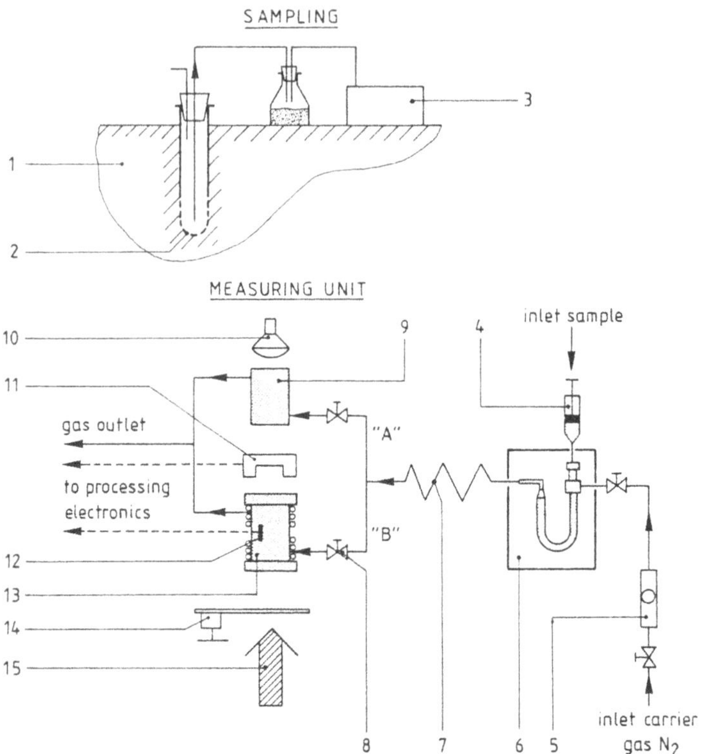

Fig. 4. Schematic diagram of PA experiment for determination of determine total organic (and inorganic) carbon content in water and soil solutions. Specimen of soil (1) is collected into a porous cup (2) by means of a forepump (3) before being injected into the combustion oven (6) and carried away by nitrogen gas streaming through the flowmeter (5). After passing the air cooled condenser (7), the gas is, by means of a two ways valve (8), directed either towards the conventional absorption cell (9) (path "A") illuminated by the infrared source (10) or to the PAHP (13) (path "B"). This latter is provided with a small microphone (12) and exposed to a chopped (14) radiation of the CO_2 laser (15). The PA signal is ratioed to the transmitted power measured behind the cell with a power detector (11).

monitoring purposes) is anticipated for the same species if strong olefin out-of-plane C-H bending vibration at 980 cm^{-1} accessible with the CO_2 laser, is excited.

For analysis of complex mixtures the PAHP, as a kind of hyphenated detector, can be used in a combination with any of well known separation techniques. This was demonstrated [11] with the CO_2 laser, gaschromatographic column and the PAHP loaded with a mixture of toluene and ethylbenzene. Recently, Jalink [13] has proposed to use a PZT crystal as a thermal detector instead of a PAHP in GC experiments. The important advantages of such experimental configuration are reasonable sensitivity of the PZT material and its ability to sustain high temperatures

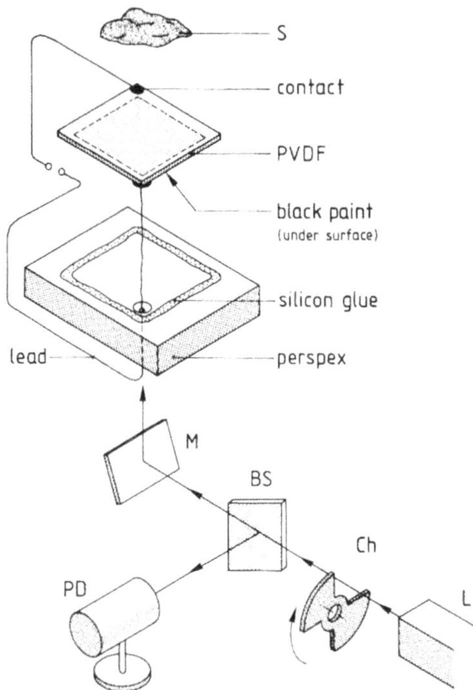

Fig. 5. A versatile instrument (based on the IPPE method) for discrimination of food samples based on their thermal effusivities. The major portion of intermittent (Ch) radiation emitted by an arbitrary heating source (here laser L) passes through the beam splitter BS, reaches the plane mirror M and is reflected towards the IPPE cell consisting of a PVDF film (its rear side painted black) attached to the perspex plate by means of the silicon epoxy. The signal at the modulation frequency is detected (with the sample S atop the PVDF foil) across the PVDF film (leads are shown) across and ratioed to the fraction of input power detected by a power detector (PD) for the purposes of normalization.

and the fact that hydrogen, normally needed to operate the conventional GC detector is no longer required.

The merrits of PT deflection methods (with inexpensive diode lasers as sources) for sensitive concentration measurements of inorganic constituents in aqueous solutions (of horticultural importance) have been discussed extensively [14]. The PA study of microbial breakdown of animal manure has been performed [15] by monitoring level of related methane concentration (with the He-Ne laser at 3.39 microns) in a liquid farm slurry. In another experiment [16] Quist et al. demonstrated 50 ng/ml (1.5 ppbv) detectivity limit for HpD (haematoporphyrine derivative, a photosensitizer in photodynamic treatment of tumour diseases) in diluted clinical solutions.

Another emerging, valuable technique is that of inverse photopyroelectric spectroscopy IPPE [17]. Relatively simple experimental set-up (Fig. 5) enabled a

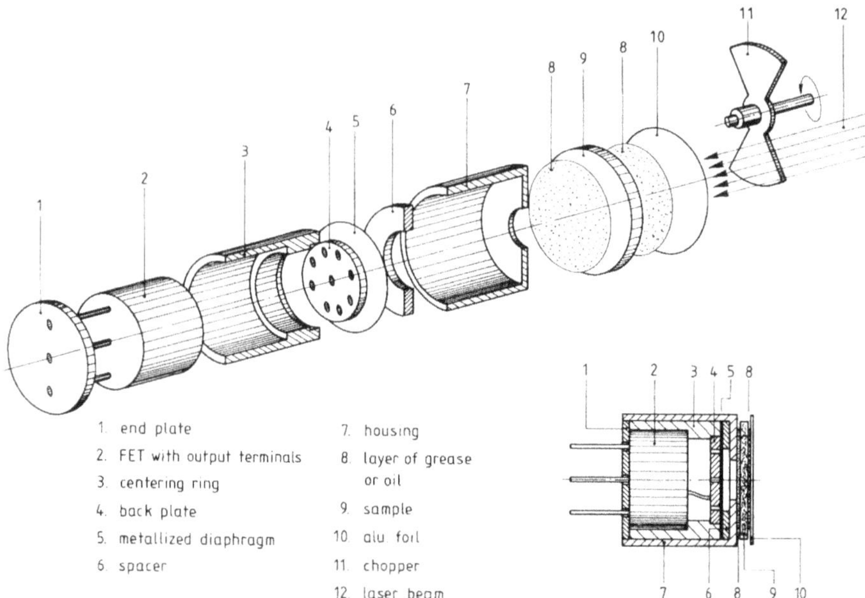

Fig. 6. The exploded wiew and the cross section of the OPC. The modulated (11) radiation originating from the heating source (12) is incident on a thin, optically opaque aluminum foil (10). The sample of interest (9) and foil are separated by means of a thin layer of grease or oil (8). The backside of the specimen is fixed to a the housing (7) that carries detecting components (1-6) in the very same way.

discrimination of various pasty food specimens and edible oils by means of their thermal effusivities [18] and could eventually serve as product's quality indicator. Preliminary experiments on food samples (ketjap an extract of soya beans and soup additives) confirmed the usability of an optothermal sensor [19] for optical characterization of these specimens in visible spectrum.

Vargas et al. proposed the concept of an open PA cell (OPC) for assessing thermal and optical data of solid samples. In a recently performed experiment the OPC detector (Fig. 6) was shown capable of providing lacking data on thermal diffusivity data of potato's periderm [20]; differences between potatoes grown in various soils were observed. The method was later applied for studies of single layer [21] and laminated [22] plastic foils used for packaging of food. Various investigations have established that PA and PT methods appear useful for studies of grain samples such as flour [23]. The construction of a novel PA instrument (for quantitative measurement of water in sand) excited by microwaves is underway. Much of other interesting work of biological and agricultural relevance is described elsewhere in these Proceedings.

In conclusion, the usability of proposed hyphenated methods with the unrivalled combination of sensitivity and specificity and of developed custom-made instrumentation for variety of applications in the field of agricultural and evironmental sciences has been demonstrated convincingly. Growing number of active research groups and manufacturers guarantees new achievements in the years to come.

References

[1] H. Sauren, Ph. D. Thesis, Wageningen Agricultural University, Wageningen, The Netherlands (1992).

[2] H. Sauren, D. Bicanic and K. van Asselt, Infr. Phys. 31,475-484 (1991).

[3] H. Sauren and D. Bicanic, Anal. Instrum. (in press).

[4] H.Jalink and D. Bicanic, Appl. Phys. Letts. 55, 1507-1510 (1989).

[5] H. Jalink, Unilever Research Vlaardingen, Report 6, (1991).

[6] H. Jalink and D. Bicanic, Meas. Scien. & Techn.(submitted).

[7] D. Bicanic, M. Chirtoc, M. Lubbers and H. Jalink , Anal.Instr. (submitted).

[8] D. Bicanic and M. Chirtoc (to be published).

[9] A.M. Robinson, Univ.of Alberta, Edmonton, Canada, Priv.Comm. (1990).

[10] J. Quist, M.Sc. Thesis (part I), Wageningen Agricultural University, Wageningen, The Netherlands (1991).

[11] D. Bicanic, H. Jalink and B. van Veldhuizen, Meas. Scien. & Techn. 1, 247-249 (1990).

[12] A. Miklós, D. Bicanic and J. de Bont, Spectr.Letts. (submitted).

[13] H.Jalink (to be published).

[14] D. Bicanic, M. Lubbers, P. Torfs and A. Tam, Acta Hortic. 304, 29-48 (1992).

[15] D. Bicanic, A. Sólyom, G. Angeli, T. Smits, Waste (submitted).

[16] J. Quist, M.Sc. Thesis (part II), Wageningen Agricultural University, Wageningen, The Netherlands (1991).

[17] M. Chirtoc, D. Dadarlat and D. Bicanic (in these Proceedings).

[18] D. Bicanic, M. Chirtoc, D. Dadarlat, P. van Bovenkamp and H. van Schayk, Appl. Spectr. (in press).

[19] P. Helander (in these Proceedings).

[20] J. Buys, D. Bicanic, J. Bontsema and H. Vargas, Anal.Instr. (in press).

[21] J. Buys, M. Sc. Thesis, Wageningen Agricultural University, Wageningen,The Netherlands (1990).

[22] A. M. Mansanares, H. Vargas, F. Galembeck, J.Buys and D. Bicanic, Jour. Appl. Phys. 70, 7046-7050 (1991).

[23] J. Buys, J.P. Favier, O. Dóka and A. Miklós (in these Proceedings).

Acknowledgements

The authors express their gratitude to Stichting Technische Wetenschappen (STW) Utrecht, Stichting FOM Utrecht, Dutch Organization for Scientific Research NWO The Hague, European Environmental Reesarch Organization Wageningen, Unilever Research Laboratory Vlaardingen and the European Community for the support received in the course of this study.

Photoacoustic In Situ Monitoring of Trace Gases in a Rural Environment

A. Thöny, B. Brand, M. Ferro-Luzzi, and M.W. Sigrist

ETH, Institute of Quantum Electronics, CH-8093 Zürich, Switzerland

Abstract. The concentrations of five different gaseous compounds in rural air have been monitored with a mobile CO_2-laser photoacoustic system. A high detection sensitivity corresponding to ppbV-concentrations has been achieved for ozone, ethene and ammonia.

1. Introduction

The sensitive and selective monitoring of gaseous air pollutants in situ or remote is of great current interest. An interdisciplinary, nationwide research program named POLLUMET has recently been launched which is aimed at a three-dimensional monitoring and modelling of the air pollution, particularly of the photosmog production in summer, over the complex topography of Switzerland. In the frame of this program we performed in situ measurements in a rural environment at Morschach (Central Switzerland) in July 1990 and in an urban environment at Dübendorf (near Zurich) in July 1991. In this paper we present results of the 1990 measurement campaign.

2. Experiments

Our measurements have been performed with the aid of our mobile CO_2 laser photoacoustic system described in detail elsewhere [1]. During a six-day period we have been monitoring the concentrations of H_2O vapor (absolute water vapor), CO_2 (carbon dioxide), NH_3 (ammonia, important component in agricultural areas), O_3 (ozone, main photosmog component) and C_2H_4 (ethene, precursor component for photosmog generation). For this purpose, the computer-controlled laser system was repetitively tuned to preselected laser transitions chosen for their characteristic absorption by the molecules to be detected. The selection of the appropriate laser transitions was based on the absorption cross sections, on minimum interfering absorptions and on good laser beam quality. As a result, nine different laser transitions were chosen for the analysis of the five compounds, namely the 10R(20) transition which is characteristic for H_2O vapor absorption, 10P(20) for CO_2, 10P(34) for NH_3, 9P(14), 9P(24) and 9P(26) for ozone, 10P(14) for ethene and, in addition, 9P(34) and 10R(34) where none of these molecules exhibits an appreciable absorption. The use of nine laser transitions yields a temporal resolution of approx. 10 minutes. The data are analyzed on the basis of calibration spectra by applying an iterative least-squares fit described previously [1] for the nine transitions considered.

Springer Series in Optical Sciences, Vol. 69 Photoacoustic and Photothermal Phenomena III
Editor: D. Bićanić

3. Results and Discussion

As a result the temporal concentration profiles of H_2O vapor, CO_2, NH_3, O_3 and C_2H_4 could be recorded simultaneously in situ. To our knowledge, it is the first time that five different gaseous compounds have been recorded simultaneously in unfiltered rural air by a single instrument. A typical example is shown in Fig. 1. In agreement with independent measurements of relative humidity and temperature, the absolute concentration of H_2O vapor varied between 0.9% and 1.8%. The CO_2 concentration varied around the expected value of 350 ppmV while the C_2H_4 concentration only exhibited single peaks below 5ppbV correlated to local traffic. In contrast, the NH_3 and O_3 concentrations showed strong variations in the ppbV-range.

It should be mentioned that the absolute concentrations of NH_3 and O_3 have not been certified yet because own calibrations for these species have not yet been performed. Ammonia which plays an important role with respect to the acidification

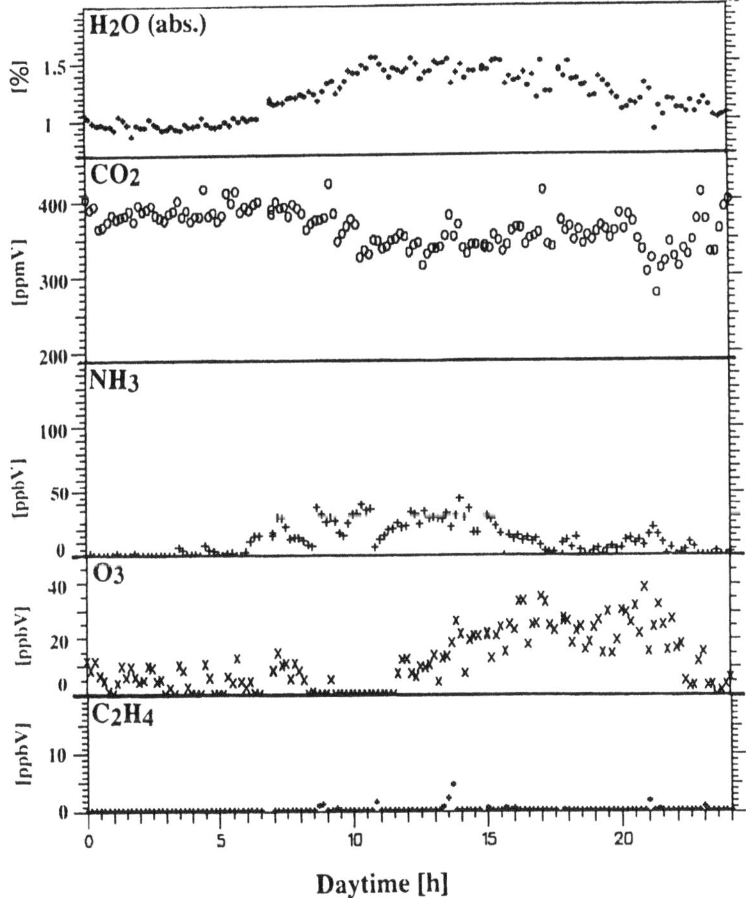

Fig. 1. Concentration profiles of H_2O vapor, CO_2, NH_3, O_3 and C_2H_4, derived from photoacoustic data. Location: Morschach/CH, Date: July 26, 1990

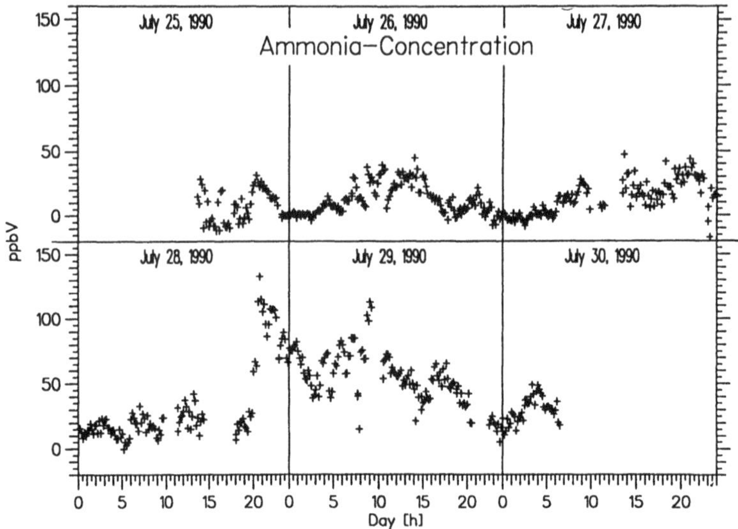

Fig. 2. Temporal variation of the NH_3 concentration in ambient air in a rural environment. Location: Morschach/CH, Date: July 25-30, 1990

of soils and water has recently been monitored photoacoustically [2, 3]. As a further result of our field studies a plot of the NH_3 concentration is presented in Fig. 2 for the period of July 25-30, 1990. The concentration varies between a few ppbV and a maximum of 135 ppbV and is directly related to local agricultural activities and meteorological parameters (wind characteristics and temperature).

In **conclusion** the successful simultaneous monitoring of five different trace gases in ambient air by a mobile CO_2 laser photoacoustic system has been demonstrated. Our 1991 measurements in urban air were performed with both $^{12}C^{16}O_2$ *and* $^{13}C^{16}O_2$ laser isotopes. The analysis includes additional hydrocarbons such as benzene, toluene and xylenes. Results will be published elsewhere.

References

1. P.L. Meyer, M.W. Sigrist: Rev. Scient. Instrum. **61**, 1779 (1990)
2. A. Olafsson, M. Hammerich, J. Bülow, J. Henningsen: Appl. Phys. **B49**, 91 (1989)
3. R.A. Rooth, A.J.L. Verhage, L.W. Wouters: Appl. Opt. **29**, 3643 (1990)

Photoacoustic Field Measurements of Methane

M. Fiedler[1], *Ch. Gölz*[2], *and U. Platt*[2]

[1]Battelle Institut, Am Römerhof 35,
W-6000 Frankfurt 90, Fed. Rep. of Germany

[2]Institut für Umweltphysik, Universität Heidelberg,
Im Neuenheimer Feld 366, W-6900 Heidelberg, Fed. Rep. of Germany

A field experiment using a nonresonant photoacoustic cell and the 3.39μm HeNe laser was carried out. The aim of the study was to monitor the diurnal variation of the methane concentration in a cowshed in order to estimate the CH_4 production rate of ruminants.

1. Introduction

Methane is one of the most important greenhouse gases and is involved in tropospheric chemistry, where the methane concentration is 1.8 ppm and increased by 1 % per year during the last hundred years [1]. The major methane sources at mid latitudes are organic waste deposition, anaerobic bacteria in the digestive system of ruminants and leakage of gas pipelines and coal mining, photoacoustic (PA) measurements offer the advantage of high dynamics, good time resolution, robust instrumentation and the achieved detection limit is appropriate for the concentration levels near methane sources. In our PA field experiment, CH_4 concentrations in a cowshed were measured in order to estimate the production rate of the ruminant source.

2. Experimental set–up

The set–up used is illustrated in Fig. 1. Based on theoretically expected PA pressure amplitude [2] a non-resonant PA cell (brass, 20 cm length, 5 mm diameter) was built. The 5 mW HeNe laser beam was modulated by a mechanical chopper at 20 Hz. A reference photoacoustic cell filled with CH_4 served to monitor the laser stability. The electret microphone signals from the measuring and reference cells were analyzed by phase sensitive lock–in procedure realized in a software.

For 60 sec integration time the microphone was the dominating noise source. The measured noise levels corresponded to a methane concentration of typically 50 ppb. The window signal (equivalent to 1–2 ppm CH_4) was consistently subtracted. The cross sensitivity to water absorption lines was measured at 25° C and 100 % rel. humidity and found to correspond to approx. 10 ppm CH_4. During the field experiment the air samples were dried by silicagel. Desorption from the cell walls induced a signal rise (memory effect) equivalent to 2 ppm CH_4 over about 10 hours at room temperature. On–line calibrations performed during the routine measurements by injecting CH_4 standards into the PA

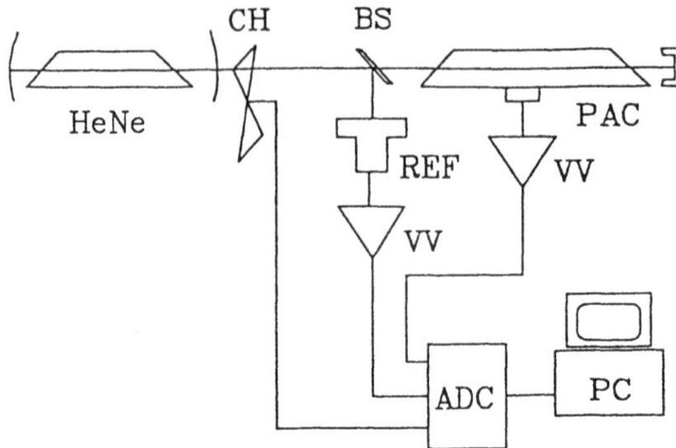

Fig. 1: Experimental setup
HeNe: 3.39 μm HeNe–Laser, CH: chopper, BS: beamsplitter, PAC: photoacoustic cell, REF: reference photoacoustic cell, VV: preamplifier, DE: detector for monitoring transmitted laser power, ADC: analog-to-digital converter, PC: personal computer.

cell showed responsivity fluctuations of about 5 %. The effects discussed above presently limit the minimum detectable CH_4 concentration to ca. 0.2 ppm.

3. Field measurement and interpretation

Figure 2 shows the results of an on–line PA measurement of the diurnal variation of the methane concentration in a cowshed with 18 cows and 10 calfs (air sampling frequency 5 minutes). The concentration varied between 36 and 192 ppm. In addition, 14 shed air samples were measured off-line in the labo-

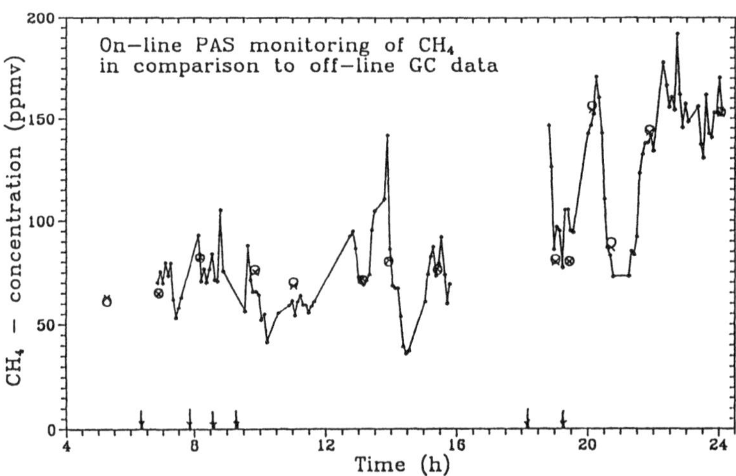

Fig. 2: Photoacoustic (PAS) on–line measurement of methane concentration in a cowshed, compared to gaschromatographic (GC) results
o : GC measurements, x : PAS crosscheck of gas samples
the arrows indicate beginning and end of the feeding times (shed doors open)

ratory both photoacoustically (crosses in Fig.2) and by gas chromatography (open circles in Fig.2) [3]. The quasi continous PAS measurements resolve short time scale fluctuations due to varying leakage and air mixing in the cowshed. A simple qualitative model [4] resulted in a lower limit of 2000 l CH_4 /24 h for the total production rate of the actual cattle population. This model assumes a constant methane production rate and fast removal of methane due to the effective air exchange processes in the shed. In the future, improved instrumentation will allow CH_4 background measurement with an accuracy of 1 %.

4. Literature

[1] Bericht der Enquete-Kommission des 11. Deutschen Bundestages, Referat Öffentlichkeitsarbeit, Bonn 1990.

[2] Rosengren, L.-G., E. Max and S.T. Eng, Infrared Phys. **13**, 109 - 121, 1973.

[3] Trapp, D., Methanisotope und Quellstärkenabschätzung bei landwirtschaftlichen Nutztieren, Diploma Thesis, Heidelberg 1990.

[4] Gölz, Ch., Photoakustische Methanmessungen, Diploma Thesis, Heidelberg 1991.

A Liquid Nitrogen Cooled CO Laser in a Photoacoustic Set-Up Monitors Low Gas Concentrations

F.G.C. Bijnen, T. Brugman, F.J.M. Harren, and J. Reuss

Department of Physics, University of Nijmegen, Toernooiveld, 6525 ED Nijmegen, The Netherlands

Introduction

Ethane is produced in plants after lipid peroxidation of the cell membrane. This process can for example be initiated by ambient SO_2 and results in the death of the cell. Ethane has proven to be a reliable indicator of this process [1]. A PA monitor of ethane is therefore constructed. The performance of two differently cooled CO-lasers (77 K and 265 K) in a PA set-up is compared. The fingerprint absorption spectrum of C_2H_6 is presented.

Experimental set-up

Our homemade liquid nitrogen cooled CO-laser [2] (see fig 1) is tuned over 200 single line transitions between 4.9 μm ($v = 4 \rightarrow 3$) and 7.7 μm ($v = 33 \rightarrow 32$). On 80 lines 1 - 3.5 Watt laser output power is obtained (see fig 2). The laser is employed in combination with an extracavity longitudinal PA cell [3]. The acoustic resonator has dimensions ⊘ 6 x 100 mm. A microphone with a sensitivity of 20 mV/Pa is used. Water has many strong absorption lines in the CO-laser wavelength region. The laser is prevented to pass through humid ambient air.

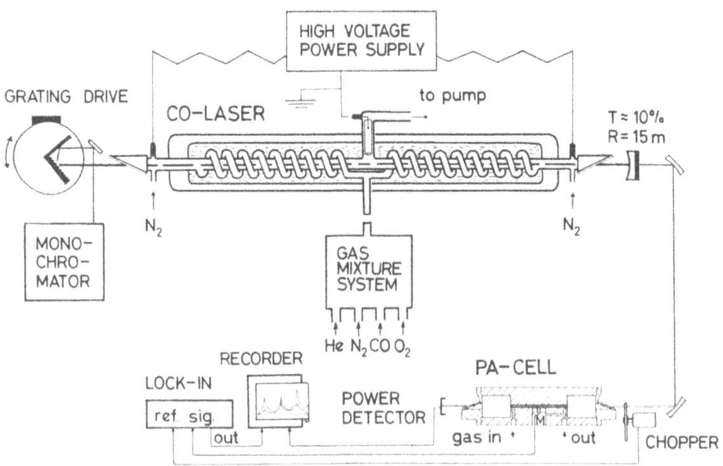

Fig 1. Diagram of the experimental set-up.

Springer Series in Optical Sciences, Vol. 69 Photoacoustic and Photothermal Phenomena III
Editor: D. Bićanić

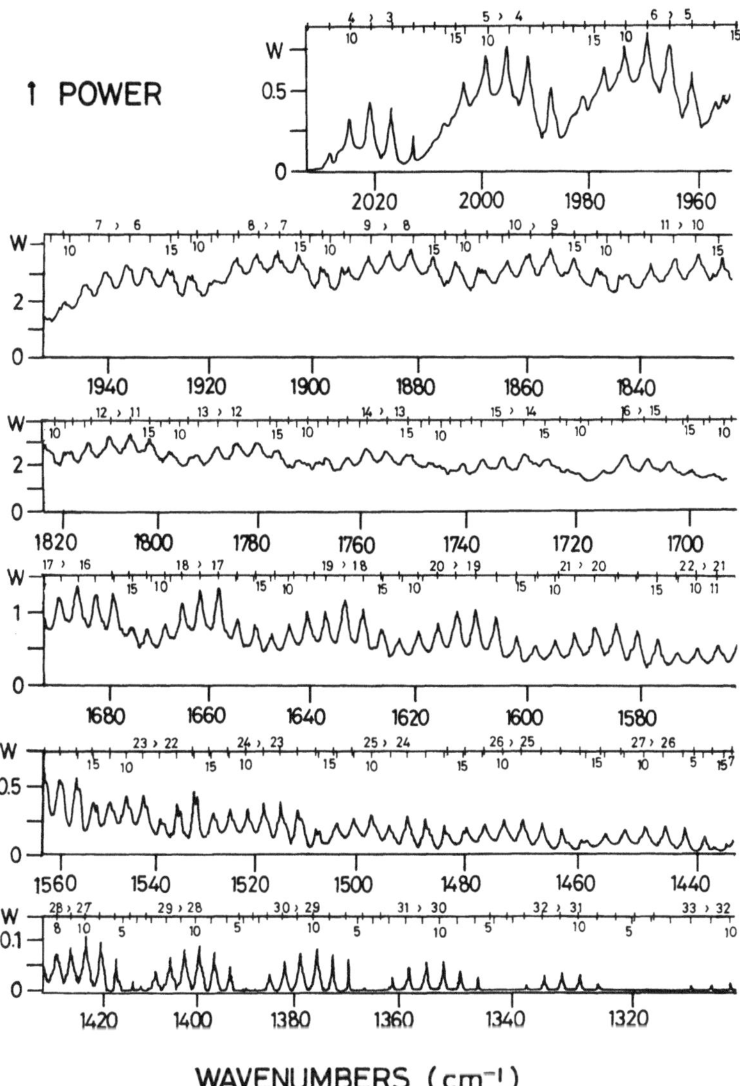

Fig 2. measured ouput of the 77 K CO-laser. The numbers above the upper line of each scan represent the $v'' \rightarrow v'$ transition. The ticks below this line represent the $P(J)$ transitions. For each scan the gasmixture/current density has been optimized.

Results and discussion

To monitor car exhausts a CW CO-laser, which was cooled down to 265 K, has been employed previously in a PA set-up [4]. A maximum output power of 1 Watt was obtained and the wavelength region from 5.1 μm to 6.4 μm was covered. Multi-line emission made a dual beam set-up with reference system necessary. Cooling the laser to 77 K yields higher population probabilities for the lower lying J-levels ($J = 8 \rightarrow 9$

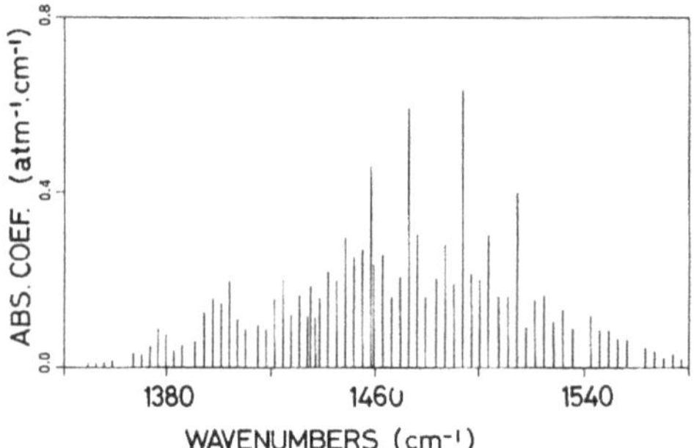

Fig 3a.
C_2H_6 absorption coefficients at CO-laser frequencies.

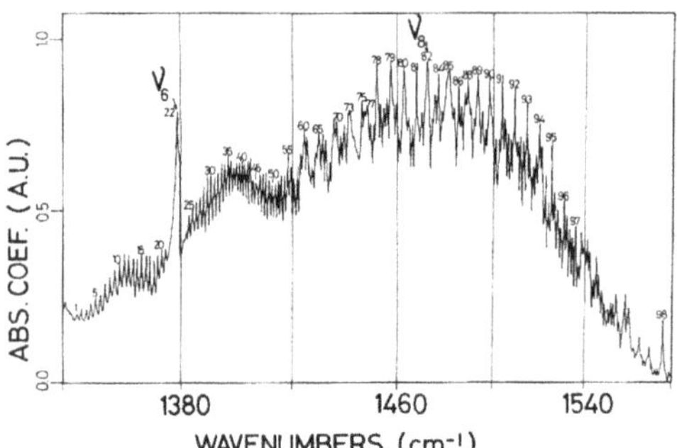

Fig 3b.
IR spectrum of fundamental ‖ band ν_6 and fundamental ⊥ band ν_8 of C_2H_6 measured with a absorption spectrometer [5]. The numbers correspond to a running index.

: $J = 14 \rightarrow 15$) causing laser action to occur on different frequencies as compared to a 265 K CO-laser ($J = 15 \rightarrow 16$: $J = 22 \rightarrow 23$). Population is distributed over fewer rotational levels. This causes less overlap between adjacent vibrational transitions. Thus the 77 K laser shows single-line performance on almost all laser lines. Since the population probability is higher for the lower lying rotational levels, the population difference between lasing levels is also larger. This means that laser action occurs even at weak vibrational transitions, yielding an extended wavelength range. The strongest laser lines we find between 5.1 and 6.2 μm. Both CO lasers cover this region; we obtain output power up to 3.5 Watt at a relatively high current density (10 mA/cm^2). This causes a rise in temperature in the laser plasma so that multi-line emission between adjacent vibrational levels occurs. Lowering the current density or putting an iris inside the cavity yields again single line performance but also a loss in power of 30 - 40 %. Since the optimum gas mixtures for laser action in the 5 μm and the 7 μm region are totally different at least two different gas mixtures are necessary to cover the whole range of 4.9 to 7.7 μm.

The absorption fingerprint of C_2H_6 has been measured (see fig 3a). It is compared with its IR absorption spectrum [5] (see fig 3b). The four strongest absorption peaks of the fingerprint are assigned (see Table 1). These four peaks all belong to the ν_8 excitation which is a CH_3 deformation mode. With this set-up we have reached a detection limit of approximately 100 ppb for C_2H_6 in pure nitrogen. With our recently constructed intracavity set-up a detection limit of 5 ppb has been accomplished.

Table 1: The assignment described in the first five columns is based on the IR absorption spectrum of Smith [5], which is reproduced in Fig 3b. The quantum number K is a projection of the angular momentum J on the molecular axis for initial state K'' and final state K'. The next four columns characterize the CO-laser transitions. The two last columns show the absorption coefficients of C_2H_6 and approximate detection limit for the corresponding transitions.
**) Peak nr. 79 is largely detuned which renders its assignment doubtful.*

C_2H_6 infrared spectrum					CO laser spectrum					
peak nr.	assignment	K''	K'	freq. (cm^{-1})	transition $v''\to v'$	$J''\to J'$	freq. (cm^{-1})	laserp. (mW)	α C_2H_6 ($atm^{-1}cm^{-1}$)	det.lim. ppb
79	ν_8	3	2	1457.91*)	27→26	6→7	1458.37	100	0.46	300
82	ν_8	0	1	1472.95	26→25	9→10	1472.91	250	0.59	100
89	ν_8	3	4	1493.68	25→24	10→11	1493.81	220	0.64	100
93	ν_8	7	8	1514.65	24→23	11→12	1514.68	400	0.40	100

[1] C.A. Riely, G. Cohen and M. Lieberman, Science 183 (1974) 208
[2] T.X. Lin, W. Rohrbeck and W. Urban, Appl. Phys. B. 26 (1981) 73
[3] F.J.M. Harren, F.G.C. Bijnen, J. Reuss, L.A.C.J. Voesenek and C.W.P.M. Blom, Appl. Phys. B. 50 (1990) 137
[4] S. Bernegger and M.W. Sigrist, Infrared Phys. 30 (1990) 375
[5] Lincoln G. Smith, J.Chem.Phys. 17,2 (1949) 139

Photothermal Detection of Trace Chemicals by Fiber-Optic Interferometric Probe

G. Vegetti[1], *M. Martinelli*[1], *L. Balconi*[2], and *F. Sigon*[2]

[1]CISE Tecnologie Innovative, P.O. Box 12081, I-20134 Milano, Italy
[2]ENEL, Italian National Electricity Power Authority, Thermal and Nuclear Research Center, Via Monfalcone 15, I-20132 Milano, Italy

Abstract. Direct measure of the refractive index change induced by photothermal effect, by an advanced extrinsic fiber-optic interferometric system is used to achieve trace level detection in liquid samples.

1. Introduction

Water chemistry is of primary concern in the conduction of large power generating plants: trace concentrations of pollutants in different points of the water-steam cycle should be monitored in order to minimize corrosion effects. On the other hand, fiber-optic interferometric sensors (FOIS) are ideally suited to be remotely operated in harsh environments. Unfortunately, chemical detection was not between the large number of possible measurements performed by this class of devices. We show that by means of sensitive interferometric systems, ruggedized to operate in hostile environments, chemical parameters enter the range of FOIS measurands. Since the early work by Stone [1], the direct detection of photothermally induced phase shifts demonstrated its usefulness in the spectroscopic assay of low-loss liquids, but suffered the environmental constraints imposed by the interferometric setup. We adopted an interferometric system which was designed to operate in harsh conditions [2], to build a fiber optic device suitable for field operation.

2. Experimental

The basic blocks of the experimental setup are shown in Fig.1. The 488 nm line of an Ar laser is launched in the fiber-optic probe to act as exciting beam. As the chemical analyte was selected an aqueous solution of Ethidium Bromide, EB, a well known dye widely used in biochemistry studies, which presents a molar extinction coefficient of 5.9×10^{3} $M^{-1}cm^{-1}$ at 480nm. The sensing beam constitutes an arm

Springer Series in Optical Sciences, Vol. 69 Photoacoustic and Photothermal Phenomena III
Editor: D. Bićanić

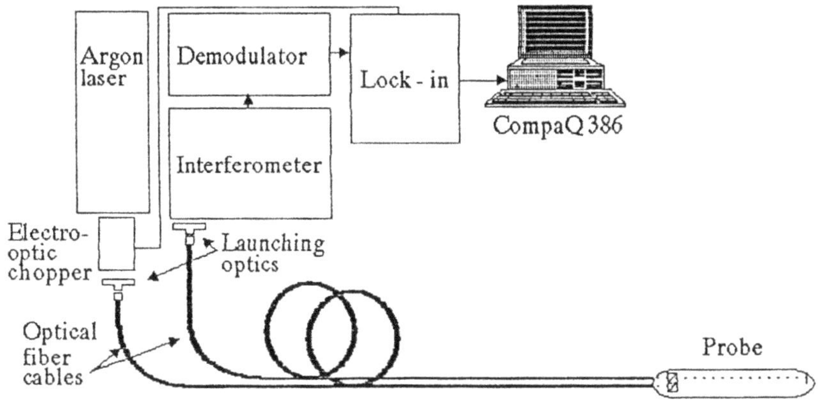

Fig. 1. Block diagram of the experimental setup.

of a remote Michelson-type fiber-optic interferometer [2] and is set collinear to exciting beam in order to achieve the maximum sensitivity. The exciting and sensing beam are integrated in the same micro-optic probe which enters the measurement area. The exciting beam is modulated at 4 KHz frequency by a chopper which also gives the reference to the lock-in amplifier for the synchronous phase detection. The Ar laser power was set at 1.5 W, whereas the He-Ne laser beam from the interferometer had a power below 0.5 mW.

3. Results and discussion

Different samples of dissolved EB, in the range 10 - 10000 μg/l (2.5×10^{-8}M - 2.5×10^{-5}M) have been prepared and the results are summarized in Fig.2, where net phase shifts, calculated as difference between actual sample phase shift and that from a water blank are plotted against dye concentration. A linear relationship appears evident. The phase-noise floor guarantees a sensitivity better than 10 ppb, which is on the other hand very close to the concentration value at which sample absorption equals that from the solvent. This first result confirms the feasibility of the approach and points out the significative prospects of integration of photothermal methods with the fiber-optic interferometric sensors family. Many improvements in the interferometric system are possible, which will results in an increase in sensitivity of this chemical sensor, even if blank absorption already emerges as the limiting value. Moreover, the chopping frequency, which was maintained in the kHz range to assess the feasibility of the measure assuming the presence of low frequency noise, can be lowered with a few minor changes to electronics, to obtain larger phase shifts without increasing exciting source power.

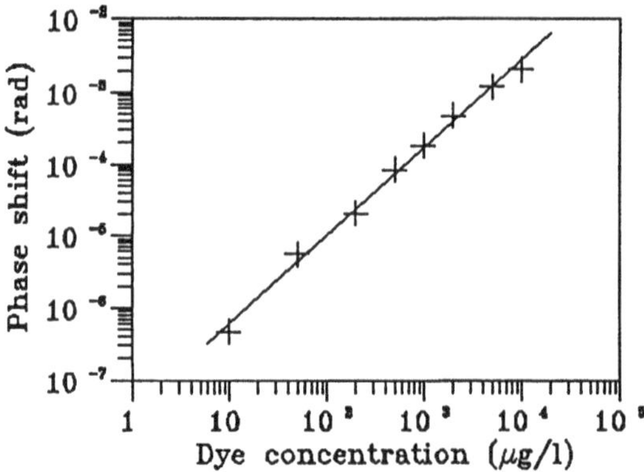

Fig. 2. Net phase shifts, measured with the interferometric probe, plotted against dye concentration for different standard dilution samples. The value measured with a solvent blank (water) was subtracted from actual samples data.

Aknowledgement

This work has been supported by the Research and Development Direction of ENEL in the framework of the CRTN-MOC contract. The Authors acknowledge CRTN Direction which authorized the present publication.

References

1. J. Stone, J. Opt. Soc. Am. **62**, 327 (1972)
2. M. Martinelli and N. C. Pistoni, Proc. SPIE **1101**, 30 (1988)

Fiber Optic Modified Laser Induced Photoacoustic Spectroscopy for the Detection of Organic Pollutants in Solutions

K. Adelhelm, W. Faubel, and H.J. Ache

Kernforschungszentrum Karlsruhe GmbH, Institut für Radiochemie, Postfach 3640, W-7500 Karlsruhe, Fed. Rep. of Germany

Abstract. The application of laser induced photoacoustic spectroscopy (LIPAS) to the in situ detection of pesticides and polycyclic aromatic compounds in liquid samples has been studied. The UV excitation laser beam (λ = 253 amd 376 nm) was transmitted to the sample cuvette through a multimode optical fiber of 600 µm diameter.

1. Introduction

The in situ monitoring of traces of organic pollutants is a very important subject in quality assurance of ground and drinking waters, but the appropriate analytical tools, which allow the detection limits required by law to be attained, are not yet available. Two main groups of environmentally relevant substances are the pesticides and the polycyclic aromatic hydrocarbons PAHs.

Laser Induced Photoacoustic Spectroscopy (LIPAS) has been proved in several works as a very sensitive analytical method of determination of special solutions, for example in nuclear waste disposal and for the speciation and detection of actinides and lanthanides in aqueous media /1,2/. In this work, we examined the application of LIPAS as analytical in line method for the determination of different organic pollutants in solution using optical fibers. Dinitrophenol herbicides show none or only little fluorescence yields and therefore they were the first samples examined. In addition, various PAHs were selected for photoacoustic measurements; they are of great importance, because a considerable number of them possesses properties like carcinogenicity and mutagenicity.

2. Experimental

The photoacoustic measurements of the substances listed were carried out with our slightly modified LIPAS apparatus described in /3/. Briefly: a dye-laser pumped by an excimer-laser generates 10 ns short laser pulses in the tuning range between 340 and 980 nm (with frequency doubling up to 240 nm). The liquid sample contained in a stainless steel cell is either excited directly by the laser pulse or guided by a 600 µm optical fiber of 2 m length (Fig. 1); the sample is in acoustic contact with a piezoelectric transducer, which detects the pressure wave resulting from the photoacoustic effect /4/. The output signal is fed, after preamplification, into a personal computer controlled boxcar integrator for further signal processing. Before collecting photoacoustic data, each sample was recorded by a Cary 2400 UV/VIS spectral photometer in order to find out the adequate absorption lines.

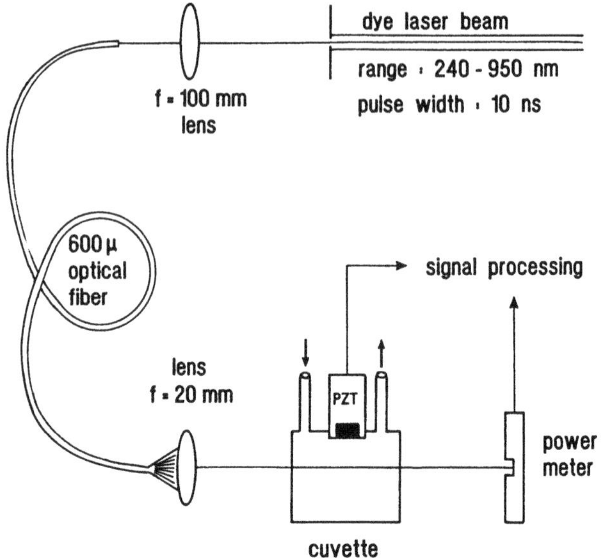

Fig. 1 Scheme of the fiber optically modified photoacoustic cell assembly.

3. Results

As a representative example of the group of dinitrophenol pesticides, the LIPAS signal of 2,4-dinitrophenol versus concentration is plotted in Figure 2. Obviously, the experimental data show a linear correlation between the intensity of the photoacoustic signal S_{PA} and the concentration which has been expected from theory /5/. The detection limit is found to reach the lower ppb level. Similar results were obtained for other herbicides.

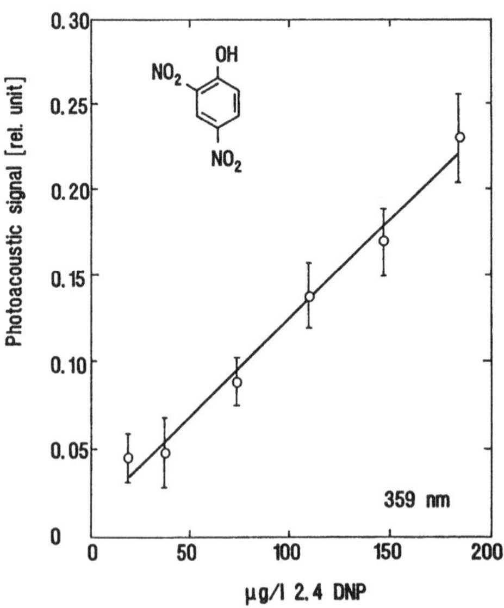

Fig. 2 Calibration curve of 2,4-dinitrophenol in H_2O.

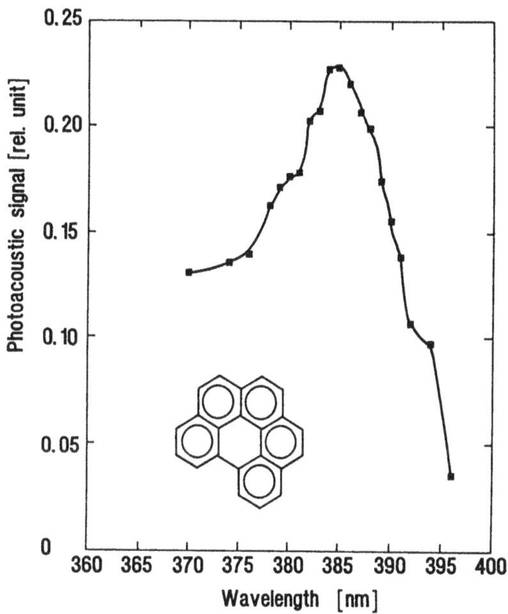

Fig. 3 LIPAS spectrum of benzo/ghi/perylene in hexane.

A difficulty in measuring other groups of pesticides and PAHs arises from their typical absorption maxima which occur in the UV-range, because for conventional laser induced excitation sources it is problematic to reach these wavelengths. We solved the problem by installing an anisotropic β-barium-borate crystal for frequency doubling in the dye laser excitation system. Through the generation of second harmonic waves as a result of non-linear optical phenomena we can reach excitation wavelengths up to 240 nm. So we are able to carry out photoacoustic measurements even in the UV-range and to collect photoacoustic data at two different wavelengths. For the selected PAHs like anthracene, photoacoustic calibration curves are plotted at two excitation wavelengths (376 and 253 nm with frequency doubling).

First photoacoustic spectra of the polycyclic aromates, anthracene and benzo/ghi/perylene were obtained. Figure 3 depicts the PAS spectrum of benzo/ghi/perylene in the range between 360 and 400 nm excitation wavelength. Contrary to the dinitrophenol pesticides, which are characterized by large absorption bands, this peak is rather small. Thus, it can be selected between different substances by collecting and processing photoacoustic spectra.

A further important step towards increasing the selectivity of LIPAS for various organic pollutants has been performed by photoacoustic measurements at several characteristic absorption wavelengths. As an example, calibration curves of the strongest absorption lines (253 and 376 nm) of anthracene are shown in Figure 4. The laser power at 376 nm was 2 mJ, whereas for the 253 nm wavelength, due to frequency doubling merely 100 µJ laser power was available. But because of the linearity of the LIPAS signal to the excitation laser power /3/ a significant increase in sensitivity should be possible with the latest generation of dye laser systems.

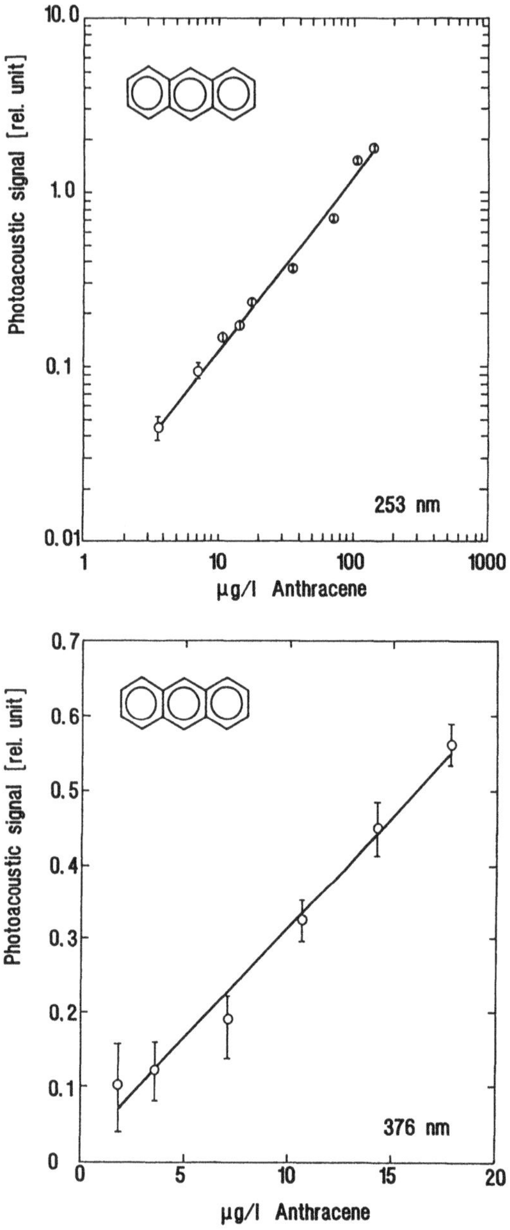

Fig. 4 Calibration curves of anthracene in hexane at 253 nm and 376 nm excitation wavelengths.

4. Conclusion

Laser induced photoacoustic spectroscopy is a promising tool for fiber optically in situ measurements which are performed to monitor continuously traces of pesticides and other organic pollutants in solutions for quality control of ground and drinking waters. But it is still necessary to improve the selectivity of the method and the performance of in situ measurements in flowing liquid media.

References

/1/ T. Sawada, S. Oda, H. Shimizu and H. Kamada, Anal. Chem. **51**, 688 (1979).

/2/ P.M. Pollard, M. Liezers, J.W. Mc Millan, G. Phillips, H.P. Thomason and F.T. Ewart, Radiochim. Acta, **44/45**, 95 (1988).

/3/ K. Adelhelm, W. Faubel, H.J. Ache, Fres. J. Anal. Chem., **338**, 259 (1990).

/4/ C.K.N. Patel, A.C. Tam, Rev. Mod. Phys., **53**, 523 (1981).

/5/ E.T. Nelson, C.K.N. Patel, Opt. Lett. **6**, 354 (1981).

Remote Photothermal Deflection Spectroscopy for Trace Analysis of Pesticides in Aqueous Solutions

B. Bohnert, W. Faubel, and H.J. Ache

Kernforschungszentrum Karlsruhe GmbH, Institut für Radiochemie, Postfach 3640, W-7500 Karlsruhe, Fed. Rep. of Germany

The transverse photothermal deflection spectroscopy (PTDS) has been developed as a trace method to determine the concentration of pesticides in water. The detection limit obtained for 2-methyl-4,6-dinitrophenol or 4,6-dinitro-ortho-cresol (DNOC) is below 400 ppt.

1. Introduction

Changes in the german regulations for the determination of pollutants in drinking water require new trace methods with lower detection limits. The transverse photothermal deflection spectroscopy meets this need. Organic pesticides which absorb in the UV region can be detected without preconcentration. Remote control is possible using optical fibres.

2. Experimental

Figure 1 shows the experimental setup for the transverse PTDS. The excitation-laser beam (UV-Ar^+, λ = 363.8 nm) was modulated by a chopper operating at 2 Hertz. It passes a 1 x 1 cm cuvette containing a solution of analyte in bidistilled water and excites the dissolved pesticides. The deexcitation is nonradiative and produces heat. Around the excitation laser beam the solution warms up periodically with the modulation frequency and changes its density and refractive index. The change of the refractive index is measured by the deflection of the He-Ne-probe-laser beam. The probe-laser beam is transmitted through a monomode optical fibre to the cuvette and crosses the excitation-laser beam at a right angle. The periodically deflection of the probe-laser beam is measured at a certain distance from the cuvette with a two dimensional position sensitive device. The deflection signal is amplified and detected by a lock-in amplifier. The whole experiment and the data processing are fully controlled by a personal computer.

Springer Series in Optical Sciences, Vol. 69 Photoacoustic and Photothermal Phenomena III
Editor: D. Bićanić

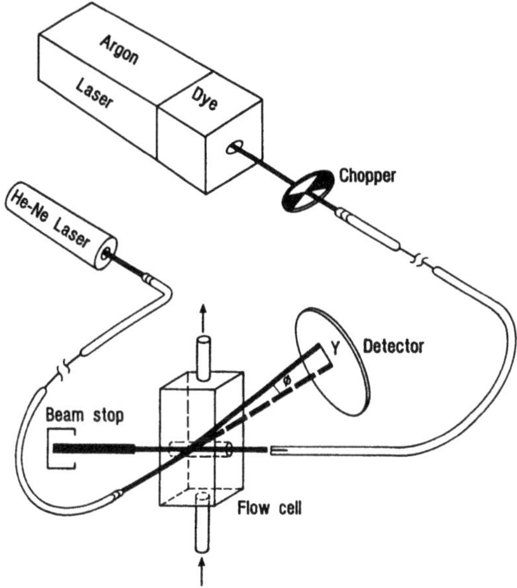

Fig. 1: Scheme of the remote controlled transverse photothermal deflection spectroscopy using optical fibers.

3. Results

Linear calibration curves are obtained for the pesticides, 2,4-dinitrophenol, 2-methyl-4,6-dinitrophenol (DNOC), 2-(sec-butyl)4,6-dinitrophenol (Dionseb) and 2-(sec-butyl)4,6-dinitrophenolacetate (Dinosebacetate), in the range of 10^{-4} up to 10^{-9} mol/l by measuring the samples successively diluted with distilled water. Figure 2 shows, as an

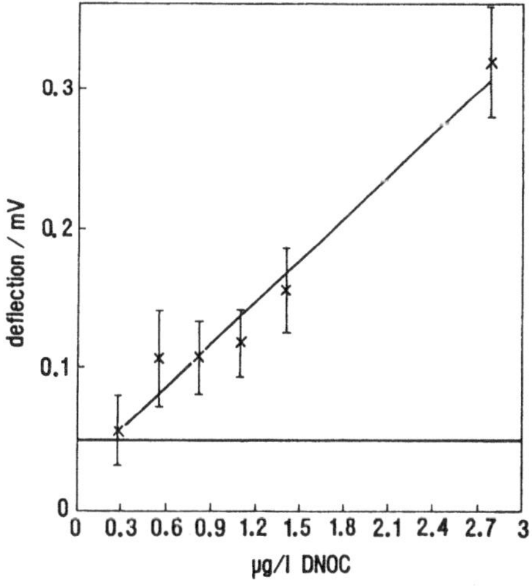

Fig. 2: Calibration curve for 2-methyl-4,6-dinitrophenol (DNOC) in bidistilled water. Error bars indicate the standard deviation.

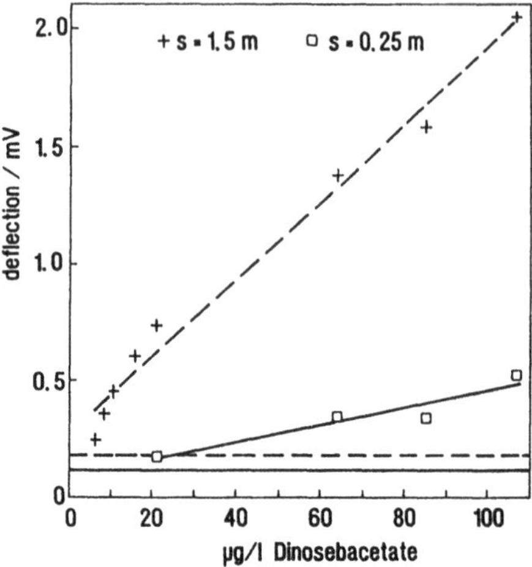

Fig. 3: Calibration curves for 2-(sec-butyl)4,6-dinitrophenolacetate (dinosebacetate) standard solutions in water for 1.5 m and 0.25 m path length of deflection.

example of all analytes used, the lower section of the calibration curve of 2-methyl-4,6-dinitrophenol (DNOC) in bidistilled water. The calibration curve is not background corrected. The limit of detection (S/N=2) is 400 pg/ml (400 ppt). For the other pesticides limits of detection have been obtained in the lower ppb range. One of the main obstacles to lower detection limits is due to the absorption of the solvent. The experiments have been carried out in distilled water, which is a optically unfavorable solvent in the IR and VIS region because of its relative high absorption [1]. The absorptivity of bidistilled water in the UV region - where our studies have been performed- is very small. In this case the limits of detection are mainly affected by the pointing stabilities of the lasers, by mechanical vibrations of the systems and by draught. For in situ measurements the He-Ne-probe-laser beam has been directed through a monomode optical fibre of 5 μm in diameter and 2 m in length. Remote control requires another optical fibre for the excitation-laser beam. These experiments are under development.

Furthermore, the dependence of the deflection signal on the path length of the deflection has been studied for distances of 1.5 m and 0.25 m between the cell and the position-sensitive detector. A decrease of the sensitivity by a factor of 4 was obtained when the distance was decreased to 0.25 m. Figure 3 shows this result for dinosebacetate in bidistilled water.

4. Conclusion

The transverse photothermal deflection spectroscopy has been proved to be a valuable ultra trace analytical method for direct monitoring of dinitrophenol pesticides in water. For future remote measurements, the argon-excitation-laser- and He-Ne-probe-laser beam will be transmitted via multimode and/or monomode optical fibres, respectively.

References

[1] A.C. Tam and C.K.N. Patel, Appl. Optics **18**, 3348 (1979).

Laser Optoacoustic Probing of Biopolymer Solutions

S.V. Egerev, O.M. Zozulya, A.E. Pashin, and O.V. Puchenkov

N.N. Andreev Acoustics Institute, Academy of Sciences,
Shvernika St. 4, 117036 Moscow, Russia

Abstract. Theoretical and experimental results concerning laser optoacoustic probing of biopolymer lipopolysaccharides (LPS) solutions of low concentration are presented. Efficient technique of LPS identification based upon investigation of OA-signal spectrum distortion is proposed.

Biopolymer solutions analysis is now a matter of great interest particularly due to the urgent need of accurate testing of polymer materials used in medicine. The presence of some types of macromolecules, i.e., bacterium membranes fragments - lipopolysaccharides(LPS), in polymers extracts is the evidence of pyrogenic activity of the material and therefore of the danger for human health. The optoacoustic(OA)-diagnostics seems to be very suitable technique for detection of ultra low LPS concentrations in weakly absorbing solutions.

In the report efficient OA-technique of LPS detection and identification is proposed. The technique is based on principles of laser pulsed dynamic OA-diagnostics described in [1]. It differs from well known OA-spectroscopy scheme developed early for small absorptions measurements(see, for example, [2]). Theoretical study of OA-effect in weakly absorbing nonhomogeneous liquids provides an opportunity to make use of extra informative features of OA-signal for solution properties exploration.

Three additional mechanisms of sound generation contribute to the acoustic response of the cell: 1 - thermal expansion of laser light heated nonhomogeneities; 2 - heat exchange between particles and liquid ; 3 - absorption of scattered light in the walls of OA-cell. The relative effectiveness of these processes essentially depends on the impurities concentration and the wavelength of laser radiation. As far as the LPS-water solutions are concerned the effect of scattering dominates in the visible region whereas the LPS fragments extra absorption contributes in the near ultraviolet region. Adequate registration of these effects has been fulfilled with the aid of specially designed OA-cell, with sensitive element being PZT-ceramics cylinder. The transfer function of PZT cylinder included two resonant peaks, the first (hoop) $f_L=c_z/\pi D$ (c_z - longitudinal sound velocity in ceramics, D-inner diameter), the second (thickness mode) $f_H=c_z/2l$ (l- thickness of the wall). To provide reliable isolation from ambient noise, the cylinder is wrapped with acoustic insulation and placed in a steel enclosure. The inner surface of the cylinder is coated by the thin polyurethane layer. Quartz cuvette with solution under study is inserted into the cylinder with immersion liquid to provide an acoustic contact. This design is expected to be optimum for transparent liquids investigations.

Springer Series in Optical Sciences, Vol. 69 Photoacoustic and Photothermal Phenomena III
Editor: D. Bićanić

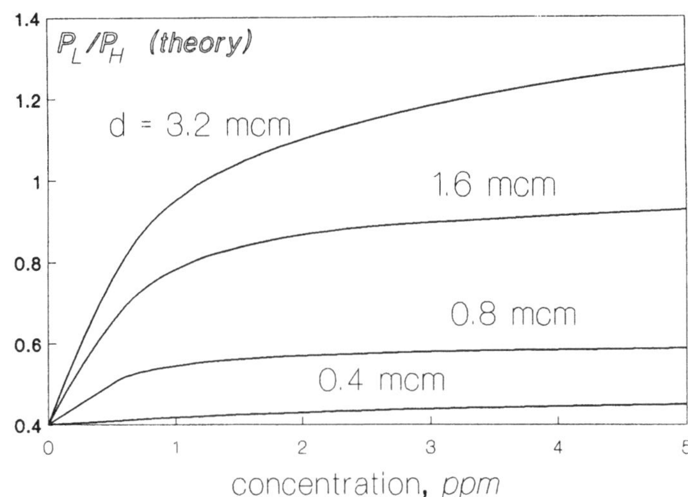

Fig.1.Ratio P_L/P_H vs concentration for various values of particle diameter (polysterene/water system).

It may be shown that for strongly absorbing particles spectrum appears to be product of well-known transfer function of cylindrical OA-array $\psi(r,\omega)$ and function $f(\omega)$ which takes into account heat exchange, adiabatic expansion of absorbing liquid and adiabatic expansion of absorbing particles (separately those effects were considered in [3],[4]):

$$P_\omega = \psi(r,\omega)f(\omega),$$

where $f(\omega)= 1 + (\mu/\xi)\cdot v\cdot \mathbf{H}(d\sqrt{\omega/2\chi})$, μ - ratio of particle and liquid absorbancies ($\mu \gg 1$), v- volume concentration, $\mathbf{H}$ - decreasing function of thermal diffusivity χ and particle diameter d, $\xi = [(\beta/\rho C_p)_{particle}/(\beta/\rho C_p)_{liquid}]$, with β-coefficient of volume expansion , ρ- density, C_p - heat capacity. After multiplying P_ω by OA-cell transfer function one obtains response spectrum with low frequency resonance (P_L) and high frequency resonance (P_H) dominating.For optically homogeneous solutions(i.e. in the case of soluble impurities) the ratio of P_L/P_H is a constant not depending on impurities concentration.Instead,in the presence of strongly absorbing particles P_L/P_H ratio varies with particles concentration v (see fig.1) and OA-cell response spectrum distortion occurs: enlargement of low frequency components due to relatively slow heat release processes.This feature may be used in diagnostics - provided known probed particles diameter their concentration would be found.

The quality of OA-experiments is characterized by the detection limit of optical energy absorbed in the cell and is equal in our case to 0.2 μJ (after 50 events averaging).Acoustic signal is amplified and then recorded by digital oscilloscope Averaging and subsequent processing of the signal is performed with the IBM-PC/AT computer.The pulsed-dye laser provides pulses at 5 pulses/sec of 30-nsec duration,pulse energy 2 mJ in the ranges 275-305 nm and 540-600 nm.

Distilled water

LPS-probe (50 ppm)

Signal spectrum

200 kHz

P_H

P_L

80 kHz

900 kHz

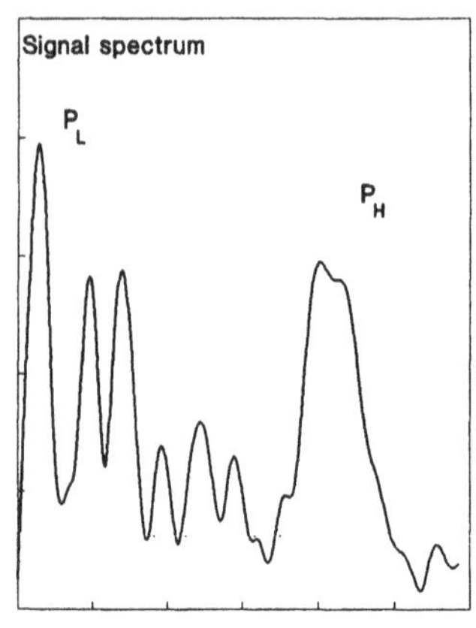

Fig.2.OA-signal profiles and spectrum components for distilled water and LPS-solution.

The test medium was prepared by dissolving dried preparation of *Escherichia coli* in hot (80^oC) double distilled deionized water with subsequent cooling. Suitable care was exercised in handling glassware and cells. For comparison, commercial medicine "*Pyrogenal*" with similar properties was tested.

There was observed slight enlargement of low frequency component during visible light probing.The value obtained was used to estimate scattered light contribution. UV probing exhibited distinct changes in OA-spectrum caused by LPS concentration changes with predicted tendencies (fig 2).Laser probing of optically homogeneous solutions (fucsin, for example) exhibited no changes of P_L/P_H as compared to distilled water response.So, spectrum changes were used for quantitative identification of LPS.The lower limit of detection (0.001-0.01 ppm) was found to be one or two orders smaller than the limit of identification. Saturation of the ratio mentioned was observed at v >0.1 ppm.

The linearity relationship between full signal amplitude and v holds only when no changes in P_L/P_H with concentration are observed. Some discrepancy with Beer's law during spectrophotometry studies are usually attributed to configurational changes that accompany dilution at extremely low concentrations of LPS. New OA-data reported may help to distinguish absorption and heat exchange processes.

REFERENCES

1.S.V.Egerev,L.M.Lyamshev and O.V.Puchenkov,Sov.Phys.Usp.,**33**, 739 (1990).

2.A.C.Tam and C.K.N.Patel, Appl.Optics,**18**,3348 (1979).

3.T.Kitamory,M.Fujii,T.Sawada and Y.Gohshi,J.Appl.Phys.**58**,1456 (1985).

4.G.J.Diebold and P.J.Westervelt,J.Acous.Soc.Am.,**84**,2246(1988).

Photoacoustic Spectroscopy of Proteins in the Visible: Human Serum Albumin

A. Figari[1], *L. Sabbioneda*[2], *and A. Zanella*[2]

[1]CISE Tecnologie Innovative, Via Reggio Emilia 39, I-20090 Segrate (Milano), Italy

[2]Centro Trasfusionale e di Immunologia dei Trapianti, Ospedale Maggiore Policlinico, Via F. Sforza 35, I-20122 Milano, Italy

ABSTRACT. Photoacoustic detection in the visible of solutions of human serum albumin is reported, down to 14.5 μM, and the molar absorptivity is given in the 360—650 nm range.

A photoacoustic (PA) technique allowed us to make the first absorption measurement of a colourless protein in the visible in 1985 /1,2,3/: this protein was spectrin. We suggested that this absorption was due to a kind of Urbach tail /4/ in a biological material, to be explained by the Davydov theory /5/ of the long-wavelength edge of the fundamental optical absorption: photons could be absorbed with energies smaller than the optical width of the forbidden band since the energy deficit $E-h\nu$ was covered by simultaneous absorption of phonons.

Unfortunately, our work passed unnoticed. However, a recent, remarkable paper by Braslavsky and coworkers /6/ has aroused new interest on this subject.

In this communication, we report on PA detection in the visible of solutions of human serum albumin (HSA). A dilution series of a 10% solution (from Sigma, 0.85% sodium chloride and 0.05% sodium azide, pH 6.7) was prepared. All the experiments were carried out at room temperature, with a cw source modulated by a chopper. The PA cell consisted in a steel tube surrounded by a piezoelectric ceramic tube /3/ and the PA signal was detected by a lock-in amplifier.

Both Figs. 1 and 2 show the PA signal (amplitude) produced by an Argon laser: $\lambda = 514.5$ nm, chopper frequency = 25 Hz.

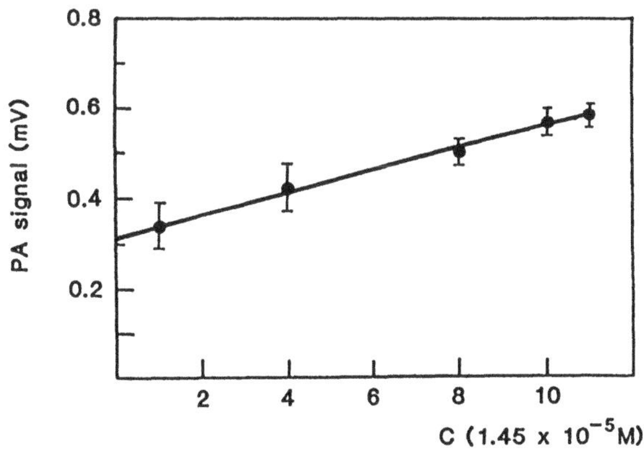

Fig.1: PA signal versus HSA concentration (laser power = 300 mW)

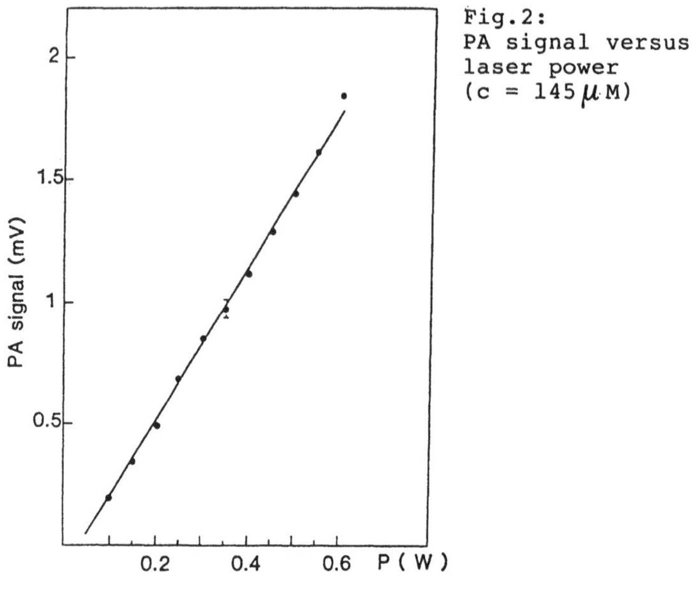

Fig.2:
PA signal versus
laser power
(c = 145 μM)

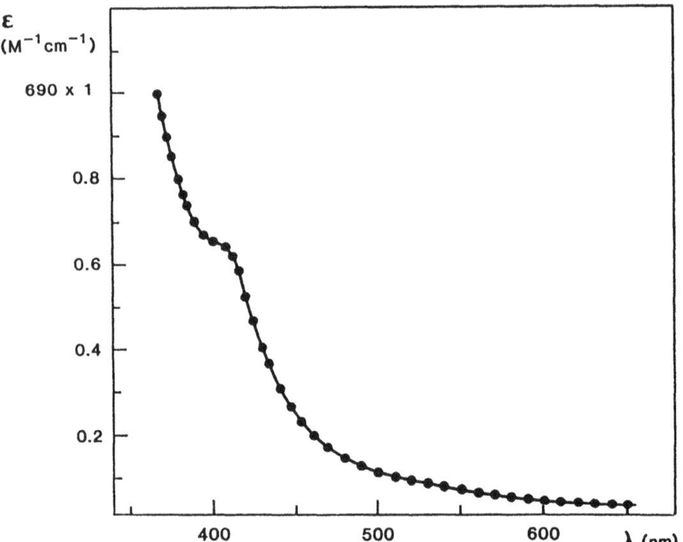

Fig.3: Molar absorptivity of HSA solution

Fig.1 shows the linear dependence of the signal on the HSA concentration c, down to 14.5 μM (laser power = 300 mW): the y—intercept gives the buffer PA signal.
Fig.2 shows the linear dependence of the signal on the laser power (c=145 μM)
Fig.3 represents the molar absorptivity in the 360—650 nm range. The source was 1kW Xenon lamp with monochromator: spectral bandwidth 2 nm. The measurements were made with

concentrations in the range from 145 μM to 1.45 mM; the coefficient of variation was 5%.
The phenomenon as a whole could be explained again as a tail of the UV absorption; another explanation could be provided by absorbing chromophores to be defined.

REFERENCES

1) A. Figari, A. Beraldo and A. Zanella, 4th Int. Topical Meeting on Photoacoustic, Thermal and Related Science. Technical Digest, Montreal (1985), postdeadline poster ThA13

2) A. Figari, A. Martinelli, A. Beraldo, M. Mariani and A. Zanella, 8th Meeting of the International Society of Haematology, Warsaw – Poland (1985), paper 286

3) A. Figari, A. Beraldo, A. Martinelli, D. Volpes and A. Zanella, Physics in environmental and biomedical research, S. Onori and E. Tabet editors, World Scientific Publishing (1986), 197

4) V.L. Bonch-Bruevich, Sov. Phys. Usp. 26 (8), 664 (1983)

5) A.S. Davydov, Phys. Stat. Sol. 27, 51 (1968)

6) O.A. Roveri, G.M. Bilmes, K. Heihoff and S.E. Braslavsky, Appl. Spectr. 44(10), (1990), 1706

7) Z.A. Yasa, W.B. Jackson and N.M. Amer, Appl. Opt. 21(1), (1982)

Photoacoustic Measurements of C_2H_4 Production and Entrapment in Plants: A Comparison with Gas-Chromatographic Results

F.J.M. Harren[1], *J. Reuss*[1], *L.A.C.J. Voesenek*[2], *R.H. Thier*[2], *and C.W.P.M. Blom*[2]

[1] Molecular and Laser Physics, and [2]Experimental Botany, University of Nijmegen, Toernooiveld, 6525 ED Nijmegen, The Netherlands

Introduction

We compare the superior performance of the photoacoustic detection method to gaschromatography in measuring the C_2H_4 production of and entrapment in *Rumex* plants during submergence.

C_2H_4 is a well-known plant hormone. Already in 1935 [1] it was shown that many plant materials produce ethylene. C_2H_4 can e.g. stimulate the ripening of fruit, the wilting of flowers or the abscission of leaves.

In our experiment we measured the C_2H_4 production and C_2H_4 release of *Rumex palustris* plants. These plants grow in a zone near rivers, where they are flooded at irregular intervals when the water level in the river rises. They respond to this inundation by an enhanced growth of the petioles, a process triggered by an increased C_2H_4 production [2].

We have investigated plants under waterlogged and submerged conditions. The amount of entrapped C_2H_4 inside the plant tissue that had been submerged for 24 hours has been measured using the photoacoustic method. A large discrepancy was found with measurements using the gaschromatographic method.

Experimental set-up

As radiation source for our photoacoustic measurements we use an infrared CO_2 waveguide laser (9-11 μm wavelength). A small resonant acoustic cell is placed inside the laser cavity between the discharge tube and the output mirror [2]. Due to the small overall volume (35 cm^3) it can be used in low flow regimes (1 l/hour) and still have a fast response time. The total set-up has an absorption sensitivity of $1.8 \cdot 10^{-10}$ cm^{-1}. For the strongest vibrational absorption of C_2H_4 on the 10P14 CO_2 laser line (949.479 cm^{-1}) we can reach a detection limit of 6 ppt (1 ppt=1:10^{12}).

To measure the C_2H_4 production of plants the photoacoustic cell is connected to a flow-through system. Air from the plant cuvettes is stripped of CO_2 by a KOH-based CO_2 scrubber to prevent interference effects of CO_2 absorption on the C_2H_4 signal.

Biological applications

In figure 1 the C_2H_4 production of a single *R. palustris* plant is compared to that of a control plant. After an acclimatization period of 18 hours (With both production levels on 3-4 nanoliter per gram dry weight per hour (nl./g DW. h.) the plant is completely inundated (W) for a 24 hours period. In this period first a decrease

Springer Series in Optical Sciences, Vol. 69 Photoacoustic and Photothermal Phenomena III
Editor: D. Bićanić

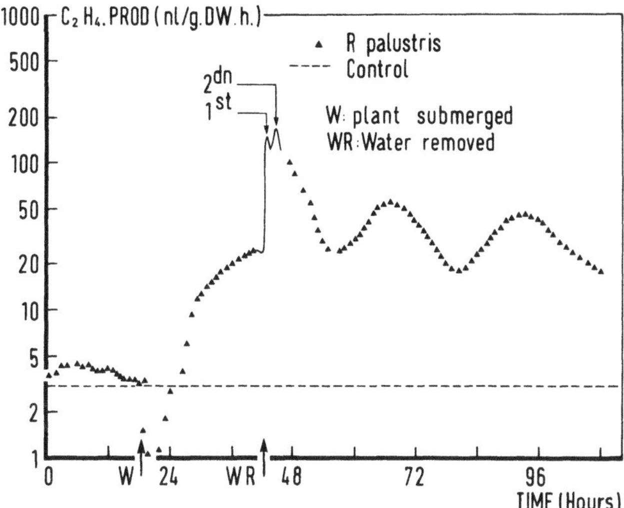

Figure 1: The C_2H_4 production of a *Rumex palustris* plant (26-30 days old) is measured during submergence (from W) and waterlogging (from WR) compared to that of a control plant. Clearly one sees a first C_2H_4 peak from the entrapped C_2H_4 and a second from the conversion from ACC. (Data are so close together that a line is drawn).

is measured of the C_2H_4 concentration in the air above the water level caused by the slow diffusion rate of C_2H_4 in water (10^4 x slower than in air). Afterwards the measured C_2H_4 level rises again due to the stress induced production of the plant.

When the waterlevel is lowered from submergence to waterlogging a fast first peak of C_2H_4 occurred within 1 hour, followed by a second, gradual one with a maximum after 3-4 hours. The first peak is attributed to the entrapped C_2H_4 inside the plant tissue during submergence. Ethylene is released immediately after lowering of the water level. The width of this peak is given by the response time of the system (a combination of flow velocity through the cuvettes and the volume of the cuvettes, tube, photoacoustic cell etc.). The second peak is connected to the onset of ACC conversion (1-aminocyclopropane-1-carboxylic acid), yielding C_2H_4. During the inundation period ACC probably accumulates inside the root tissue, but cannot be converted into C_2H_4 since this conversion needs oxygen . After this second peak the C_2H_4 production remained higher for several days. Inspite of the continuous light and temperature regime a diurnal rhythm in the ethylene production is observed.

From the integrated first peak we calculate the total amount of C_2H_4 entrapped: for this specific plant 12 ppm. or 12 nanoliter per milliliter internal air volume. (The average internal air volume amounts to 0.35 ml.) This concentration is in contradiction to the results of gaschromatographic measurements which yield an average concentration of 0.5 ppm C_2H_4 inside the internal air volume of the plant[3]. Because concentrations of exogenous C_2H_4 become effective on the growth of petioles at a level of 5ppm [3] the gaschromatographic result appear to be doubtful.

The discrepancy between the two measurements can be explained by the difference in plant treatment. Before the C_2H_4 gas sample is injected into the gaschro-

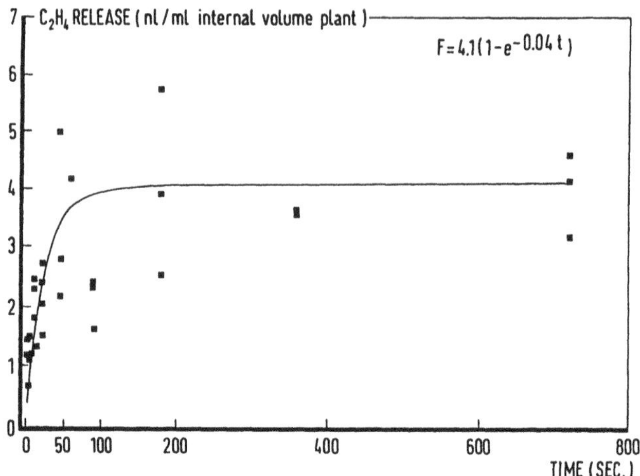

Figure 2: The C_2H_4 release of *R. palustris* plants is plotted against the exposure time to the atmosphere after a 24 hour period of submergence. The amount of C_2H_4 is normalized for its production rate and its internal volume. C_2H_4 release = 4.1(1-e$^{-0.04t}$) nanoliter/plant.

matograph the plant is removed from the water, cut into pieces and put into an ammonium sulphate solution from where the internal air volume of the plant is collected by vacuum extraction[3]. During the treatment (which can take a few minutes) a large amount of C_2H_4 can escape from the plant.

In a separate photoacoustic experiment we have investigated the C_2H_4 release rate, lifting the plant above the water surface during a limited period of time. (fig 2.) and afterwards submerging it again. Just before the lifting of the plant the ethylene production rate (nl / plant h.) was determined. The ratio of the mean C_2H_4 production rate (n=36) divided by each individual production rate was used to correct the ethylene release after de-submergence. High ethylene production rates will lead to high internal ethylene concentrations during submergence and thus to high release peaks after de-submergence. The ratio corrects for inter plant variation in ethylene production under water. The air exposure time varied from 2 till 360 seconds. From the concentrations in the air flow we calculated the total amount of C_2H_4 that was released during this period; corrected for the production ratio it is plotted in figure 2. From this figure we obtained a mean internal concentration of 4.1 ± 0.3 ppm C_2H_4 in the plant tissue just before de-submergence. The $^1/e$ release time is 25 seconds. This demonstrates that, unless the plants are handled very carefully the photoacoustic method is far more suitable for reliable measurements of C_2H_4 production and release than the gaschromatographic method.

[1] W. Crocker, A.E. Hitchcock and P.W. Zimmermann, Contrib. Boyce Thompson Inst. 4 (1935) 177-218

[2] F.J.M. Harren, F.G.C. Bijnen, J. Reuss, L.A.C.J. Voesenek and C.W.P.M. Blom, Appl. Phys. B 50 (1990) 137-144

[3] L.A.C.J. Voesenek and C.W.P.M. Blom, Plant, Cell and Environment 12 (1989) 433-439

Interpretation of the Photoacoustic Effect on Leaves by Evolution and Transport of Heat and Oxygen

P. Korpiun[1] *and S. Malkin*[2]

[1]Physik-Department E13, Technische Universität München, James-Franck-Strasse, W-8046 Garching, Fed. Rep. of Germany
[2]Biochemistry Department, The Weizmann Institute of Science, Rehovot 76100, Israel

A model of the photoacoustic effect observed on intact leaves was developed several years ago by Malkin, Cahen and their coworkers /1,2/. Following these authors, the PA signal created by the absorption of light in the chlorophyll of a leaf consists of two components. The first one comes from the heat accompanying the radiationless transitions, the second one from the photosynthetically evolved oxygen. The latter is commonly known as photobaric component of the PA signal.

In the present work we make an attempt to quantitatively relate the PA signal to the diffusive transport of heat and oxygen from the chloroplast across the mesophyll to the intercellular gas (Fig. 1). The heat wave changes the enthalpy of the intercellular gas in the leaf and of the transducer gas surrounding the leaf by the amount Q(t). The periodically evolved oxygen entering the intercellular gas varies its total molar concentration by $\psi(t)$. The total pressure variation p(t) detected by a microphone is then

$$p(t) = \gamma\left[\frac{Q(t)P_o}{\rho\, c_p T_o} + RT_o\psi(t)\right] \tag{1}$$

where ρ and c_p are the density and the isobaric specific heat capacity of the intercellular gas (air), respectively, P_o is the mean pressure, T_o the mean temperature of the gas, and $\gamma = c_p/c_v$ (c_v: isochoric specific heat capacity of the gas).

The heat wave generated by radiationless transitions in the chlorophyll is described by solutions of the heat conduction equation in the entire mesophyll cell including the vacuole. For frequencies > 10 Hz this cell with a diameter > 30 μm can be assumed to be thermally thick. One finds that the behavior of enthalpy Q(t) with time is shifted in phase angle by 90^0 with respect to the modulated light intensity $J_o \cos \omega t$, i.e. $Q(t) \propto J_o \cos(\omega t - \pi/2)$.

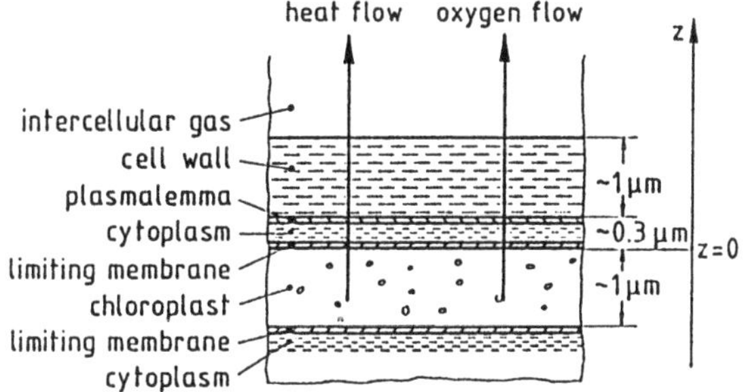

Fig. 1: Simplified model of mesophyll cell used to calculate O_2 concentration waves

Oxygen is evolved after a sequence of reactions. We assume for simplicity that these are first order reactions characterized by a rate constant k /1,3/. The transport of O_2 in the chloroplast, the cytoplasm, the cell wall, and the gas is described by one–dimensional diffusion equations. The diffusion equation for the concentration C_1 in the chloroplasts with diffusion coefficient D_1 in z–direction is

$$D_1 \frac{\partial^2 C_1}{\partial z^2} - \frac{dC_1}{dt} = - \frac{\phi J_0 \beta k \exp(\beta z)}{2N (k^2+\omega^2)^{1/2}} \cos\left[\omega t - tg^{-1}(\omega/k) - \pi/2\right]. \tag{2}$$

The right hand side of this equation is the source term, where β is the optical absorption coefficient, N the density of photochemical sources and ϕ the quantum efficiency. The diffusion equation in the cytoplasm underneath (i=0) and above (i=2) the chloroplast, the cell wall (i = 3) and the gas (i = 4) is $D_i \partial^2 C_i / \partial z^2 - \partial C_i / \partial t = 0$ respectively. The time dependent concentration distributions in the regions i = 0 to 4 are superpositions of concentration waves $C_i^{\pm} = A_i^{\pm} \cos(\omega t \pm z/\kappa_i)$ running in +z and –z direction with amplitudes $A_i^{\pm}$. The frequency dependent quantity $\kappa_i = \sqrt{D_i/\pi f}$ is the oxygen diffusion length in the region labelled by i. The amplitudes $A_i^{\pm}$ are determined from the boundary conditions for the concentration and the oxygen fluxes at the different $z = l_i$ also considering the solubility or the partition coefficients of oxygen in the various regions of the mesophyll cell. Traversing the chloroplast, the cytoplasm and the cell wall with thicknesses l_i, the phase of the concentration wave is shifted back by the approximate amount of $\sim l_i/\kappa_i$. The thickness of the limiting membrane and the plasmalemma (Fig. 1) is much smaller than the concentration diffusion lengths κ_i. Thus the influence of these membranes on the oxygen concentration wave is taken into account as permeabilities /4/. They reduce the amplitude of the wave and also shift its phase angle. The phase angle of the contribution of the O_2 concentration variation $\psi(t) \propto J_0 \cos(\omega t + \varphi - \pi/2)$ to the pressure oscillation p(t) in eq. (1) is finally shifted by the amount

$$\varphi = - \left[tg^{-1}(\omega/k) + \sum_{i=1}^{3} \varphi_i(\omega) \right] \tag{3}$$

relative to the enthalpy oscillation Q(t). It can be shown that the phase angle of the concentration wave in contrast to that of the heat wave strongly depends on the modulation frequency. This was found experimentally several years ago by Poulet et al. /2/. Their data for the difference in the phase angle between the thermal and photobaric component of the signal measured on a tobacco leaf are plotted in Fig. 2. In this figure are also shown data measured by Geel and Snel /5/ on a spinach leaf.

We have calculated the thermal and photobaric pressure variation in the intercellular gas assuming that the solubility of oxygen in the cell wall is 0.026 $cm^3(O_2\text{–gas})/cm^3$ (H_2O) /6/ and that the diffusion coefficients D_i and the partition coefficients have the same value in the chloroplasts, the cytoplasm and the cell wall. Some results for the phase shift φ of the photobaric component relative to the enthalpy oscillation are shown in Fig. 2. The data of the thicknesses of the chloroplasts, the diffusion coefficient, the permeabilities of the membranes and the rate constant used to calculate the plots in Fig. 2 are listed in Table 1. Because those data are not exactly known for the leaves investigated by Poulet et al. /2/ and Geel and Snel /5/ we selected reasonable values from literature /1,2,3,6,7/ in order to fit the experimental data.

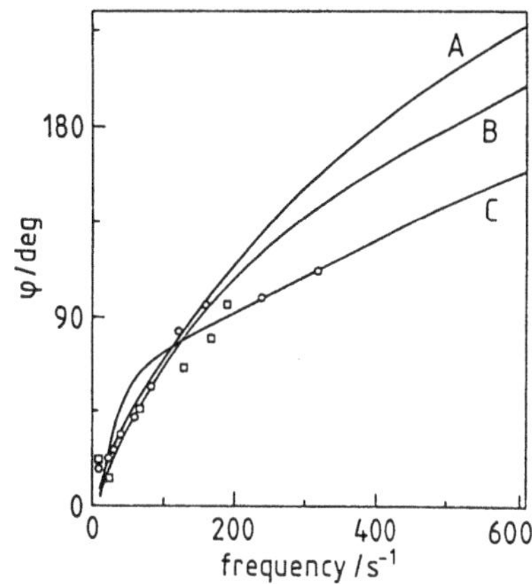

Fig. 2:
Phase angle difference φ between thermal and photobaric part of the PA signal. Solid line: calculated; circles: expt./2/, squares: expt./5/.

Table 1: Chloroplast parameters

Plot	Thickness of Chloroplasts (μm)	Permeabilities (cm/s) limiting membrane	Plasmalemma	D(cm^2/s)	k(1/s)
A	1	1	0.1	$3.7 \cdot 10^{-6}$	$2 \cdot 10^3$
B	2	0.1	0.1	$3.7 \cdot 10^{-6}$	$3 \cdot 10^5$
C	1	0.15	0.1	$2.5 \cdot 10^{-5}$	700

In conclusion our calculations confirm the experimental observations that the frequency dependent phase angle of the PA signal can be related to the evolution of oxygen and its transport across the mesophyll cell. The shape of the plot of the phase angle versus modulation frequency may appreciably vary, since it depends on the exact structure of the cell and numerous transport coefficients.

/1/ G. Bults, B.A. Horwitz, S. Malkin, D. Cahen, Biochim.Biophys.Acta 679, 452 (1982)
/2/ P. Poulet, D. Cahen, S. Malkin, Biochim.Biophys.Acta 724, 433 (1983)
/3/ P. Joliot, M. Hofnung, R. Chabaud, J.Chim.Phys. 63, 1423 (1966)
/4/ R. Osiander, B. Helm, P. Korpiun in Photoacoustic and Photothermal Phenomena II, Springer Series in Optical Sciences; J.C. Murphy, J.W. Maclachlan–Spicer, L.C. Aamondt, and B.S.H. Royce eds., Springer Heidelberg, New York 1990, p.107
/5/ C. Geel, J.F.H. Snel (unpublished data) (1990)
/6/ Landolt–Börnstein, Zahlenwerte und Funktionen, Vol.2, Part 2b, K. Schäfer, E. Lax eds. (Springer, Berlin, Göttingen, Heidelberg, 1962) p. 1–127
/7/ P.S. Nobel: Introduction to Biophysical Plant Physiology (W.H. Freeman and Co., San Francisco, 1973)

Photoacoustically Monitored Energy Dissipation and Xanthophyll Cycle Compared in Two Durum Wheat Varieties Exposed to a High-Light Treatment

M. Eyletters and R. Lannoye

Université Libre de Bruxelles, Laboratoire de Physiologie Végétale, Av. Paul Héger, 28 CP169, B-1050 Bruxelles, Belgium

In vivo thermal energy dissipation was photoacoustically monitored in two durum wheat varieties before and after a high-light treatment. Concomitant with the conversion of carotenoid violaxanthin into zeaxanthin, an increase in the heat emission signal was observed in the light-stressed leaves. This increase was more marked in the variety sensitive to drought conditions.
When the xanthophyll cycle was blocked by dithiothreitol, the photothermal signal did not increase indicating a relationship between these 2 phenomena.

Prolonged exposure of leaves to a high-light treatment induces an inhibition of leaf photosynthesis. This process is called "photoinhibition". It essentially acts on photosystem II [1]. The excess of light energy induces a structural change in the D1 protein of the reaction center of PSII. The D1 protein becomes inactive for the reactions of the electron transport system [2].
It has been suggested that photoinhibition is usually accompanied by an increase in the rate constant of thermal deactivation of PSII antenna-chlorophyll.
Photoacoustic studies now comfirm this idea. Demmig and al. [3] have tried to link this phenomenon with the functionning of the xanthophyll cycle (the light induced enzymatic de-epoxidation of violaxanthin in zeaxanthin). They concluded that the function of zeaxanthin is to act as an alternative quencher competing with fluorescence emission and photochemistry for excitation energy, thus leading to increased heat emission.

In this study, we have used two varieties of durum wheat: one sensitive and the other resistant to high temperature, water stress and high light. We have compared the response of these varieties in order to show genotypic differences.

Springer Series in Optical Sciences, Vol. 69 Photoacoustic and Photothermal Phenomena III
Editor: D. Bićanić

Table 1: Effects of photoinhibitory treatment (1H at 2300 $\mu mol.m^{-2}.s^{-1}$) on the relative amplitude of the photothermal (Apt) and O_2 evolution (Aox) components of the *in vivo* photoacoustic signal of two durum wheat varieties (sensitive and resistant varieties) infiltrated or not with buffer containing 2 mM DTT. C, P are respectively the control and the photoinhibited plants. % indicates data expressed in % of the control Apt, Aox and PL before photoinhibition.

A) Without infiltration with DTT

	Aox		Apt	
	sensitive	resistant	sensitive	resistant
C	41.5	43.1	42.1	34.2
P	7.1	5.15	65.7	39.6
%	-83	-88	+56	+16

B) Infiltration buffer + DTT

	Aox		Apt	
	sensitive	resistant	sensitive	resistant
C	37.05	37.07	45	46.7
P	5	6.33	42	43.3
%	-86	-83	-6.6	-8

A high-light treatment of 2300 $\mu mol.m^{-2}.s^{-1}$ was applied during 1 hour. A drastic reduction of the photosynthetic oxygen evolution component (Aox) of the *in vivo* photoacoustic signal was observed (table 1), indicating an alteration of the photosynthetic electron transport system.

In table 1A, it can been seen that, after a high-light treatment, heat emission substantially increased. This increase was more marked in the sensitive variety than in the resistant one.

As shown in the table 1B, no rise in the heat emission was observed when the xanthophyll cycle is blocked by dithiotreitol (DTT).

Conclusion: High-light treatment inhibits the photosynthetic electron transport system and favors thermal deactivation of excited pigments. A relationship exists between these phenomena.

Acknowledgements: M. Eyletters acknowledges the receipt of a research grant from IRSIA (n°890667)

References:

[1] Krause GH., 1988 Physiol. Plant.74: 566-574
[2] Ohad I., Kyle D.J. and Hirschberg J., 1985 EMBO J. 4:1655-1659
[3] Demmig B., Winter K., Krüger A. and Czygan F. Plant Physiol. 1987, 84: 218-224

Deconvolution of Photobaric and Photothermal Signals from Spinach Leaves

J.F.H. Snel, M.W. Polm, W.F. Buurmeijer, and W.J. Vredenberg

Department of Plant Physiological Research,
Wageningen Agricultural University, Gen. Foulkesweg 72,
6703 BW Wageningen, The Netherlands

Abstract. The kinetics of the photobaric and the photothermal components of the photoacoustic signal of spinach leaf-discs were measured simultaneously at two frequencies in one single experiment. This method avoids some of the problems associated with conventional single-frequency modulation, *e.g.* biological variability and repeatability. This communication shows that at modulation frequencies from 35 to 222 Hz the phase of the photobaric component remains constant during photosynthetic induction. This indicates that the photobaric component of the PA signal can be treated as a single component under the conditions used here.

Abbreviations. LED - light emitting diode; LIA - lock-in amplifier; PA - photoacoustic.

1. Introduction

At low modulation frequency the PA signal of plant leaves consists of a photobaric and a photothermal component [1]. The photobaric component is related to changes in gas concentrations, mainly oxygen, associated with the photosynthetic process. The photobaric process is relatively slow ($t_{½}$≈10ms), which causes an additional damping at high modulation frequencies. The phase and the amplitude of the photothermal and the photobaric component can be determined by a procedure first described in [2] and briefly summarized here. The photobaric component is eliminated by saturating photosynthesis with high-intensity non-modulated light. Saturating light, however, increases the thermal losses in photosynthesis. This means that the amplitude of the photothermal component has to be measured separately. This can be done at a frequency of 500 Hz or higher, where the photobaric component is virtually absent and the PA signal therefore is mainly of photothermal origin [2]. Knowing the phase and the amplitude of the photothermal component, the phase and the amplitude of the photobaric component can be determined.

In steady-state [*e.g.*, 2] and during dark-light transients [3,4] the magnitude and the phase of the photothermal and the photobaric components have been determined by extrapolation of the magnitude of the photothermal component measured at high modulation frequency in a separate experiment. For the measurement of the kinetics of the photobaric and the photothermal components the kinetics of the PA signal have to be measured at two frequencies with the same material under identical conditions, which is quite problematic due to biological variability and sample history [3,4]. This problem can be avoided by using dual-frequency modulation to measure the photothermal signal at high frequency and the photobaric signal at low frequency simultaneously in one experiment.

The results show that the phase of the photobaric component of the PA signal is frequency dependent as in [2]. At a given frequency, however, the phase remains constant during illumination at the conditions used in our experiments.

2. Methods

Leaf-discs of spinach (*Spinacia oleracea* cv. Amsterdams breedblad) plants were grown as described before [4]. Leaf-discs (Ø 12 mm) were cut at least 60 min. before the start of the experiment and kept in darkness between moist filter paper. All experiments were carried out at room temperature (20-25 °C).

The setup and methods were essentially as in [4] with the exception that a new PA-cell (Fig.1) was designed to achieve a higher signal to noise ratio. The microphones (M40, Microtel) are used in the current source mode. The currents of the individual microphones are summed across a single resistor to provide a single voltage output. Instead of the light-source used before [4], nine LED's driven by a transconductance amplifier [5], were placed above the upper window of the PA cell to provide modulated PA measuring light (660 nm, 20 nm half bandwidth). The modulation signal applied to the transconductance amplifier resulted in modulated illumination of the form $I(t)=30(1+0.5\sin 2\pi f_1 t+0.5\sin 2\pi f_2 t)$ $W.m^{-2}$. The high-frequency signal was measured as in [4] using the magnitude output of the 2-phase LIA (EG&G/PAR 5204) and the PM 5217 function generator as the modulation source for f_1. The low-frequency modulation signal was obtained using a second 2-phase LIA (EG&G 5210) with its internal oscillator as the modulation signal for f_2. Saturating non-modulated light pulses were given as described [4]. The duration was 500 ms and the repetition rate 0.1 s^{-1}. Data-acquisition and process control was performed by the program PHOTACOS (author W.F. Buurmeijer) running on a PC equipped with an intelligent interface (CED 1401, Cambridge Electronic Design Ltd.). The phase and the amplitude of the photobaric and the photothermal signal were measured as in [2].

At the start of the experiment the phase of the EG&G 5210 LIA was set to the phase of the photothermal component. After a dark adaptation period of 300 s the modulation voltage was applied to the transconductance amplifier by means of an electronic switch iniating the dark-to-light transition and after another 1 s the saturating pulse train was started.

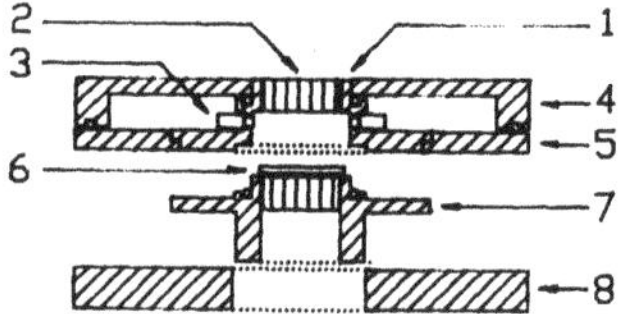

Figure 1: Schematic drawing of the photoacoustic cell. The cell is made of an aluminium body (1) with a cylindrical measuring chamber (Ø 12 mm) bounded by a perspex window (2) which has been pressed into the chamber. Nine miniature electret microphones (3) are mounted radially onto small holes in the wall the wall of the measuring chamber. Leaks in the junction are prevented by a silicon based seal (Xantopren, BAYER). The cell is clamped in a rectangular aluminium housing (4) by a cover (5) to suppress external noise. The leaf disc (6) rests on the sample holder (7). The aluminium sample holder with its perspex window is tightly fixed onto the cell by the base plate (8) which is attached to the housing.

3. Results

Fig.2 shows the kinetics of the phase difference between the photothermal and the photobaric signal measured at three different modulation frequencies using dark adapted spinach leaf-discs. As the phase of the photothermal component remains constant during dark-to-light transients [Snel, unpublished data], any change in the phase of the photobaric component should be reflected in the phase difference between the photothermal and the photobaric component. Such a change is not apparent in Fig.2. This means that the phase of the photobaric signal is constant as well during dark-to-light transitions and that, at least under our conditions, there are no indications for a light induced oxygen uptake

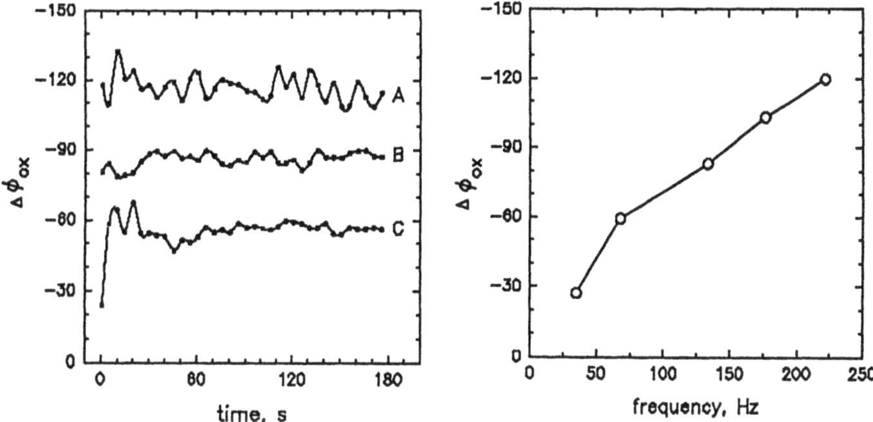

Figure 2 (left): Kinetics of the relative phase of the photobaric component of the PA signal in dark adapted spinach leaf discs. Conditions as described in section 2. Modulation frequency (f_2) was 68 (A), 134 (B) and 222 Hz and high-frequency modulation (f_1) was at 600 Hz (A,B,C). The timeconstant was 60 ms.

Figure 3 (right): Frequency dependence of the steady state relative phase of the photobaric component of the PA signal. Experimental data taken from Fig.2.

resulting in phase changes of the photobaric component as observed by M. Havaux and S. Malkin (unpublished results).

The steady-state phase difference between the photothermal and the photobaric component is shown in Fig.3 as a function of the low-frequency modulation f_2. The phase difference increases from -30 degrees to -117 degrees when the frequency increases from 35 to 222 Hz. This result is in approximate agreement with the results for Tobacco (*Nicotiana tabacum* var. Xanthii) leaves [2]. Above 150 Hz however, the phase of the photobaric component does not seem to increase very much in Tobacco leaves [2] while the phase shown in Fig.3 is still increasing at 222 Hz. This different behavior might be related to different diffusion properties resulting from structural differences in the leaves of the two plant species.

4. Discussion

The dual-frequency modulation technique enables deconvolution of the photothermal and photobaric components of the PA signal from plant leaves. The data presented here show that the phase of the photobaric and the photothermal component do not change significantly during dark-to-light transitions. Light-induced O_2-uptake in leaves has been suggested to occur in the ms range [7,8] and could therefore easily lead to a negative photobaric component. This additional photobaric component is characterized by a delay of 1-2 ms using a pulse-induced PA technique [8] and has also been inferred from phase changes of the photobaric component observed using conventional LIA techniques (M. Havaux and S. Malkin, unpublished results). Apparently such a component is not present under our conditions, since it would have introduced a change of the phase of the photobaric component at the higher f_2 frequencies.

We have carefully avoided to keep the leaf-discs too long in the dark in the PA cell. As the volume of the cell is necessarily small, respiration of the leaf could easily increase the

CO_2 concentration in the cell to a high level which could lower the pH of the chloroplast stroma and consequently lead to a stimulated O_2-uptake in the Mehler reaction [6].

5. References

1 G. Bults, B.A. Horwitz, S. Malkin and D. Cahen: Biochim. Biophys. Acta 679, 452 (1982)
2 P. Poulet, D. Cahen and S. Malkin: Biochim. Biophys. Acta 724, 433 (1983)
3 D. Bicanic, F. Harren, J. Reuss, E. Woltering, J.F.J. Snel, L.A.C.J. Voesenek, B. Zuidberg, H. Jalink, F. Bijnen, C.W.P.M. Blom, H. Sauren, M. Kooijman, L.W.A. van Hove and W.J.M. Tonk: In *Photoacoustic, Photothermal and Photochemical Processes in Gases*, ed. by P. Hess, Springer Ser. Top. Curr. Phys., vol.46, (Springer, Berlin Heidelberg 1989) p. 213
4 J.F.H. Snel, M. Kooijman and W.J. Vredenberg: Photosynth. Res. 25, 259 (1990)
5 J.F.H. Snel: In *Applications of Chlorophyll Fluorescence*, ed. by H.K. Lichtenthaler, (Kluwer Academic Publishers, Dordrecht 1989) p. 93
6 U. Schreiber, H. Reising and C. Neubauer: Z. Naturforsch. in press (1991)
7 S. Malkin: Planta 171, 65 (1987)
8 J. Kolbowski, H. Reising and U. Schreiber: Photosynth. Res. 25, 309 (1990)

In Vivo State Transitions as Followed by Photoacoustically Measured Energy Storage

M. Charland, K. Veeranjaneyulu, D. Charlebois, and R.M. Leblanc

Centre de Recherche en Photobiophysique, Université du Québec à Trois-Rivières, C.P. 500, Trois-Rivières, Québec, Canada G9A 5H7

Abstract. State 1-state 2 transitions in intact sugar maple leaves were monitored by following changes in energy storage (ES) of PSI (ES_{PSI}) and PSII (ES_{PSII}). ES of both photosystems (ES_T) was recorded under saturating background white light, while ES_{PSI} was determined under saturating far-red light. The difference between ES_T and ES_{PSI} represented ES_{PSII}. Adaptation of the leaf to far-red light (light 1, λ > 715 nm) and red light (light 2, 650 nm) led to state 1 where light 2 was distributed mainly to PSII. Exposure of the leaf to state 2 conditions (light 2 alone) reversed the phenomenon. These state transitions needed about 20 min to occur. From our experiments on light to dark transitions, and vice-versa, we suggest that the dark state was state 2. We also studied the effect of NaF, a known inhibitor of state 2 to state 1 transition. Our results support the concept of phosphorylationand migration of light harvesting complexes, as responsible for the light energy distribution between both photosystems in vivo.

1. Introduction

Plants and algae possess a regulatory mechanism which permit the balanced excitation of both photosystems under different light conditions. This mechanism, known as state transitions, insure a maximal efficiency of the photosynthetic apparatus ¦1¦. State transitions were already explained by following the changes in fluorescence emission and photoacoustic O_2 evolution ¦2¦. We report in the present communication a more direct method, based on energy storage measurements, to follow state transitions.

2. Materials and Methods

The experimental set-up used was similar to that previously described ¦3¦. The leaf disk (1.8 cm diameter of sugar maple (Acer saccharum Marsh.)) was driven to state 1 by illumination with both light 1 (λ > 715 nm, 17 W m^{-2}) and light 2 (650 nm, 7 W m^{-2}, 80 Hz) or to state 2 by shining with light 2 for 25 min. We used the vectorial analysis of signal developped by Poulet et al. ¦4¦ to insure that no O_2 was contributing to the signal at 80 Hz, which permit an accurate measurement of ES. The saturation of both photosystems was done by illuminating with a background white light (160 W m^{-2}), which gave ES_T, the energy storage of both photosystems ¦4¦. According to Malkin et al. ¦5¦, a saturating far-red light permits the measurement of ES of PSI (ES_{PSI}). We used a 150 W m^{-2} far-red light provided by the background light (with Schott GG427, OG550, RG645 and RG715 filters) for this purpose. ES_{PSII} represents the difference between ES_T and ES_{PSI} |6|.

3. Results and Discussion

The most critical step in our methodology is to insure that the intensity of far-red light used is sufficient to saturate PSI photochemistry. This was checked and results are presented in figure 1 up. This far-red light was not absorbed by PSII as shown by the fact that any O_2 evolution has not been observed when far-red light is modulated. It is then possible to determine ES_T, ES_{PSI} and ES_{PSII} following adaptation under different light conditions (Figs 1 center and bottom).

The variation of ES_{PSII}/ES_{PSI} is found to be a good indicator of state transitions. The value of this ratio is found to be dependent of the age and the status of the leaf. This ratio changes during state transitions and the transition is fully reversible (Fig. 2). The completion of the state transitions occurs in about 20 minutes.

Under strong white light illumination, the photosynthetic apparatus is in state 1 conditions, as revealed by the constant ES_{PSII}/ES_{PSI} when state 1 light conditions are provided, and a decrease of this ratio under state 2 conditions. This state may be dependent of white light intensity and/or quality since seedlings grown under weak light presented state 2 behavior. During a dark to light transition, the ratio ES_{PSII}/ES_{PSI} is close to the one obtained under state 2 conditions, suggesting that dark state is state 2.

To insure that we were studying state transitions, we vacuum-infiltrated the leaves with NaF, a known inhibitor of phosphatases, enzymes responsible of state 2 to state 1 transition. Table 1 shows that this transition was almost inhibited in NaF-treated leaves.

Table 1

ES_{PSII}/ES_{PSI} following NaCl or NaF infiltration of sugar maple seedlings leaves. (n= 6 samples, mean ± standard deviation)

	state 2	state 1
NaCl (0.125 M)	1.9 ± 0.4	5.8 ± 1.4
NaF (0.125 M)	1.4 ± 0.6	2.2 ± 1.4

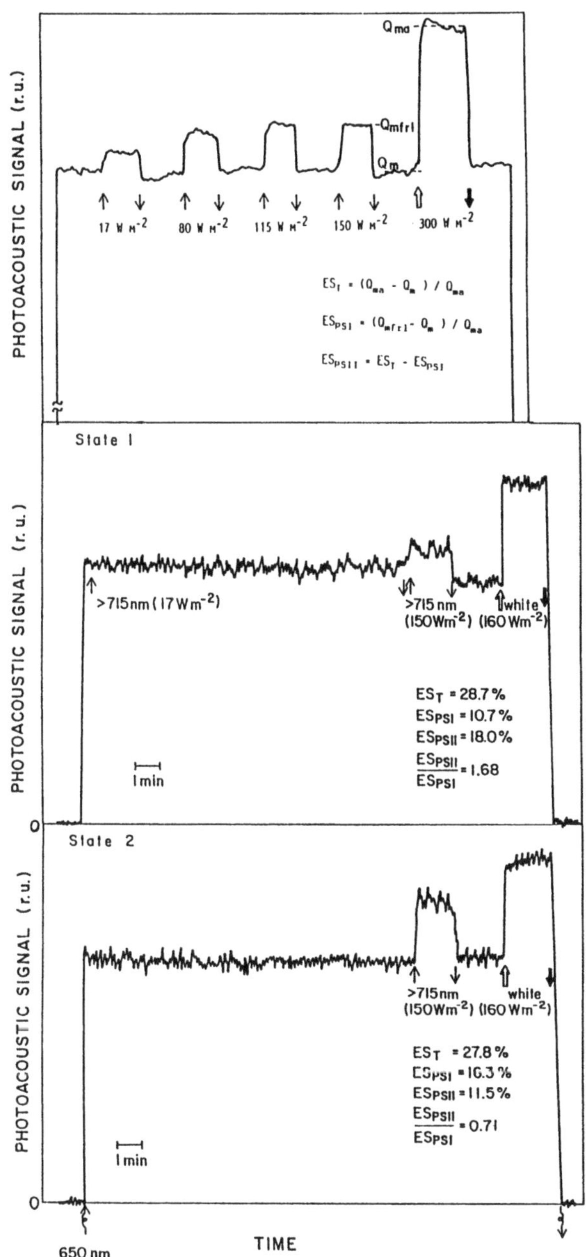

Figure 1: Example of raw data to describe ES_T, ES_{PSI} and ES_{PSII} (up). Photoacoustic signal of a sugar maple sapling leaf in course of time during acclimatation to state 1 (center), and state 2 (bottom).
(⇞,⇟) light 2 on and off, respectively;
(↑,↓) background far-red light on and off, respectively;
(⇧,⬇) background white light on and off, respectively.

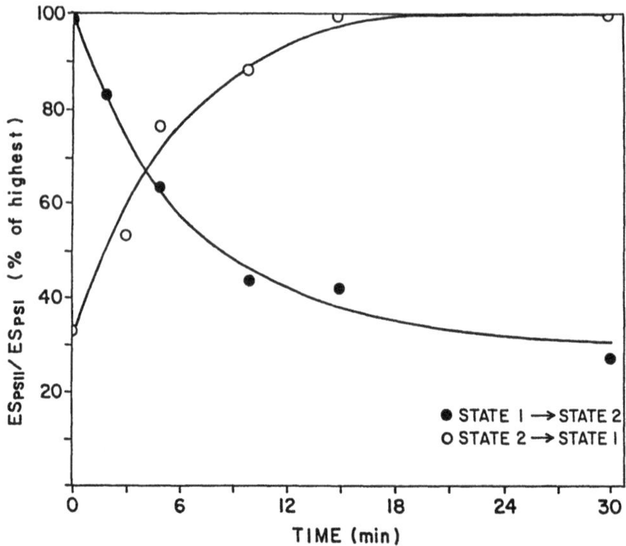

Figure 2: Variation of the ratio ES_{PSII} to ES_{PSI} in course of time during state transitions.

We thank the Natural Sciences and Engineering Research Council of Canada and the Fonds FCAR du Québec for providing grants.

4. References

|1| D.C. Fork and K. Satoh, Annu. Rev. Plant Physiol. **37**, 335 (1986).

|2| O. Canaani, J. Barber and S. Malkin, Proc. Natl Acad. Sci. USA **81**, 1614 (1984).

|3| R. Carpentier, B. LaRue and R.M. Leblanc, Arch. Biochem. Biophys. **222**, 403 (1983).

|4| P. Poulet, D. Cahen and S. Malkin, Biochim. Biophys. Acta **724**, 433 (1983).

|5| S. Malkin, S.K. Herbert and D.C. Fork, Biochim. Biophys. Acta **1016**, 177 (1990).

|6| K. Veeranjaneyulu, M. Charland, D. Charlebois and R.M. Leblanc, Plant Physiol. **97**, 330 (1991).

Laser-Induced Optoacoustic Spectroscopy (LIOAS) of Bacteriorhodopsin (BR): A Mathematical Model and Temperature Studies

M. Rohr, W. Gärtner, and S.E. Braslavsky

Max-Planck-Institut für Strahlenchemie, Stiftstr. 34–36, W-4330 Mülheim/Ruhr, Fed. Rep. of Germany

Excitation of BR with ns pulses induces a photochromic equilibrium between BR and its early intermediate K, produced in less than 5 ps [1] and living ca. 2 μs [2]. During this equilibrium the five species present (BR, BR*, J, K and K*) relax by various processes, delivering heat to the medium. With a LIOAS experiment this heat is integrated over a time which depends on the acoustic transit time of the sound wave, τ_a', travelling through the laser beam cross section [3]. Although the absorption spectra of BR and K strongly overlap [1], the small differences in their absorption coefficients, ε_{BR} and ε_K, permit the determination of the heat evolved in the system as a function of the excitation wavelength, λ_{exc}, and excitation energy, E_{exc}. In addition, fs-LIOAS affords the quantum yield of the primary photoreaction BR → K without the need to consider the spectral properties of the photoproduct and serves as basis for the interpretation of the ns-LIOAS data. A mathematical model describing the kinetic behavior of the species involved (Fig. 1) and the time behavior of the heat evolved was developed, similar to that in ref. [4]
In contrast to the linear energy dependence found for the fs-experiments (not shown), ns-excitation yields a nonlinear dependence on E_{exc} (Fig. 2) due to the establishment of the photoequilibrium BR ⇌ K. In order to describe the system, five differential equations are required:

$$\frac{dN_0}{dt} = -I_{(t)}\sigma_\lambda^{BR}N_0 + k_1N_1 + k_5N_4 \tag{1}$$

$$\frac{dN_1}{dt} = I_{(t)}\sigma_\lambda^{BR}N_0 - (k_1+k_2)N_1 \tag{2}$$

$$\frac{dN_2}{dt} = k_2N_1 - k_3N_2 \tag{3}$$

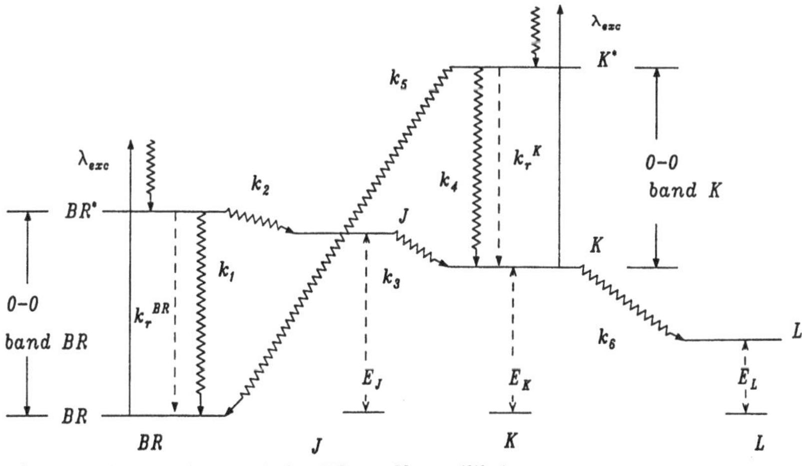

Fig 1 : Kinetic sheme of the BR ⇌ K equilibrium

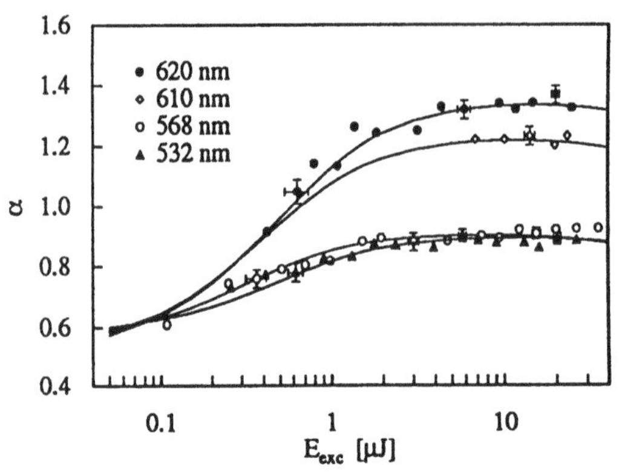

Fig. 2 : Semilogarithmic plot of α *vs* E_{exc}, for various excitation wavelengths

$$\frac{dN_3}{dt} = -I_{(t)}\sigma_\lambda^K N_3 + k_3 N_2 + k_4 N_4 - k_6 N_3 \tag{4}$$

$$\frac{dN_4}{dt} = I_{(t)}\sigma_\lambda^K N_3 - (k_4 + k_5)N_4 \tag{5}$$

(σ_λ^{BR}, σ_λ^K : absorption cross sections of BR and K at the various wavelengths; $I_{(t)}$: photon fluence of the incident laser, the time profile of the laser pulse was measured with an ultrafast photodiode and was approximated by a sum of two Gaußian-functions; $\Phi_f^{BR,K} < 10^{-3} \Rightarrow k_r^{BR} \ll k_1 + k_2$ and $k_r^K \ll k_3 + k_4$). The intermediate J is only included as ground state, due to its very short lifetime. The linear approximation of the Lambert Beer law was used (low absorbances). Relationships between rate constants, lifetimes and quantum yields are given by :

$$\tau_{BR^*} = \frac{1}{k_1 + k_2 + k_r^{BR}} \quad , \quad \tau_{K^*} = \frac{1}{k_4 + k_5 + k_r^K} \tag{6}$$

$$\Phi_{BR\to K} = \frac{k_2}{k_1 + k_2 + k_r^{BR}} \quad , \quad \Phi_{K\to BR} = \frac{k_5}{k_4 + k_5 + k_r^K} \tag{7}$$

The differential heat evolution within the acoustic transit time is given by :

$$\frac{dQ^S}{dt} = N_0 I_{(t)}\sigma_\lambda^{BR}\left(\frac{hc}{\lambda_{exc}} - \frac{hc}{\lambda_{BR}^{00}}\right) + N_1 k_1 \frac{hc}{\lambda_{BR}^{00}} + N_1 k_2\left(\frac{hc}{\lambda_{BR}^{00}} - E_J\right) + N_2 k_3 (E_J - E_K) + N_3 I_{(t)}\sigma_\lambda^K\left(\frac{hc}{\lambda_{exc}} - \frac{hc}{\lambda_K^{00}}\right) + N_3 k_6 (E_K - E_L) + N_4 k_4 \frac{hc}{\lambda_K^{00}} + N_4 k_5\left(\frac{hc}{\lambda_K^{00}} + E_K\right) \tag{8}$$

(Q : integrated heat over τ_a'; λ_{BR}^{00}, λ_K^{00} : wavelengths for the 0-0 bands for BR and K, λ_{BR}^{00} = 656 nm [1], λ_K^{00} = 656 nm [2] were estimated from the crossing of the absorption and emission bands of BR and K.) With a similar differential equation for the calorimetric reference a calculated α results :

$$\alpha_{calc} = \int_0^{\tau_a'} \frac{dQ^S}{dt}\, dt \cdot \left(\int_0^{\tau_a'} \frac{dQ^R}{dt}\, dt \right)^{-1} \tag{9}$$

Values of σ_λ^K were adopted from ref. [5] in order to fit τ_{K^*} and the quantum yield $\Phi_{K\to BR}$ for the photoisomerization K → BR. The value of τ_K and $\Phi_{BR\to K}$ were those determined with LIOAS fs-experiments ($\Phi_{BR\to K} > 0.5$, $\tau_K = 1.5\ \mu s$, oral presentation Rohr et al.), $\tau_{BR^*} = 0.5$ and $\tau_J = 3.0$ ps [1, 6]. Applying a gradient method, the parameters were fitted with a global target analysis to the experimental data (lines in Fig. 2). The best fit was

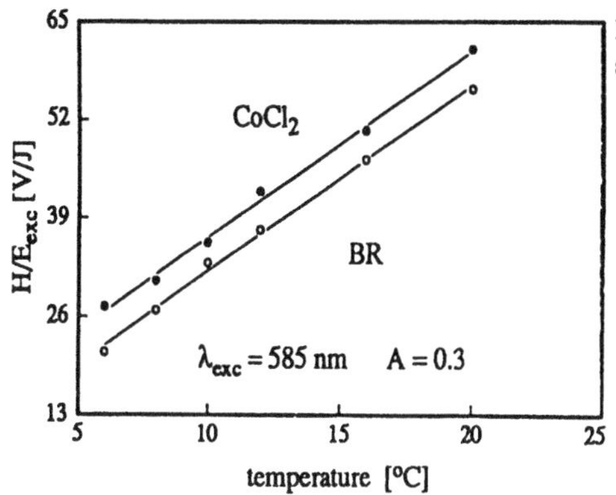

Fig 3 : Temperature dependence of H for sample and reference

obtained for $\Phi_{K \rightarrow BR} = 0.6 \pm 0.05$, $\tau_{K^\bullet} = 4$ - 10 ps, and $\Phi_{BR \rightarrow K} = 0.6$, which requires $E_K - E_L = 30$ kJ/mol.

The procedure of combining fs- and ns-LIOAS experiments is of general applicability to obtain kinetic and thermodynamic parameters for intermediates in cases of relatively rapid establishment of photochromic equilibrfiia.

The optoacoustic signal can be divided into a part due to thermal effects, ΔV_{th}, induced in the sample by the heat pulse, and a part due to possible molecular (perhaps conformational) changes, ΔV_{con}, concomitant with the photochemical reaction [7, 8].

$$H^S = k' \left[\underbrace{\alpha \frac{\beta}{C_p \varrho} (1 - 10^{-A}) E_{exc}}_{\Delta V_{th}:\ thermal} + \underbrace{\frac{1}{C_p} \Delta V_{con} \frac{1}{C_p}}_{molecular\ changes} \right] \tag{10}$$

(C_p : heat capacity, ϱ : density, β : cubic expansion coefficient, k' : instrumental constant). In water, ΔV_{th} is strongly temperature dependent, through the cubic expansion coefficient, β. For dilute solutions the parameters C_p, ϱ and β are those for the solvent, H_2O in our case. This was checked by the linearity of H *vs.* T (Fig. 3).

Using a calorimetric reference (signal only arises from thermal effects), a simple linear dependence on $1/\beta$ results for the ratio H^S/H^R normalised for the absorbed energy, with slope = ΔV_{con} :

$$\frac{H^S/_{(1-10^{-A^S})E^S_{exc}}}{H^R/_{(1-10^{-A^R})E^R_{exc}}} = \alpha + \frac{\Delta V_{con}}{(1 - 10^{-A^S}) E^S_{exc}} \frac{C_p \varrho}{\beta} . \tag{11}$$

The results of temperature dependent studies at various absorbances and various excitation conditions, fs- and ns-pulses (Fig. 4) are summarized in the table.

Parameter	Absorbance	ΔV_{con} 10^{-6} $[\frac{cm^3}{J}]$	ΔV_{con} 10^{-31} $[\frac{m^3}{molecule}]$	$\sqrt[3]{\Delta V_{con}}$ $[pm]$
fs-experiment	0.325	2.87	9.75	100
ns-experiment	0.090	1.66	5.9	84
ns-experiment	0.170	2.15	7.6	91
ns-experiment	0.245	2.98	10.6	102

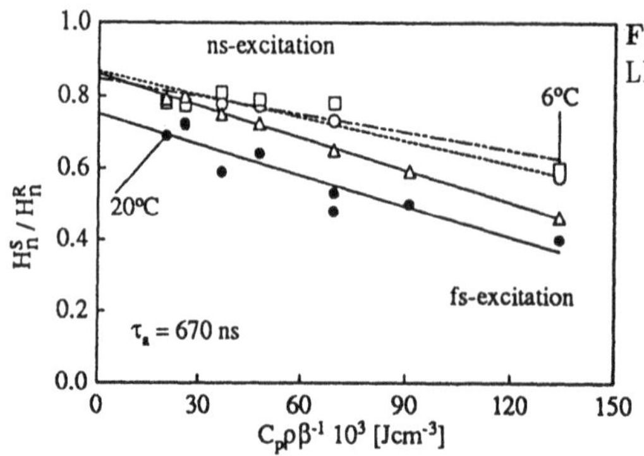

Fig. 4 : Ratio of energy normalized LIOAS signals *vs* $C_p\varrho/\beta$

After fs excitation only the photoisomerization BR $\rightarrow$ K occurs, and no back reaction should contribute to ΔV_{con}. Thus, a different volume change should be expected than that for ns excitation. However, the value $\sqrt[3]{\Delta V_{con}} = 100$ pm is in the same range for both sets of experiments. Since the data for ns-excitation show a dependence on the absorbance, after normalising for the absorbed energy, the reason for this additional dependence should be understood before conclusions are drawn from a comparison of ns and fs data.

Acknowledgments : We are indebted to of Professor K. Schaffner for his support and interest. M.R. is a recipient of a fellowship from the Alfried Krupp von Bohlen und Halbach Foundation. We thank S. Griebenow, D. Lenk, and A. Kessel for their conscientious technical assistance.

References

[1] Polland, H.J.; Franz, M.A.; Zinth, W.; Kaiser, W.; Oesterhelt, D. *Biochim. Biophys. Acta*, **1986**, 851, 407–415.

[2] Atkinson, G.H.; Blanchard, D.; Lemaire, H.; Brack; T.L.; Hayashi, H. *Biophys. J.*, **1989**, 55, 263–274.

[3] Braslavsky, S.E. Heihoff, K. In Scaiano, J.C. , *Handbook of Organic Photochemistry I*, CRC Press Inc. Boca Raton, Florida, **1989**, pp 327–356.

[4] Bilmes, G.M.; Tocho, J.O.; Braslavsky, S.E. *J. Phys. Chem.*, **1988**, 92, 5958–5963.

[5] Atkinson, G.H.; Brack, T.L.; Blanchard, D.; Rumbles, G. *Chem. Phys.*, **1989**, 131, 1–15.

[6] Sharkov, A.V.; Pakulev, A.V.; Chekalin, S.V.; Matveetz, Y.A. *Biochim. Biophys. Acta*, **1985**, 808, 94–102.

[7] Ort, D.R. Parson, W.W. *Biophys.J.*, **1979**, 25, 355–364.

[8] Westrick, J.A.; Goodman, J.L.; Peters, K.S. *Biochemistry*, **1987**, 26, 8313–8318.

Sub-ps and ns Laser-Induced Optoacoustic Spectroscopy (LIOAS) of Bacteriorhodopsin (BR)

M. Rohr, W. Gärtner, G. Schweitzer, A.R. Holzwarth, and S.E. Braslavsky

Max-Planck-Institut für Strahlenchemie, Stiftstr. 34–36, W-4330 Mülheim/Ruhr, Fed. Rep. of Germany

One of the challenging problems in the BR cycle is the accurate determination of the photochemical quantum yield for the reaction of BR yielding its early intermediates, J and K. Due to the strong overlap between the absorption spectra of BR and K [1], it is difficult to obtain this yield through optical methods. Sub-ps LIOAS circumvents this problem by measuring the heat stored after excitation with a laser pulse, affording thus the quantum yield of the species storing the energy for a time longer than the heat integration time of the experiment.

Since K is formed in less than 5 ps [2], a photochromic equilibrium between BR and K is established already after some ps of excitation of the system, with ns pulses. 585 nm pulses of < 500 fs were used in order to avoid this problem and solely excite BR. The minimal repetition rate of the fs-laser system was 1 kHz. In order to reduce this rate a mechanical chopper synchronized at 32 Hz was introduced and allowed the BR cycle to come to completion before the next pulse reached the sample. A lead zirconium titanate detector (PZT) served to detect the signal (Fig. 1).
A linear energy dependence of the optoacoustic signal, H, results for BR (Fig. 2). Comparing this with the slope of the line obtained from a calorimetric reference, $CoCl_2$, which converts promptly all the absorbed energy into heat ($\alpha = 1$), k is eliminated. The fraction of heat delivered promptly by the sample, α, for the particular solvent, absorbance, and geometrical arrangement results from the ratio of the energy normalised amplitudes of the signal for the sample, H^S, and reference, H^R, *i.e.*, $\alpha = H_n^S/H_n^R$ (slopes of Fig. 2) $= 0.71 \pm 0.07$ is obtained. Since a heat integration time, $\tau_a' = 670$ ns (acoustic transit time) was used in this experiment, the optoacoustic signal is in part due to the decay of K to the subsequent intermediate L (lifetime of K-groundstate, $\tau_K \approx$ 1-2 μs [3, 4]). In order to determine this contribution to the prompt heat, the integration time for the heat

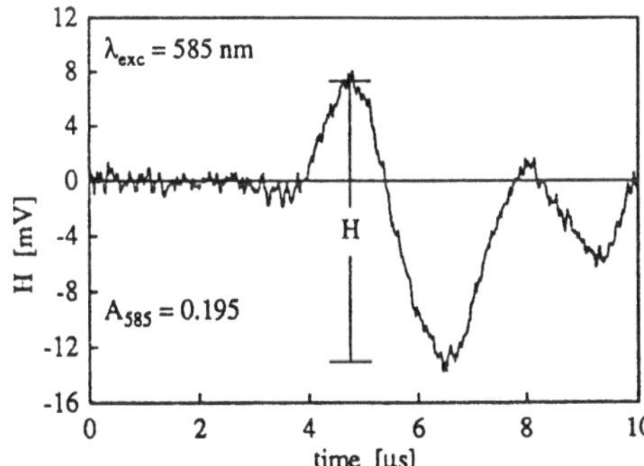

$$H = k \;\; \alpha \;\; E_{exc} \;\; (1 - 10^{-A})$$

k : thermoelastic and geometrical constant, α : prompt heat, E_{exc} : excitation energy, A : absorbance.

Fig.1 : Optoacoustic signal of BR after fs laser excitation (average of 200 shots)

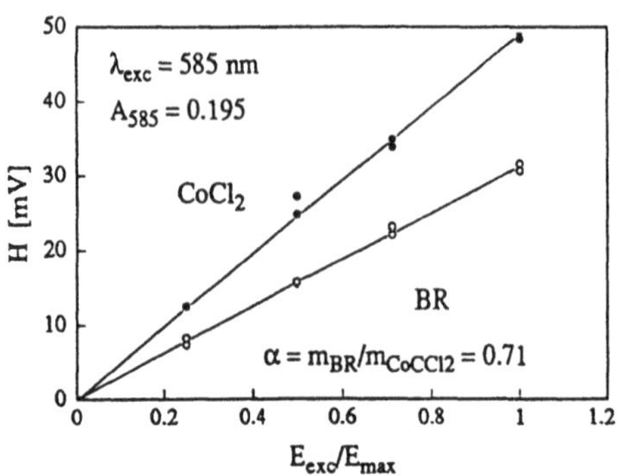

Fig 2 : Energy dependence of H for sample and reference

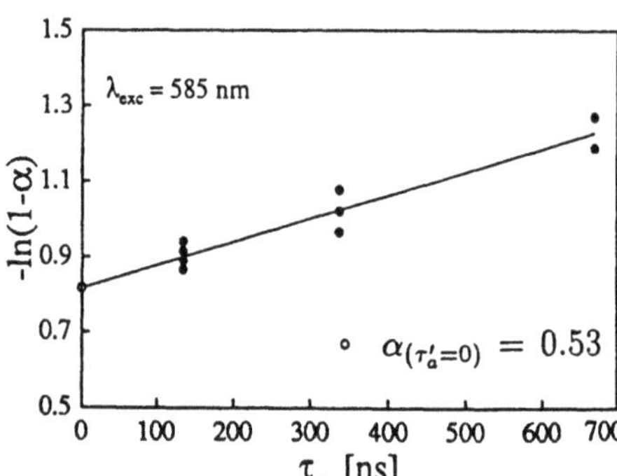

Fig. 3 : Semilogarithmic plot of the heat stored (1-α) by the excited BR, K and L ground states *vs* the acoustic transit time, τ_a'

detection was varied by changing the laser beam diameter [5]. Measurements with beam diameters of 1, 0.5 and 0.2 mm, corresponding to τ_a' between 670 and 135 ns, respectively, are shown in Fig. 3.

Upon excitation, the energy is distributed between the various processes: prompt heat, including the energy released during the forward and backward photoreactions, slow energy release during the thermal conversion from K to the intermediate L, energy storage by L, and fluorescence. Assuming unimolecular reactions the integrated form of the energy balance is given by equation 1 :

$$E_{abs} = \underbrace{\alpha E_{abs}}_{prompt\ heat} + \underbrace{\Phi_{BR\to K}(E_K - E_L)e^{-\frac{\tau_a}{\tau_K}}}_{decay\ K\to L} + \underbrace{\eta_L E_L(1 - e^{-\frac{\tau_a}{\tau_K}})}_{stored\ heat\ in\ L} + \underbrace{\Phi_f E_f}_{fluorescence\ \approx\ 0} \qquad (1)$$

Since with fs-excitation no photochemical repopulation of the BR ground state from the K intermediate needs to be considered, with the assumption of mono-exponential decays, an efficiency $\eta_L = 1$, and the verification of a linear energy dependence ($E_{abs} \Leftrightarrow E_{exc}$, see Fig. 2) the energy balance is reduced to :

$$1 - \alpha = (\Phi_{BR}\frac{(E_K - E_L)}{E_{exc}} - \frac{E_L}{E_{exc}})e^{-\frac{\tau_a}{\tau_K}} + \frac{E_L}{E_{exc}} \qquad (2)$$

Linearity, as shown in Fig. 3, can only be explained from eq. 2 if $E_L/E_{exc} \ll (1-\alpha)$ and thus $E_L \ll E_{exc}$. A simplified energy balance equation results :

$$1 - \alpha = \Phi_{BR\to K}\frac{(E_K - E_L)}{E_{exc}}\ e^{-\frac{\tau_a'}{\tau_K}} \qquad (3)$$

With the alternative approximation of a small energy step K $\rightarrow$ L ($E_K - E_L \approx 0$) a non linear relationship for -ln(1-α) *vs.* τ'_a would result.

In order to withdraw the fraction of α which is due to the decay K $\rightarrow$ L, the plot of the time-resolved measurements was extrapolated to zero time (see Fig. 3) yielding a value for $\alpha_{(\tau'_a=0)} = 0.53$. Since the shortest heat integration time was 135 ns, no contribution of a vibrationally excited ground state of K had to be taken into account.

For the very short transit times the intermediate K is the only energy storing species, and storage by L does not need to be considered. Thus the energy balance equation is further reduced to :

$$1 - \alpha \quad = \quad \Phi_{BR \rightarrow K} \; \frac{E_K}{E_{exc}} \tag{4}$$

Since the energy content of K, E_K, cannot be precisely determined, assuming the photocycle of BR to be exothermic, an upper limit for E_K is evaluated using the 0-0 energy of BR derived from the crossing point of the absorption and emission spectra of BR (628 nm $\Leftrightarrow$ 190 kJ/mol) [2]. The value of $\alpha_{(\tau'_a=0)}$, together with the assumed upper limit for E_K allows the calculatfiion of

$$\Phi_{BR \rightarrow K} \quad > \quad 0.5$$

as a lower limit for the quantum yield of the primary photoisomerization of BR. The principle limit for the quantum yield $0 \leq \Phi \leq 1$ yields also a lower limit for E_K of $E_K \geq$ 96 kJ/mol.

The slope of the straight line in Fig. 3, affords $\tau_K = 1.5 \pm 0.2\ \mu s$, in agreement with literature data.

LIOAS experiments were performed with 8 ns pulses in order to obtain the quantum yield for the back photoreaction K $\rightarrow$ BR. The experiments were carried out as previously described with a PZT detector [5]. In order to take into account the photochromic equilibrium between BR and K established within the pulse duration, a mathematical model was applied with five differential equations describing the kinetic behaviour of all species present in the system (BR, BR*, J, K and K*) and an equation describing the heat evolution within the acoustic transit time [6]. With this model the energy dependencè of the LIOAS signal for six excitation wavelength (532 - 630 nm; $\varepsilon_K/\varepsilon_{BR}$ varied between 0.7 and 2.7) was fitted with a global target analysis. The best fit was obtained for

$$\Phi_{K \rightarrow BR} \quad \geq \quad 0.6 \pm 0.05 \qquad and \qquad \tau_{K^*} = 4 - 10 \quad ps$$

In addition, a better fit was obtained for $\Phi_{BR \rightarrow K} = 0.6$ with $E_K - E_L = 30$ kJ/mol. For τ_{BR^*} a value of 0.5 ps is given in literature [4, 7]. The relation $\tau_{K^*} > \tau_{BR^*}$ is consistent with the higher fluorescence and the smaller Stokes shift reported for this intermediate, when compared to BR [8].

The LIOAS measurements with fs-pulse excitation constitute the first example of this type of experimentes with biological samples and are a general method to obtain kinetic and thermodynamic parameters for intermediates living longer than the acoustic transit time. Combination with ns-LIOAS yields additional information, in particular when photochromic equilibria are established within the longer laser pulses.

Acknowledgments :

We are indebted to Professor K. Schaffner for his support and interest. M.R. is a recipient of a fellowship from the Alfried Krupp von Bohlen und Halbach Foundation. We thank S. Griebenow, D. Lenk, and A. Kessel for their conscientious technical assistance.

References

[1] Dioumaev, A.K.; Savransky, V.V.; Tkachenko, N.V.; Chukharev, V.I. *J. Photochem. Photobiol. B.*, **1989**, 3, 397–410.

[2] Polland, H.-J.; Franz, M.A.; Zinth, W.; Kaiser, W.; Koelling, E.; Oesterhelt, D. *Biophys. J.*, **1986**, 49, 651–662.

[3] Atkinson, G.H.; Brack, T.L.; Blanchard, D.; Rumbles, G. *Chem. Phys.*, **1989**, 131, 1–15.

[4] Polland, H.J.; Franz, M.A.; Zinth, W.; Kaiser, W.; Oesterhelt, D. *Biochim. Biophys. Acta*, **1986**, 851, 407–415.

[5] Braslavsky, S.E. Heihoff, K. In Scaiano, J.C. , *Handbook of Organic Photochemistry I*, CRC Press Inc. Boca Raton, Florida, **1989**, pp 327–356.

[6] Rohr, M.; Gärtner, W.; Braslavsky, S.E. *this book*, **1992**.

[7] Sharkov, A.V.; Pakulev, A.V.; Chekalin, S.V.; Matveetz, Y.A. *Biochim. Biophys. Acta*, **1985**, 808, 94–102.

[8] Atkinson, G.H.; Blanchard, D.; Lemaire, H.; Brack; T.L.; Hayashi, H. *Biophys. J.*, **1989**, 55, 263–274.

Fourier Transform Infrared Photoacoustic Spectra of Bacteria

M. Ardeleanu[1], *R. Morisset*[1], *and L. Bertrand*[2]

[1]Hôpital de l'Hôtel-Dieu, Département de Microbiologie,
[2]Département de Génie Physique, Ecole Polytechnique de Montréal,
C.P. 6079, Succ. "A", Montréal, Québec, Canada H3C 3A7

Abstract. The development of rapid bacterial identification methods is an important issue in modern medicine. The photoacoustic spectra of bacteria enable us to discriminate between different species. We discuss here preliminary results, which strongly suggest that, by taking advantage of the Fourier Transform Infrared Photoacoustic Spectroscopy, a rapid identification method could be developed.

1. Introduction

Bacterial identification is an applied discipline of microbiology with a great practical significance. The conventional identification methods used in clinical laboratories are based on a combination of morphological, chemical and physiological analysis. For non fastidious organisms, these methods usually cannot provide the identification earlier than 36 or 48 hours. The immediate administration of a specific antibiotic therapy is of crucial importance for the outcome of a patient with severe acute bacterial infection. The choice of the antibiotic is based only on clinical and epidemiological data and on physician's own experience. A few days later, when the pathogen bacterium has been identified, it may prove necessary to readjust the therapy. Thus, the development of rapid identification methods is of great practical interest since it improves the accuracy of the medical decision and thereupon the patient's clinical outcome.

Physical methods are rapid, but none of the attempts made during the last decade resulted in a practical solution [1]. Promising results were obtained by Naumann, who used the Fourier Transform Infrared Spectroscopy (FTIR). Mathematical analysis of the infrared (IR) absorption spectra showed that one could identify a specific fingerprint-like bacterial IR pattern [2,3].

The Fourier Transform Infrared Photoacoustic Spectroscopy (FTIR/PAS) is a similar approach. The same FTIR advantages are present coupled with the reliability of the photoacoustic (PA) technique: increased sensitivity for low absorption measurements, non destructive analysis and the possibility to analize IR opaque samples. All of these properties suggest that FTIR/PAS may be more appropriate for rapid bacterial identification. We report here preliminary results of PA spectra of bacteria, which strongly suggest that, by taking advantage of the FTIR/PAS technique improved at the Ecole Polytechnique [4], a rapid identification method could be developed.

2. Methods

The following bacteria from the American Type Culture Collection were studied: Staphylococcus aureus ATCC 25923, Streptococcus pyogenes ATCC 19615, Escherichia coli ATCC 25922 and Proteus vulgaris ATCC 13315.

Isolated colonies of the bacteria grown at 37°C for 18 ± 2 hours were taken from the solid culture plates (blood - agar) using calibrated platinum loops and were suspended in 3 μl of isotonic saline solution, directly in the central cavity of the sample holder. After a few minutes of drying at room temperature in order to obtain a slow sedimentation and a uniform layer of bacteria, the sample was introduced in the PA cell and further drying was achieved by applying a vacuum of 5 ± 2 Torr for 10 minutes. The dried samples' weight was about 10 ± 2 mg. Helium was introduced in the PA cell up to 760 Torr.

FTIR/PAS measurements were performed on a BOMEM DA3 interferometer. Globar source and KBr beamsplitter were used with a home-made photoacoustic cell, not specifically designed for this experiment [5]. The mirror speed was 0.1 cm/s corresponding to a thermal diffusion length much longer than the sample thickness. Optical resolution was 4 cm^{-1} and the scan number was 200, corresponding to approximately 5 min scanning time for acquiring one spectrum.

3. Results and discussion

Two typical PA spectra from idependently-grown bacteria of the same species are shown in Fig. 1. The broad bands dominating the spectra are very likely the result of multiple superimposed bands, in accord to the known complexity of any biological material. By analogy with the FTIR transmission spectra described in literature [6], some of the bands can be assigned to the major biochemical components of the microbial cells. The bands between 900 and 1200 cm^{-1} are mainly due to carbohydrates present within the cell wall and capsule, especially polysaccharides. The range between 1200 and 1500 is less well defined, this representing a mixed region containing information from proteins, fatty acid and phosphate-carrying compounds. The intense Amide II near 1547 cm^{-1} and Amide I bands near 1656 cm^{-1} are due to the high content in proteins and peptides. Bands near 1735-1745 cm^{-1} (> C=O stretching of ester groups) mostly reflect the relative amount of O-acetyl groups bound to the C_6-position of muramic acid of peptidoglycan, a main constituent of the bacterial cell wall. The region between 2800 and 3000 cm^{-1} is dominated by the $-CH_3$, $> CH_2$ and $= CH$ stretching vibrations of the functional groups usually present in the fatty acids.

Speaking in terms of broad bands and general contour, the FTIR/PA spectra were similar to the FTIR transmission spectra published by Naumann et al.

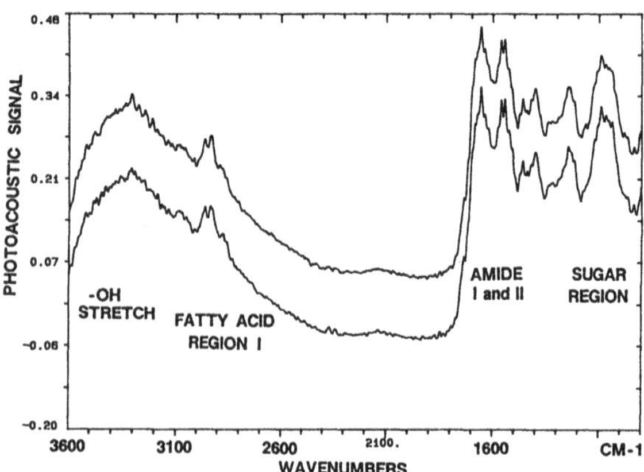

Fig. 1: Photoacoustic spectra of two different samples of Staphylococcus aureus.

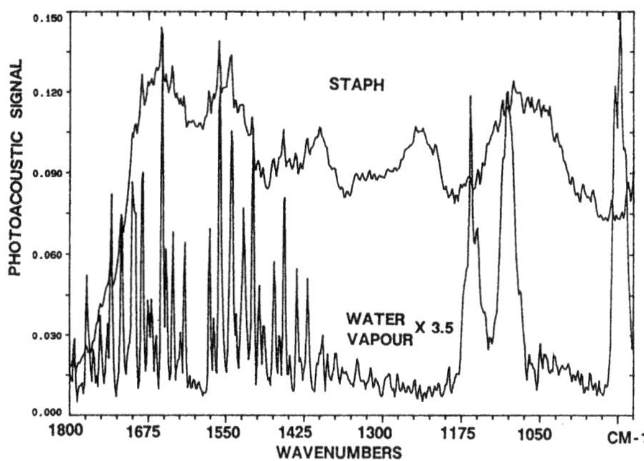

Fig. 2: Water vapours and Staphylococcus aureus PA spectra.

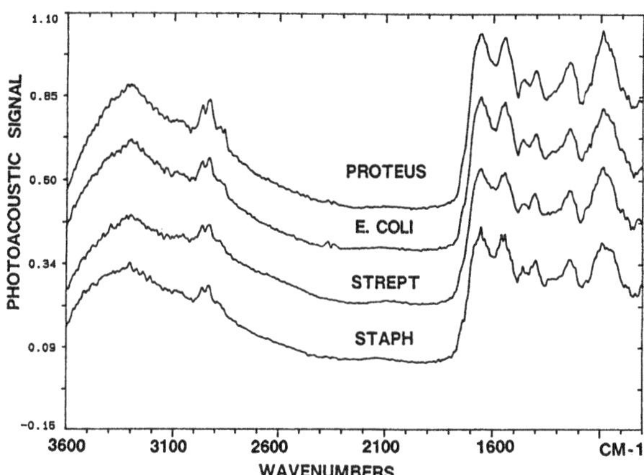

Fig 3: PA spectra of different bacteria. proteus: Proteus vulgaris; e. coli: Escherichia coli; strept: Streptococcus piogenes; staph: Staphylococcus aureus.

More details, i.e. as yet unassigned small narrow peaks, can be observed on our PA spectra, which confirms the higher sensitivity expected with the FTIR/PAS. The samples were only partially dried in order to avoid water condensation on the PA cell's window, so we assumed that at least some of these peaks correspond to water bound to hydrophilic molecules. This appeared to be true when we compared these spectra with a water vapour spectrum, as shown in Fig. 2. Further drying by vacuum only slightly reduced the intensity of these "water peaks". Furthermore, the intensity of the "water peaks" varied from one species of bacteria to another. Therefore we suppose that the intensity of the "water peaks" reflects the abundance of hydrophilic molecules and may be taken into consideration as characteristic of a particular species.

In the spectral domain that we investigated, one can recognize the features common to all living cells. However, when analyzing the number of bands, the peak positions, the half width, the band contours and the relative peak intensities, distinctive details may be observed even with the naked eye, as shown in Fig. 3. These details, interpreted as part of a

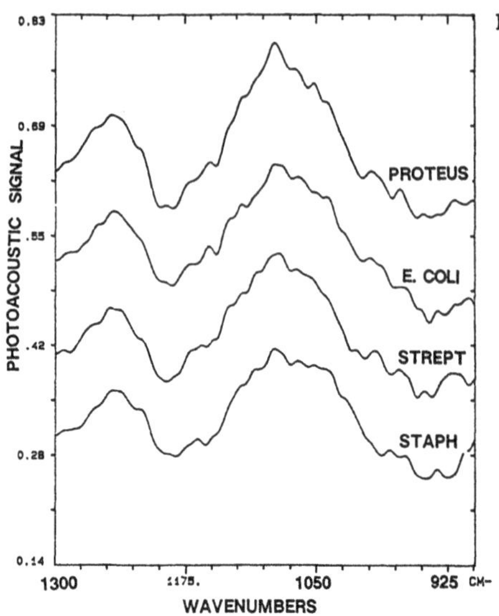

Fig 4: Sugar region of spectra of the four bacteria shown in Fig. 3.

The spectra are drawn on different scales.

specific spectral signature, may be seen in Fig. 4, where the "sugar region" of the same previous four spectra is shown.

4. Conclusions

Although preliminary, we consider these results very encouraging. We already obtained good quality, reproducible spectra in conditions not entirely adapted for such measurements. We are now trying to define better experimental conditions, regarding the technique itself (cell design, background noise, resolution, scan number, mirror speed, etc.) and the sample (dried bacteria or bacteria in suspension, optimal quantity, culture media, etc.).

If we succeed to overcome these difficulties, the simplicity and rapidity of the technique are good reasons to make us believe that the FTIR/PAS might open a new path in the methodology of bacterial identification. The information provided by PA spectra may enable us to discriminate between different, phylogenetically close species and even between different strains.

Acknowledgment: We want to thank A. Domingue and P.A. Dion for their technical assistance.

References:

1. Salzman, G.C., Gregg, C.T., K. Grace, 83rd Annual Meeting of the American Society for Microbiology, (1983), 353.
2. Naumann, D., Fijala, V., Labischinski, H., Giesbrecht, P., J. Mol. Struc. 174, (1988), 165.
3. Naumann, D., Helm, D. and Labischinski, H., Nature, 351, (1991), 81.
4. Bertrand, L. Applied Spectroscopy, 42, (1988), 134.
5. Rousset, G., Monchalin, J.-P. and Bertrand, L., J. de Phys., Coll. C6, (1983), 165.
6. Helm, D., Labischinski, H., Schallehn, G. and Naumann, D., J. Gen. Microbiology, 137, (1991), 69.

New and Versatile Method for Thermophysical Characterization of Materials: The Inverse Photopyroelectric Technique

M. Chirtoc[1], *D. Bićanić*[2], *D. Dadarlat*[1], *and I. Chirtoc*[1]

[1]Institute of Isotopic and Molecular Technology, P.O. Box 700, Cluj-Napoca 5, Romania

[2]Department of Agricultural Engineering and Physics, Agrotechnion, Wageningen Agricultural University, Bomenweg 4, 6703 HD Wageningen, The Netherlands

Abstract. The new Inverse Photopyroelectric (IPPE) technique for thermal characterization of materials is presented. The theoretical principles are derived and a number of applications concerning the thermal properties evaluation of metals, food industry products, binary liquid mixtures, as well as the assessment of a sensor for monitoring the condensation of atmospheric water vapour are reported.

1. Introduction

The conventional PPE technique [1] is based on the measurement of the 2emperature variation of a sample irradiated by modulated light. This is achieved by a pyroelectric sensor placed in thermal contact with the rear side of the sample. The symmetry arguments of the theoretical approach [2] suggest another configuration for the PPE cell, in which the sample is placed behind, and not in front of the pyroelectric. We call this arrangement illustrated in Fig.1 an Inverse Photopyroelectric (IPPE) cell [3]. The physics of the IPPE detection process is as follows: the sample acts as a heat sink for the irradiated pyroelectric and therefore the signal amplitude will be reduced when compared to the situation where no sample is being present.

2. Theory

Starting from the general theory [2] it can be shown that as far as the sample-related parameters are concerned, and for a periodic excitation with angular frequency ω, the signal amplitude V is proportional to a dimensionless factor Γ:

$$\Gamma=1-(S_+ + S_-)/[S_+\exp(\sigma_p L_p) + S_-\exp(-\sigma_p L_p)] \quad (1)$$

with the following notations:

$$S_\pm=(b_{sp}\pm1)\exp(\sigma_s L_s)+(b_{sp}\mp1)\exp(-\sigma_s L_s)$$
$$b_{sp}=k_s a_s/k_p a_p \quad (2)$$
$$\sigma_{s,p}=(1+i)a_{s,p} \quad ; \quad a_{s,p}=(\omega C_{s,p}/2k_{s,p})^{1/2},$$

where $L_{p,s}$ are the thicknesses of the pyroelectric and sample, b_{sp} is the ratio of the thermal effusivities with $k_{s,p}$ being the thermal conductivities, $a_{s,p}$ the thermal diffusion coefficients and $C_{s,p}$ the volume specific heats, respectively.

An obvious requirement is that the pyroelectric should be thermally thin, i.e. $a_p L_p<1$. Then there are two cases of interest, namely a thermally thick sample ($a_s L_s>1$), yielding:

$$\Gamma=\sqrt{2}b_{ps}a_p L_p\exp(i\pi/4) \quad (3)$$

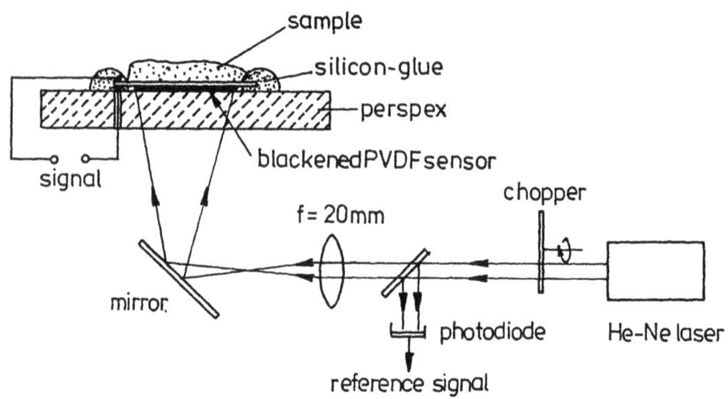

Fig. 1 Schematic diagram of the apparatus including the IPPE cell used for the measurement of the thermal effusivity of fluid samples.

and a thermally thin sample ($b_{ps}a_pL_p<a_sL_s<1$), for which:

$$\Gamma=C_pL_p/C_sL_s. \tag{4}$$

Equations (3,4) show that the IPPE configuration allows for the measurement of the thermal effusivity or the specific heat of the sample. There exists a third case when Γ is proportional to k_s^{-1}, if the backing instead of air is a good thermal conductor.

3. Applications

The above principles were applied in the experimental conditions of Eq.(3) to several classes of materials. In the case of metals, we have derived correct values of k_s for Ni and Fe after calibration of the cell with Al [4]. In the case of liquids, we have shown that the concentration of alcoholic beverages can be estimated based on their thermal effusivity values [5]. Alternatively, with precisely prepared ethanol-water mixtures, a calibration curve for the IPPE instrument was generated [6] in order to measure the absolute thermal effusivity values of fluid samples, (Fig.2). The reference sample was water. By using this method, it was possible to evaluate the fat/water content of industrial food products like mayonnaise, shortening, edible oil [6] yoghurt, custard and margarine [7]. For solid foodstuffs, the method enabled the homogeneity analysis and monitoring of a drying process in cheese [7].

Due to the large thermal effusivity difference between water and air, the IPPE cell proved to be a sensitive instrument for monitoring the condensation of atmospheric water vapour [8]. Similarly, the thickness of thin liquid layers in the μm range can be accurately measured [8].

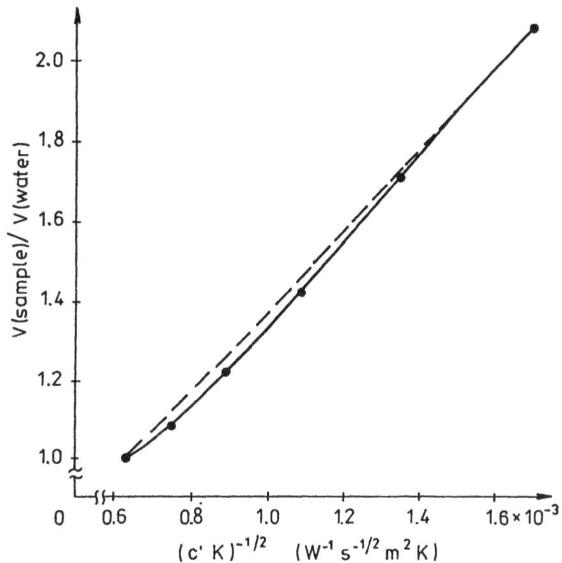

Fig.2 The magnitude of IPPE signal vs. the inverse of thermal effusivity, normalized to water as reference substance: theory (dashed curve); experimental calibration with ethanol-water mixtures (solid curve).

4. Conclusions

The IPPE method is a simple, inexpensive and easy to apply photothermal technique yet capable of quantitative and qualitative characterization of a wide range of materials by means of their thermal effusivity. When appropriately calibrated, it has advantages above the conventional PPE method since the optical properties of the sample do not alter the results. Also its geometry and shape are irrelevant as long as the sample is thermally thick. The intrinsic potentiality of the IPPE method to measure the specific heat of the sample might be exploited for phase transition investigations.

References

[1] A.Mandelis and M.M.Zver, J.Appl.Phys. **57** 4421 (1985).

[2] M.Chirtoc and G.Mihailescu, Phys.Rev.**B 40** 9606 (1989).

[3] M.Chirtoc, D.Dadarlat and D.Bicanic, in this Conference.

[4] D.Dadarlat, M.Chirtoc, C.Neamtu, R.M.Candea and D.Bicanic, Phys.Stat.Sol.(a) **K121** K231 (1990).

[5] M.Chirtoc, V.Tosa, D.Bicanic and P.Torfs, Ber. Bunsenges. Phys. Chem. **95** 766 (1991).

[6] D.Bicanic, M.Chirtoc, D.Dadarlat, P.van Bovenkamp and H.van Schayk, Appl.Spectroscopy, (1992), in press.

[7] D.Bicanic, M.Chirtoc and D.Dadarlat, Dairy Sci. (1991) (submitted).

[8] M.Chirtoc, D.Bicanic and V.Tosa, Rev.Sci.Instrum. **62** 2257 (1991).

Use of an Open Photoacoustic Cell for some Applications in Agriculture

A.M. Sólyom[1], *D. Bićanić*[2], *Gy.Z. Angeli*[1], *A. Miklós*[3], and *M. Lubbers*[2]

[1]Analtron Applied Research Company, P.O. Box 63, H-1311 Budapest, Hungary

[2]Department of Agricultural Engineering and Physics, Agrotechnion, Wageningen Agricultural University, Bomenweg 4, 6703 HD Wageningen, The Netherlands

[3]Institute of Isotopes, Hungarian Academy of Sciences, P.O. Box 77, H-1525 Budapest, Hungary

Abstract. The photoacoustic spectrometer consisting of a tunable CO_2 laser and windowless resonant cell was used for detection of ammonia and ethylene.

The potential of CO_2 laser photoacoustic spectroscopy (PAS) and the traditional indophenol blue colorimetric method detecting the gaseous ammonia have been compared. The results obtained with the two independent techniques are comparable in the range of concentrations studied.

In another experiment the spectra of volatile products produced by UV radiation in different types of polymer tubing were investigated. The spectra of transparent polyethylene tubing was dominated by an absorption peak at 10P(14) laser line. For PTFE and colored polyethylene tubing no signals were measured.

1. Introduction

The measurement of low concentration ethylene and ammonia has great importance in agriculture. In recent years numerous studies demonstrated the feasibility of infrared-laser photoacoustic spectroscopy for detecting these gases in the variety of environmental and agricultural applications [1].

The commonly recommended standard procedure for detection of ammonia is the modified indophenol blue method; the ammonia is determined as an emerald green color complex formed by the reaction of ammonia with salicylate and phenol-sodium-hypochlorite solution and catalyzed by sodium-nitroprusside in a buffered alkaline medium at pH of 12.8-13 [2]. The detection limit of this method is approximately 1.3 ppb NH_3 at 760 torr and 25 °C [3]. Our aim was to investigate whether or not the photoacoustic spectroscopy provides the same analytical results as the colorimetric method.

In a conventional on-line monitoring (such as for example atmospheric pollution or metabolic changes in stressed ornamental flowers) one often draws the gaseous sample into the photoacoustic cell by means of a pump and polymer tubing. On the other hand it is known that exposure of tubing to the UV light might induce generation of volatile products in tubing on account of such interaction. In this experiment the extent of such

Springer Series in Optical Sciences, Vol. 69 Photoacoustic and Photothermal Phenomena III
Editor: D. Bićanić

effect (that if neglected posses a potential danger especially when dealing with ppbv levels) has been determined for several common tubings using PA spectroscopy.

2. Experimental

A step tunable CO_2 laser radiation source in conjunction with a windowless (open) resonant cell (with a high acoustic quality factor) was used for trace analysis of ambient air. The photoacoustic cell manufactured from brass was provided with an electronic resonance locking circuitry [4].

The colorimetric measurements were carried out at 660 nm using Technicon Auto-Analyzer II Continuous Flow System. Before the colorimetric measurement the ammonia was preconcentrated by bubbling of the gaseous mixture in a dilute sulphuric acid in a midget impinger to form ammonium sulphate.

Gas mixtures were prepared from reagent grade gases (5.0 nitrogen, 100 ppmv of ammonia and 136 ppm of ethylene) and used without further purification. The total flow rate of the mixtures was kept below 1 l/min in order to reduce the acoustic noise generated by the streaming gas.

3. Results and discussion

Ammonia measurement. The ammonia was measured using seven different samples within the 4.9-17.8 ppmv concentration range. The samples were measured by both, the photoacoustic and the colorimetric method, and the analytical results were compared.

The dependence of the PA signal on the concentration of ammonia at the 9R(30) laser line was measured in the 1.5 ppm to 15.7 ppm range. A plot of measured absorbance (Figure 1) versus the ammonia concentration for the data is linear (R coefficient 0.998) and the extrapolated detection limit corresponding to a signal to noise ratio of unity is 0.43 ppb.

The analytical results obtained by the PAS and indophenol-blue measurements are compared in Table 1.

Table 1.:The comparison of colorimetric and photoacoustic methods for detection of ammonia

Sample concentration ppmv		4.9	5.7	6.3	6.9	10.1	12.7	17.8
Measured conc. ppmv	Indoph.-blue	5.1	5.8	6.4	7.6	9.9	10.5	18.2
	PAS	4.8	5.6	6.2	6.8	9.8	12.3	17.3

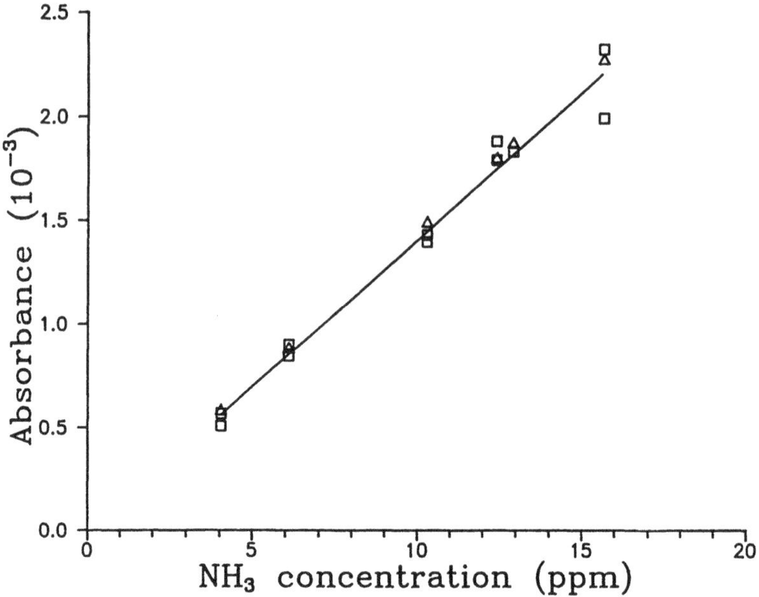

Figure 1.
Absorbance versus concentration of ammonia at 9R(30) laser line
(□ : measured points, ▵ : calculated points)

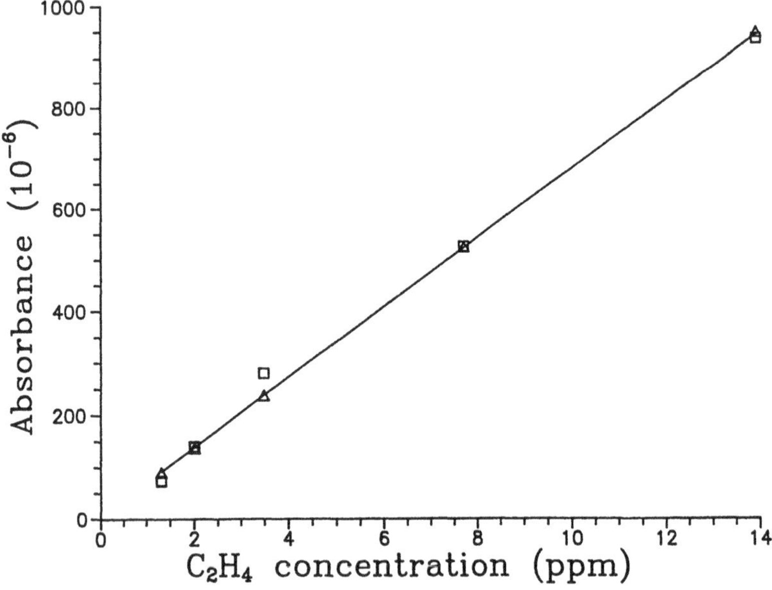

Figure 2.
Absorbance versus concentration of ethylene at 10P(14) laser line
(□ : measured points, ▵ : calculated points)

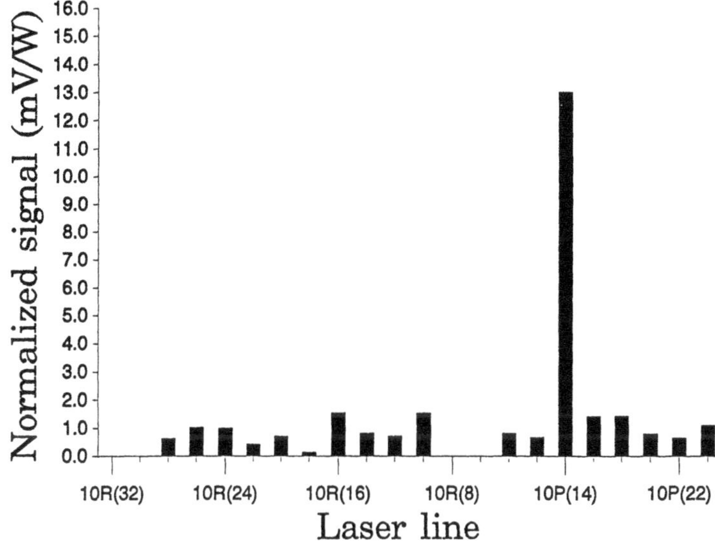

Figure 3.
Measured photoacoustic spectrum of clear polyethylene tubing

Ethylene measurement. In order to calibrate our spectrometer for ethylene measurements the calibration curve of ethylene was determined. Measurements were made with standard samples the concentration of which varied between 1.3 and 13.9 ppm ethylene in nitrogen. As expected, the plot of the measured absorbance at selected laser line 10P(14) versus ethylene concentrations (shown in Figure 2) exhibits a linear dependence with the correlation coefficient of 0.998 and the derived detection limit of 1.3 ppb.

Finally the spectrum of volatile products present in the clear polyethylene tubing irradiated by UV light while being continuously flushed by pure nitrogen is shown in Figure 3. The absorption peak detected at 10P(14) line corresponds to about 70 ppb ethylene emission suggesting a need for shielding of a tubing in particular at low concentration.

No signals were obtained from PTFE and coloured polyethylene tubing.

4. References

[1] *Photoacoustic, Photothermal and Photochemical Processes in Gases* ed. Hess, P. (Topics in Current Physics, Springer-Verlag, Berlin, 1989)

[2] M.W. Weatherburn: Anal. Chem. **39**, 971. (1967)

[3] *Handbook of Air Pollution Analysis*, ed. Perry, R. and Young, R. J., pp. 274-277. (Chapman and Hall, London, 1977)

[4] Gy.Z. Angeli, Z. Bozóki, A. Miklós, A. Lőrincz, A. Thöny, M. W. Sigrist: Rev. Sci. Instrum. **62**, 810-813 (1991)

In Vivo Evaluation of the Human Skin Permeation of Topically Applied Light Absorbing Agents by a Mirage Effect Spectrometer

K. Plamann[1], K. Giese[1], B. Sennhenn[1], N. Harendt[1], and K. Kölmel[1]

[1]Institut für Medizinische Physik und Biophysik, Georg-August-Universität, W-3400 Göttingen, Fed. Rep. of Germany
[2]Hautklinik, Georg-August-Universität, W-3400 Göttingen, Fed. Rep. of Germany

Photothermal spectroscopy allows to determine the light absorption within a sample depth-resolved because of the thermal diffusion length depending on the chopping frequency. Therefore, it can be used to evaluate concentration profiles of topically applied light absorbing drugs in the upper layers of human skin. A new experimental setup with Mirage-Effect detection for in vivo-measurements is described which considerably increases the measuring depth and the accuracy of the measurements at low chopping frequencies.

Compared to other spectroscopic methods, photothermal spectroscopy offers the advantage of an adjustable measuring depth as the thermal diffusion length depends on chopping frequency. Therefore it is an adequate method for determining the penetration behaviour of light absorbing drugs topically applied to human skin.

Since several years, we use a photoacoustic setup for photothermal in vivo-measurements on human skin [1]. The available frequency range down to 180 Hz corresponds to a maximum measuring depth of about 7 μm, which is about half the thickness of the stratum corneum. At lower frequencies, below the skin resonance, the sensitivity of the open-ended photoacoustic cell and the reproducibility of the measurements become poor [1]. To reach the viable epidermis by photothermal measurements, an in vivo-spectrometer with Mirage-Effect detector [2,3] has been developed which extends the usable frequency range down to 10 Hz, corresponding to a maximum measuring depth in human skin of about 30 μm. This low frequency compares well with frequencies usable by the recently introduced technique of optothermal infrared spectroscopy [4]. That method, however, requires the skin area of interest to be brought into contact to a thermoacoustic detector.

In our experimental arrangement the forearm of a subject is used for in vivo-measurements on skin. The forearm lying on an arm rest adjustable for height is locked in place by a metal clamp shown in fig. 1 to provide a flat surface. Along the clamp is a groove, the laser beam is running through. For illumination of the skin the groove has a gap of 2 cm in length and 2 mm in width. The skin is pushed through the gap as shown in fig. 1 so that the laser beam can be directed directly across this illuminated area. The automatic fine adjusting of the distance laser beam/skin is achieved by use of computer-controlled step motors, the laser and the postion detector are mounted on. The position "zero" is defined as the position where 97% of the light intensity of the laser beam still hits the position detector. The axis of the laser beam then runs in a distance of about 125 μm of the skin surface. This position turned out to be optimal for low-frequency in vivo-measurements because at a position closer to the skin surface, the measurement is influenced by unavoidable skin vibrations, whereas a larger distance yields a poor signal-to-noise ratio. As light source we use a xenon arc lamp emitting a light intensity from 2 mW/cm^2 at 250 nm up to 25 mW/cm^2 at 500 nm.

The frequency range usable for photothermal in vivo-measurements on human skin with Mirage-Effect detection extends from 10 to 200 Hz, that with acoustic detection

Springer Series in Optical Sciences, Vol. 69 Photoacoustic and Photothermal Phenomena III
Editor: D. Bićanić

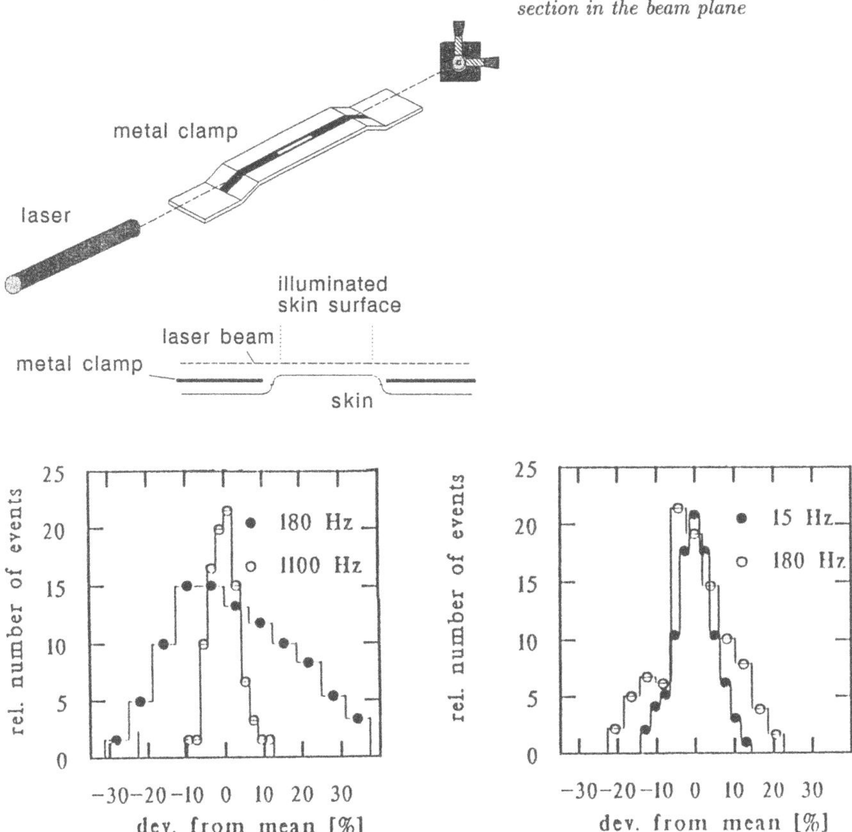

Figure 1: Metal clamp with schematic section in the beam plane

Figure 2: Reproducibility of photoacoustic measurements (1100 and 180 Hz, left) and Mirage-Effect measurements (180 and 15 Hz, right)

from 150 Hz to 2000 Hz. A comparision of the reproducibility of photoacoustic (1100 and 180 Hz, left) and beam deflection (180 and 15 Hz, right) measurements is depicted in fig. 2. Each of the plotted histograms results from a set of about 100 in vivo measurements. The PA–measurements at 1100 Hz show a satisfactory accuracy with respect to biological samples, those at 180 Hz have a reduced reproducibility because of the reasons mentioned above. In this frequency range, the Mirage-Effect spectrometer already offers a better accuracy. The best results, however, are obtained by using the spectrometer at low frequencies, as demonstrated by the measurements at 15 Hz.

The Mirage-Effect spectrometer has been used to obtain spectra of human skin. Fig. 3 left shows the spectrum of skin treated with an UV-A-absorbant (Eusolex 8020, E. Merck, Darmstadt) solved in Isopropylmyristate, a substance of oily consistence, and the spectrum of stratum corneum treated with the pure solvent. The difference spectrum is depicted in fig. 3 right. A chopping frequency of 30 Hz was used, corresponding to a measuring depth containing the whole horny layer and parts of the underlying viable epidermis. The difference spectrum yields information about the amount of drug present within the photothermal measuring depth. Its time dependence reflects the diffusive transport of drug to skin layers underneath the measuring depth. At visible wavelengths

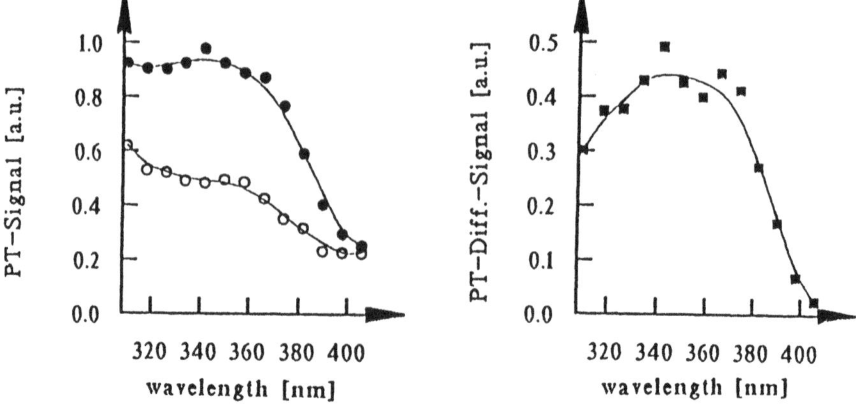

Figure 3: Photothermal in vivo-spectra of skin treated with Isoprophylmyristate with (upper curve, left) and without (lower curve, left) 3 % of the UV-A absorbant Eusolex 8020 obtained at a chopping frequency of 30 Hz. The right curve depicts the difference of both spectra.

the optical penetration length of untreated skin is much larger than the photothermal measuring depth under consideration here. In the UV–spectral range, the optical penetration length decreases with decreasing wavelength [5]. In case of UV–A–absorbants of suitable concentrations, as considered in fig. 3, photothermal signal saturation is of low influence to the difference signal. At UV–B–wavelengths, saturation effects at low chopping frequencies prevent a sensitive evaluation of drug diffusion into deeper skin layers.

At present, the spectral range accessible to in vivo–measurements on skin is limited to wavelengths above about 315 nm from another reason. Due to the biological effects of UV–radiation, the exposure has to be small. E. g. at a wavelength of 270 nm the exposure limit is 3 mWs/cm^2 [6], which with the given light intensity of the xenon arc lamp corresponds to a maximum exposure time of about 1 s. Reducing the light intensity is possible, however, not practicable at present because of the too low signal-to-noise ratio.

References

[1] Giese K, Nicolaus A, Sennhenn B, Kölmel K: Can. J. Phys. **64** (1986) 1139 - 1141

[2] Murphy J C, Aamodt L C: J. Appl. Phys. **51** (1980) 4580 - 4588

[3] Boccara A C, Fournier D, Badoz J: Appl. Phys. Lett. **36** (1980) 130 - 132

[4] Frödin T, Helander P, Molin L, Skogh M: Acta Derm. Venereol. **68** (1988) 461 - 467

[5] Kölmel K, Sennhenn B, Giese K: Br. J. Dermatol. **122** (1990) 209 - 216

[6] International Non-Ionizing Radiation Committee of the International Radiation Protection Association: Health Physics **49** (1985) 331 - 340

Optothermal Characterisation of Human Skin

R.M.S. Bindra[1], *G.M. Eccleston*[2], *R.E. Imhof*[1], *and D.J.S. Birch*[1]

[1]Department of Physics and Applied Physics, and
[2]Department of Pharmacy, Strathclyde University,
Glasgow G4 0NG, UK

Abstract. The non-invasive nature of opto-thermal transient emission radiometry (*OTTER*) facilitates *in-vivo* measurements on human skin. We report experiments to characterise different skin types and conditions, using a range of excitation and thermal emission wavelengths.

1. Introduction

Opto-thermal transient emission radiometry lends itself to the study of biological subjects, such as skin *in-vivo*, because it is non-invasive and insensitive to small movements of the sample. It has potential for monitoring skin condition, and for sensing the presence and disappearance of topically applied substances such as sunscreens [1], industrial chemicals, cosmetics and trans-dermal drug preparations. All such measurements rely on a knowledge and understanding of the opto-thermal properties of different skin types and sites, and their natural variations. This study, restricted as it was to the fingers of several volunteers, nevertheless goes some way towards satisfying this need.

No satisfactory methods for extracting physical parameters from opto-thermal measurements on systems as inhomogeneous as skin are available at present. Whilst mathematical modelling research is ongoing, our current approach to data analysis is two-fold. First, we use least-squares analysis to assess the extent to which a given skin measurement deviates from the theoretical impulse response function of a homogeneous, semi-infinite sample [2]. Surprisingly, this model has been found to give reasonable fits in some instances. When it is inappropriate, we are reduced to least-squares fitting of empirical expressions, such as

$$I(t) = A_1 \exp(t/\tau_1)\, erfc(\sqrt{t/\tau_1}\,) + A_2 \exp(-t/\tau_2) \tag{1}$$

where A_1, A_2 τ_1 and τ_2 are adjustable parameters. The expression in A_1 and τ_1 is the impulse response of a semi-infinite, homogeneous sample; the exponential term in A_2 and τ_2 is added to represent deviations commonly observed in skin measurements. A loose interpretation would identify τ_1 with the opto-thermal decay time of a homogeneous material ($\tau_1=\alpha^{-2}D^{-1}$ where α=optical absorbance and D=thermal diffusivity), A_2/A_1 and τ_2 as inhomogeneity indicators.

Springer Series in Optical Sciences, Vol. 69 Photoacoustic and Photothermal Phenomena III
Editor: D. Bićanić

2. Results and Discussion

The fingers of several volunteers were washed thoroughly with soap and water prior to measurement, in order to establish a common starting point. Opto-thermal impulse response curves were then measured by averaging typically 100 shots of the laser for each of several excitation and thermal emission wavelengths.

a) Opto-thermal Impulse Response Spectra

Opto-thermal impulse response curves depend on the depth of skin probed, and therefore on excitation and thermal emission absorption depths. Typical values for excitation at 266, 355 and 532nm are ~2, 60 and 400μm respectively. Absorption depths for thermal emissions were studied by measuring opto-thermal impulse response spectra using interference filters of ~10% bandwidth, in the range 6-13μm, as shown in figure 1. The ordinate uses the fitted decay time, $\tau_1^{-1/2}$ of equation (1) which, in homogeneous materials, is proportional to the lesser of excitation or thermal emission absorbance [3,4].

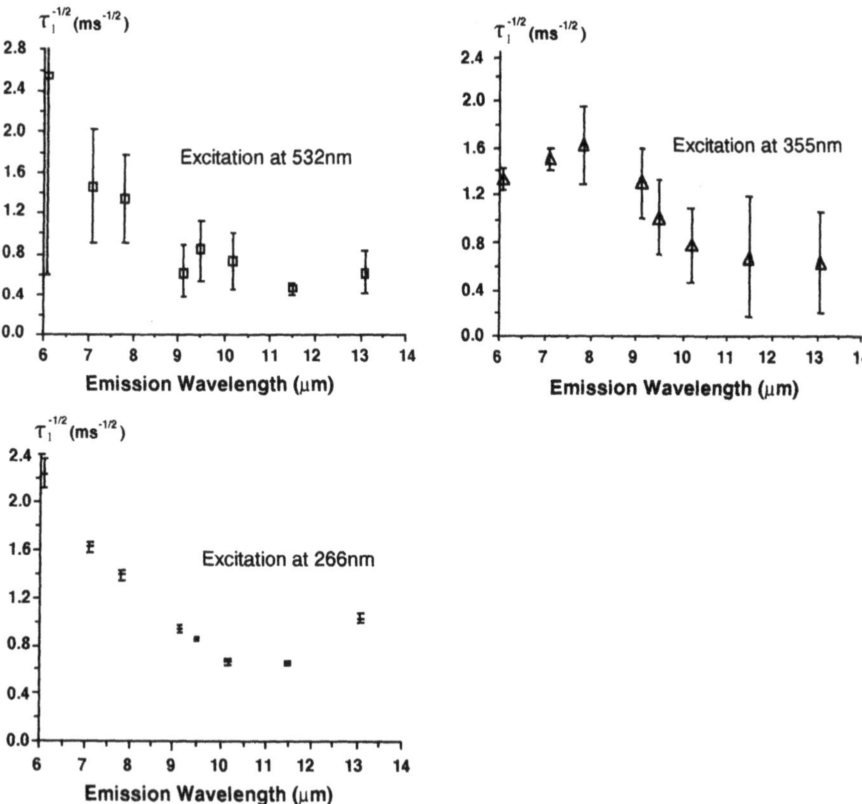

Figure 1

Opto-thermal impulse response spectra of *in-vivo* skin excited using 532, 355 and 266nm wavelength radiations.

All three excitation wavelengths were found to produce unstructured spectra, with signal strengths varying in proportion to absorbance for similar fluences. The minimum near 10μm in the spectrum excited with 266nm radiation is a feature of the water spectrum. The spectrum excited with 355nm radiation deviates from this, with a peak near 8μm. With 532nm excitation, signal strengths were too low for reliable parametrisation, especially at short emission wavelengths.

The absorption depths calculated from the decay times of figure 1, using a value of $D \approx 6\times10^{-8}m^2s^{-1}$ [5], are in the range 5-10μm. Whilst these values might be plausible for 266nm excitation, they are irreconcilable with the much larger absorption depths for excitation at 355nm and especially 532nm, if considerations of photothermal saturation in homogeneous materials were valid [4]. We therefore conclude that the opto-thermal decay characteristics of *in-vivo* skin are strongly influenced by inhomogeneity, and further progress in theoretical modelling is required, before such measurements can be used to calculate physiologically meaningful parameters.

Similar investigations with different individuals and skin sites have revealed wide variations in opto-thermal decay behaviour, including delayed thermal waves from deep layers in some instances. Of particular interest was the observation that dormant dermatitis, indiscernible by eye, results in faster than normal decay times.

b) Externally Applied Chemicals

Chemicals applied to skin can be detected either through their chromophores or through their effects on skin condition. Sunscreens are examples of the first kind. Strong UV absorption marks these chemicals out from normal skin. Their presence and disappearance can therefore be monitored straightforwardly [1], although the effect on underlying skin condition can never be discounted, especially in commercial preparations that contain condition altering ingredients.

A fundamentally important example of a condition altering chemical is water. Figure 2 shows how the opto-thermal impulse response (355nm excitation, 6-13μm emission) from a representative individual, whose finger was soaked in tap water until wrinkled (~45 minutes), changed with hydration. Recovery to normal, dry skin response occurred in ~15 minutes. Similar results were obtained with 532nm excitation and other volunteers.

The observed change in shape of the impulse response, from one approximating a semi-infinite, homogeneous response for normal skin, to one which deviates strongly from that model and from equation (1) for hydrated skin, is contrary to expectation. The stratum corneum is known to swell almost ten-fold through hydration, which might suggest that a more homogeneous impulse response would result from hydrated skin. Our current interpretation is that two different mechanisms are at work in the early and late parts of the signal. The increased initial decay rate may be caused by increases of thermal diffusivity and infrared absorbance with hydration of near-surface skin. The languid tail may be caused by a superposition of delayed thermal waves from a range of depths beneath the surface. These are likely to be present in normal as well as hydrated skin, but become more visible in the latter case because of the postulated higher opacity of hydrated skin to thermal radiations. Further work with dispersed infrared detection and shorter excitation

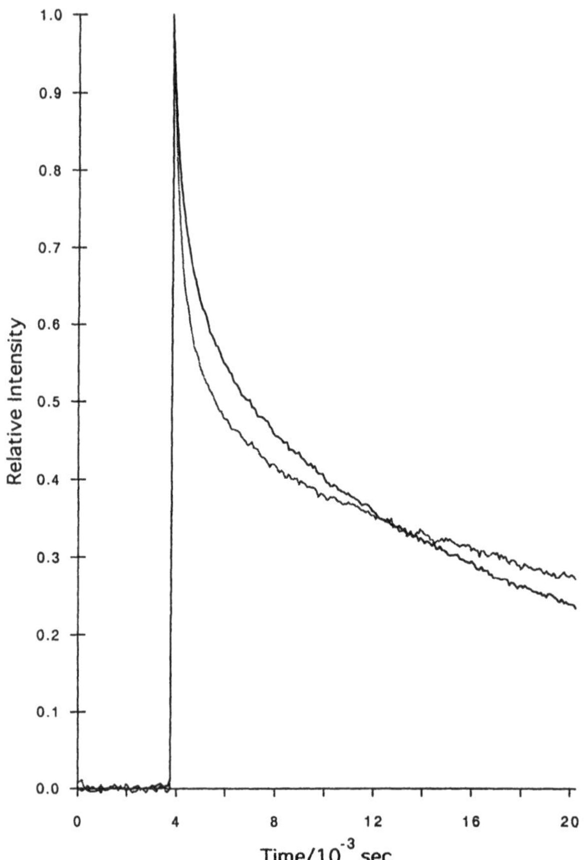

Figure 2
Opto-thermal impulse response curves for normal (line) and hydrated (points) skin, measured using 355nm excitation and broadband thermal emission.

wavelengths, where spectral shifts with hydration have been reported [6], are planned to clarify the relationships between the physiological and opto-thermal effects of hydration.

3. Conclusions

These results form the basis against which changes in skin condition, caused by disease or externally applied chemicals for example, can be understood. The experimental work is now being extended to include all relevant skin sites and other wavelengths. Theoretical work is in progress to model the complex optical and thermal properties of skin, in order to facilitate the physiological interpretation of opto-thermal impulse response curves.

Acknowledgments

We thank SERC, the Health and Safety Executive and Unilever Research for financial support.

References

1. R.E. Imhof, C.J. Whitters and D.J.S. Birch, Phys. Med. Biol. 35(1), 95-102, (1990).
2. R.E. Imhof, D.J.S. Birch, F.R. Thornley, J.R. Gilchrist and T.A. Strivens, J Phys E: Sci. Instrum. 17, 521-5, (1984).
3. R.E. Imhof, C.J. Whitters, D.J.S. Birch and F.R. Thornley, J Phys E: Sci. Instrum. 21, 115-7, (1988).
4. R.E. Imhof, C.J. Whitters and D.J.S. Birch, *Photoacoustic and Photothermal Phenomena*, (edited by J.C. Murphy, J.W. Maclachlan Spicer, L.C. Aamodt and B.S.H. Royce), Springer Ser. Opt. Sci. 62, 46-54, (1990).
5. A.J.H. Vendrick and J.J. Voss, J. Appl. Physiol. 2(2), 211-5, (1957).
6. B. Sennhenn, PhD Dissertation, 73, Göttingen (1990).

Part II

Spectroscopy

Standard Versus New Photopyroelectric (PPE) Techniques: Reflection (RPPE) and Inverse (IPPE) Configurations

M. Chirtoc[1], *D. Dadarlat*[1], *and D. Bićanić*[2]

[1]Institute of Isotopic and Molecular Technology, P.O. Box 700, R-3400 Cluj-Napoca 5, Romania

[2]Department of Agricultural Engineering and Physics, Agrotechnion, Wageningen Agricultural University, Bomenweg 4, 6703 HD Wageningen, The Netherlands

Abstract. The paper presents the salient features of the standard photopyroelectric (PPE) method for optical and thermal characterization of materials as compared to the new reflection (RPPE) and inverse (IPPE) configurations, which are shown to be better suited for spectroscopic and thermophysical investigations, respectively. A number of applications are reviewed in order to support the theoretical arguments.

1. Introduction

Among the photothermal methods for optical and thermophysical characterization of materials, the calorimetric techniques alone are based on the direct detection of the temperature field [1], therefore having a number of advantages over other detection schemes involving secondary mechanisms. Pyroelectric sensors have a temperature-measurement resolution of 10^{-6}K [2] and when combined with periodic excitation and lock-in signal processing, this limit can be lowered even further. Their use as temperature sensors in the calorimetric technique applied to spectroscopic investigations of solids and liquid samples has marked the appearance of the photopyroelectric (PPE) spectroscopy [3-5].

2. Basic configurations

As it is well known, in the PPE technique the pyroelectric transducer is placed in thermal contact with the rear side of the sample slab or layer, thus measuring the rear temperature variation when the sample front side is being irradiated by pulsed, single-frequency or wide band modulated radiation. The following discussion will consider only single-frequency, periodic modulation.

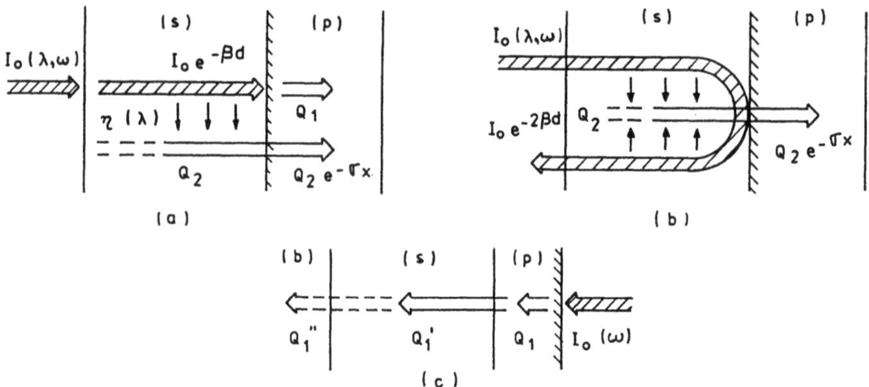

Fig.1. Basic PPE cells: (a) standard - (PPE); (b) reflection - (RPPE); (c) inverse - (IPPE) configurations, with (p), pyroelectric; (s), sample; (b), backing. Q_1 is the heat developed in (p) and Q_2 in (s).

The basic phenomena behind the PPE detection mechanism are shown in Fig.1a. The overall PPE signal is produced by the vector sum of two periodic heat sources. The source Q_1, generated upon absorption in the opaque pyroelectric, is proportional to the fraction of radiation transmitted through the sample. The detector acts as a radiation detector and produces a signal analogous to the conventional transmission spectrum. The source Q_2, on the other hand, arises due to the radiation absorption within the partially transparent sample and it contains three types of information: optical transmission into the sample depth, optical absorption at various depths d-x and the radiation-to-heat conversion quantum efficiency $\eta(\lambda)$. Before reaching the pyroelectric, the thermal wave is also influenced by the thermal properties of the sample and of the adjacent layers, being attenuated and dephased. For this signal component the pyroelectric sensor is a true thermal wave sensor, yielding a combined transmission-absorption spectrum. The quantitative description of this complicated situation was given by Mandelis and Zver [6]. It was shown that the particularization of the general PPE response leads to six special cases denoted by 1a-c, 2a-c. The selected specific case is completely defined by only two independent, sample-related parameters, i.e. the optical thickness βd and the thermal thickness ad [7], where d is the sample thickness, β the bulk absorption coefficient, and $a^{-1}=(2\alpha/\omega)^{1/2}$ is the thermal diffusion length with α the thermal diffusivity and ω the angular modulation frequency of light.

In the standard configuration of Fig.1a the signal is sometimes saturated due to the dominance of the component Q_1, especially for transparent samples. It should be expected that by suppressing it, the PPE signal component left, Q_2, would be richer in spectroscopic information on the sample, Fig.1b. Experimentally we have achieved this [8] by using a pyroelectric sensor provided with a highly reflecting front electrode, thus sending back out of the cell the transmitted radiation. We have developed the corresponding theory of the reflection PPE (RPPE) spectroscopy [9,10], which predicts the following spectroscopically relevant factors in the RPPE signal expressions, by comparison with the PPE ones:

case:		PPE (R=0)	RPPE (R=1)
1a,b	($\beta d>1$)	$\exp(-\sigma d)$	$\exp(-\sigma d)$
1c,2c	($ad>1$)	$\exp(-\beta d)$	$\exp(-\beta d)[1-\exp(-2\beta/\sigma)]$
2a,b	($ad,\beta d<1$)	1	$1-\exp(-2\beta d)\approx 2\beta d$

The dependence of the signals is obviously different in the two configurations, except for opaque samples, cases (1a,b), when the signal is saturated with respect to β and is attenuated and dephased $[\sigma=(1+i)a]$ with increasing sample thickness. For thermally thick samples, cases (1c,2c), the standard PPE method produces a transmission spectrum whereas the RPPE spectrum is either a transmission one, for high β, or an absorption one, for low β. Finally, the PPE spectrum of transparent and thermally thin materials (cases 2a,b) is saturated, but the corresponding RPPE absorption signal is proportional to β.

The general expression of the complex PPE signal can be best represented as amplitude and phase surfaces as a function of the two dimensionless ad and βd parameters [10]. In Fig.2 the amplitude values are proportional to the local heights of the surfaces. The special cases 1a-2c are assigned to distinct regions on these surfaces and the search for spectroscopically interesting cases reduces to the identification of the regions with positive or negative slope with respect to βd, corresponding to absorption or transmission spectroscopy, respectively. It is easy to conclude from Fig.2a that an optical transmission experiment should be designed in the cases (1c,2c). Fig.2b displays more experimental opportunities for the β measurement, via optical absorption (cases 2a-c) or transmission (case 1c). Concerning the phase surfaces, they contain less spectroscopic information on the sample, except for a steep discontinuity at the boundary between cases (1b)-(1c).

Fig.1a tells that if one is interested in the thermal properties of various samples, the wavelength dependence of the

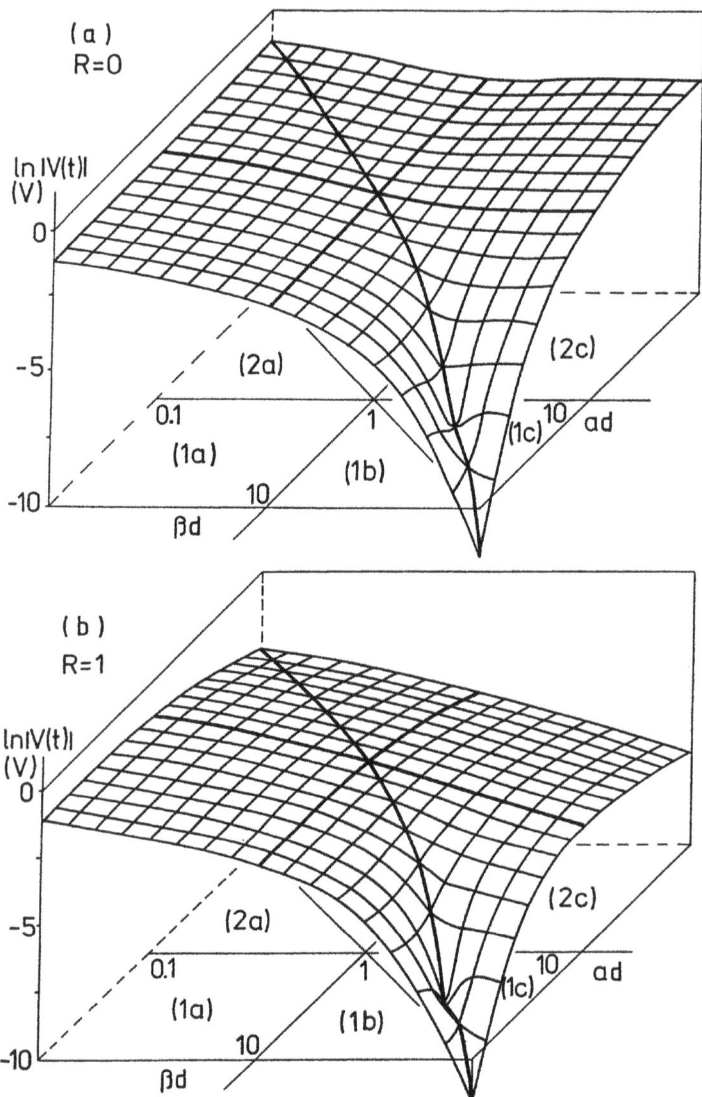

Fig.2. PPE (a) and RPPE (b) signal amplitude vs sample thermal and optical thicknesses.

PPE signal should be avoided. This can be done in the standard configuration by covering them with a thin black absorber, but this is not always possible and convenient. In the ideal case, the signal component Q_2 would be suppressed and the sample would not come into direct interaction with the radiation. We found the solution to these requirements in another new configuration, the inverse PPE (IPPE) technique [11], Fig.1c. Contrary to the conventional PPE, in the IPPE approach it is the thin sensor which faces the impinging radiation. The thermal wave propagating inwards from the blackened front face across the sensor reaches eventually the sample placed at its rear face, so that a fraction of the developed heat will leak into the sample acting like a heat-sink for the sensor. The average temperature variation, and hence the IPPE signal will be reduced when compared to the situation where no sample is being present. The same theory of Ref.10 with some symmetry considerations predicts various special cases in which it is possible to measure the thermal effusivity, the specific heat or the thermal conductivity of the sample. The IPPE technique is based on thermal wave absorption, and the PPE and RPPE ones are based on thermal wave transmission.

3. Applications

In the preceding section the radiation reflected from the sample surface was disregarded. It influences the signal amplitude in all the special cases by the constant factor $1-R_s$. It follows that in the otherwise saturated case (1a) it is possible to measure the surface absorptance of opaque materials. We have thus identified in the PPE spectra of anodized Al the presence of acid molecules adsorbed during the anodization process [5]. A more detailed analysis reveals that in fact it is the surface absorption modulated by the factor $\eta(\lambda)$ that is being measured. The ratio of the PPE signal to the quantity $1-R_s$, which can be derived from a black pyroelectric detector placed in front of the cell, would give directly the quantum efficiency spectrum. By using the set-up of Fig.3a, various fundamental processes such as optical transitions of polarons in mixed-valence compounds (Fig.3b) and in Pd-hydride have been investigated [12], or excitonic processes in La_2CuO_4 type superconductors.

For rather restrictive conditions, in the same case (1a) the PPE signal is proportional to the inverse specific heat, whereby the possibility to measure phase transitions [13]. We have observed the phase transitions of ferroelectric [14,15], ferromagnetic and itinerant electron antiferromagnetic [16], and superconducting [17] materials. It is worth mentioning a remarkable feature of this method if the sample is almost thermally thick. Then the signal amplitude decreases, but the relative sensitivity to the specific heat discontinuity increases compared to case (1a, $ad<1$), due to the exponential

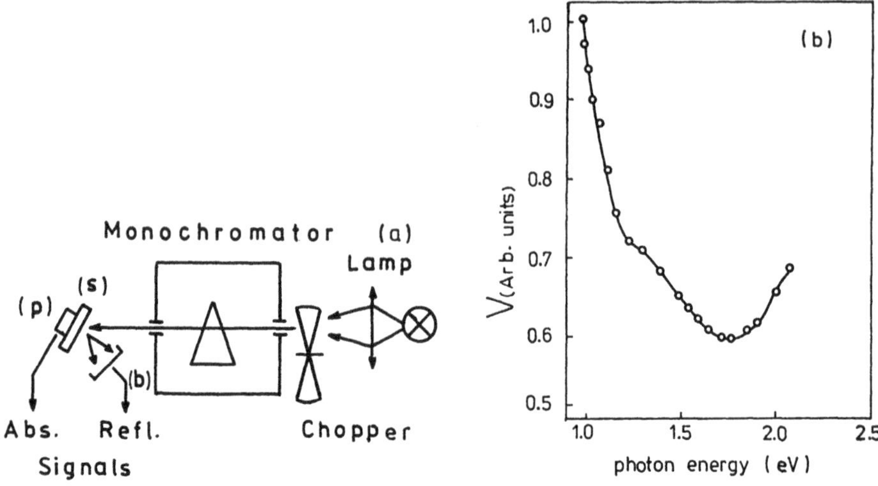

Fig.3. The set-up (a) for PPE radiation-to-heat conversion spectroscopy, and the spectral dip (b) showing non-thermal polaron transitions in KFe(III)Fe(II)$(CN)_4$, with: (p), pyroelectric; (s), sample; (b), black pyroelectric detector.

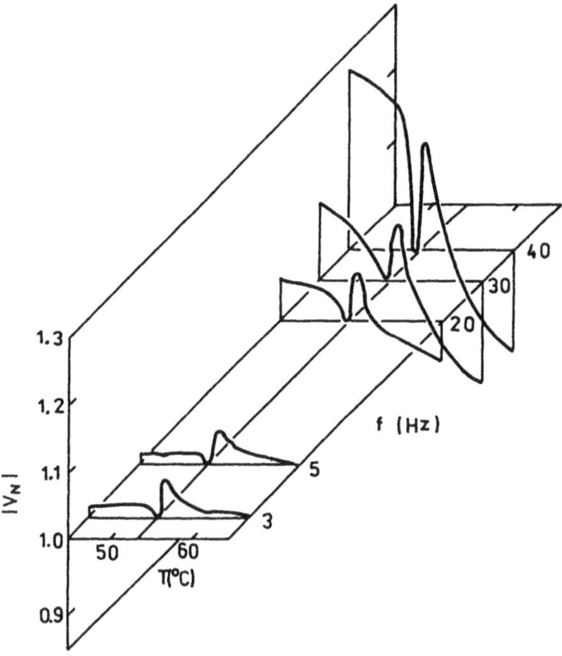

Fig.4. Normalized PPE signal amplitude exhibiting discontinuity enhancement upon ferro- to paramagnetic phase transition of $Ni_{70}Cu_{30}$ with increasing modulation frequency.

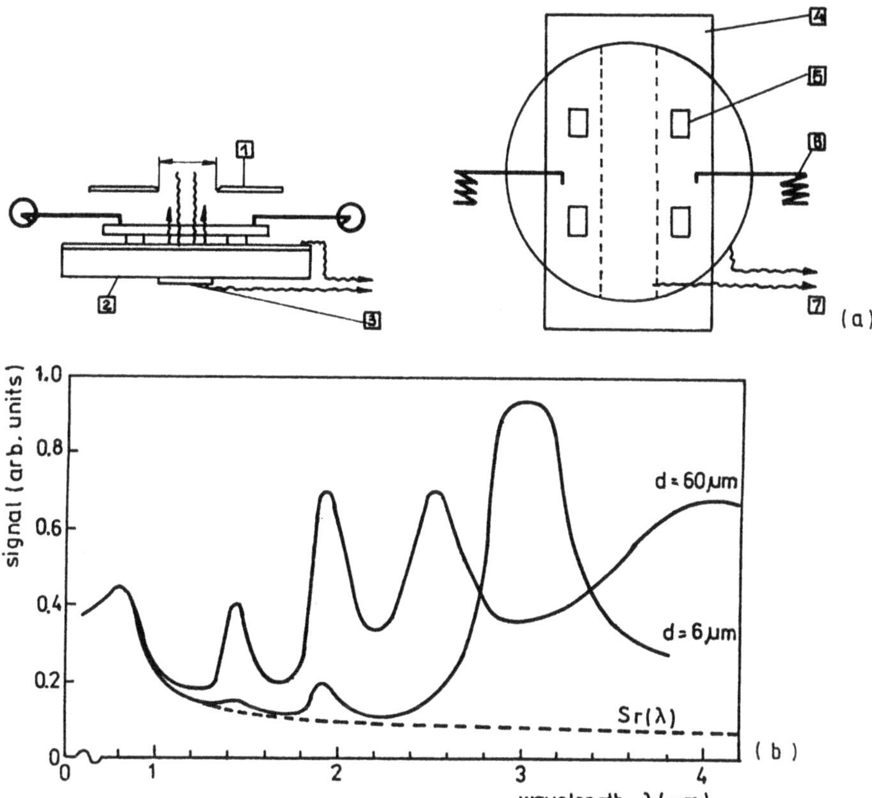

Fig.5. RPPE cell for liquids (a) and the spectra obtained for water (b). Sr is the residual cell absorption. 1, monochromator exit slit; 2, PZT sensor; 3, rear electrode; 4, window; 5, spacers; 6, joining spring; 7, signal.

dependence of case (1b, ad>1). This effect is illustrated in Fig.4 by the series of curves obtained at different chopping frequencies.

A compact RPPE cell for liquids and the obtained test spectrum of water are shown in Fig.5 [8]. For a 6μm sample thickness (low βd), the correct absorption spectrum is observed, but for 60μm, the strong absorption peak at λ=3μm, with βd=17, appears inverted, as a transmission peak. Such spectral inversion effects are discussed in Ref.9. The coaxial cell design of Fig.6 avoids also direct absorption of radiation in the pyroelectric and is intended for very low absorption measurements of He-Ne laser radiation in water, in order to detect sub ppm levels of phosphate pollutants.

We have applied the IPPE technique [11] for thermal effusivity measurement of metals [18] and for thermophysical

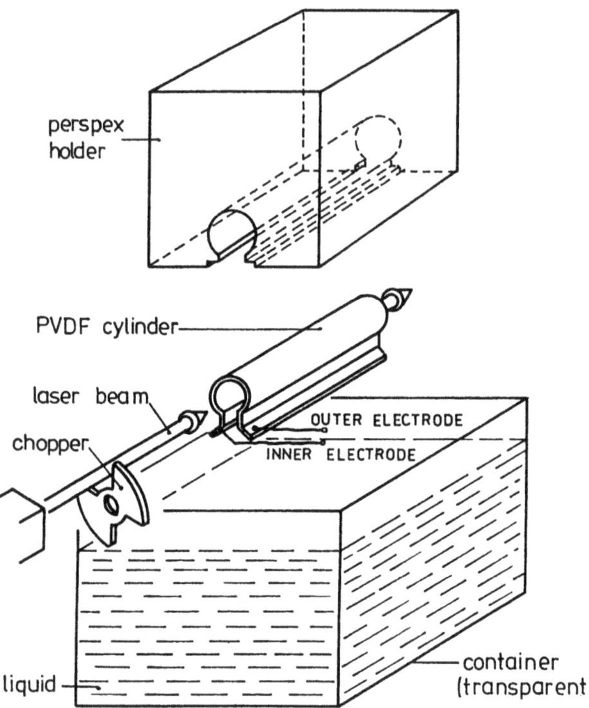

Fig.6. Coaxial configuration PPE cell for absorption spectroscopy of transparent liquids.

evaluation of food industry products, like cheese, yoghurt, margarine, pudding [19], mayonnaise, shortening, edible oil [20] and alcoholic mixtures [21]. The sensitivity of the PPE cell to thin layers of materials in the μm range has suggested its application as atmospheric water vapour condensation sensor [22], Fig.7. The IPPE technique has the advantage to compare materials regardless of their optical properties and sample geometry or size, since the excitation and the temperature measurement are made indirectly and from the same side.

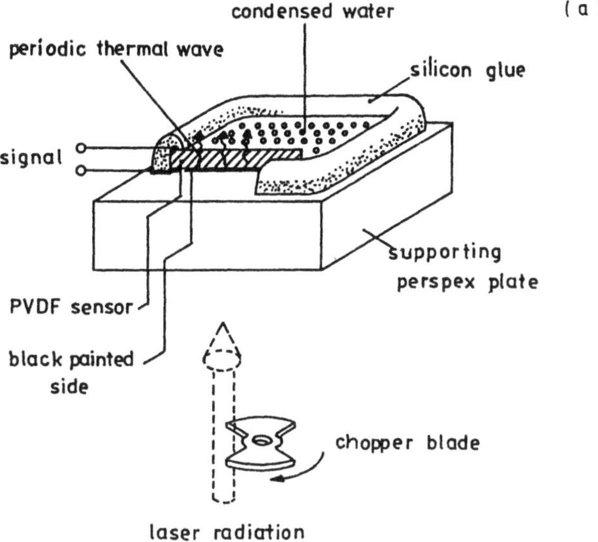

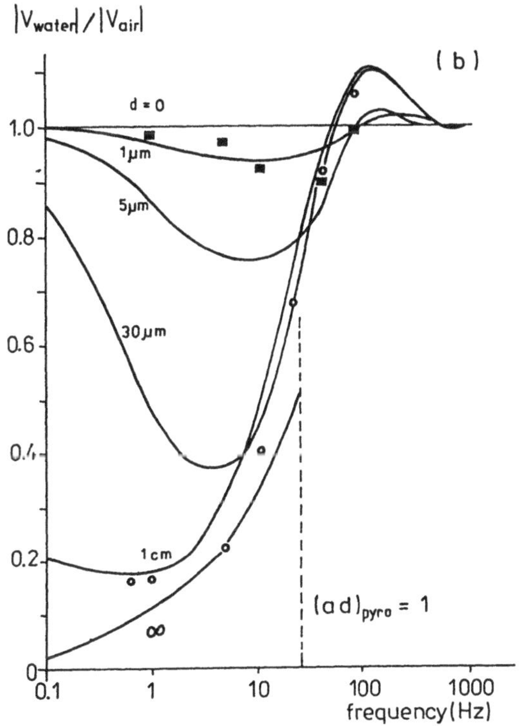

Fig.7. IPPE water vapour condensation monitor, (a) and the calculated and experimental sensitivity, (b).

4. Conclusion

As a general conclusion, the PPE methods are remarkably versatile for thermal and optical materials characterization. The various configurations are complementary and should be chosen according to the specific experimental tasks.

References

[1] A.C.Tam, Rev.Mod.Phys. **58**, 381 (1986).

[2] S.B.Lang and F.Steckel, Rev.Sci.Instrum. **36**, 1817 (1965).

[3] H.Coufal, Appl.Phys.Lett. **44**, 59 (1984).

[4] A.Mandelis, Chem.Phys.Lett. **108**, 388 (1984).

[5] D.Dadarlat, M.Chirtoc, R.M.Candea and I.Bratu,Infrared Phys. **24**, 469 (1984).

[6] A.Mandelis and M.M.Zver, J.Appl.Phys. **57**, 4421 (1985).

[7] M.Chirtoc, G.Mihailescu and D.D.Bicanic, in "Photoacoustic and Photothermal Phenomena II", J.C.Murphy, J.W. Maclachlan-Spicer, L.Aamodt and B.S.N.Royce, Editors, Springer Series in Optical Sciences Vol.62; A.L. Schawlow, Series Editor Springer Verlag, Berlin, 1990, pp 537.

[8] M.Chirtoc, D.Dadarlat, I.Chirtoc and D.D.Bicanic, Spectroscopy Letters **21**, 413 (1988).

[9] M.Chirtoc and I.Chirtoc, Infrared Phys. **29**, 847 (1989).

[10] M.Chirtoc and G.Mihailescu, Phys.Rev. **B40**, 9606 (1989).

[11] M.Chirtoc, D.D.Bicanic, D.Dadarlat and I.Chirtoc, in this Conference.

[12] D.Lupu, M.Chirtoc, D.Dadarlat, R.M.Candea, I.Bratu and V.Mecea, Phys.Stat.Sol. (b) 163, 519 (1991).

[13] A.Mandelis, F.Care, K.K.Chan and L.C.M.Miranda, Appl.Phys. **A38**, 117 (1985).

[14] D.Dadarlat, M.Chirtoc and D.D.Bicanic, same as Ref.7, pp.300.

[15] D.Dadarlat, M.Chirtoc and D.D.Bicanic, Appl.Phys. **A50**, 357 (1990).

[16] D.Dadarlat, M.Chirtoc, C.Neamtu and D.D.Bicanic, J.Phys. Chem.Sol. **51**, 1369 (1990).

[17] D.Dadarlat, R.M.Candea, A.Frandas, M.Chirtoc and D.D.Bicanic, in this Conference.

[18] D.Dadarlat, M.Chirtoc, C.Neamtu, R.M.Candea and D.D.Bicanic, Phys.Stat.Sol. (a) **K121**, K231 (1990).

[19] D.D.Bicanic, M.Chirtoc, D.Dadarlat and I.Chirtoc, Dairy Science, (1991) (submitted).

[20] D.D.Bicanic, M.Chirtoc, D.Dadarlat, P.van Bovenkamp and H. van Schayk, Appl.Spectroscopy, (1991) (in press).

[21] M.Chirtoc, V.Tosa, D.D.Bicanic and P.Torfs, Ber.Bunsen Ges.Phys.Chem., (1991) (in press).

[22] M.Chirtoc, D.D.Bicanic and V.Tosa, Rev.Sci.Instrum., (1991) (in press).

A Tutorial on the State-of-the Art of FTIR Photoacoustic Spectroscopy

J.F. McClelland[1,2,3], *S. Luo*[1], *R.W. Jones*[1,2], *and L.M. Seaverson*[3]

[1]Ames Laboratory–USDOE, Spedding Hall and
[2]Center for Advanced Technology Development,
Iowa State University, Ames, IA 50011, USA
[3]MTEC Photoacoustics, Inc., P.O. Box 1095, Ames, IA 50010, USA

Abstract. The current status and capabilities of FTIR PAS are discussed including historical background, signal generation, FTIR spectrometer and photoacoustic detector technology and use, and qualitative and quantitative applications that demonstrate the state-of-the-art analytical capabilities of the technique.

1. Introduction

The photoacoustic effect, in which light absorption produces heat and consequently a thermal expansion driven acoustic response, was discovered in 1880 by A. G. Bell[1] during his research on the photophone, a device that transmitted the human voice via an intensity modulated light beam. Bell also investigated photoacoustic spectroscopy (PAS) as a means to measure the absorption spectrum of solid materials. A PAS renaissance occurred in 1973 stimulated by the work of Rosencwaig then at Bell Laboratories[2]. In 1978 Busse performed the first PAS with an FTIR spectrometer when he measured the absorption spectrum of gases[3]. In 1980 Rockley[4] and Vidrine[5] independently published the first FTIR PAS measurements on solids.

The main factor behind the interest in PAS is that the spectra of opaque materials can be measured directly because the sampling depth in PAS is defined by the material's thermal and optical properties rather than by physical thickness. In transmission based spectroscopies (which include diffuse reflection[6]) the physical sample thickness defines the sampling depth. Since most materials are opaque as they naturally exist, sample preparation is required consisting of thinning and/or dilution in order to allow light to be transmitted. Such preparations are often not desireable due to time, effort, and sample alteration considerations.

Although PAS enables spectra to be measured of opaque materials, two factors have retarded the rapid utilization of the technique. First, heat transfer is very inefficient for light absorption generated heat from solid materials into an adjacent gas where thermal expansion causes the acoustic signal. This inefficiency led to a poor signal-to-noise ratio for early photoacoustic spectra and consequent difficulties for computer processing of spectra to extract analytical information. Secondly, the strong bands of photoacoutic spectra are truncated due to signal saturation which is discussed in the next section. This spectral distortion and nonlinearity problem was a barrier to qualitative and quantitative analyses until the signal-to-noise ratio of FTIR PAS spectra improved to the point that advanced computer data reduction methods could be employed. Only at that point did it become apparent that FTIR PAS spectra have the information content necessary for accurate qualitative and quantitative analyses.

2. Signal Generation[7]

The sequence of PA signal generation shown in Figure 1 involves a fraction, R, of the incident light beam, P_o, being reflected at the sample surface and the fraction (1-R) passing into the sample and experiencing an exponential decay due to the sample's absorption coefficient, β. Some, if not all of the absorbed radiation is converted into heat which causes temperature oscillations in the sample if the intensity of the beam is modulated at a frequency, $f = \omega/2\pi$. The amplitude of the temperature oscillations in a layer dx at depth x is proportional to $P_o(1\text{-}R)\beta e^{-\beta x}dx$.

The temperature oscillations generate thermal-waves[8] which propagate with a decay coefficient, $a_s = (\pi f/D)^{1/2}$, where D is the sample's thermal diffusivity. The thermal-waves generated at a depth x in the sample decay to an amplitude proportional to $\beta e^{-(\beta+a_s)x}$ at the surface on which the light beam is incident and from which heat is transferred into an adjacent gas layer as shown in Figure 2. Only a small fraction of the thermal-wave is actually transmitted into the gas due to thermal-wave reflection at the sample surface. This transmitted heat causes expansion of the gas which is the PA signal detected by a sensitive microphone. Note that when $\beta \ll a_s$, the PA signal is directly proportional to the spectroscopic quantity of interest, β. This is not the case in transmittance and reflectance spectroscopy where absorbance is not measured directly and must be calculated after using simplifying assumptions that may not be realistic.

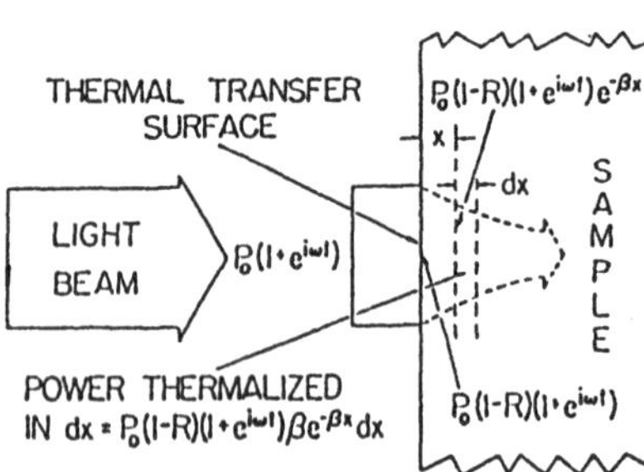

Figure 1. Illustration of the optical power deposition in a sample with incident power P_o, sample reflectivity R, and absorption coefficient, β.

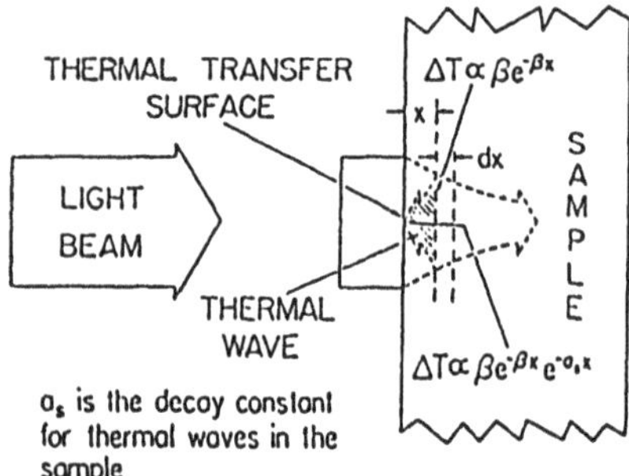

Figure 2. Illustration of the thermal-wave propagating from a layer at x toward the thermal transfer surface. ΔT, the wave amplitude, is dependent on the optical and thermal properties of the sample.

Both the optical and thermal-wave decay coefficients, β and a_s respectively, play an important role in PA signal generation. When $\beta<<a_s$, a_s controls the depth from which absorbed light can contribute to the PA signal. 63% of the PA signal comes from a depth within one thermal-wave decay length, L, given by $L = 1/a_s$. Hence, when $\beta<<a_s$, the effective PA sample thickness or sampling depth of an FTIR PA measurement is approximately $L = (D/\pi v\nu)^{1/2}$ where the substitution $f = v\nu$ has been made since the FTIR interferometer optical path difference (OPD) mirror velocity, v, and the wavenumber, ν, determine the modulation frequency in FTIR spectrometers where the mirror moves with constant velocity. Note that the sampling depth increases with decreasing wavenumber across the spectrum of a sample. In recently introduced step-scan FTIR spectrometers a constant modulation frequency and consequently sampling depth, $L = (D/\pi f)^{1/2}$, can be achieved across the spectrum by independently controlled phase or chopper modulation methods.

The PA signal increases linearly with β when $\beta<<a_s$ but loses linearity (often called the onset of signal saturation) with increasing β as the optical absorption induced heating occurs more and more within the effective PA sample thickness, L. At very high absorption, heating occurs very close to the sample surface and the PA signal becomes insensitive to increases in absorption coefficient (often called full signal saturation) although some spectral features may still occur due to the (1-R) term in the PA signal expression.

The weaker bands of most FTIR PA spectra are in the range where the PA signal is linear with absorption. Stronger bands are usually truncated to some degree due to the onset of saturation having occured but spectral features seldom appear due to reflectivity effects associated with full saturation. The effective PA sample thickness or sampling depth, L, can be decreased in order to move the onset of saturation to higher absorbance by measuring spectra at higher modulation frequencies as controlled by the FTIR mirror velocity or independently controlled modulator.

In the analysis of samples with concentration gradients or layers it is particularly desireable to be able to vary the sampling depth. The most straight forward situation is one where the thermal diffusivity is homogeneous and where weak absorption bands (such that the peak β values are $<<a_s$) can be used both for monitoring the depth varying concentration and for an internal standard associated with a constant concentration with depth. Furthermore, the sample's thermal diffusivity, D, and the modulation frequency, f, must allow the sampling depth, $L = (D/\pi f)^{1/2}$, to be adjusted over the range of interest by setting f via the mirror velocity or the external modulator. Currently the most advanced FTIR interferometer (Bio Rad 896[9]) can generate mirror velocities from 2.4 cm/s to 2.4×10^{-7} cm/s which corresponds to a very wide range of modulation frequencies over the UV to far IR spectral range. Currently the photoacoustic detector (MTEC 200) frequency range is approximately 5 Hz to 6 kHz so not all of the modulation

frequency range of the Bio Rad 896 is useable. In externally modulated operation the Bio Rad 896 covers the range 5 Hz to 400 Hz.

Nonhomogeneous samples are more difficult to analyze when the thermal diffusivity varies significantly with depth and when the absorption coefficient is not small relative to the thermal-wave decay coefficient. Variations in the thermal diffusivity between thin layers of a material result in variations in sampling depth within the sample and in thermal-wave reflection effects which make interpretation of results less obvious. Analyses based on stronger absorption bands also can lead to ambiguities since the sampling depth is now determined by the decay of both the infrared beam and thermal-waves within the sample. Furthermore when the sampling depth is reduced by increasing the mirror velocity and consequently the modulation frequency, some reduction in saturation induced truncation of strong bands occurs. This observation can be incorrectly interpreted as an increase in species concentration near the surface.

Table 1 gives sampling depths for typical values of thermal diffusivities and modulation frequencies. The thermal-wave decay coefficient, which should be larger than the analysis peak absorption coefficient value, is just the reciprocal of the sampling depth.

Table 1. Sampling depths in μm for typical thermal diffusivities from 0.1 to 0.001 cm^2/s and modulation frequencies from 10 to 10^4 Hz.

$D(cm^2/s)$	10 Hz	100 Hz	1000 Hz	10000 Hz
0.1	600	200	60	20
0.01	200	60	20	6
0.001	60	20	6	2

3. Experimental

The overall signal-to-noise ratio improvement in FTIR PAS spectra is due to a number of improvements in FTIR and PAS instrument performance and operating parameters. A complete discussion of these isn't practical within the scope of this paper which focuses only on the key points. Typical FTIR operating parameters are 8 cm^{-1} spectral resolution, 400-4000 cm^{-1} spectral range, 0.10 cm/s mirror velocity (40 Hz at 400 cm^{-1} and 400 Hz at 4000 cm^{-1}) and 32 co-added interferometer scans (the signal-to-noise increases with the square root of the scan number).

The photoacoustic detector microphone sensitivity should be 50mV/Pa (Bruel and Kjaer 4176) to achieve a high signal-to-noise ratio. The detector should operate with a helium gas atmosphere and with a vibration isolation mounting to reduce structure borne noise. The sample-to-detector window distance should be about 1 mm to allow generation of the photoacoustic signal in the gas, but any additional gas volume should be filled by spacers to increase the signal level. If moisture is associated with the sample, magnesium perchlorate desiccant should be used in the sample chamber to avoid spectral interference and signal phase distortion due to vapor species absorption.

FTIR PAS spectra should be normalized to account for spectral variations in the FTIR source and optics and for sample-to-sample volume variations in the quantitative analyses. Spectral normalization is done by ratioing the sample spectrum by the spectrum of carbon black whereas volume normalization is based usually on an internal standard (an absorbance band that doesn't vary with concentration) or on weighing.

4. Applications

In order to illustrate the state-of-the-art analytical capabilities of FTIR PAS, three applications are discussed in this section including a qualitative forensic analysis to identify the chemical compound type of single fibers, quantitative analysis of polymer concentrations in copolymers and minerals in coal, and depth profiling of a multilayer polymer material.

Qualitative Analysis

Photoacoustic spectra for forensic single fiber identification were measured (at an OPD velocity of 0.05 cm/s) directly on single fibers without the pressing operation to reduce optical density common to FTIR microscopy. Figures 3 & 4 show spectra of nylons and polyesters, respectively, measured by FTIR photoacoustic spectroscopy (top group) and FTIR microscopy (bottom group). The strong bands of the photoacoustic spectra are truncated due to signal saturation while weak bands at low wavenumber are enhanced due to the multiple heat transferring reflections of low frequency thermal-waves which occur in photoacoustic signal generation with microsamples. These two effects result in spectra with more distinctive information over a wider spectral range because the limitation of the narrow band MCT detector common to FTIR microscopy is avoided.

Figure 5 shows FTIR photoacoustic spectra of nylon 6,6 and PET fibers of different diameters and cross-sections. The peak locations remain rather well fixed from spectrum to spectrum although relative peak heights vary. Due to this variation and the discussion related to Figure 3, it is important to base computer searches on peak locations only (not on heights) and to use photoacoustic spectra in the search library. When these conditions are met 100% search success has been achieved on over 20 unknowns using a library of 43 fibers[10] of different polymer compositions with the Perkin-Elmer Search routine and using an appropriate search threshold.

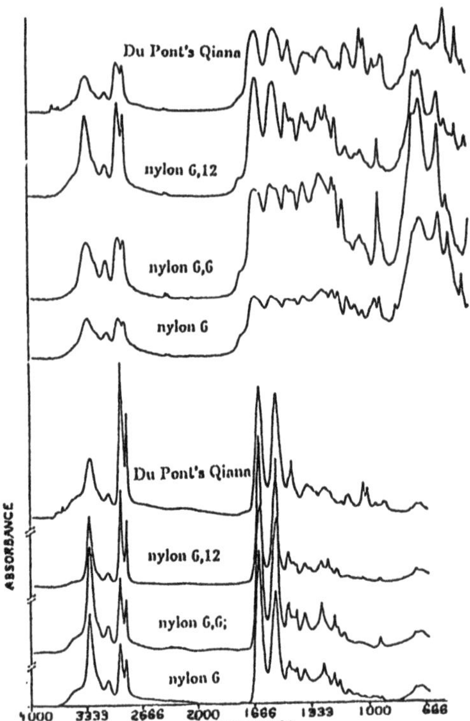

Figure 3. Photoacoustic (top four) and FTIR microscope (bottom four) spectra of various nylon single fibers. The microscope spectra are from Ref. 10.

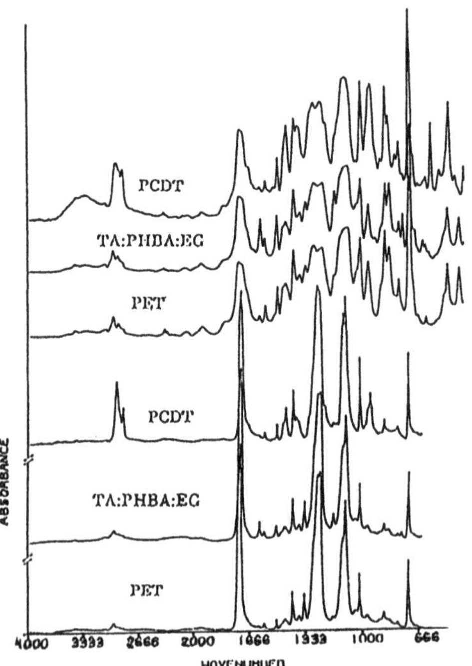

Figure 4.
Photoacoustic (top three) and FTIR microscope (bottom three) spectra of various polyester single fibers. The microscope spectra are from Ref. 10.

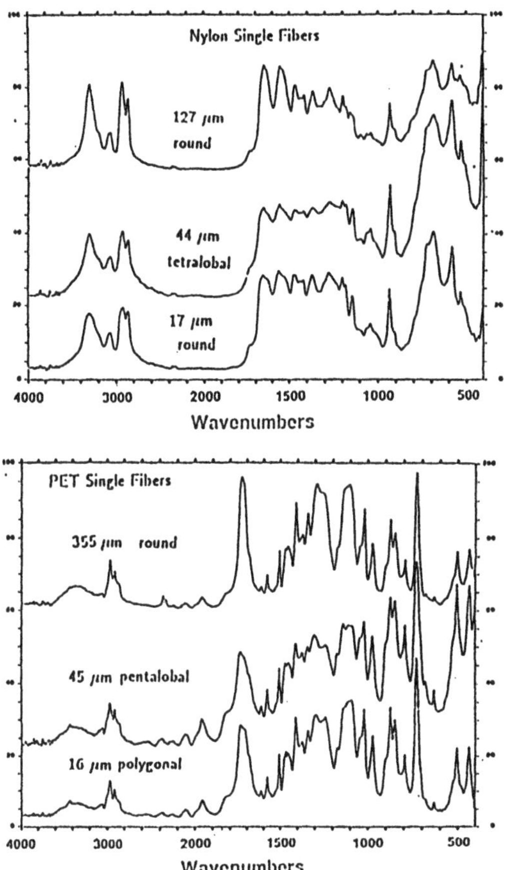

Figure 5. Spectra of nylon 6,6 and PET single fibers with different diameters and cross-section geometries

Quantitative Analysis

Copolymers of vinyl acetate in polyethylene were quantitatively analyzed with material in both molding pellet and sheet form. Thirteen pellet and fourteen sheet samples of different concentrations were measured. Spectra of three pellet samples are shown in Figure 6. In order to account for signal level differences due to varying sample volume all spectra were scaled by the C-H band area (3120-2750 cm^{-1}) The spectra were processed to determine mass % vinyl acetate concentration by the Perkin-Elmer CIRCOM factor analysis program set to include these spectral ranges: 3120-2750 cm^{-1}, 1900-1630 cm^{-1}, 1490-578 cm^{-1}. Standard errors of prediction (SEP) were calculated by successively treating one sample as an unknown and the others as the learning set. A SEP of 0.64% vinyl acetate was calculated using spectra of both pellet and sheet stock, whereas SEPs of 0.70% and 0.48% were found for pellet and sheet stock spectra alone, respectively. Figure 7 shows a plot of vinyl acetate concentration predicted from the spectra vs. the concentration measured by titration.

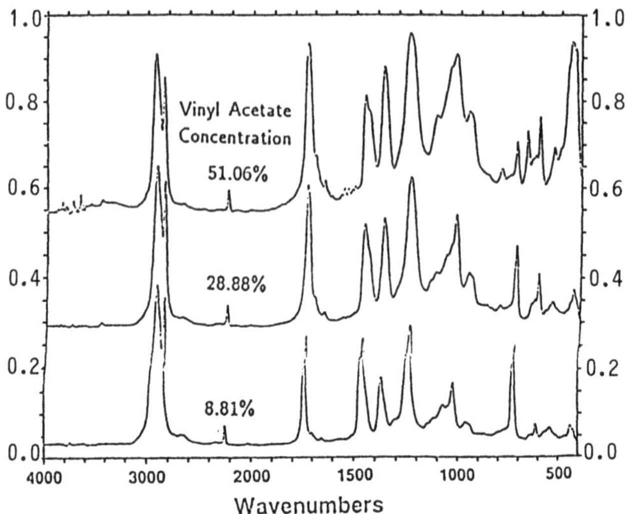

Figure 6. FTIR photoacoustic spectra of copolymer molding compound beads with three different concentrations of vinyl acetate.

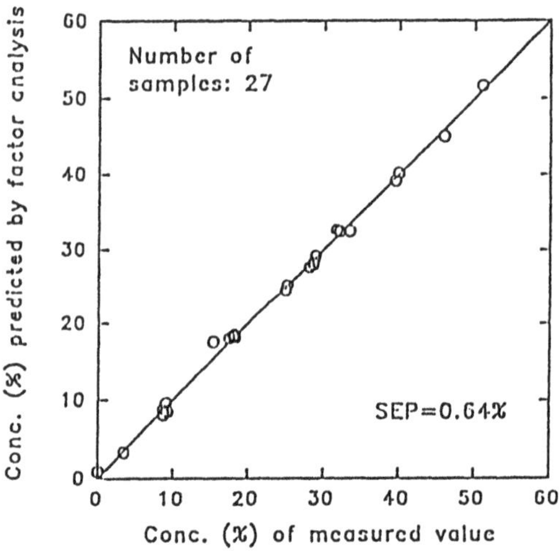

Figure 7. Plot of vinyl acetate concentration of measured value against predicted concentration by factor analysis for both pellet and sheet copolymer samples.

A four component system was analyzed to demonstrate the capability of PAS on a multicomponent system. It consisted of coal and the commonly coal associated minerals kaolinite, pyrite, and quartz. The four components were ground to -100 mesh and mixed into 15 samples of various compositions with each mineral concentration varying from 0% to 30% and a total mineral concentration for each sample of about 30% by weight. FTIR photoacoustic spectra were measured and

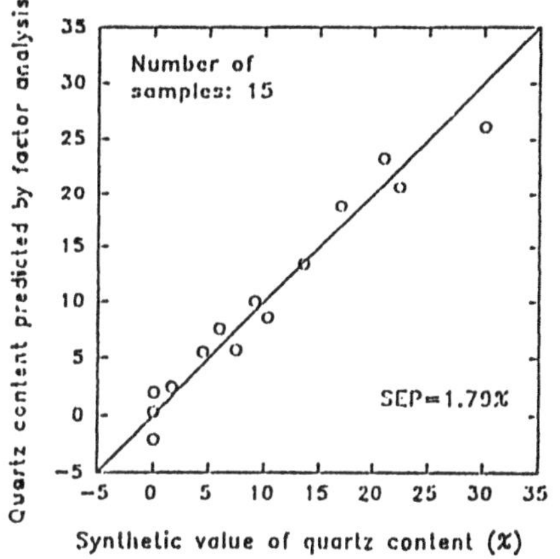

Figure 8. Plot of the factor analysis vs. the synthetic values of quartz content

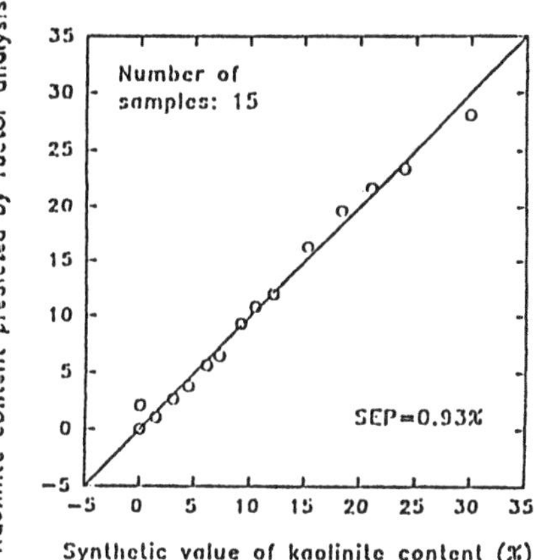

Figure 9. Plot of the factor analysis vs. the synthetic value of kaolinite content

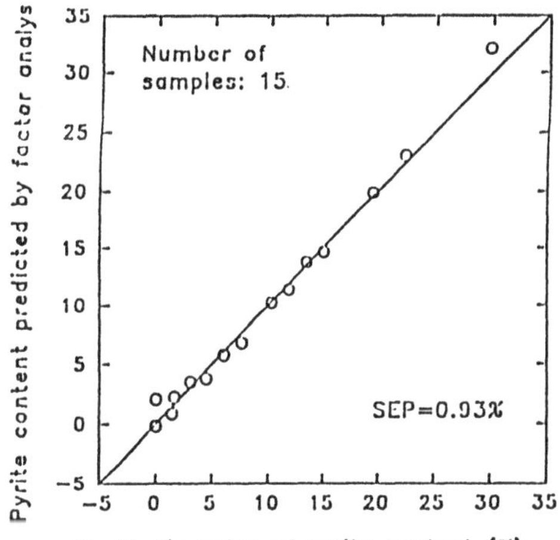

Figure 10. Plot of the factor analysis vs. the synthetic values of pyrite content

processed by the Perkin-Elmer CIRCOM program using the spectral ranges of 3780-3580 cm^{-1}, 2130-1830 cm^{-1}, and 1160-440 cm^{-1} after normalization by the area of the 1600 cm^{-1} band (1790-1532 cm^{-1}). SEP's for quartz, pyrite, and kaolinite of 1.79%, 0.93%, and 0.93%, respectively, were calculated when a single sample was successively treated as an unknown and the others as the learning set. Figures 8, 9, and 10 show the mineral concentrations predicted from the infrared spectra vs. the values known from synthesis.

Depth Profiling

Multilayered samples are an important class of materials for infrared analysis. FTIR PAS allows the sampling depth of spectra to be varied based on a material's

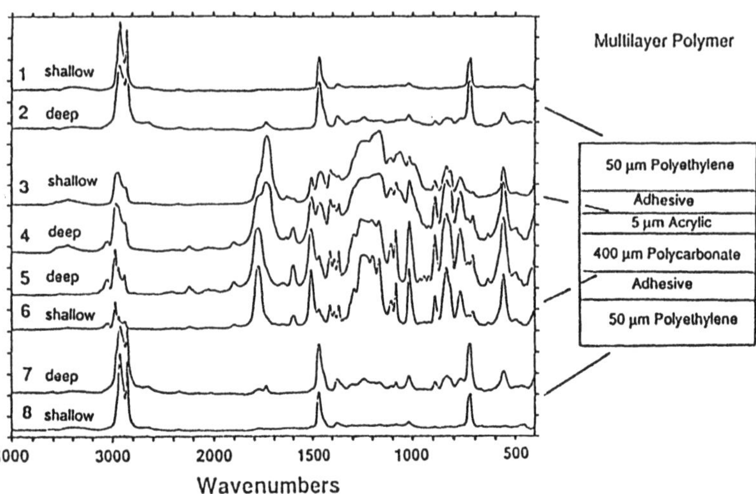

Figure 11. Multilayer polymer spectra

thermal response at its surface which receives infrared absorption generated heat from deeper into the material as the infrared beam modulation frequency imposed by the FTIR mirror velocity decreases. Figure 11 shows spectra of a multilayer polymer measured with shallow and deep sampling depths corresponding to optical path difference velocities of 0.50 cm/s and 0.05 cm/s, respectively. Spectrum 1 has the bands of the polyethylene surface layer alone whereas the deeper probing spectrum 2 shows bands from the adhesive and acrylic layers. Spectra 3 and 4, measured after stripping off the polyethylene/adhesive layers, show bands of the acrylic surface layer and polycarbonate substrate with the latter's bands being more prominent in the deep sampling spectrum. Spectra 5 and 6, also measured after stripping, show only bands of polycarbonate due to the 400 μm thickness of the substrate. The differences that do appear between spectra 5 and 6 are due to truncation of strong bands associated with the deeper sampling of spectrum 5 relative to 6. Spectra 7 and 8 are similar to Spectra 1 and 2 except that in the deep sampling spectra some polycarbonate rather than acrylic bands show up.

In Figure 12 spectra are plotted of the bare and adhesive sides of the 50 μm polyethylene film which composed parts of the multilayer polymer of Figure 11. Spectral subtraction allows the adhesive spectrum to be isolated and the adhesive band positions in spectra 2 and 7 of Figure 11 to be located.

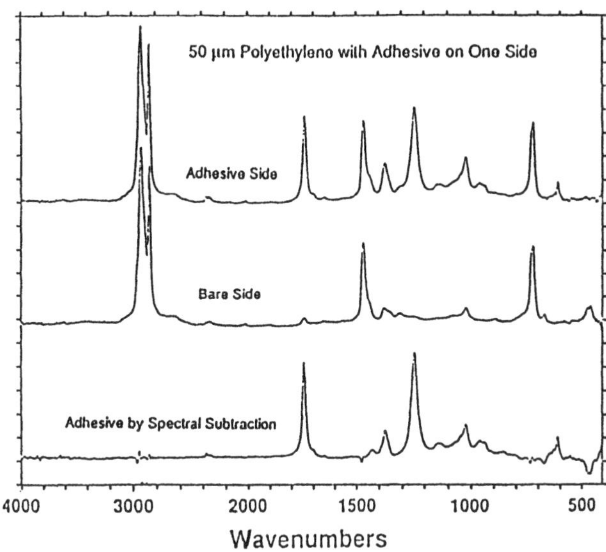

Figure 12. Spectral subtraction allows the adhesive of the multilayer polymer sample (Figure 11) to be isolated

5. Conclusions

The applications studies show that the information content in FTIR PAS spectra is sufficient to perform accurate qualitative and quantitative analyses of solid materials. These applications demonstrate that the "no sample preparation" capability of FTIR PAS can now be exploited to great advantage. Futhermore, FTIR PAS is applicable both to microsamples such as single fibers and to macrosamples such as molding compound pellets. Finally, the depth varying sampling capability of FTIR PAS is demonstrated to be useful in the analysis of multilayered polymer systems.

Acknowledgements

Mary Tungol and Ed Bartick (FBI Academy, Forensic Science Research Unit) provided fiber samples and information.

Trevor Matheson (Coal Research Association of New Zealand, Inc.) supplied samples of New Zealand Pyramide No. 2 coal.

This work was funded by MTEC Photoacoustics, Inc. and by the Center for Advanced Technology Development, which is operated for the U.S. Department of Commerce by Iowa State University under Grant No. ITA 87-02, and by Ames Laboratory, which is operated for the U.S. Department of Energy by Iowa State University under Contract No. W-7405-ENG-82, supported by the Assistant Secretary for Fossil Energy.

References

1. A. G. Bell, Proc. Amer. Ass. Advan. Sci. 29, 115 (1880).

2. A. Rosencwaig, Opt. Commun. 7, 305 (1973).

3. G. Busse and B. Bullemer, Infrared Phys. 18, 225 (1978).

4. M. G. Rockley, Appl. Spectrosc. 34, 405 (1980).

5. D. W. Vidrine, Appl. Spectrosc. 34, 314 (1980).

6. The spectral information in a diffuse reflectance measurement is primarily encoded by transmission of light through powder particles as evidenced by negative going peaks corresponding to absorbance bands.

7. This section has been updated to include discussion of step-scan FTIR PAS spectroscopy. In its earlier form it was part of a presentation at and proceedings of the International Workshop on Fourier Transform Infrared Spectroscopy, Sept. 19-21, 1990, Antwerp, Belgium.

8. F. A. McDonald, American Journal of Physics. 48, 41 (1980).

9. R. A. Crocombe and S. V. Compton, The Design Performance and Applications of a Dynamically-Aligned Step-Scan Interferometer, FTS/IR Notes No. 82, Bio-Rad, Digilab Div., 237 Putnam Ave., Cambridge, MA 02139 USA (June 1991).

10. M. W. Tungol, E. G. Bartick, and A. Montaser, Appl. Spectrosc. 44, 543 (1990).

Transient Infrared Spectroscopy

R.W. Jones and J.F. McClelland

Center for Advanced Technology Development
and Ames Laboratory–USDOE, Spedding Hall,
Iowa State University, Ames, IA 50011, USA

Abstract. Transient infrared spectroscopy is a pair of closely related techniques that allow measurement of useful spectra from opaque, moving materials. In the techniques a transient thermal gradient is used to spectroscopically isolate a thin surface layer from the bulk of the sample. The basic elements of the techniques and some industrially relevant applications are described.

1. Introduction

The real-time, on-line analysis of materials is increasingly important for process and quality control. Although infrared spectroscopy could supply much of the desired information, many materials can not be analyzed by conventional techniques without sample preparation. Conventional infrared (thermal) emission spectroscopy is a single ended, noncontact method that could be used on line except that it has a sample opacity limitation similar to that of transmission spectroscopy. The emission from deep layers of optically thick materials is absorbed by overlying layers. This self-absorption results in a featureless, blackbody-like emission spectrum [1].

Transient infrared emission spectroscopy (TIRES) [2] and transient infrared transmission spectroscopy (TIRTS) [3] are two closely related methods which are based on thermal emission, but avoid the self-absorption problem and produce useful spectra from moving materials, even if they are opaque. In both methods a transient thermal gradient is induced at the sample surface and the emission from the surface is analyzed. Figure 1 shows the general scheme. In TIRES a jet of hot gas and in TIRTS a jet of cold gas is aimed onto the moving sample within the field of view of a spectrometer. Although the heated or cooled layer thickens quickly, sample motion carries it out of

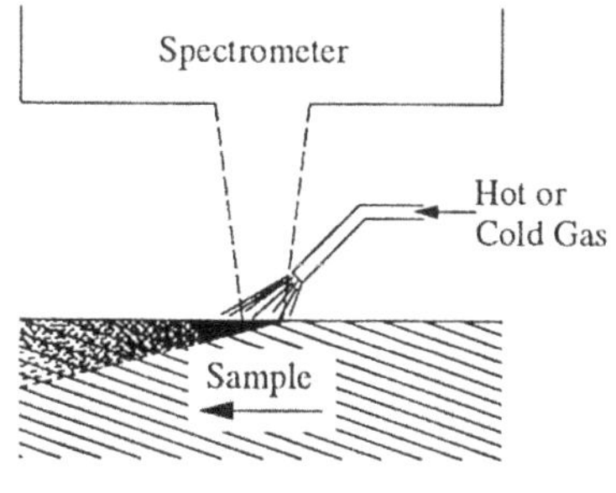

Figure 1. Schematic of TIRES and TIRTS methods. A jet of hot or cold gas aimed on the moving sample surface within the field of view of the spectrometer (dashed lines) produces a thin hot or cold layer from which the spectrum arises.

Springer Series in Optical Sciences, Vol. 69 Photoacoustic and Photothermal Phenomena III
Editor: D. Bićanić

the field of view, so the spectrometer always sees a thin, hot or cold layer. In TIRES the heated layer is an emission source and the TIRES spectrum is equivalent to the emission spectrum of an optically thin sample. In TIRTS the cold layer acts as a thin transmission sample. Thermal (blackbody-like) emission from the uncooled bulk of the sample undergoes selective absorption as it passes through the cooled layer on its way to the spectrometer.

2. Experimental Section

The details of our TIRES and TIRTS apparatus have appeared elsewhere [2,3], so only a brief description is given here. The sample was mounted where the source for our FTIR spectrometer is usually located. The sample was formed into or attached to a disk and mounted on the shaft of a motor so that it could be spun through the spectrometer field of view. The hot jet consisted of nitrogen passed over a resistive heating element, and the cold jet was helium passed through a liquid-nitrogen bath. The spectrometer operated at 8 cm^{-1} resolution with an MCT detector.

Except for structure, the raw TIRES and TIRTS spectra vary in intensity with wavenumber as blackbody spectra do. The spectra are normalized (that is, converted to emittance and transmittance) by ratioing them against blackbody spectra.

3. Results and Discussion

Figure 2 shows spectra taken on a 3-mm-thick sample of polycarbonate. The conventional emission spectrum in Figure 2a was observed with the whole sample heated to 91 °C. Directly below it is a blackbody spectrum from carbon black. There is very little difference between these two. In contrast, the TIRES and TIRTS spectra have the same structure one would find by conventional means with a properly prepared (thinned) sample. These same polycarbonate spectra are displayed in Figure 2b after

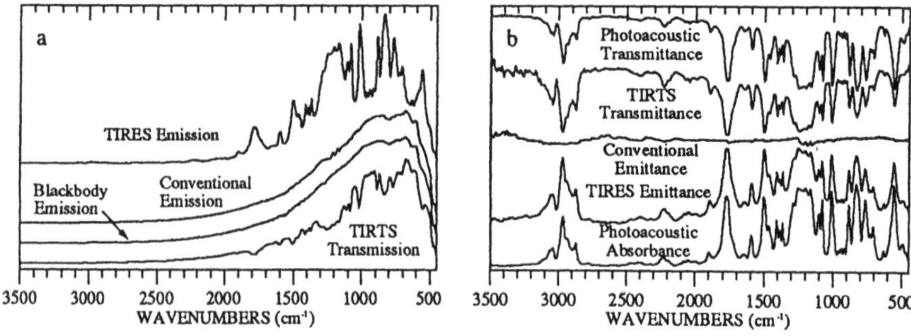

Figure 2. (a) Raw and (b) normalized spectra of polycarbonate. The TIRES and TIRTS spectra were taken at a sample speed of 41 cm/s.

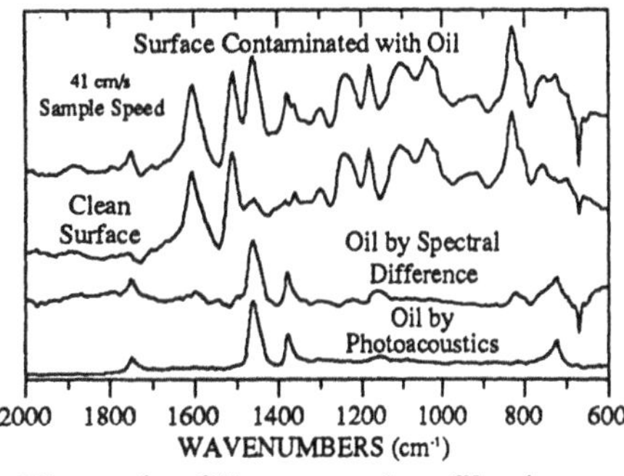

Figure 3. Oil on a carbon-fiber/epoxy composite observed using TIRES.

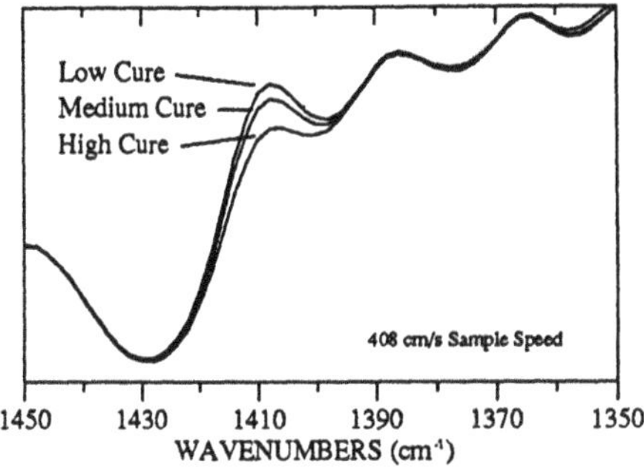

Figure 4. Cure of an acrylic coating on polycarbonate observed using TIRES.

normalization to remove the intensity variation. An absorbance spectrum of the polycarbonate taken with a photoacoustic detector is shown at the bottom of the figure for reference. This same spectrum, converted mathematically to transmittance form, is shown at the top. The normalized conventional spectrum is nearly flat, while the TIRES and TIRTS spectra duplicate the structure seen in the photoacoustic spectra. The signal-to-noise ratios of the TIRES, TIRTS, and conventional spectra all degrade on the left side, but this is the direct result of the fall off in intensities visible in Figure 2a. Within the fingerprint region the signal-to-noise ratios are good.

Our early qualitative and quantitative analyses of a range of materials using TIRES and TIRTS are reported elsewhere [2,3]. Most recently we have been transferring the technology from the laboratory to the industrial plant. As part of that process we have been examining analysis problems brought to us by industry. Figures 3 and 4 show examples of these. One company needed to detect and identify contamination on carbon-fiber/epoxy composite surfaces. The composite is a hard, black material not amenable to spectroscopic analysis. Figure 3 shows normalized TIRES spectra of the clean and oil-contaminated composite, along with a spectrum of their difference. This difference spectrum compares well with the reference spectrum of the pure oil taken photoacoustically. Another company wished to monitor the cure of a thin acrylic coating on polycarbonate sheet. Figure 4 shows a portion of the raw (not normalized) TIRES spectra of three different samples. The variation in cure is readily apparent.

These spectra and others [2,3] show that TIRES and TIRTS are capable of analyzing moving materials in real time, even if those materials when stationary are not readily analyzed by most other means.

4. References

1. Peter R. Griffiths: Appl. Spectrosc. **26**, 73 (1972)
2. Roger W. Jones, John F. McClelland: Anal. Chem. **62**, 2074 (1990)
3. Roger W. Jones, John F. McClelland: Anal. Chem. **62**, 2247 (1990)

Photoacoustic Spectra of Polymer Fibres and Textile Materials

M. Kaplanová

Department of Chemical Technology of Fibrous Materials,
University of Chemical Technology, Čs. Legií 565,
CS-532 10 Pardubice, Czechoslovakia

Abstract. Photoacoustic spectra and calibration curves of PET fabrics dyed with different disperse dyes can be used for the identification of these dyes and determination of their concentrations in the volume and on the surface of fibres. The linear relationship between the PAS signal of the dyed PET fabrics, measured at the absorption maxima of dyes at a fixed chopping frequency, and the optical absorption coefficients of samples enables the evaluation of the thermal parameters of the PET textile material.

1. Introduction

As photoacoustic spectroscopy offers the ability to measure non-radiative de-excitation processes, the results should be comparable with absorption spectra. Consequently photoacoustic spectroscopy might be expected either to replace or to complement conventional reflectance measurements. We have therefore compared the two types of optical spectra measurement on different dyed polyester fabrics to determine whether photoacoustic spectrometry would be useful for the estimation of dyes on textiles and the determination of the dyestuff concentrations.

Whilst colour and its intensity can be studied in solution, it is also valuable to assess these properties under conditions in which the dye will be used in practice. The way in which a dye's reflectance is modified by its presence in the polyester fibers is of importance, because it is a factor in determining the amount of dye that must be used in order to produce a particular colour. Studies of the adsorption of dyes on wool [2] and cotton [3] fabrics showed, that photoacoustic spectroscopy may be of value for the analysis of two-dye and three-dye combinations and very dark colours.

2. Results and Discussion

Polyester fabrics were dyed in dyebaths with different concentrations of disperse dyes: C.I. Disperse Blue 56, Disperse Yellow 211, Disperse Red 50 and Disperse Red 72. Reflectance spectra were obtained using the Zeiss Spekol reflectance spectrophotometer with barium sulphate as reference.

Photoacoustic spectra were measured using the home-built single-beam photoacoustic spectrometer with microcomputer system attached to the spectrometer [1]. The light source was a 100 W Xenon arc lamp, the light modulation frequency being 18 Hz or

Springer Series in Optical Sciences, Vol. 69 Photoacoustic and Photothermal Phenomena III
Editor: D. Bićanić

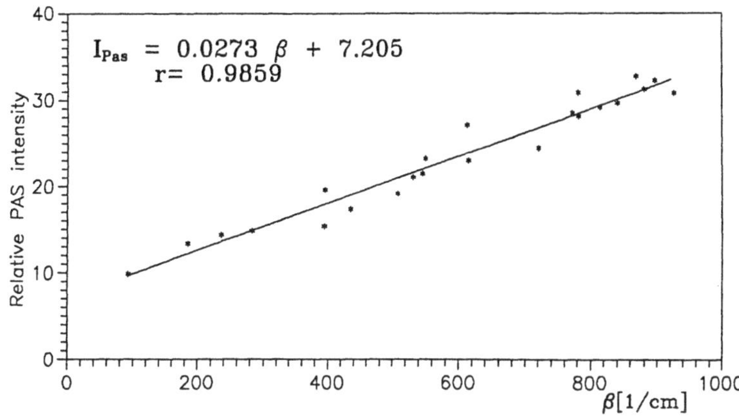

Fig 1. Dependence of PAS signal on the optical absorption coefficients of the dyed PET fabrics.

gradually changed from 15 Hz to 90 Hz. The photoacoustic spectra were normalized and corrected for the absorption of the undyed polyester fabric.

The absorption spectra of dye solutions, and the reflectance and photoacoustic spectra of the dyed polyester fabrics have been carefully compared. The general agreement between these spectra was very good. Photoacoustic absorption peaks are slightly shifted to the short-wavelengths region as compared with the absorption spectra of dye solutions in acetone. Dyed polyester fabrics are optically opaque and thermally thick samples. A thermal diffusion length of 15 μm at the modulation frequency of 18 Hz, calculated on the basis of the published data of thermal parameters of polyester fibers (thermal conductivity defined along the fiber axis [4]), is small in comparison with the thickness of fabrics (~ 380 μm) or yarns (~ 195 - 290 μm) and very close to the fiber diameter of 13 μm.

The photoacoustic signal measured at the fixed modulation chopping frequency depends on the amount of disperse dye on the PET fabric and consequently on the absorption coefficients of samples . A log-log plots of the PAS signal versus chopping frequency show a frequency dependence that varies from close to ω^{-1} when $\mu > l_\beta$ to $\omega^{-3/2}$ when $\mu < l_\beta$ ($l_\beta = 1/\beta$). The photoacoustic signal, measured at the dye absorption maxima of the dyed samples (the same fabric substrate and the same type of dyes and mode of the dye application) is proportional to the optical absorption coefficient (Fig.1) up to $\beta \sim 1000 cm^{-1}$. A least-squares fit of these data was used to determine the linear regression constants of that empirical relation and made possible the quantitative analysis of the dyed polyester fabrics. The saturation of the reflectance spectra have been observed for relatively low concentrations of dyes (for $\beta \sim 500 cm^{-1}$). Photoacoustic spectroscopy gives better resolution than can be obtained from the reflectance spectra measurement.

We studied also photoacoustic spectra of partially decolourized samples from which were gradually extracted dyes. We made the depth-profile analysis of the optical properties of samples at the different stages of bleaching of dyed PET fabrics from the measurement of the intensity of the photoacoustic absorption maxima at different modulation frequencies. The results showed that photoacoustic spectroscopy is capable of detection of the differences in the spectral properties of dyes on polyester, which appear to be similar in color visually.

References

1. M. Kaplanová, M. Trojan, P. Sladký , Czech. J. Phys. **B38**, 104 (1988).
2. R.S. Davidson, D. King , P.A. Duffield, D.M. Lewis, JSDC **V99**, 123, (1983).
3. Xiao-Li, K.-H. Brücher, H.-H. Perkampus, Fresenius Z. Anal. Chem. **319**,63, (1984).
4. Polymer Handbook, eds. J. Brandrup, E.H. Immergut, J. Wiley & Sons, (1975).

Study of Absorption Properties of Dark Coloured Textile Fabrics

A. Eickmeier and E. Schollmeyer

Deutsches Textilforschungszentrum Nord-West e.V., Frankenring 2, W-4150 Krefeld, Fed. Rep. of Germany

Abstract. The characterisation of dyed textile fabrics by means of photoacoustic (PA) spectroscopy is investigated. Special attention is payed to the influence of straylight and textile structure.

1. Introduction

Reflection spectroscopy is usually applied to determine the dyestuff concentration in textile analysis. In that case the semi empirical Kubelka-Munk function $k/s=(1-R)^2/2R$ gives a relationship between a concentration dependent coefficient k, the scattering coefficient s and the reflection coefficient R. In most cases this method is unsuitable for dark and saturated colourations because the reflection coefficient R of such sample is extremely small. In contrast, the photoacoustic spectroscopy (PA) is a powerful tool to measure the optical properties especially of strongly absorbing materials [1,2].

Application to textile analysis is studied here systematically. Attention has been paid to the complex geometric structure of textile fabrics which influences the detectable PA amplitude and phase by characteristic stray-light effects and by the effective thermophysical properties resulting from roughness and porosity [3].

2. Experimental Results and Discussion

Samples were fitted to the quartz glass backing of the PAS-cell. In such way the interaction between transmitted light and the cell walls could be minimized. The periodical heating of the sample was achieved by a HeNe-laser (wavelength 632.8 nm). Fabrics of poly(ethylene terephthalate) dyed with disperse blue 79 served as samples. The dyestuff has an absorption maximum at 620 nm.

As a test of the method's sensitivity a neutral density filter (D=4) and a quartz window blackened with carbon-black were compared with a dark, saturated dyed fabric. The resulting PA amplitudes are plotted versus the modulation frequency in Fig.1. In reflection spectroscopy it is almost impossible to distinguish between the samples having a reflection coefficient $R < 0.02$. The PA amplitudes however yield differences of roughly one order of magnitude between the three samples. This considerable enhancement of resolution demonstrates the advantage of PAS for textile analysis.

In a further experiment, fabrics were dyed with dyestuff of different concentrations. For each sample the PA amplitudes of

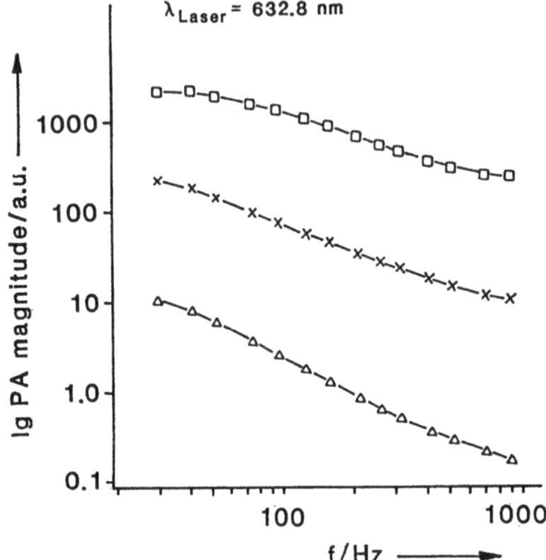

Fig.1. Photoacoustic amplitudes of carbon-black coated quartz (+++), blue dyed fabric (xxx), and neutral density glass (ΔΔΔ)

the fabrics and the stray light contributions were measured as a function of the modulation frequency. To measure the stray light response the cell was closed with a quartz glass window and the sample fixed to the back side of the window. Photoacoustic amplitudes increasing with the dyestuff concentration were obtained up to 2% dyestuff (relative to fabric weight) and are shown in Fig. 2. With dyestuff concentrations up to 6% and 8% the values converged to the saturation limit, differing only marginally. A stray light correction, however, contibutes to seperate the 6% colouration from the darker 8% colouration.

3. Theory

Under the assumption of an absolutely black reference material ($\sqrt{4\pi f/\alpha_r \ \beta_r^2(\lambda)} \ll 1$) the normalized photoacoustic amplitude for a homogenous compact sample is given by [4]

$$\frac{S_t}{S_r} = \frac{(1-R_t)}{(1-R_r)} \frac{(k\rho c)_r^{\frac{1}{2}}}{(k\rho c)_t^{\frac{1}{2}}} \left(1 + \left(\frac{4\pi f}{\alpha_t \beta_t^2}\right)^{\frac{1}{2}} + \frac{2\pi f}{\alpha_t \beta_t^2}\right)^{-\frac{1}{2}} \qquad (1)$$

where t refers to the textile and r to the reference material. The parameters are: The reflectivity R_t the relevant parameter in colour analysis, the effusivity $(k \rho c)^{1/2}$, the thermal diffusivity α and the optical absorption constant β. To

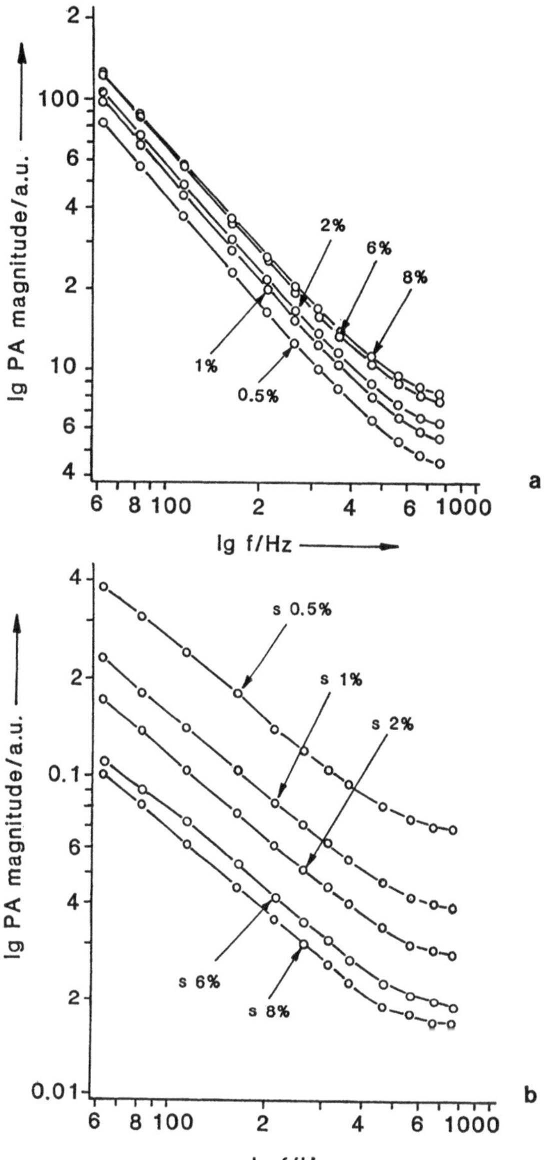

Fig.2. a) Photoacoustic amplitudes; b) straylight (s) for different dyestuff concentrations

determine R_s from the amplitude measurement it is necessary to have an information about the factor $\sqrt{\alpha}\,\beta$. This can be obtained by measuring the phase shift of the PA signal. For a textile sample normalized with a carbon-black coated quartz glass one can set $(\alpha_t/\alpha_r)^{\frac{1}{2}} \cdot \beta_t/\beta_r \ll 1$ and with this approximation the solution for the phase shift has the simple form [4]

$$f^{\frac{1}{2}}\ [\cot(\Phi_t-\Phi_r)-1] = \sqrt{\alpha_t/\pi}\,\beta_t \tag{2}$$

This result, together with eq. 1 and the asumption that $R_r \approx 0$ lead to a solution for the reflection coefficient in the form

$$R_t = 1 - \frac{S_t}{S_r} \frac{(k\rho c)^{\frac{1}{2}}_t}{(k\rho c)^{\frac{1}{2}}_r} [1+2(\cot(\Phi_t - \Phi_r)-1)^{-1}+2(\cot(\Phi_t-\Phi_r)-1)^{-\frac{1}{2}}]^{\frac{1}{2}}. \quad (3)$$

Using this equation PA data taken from a colouration series of a textile fabric can be compared to results obtained by means of reflection spectroscpy.

4. Acknowledgements

The authors like to acknowledge the financial support received by the Forschungsgemeinschaft Gesamttextil for this work (AIF-No. 8157).

5. References

[1] A. Rosencwaig, and A. Gersho, J.Appl.Phys. **47**, 64 (1976)
[2] H. Coufal, Fres. Z. Anal. Chem. **324**, 456 (1986)
[3] S. Krueger et al., J.Appl.Phys. 62, 55 (1987)
[4] B.K. Bein et al., Can. J. Phys. **64**, 1208 (1986)

Photoacoustic Response to X-Ray Absorption in CuO and Cu_2O

T. Toyoda[1], *T. Masujima*[2], *H. Shiwaku*[3], *A. Iida*[3,4], *and M. Ando*[3,4]

[1]Department of Applied Physics and Chemistry, The University of Electro-Communications, Chofu, Tokyo 182, Japan
[2]Institute of Pharmaceutical Sciences, Hiroshima University, School of Medicine, Minami-ku, Hiroshima 734, Japan
[3]Department of Synchrotron Radiation Science, The Graduate University for Advanced Studies, Tsukuba, Ibaraki 305, Japan
[4]Photon Factory, National Laboratory for High Energy Physics (KEK), Tsukuba, Ibaraki 305, Japan

Abstract : The hard X-ray spectra of CuO and Cu_2O have been measured at XANES(X-ray absorption near-edge structure) regions with a photoacoustic detector using synchrotron radiation. It is shown that the information derived from XANES is also included in the X-ray photoacoustic spectrum which reflects the heat production processes in those copper oxides. However, the results showed that the increases and the changes of the photoacoustic signal were different from those of the X-ray absorption coefficient at XANES regions.

1. Introduction

Hard X-ray absorption is concerned with the study of electronic transitions from atomic inner shells to unoccupied states, and X-ray absorption spectroscopy is now widely used in the characterization of materials. Although heat production is associated with X-ray absorption, there have been few investigations into its photon energy dependence. Recently, the heat production by X-ray absorption has been detected by the photoacoustic method using synchrotron radiation/1/. Further spectroscopic study showed that the spectra of the heat production also included fine structure in the K-edge absorption regions of Cu and Ni, i.e., EXAFS(extended X-ray absorption fine structure) comparable to the usual absorption spectra carried out simultaneously/2-4/. The purpose of this paper is to present detailed experimental results concerning the hard X-ray absorption spectra with a photoacoustic detector for copper oxides, CuO and Cu_2O, and to compare them with the usual X-ray absorption character at or near Cu K-edge regions. CuO and Cu_2O show monoclinic and cubic structure, respectively. Recently, antiferromagnetic ordering and its possible relations to high-Tc superconductivity have prompted studies of the magnetic properties of CuO by neutron diffraction/5/ and Mössbauer spectroscopy/6/. The results presented here should provide a basis for a better understanding of the mechanisn of heat production by X-ray absorption because it is related to the inelastic interaction between the photoelectrons or Auger electrons and the valence electrons. Also, they should give useful information for the use of X-ray photoacoustic spectroscopy as a new experimental tool.

2. Experimental Procedure

Monochromatic X-rays were obtained from a Si(111) double crystal monochromator using the white X-rays. In order to measure the absorption spectrum simultaneously, the ionization chambers were set at both sides of the photoacoustic cell. The X-ray beam intensity was modulated by a lead plate chopper with a frequency of 9 Hz. The cylindrical cell has a sample chamber with a volume of 0.16 cm^3 at the center and two Be windows together with an electret michrophone. The signal intensities for photoacoustic and ionization chamber current were proportional to the storage ring current which was the measure of the photon flux, so that the photoacoustic and ionization chamber current responses are in the linear regions. The photoacoustic signal intensity was divided by the first ionization chamber current for normalization against the photon flux. CuO and Cu_2O powders(commercial) were used as specimens.

3. Results

Fig. 1 shows the photon energy dependence of the X-ray photoacoustic (PAS) and optical density($\log(I_0/I_t)$) signals for (a)CuO and (b)Cu_2O at Cu K-edge regions. In Fig. 1, the fine oscillation structures shown in the X-ray photoacoustic spectra for each case correspond to those in the optical density spectra. Energy peak values derived from the X-ray photoacoustic spectra agree with those derived from the optical density ones for each case. The results suggest that the information in XANES is also included in the X-ray photoacoustic spectra, so that the heat production processes in CuO and Cu_2O are reflected in XANES. The first energy peak value for CuO is the same as that for Cu_2O, and they are 9eV smaller than that for pure Cu metal, indicating a chemical shift. Although the second energy peak value for CuO is the same as

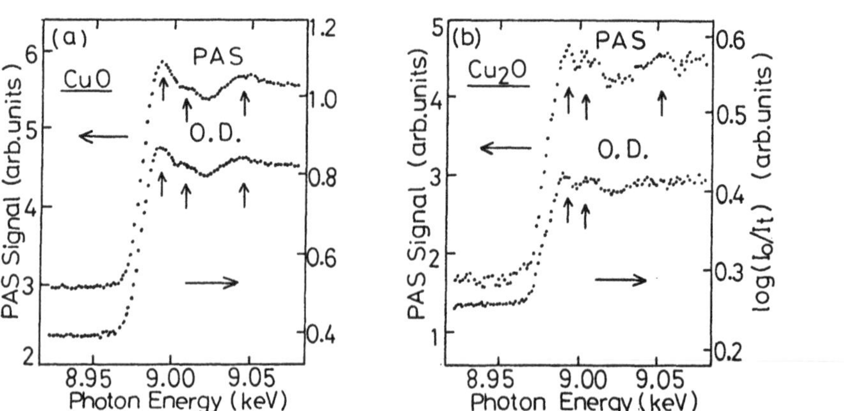

Fig. 1 X-ray photoacoustic(PAS) and optical density($\log(I_0/I_t)$) spectra for (a)CuO and (b)Cu_2O at Cu K-edge regions.

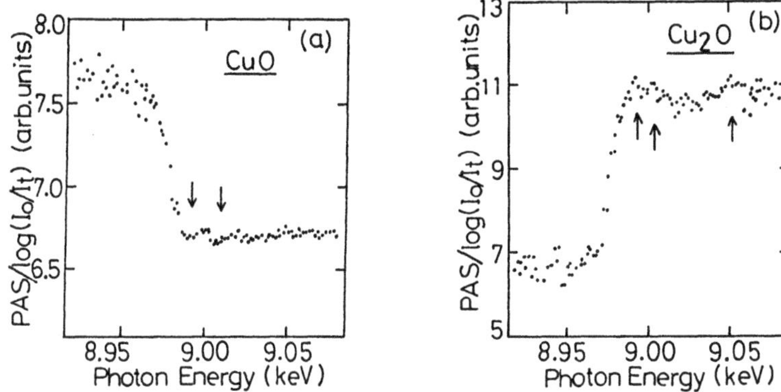

Fig. 2 PAS/log(I_0/I_t) spectra for (a)CuO and (b)Cu_2O at Cu K-edge regions.

that for pure Cu, that for Cu_2O is 5eV smaller than that for pure Cu. Fig. 2 shows the PAS/log(I_0/I_t) spectra(ratio of photoacoustic signal to optical density) for (a)CuO and (b)Cu_2O. The results indicate that the increases and the changes of the X-ray photoacoustic signal are different from those of the X-ray absorption coefficient at the XANES regions. As photon energy approaches the K-edge regions, it is found that PAS/log(I_0/I_t)decreases rapidly in the case of CuO and increases rapidly in the case of Cu_2O.At XANES regions, energy peaks in the X-ray photoacoustic spectra(↑marks in Fig.1(a)) correspond to the energy bottom of PAS/log(I_0/I_t) spectra(↓ marks in Fig.2(a)) for CuO that is similar to the case of pure Cu and brass/8/. Energy peaks of Cu_2O in the photoacoustic signal(↑marks in Fig.1(b)) correspond to the energy peaks of PAS/log(I_0/I_t) spectra(↑marks in Fig.2(b)).

References

1. T. Masujima et al., Chem. Lett. 1987, 973 (1987)
2. T. Masujima, Photoacoustic and Photothermal Phenomena(Springer, 1988) p. 19.
3. T. Masujima et al., Japan. J. Appl. Phys. 28, L513 (1989).
4. T. Masujima, Photoacoustic and Photothermal Phenomena II (Springer, 1990), p. 222.
5. B. X. Yang et al., Phys. Rev. B38, 174 (1988).
6. Ch. Niedermayer et al., Phys. Rev. B38, 2836 (1988).
7. T. Masujima et al., Rev. Sci. Instrum. 60, 2522 (1989).
8. T. Toyoda et al., Japan. J. Appl. Phys. 29, L1723 (1990).

Photoacoustic Response to X-Ray Absorption in Copper, Phosphor Bronze, and Brass

T. Toyoda[1], *T. Masujima*[2], *H. Shiwaku*[3], *A. Iida*[3,4], and *M. Ando*[3,4]

[1]Department of Applied Physics and Chemistry, The University of Electro-Communications, Chofu, Tokyo 182, Japan
[2]Institute of Pharmaceutical Sciences, Hiroshima University, School of Medicine, Minami-ku, Hiroshima 734, Japan
[3]Department of Synchrotron Radiation Science, The Graduate University for Advanced Studies, Tsukuba, Ibaraki 305, Japan
[4]Photon Factory, National Laboratory for High Energy Physics (KEK), Tsukuba, Ibaraki 305, Japan

Abstract: Although heat production is associated with X-ray absorption, there have been few investigations into the photon energy dependence of them. The purpose of this paper is to present detailed experimental results concerning the X-ray photoacoustic signal for copper, phosphor bronze, and brass and to compare them with the usual X-ray absorption character at near Cu K-edge regions. The results suggest that the information in XANES (X-ray absorption near-edge structure) is also included in the heat production in those materials.

1. Introduction

When materials are irradiated with X-rays, several effects such as absorption, fluorescence, electron release, and heat production occur. In these effects, X-ray absorption concerns the study of electronic transitions from atomic inner shells to unoccupied states. Electronic and structural information on selected sites can be obtained in complex systems, and X-ray absorption spectroscopy is now widely used in the characterization of materials. Although heat production is associated with X-ray absorption, there have been few investigations into the photon energy dependence of the heat production. Until recently, main experimental difficulty in detecting the heat production was the lack of sufficient X-ray flux. However, very intense sources have become available with the advent of synchrotrons for X-ray research. Recently, the heat production by X-ray absorption has been detected by the photoacoustic method using synchrotron radiation/1/. Further spectroscopic study showed that it also included the EXAFS (extended X-ray absorption fine structure) in the K-edge absorption region of Cu and Ni. The purpose of this paper is to present detailed experimental results concerning the X-ray photoacoustic signal for pure copper, phosphor bronze (94wt.%Cu, 6wt.%Sn), and brass (65wt.%Cu, 35wt.%Zn) and to compare them with the usual X-ray absorption character at near K-edge regions. In order to compare the difference between the heat production by Cu atoms and that by Zn atoms in cubic structure, the X-ray photoacoustic spectra at both Cu and Zn K-edge regions have been measured in brass.

Springer Series in Optical Sciences, Vol. 69 Photoacoustic and Photothermal Phenomena III
Editor: D. Bićanić

2. Experimental Procedure/2/

Monochromatic X-rays were obtained from a Si(111) double crystal monochromator using the white X-rays. In order to measure the absorption spectrum simultaneously, the ionization chambers were set at both sides of the photoacoustic cell. The X-ray beam intensity was modulated by a lead plate chopper with a frequency of 9 Hz. The cylindrical cell has a sample chamber with a volume of 0.16 cm^3 at the center and two Be windows together with an electret microphone. The photoacoustic signal intensity was divided by the first ionization chamber current for normalization against the photon flux. Pure copper, phosphor bronze, and brass foils of 10 μm thickness were used as specimens.

3. Results

Fig. 1 shows the X-ray photoacoustic and optical density spectra for pure copper and phosphor bronze at Cu K-edge regions. The fine structures shown in the X-ray photoacoustic spectra correspond to those in the optical density spectra. PAS/log(I_0/I_t) spectra (ratio of photoacoustic signal to optical density) indicate that the increases and changes of the X-ray photoacoustic signal are different from those of X-ray absorption coefficient at the XANES regions. Fig. 2 shows the X-

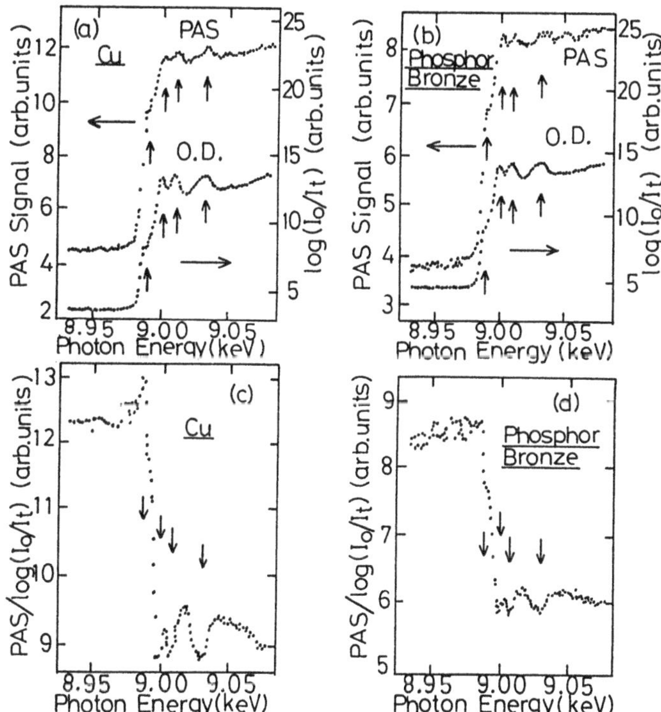

Fig. 1 X-ray photoacoustic(PAS) and optical density(log(I_0/I_t)) spectra for (a)pure copper and (b)phosphor bronze, and PAS/log(I_0/I_t) spectra for (c)pure copper and (d)phosphor bronze.

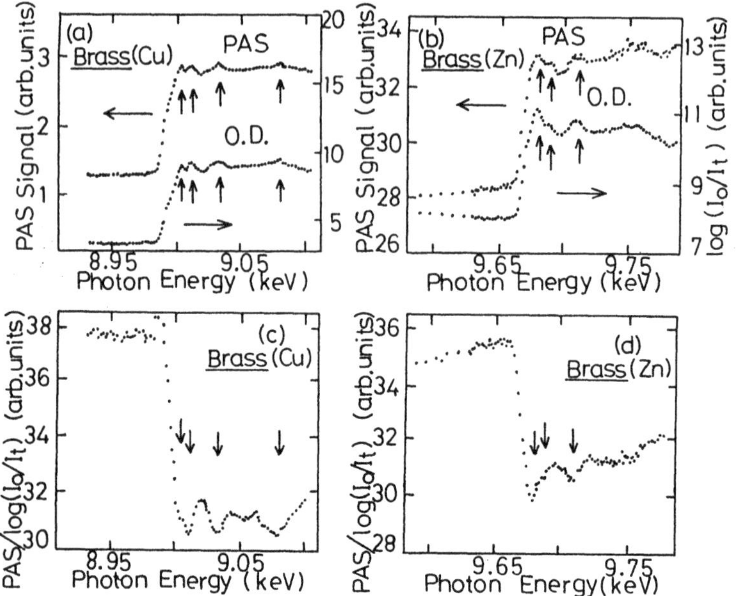

Fig.2 X-ray photoacoustic(PAS) and optical density(log(I_0/I_t)) spectra for brass at (a)Cu K-edge and (b)Zn K-edge regions, and PAS/log(I_0/I_t) spectra of them((c) and (d)).

ray photoacoustic and optical density spectra for brass at (a)Cu K-edge and (b)Zn K-edge regions. The fine structures shown in the X-ray photoacoustic spectra correspond to those in the optical density spectra. The results suggest that the information in XANES is also included in the heat production of brass. Energy peak values of brass at Cu K-edge region show the same values as for pure copper, indicating that no chemical shift is seen in spite of the difference of coordination numbers. PAS/log(I_0/I_t) spectra of Cu and Zn parts indicate that the increases and changes of the X-ray photoacoustic signal are different from those of X-ray absorption coefficient.

References

1. T.Masujima et al., Chem.Lett. 1987, 973 (1987).
2. T.Masujima et al., Rev.Sci.Instrum. 60, 2522 (1989).

Theory of the Photoacoustic Effect in Photoactive Samples

P. Poulet, M. Ouzafe, and J. Chambron

Unité Associée CNRS UA 1173, Institut de Physique Biologique,
Faculté de Médecine, 4 rue Kirschleger, F-67085 Strasbourg, France

Abstract. A theory of the photoacoustic (PA) effect in photoactive samples is presented. It takes into account two heat sources corresponding to the radiationless relaxation processes of two states with different lifetimes and heat diffusion across the sample. The amplitude and the phase of the PA signal depend on the lifetime of the long-lived excited state and on the product of the energy of this state times the quantum yield of its production. Simulations of PA amplitude and phase variations versus light modulation frequency exhibit new features of the PA signal.

1. Theory

We have developed a theoretical model of the PA effect while keeping in mind : a) the heat produced by non radiative transitions from two excited states, the first with a short lifetime compared to the light modulation period, and the second with a lifetime of the same order of magnitude, b) the heat diffusion from the production sites to the surface of the thermally thick sample. We hereto consider that only variations in the sample's surface temperature can generate a measurable signal. The theoretical model developed is therefore valid for surface temperature detection methods. Using the Jablonski energy diagram, the heats released per absorbed photon, Q_e from the singlet excited state and Q'_m from the metastable state, are given by :

$$Q_e = E_{ex} - \gamma_f E_f - \gamma_2 E_2$$

$$Q'_m = \gamma_2 (E_2 - \gamma_p E_p - \gamma_4 E_4)$$

With E_{ex} the energy of incident photons, γ_f the quantum yield of fluorescence, E_f the mean energy of a fluorescence photon, γ_2 the quantum yield of crossing to the long-lived state (in practice a triplet or singlet oxygen), E_2 its energy, γ_p the quantum yield of phosphorescence, E_p the mean energy of a phosphorescence photon, γ_4 the quantum yield of energy transfer to another molecule (photochemistry), E_4 the energy of the final state.

Writing R the ratio Qe/Q'_m and τ_m the lifetime of the long-lived state, the amplitude ΔP_0 and phase $\Phi(\omega)$ of the PA signal generated by these two relaxation processes are :

$$\Delta P_0 = \frac{K\, I_0\, Q_e\, \beta\mu_s}{2\omega\sqrt{(k\rho c)_s}\, E_{ex}\sqrt{(1+(\beta\mu_s+1)^2)}} \left\{ 1 + \frac{2R+1}{R^2(1+(\omega\tau_m)^2)} \right\}^{1/2}$$

$$\Phi(\omega) = -\pi - \Psi(\omega) + tg^{-1}(\beta\mu_s+1) - \left\{ tg^{-1} \frac{\omega\tau_m}{1+R(1+(\omega\tau_m)^2)} \right\}$$

With the usual PA notations : I_0 is the light intensity, β the absorption coefficient, μ_s the thermal diffusion length, $\omega/2\pi$ the modulation frequency, $(k\rho c)_s$ the thermal effusivity. K is a proportionality factor and $\Psi(\omega)$ a phase shift; both are dependent variables of the properties of the gas as well as of periodically vaporized and/or condensed molecules.

The terms written outside of the brackets represent the signal generated by a thermically thick sample in which the energy absorbed is converted into heat by fast relaxation processes. They are equal to the terms calculated by Poulet et al. [1]. We will call such a sample a photoinactive sample. The terms inside the brackets show the contribution to the PA signal of the metastable state; the phase is identical to the result of Hunter et al [2] as used by Moore et al [3].

The sample's characteristic parameters which determine the amplitude and phase of the PA signal : $B = \beta\mu_s\sqrt{\omega}$, τ_m and R, can be measured from the study of the variations of the PA signal versus ω. R is a PA parameter that describes the relaxation mechanisms. Usual photophysical parameters can be calculated from R. The product of the intersystem crossing yield times the energy of the metastable state or the product of the yield of photochemical pathways times the energy of the final state can be measured by PA experiments.

2. Simulations

The variations of the PA amplitude times frequency product (Fig. 1A) and of the phase (Fig. 1B) are studied as a dependent variable of the modulation frequency for various conditions. The represented phases are the differences between the

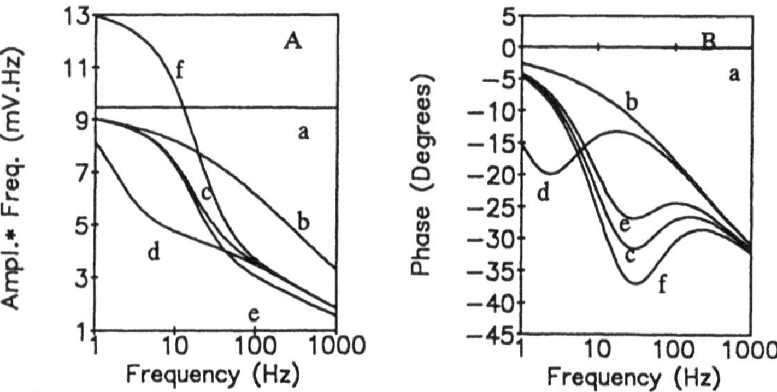

Figure 1 : Variations of the amplitude times frequency product (A) and of the phase (B) versus modulation frequency.

phase of the studied samples and that of a reference sample, the last being equal to $-\pi/2 - \Psi(\omega)$.

The results correspond to various aqueous solutions: (a) to a reference sample producing a fully saturated PA signal; all other samples have $\beta = 10^3$ cm^{-1}. (b) is a photoinactive sample : $\gamma_2 = 0$. The effect of γ_2 and τ_m are shown on other curves : (c) corresponds to $\tau_m = 10^{-2}$ s, $\gamma_2 = 0.75$; (d) to $\tau_m = 10^{-1}$ s, $\gamma_2 = 0.75$; (e) to $\tau_m = 10^{-1}$ s, $\gamma_2 = 0.90$. The final state reached through relaxation of the long-lived excited state is the ground state with 0 energy. Curve (f) corresponds to $\tau_m = 10^{-2}$ s, $\gamma_2 = 0.75$; the molecules at the metastable state are submitted to photochemical reactions ($\gamma_4 = 1$), leading to the production of a stable compound whose energy is equal to $E_4 = -1.2$ eV.

3. Conclusion

The described theoretical analysis of the PA effect demonstrates the sensitivity of PA detection to the existence of a long-lived excited state. The fast decrease ($> \omega^{-3/2}$) and the point of inflexion on the plot of the amplitude against ω, as well as the phase minimum or maximum, are characteristic of such a state. The lifetime, the quantum yield and the energy of this state can be calculated from experiments using PA detection. Experimental verifications have been done under different conditions, they are presented elsewhere [4].

References

1. P. Poulet, J. Chambron, R. Unterreiner : J. Appl. Phys. **51**, 1738-1742 (1980)
2. T.F. Hunter, D. Rumbles, M. Stock : J. Chem. Soc. Faraday Trans. II **70**, 1010-1021 (1974)
3. T.A. Moore, D. Benin, R. Tom : J. Am. Chem. Soc. **104**, 7356-7357, (1982)
4. M. Ouzafe, P. Poulet, J. Chambron : this conference

Photoacoustic Detection of Triplet State and Singlet Oxygen

M. Ouzafe, P. Poulet, and J. Chambron

Unité Associée CNRS UA 1173, Institut de Physique Biologique, Faculté de Médecine, 4 rue Kirschleger, F-67085 Strasbourg, France

Abstract. The amplitude and phase variations of the photoacoustic (PA) signal against modulation frequency are used to measure the lifetime and conversion yield of a long-lived excited state which can be a triplet state or singlet oxygen, $O_2(^1\Delta_g)$. The addition of an acceptor, specific to $O_2(^1\Delta_g)$, induces changes in the amplitude of the PA signal which can be used to study the production and deactivation of this excited form of oxygen.

1. Introduction

The amplitude and phase of the PA signal generated by a photoactive, condensed sample, which relaxes through a long-lived excited state, have been calculated by Poulet et al. [1]. We present preliminary experimental verifications of this model. The PA signal is characterized by three parameters: $\beta\mu_s$ the absorption coefficent times the thermal diffusion length, τ_m the lifetime of the long-lived excited state and R the ratio of the heats produced by relaxation of the short and of the long-lived state. The study of the PA signal variations as a dependent variable of light's modulation frequency $\omega/2\pi$ allows us to determine these parameters. The phase variations have already been used [2,3] for this purpose, the amplitude variations analysis have been presented for the first time.

2. Materials and methods

Built with a closed cell, the PA spectrometer uses a microphone detector. The cell is equiped with gas entry and exit valves which allow to modify the nature of the gas. PA signals are analyzed by a double-phased lock-in amplifier and recorded by a micro-computer. The adjustment of the experimental results to the theoretical model, to determine the parameters $\beta\mu_s/\omega$, τ_m and R, is carried out with a non-linear algorithm (Marquardt).

The following dyes were studied : acridine orange (AO), fluorescein isothiocyanate (FL), methylene blue (MB), hematoporphyrin IX (Hp). All dyes were used without any additional purification. They are dissolved to 5.10^{-3} M in the following solvents : ethanol, methanol or water. The solutions are equilibrated under nitrogen or oxygen during at least ten minutes before the manipulations. Some dyes are fixed on agarose 1% gels to study the effect of the

Springer Series in Optical Sciences, Vol. 69 Photoacoustic and Photothermal Phenomena III
Editor: D. Bićanić

lifetime of the triplet state. Different chemicals are added to solutions in order to modify the relaxation pathways. Potassium iodide, KI, is used as a fluorescence quencher, with a concentration of 3 M. A specific receiver of $O_2(^1\Delta_g)$, tetramethyl ethylene (TME) is used at a 0.2 N concentration.

3. Results and discussion

The PA signal's amplitude (Fig. 1A) and phase (Fig. 1B) variations are measured on AO in aqueous solutions (5mM) and in agarose gel after having purged the sample with nitrogen. The triplet state's lifetime rises from 1 to 5 ms when the radiationless deactivation is reduced by incorporation of the dye into the gel.The R ratios measured on nitrogen equilibrated solutions through amplitude and phase studies are consistent, around 1.6. This ratio increases up to 2.4 in the gel. R can be associated to the product of the γ_T intersystem crossing yield times the E_T energy of the triplet state. Using a fluorescence quantum yield $\gamma_f = 0.46$ and a mean energy of the emitted photons $E_f = 2.18$ eV, the measured $\gamma_T E_T$ products are around 0.6 eV in solution and 0.46 eV in gel.

PA signals are measured on FL on gels with and without KI. Fig. 1A shows the amplitude times frequency and Fig.1B the phase variations versus the modulation frequency. Below 80 Hz, the PA signal produced by FL + KI is greater than the one generated by FL alone. This amplification is due to an increase of the non-radiative rate as usually verified on PA experiments [4]. At modulation frequencies higher than 80 Hz, the signal produced by FL + KI is lower than the one generated by FL alone. This phenomenon is due to the existence of a triplet state with a long lifetime. The $\gamma_T E_T$ product has been calculated thanks to the R ratio. Using $\gamma_f = 0.92$ and $E_f = 2.33$ eV for FL [4], the measured $\gamma_T E_T$ products are around 0.12 eV for FL and 1.6 eV for FL + KI.

TME is an acceptor, specific to $O_2(^1\Delta_g)$, quenching all photoproduced $O_2(^1\Delta_g)$. The addition of TME induces variations of the PA signal amplitude. The phase

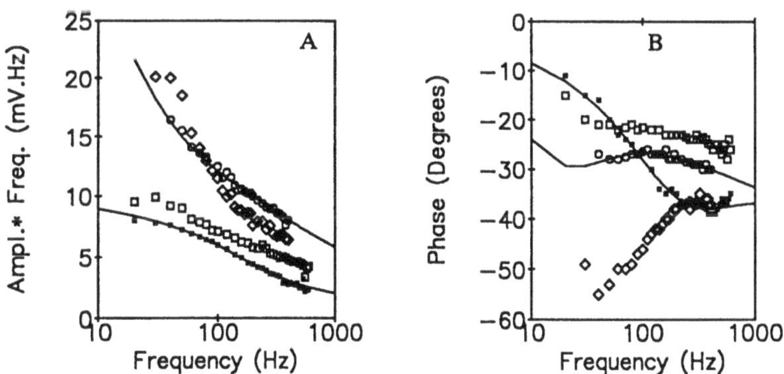

Figure 1: Amplitude times frequency product (A) and phase (B) of PA signals measured on AO, under N_2, in water (-■-■) and in gel (□ □) and measured on FL without (-o-o-) and with (◊ ◊) KI.

remains unaltered. In the absence of the acceptor, all $O_2(^1\Delta_g)$ is deactivated by collisions and its E_Δ energy is converted into heat. In the presence of TME, the released heat is $Q_p = E_\Delta - E_4$: E_4 being the energy of the final state. Experiments were performed on various dyes. The triplet state is quenched by oxygen and its lifetime is too short to be measured by PA spectroscopy. When TME is added to MB in methanol, the signal decreases , the amplitude times frequency product is multiplied by 0.60. When added to Hp solutions, TME produces an amplification of the PA signal, the amplitude times frequency product is multiplied by 1.31 in methanol and by 1.23 in ethanol. The $\gamma_\Delta E_4$ product is calculated from this factor. If the E_4 energy is known, γ_Δ is the result of such PA measurements. Measured γ_Δ for Hp, 0.64 in ethanol and 0.86 in methanol, agree with known values [5].

4. Conclusion

PA detection of the relaxation processes producing fast and slow heat sources allows to measure the lifetime of a long-lived excited state, in the range 10^{-4} to 1 s, and the R ratio of these two heat sources. R is related to parameters such as intersystem crossing yield, singlet oxygen production yield and deactivation, and photochemical reaction thermodynamics. Therefore PA appears as a suitable technique to measure the photophysical properties of dyes in highly absorbing media.

References

1. P. Poulet, M. Ouzafe, J. Chambron : this conference
2. T.F. Hunter, D. Rumbles, M. Stock : J. Chem. Soc. Faraday Trans. II **70**, 1010-1021 (1974)
3. T.A. Moore: *Photochemical and Photobiological Reviews,* **7**, ed by K.C. Smith (Plenum Press, New York 1983) pp 187-221
4. D. Cahen, H. Garty, R.S. Becker : J. Chem. Phys. **84**, 3384-3389 (1980)
5. T.G. Truscott : *Primary Photoprocesses in Biology and Medicine,* ed by R.V. Bensasson, G. Jori, E.J. Land, T.G. Truscott (Plenum Press, New York 1985) pp. 309-320

Origin of Photoacoustic Signals in the Visible Spectrum of I_2 Vapour

K. Narayanan and S.N. Thakur

Laser and Spectroscopy Laboratory, Department of Physics, Banaras Hindu University, Varanasi 221 005, India

Abstract : Vibronically resolved laser photoacoustic spectrum of I_2 vapour at room temperature in presence of air at atmospheric pressure has been recorded in the region 14700-20300 cm^{-1} for the first time. The bands are identified with the $B^3\pi\ (O\,u^+) \leftarrow X\ ^1\Sigma_g^+$ transition and evidence of energy transfer to atmospheric oxygen around 19700 cm^{-1} region as well as predissociation around v' ~ 23-27 have been observed. The relative intensities of nonradiative transitions have been estimated in the light of the Frank-Condon Factors (FCF's) from optical absorption spectra and Boltzmann factors of the ground state vibrational levels. The entire spectrum has been divided into four parts on the basis of radiationless processes leading to the photoacoustic signals.

1. Introduction:

Venkateswarlu et al.[1] reported the vibronically resolved laser photoacoustic spectrum of I_2 vapour between 17513-15155 cm^{-1}, but no definite conclusions were given regarding the origin of photoacoustic signals. In the present work the spectrum of I_2 in the region 14700-20300 cm^{-1} was recorded with a pulsed Nd- YAG laser- pumped dye laser and a photoacoustic cell filled with air at atmospheric pressure, as well as at lower pressures inorder to understand the mechanism of radiationless transitions in I_2 .

2. Experimental:

Nd-YAG laser pumped dye laser (Quanta Ray DCR-III,PDL II) of energy ~3 mJ/p and 0.05 nm bandwidth is split into two parts of which the main part is focussed into the photoacoustic cell using a lens of 15 cm focal length. The weaker part of the dye laser beam is further split into two equally intense beams of which one is sent to a Scientech power meter for recording the dye laser power profile, and the other is directed into a Hamamatsu Fe-Ne hollow cathode lamp to obtain the optogal vanic lines for wavelength calibration.The photoacoustic signals are picked up with a Sennheiser KE-4-211-2 microphone, whose output is amplified 300 times and fed to a boxcar (EG&G)for processing. The processed signal is plotted along with the calibration lines by scanninng the dye laser wavelength. The experimental setup is shown in Fig.1. The experiments have been carried out at atmospheric pressure as well as with saturated vapour of I_2 at 300 K in the presence of air at 15 torr. The photoacoustic spectra are then normalised on a point -to- point basis to remove spectral intensity variations of the source,and these are shown in Figs. 2(a) and 2(b) respectively.

3.Results and Discussion:

The I_2 spectrum in the presence of air at atmospheric pressure shows no bands in the vicinity of 20000 cm^{-1} and is marked by a region of minimum signals around 19840 cm^{-1}. However, discrete bands appear at lower frequencies superimposed on a background continuum which reaches a maximum near 19225 cm^{-1}. Laser light absorption by I_2 leads the $B^3\pi$ (Ou^+) state to dissociate into an

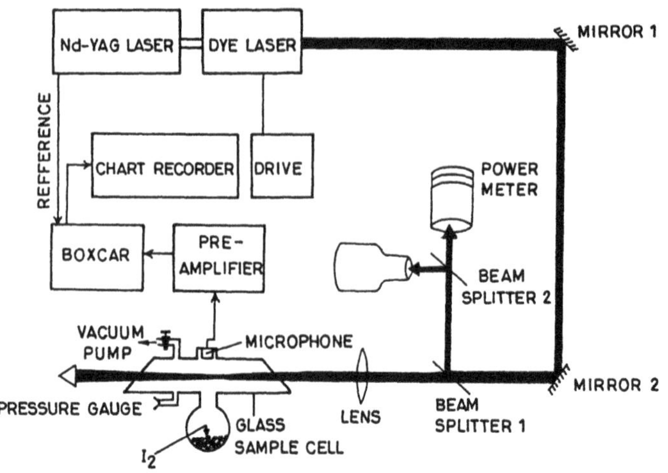

Fig. 1. Experimental setup used for recording the photoacoustic spectrum of I_2 vapour.

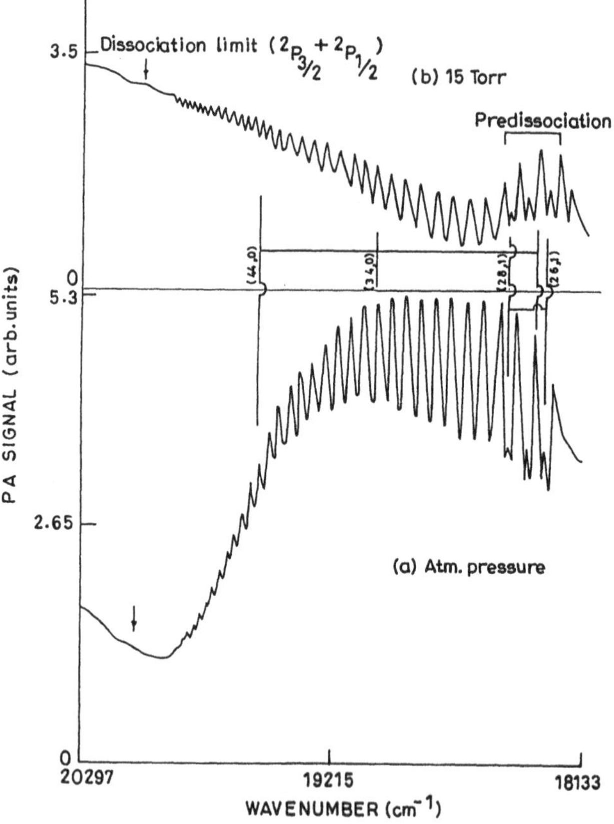

Fig. 2. Normalised photoacoustic spectrum of I_2 vapour.
(a) at atmospheric pressure and
(b) at 15 torr pressure.

excited $^2P_{1/2}$ atom and a ground state $^2P_{3/2}$ atom and the observed decrease in the photoacoustic intensity here is primarily due to energy transfer to O_2 molecules according to the following process [2]

$$I\,(^2P_{1/2}) + O_2\,(^3\Sigma^-_g) + 284\ \mathrm{cm^{-1}} \rightarrow I(^2P_{3/2}) + O_2\,(^1\Delta) \quad (1)$$

O_2 molecules in the $^1\Delta$ state are long lived with a life time of 0.1seconds[3] and their radiationless decay is not likely to be detected in our experiments,which have a time delay of the order of 800 microseconds with respect to the exciting laser pulse. This is also confirmed by recording the photoacoustic spectrum at a low pressure where the photoacoustic signals show an increasing trend, and the background continuum is shifted to higher frequencies near the dissociation limit of I_2 at 20043 cm^{-1}.

The optical absorption intensities of the various bands has been calculated from the relation,

$$I_{OA}\,(v', v'') = Q\,(v', v'')\,\exp\,[\,-\Delta E\,(v'')\,/\,kT\,] \quad (2)$$

where Q (v', v'') is the Frank-Condon factors for the vibrational bands (v',v''), ΔE (v'') is the vibrational energy in the ground electronic state relative to that of v''=0, k the Boltzmann constant and T is the ambient temperature (300k) of the photoacoustic cell. Values of Q(v', v'') were taken from the very accurate work of Tellinghuisen [4],the Boltzmann depended factors were calculated from our experimental conditions and the vibrational energy seperations for v''=1,2,3 etc. were taken from Luc's [5] work. On the basis of a comparison between the relative photoacoustic intensities and the calculated optical absorption intensities, the mechanism of heat generation following the optical excitation in the visible spectrum of I_2 vapour may be divided into four different categories in terms of the dominating radiationless processes.

(i) For excitation frequencies greater than 20043 cm^{-1} the major contribution to photoacoustic signals is from the photodissociation of the repulsive $^1\pi$ (1u) state into two ground state $^2P_{3/2}$ iodine atoms. **(ii)** Photodissociation of the $B^3\pi\,(Ou^+)$ state is the major cause of photoacoustic signals in the range 19225-20043 cm^{-1}, with some contribution also from the photodissociation of the $^1\pi$ (1u) state. **(iii)** At 18513 cm^{-1} , the optical extinction coefficients [6] for excitation into $B^3\pi\,(Ou^+)$, $^1\pi$ (1u) and $^3\pi$ (1u) states are 638, 81 and 0.9 respectively and their corresponding values at 16662 cm^{-1} are 133, 7.4 and 14.5 respectively. The major contribution to photoacoustic signals in the region 18862-16662 cm^{-1} comes from the radiationless decay of vibronically excited I_2 molecules in the $B^3\pi\,(Ou^+)$ state. **(iv)** For excitatin frequencies smaller than 16662 cm^{-1}, the main contribution to photoacoustic signals is fom the photodissociation of the weakly bound $^3\pi$ (1u) alongwith smaller contribution form vibronically excited molecules in the $B^3\pi\,(Ou^+)$ state.

If we assume that the average heat energy released by nonradiative deexcitation from the molecular vapour does not change appreciably in the spectral region under investigation, we may express the relative photoacoustic intensity at each frequency as a product of optical absorption intensity (I_{OA}) and the relative nonradiative deecxcitation rate (A_{nr}) at that frequency as

$$I_{PA} = I_{OA} \times A_{nr} \quad (3)$$

The relative nonradiative deexcitation rates (Anr) thus obtained for each band of the B — X system are plotted in Fig.3(b) as a function of v' along with the rates of radiative decay [6]. (see Fig.3(a)). The radiationless rate is found to increase significantly from its general trend for the vibrational levels v' ~ 23-27, which suggests that there must be some additioinal channel of heat generation in this region apart from that described under category (iii) above. It is known that the potential curve of $B^3\pi\,(Ou^+)$ state is crossed by that of the repulsive $^1\pi$ (1u) around v'~25 leading to a predissociation [7] into two ground state $^2P_{3/2}$ iodine atoms,and this has been identified as the cause of sudden increase of radiationless decay in the present case.

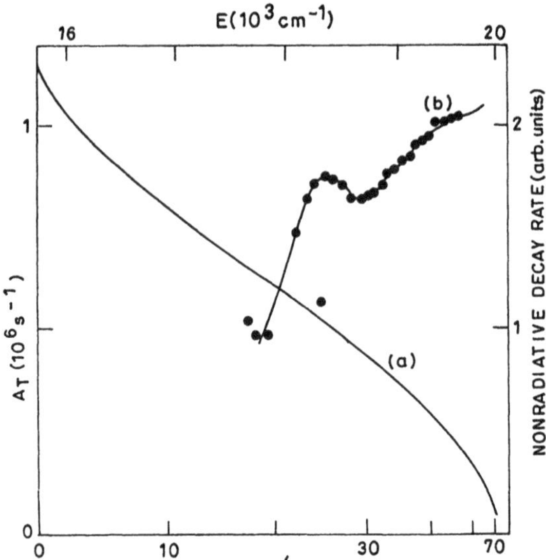

Fig. 3. Variation of (a) radiative and (b) nonradiative decay rates.

Acknowledgements : This work has been partially supported by Indo - US project on "Chemical Dynamics and Laser Spectroscopy". KN is thankful to CSIR ,New Delhi for a senior research fellowship.

References

[1] P.Venkateswarlu, D.Kumar and S.P.Mc Glynn, Lasers and Applications (Eds. H. D. Bhist and J.S. Goela) Tata McGraw Hill,New Delhi, p-89 (1984)

[2] W.R. Harshbarger and M.B. Robin, Chem. Phys. Lett.; **21**, 462 (1973).

[3] I.D. Clark and R.P. Wayne, Chem. Phys. Lett; **3**, 93 (1969).

[4] J.Tellinghuisen, J. Quant. Spectrosc. Radiat. Transfer, **19**, 149 (1978).

[5] P. Luc, J. Mol. Spectrosc., **80**. 41 (1980).

[6] J. Tellinghuisen, J. Chem. Phys., **76**, 4736 (1982).

[7] Ara Chutjian, J. Chem. Phys., **51**, 5413 (1969).

Fourier Transform Infrared Photoacoustic Spectroscopy of Amorphous Carbon Films

A. Raveh[1], *L. Martinu*[2], *A. Domingue*[2], *M.R. Wertheimer*[2], *and L. Bertrand*[2]

[1]Nuclear Research Center – Negev, P.O. Box 9001, Beer-Sheva 84190, Israel

[2]"Groupe des Couches Minces" (GCM) and Department of Engineering Physics, Ecole Polytechnique, Montréal, Québec, Canada H3C 3A7

Abstract. Phase Fourier Transform Infrared Photoacoustic Spectroscopy (PHAS-FTIR/PAS) can distinguish between surface and bulk chemical structure. In the present work, PHAS-FTIR/PAS is applied to the study of hydrogenated amorphous carbon (a-C:H) films deposited in a dual microwave/radio-frequency plasma. It is shown that spectra correlate well with physico-chemical properties of a-C:H films.

1. Introduction

Phase Fourier Transform Infrared Photoacoustic Spectroscopy (PHAS-FTIR/PAS) is a powerful tool for analyzing and characterizing various materials [1]. Excellent candidates among such materials are amorphous hydrogenated carbon (a-C:H) and diamond-like carbon (DLC) films, because of their unusual structure and properties [2,3], ranging from graphite, polymer-like hydrocarbon, to diamond-like. Since such layers are generally obtained by decomposition of hydrocarbons such as methane, butane and acetylene, they contain hydrogen (between 20 and 50 at.%) which can be bonded or "free" (i.r. inactive). We use the PHAS-FTIR/PAS technique to distinguish different types of chemical bonding in a-C:H films with the aim to correlate and eventually qualitatively predict physical properties such as density and microhardness.

2. Experimental

In this study a-C:H films have been grown on silicon and titanium substrates from low pressure methane plasma under different discharge modes, namely microwave (MW), radio-frequency (RF), and mixed-frequency MW/RF plasma. The MW/RF plasma system and the deposition procedure are described in detail elsewhere [4,5]. Microstructure of the a-C:H films was studied using a Bomem DA3 FTIR spectrometer equipped with a globar source and a KBr beamsplitter, in a spectral range between 500 and 4000 cm^{-1}, and with a 4 cm^{-1} resolution. A deuterated triglycine sulphate detector (DTGS) is used in transmission FTIR, whereas a homemade photoacoustic cell, and data acquisition and manipulation developped in this laboratory [6], are used in PHAS-FTIR/PAS. The mirror speed is 0.1 cm s^{-1}, corresponding to a thermal diffusion length λ between 100 μm and 300 μm for polymeric carbon [7], and between 140 μm and 400 μm for diamond-like materials [8].

3. Results and Discussion

Figure 1 represents two FTIR spectra of films prepared from methane plasma in the MW/RF and MW modes of operation. The MW/RF spectrum shows two important regions of stretching vibrations: the broad absorption band around 2900 cm^{-1}, and the peaks between

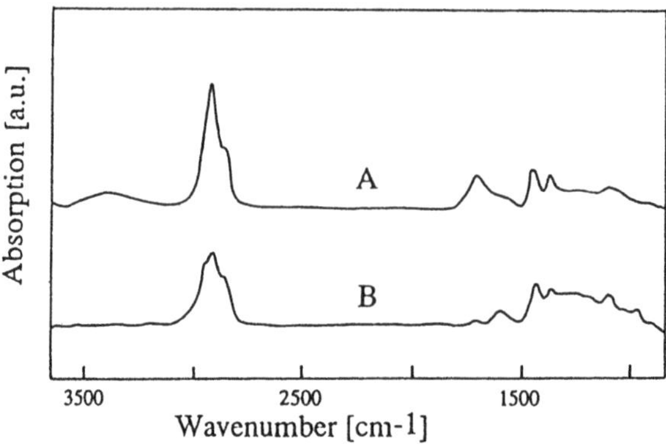

Figure 1: Fourier Transform Infrared spectra of carbon films: MW plasma in CH_4 (A), and MW/RF plasma in CH_4 (B).

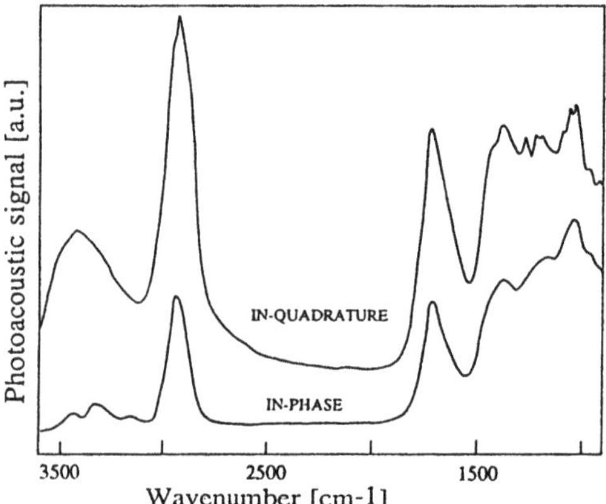

Figure 2: IN-QUAD and IN-PHASE photoacoustic spectra of a MW plasma grown a-C:H film.

1700 and 800 cm^{-1}. The band at 2920 cm^{-1} is due to CH_2 stretch vibration (sp^3 carbon), whereas the peaks at 1440 cm^{-1} and 1375 cm^{-1} correspond to deformation of sp^3CH_2 and sp^3CH, respectively. The intensities and shapes of these peaks give information about the sp^3/sp^2 ratio. Thus, it is observed that the ratio of intensities of the band at 2920 cm^{-1} to that at 3045 cm^{-1} (sp^2CH) increases, indicating an increase of the sp^3 component, while the intensity ratio of peaks at 1440 cm^{-1} and at 1375 cm^{-1} decreases; this, in turn, is semi-quantitatively correlated with the hardness and total bonded hydrogen, as shown and discussed in detail elsewhere [9]. Additional peaks near 3400 cm^{-1} and 1700 cm^{-1} in the MW film spectrum indicate C=O stretch vibrations. This film has much lower hardness and high hydrogen content, ~50 at.%. Based on the spectral analysis, the MW/RF films (curve B) can be classified as dense and hard a-C:H, while the MW films (curve A) have a predominantly polymer-like character.

Using PHAS-FTIR/PAS analysis, the IN-PHAS and IN-QUAD spectra have been found to contain all the absorption bands corresponding to the thermal diffusion length λ (see Fig. 2). However, the surface absorption bands (within the depth < 1/10 λ) are amplified in the IN-QUAD spectra [6]. Since the OH and C=O absorptions appear mainly in the latter, it follows that the bonded oxygen occurs mostly near the film surface.

The photoacoustic spectra obtained with our data manipulation show higher sensitivity than those obtained in transmission. This is exemplified in Fig. 3 for MW/RF films deposited from

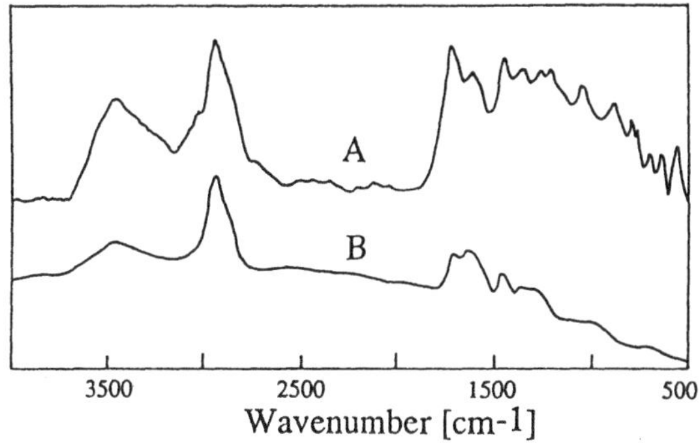

Figure 3: Fourier Transform Infrared spectra of MW/RF plasma a-C:H films: Photoacoustic (A) and transmission (B) spectra.

CH_4 plasma. The photoacoustic spectrum (curve A) exhibits more resolved and more intense peaks of C=O (at 1700 cm^{-1}), C-CH_3 (1375 cm^{-1}), sp^3CH_2 (1030 cm^{-1}), sp^3C-C (885 cm^{-1}) and sp^2CH (800 cm^{-1}) than the transmission spectrum (curve B). Here, too, the amplitudes and shapes bear witness to the sp^3/sp^2 ratio and to the polymer-like versus hard a-C:H structural character. However, the broad absorption band near 2900 cm^{-1} is the same for photoacoustic and transmission measurements. The higher sensitivity in the former mode can be explained by the optical diffusion at the rough sample surface which decreases the light intensity in transmission FTIR, whereas our photoacoustic PHAS-FTIR/PAS is insensitive to this optical diffusion.

Variations of FTIR spectra with methane flow rate during deposition were also found to correlate with mechanical and compositional properties measured by other methods [9]. This confirms that MW/RF films are characterized by a higher sp^3/sp^2 ratio, higher microhardness (~51 GPa), and a higher density (1.81 g/cm^3) than for MW or RF deposited films. This correlation, too, is discussed in more detail elsewhere [9].

4. Conclusions

Amorphous carbon films have been grown on silicon and titanium substrates from low pressure methane plasma under different discharge modes, namely microwave (MW) and mixed-frequency (MW/RF) plasma. PHAS-FTIR/PAS analysis confirms that simultaneous application of MW and RF power yields very dense, hard a-C:H films, compared with those obtained by MW plasma. These MW/RF films also exhibit higher concentrations of sp^3 features than those obtained in RF discharges, while MW films show a polymer-like character.

References

1. Bertrand, L., Applied Spectroscopy, 42 (1988) 134.
2. Dischler, B., in "Amorphous Hydrogenated Carbon Films", Koidl, P., and Oelhafen, P., eds. Proc. E-MRS, Vol. 17, Les Editions de Physique, Paris 1987, p. 189.
3. Angus, J.C., Koidl, P., and Domitz, S., Chapter IV, in "Plasma Deposited Thin Films", Mort, J., and Jansen, F., eds. (CRC Press, Boca Raton, 1966).
4. Martinu, L. , in: "Diamond and Diamond-like Films and Coatings", Angus, J.C., Clausing, R., Koidl, P., and Horton, L.L., eds., Plenum Press, New York, 1991.
5. Martinu, L., Raveh, A., Domingue, A., Bertrand, L., Klemberg-Sapieha, J.E., Gujrathi, S.C., and Wertheimer, M.R., Thin Solid Films (in press).

6. Rousset, G., Monchalin, J.P., and Bertrand, L., J. de Phys., Coll. C6, 44 (1983) 165.
7. Frederikse, H.P.R., Feldmann, A., and Ying, X.T., in "Photoacoustic and Photothermal Phenomena II", Murphy, J.C., Maclachlan-Spicer, J.W., Aamodt, L.C., and Royce, B.S.H., eds., Springer Verlag, N.Y. 1990.
8. Touloukian, Y.S., Powelle, R.W., Yo, C.Y., and Nicolaou, M.C., in "Thermal diffusivity", Plenum Press, N.Y. 1970.
9. Raveh, A., Martinu, L., Klemberg-Sapieha, J.E., and Wertheimer, M.R, 38th Annual AVS (Amer. Vacuum Soc.) Symp. and Topical Conf., Seattle, Washington, Nov. 11-15, 1991.

An Excitation Energy Transfer Between Monomer and Dimer of Rhodamine 6G in a Viscous Solution

J. Szurkowski

Institute of Experimental Physics, University of Gdansk,
Wita Stwosza 57, 80-952 Gdansk, Poland

The deactivation of dimers and monomers of Rhodamine 6G in a polymer matrix (PVA) for different dye concentrations was studied using absorption, fluorescence spectra as well as photoacoustic and fluorescence excitation spectra. The analysis of the shape of the spectra leads to understanding the energy balance in the sample [1,2].

At the lowest concentration /$4*10^{-3}$M/, the dye is found practically as monomers, while at the highest one /$1*10^{-1}$M/ 25% of the molecules are in dimer form as computed from the data of [3]. In order to omit corrections related to reabsorption of luminescence and to keep on the linear dependence between the energy and registered signal, measurements were performed for very thin samples (the thickness of PVA layer on a glass plate does not exceed 25 μm), i.e., a thermally thin sample for a 20 Hz modulation frequency.

Fig.1 presents normalized photoacoustic and fluorescence excitation spectra for lowest and highest concentrations. The concentration change affects the shape of absorption spectra in the smallest extend. The largest concentration-dependent changes in the shapes occur for photoacoustic and fluorescence excitation spectra. That is a result of far higher thermal deactivation yield in the case of dimers and the monomer-dimer energy exchange process.

If the process of energy transfer does not take place in the sample, the sum of thermal deactivation energy and that given up by radiative transitions should be equal to the energy absorbed by the sample for each wavelength. By adding the excitation and photoacoustic spectra with appropriate corrective coefficients one can obtain the sample absorption spectrum. In contrast, if energy transfer between Rhodamine 6G dimer and monomer forms takes place the simple operation will not lead to the absorption spectrum as expected at high concentrations. Fig.2 presents the comparison of the best-fit curve originated from the sum of the thermal and radiative deactivation spectra to the measured absorption spectrum for Rhodamine 6G at concentrations; $4*10^{-3}$ and $1*10^{-1}$M. For the lower concentration, one can observe good agreement between the both curves /Fig.2a/. Slight discrepancies may result from the presence of small amount of dye in an aggregate form at this concentration. At the higher concentration /Fig.2b/ it was not possible to achieve such agreement with any values of the best-fit approximation coefficients. Taking into account the fluorescence yield for Rhodamine 6G in PVA known from Ref.[1], for this concentration, makes the agreement between the both curves even worse.

Fig.3 presents differences between the measured absorption spectra and these best-fitted sums of photoacoustic and fluorescence excitation spectra. The

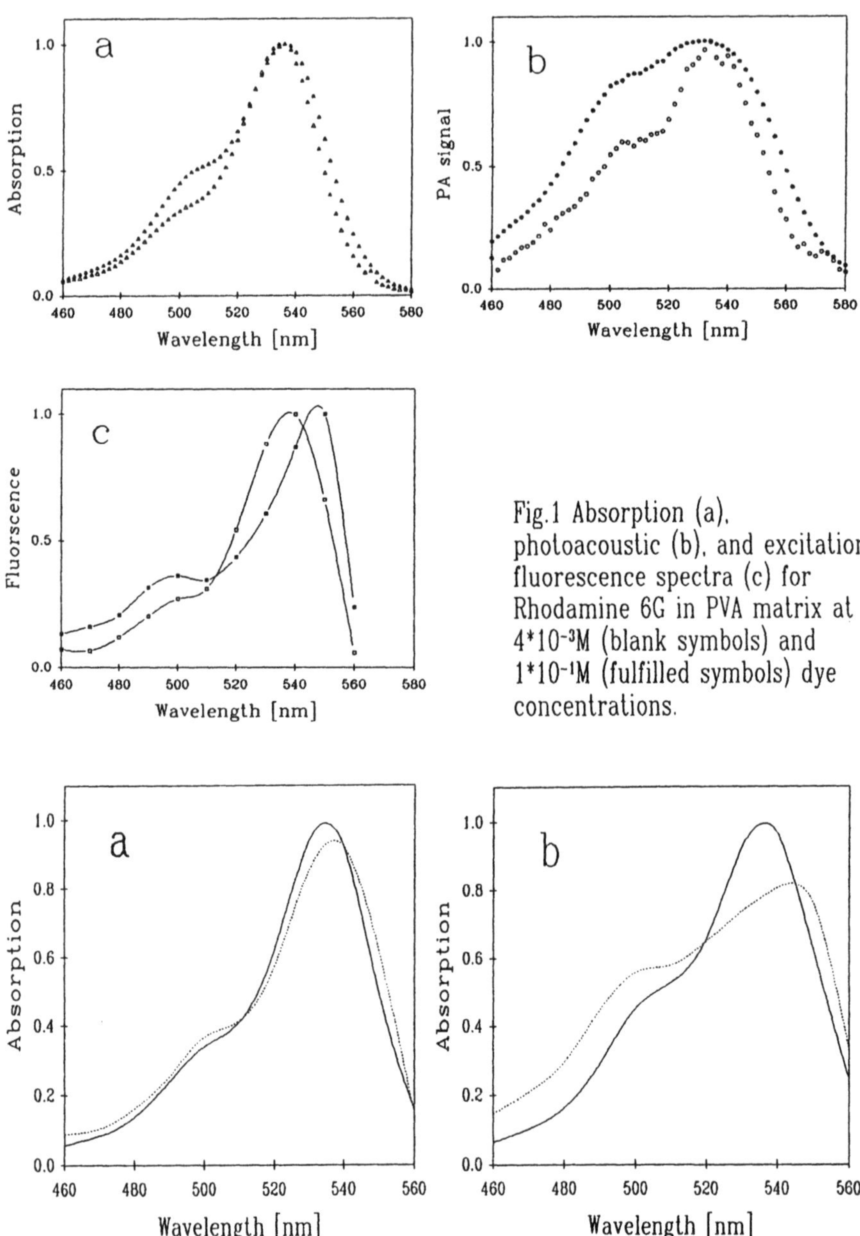

Fig.1 Absorption (a), photoacoustic (b), and excitation fluorescence spectra (c) for Rhodamine 6G in PVA matrix at 4*10^{-3}M (blank symbols) and 1*10^{-1}M (fulfilled symbols) dye concentrations.

Fig.2 Results of the best-fit-mean-square method of the photoacoustic and excitation fluorescence spectra sum with reference to the measured absorption spectra at (a) 4*10^{-3}M and (b) 1*10^{-1}M concentrations. (__) absorption spectra and (...) their best-fit approximations.

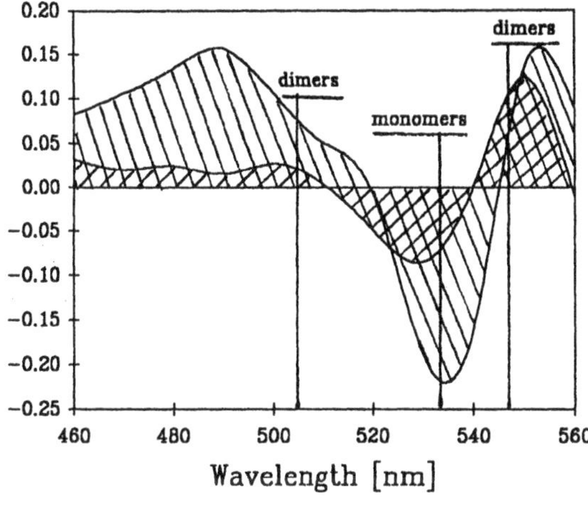

Fig.3 Difference between the measured and calculated (best-fit) absorption spectra from Fig.2 for concentration;

1*10⁻¹M.
4*10⁻³M.

characteristic wavelengths for the absorption maxima of monomers and dimers are also indicated. The results point to an energy transfer process from monomers to dimers of Rhodamine 6G taking place in the sample.

A method for determining the energy transfer within the dye sample reported in this paper is of simplified character and does not take into account several important relations (factors) affecting the result like the PA spectrum shape versus sample thickness dependence, lack of linearity between the absorbance and energy absorbed etc. It seems that in the case of "thin" samples the method can be a good first approximation.

Acknowledgments

I wish to thank Professor C.Bojarski for suggesting these measurements and Dr. G.Zurkowska for preparing the samples.

References

[1] Szurkowski J., Studia Biophysica 125, 63-69 (1988)
[2] Frackowiak D., Balter A., Szurkowski J., Lorrain L., Szych B., Hotchandani S., Acta Phys. Pol. A74, 125-138 (1988)
[3] Bojarski C., Obermueller G., Acta Phys. Pol. A50, 389-411 (1976)

Vibrational Spectroscopy of Monolayers by Pulsed Photothermal Beam Deflection (PBD)

C.D. Bain, P.B. Davies, and T.H. Ong

Department of Chemistry, University of Cambridge,
Lensfield Road, Cambridge CB2 1EW, UK

Abstract: The vibrational spectra of organic monolayers on both smooth and rough solid substrates were obtained by pulsed Photothermal Beam Deflection (PBD). The system used as a prototype was octadecyltrichlorosilane (OTS) adsorbed on silicon. The spectra, recorded both in air and under liquids, illustrate that sub-monolayer sensitivity can be achieved with pulsed lasers with short acquisition times.

1. Introduction

Photothermal Beam Deflection (PBD) exploits the thermally induced refractive index gradient near a surface to measure the amount of radiation absorbed by the surface. In previous studies, a modulated, continuous wave pump source was focussed onto a surface, and the amount of radiation absorbed measured by monitoring the deflection of a probe laser in the medium adjacent to the surface. The absorption spectrum of the surface was obtained by scanning the frequency of the incident light. These studies included UV-vis spectroscopy of thin organic films on solid substrates[1], trace gas detection[2] and the characterisation of the electrode-electrolyte interface[3]. We now report the vibrational spectra of organic monolayers on solid surfaces obtained by photothermal beam deflection with a nanosecond pulsed infrared laser.

2. Experimental Section

The experimental arrangement is shown schematically in Figure 1. In this experiment we employed 700 μJ of infrared in a 4 ns pulse, generated by difference-frequency mixing of the output from a tunable dye laser and a Nd:YAG laser in a lithium niobate crystal. The pump beam was focussed to a 500 μm diameter spot on the surface by a CaF_2 lens. The 0.8 mW He-Ne probe laser was tightly focussed at the surface and its deflection detected by a bis-photodiode. The output from the photodiodes was fed into a differential amplifier and spectra were acquired in digital form using a boxcar and analogue to digital converter.

Springer Series in Optical Sciences, Vol. 69 Photoacoustic and Photothermal Phenomena III
Editor: D. Bićanić

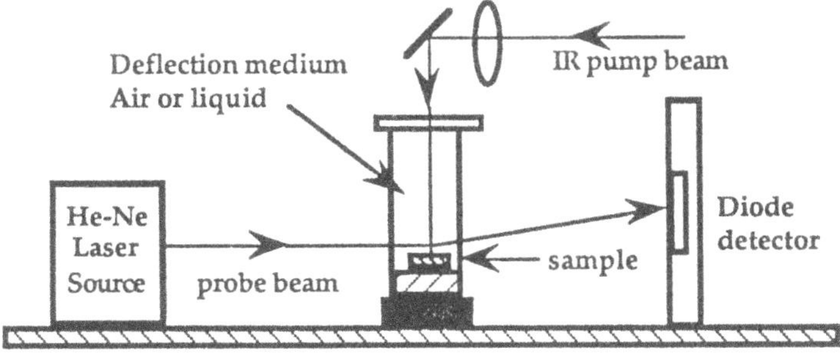

Figure 1: Schematic Diagram of Experimental Arrangement for PBD

3. Results

Figure 2a shows a photothermal spectrum of a monolayer of octadecyltrichlorosilane ($CH_3(CH_2)_{17}SiCl_3$, OTS) on silicon recorded in air. The acquisition time was 30 min. The monolayer was prepared by immersing a clean polished (111) silicon substrate (1 mm x 10 mm) into a 1mM solution of OTS in hexadecane for about 1 min. The self-assembled monolayer[4] forms by the spontaneous reaction of the surface silanol groups with the Si-Cl bonds to form a stable cross-linked sheet of Si-O-Si links. The spectrum shows two strong absorption peaks at 2850 cm^{-1} and 2918 cm^{-1}, together with weak features around 2876 cm^{-1} and 2955 cm^{-1}. It agrees well with the optoacoustic spectrum of LB films of cadmium stearate on sapphire recorded by Rothberg et al[5], in terms of the peak positions, relative peak intensities and line widths. The two strong modes observed can be assigned to the symmetric (d^+) and asymmetric (d^-) stretches, respectively, of the methylene groups in the hydrocarbon chain. The weak features arise from the terminal methyl stretches. A third methyl peak expected at 2935 cm^{-1} is obscured by the methylene absorption. Figure 2b shows that similar results are obtained on unpolished etched silicon substrate.

Spectra of the same monolayer system under a liquid deflection medium are shown in figure 2c and 2d for perfluorodecalin and CCl_4, respectively. Since the temperature dependence of the refractive index of liquids is greater than that of gases, the presence of the liquid amplifies the beam deflection significantly, giving improved signal to noise ratios and rapid acquisition times (≈ 5 mins). The peak positions of the methylene stretches show no significant shifts under these liquids. This agrees well with the results reported by Porter for the infrared spectra of a monolayer of octadecanethiol on gold under various liquids[6].

The chief drawback of PBD as a vibrational probe for monolayer studies is solvent absorption. The silicon wafer acts as an efficient heat sink due to its high thermal conductivity. Absorption of the infrared beam by the solvent creates a thermal gradient, and hence a refractive index gradient, opposite to that caused by the monolayer. Consequently, the probe beam is deflected in the opposite direction, giving rise to negative dips in the spectra. For weak solvent absorption, these can be subtracted using a background reference sample. We have found that 0.5% of methanol in CCl_4 can be tolerated (Figure 2e).

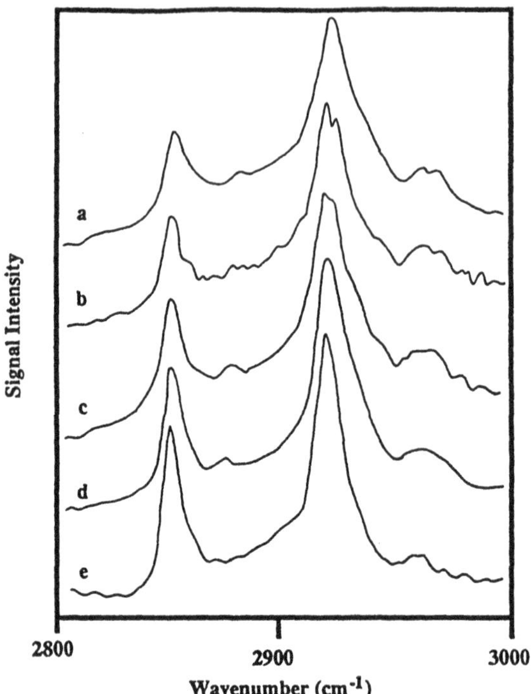

Figure 2: PBD spectra of a monolayer of octadecyltrichlorosilane a) on a polished silicon substrate under air, b) on an unpolished silicon substrate under air, c) on a polished substrate under perfluorodecalin, d) under CCl_4 and e) under CCl_4 with 0.5% methanol, after subtraction of solvent peaks. The relative signal intensities in spectra a – e are 1: 0.5 : 6 : 30 : 30.

This study illustrates the high sensitivity of pulsed photothermal beam deflection as a vibrational probe. Good signal to noise can be obtained on monolayer films in relatively short times.

We thank the SERC and Unilever Research Port Sunlight Laboratories for support for this work.

4. References:

1. A.Skumanich & J.C.Scott, *Mol. Cryst. Liq. Cryst.*, **183**, 365 (1990)
2. D.Fournier, A.C.Boccara, N.M.Amer & R.Gerlach, *Appl. Phys. Lett.*, **37**, 6 (1980)
3. R.A.Palmer & M.J.Smith, *Can. J. Phys.*, **64**, 1086 (1986)
4. S.R.Wasserman, G.M.Whitesides, I.M.Tidswell, B.M.Ocko, P.S.Pershan & I.D.Axe, *J. Am. Chem. Soc.* **111**, 5852 (1989)
5. L.Rothberg, G.S.Higashi & D.L.Allara, *Chem. Phys. Lett.*, **133**, 67 (1987)
6. S.M.Stole & M.D.Porter, *Langmuir*, **6**, 1199 (1990)

Reverse Mirage Photothermal Beam Deflection: Theory and Experiment

C.J. Manning[1], *F. Charbonnier*[2], *J.L. Chao*[3], *and R.A. Palmer*[1]

[1]Department of Chemistry, Duke University, Durham, NC 27706, USA
[2]ESPCI, 10, rue Vauquelin, F-75005 Paris, France
[3]IBM Corporation, Research Triangle Park, NC 27705, USA

Abstract. A relatively simple but complete theoretical model of the magnitude and phase of the PBD signal for the reverse mirage geometry in the presence of absorption in the deflection medium, and including the finite probe beam dimension and intensity distribution, has been developed. Numerical solutions yield results consistent with previous models. Step-scan FT-IR reverse mirage spectra of liquid/deflection media have been used to determine absolute absorptivities. Simulated spectra agree in both magnitude and phase with experimental data. Acetonitrile is used as an example.

1. Introduction

The transverse photothermal beam deflection signal in the reverse mirage geometry, i.e. with the probe beam on the non-illuminated side of the interface, (Figure 1) includes contributions from absorption localized at (or within a distance μ of) the interface, as well as absorption within the deflection medium further from the interface. The latter contribution is not present in data collected in the conventional mirage geometry, since radiation absorbed in the deflection medium is extinguished far from the sample. Thus, the theory appropriate for modeling the result of conventional mirage experiments [1] does not consider absorption in the deflection medium and is inadequate to describe the reverse mirage result. The reverse mirage geometry has previously been used to measure the propagation of charge carriers in silicon [2], and also to measure mid-IR absorption spectra of liquids [3,4]. In the later case, the absorbing/deflecting medium is a liquid at the interface with an IR-transparent solid (Figure 1).

Theoretical models for the reverse mirage effect, which include deflection in the absorption medium and, which give basic agreement with experiment [3,4], have been proposed. However, these have not been used to describe the phase of the signal, nor considered the finite dimension of the probe beam. In this paper we present a relatively simple but more complete theoretical treatment of reverse mirage (PBD), along with experimental results obtained by use of step-scan FT-IR, which are in quantitative agreement with the theory.

The experimental reverse mirage data used for comparison to the theoretical calculations have been obtained by use of an IBM Instruments IR44 FT-IR spectrometer, modified for optional step-scan operation [5]. In contrast to the use of single frequency or multiple discrete frequency lasers, FT-IR allows continuous

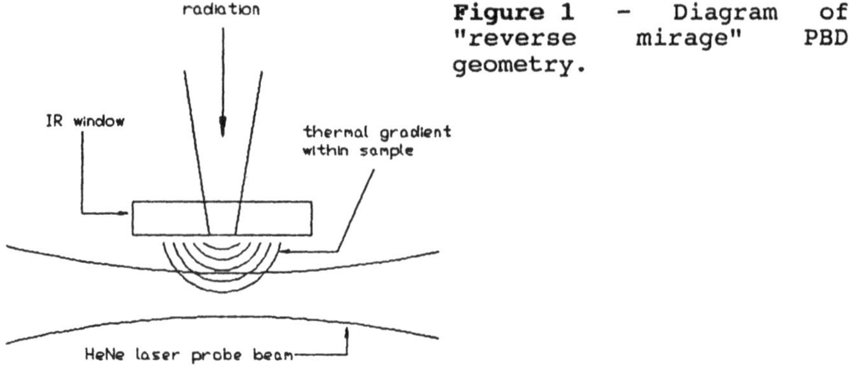

Figure 1 - Diagram of "reverse mirage" PBD geometry.

coverage of wide regions of the spectrum. The step-scan mode of data collection in FT-IR is particularly useful for photothermal spectroscopies. A single modulation frequency can be used (which applies to all wavelengths), permitting a lock-in amplifier to be used for demodulation and for the associated easy extraction of the signal phase. In the step-scan mode the PA phase is completely separated from the beamsplitter phase [6], in contrast to conventional continuous-scan (rapid-scan) FT-IR PA/PT spectroscopy. Values of absolute absorptivities of the various organic liquids for comparison to those determined by PBD were estimated from ATR data obtained with a Circle Cell.

2. Results and Discussion

The essence of the theory is seen in Equations 1 and 2. Equation 1 is the solution of the Rosencwaig-Gersho model for the case of absorption in the deflection medium. Numerical solution of equation 1 can be used to generate theoretical phase and magnitude spectra.

$$\theta(x) = E \left(\frac{(\sigma_s + \beta B) e^{\sigma_s x}}{(1 + B)} - \beta e^{\beta x} \right), \tag{1}$$

$$\text{where} \quad E = \frac{A}{(\beta^2 - \sigma_s^2)} \tag{2}$$

Examples of comparison of such calculated spectra to experiment are shown in Figures 2 and 3. In these spectra the transparent solid is zinc selenide (ZnSe) and the absorbing/deflecting medium is acetonitrile, CH_3CN.

Plots of the logarithm of the experimental PBD magnitude vs. effective probe beam distance yield linear plots, the slopes of which are the absolute absorptivity, β [4]. An example of such a plot for the 2944 cm^{-1} band of CH_3CN is shown in Figure 4. Data for this band and ten others of CH_3CN were used to calculate the

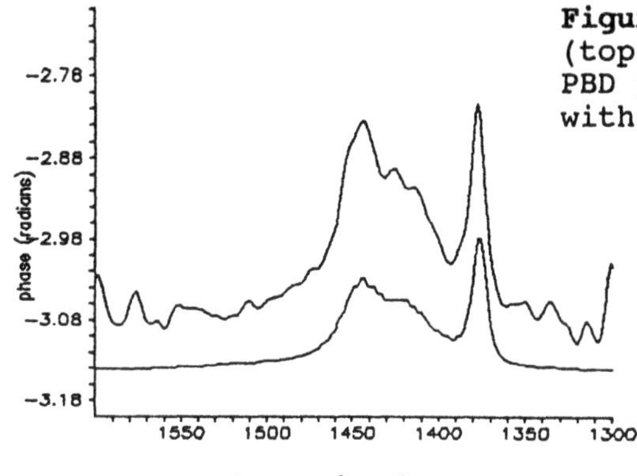

Figure 2 - Experimental (top) and model (bottom) PBD phase for acetonitrile with 70 μm probe depth.

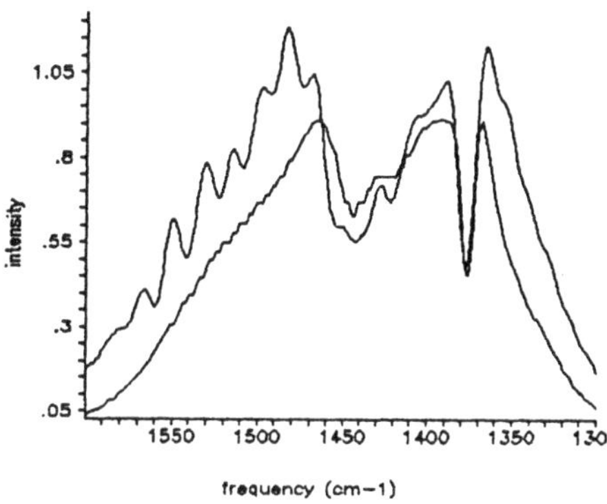

Figure 3 - Experimental (top) and model (bottom) PBD magnitude for acetonitrile with 70 μm probe depth.

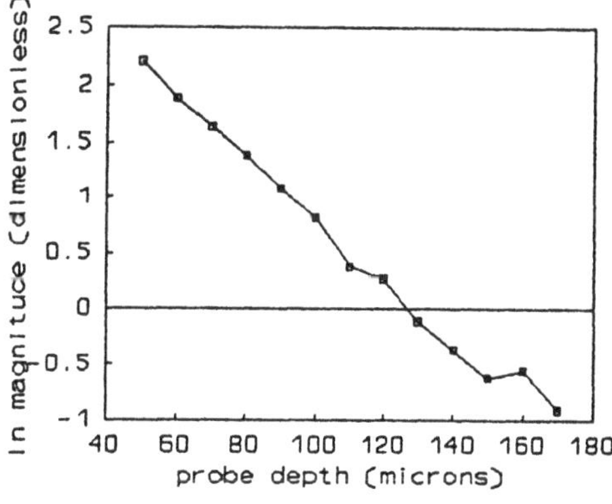

Figure 4 - Plot of log experimental PBD magnitude vs. probe depth for the 2944 cm^{-1} band of acetonitrile.

absorptivities given in Table I, where the PBD data are compared to approximate data obtained by ATR. Both qualitative and quantitative agreement between theory and experiment is good. The residual differences are probably due to uncorrected ATR artefacts.

Table I - Values of absorptivity, ß, measured by ATR and PBD (reverse mirage) for acetonitrile

Peak Location (wavenumbers)	ß-ATR (cm^{-1})	ß-PBD (cm^{-1})	correlation coefficient	#obs.
3164	82.7	201	0.996	11
3006	163	298	0.995	13
2944	154	286	0.997	11
2628	xxx	73.9	0.930	11
2292	215	348	0.996	10
2254	862	855	0.992	3
1443	456	675	0.999	7
1377	503	621	0.999	7
1038	296	376	0.985	9
918	201	285	0.989	11
748	107	180	0.980	13

1. W. B. Jackson, N. M. Amer, A. C. Boccara and D. Fournier, Appl. Opt. **20**, 1333 (1981).
2. D. Fournier, C. Boccara, A. Skumanich, and N.M., Amer. J. Appl. Phys. **59**, 787 (1986).
3. M. J. Smith and R. A. Palmer, Appl. Spectrosc. **41**, 1106 (1987).
4. D. Bicanic, S. Krüger, P. Torfs, B. Bein, and F. Harren, Appl. Spectrosc. **43**, 148 (1989).
5. C. J. Manning, J. L. Chao, and R. A. Palmer, Rev. Sci. Instrum. **62**, 1219 (1991).
6. C. J. Manning, F. Charbonnier, J. L. Chao, and R. A. Palmer, Infrared Phys., in press.

Piezophotoacoustic Spectroscopy of Gyrotropic Layered Structures

G.S. Mityurich, V.P. Zelyony, and A.N. Serdyukov

Gomel State University, Gomel 246699, Belorussia

Abstract. The peculiarities of thermooptical sound generation in gyrotropic 2-layered piezoelectric samples was studied. The resonance frequencies of the layered structure was determined. It was established, that the multibeam interference of light waves within layers essentially change the mechanism of PA signal formation.

The photoacoustic (PA) method has been widely adopted for studying multilayered media, thin films deposited on dielectric and conductive basing materials.

The investigation of the layered samples using the PA method has been conducted till now without taking into account gyrotropy which may cause a significant change of the mechanism of PA signal formation.

In this paper we investigate PA transformation in gyrotropic 2-layered piezoelectric samples taking into account multibeam interference of light waves within layers. We assume that the layered sample under consideration is simultaneously the PA detector and comprises absorbing cubic crystals of 23 or $\bar{4}3m$ symmetry, cut along the direction [110], parallel to the propagation direction of amplitude-modulated electromagnetic radiation. Based on the the Dyugamel- Neuman relations

$$\sigma_{ij}=c^{E}_{ijkl}u_{kl} - e^{T}_{ijm}E_{m} - \lambda^{E}_{ij}\theta,$$

$$D_{i}=e^{T}_{ijk}u_{jk} + \varepsilon^{S}_{ij}E_{j} + P^{S}_{i}\theta,$$

and using the expressions for the energy dissipation [1] in the 1-st and 2-nd layers

$$Q^{\pm}_{1} = A^{(1)}_{\pm}\exp(-\alpha_{\pm}x) + A^{(2)}_{\pm}\exp(-\alpha_{\mp}x),$$

$$Q^{\pm}_{2} = B^{(1)}_{\pm}\exp(-\beta_{\pm}x) + B^{(2)}_{\pm}\exp(-\beta_{\mp}x)$$

the voltage drop in the piezostructure under different boundary conditions - free (a) ($\sigma_1(0) = 0$, $\sigma_2(d_1+d_2) = 0$), clamped (d) ($u_1(0) = 0$, $u_2(d_1+d_2) = 0$) and clamped by turns (b) ($\sigma_1(0) = 0$, $u_2(d_1+d_2) = 0$), (c) ($u_1(0) = 0$, $\sigma_2(d_1+d_2) = 0$) the sample boundaries - is given below (see Table 1.). In the expressions for the energy dissipation we used the following designations $\alpha_{\pm} = 4\pi/\lambda(\varepsilon''_1/2\sqrt{\varepsilon'_1} \pm \gamma''_1)$, $\beta_{\pm} = 4\pi/\lambda(\varepsilon''_2/2\sqrt{\varepsilon'_2} \pm \gamma''_2)$, $A^{(1,2)}_{\pm} =$

Springer Series in Optical Sciences, Vol. 69 Photoacoustic and Photothermal Phenomena III
Editor: D. Bićanić

$= A_0^{(1,2)}\alpha_\pm(1 \pm \tau)^2/(1 + \tau^2)$, $B_\pm^{(1,2)} = B_0^{(1,2)}\beta_\pm(1 \pm \tau)^2/(1 + \tau^2)$, $\varepsilon = \varepsilon' + i\varepsilon''$ - the complex tensor of the dielectric permeability $\gamma = \gamma' + i\gamma''$ is the complex pseudoscalar gyration parameter, the imaginary part of which γ'' describes the circulary dichroism, τ, λ - the incident light wave ellipticity and wave length. A_0, B_0 - the constants, determined by optical parameters of the crystal and the incident light beam characteristics.

Table 1. The expressions for the voltage drop in the piezostructure under different boundary conditions

boundary conditions	
a	$V_1 = \dfrac{h_1 V_1^{(1)}(\cos k_1 d_1 - 1) + h_2 V_2^{(1)}(\cos k_2 d_2 - 1) + V_3^{(1)}}{k_2 c_{02}\cos(k_1 d_1)\sin(k_2 d_2) + k_1 c_{01}\sin(k_1 d_1)\cos(k_2 d_2)} + V_0$
b	$V_2 = \dfrac{h_1 V_1^{(2)}(\cos k_1 d_1 - 1) + h_2 V_2^{(2)}\sin k_2 d_2 + V_3^{(2)}}{k_2 c_{02}\cos(k_1 d_1)\cos(k_2 d_2) + k_1 c_{01}\sin(k_1 d_1)\sin(k_2 d_2)} + V_0$
c	$V_3 = \dfrac{h_1 V_1^{(3)}\sin k_1 d_1 + h_2 V_2^{(3)}(\cos k_2 d_2 - 1) + V_3^{(3)}}{k_2 c_{02}\sin(k_1 d_1)\sin(k_2 d_2) - k_1 c_{01}\cos(k_1 d_1)\cos(k_2 d_2)} + V_0$
d	$V_4 = \dfrac{h_1 V_1^{(4)}\sin k_1 d_1 + h_2 V_2^{(4)}\sin k_2 d_2 + V_3^{(4)}}{k_2 c_{02}\sin(k_1 d_1)\cos(k_2 d_2) - k_1 c_{01}\cos(k_1 d_1)\sin(k_2 d_2)} + V_0$

The following designations are used in the table: $h_{1,2} = e_{14}^T/\varepsilon_{11}^S$, $c_{01,2} = c_{44}^E + h_{1,2}e_{14}^T$, $k_{1,2}^2 = \rho\Omega^2/c_{01,2}$, ρ - the density, Ω - the modulation frequency, c_{44}^E, e_{14}^T - elasticity coefficient and piezoelectric constants tensors. Indexes "1" and "2" correspond to the 1-st and the 2-nd layer. The values V_0, $V_{1,2,3}^{(1,2,3)}$ is functions of optical and thermal parameters of layered sample.

We made the numerical analysis of the derived expressions, from which we conclude that the PA - amplitude depends on relation between acoustic, thermal and optical parameters of the gyrotropic piezostructure in the complex way. In the range of the layer thickness $k_{1,2}d_{1,2} \leqslant 1$, $k_{1,2} = 2\pi n_{1,2}/\lambda$ the major contribution in the detected PA signal is related to the sample optical characteristics. Multibeam interference of light waves in gyrotropic layers leads to the heavy oscillating dependence

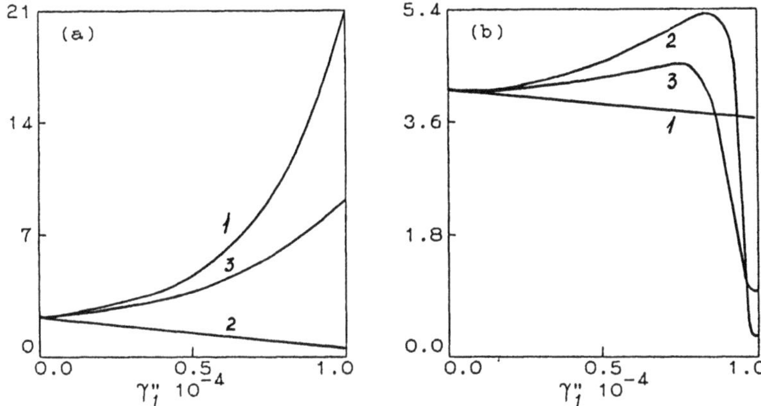

Fig.1. The dependence of PA signal amplitude on the circular dichroism parameter of the 1-st layer (γ'') for different polarizations of the incident radiation; 1 - τ=+1, 2 - τ=-1, 3 - τ=0, a) taking into account the light waves reflection within the layer thickness of $d_1 = 5\cdot 10^{-6}$m, b) without reflection of the light waves.

of the PA response amplitude on the layer thickness. The reflection of light waves within layer thickness of $d_1 = 5\cdot 10^{-6}$m essentially influences on the behaviour PA signal amplitude and phase characteristics for different polarizations of the incident radiation (see Fig.1(a,b)). The resonances in the layered structure are detected; the resonance frequencies depend upon specific boundary conditions. For example, for the case of free and clamped sample boundaries:

a) $\Omega_{рез} = \left(\frac{\pi}{2} + \pi k \right) \frac{1}{d} \sqrt{\frac{c_{01}}{\rho_1}}$, k=0,1..., $d_2 = \sqrt{\frac{c_{02}\,\rho_1}{c_{01}\,\rho_2}}\; d_1$

b) and for the clamped by turns boundaries

$$\Omega_{рез} = \left(\frac{\pi}{2} + \pi k \right) \frac{1}{d_1} \sqrt{\frac{c_{01}}{\rho_1}} \text{ , k=0,1...,}$$

$$d_2 = \sqrt{\frac{c_{02}\,\rho_1}{c_{01}\,\rho_2}}\; \frac{\pi + \pi k}{\pi/2 + \pi k}\; d_1 .$$

Choosing modulation frequencies low enough $\Omega \leqslant 10^3$ Hz up to $d_{1,2} = 10^{-4}$ m thickness we can neglect the two-layered sample acoustic characteristics. In this case the PA signal is defined by the optical and thermal characteristics of the crystals. It is easy to show that measuring the difference between PA signals amplitudes corresponding to clockwise (τ = +1) and anticlockwise (τ = -1) polarization of the incident radiation gives us the possibility to determine the circular dichroism parameter of the gyrotropic medium.

References

1. G.S. Mityurich, V.V. Sviridova, A.N. Serdyukov: J. Appl. Spectr. 53, 611(1990).

Photoacoustic Signal Generation and Spectra of Powders

J. Buijs[1], *J.P. Favier*[2], *O. Dóka*[3], and *A. Miklós*[3]

[1]Department of Physical and Colloid Chemistry,
Wageningen Agricultural University, Dreijenplein 6,
6703 HB Wageningen, The Netherlands
[2]Department of Agricultural Engineering and Physics, Agrotechnion,
Wageningen Agricultural University, Bomenweg 4,
6703 HD Wageningen, The Netherlands
[3]Department of Photophysics, Radiation Research Laboratory,
Institute of Isotopes, P.O. Box 77, H-1525 Budapest, Hungary

Abstract. Photoacoustics is used to obtain thermal excitation spectra in the visible region of powdered samples. The process of signal generation and as application a few spectra of different types of flours are described.

1 Photoacoustic signal generation in powders

Photoacoustics opens the possibility to obtain thermal excitation spectra of inhomogeneous and light scattering samples, such as powders. In this case photoacoustic spectroscopy is used for measuring the acoustic signals produced by powders (irradiated by a modulated light source) as a function of the optical wavelength, ranging from 300 to 700 nm. For most of the samples the deexcitation proces is non-radiative and therefore the thermal excitation spectrum correlates well with the absorption spectrum of the sample. The set-up is shown in Fig. 1.

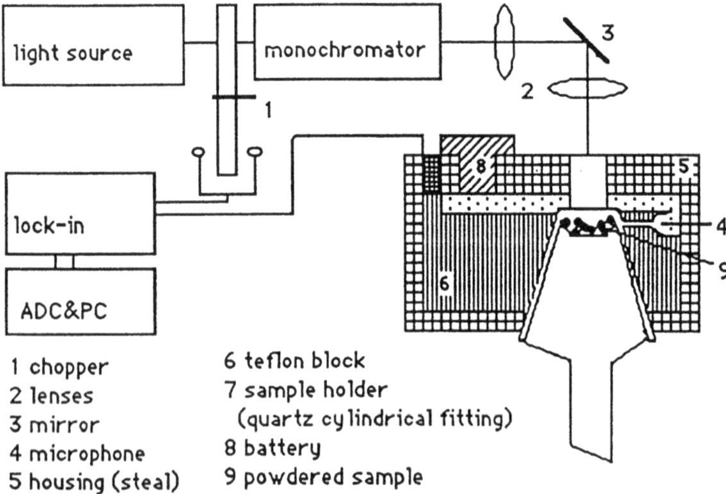

Figure 1. Set-up used for measuring thermal absorption spectra of powders.

Springer Series in Optical Sciences, Vol. 69 Photoacoustic and Photothermal Phenomena III
Editor: D. Bićanić

Compared to solid plate shaped samples the light intensity distribution and the heat diffusion in the sample needs a more detailed describtion. A one dimensional model for signal generation can be formulated, using general equations and correcting them for air gaps and reflections inside the sample [1]. A cylindrical symmetric sample with thickness l_s is considered, where the light incident on the sample surface is at z=0. Due to the large thermal diffusion coefficient of air, one can assume that the air inside the sample has the same temperature depth profile as the surface of the grains. Taking this into account, all the physical parameters used in the equations below are the effective parameters, corrected for the influence of air inside the sample. Besides the surface/volume ratio of the grains also some effective parameters, like the thermal conductivity, will be highly affected by the grain size. From experimental data [2] it was shown that the transmittance can be neglected, even for optical transparant samples.

More insight in the temperature distribution in the sample is gained if the general solution is divided into two major cases, based on the optical absorption length compared to the grain size, i.e. optical transparant and optical opaque samples.

For optically transparant samples, the optical absorption length is larger than the grain size and light will penetrate through the grains. If the grains are relative small compared to the whole sample and if the thermal diffusion between the grains can be neglected it is a reasonable assumption that there is almost no heat diffusion inside the sample. Therefore the temperature distribution is proportional to the absorption and thus to the light intensity distribution P_o in the sample. The light distribution in the sample depends on the reflectance- and the absorption coefficient, R and A respectively, and it can be shown that the signal S in this case is given by

$$S = T\frac{\beta_g}{VK_T}\left[\frac{A}{\sigma_s\sigma_g\left(\kappa_s\sigma_s+\kappa_g\sigma_g\right)}\left(1+\frac{2(1+r_o))}{(1-r_o}\right)+\frac{\beta_s}{\beta_g\kappa_s\sigma_s^2}(1-R)\right]$$

with T the responsivity of the microphone, V the effective cell volume, β the thermal expansion coefficient, K_T the thermal compressibility, κ the thermal conductivity, σ the complex thermal diffusion and r_0 the effective reflection coefficient for internally incident diffuse light at the sample gas boundary. The first term is the signal contribution from the heat diffused into the gas, which is proportional to the optical absorption coefficient and increased by a factor $[1+2R(1+r_0)/(1-r_0)]$ due to scattering. The second term is the signal due to thermal expansion of the sample.

For optically opaque samples one can assume that most of the energy is deposited in the upper layer of the sample. Further the thermal diffusion will be strongly blocked by the air gaps in the sample. In this case the PA signal can be described by

$$S = T\frac{\beta_g}{VK_T}\frac{P_o}{\kappa_s\sigma_s\sigma_g}\left[F+\frac{\beta_s\sigma_g}{\beta_g\sigma_s}\left(1-\frac{\kappa_g\sigma_g}{\kappa_s\sigma_s}F\right)\right](1-R)$$

where

$$F=\frac{(1+b)e^{\sigma_s l_s}+((1-b)e^{-\sigma_s l_s}}{(1+b)(1+g)e^{\sigma_s l_s}-(1-g)(1-b)e^{-\sigma_s l_s}}$$

with $g=\kappa_g\sigma_g/\kappa_s\sigma_s$ and $b=\kappa_b\sigma_b/\kappa_s\sigma_s$. The first and second term in parentheses are the factors for the relative signal magnitude due to thermal diffusion and thermal expansion respectively.

Normalization of the signal has been carried out by using an intensity spectrum, produced by measuring the intensity of the incident light each 5 nm and than fitting these points with a carbon black spectrum into a normalization spectrum.

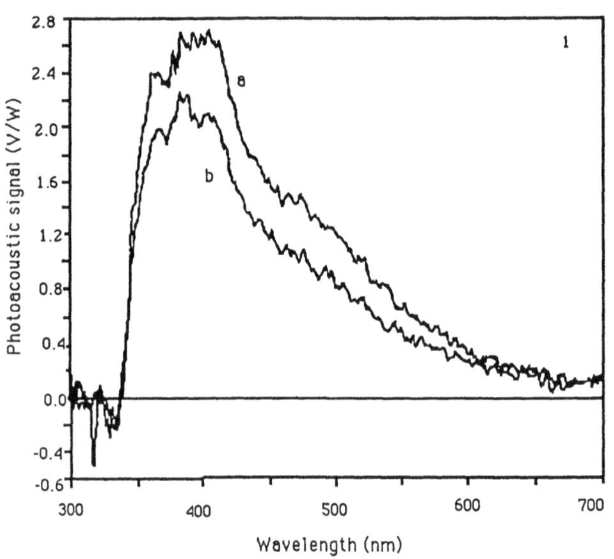

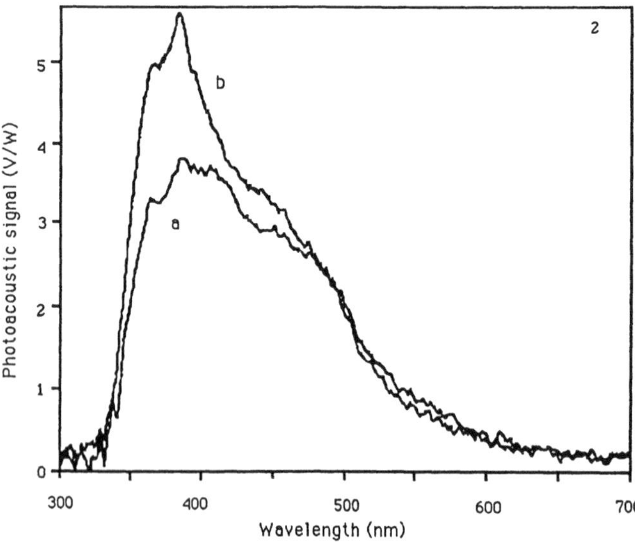

Fig 2. Photoacoustic powder spectra of flours; (1a) bread flour, (1b) fine bread flour, (2a) doughnut flour, (2b) semolina flour

2 Applications

The main interest lays in the field of biological samples and this technique has been used in order to study all kinds of flours. With the measured spectra, discrimination can be made between all different kinds of flours based on the colour and the grain size of the sample. As an example, four spectra of the flours are shown in Fig. 2. Note that for wavelengths below 350 nm most of the intensity is absorbed by the lenses and mirror, resulting in an inaccurate calibration of the signal.

References

1 A.Y. Zafer, W.B. Jackson and N.M. Amer, Appl. Opt., (1982) , 21, 21-31 .

2 J. Buijs and J.P. Favier; Photoacoustic measurements of powdered samples, Msc thesis, Dept. of Agro and system Physica, Agicultural University Wageningen, The Netherlands, (1991).

Part III

Non-Destructive Evaluation and Material Characterization

Recent Developments in the Thermal Diffusivity Measurements of Diamond Materials

P.K. Kuo, L. Wei, and R.L. Thomas

Department of Physics and Institute for Manufacturing Research, Wayne State University, Detroit, MI 48202, USA

Abstract. The different methods used for the determination of thermal diffusivity or thermal conductivity of diamond crystals, polycrystalline diamond films and diamond-like-carbon films are reviewed. Recent results of the thermal wave methods are discussed.

1. INTRODUCTION - WHY DIAMOND?

Diamond as a material has many outstanding physical properties. For example, its thermal conductivity at room temperature is about 6 times that of copper, and its thermal expansion coefficient is less than that of invar. These properties, together with its wide band gap (5.45 eV), breakdown voltage (10^7 V/cm) and saturation velocity (2.7 x 10^7 cm/s), combine to make it a very attractive semiconductor and optical material. Historically, its popularity as a gem stone has sustained a high market value and has limited its application as an industrial material. The study of its properties has remained an academic exercise, mainly for the verification of theories of solid state physics. However, the situation has changed drastically in recent years. First, some industrial laboratories have been making steady progress in the synthesis of single crystal diamond. Both General Electric in the US and Sumitomo in Japan have been able to grow high quality crystals up to several carats in weight. Secondly, low pressure vapor growth of diamond films has been carried out with moderate success since the late forties, and in recent years there have been dramatic increases in both the growth rates (polycrystalline diamond films can now be produced at rates of tens to hundreds of micrometers per hour) and the quality of the films, which have made possible the first practical applications of this technology. A useful recent review of the field by Angus and Hayman [1] describes both the technological and the scientific development, and provides a comprehensive list of references. About a year ago (July '90), General Electric announced the synthesis of isotopically enriched diamond single crystals which have thermal conductance almost 50% higher than the best natural diamonds. All these developments have made diamond one of the most attractive new materials in the nineties.

2. THERMAL CONDUCTANCE AND THERMAL DIFFUSIVITY

The measurement of thermal properties of diamond presents a special challenge to the material scientists and engineers. This is so because of two factors. The first is the fact that diamond is a better thermal conductor at room temperature than any other material by a wide margin. Any attempt to attach heaters or thermal probes is likely introduce thermal resistance which is substantial compared to the intrinsic thermal resistance of the diamond specimen being measured. The second factor is the preciousness of diamond, which makes availability of large samples unlikely. Very often, the geometry of the sample is dictated by the application. In the case of CVD (chemical vapor deposition) diamond or diamond-like-carbon, the samples are often in the form of thin films, either attached or removed from substrate. The latter fact only makes the task even more difficult.

The measurement of thermal conductance, which describes how much heat passes through a given sample of unit area and unit thickness for given temperature gradient, requires calorimetry, a cumbersome experimental procedure. Often one makes use of the heat capacitance of the sample as the medium of calorimetry. In essence, one is measuring the thermal diffusivity of the material, which is related to thermal conductivity by:

$$\alpha = \kappa/c\rho$$

where κ is the thermal conductivity, c the thermal capacity, and ρ is the density of the material. Since only the temperature change over a distance needs to be measured as a function of time, it is the favored mode of operation in many of the methods described below. The product, $c\rho$, is simply the thermal capacitance per unit volume of the sample. It is not expected to change much for different sample geometries, or even different isotopic contents. Thus once the thermal diffusivity is measured, one can compute the thermal conductance by using the value of $c\rho$ from bulk samples. As a diffusion coefficient, the thermal diffusivity actually describes the ability of the material to disperse heat, a relevent property when the material is used as a heat sink or heat spreader. At the interface between two different materials, the conservation of heat flux brings in another quantity into play, the effusivity:

$$\eta = \sqrt{(\kappa c\rho)}$$

Crudely speaking, effusivity describes the "branching ratio" of heat at an interface between two materials. Because it is derivable from κ and $c\rho$ it is not necessary to make a separate measurement for this quantity. However in an experiment involving at least two dissimilar materials, both α and η are needed for a proper modeling of the data.

3. METHODS OF MEASUREMENT

There are no less than five different approaches (and their variations) to the measurement of thermal diffusivity or conductivity of diamond materials. We give each a short description and comment below on the relative pros and cons.

a) The conventional method

It should be clear from the above discussion that the conventional method is not advantageous for measuring the thermal conductivity of diamond materials. However, it is still employed today, and is the method every other method must compare with to gain acceptance by the material science community. In this method, the sample is shaped into a rectangular bar. One end of the bar is attached to a heating block and the other end to a heat sink. Two thermocouples are then attached to the sample bar at a known distance from each other. The heater block and the heat sink must be designed to have sufficient thermal mass to provide a steady state temperature gradient along the bar. The thermocouples, on the other hand, must be small enough to not have a thermal gradient of their own so that they reflect accurately the temperature of the sample. The thermocouples provide the information on the temperature gradient established in the sample. The heat flux through the sample is obtained by knowing the electrical power input to the heater, after all the revenues of heat loss have been quantified. The whole assembly is usually enclosed in a radiation shield which is heated to have a temperature profile matching that of the steady state temperature profile of the sample, to minimize unwanted heat transfer through radiation. With a careful design, the main source of uncertainty of this method is the uncertainty in the positioning of the thermocouples. The latter, in turn, is limited by the physical size of the thermocouples.

An alternative approach is to use sinusoidal instead of steady state heat flow. Then a propagating thermal wave would be detected by the thermocouples. Either the change in amplitude or the phase of the wave would give the diffusivity of the sample material. More than a century ago (1863), Angström, using hot and cold running water, performed perhaps the very first thermal wave experiment, with frequencies in the millihertz range.[2] More modern versions of his experiment were reported in 1961.[3]

b) The flash method[4]

This method is basically a variation of the above method where the steady state heating is replaced by transient heating. Typically, a sample in the shape of a flat slab is used. A pulse of heat is applied to one side of the sample from a laser beam or a flash lamp. The temperature of the opposite side is monitored with an infrared detector. Either the time delay for the temperature to reach the peak value or one-half of its peak value is measured to determine the diffusivity of the material. Note that since no calorimetry is involved, only the thermal diffusivity, not the thermal conductivity, is determined. For a sample which is transparent either in the visible or infrared regions, an opaque film must be placed over the sample surfaces to insure that the conversion of light energy to heat occurs at the surface and that the observed infrared radiation originates only at the surface. Given the right sample, this method is probably the fastest among all methods. Its drawback is in the restrictive sample geometry.

c) The thermal wave/mirage effect method[5]

This method uses a focused, modulated laser beam as the source of heat. The laser beam launches a thermal wave in the sample. Detection is through the mirage effect, which utilizes a second laser beam to traverse the heated air above the heated sample. The time varying index of refraction in the air causes the probe laser beam to be refracted, just like what happens to light rays in the familiar mirage effect; hence the name. This method is described in more detail in the accompanying poster paper in this conference. The shortcoming of this method is in the requirement of intensive data analysis. The mirage signal is dependent in an involved manner on the diffusivity of the sample, along with several other experimental parameters, such as the radii of both the heating and probing beams, the offset distance between beams, etc. However, it does have the least restrictive requirement of sample geometry. Hence it is equally applicable to thin films (free-standing or on-substrate) and multilayered structures. Except for the fact that an opaque film has to be deposited on transparent sample surface to facilitate light absorption, this method is contactless. It also requires only one surface of the sample to be accessable.

d) The convergent thermal wave method[6]

This is a variation of the flash method described above. The heating source is in the form of laser beam, focused into a ring through a lexicon lens. In the interior of the ring, the thermal wave converges toward the center with undiminished magnitude, where it is monitored with an infrared detector. This method is just as fast as the flash method, with improved signal-to-noise ratio. It also shares the same drawback of restrictive sample geometry. In its latest implementation, it is only applicable to free standing diamond films with a diameter of 4 cm or larger.[7]

e) The three omega (3ω) method[8]

In this method a thin film is deposited on a flat surface of the sample. The thin film serves both as the heating element and the detector. A sinusoidal current of frequency ω (and high harmonic purity) is sent through the resistor, launching a thermal wave of frequency 2ω in the sample. This time varying temperature fluctuation in turn induces a resistance fluctuation (through the temperature coefficient of the resistivity) at the same 2ω frequency. If a lock-in amplifier is used to monitor the voltage across the thin film resistor, and if the reference frequency of the lock-in amplifier is tuned to 3ω (hence the name of this method), it detects a signal which is proportional to the surface thermal impedance of the sample. This method is conceptually very similar to the thermal wave/mirage effect method above, except using a different detection mechanism. It requires somewhat less intensive data analysis, and is in principle applicable to thin films and multilayered structures. Its success depends largely on the fine art of depositing a thin film resistor and making connections to it.

4. RECENT RESULTS

There are mainly two areas of experimental activity to be described: a) Synthetic, type IIA diamond crystals with varying isotopic contents, and b) Polycrystalline diamond films.

Table I. Room-temperature thermal diffusivity and conductivity for Type IIA Diamond crystals with different isotopic contents.

Isotopic content (C^{13}%)	Diffusivity (cm^2/s)	Conductivity* (W/cm-K)
0.07	17.5	31.9
0.5	14.5	26.4
1.0	12.4	22.6
natural diamond	12.2	22.2
4.54	7.75	14.1
36**	4.65	8.46
68**	4.20	7.64
99**	12.0	21.8

*Based on $c\rho = 1.82$ J/cm^3-K.
**Isotopic content is that of feed stock CH_4.

Table II. Thermal Conductivity of Isotopically Pure Diamond.

Investigator	Technique	Conductivity(W/cm-K)
Wayne State U P.K. Kuo, R.L. Thomas, and Lanhua Wei	Thermal Wave	30-32
U. of Delaware David Onn and Adam Witek	Xenon Photo Flash	35
General Motors Don Morelli and Joseph Heremans	Longitudinal Bar	22-25
U. of Illinois A.C. Anderson and Daming-Zhu	Longitudinal Bar	22-25
Stanford University	Photo Flash	In Progress
Cornell University Jeff Olson and Robert Pohl	3ω	In Progress
Brown University Humphrey Maris	Laser pump and probe	In Progress

Diamond-like-carbon film, although a member of the diamond family, turns out to have thermal conductivity very small, about 2 to 3 orders of magnitude smaller than those of diamond.

It is a year after the publication of the initial results of the isotopically enriched diamond crystals[9]. Although much more data are now available, the experimental situation is still somewhat unsettled. The General Electric-Wayne State University collaboration has produced more synthetic diamond crystals and have performed thermal diffusivity measurements on them. The results are summarized in Table I. These results are preliminary in the sense that the isotopical contents of the individual crystals have not all been measured. The listed values, where noted, are those of the methane gas which were used as the raw material in growing the crystals. The data look sufficiently consistent among themselves, but the dependence on isotopic content seems stronger than those of He, Li and Ge. Theorists are hard at work to reach a quantitative understanding. There are also more independent experimental groups taking part in the thermal diffusivity measurement of the isotopically enriched diamonds crystals. Several different methods were used, and the discrepancies are larger than the estimated combined uncertainties in measurements. The results are summarized in Table II.

On the polycrystalline diamond films, because of the crystalline boundaries and graphitic contamination, it is not expected that they will have the same high thermal diffusivity as single crystals. Results differ widely from process to process and from manufacturer to manufacturer. Some reported values are very high, to more than 50% of the single crystal values.[7] Because of the great expected potential in application as heat sinks or heat spreaders, the competition is very fierce. Consequently, there are not many published results. It can be reported that the art of growing these films with microwave assisted chemical vapor deposition has progressed to the point that very high quality, thick (up to hundreds of micrometers) and large size (up to a few inches) films can be obtained routinely. These films contain microcrystals of several micrometers to tens of micrometers, are transparent after polishing and have thermal conductivities up to about half that of the highest quality single crystals. They can be produced with sufficient growth rates that they are strong candidates for heat sinks/spreaders and optical windows. The thermal wave methods described above work quite well with these films with very little sample preparation. They are being used routinely as tools for quality control.

5. SUMMARY

Modern thermal wave methods for the characterization of thermal properties of diamond materials have been shown to be capable of meeting the challenge from the advancement of manufacturing skills. These methods are non-destructive, fast, accurate, and requires very little sample preparation. Although some disagreement still lingers, the general picture is that they *do* help to make better materials.

ACKNOWLEDGEMENTS

This work was sponsored by ARO under Contract No. DAAL03-88-K-0089; and by the Institute for Manufacturing Research, Wayne State University.

REFERENCE

1. John C. Angus and Cliff C. Hayman, Science **241**, 913 (1988).
2. M.A.J. Angström, Philo. Mag. **25**, 181, (1863).
3. A. Hirschman, J. Dennis, W. Derken, and T. Monahan, in *International Developments in Heat Transfer*, Part IV, p. 863, ASME, New York, (1961).
4. W.J. Parker, R.J. Jenkins, C.P. Butler, and G.L. Abbot, J. Appl. Phys., *32*, 1679, (1961).
5. P.K. Kuo, M.J. Lin, C.B. Reyes, L.D. Favro, R.L. Thomas, D.S. Kim, Shu-yi Zhang, L.J. Inglehart, D. Fournier, A.C. Boccara, and N. Yacoubi, Can. J. Phys., **64**, 1165, (1986); P.K. Kuo, E.D. Sandler, L.D. Favro, and R.L. Thomas, Can. J. Phys., **64**, 1168, (1986).
6. P. Cielo, L.A. Utracki, and M. Lamontagne, Can. J. Phys., **64**, 1172, (1986); F. Enguehard, D. Boscher, A. Deom, and D. Balageas, Mater. Sci. Eng. **B5**, 127, (1990).
7. G. Lu, and W.T. Swann, First International Conference on the Applications of Diamond Films and Related Materials, August 17-22, 1991, Auburn, Alabama.
8. D.G. Cahill, and R.O. Pohl, Phys. Rev. **B35**, 4067, (1987).
9. T.R. Anthony, W.F. Banholzer, J.F. Fleischer, Lanhua Wei, P.K. Kuo, R.L. Thomas, and R.W. Pryor, Phys. Rev. **B42**, 1104 (1990).

Photoacoustic Study of the Thermal Effusivity of Cellulose and Paper

M. Kaplanová[1] *and G. Katuščáková*[2]

[1]University of Chemical Technology, Čs. Legií 565, CS-53210 Pardubice, Czechoslovakia

[2]Research Institute of Cellulose and Paper, Lamačská 3, CS-81520 Bratislava, Czechoslovakia

Abstract. Three different types of paper samples were studied by photoacoustics and photoacoustic spectroscopy: cotton linters, coating papers and Xerox papers. The presence of lignin fibres in mostly cellulosic papers were demonstrated by expressive absorption in the UV region of PA spectra. The thermal effusivities of the samples were evaluated from the frequency dependences of the PA signal at the saturation light conditions.

1. Introduction

The photoacoustics can be an effective method for determining of the thermal parameters of various materials when the photoacoustic signal is measured as a function of chopping frequency for thermally thin samples. The thermal conductivity is an important physical parameter which is often very difficult to measure, particularly on powders and porous materials. Cellulose and related filamentary materials are generally available in a form not convenient for conventional spectroscopic methods while the photoacoustic technique ideally suits as a means to obtain optical and thermal parameters. Thermal parameters as the thermal effusivity, can be used to monitor inner structural changes of materials.

2. Results and Discussion

We investigated several kinds of cellulosic materials (cotton linters, lignocellulosic samples and special reprographic papers) by means of the photoacoustic as well as by other physical experimental methods (structure - electron microscopy, measurement of the porosity and apparent volume density) to find relations between the thermal parameters and the structure of the porous paper samples. The samples were in the form of sheets, the thickness being from 70 to 120 μm. Photoacoustic spectra of the samples, containing lignin fibres reveal a shoulder near 380 nm, which is not observable in the spectra of pure cellulosic samples and relatively high PAS intensity at wavelengths shorter than 350 nm. This observation is in accordance with the results of the measurement of the photoacoustic spectra of pure lignin [1] and with our measurement of the photoacoustic spectra of the plate of beech wood. We measured photoacoustic spectra using single-beam photoacoustic spectrometer with microcomputer system attached to the home-built spectrometer [2].

The photoacoustic signal of the optically opaque and thermally thin samples depends on the thermal effusivity of the

Springer Series in Optical Sciences, Vol. 69 Photoacoustic and Photothermal Phenomena III
Editor: D. Bićanić

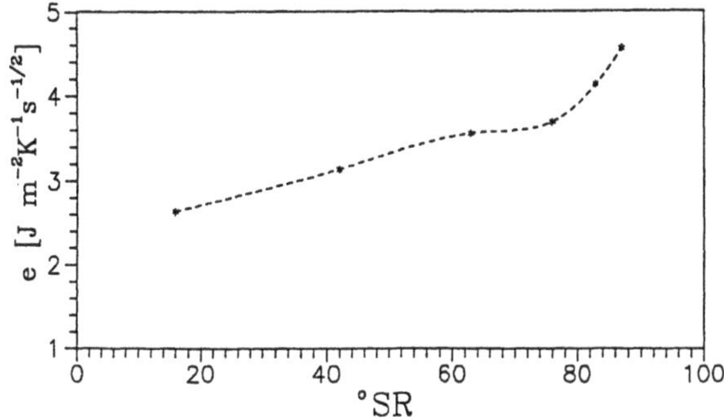

Fig. 1. Dependence of the thermal effusivity of cotton linters on the degree of beating.

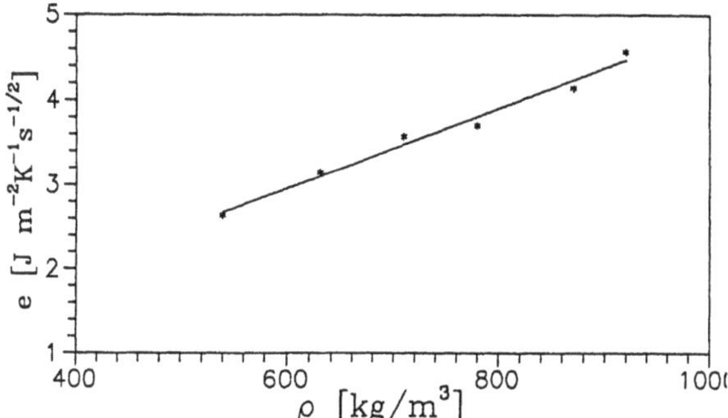

Fig. 2. Dependence of the thermal effusivity of cotton linters on the apparent volume densities.

sample and varies with f^{-1}. The thermal effusivity e is determined by the relation

$$e = (k \varrho C)^{1/2},$$

where k is the thermal conductivity, ϱ the density and C is the specific heat of the sample. The intensity of white light from a 100 W Xenon lamp was modulated by a variable frequency chopper. The frequency of the modulation varied from 10 to 120 Hz.

From the log-log dependence of the photoacoustic signal on the modulation frequency we evaluated the effusivity of the sample, multiplied by the constant factor, which depends on the experimental apparatus and conditions of measurement. We determined this constant factor by the calibration measurements (at the same experimental conditions) of a poly(methylmethacrylate) thin plate (thickness 1 mm) with known thermal parameters. Using the relatively low modulation frequencies, at which the thermal diffusion length is comparable or greater than the thickness of the sheet, the effusivity obtained is parameter, which characterizes the whole porous sample.The effusivities of sheets of cotton linters prepared from the fibres milled to the different degree (from the beating degree 16°SR to 87°SR, Fig.1). depend on the structure of sample and are proportional to the apparent volume density of paper sheets as illustrate Fig.2. The effusivity values from 2.64 to 4.56 $Jm^{-2}K^{-1}s^{-1/2}$ are in agreement with the values of thermal effusivities from 4.24 to 14.33$Jm^{-2}K^{-1}s^{-1/2}$of cellulosic papers determined by a conventional thermal transport experiments [3].

References:

1. J.R. Harbour, M.A. Hopper, R.H. Marchessault, C.J. Dobbin, E. Anczurowski, J. Pulp and Paper Sci. **V11**, 142, (1985)
2. M. Kaplanová, M. Trojan, P.Sladky, Czech. J. Phys. **B38**, 104, (1988)
3. Polymer Handbook, eds. J. Brandrup, E.H. Immergut, J. Wiley & Sons,(1975)

Photoacoustic Characterization of Calcinated and Non-Calcinated Zeolitic Chromosilicate

O. Nakamura[1], *A.M. Mansanares*[1], *H. Vargas*[1], *H.O. Pastore*[2], *E.J.S. Vichi*[2], *N.F. Leite*[3], *and L.C.M. Miranda*[3,4]

[1]Instituto de Física, Universidade Estadual de Campinas, 13081 Campinas, SP, Brazil
[2]Instituto de Química, Universidade Estadual de Campinas, 13081 Campinas, SP, Brazil
[3]Instituto Nacional de Pesquisas Espaciais, 12201 São José dos Campos, SP, Brazil
[4]Instituto Politécnico Rio de Janeiro, 28630 N. Friburgo, RJ, Brazil

Abstract. In this paper we confirm, using the photoacoustic spectroscopy, that the incorporation of oxygen to crytalline zeolitic chromosilicates promotes the oxidation of anchored $Cr(III)$ to non - interacting $Cr(VI)$ species. The thermal diffusivity, the non - radiative relaxation time and the optical absorption coefficients of both calcinated and non - calcinated chromosilicate samples are also reported.

The study of chromium ion states in high - content silica zeolites is of great interest since new catalysts can be designed with these systems [1]. The preparation of highly crystalline zeolitic chromosilicate containing $Cr(III)$ both substituting silicon atoms at the lattice sites, and outside the silicon skeleton, probably linked to oxygen atoms to form oxides like Cr_2O_3, has been recently reported [2]. Submitting the chromosilicate samples to a calcination process, $Cr(VI)$ species become evident. In this paper, we report on the photoacoustic characterization of the optical and thermal properties of calcinated and non - calcinated chromosilicate samples.

In Fig.1a we show the PA spectrum of a freshly prepared chromosilicate sample exhibiting the characteristic ligand field bands of $Cr(III)$ chromosilicate sample [3]. This sample was obtained by heating a freshly prepared chromosilicate sample at 693K for 24 hours, in a shallow bed, under oxygen flow. During this process the sample changes its colour from light green to yellow. Figure 1b exhibits the strong 370nm charge transfer absorption band characteristic of $Cr(VI)$.

The thermal diffusivity α, the non - radiative relaxation time τ and the optical absorption coefficient $\beta(\lambda)$ were obtained from the PA signal data as a function of modulation frequency. We first carried out a PA signal modulation frequency scanning to identify the mechanism respon-

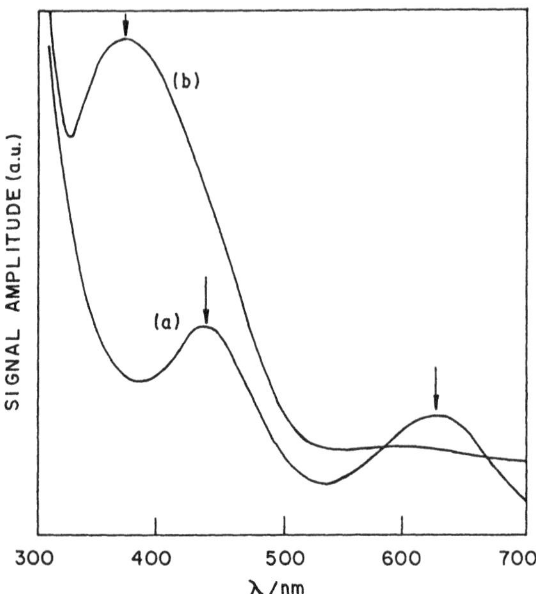

Fig. 1. PA spectrum of a non-calcinated (a) and calcinated (b) chromosilicate samples.

sible for the production of the PA signal. It was found that sample's thermoelastic contribution was a dominant mechanism as in the case of doped glasses [4]. Having identified the mechanism we have next carried out a PA signal data fitting to the corresponding theoretical expression of the PA signal and calculated α, τ and β.

The thermal diffusivities were measured using the PA data fitting to the corresponding expression for the thermoelastic phase contribution of an opaque and thermally thick sample. We got these conditions using a rear-illuimination PA configuration by putting an Al foil in contact with the sample and the rear side of this Al foil in contact with the outside air and exposed to light beam [5]. The non-radiative relaxation times τ and the optical absorption coefficients β at the corresponding absorption band peaks, were measured using the PA phase data fitting to the corresponding expression for the thermoelastic expansion contribution of an optically transparent and thermally thick sample [4].

The values obtained for thermal diffusivities were 0.0382 ± 0.0014 cm^2/s and 0.0415 ± 0.0023 cm^2/s for the calcinated and non - calcinated samples, respectively. The values for the non- -radiative relaxation time and the optical absorption coefficient for the calcinated and non - calcinated samples are shown in Table 1.

Table 1. Values of the non-radiative relaxation time and optical absorption coefficient for the calcinated and non-calcinated chromosilicate samples.

	Non - Calcinated		Calcinated
$\lambda(nm)$	440	640	380
$\tau(ms)$	2.70± 0.35	2.17± 0.12	2.26±0.09
$\beta(cm^{-1})$	230± 19	211± 15	233±9

References

1. R. B. Borade, A. B. Halgeri and T. S. R. Rao: *Advances in Catalysis Science and Technology* (Wiley, New Delhi, 1985) p.389.
2. H. O. Pastore, E. Stein, C. U. Davanzo, E. J. S. Vichi, O. Nakamura, M. Baesso, E. C. Silva and H. Vargas: J. Chem. Soc.: Chem. Commun. 772 (1990).
3. T. Bates, *Ligand Field Theory and Absorption spectra of Transition - Metal Inos in Glasses: Modern Aspects of the Vitreous States*, ed. Mackenzie J. D. (Butterworths, London, 1962) vol. 2 p.195.
4. M. L. Baesso, A. M. Mansanares, E. C. Silva, H. Vargas and L. C. M. Miranda: Phys. Rev. (B) **40** 1880 (1989).
5. N. F. Leite, N. Cella, H. Vargas and L. C. M. Miranda: J. Appl. Phys. **61** 3025 (1987).

Photoacoustic Depth Profiling of Continuously Inhomogeneous Condensed Phases and Application to the Liquid Crystal Octylcyanobiphenyl (8 CB)

A. Mandelis[1], *E. Schoubs*[2], *S.B. Peralta*[1], *and J. Thoen*[2]

[1]Photoacoustic and Photothermal Sciences Laboratory and Ontario Laser and Lightwave Research Center, Department of Mechanical Engineering, University of Toronto, Toronto, Canada M5S 1A4

[2]Laboratorium voor Akoestiek en Warmtegeleiding, Department of Physics, K.U. Leuven, B-3001 Leuven, Belgium

Abstract. An application is presented of the Hamilton-Jacobi formulation of thermal-wave physics [A. Mandelis, J. Math. Phys. **26**, 2676 (1985)] to the problem of photoacoustic depth profiling of inhomogeneous condensed phases with arbitrary, continuously varying thermal diffusivity profiles, which exhibits excellent profile reconstruction fidelity. The method was used to obtain quantitative depth profiles of thermal diffusivity decreases extending 20-30 μm below the surface of the liquid crystal 8 CB, as a result of observed changes in the photoacoustic signal frequency response upon the application of a transverse magnetic field in the nematic phase at 37°C. These decreasing profiles are a measure of the extent of bulk reorientational effects due to the magnetic field, as well as the extent of the influence of the surface as a domain reorientation inhibitor.

We have addressed the depth-profiling thermal-wave problem for continuously thermally varying condensed phases PAS. Unlike earlier treatments, [1-3] well-posed, direct, simple and convenient expressions for the frequency dependence of the PA signal are obtained based on the classical mechanical concept of the Hamilton-Jacobi thermal harmonic oscillator (THO) [4]. It can be shown that the solution to the thermal-wave inhomogeneous problem

$$\frac{d}{dx}\left[k(x)\frac{d}{dx}T(x)\right] - i\omega\rho(x)c(x)T(x) = -\frac{1}{2}\eta\beta I_0 e^{\beta x}; \qquad x \le 0 \tag{1}$$

with the well-known quantities T (ac temperature); η (nonradiative efficiency); β (absorption coefficient); I_0 (incident irradiance) in a semi-infinite condensed phase medium with continuously variable thermal conductivity, $k(x)$; density, $\rho(x)$; and specific heat $c(x)$, gives rise to a THO field. In the limit of photoacoustically saturated media and assuming a specific convenient thermal diffusivity profile

$$\alpha_s(x) = \alpha_0\left[\frac{1+\Delta e^{-qx}}{1+\Delta}\right]^2; \; \Delta \equiv \left[\frac{\alpha_0}{\alpha_\infty}\right]^{1/2} - 1, \tag{2}$$

the normalized PA signal (amplitude, $|M(\omega)|$, and phase, $\Delta\phi(\omega)$) becomes

Springer Series in Optical Sciences, Vol. 69 Photoacoustic and Photothermal Phenomena III
Editor: D. Bićanić

$$|M(\omega)|e^{i\Delta\phi(\omega)} = 1 + \frac{1}{4} R^{1/2}(\infty) \exp\left[-\frac{(1+i)\sqrt{\omega}}{2\sqrt{2}q} \frac{1}{\sqrt{\alpha_\infty}} \ln\left[\frac{\alpha_0}{\alpha_\infty}\right]\right] \tag{3}$$

where as a reference we assumed a homogeneous material with constant thermal diffusivity α_0. In the bulk of the inhomogeneous sample the thermal diffusivity $\alpha_s(x)$ assumes an arbitrary value $\alpha_s(\infty) = \alpha_\infty$. $R(x)$ is the ratio of thermal effusivities:

$$R(x) \equiv e_s(0)/e_s(x) \tag{4}$$

The essence and versatility of the technique lie in its treatment of differential subsurface layers corresponding to a measurement at modulation frequency $f_j = \omega_j/2\pi$ with local $\alpha_s(x;f_j)$ values determined by redefining (updating) the values for q and $R(\infty)$ found at a neighboring frequency, f_{j+1}, differing from f_j by $\delta f = f_{j+1} - f_j << f_j$. Thus, local variations of $\alpha_s(x)$ are accounted for, as manifested by normalized signal differences between neighboring frequencies, and the original single exponential decay for $\alpha_s(x)$, Eq. (2), becomes irrelevant in the determination of the global depth profile: this is determined as an inverse problem from successive values of the entire set of the experimental data dependences on ω_j and can be any function of x, numerical or in closed form. A test $\alpha_s(x)$ profile input to Eq. (3) according to Eq. (2) was used. Excellent reconstruction of the same profile from the simulated amplitude ratio and phase difference data with a homogeneous reference sample with $\alpha_s = \alpha_0$, was thus obtained.

The experimental apparatus was described elsewhere. [5] An 8 CB sample without an applied transverse magnetic field was the reference of the same sample in the presence of a magnetic field **B**. A 3.39 μm He-Ne laser was chosen, a radiation strongly absorbed by 8 CB, so that the sample was photoacoustically saturated. Figures 1a,b show the normalized PA frequency amplitude and phase response with B = 1.65 kG at T = 37.5°C. Figure 1c shows the reconstructed thermal diffusivity depth profile $\alpha_s(x_j)$, with [6] $\alpha_s(0) = \alpha_0 = 1 \times 10^{-7}\ m^2/s$.

The results indicate that the application of the transverse magnetic field across the 8 CB nematic liquid crystal generates a region in the bulk with lower thermal transport properties than near the surface, the "bulk" behavior commencing at and below 26-28 μm from the surface. Although this is the first time that such a variation is identified, yet it is consistent with the lowering of the nematic values of k (T) in the presence of a transverse magnetic field in 8 CB, when the conductivity is measured photoacoustically [7] at low f. The depth profile of Fig. 1c is considered to be clear evidence of an interface-induced alignment of the long molecular axes perpendicular to the surface as a long-range phenomenon (tens of microns). This novel effect was shown to be completely removed at 43°C (isotropic phase). The present results may be significant for future studies of the long-range positional order of liquid crystal molecules due to interface-molecule interactions and of the effect of surfaces in phase transition behavior. This work also shows the rich potential of the THO concept in the study of depth dependent thermo-physical phenomena.

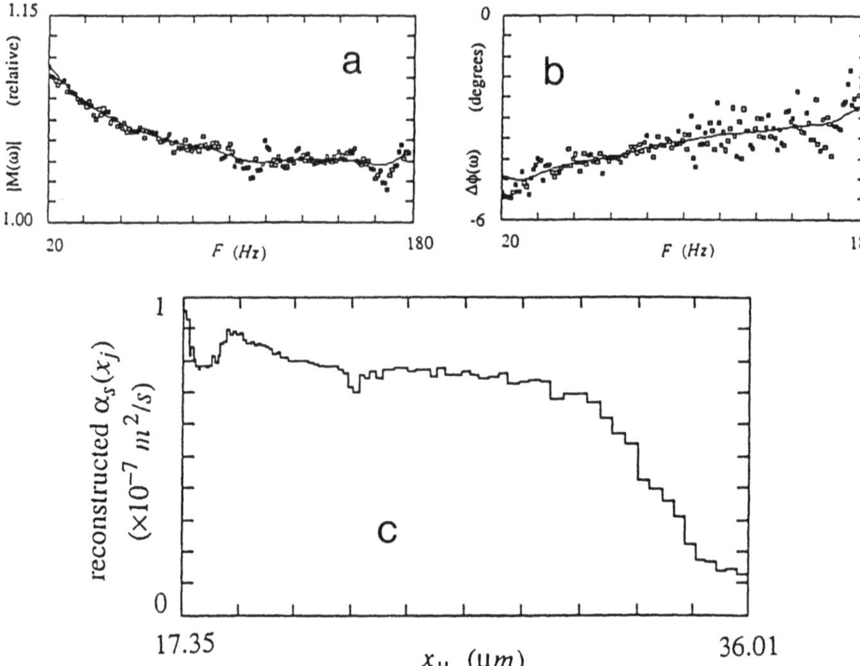

Fig. 1 Amplitude ratio (a) and phase difference (b) of averaged data between a nematic 8 CB sample with an applied transverse magnetic field, and the same sample with B = 0 ; (c) Reconstructed thermal diffusivity profile, using $\delta f = 1 Hz$.

References

[1] J. Opsal and A. Rosencwaig, J. Appl. Phys. **53**, 4240 (1982).

[2] A.K.S. Thakur, Lett. Heat Mass Transfer **9**, 385 (1982).

[3] H.J. Vidberg, J. Jaarinen, and D.O. Riska, Can. J. Phys. **64**, 1178 (1986).

[4] A. Mandelis, J. Math. Phys. **26**, 2676 (1985).

[5] C. Glorieux, E. Schoubs, and J. Thoen, Mater. Sci. Eng. **A122**, 87 (1989).

[6] T. Akahane, M. Kondoh, K. Hashimoto, and M. Nagakawa, Jpn. J. Appl. Phys. **26**, L1000 (1987).

[7] J. Thoen, E. Schoubs, and V. Fagard, Proc. Symp. Phys. Acoustics, Kortrijk Belgium (1990).

Computer Simulation of the Thermal Diffusivity Measurements of Oriented Polymer Samples Using the Mirage Effect

J. Rantala[1], *J. Jaarinen*[2], *and P.K. Kuo*[3]

[1]Department of Physics, University of Helsinki, Siltavuorenpenger 20D, SF-00170 Helsinki, Finland
[2]Neste, Corporate R&D, P.O. Box 356, SF-02151 Espoo, Finland
[3]Department of Physics, Wayne State University, Detroit, MI 48202, USA

Abstract. The effect of the measurement parameters on the mirage signal obtained when determining the thermal diffusivity of oriented polymer samples has been studied in order to clarify the difficulties arising from the low diffusivity of the samples and to find out the best measurement conditions. Because the mirage data obtained from these samples do not include any single feature to determine the thermal diffusivity, it is concluded that statistical methods applied to the entire measurement data should be used for this purpose.

1. Introduction

When the thermal properties of oriented polymers are determined, two main factors have to be considered. Firstly, the diffusivity of polymers is low when compared to the fluid surrounding the sample, such that the effect of the fluid cannot be neglected. Secondly, the orientation of the polymer fibres causes anisotropy in thermal conductivity, which can even be 3-dimensional. Therefore, the standard methods to determine the thermal diffusivity of foils, like the flash method, do not give the needed information of the diffusivity parallel to the foil surface.

However, the mirage effect offers a non-contact method for thermal diffusivity measurements in the surface plane of the sample. The method has been successfully applied to samples with higher diffusivities than those of polymers. Unfortunately,

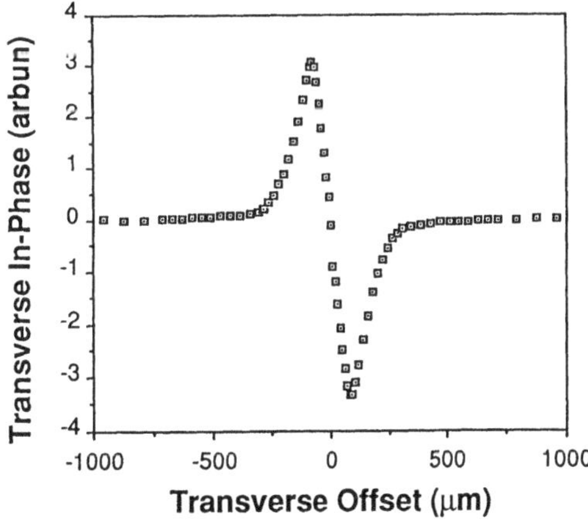

Fig. 1. Experimental data of a transverse in-phase mirage signal obtained from an ethyl vinyl alcohol polymer sample.

the most widely used zero-crossing technique [1] cannot be applied to polymers, because when low diffusivity samples are measured, either no zero-crossing exists or the signal level at the zero-crossing point is so small that noise makes its determination impossible. As an example of this, experimental data of polymer diffusivity measurement are presented in the Fig. 1. Here the signal approaches asymptotically zero but does not cross it.

Because zero-crossing or any other simple feature for the diffusivity determination can not be found, the diffusivity has to be calculated from the entire mirage measurement data. Therefore, it is important to understand the effect of different measurement parameters on the signal obtained. This is the subject of this paper, which continues our earlier published studies [2].

2. Some Examples of the Results

The data presented in this paper are calculated for opaque samples totally surrounded by a fluid and using the bouncing technique as the probing method. The simulation model is described elsewhere [3]. The parameters under study here are the diffusivity of the fluid surrounding the sample, the modulation frequency, the widths of both the heating and the probe beams, and the height of the probe beam.

Because thermal waves are more damped the lower the diffusivity of the fluid is, the mirage signal obtained is dominated by the thermal properties of the low-diffusivity sample only near the heating point. Outside the heated area the signal is due to the less-attenuated thermal wave propagating in the fluid. This causes the temperature above the sample not to fall monotonically as a function of the distance from the heating beam, resulting in a second maximum in the magnitude signal, which is shown in the Fig. 2. This also causes the non-existence of zero-crossings that otherwise could be used for diffusivity determination [4,5].

The local maximum becomes higher when the thermal diffusivity of the fluid increases. Therefore, by selecting a fluid with a low thermal diffusivity, the effect of the fluid can be decreased and the curves obtained from different sample diffusivities

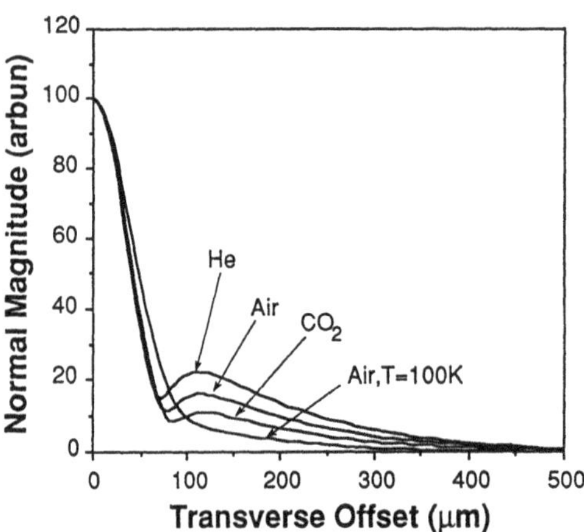

Fig. 2. The normalized magnitude of the normal deflection mirage signal obtained using gases with diffusivities.of 0.025, 0.10, 0.22, and 1.7 cm^2/s representing respectively air at the temperature of 100 K, and CO_2, air, and He at room temperature. The thermal diffusivity of the sample is simulated to be 0.03 cm^2/s corresponding to a plastic sample of high draw ratio. The modulation frequency of the heating beam is 100 Hz.

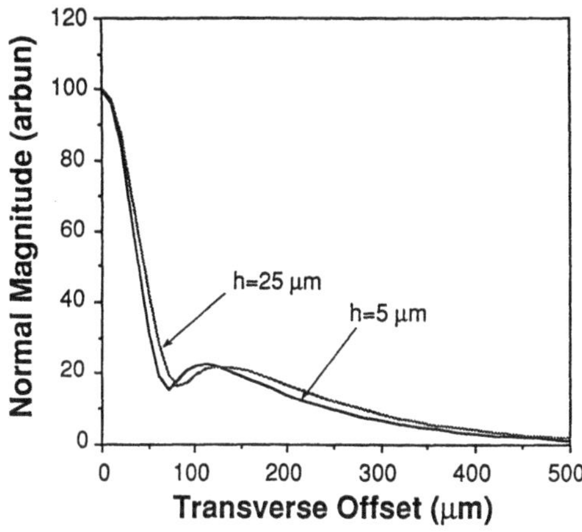

Fig. 3. The normalized magnitude of the normal deflection mirage diagrams obtained with probe beam heights of 5 and 25 μm. The diffusivity of the sample is 0.001 cm^2/s and that of the gas is 0.10 cm^2/s, corresponding respectively to a plastic sample of low draw ratio and CO_2 The modulation frequency is 3 Hz.

will be better distinguished from each other. Because CO_2 is easy to obtain and handle, the measurements should be done in that. When low temperature behavior is studied, even better results are to be expected because of the decrease of the thermal diffusivity of the fluid.

When the sample diffusivity is determined from the measurement data, the values of all the parameters affecting the data should be known as accurately as possible. Unfortunately, the height of the probe beam is difficult to measure exactly, and therefore one should know how much the data is affected by small changes of the height. This is clarified in the Fig. 3 that shows the effect of the height being small. Therefore the awareness of the exact value of the height is not essential, and an approximate value can be used instead. Anyhow, in practical measurements the height should be kept small to keep the signal to noise ratio good.

3. Conclusions

The low thermal diffusivity of the sample causes difficulties for measurements because the thermal properties of the fluid surrounding the sample have a strong contribution to the mirage signal. Because of this, the mirage data do not include any simple feature (like zero crossings, slope of the phase curve, etc.) from which the diffusivity could be directly measured. Instead, the data contain information of all the parameters affecting the measurement result, which are the thermal properties of the fluid and the sample, the radius and the modulation frequency of the heating beam, and the height and the radius of the probe beam. Also, the diffusivity information is included in the entire data from the whole measurement area, and it can not be determined from the behavior of a single measurement point, as in the zero-crossing method. To extract the thermal diffusivity of the sample from all these data, one has to use either multi-parameter fitting or other kind of statistical analysis. The use of these methods is described in another papers [6, 7] as well as the comparison of this computer simulation study to experimental data [7].

To make the signals obtained from different low-diffusivity samples more easily distinguished from each other, the fluid surrounding the sample should have as low diffusivity as possible. Because of the combination of ease of use and the low diffusivity, CO_2 is best for this purpose. Also, the heating and probing parameters are to be optimized. The modulation frequency is to be chosen as low as the measurements can be made without the noise level increasing too high, which in practice means about 3 Hz. The widening of the heating beam radius also helps to maximize the differences between the magnitude signals from samples of different diffusivities. However, the probe beam should not be too wide because this deteriorates the phase measurement. About 30 – 60 μm seems to be suitable for the both beams. On the other hand, the height of the probe beam does not have a significant effect as long as it is low enough to keep the signal to noise ratio acceptable.

References

[1]P. K. Kuo, E. D. Sendler, L. D. Favro, and R. L. Thomas: Can. J. Phys. **64**, 1168 (1986)
[2]J. Rantala, J. Jaarinen, L. Wei, P. K. Kuo, and R. L. Thomas: *Review of Progress in Quantitative NDE, Vol 10B* (edited by D. O. Thompson and D. E. Chimenti, Plenum Press, New York, 1991) p. 2165
[3]J. Rantala, J. Jaarinen, and P.K. Kuo: submitted for publication
[4]C.B. Reyes: Ph.D. Thesis, Wayne State University, 1988
[5]A. Salazar, A. Sánchez–Lavega, and J. Fernández: J. Appl. Phys., **69**, 1216, (1991)
[6]B. Skagerberg, J. Jaarinen, and J. Rantala: submitted for publication to *Photoacoustic and Photothermal Phenomena III* (Springer–Verlag, 1992)
[7]J. Rantala, L. Wei, P.K. Kuo, J. Jaarinen, and R.L. Thomas: to be published

The Determination of Thermal Diffusivity of Oriented Polymer Films by Means of Optical Beam Deflection

B. Skagerberg [1], *J. Jaarinen* [2], *and J. Rantala* [3]

[1] Neste, Scientific Services, P.O. Box 310, SF-06101 Porvoo, Finland
[2] Neste, Corporate R&D, P.O. Box 356, SF-02151 Espoo, Finland
[3] Department of Physics, University of Helsinki, Siltavuorenpenger 20D, SF-00170 Helsinki, Finland

Abstract. The understanding of the structure-properties relationships in polymers is of increasing importance in the development of new engineering polymers. This places requirements to the characterization of physical properties of differently treated polymers. In this work we present a method for determination of the thermal diffusivity of oriented polymer films using the mirage (optical beam deflection,OBD) technique. We present here the application of chemometrics, in the form of multivariate calibration, to obtain the thermal diffusivity from the mirage data.

1. Introduction

It is well known that the thermal conductivity of an oriented polymer film can be much greater in the direction of orientation than perpendicular. The macroscopic thermal conductivity is typically 2-20 times larger in the direction of the orientation than perpendicular. The largest reported anisotropy factor is about 100 [1]. There is no well established technique to measure the thermal conductivity or thermal diffusivity of polymer films in the direction of the film. The mirage technique provides a means to generate thermal waves propagating in the direction of the film. However, the application of the mirage technique to polymers has been difficult due to the low thermal diffusivity of polymers, which causes the thermal waves propagating in the air and in the sample to interfere strongly thus making the determination of the sample thermal diffusivity difficult. In this case the two data analysis techniques, the "zero-crossing" technique [2] and the least squares analysis [3], used for extracting the diffusivity value from the data fail.

In this paper we present a chemometric approach to the analysis of the mirage signal data. We apply a projection method, partial least squares regression (PLS) [4], to build a predictive model for the determination of thermal diffusivity.

2. The model

Partial least squares regression, PLS, is a data analysis technique based on projection techniques. The objective of the technique is simultaneously to model a matrix X, the mirage data, and predict a matrix Y, the physical parameters to be determined. In linear PLS two blocks of data, X and Y, are projected down to their latent structures

Springer Series in Optical Sciences, Vol. 69 Photoacoustic and Photothermal Phenomena III
Editor: D. Bićanić

[5]. The predictive power of the model is determined by cross validation of the data in the training set. Each modelling sample is left out once of the data and the sum of the squares of the prediction errors is minimized. It is to be noted that the model for prediction is not identical with the optimal fitting model.

3. The case study

As a case study we have applied the PLS technique to the prediction of thermal diffusivity and probe beam height from simulated mirage data. The typical mirage data consists of four spatial probe beam deflection spectra. These are the deflections parallel and normal to the sample both of which consist of the inphase and quadrature components. The spatial coordinate is the distance between the probe beam and heating beam axis, ie. the transverse offset. The calculated mirage signals, 60 offsets altogether 240 variables, used for building the model consisted of five diffusivities between 10^{-5} - 10^{-7} m^2/s, corresponding to the typical range of diffusivities for polymers, and six probe beam heights between 0 to 25 microns, resulting 30 calibration samples. The mirage signal was simulated by using the model by Kuo and co-workers.

The mirage signal shows clear changes from sample to sample, but the changes are global by character thus no single feature can be found that corresponds to the sample diffusivity.

4. Results

The model based on the training set of 30 simulated samples was tested with 14 simulated samples (test set), whose diffusivities were to be predicted. In some of the samples gaussian noise was added in order to test the validity of the model in a more realistic case. It turned out that the height of the probe beam was difficult to predict the predictive power of the model was only about 55 %. Especially in the case of added noise the height could not be predicted well. Apparently the characteristic changes due to height difference are small and typical experimental noise suppresses the height information in the spectra.

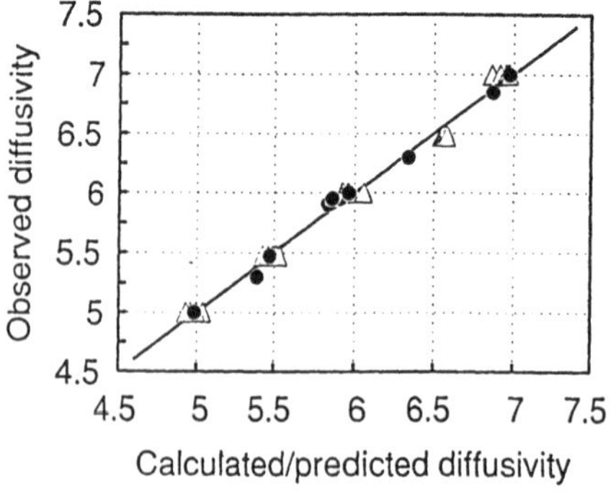

Figure 1. Observed (open triangles, the training set) versus predicted (filled circles, the test set) thermal diffusivities.

The prediction of the thermal diffusivity by the model is presented in Figure 1. The predictive power for diffusivity of the PLS model is 99 %. Even in the case of added noise the thermal diffusivity of the test samples are predicted excellently.

5. Conclusion

The PLS technique to analyze the mirage data has been shown to be capable to predict the thermal diffusivity of polymeric samples. Due to the two-dimensional character of mirage detection the technique is applicable to characterize the anisotrophy of oriented polymer foils. We are now in the process to apply the technique to real experimental model building and to predict the thermal diffusivity from real experiments.

Acknowledgement

Professor P.K. Kuo at Wayne State University is greatly acknowledged for placing the mirage signal simulation theory and program at our disposal.

References

1. C.L. Choy, W. H. Luk and F.C. Chen, Polymer, **19**, 155 (1978)
2. P.K. Kuo et al., Can.J. Phys., **64**(9), 1165 (1986)
3. T.R. Anthony et al., Phys. Rew., B42, 1104 (1990)
4. S. Wold et al., In Chemometrics: Mathematics and statistics in chemometry, B. Kowalsky ed., Reidel Publ. Co, Dordrecht NL (1984), pp 17-95
5. B. Skagerberg, J. Jaarinen, and J. Rantala, submitted for publication

Photothermal Measurement of Temperature Dependent Absorbance of Infrared Components

P. Eckardt, H.G. Walther, and W. Richter

Physikalisch-Astronomische Fakultät, Friedrich-Schiller-Universität, Max-Wien-Platz 1, O-6900 Jena, Fed. Rep. of Germany

Abstract. The absorbance of an Au/ZnS/Ge/ thin film system and of Au/ZnS and Au/BaF_2 film combinations have been measured as a function of increasing temperature in the infrared spectral range with a photothermal beam deflection arrangement. At an Au/Ge thin film combination without diffusion barrier the beginning of the diffusion has been observed.

Mirrors and transmission optics used in high power CO_2-laser applications must be minimized with respect to their absorption losses. The absorption induced sample heating can produce misalignment, especially because of the run-away-effect. For the study of the commonly very low but temperature dependent absorbance of coated IR-films we used a PBD-arrangement described in /1/. The periodical deformation of the sample generated by the CO_2-laser absorption was detected by the beam deflection of a HeNe-laser probe which was focused onto the measuring slit. By an electric heater contacted to the sample its temperature could be increased up to 300°C. Of great interest are the conditions for using the high refractive Ge as a coating material.

The photothermal signal of a sample with surface absorbance A and bulk absorption coefficient $\beta(cm^{-1})$ can be expressed as a function of sample temperature T by

$$S(T) = c_1(T) \cdot A(T) + c_2(T) \cdot \beta(T).$$

where c_1, c_2 are calculable complex transfer coefficients whose temperature dependences are determined by the temperature trends of the thermal and elastical properties of the sample. The analysis of the phase shift of the photothermal signal permits a conclusion about the depth where the laser absorption takes place /2/. Fig.1 shows the absorbance of bulk Ge consisting of two contributions. One of them is the surface absorbance, which is widely temperature independent and dominates in case of the examined sample at room temperature, while the other one is the bulk absorption which increases exponentially accordingly to exp(-const/T) and prevails at temperatures above 60°C. The identification of bulk or surface absorptions succeeds in considering the phase of the signal which is shifted by about 55° at elevated temperatures with respect to room temperature.
The temperature dependent refractive index of thin film material was calculated from the measured temperature dependence of the photothermal signal taken into account the transfer coefficients c_1 and c_2. For ZnS and BaF_2 it could not be observed a noticable increase of the absorbance within the considered temperature range what may be caused by the energy gap of 3.6 eV of ZnS.

It is possible to reduce the absorption of Au-mirrors by means of dielectric coating. This fact is demonstrated in Fig.2

Springer Series in Optical Sciences, Vol. 69 Photoacoustic and Photothermal Phenomena III
Editor: D. Bićanić

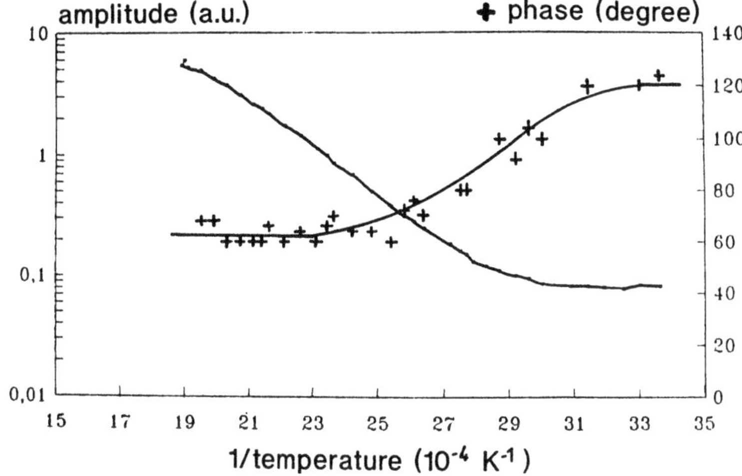

Fig.1 PBD-signal amplitude and phase vs. temperature

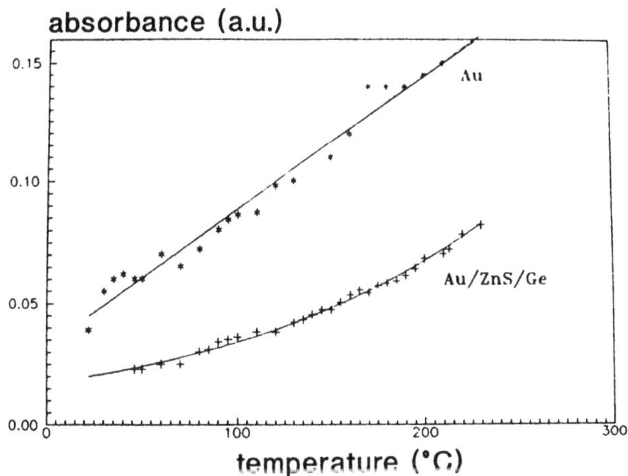

Fig.2 PBD-signal amplitude of pure Au-mirror and ZnS/Ge-coated Au-mirror vs. temperature

at a ZnS/Ge two-layer coating deposited on gilded Si which absorbs less than half the value of the pure gilded Si at room temperature. However, at elevated temperatures the absorbance increases exponentially because of the run-away-effect.

Using the well known matrix formalism /3/ it is possible to calculate the absorbance of the complete thin film system with given real and imaginary refractive indices n and k, respectively. Taking the data of ZnS n=1.2 and assuming a temperature independent $k=1.10^{-3}$, and of Ge n=4, and reading out from Fig.2 the value of the temperature dependent absorbance of the complete thin film system, one can derive the temperature dependent extinction coefficient k of Ge. An Arrhenius-representation enables one to determine the energy gap of thin film Ge of 0.31 eV which is half the value of the crystalline Ge and seems to be

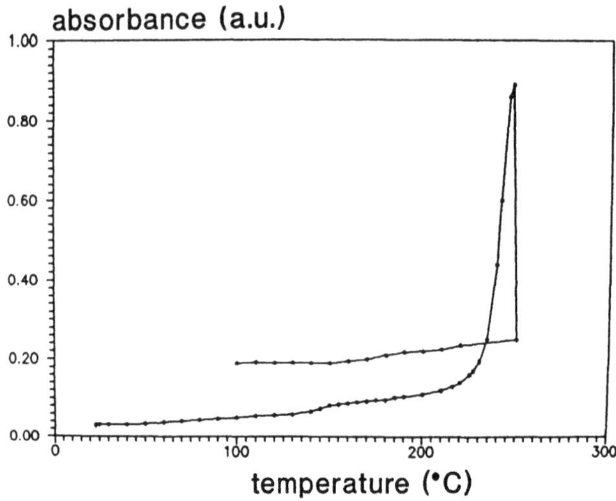

Fig.3 PBD-signal amplitude of an Au/Ge thin film system without diffusion barrier vs. temperature

trustworthy. From these results it can be evaluated the high power laser resistivity of boosted Au-mirrors.

The reported photothermal arrangement enables an instantaneous and simple measurement of the temperature dependence of the absorbance. The calibration of the room temperature absorbance was carried out by laser calorimetric measurements. Differing from laser calorimetric measurements it is not necessary to stabilize the temperature with the photothermal arrangement because of its immediate signal response.

Another advantage of this arrangement based on the instantaneous signal response is given by the possibility to study short time change of the temperature dependent absorbance. Fig.3 shows the absorbance versus temperature of a sample which consits of a thin Ge-film deposited on an Au-film. While at first the deposited heat is conducted by the metallic film, obviously, a diffusion process is starting at about 200°C, which is connected with a strong increase of the absorbance. It culminates in the destruction of the smooth sample surface. The signal drops down because the probe beam is scattered at the roughened sample surface. On this way one can observe the beginning of the diffusion process just before anything is noticed visually.

References

1. H.G. Walther, E. Welsch, in Photoacoustic and Photothermal Phenomena II, Springer Series in Optical Sciences 62, 319 Springer-Verlag, Heidelberg(1990)
2. E. Welsch, H.G. Walther, K. Friedrich, P. Eckardt, J.Appl.Phys.67, 6575(1990)
3. F. Abeles, J.Phys.Rad.II, 310(1950)

Numerical Analysis of Simultaneous Surface and Subsurface Features in Homogeneous Solids and in Coatings by the Mirage Technique

A. Salazar, A. Sánchez-Lavega, and J. Fernández

Dpto. Física Aplicada I, E.T.S. Ingenieros Industriales y de Telecomunicación, Universidad del País Vasco, Alda. Urquijo s/n, E-48013 Bilbao, Spain

1. INTRODUCTION

The mirage technique has been proved to be very performant for nondestructive evaluation of materials and layered systems, and in particular, for the analysis of close subsurface defects[1, 2]. In this communication we evaluate the possibility to detect and measure subsurface defects when surface inhomogeneities are simultaneously present, both in opaque homogeneous materials and in opaque coatings. A fully 3D theoretical treatement which incorporates the effects of the mirage parameters (a=exciting beam radius, b=probe beam radius, z_o=probe beam height) has been employed. The surface feature is modeled as a horizontal parallel step of given height, while the subsurface one is a thin horizontal air slab. We have computed the amplitude and phase of the normal component of the mirage deflection, which is sensitive to the normal heat flow. We have calculated this component in two regions: One above the defect (Φ_n) and the other one in a region free of defect (Φ_{no}), using probe and pump beams superimposed. Then, we have evaluated the normalized amplitude (Φ_n/Φ_{no}) and the phase lag [$\Delta\Psi(\Phi_n) = \Psi(\Phi_n) - \Psi(\Phi_{no})$] as a function of the modulation frequency (f).

2. OPAQUE HOMOGENEOUS SOLIDS

In Figure 1 we have drawn schematically our numerical analysis results for opaque homogeneous samples. As shown, different geometrical configurations for the surface and subsurface defects have been studied.

Our main results are [3] : (i) Surface and subsurface defects can be detected and discriminated by making use of the frequency dependence of Φ_n and $\Psi(\Phi_n)$; (II) The amplitude and phase contrast rise as a and z_o increase; (iii) The influence of the probe beam size can be neglected when it is focused. A reasonable theoretical limit is $b < 2\sqrt{2}\,\mu_g$ (being μ_g the thermal diffusion length of the air); (iv) Subsurface defects with a thickness (l_2) as low as 0.1 μm and located down to a depth (l_1) of $0.3\sqrt{k_s}$ (k_s is the thermal diffusivity of the solid), can be detected; (v) The determination of l_1 and l_2 can be made through a model fitting.

3. OPAQUE COATINGS

In Figure 2 are drawn the set of geometric configurations we have studied to simulate defects (inhomogeneities) in the substrate, in the coating and in their interface.

Our numerical results for the dependence of the normalized amplitude and phase lag, against frequency are shown in Figure 3. As in section 2, it is evident that it is not only possible to detect such a kind of features, but also to discriminate between them

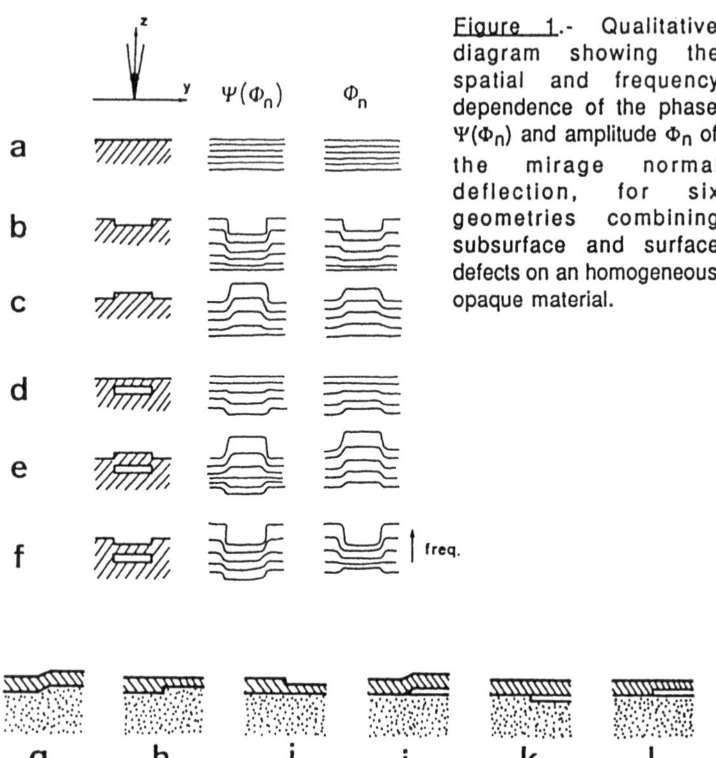

Figure 1.- Qualitative diagram showing the spatial and frequency dependence of the phase $\Psi(\Phi_n)$ and amplitude Φ_n of the mirage normal deflection, for six geometries combining subsurface and surface defects on an homogeneous opaque material.

Figure 2.- Geometric configurations of the surface and subsurface features simulating coating/substrate defects

using frequency. Moreover, through a model fitting, it is possible to measure the height of the step defect and the delamination depth and thickness. In these calculations we assume that the coating thickness and the thermooptical properties of both, coating and substrate, are known. The effect of the mirage parameters (a, b, z_o) has the same behavior as mentioned previously for homogeneous samples.

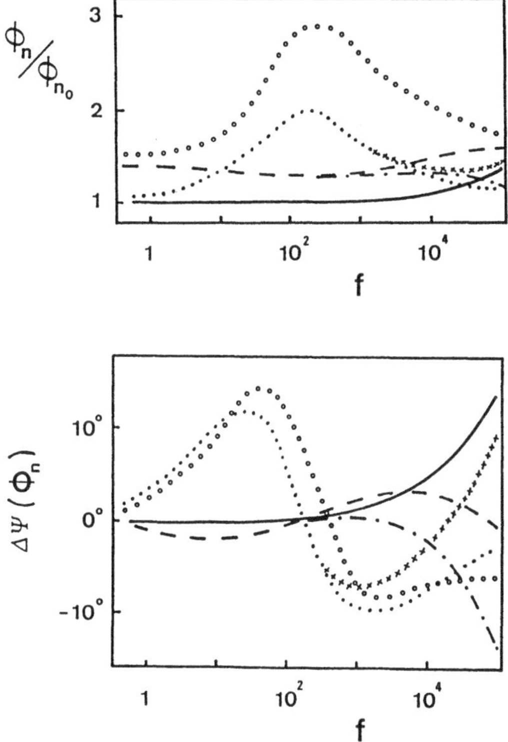

Figure 3.- Normalized amplitude and phase lag corresponding to opaque coatings: g (———), h (— — —), i (— · — · —), j (x x x x), k (• • • • • • • •), l (o o o o o). Calculations have been made for a coating of Copper, 5 μm thick, onto a substrate of crown glass. The air slab thickness is 2 μm and the coating inhomogeneity (step) is 2 μm. Mirage parameters are $a = z_0 = 100$ μm, $b = 0$.

This work was developed through grants from Gobierno Vasco PGV 0036/89 and from Universidad del País Vasco UPV E174/90.

REFERENCES

1. P.K. Kuo, L.D. Favro, R.L. Thomas, in Photothermal Investigation of Solids and Fluids, ed. J.A. Sell, pp 191-212. Academic Press, 1989.
2. J.P. Roger, D. Fournier, A.C. Boccara, F. Lepoutre, J. Physique, Col.C5, 295-310, 1989.
3. A. Salazar, A. Sánchez-lavega, J. Fernández, J. Appl. Phys. , 70, 3031-3037, 1991.

Dichroic Ratio and Thermal Diffusivity Measurements on Polydiacetylene Single Crystal Films by Photothermal Probe Beam Deflection

X. Quelin[1], *C. Lapersonne-Meyer*[2], *J. Berrehar*[2], *G. Louis*[1], *and P. Peretti*[1]

[1]Département de Recherches Physiques (URA 71 CNRS), Université Paris 6, Tour 22, B.P. 136, 4 Place Jussieu, F-75252 Paris Cedex 05, France

[2]Groupe de Physique des Solides (URA 17 CNRS), Universités Paris 7 et Paris 6, 2 Place Jussieu, F-75251 Paris Cedex 05, France

Abstract : Substituted diacetylenes have been polymerized by low energy electron irradiation, leading to single crystal polymers. Films thicknesses can be controlled by adjusting the incident electron energy. The work reported here concerns TS-6 single crystals. Using photothermal probe beam deflection technique, we have measured the ratios of anisotropic properties : optical dichroism and thermal diffusivity.

1. Introduction

It is well known [1] that many molecular crystals of substituted diacetylenes can go through a solid state polymerization under UV, X-ray or electron irradiation according to the reaction

$$n\,(R - C \equiv C - C \equiv C - R') \;\rightarrow\; (= \overset{\displaystyle R}{\overset{|}{C}} - C \equiv C - \underset{\displaystyle R'}{\underset{|}{C}} =)\,n$$

The polymerization process thus results in single crystals of conjugated, straight, unbranched polymer chains, showing a large degree of anisotropy in their properties.The work reported here concerns (p - toluenesulfonate) of 2,4-hexadiyne-1,6- diol (abbreviated designation TS-6) for which the side groups R and R' are identical

$$CH_2\,O - SO_2 - \phi - CH_3$$

Using phototermal probe beam deflection, the dichroic ratio of 250 and 440 nm thick samples have been measured and the thermal diffusivity constants parallel (//) and perpendicular (⊥) to the polymeric chains direction could be determined for a 100 μm thick sample.

2. Film preparation

This crystal is monoclinic, $P2_1/\,c$, with two polymer chains per unit cell extending in the crystallographic **b** direction. Polymerization is achieved by irradiating the surface of a *c.a.* 100

Springer Series in Optical Sciences, Vol. 69 **Photoacoustic and Photothermal Phenomena III**
Editor: D. Bićanić

μm thick monomer single crystal with low energy monokinetic electrons (electron energies $\lesssim$ 5 kev) [2]. If the polymeric chains grow parallel to the sample surface, which is the case of TS-6 cleaved along the b-c plane, the polymer film thickness is of the order of the penetration depth of the electrons and thus directly related to the incident electron energy.

After polymerization, the film remains on the monomer substrate and is separated from it by selective dissolving of the monomer and transferred onto a pyrex glass substrate.

3. Dichroic ratio measurements

The optical properties of these monocrystalline films - absorption and resonance Raman scattering spectra - have been reported elsewhere [3].

Boccara *et al* [4] have shown that, when the factor $\exp(-\beta l)$ is $<<1$ (where β and l respectively stand for the light absorption coefficient and the sample thickness), the deflection amplitude is proportional to β. Thakur et al [5] have studied for a 4 μm thick sample of TS-6, the low non resonant absorption coefficient in the wavelength range {650 nm - 1300 nm}.

We present here measurements of the dichroic ratio $R_\beta = \beta_{//} / \beta_\perp$ for two wavelengths : 514 nm and 488 nm. In the rear configuration (which means that the pump beam is focused on one side of the sample, whereas the probe beam runs along the other one), the plot of the normal deflection Φ_n versus the angle between the pump beam polarization and the polymer chains axis gives the following results : for 514 nm and 488 nm excitation wavelengths, R_β is found to be equal to 5 and 10 in the case of 255 nm and 440 nm thick samples respectively.

4. Thermal diffusivity measurements

It is well know [6] that oriented semicrystalline polymers exhibits large anisotropic properties of the thermal conductivity.

Kuo *et al* [7, 8] have shown that there is a linear relation between the first non central zeros crossing of $Re(\phi_t)$, denoted D, and the inverse square root frequency. The slope of the line is then directely related to the thermal diffusivity coefficient of the sample. We have used this method to measure the thermal diffusivities $\alpha_{//}$ and $\alpha_\perp$ of TS-6 100 μm thick samples.

The experimental set up is the same as in ref [9], except that the sample is enclosed in a cell where the surrounding air gas is replaced by propan gas which provides a noticeable enhancement of the photothermal signal. We have checked that no adsorption of this gas on the sample occurs.

According to Salazar *et al* [10], in the case of an optically thick and thermaly thin sample (the thermal diffusion length is about 200 μm compared to the 100 μm sample thickness), the slopes of the lines (a) and (b) on the figure 1 are respectively equal to $(\pi \alpha_{//})^{1/2}$ and $(\pi \alpha_\perp)^{1/2}$ so that we find $\alpha_{//} = 1.5\ 10^{-5}\ m^2/s$ and $\alpha_\perp = 5.4\ 10^{-6}\ m^2/s$ at room temperature.

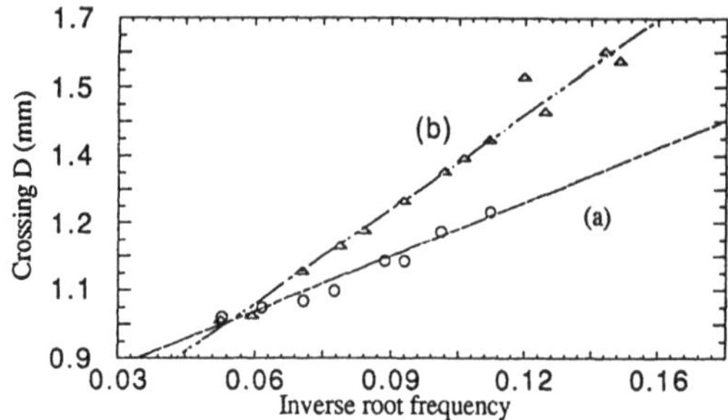

Figure 1 : Experimental plot of D versus the inverse of the square root $f^{-1/2}$ of the modulation frequency : (a) normal and (b) parallel to the chains axis. The TS-6 sample is 100 µm thick.

5. Conclusion

We have measured the dichroïc ratio R_β on thin layers of TS-6 samples and found no noticeable change of R_β with the incident wavelengh, in the range 488 nm - 514 nm. The value of $R_\beta = 10$ for the 440 nm thick sample compares with other determinations [2] reported elsewhere. The lower value obtained for the 255 nm thick sample may be attributed to the presence of a certain amount of dioriented chains at the sample surface.

Quantitative thermal diffusivity anisotropy of a 100 µm thick TS-6 polymer sample is given for the first time.

References

1. D. Bloor and R.R. Chance (Eds), Polydiacetylenes : Synthesis, Structure and Electronic Properties, Martines Nijhoff, Dordrecht, Netherlands (1985).
2. J. Berrehar, C. Lapersonne-Meyer and M. Schott, Appl. Phys. Lett., **48,** 630 (1986).
3. J. Berrehar, P. Hassenforder, C. Lapersonne-Meyer and M. Schott, Thin Solid Films, **190**, 181 (1990).
4. A.C. Boccara, D. Fournier, W. Jackson and N.M. Amer, Opt. Lett.,5, 377 (1980).
5. M. Thakur, R.C. Frye and B.I. Greene, Appl. Phys. Lett., **56**, 1187 (1990).
6. C. L. Choy, Polymer, 18, 984 (1977).
7. P. K. Kuo, M.J. Lin, L.D. Favro, R.L. Thomas, D.K. Kim, Shu Yi Zhang, L.J. Inglehart, D. Fournier, A.C. Boccara and N. Yacoubi, Can. J. Phys., **64**, 1165 (1986).
8. P.K. Kuo, E.D. Sendler, L.D. Favro and R.L. Thomas, Can. J. Phys., **64**, 1158 (1986).
9. M. Gharbia, A. Hadj-Sahraoui, G. Louis, A. Gharbi and P. Peretti, in Photoacoustic and Photothermal Phenomena II, J.C. Murphy, J.W. Maclachlan Spice, L.C. Aamodt Editors (Eds Springer Verlag), **62**, 306 (1989).
10. A. Salazar, A. Sanchez-Lavega and J. Fernandez, J. Appl. Phys., **65**, 4150 (1989).

Photothermal Analysis of Layered Materials Under Random Excitation

P. Egée[1], *E. Mérienne*[2], *K. Hakem*[2], *M. Heuret*[2], and *M. Egée*[2]

[1]CEA-CESTA, B.P. 2, F-33114 Le Barp, France

[2]Laboratoire d'Energétique et d'Optique, Faculté des Sciences de Reims, B.P. 347, F-51062 Reims Cedex, France

Abstract. We have developed a photothermal analysis device using randomly modulated local excitation. This method, which primarily makes it possible to obtain the pulse response, then the harmonic response from the sample, has a number of advantages over the other systems available up to now: lower thermal stress than in the pulsed regime, faster acquisition and lower frequency band than with sinusoidal excitation. These advantages can be made use of in studying different types of samples over a wide thickness range. The method and the device were first validated by comparing the experimental results with the theoretical results obtained using a 3D model for dielectric and metallic samples of known thermophysical properties. The method was then tried on sprayed ceramic coatings on metal substrates with intentional adhesion defects. The experimental results were compared with those obtained using a 1D theoretical model suitable for semi-transparent and scattering materials.

1. Photothermal radiometry under random excitation

In photothermal radiometry, two different techniques are normally used depending on the sample thickness range.

Photothermal thermography [1, 2] is suitable for materials with a thickness of 1 millimetre or more. Use is generally made of extended pulse excitation and, for detection, digital infrared thermography. This technique has been the subject of industrial developments [3] essentially intended for inspection of composite materials and bonded structures.

For materials with thicknesses between a few micrometers and one millimetre, focused excitation with a local detector is used. The choice of the excitation mode, which may be pulsed [4] or sinusoidal [5], depends on the constraints on analysis (structure and thermophysical properties of the sample, time available for measurement), and possibly on other considerations (experimental apparatus, models available, etc.).

These methods have a number of drawbacks. For instance, under sinusoidal excitation , the exploration time is relatively long if it is required to make measurements over a wide frequency band. Also, under single pulse excitation, it is necessary to deposit a substantial amount of energy to obtain an adequate response and to explore a substancial depth.

Other excitation systems have been utilised, for photoacoustic devices and for a mirage effect device. Mandelis [6] has developed a mirage effect device using pseudo-random excitation, thus obtaining a pulse response from the sample or its harmonic response.

The device which we describe here addresses some of these problems and has a number of features which we consider to be important: relative flexibility as regards the frequencies involved and the thicknesses inspected (particularly for higher thicknesses), speed of multi-frequency acquisition, less thermal stress than in the single pulse regime, the existence of a self-calibration procedure for the photothermal signal phase, and the possibility of displaying the results in the harmonic pulse response form.

The method uses pseudo-random binary excitation equivalent to "white noise", which in our device is generated by a computer controlled optronic device sequentially modulating a continuous laser beam. This excitation sequence consists of 511 steps. The intercorrelation product of the photothermal signal and the excitation sequence gives the pulse response. The Fourier transformation of this pulse response, provided certain precautions are taken, then gives the harmonic response of the sample.

The sample subjected to laser excitation is heated and only reaches its equilibrium temperature after a transient regime, and it is not until this point has been reached that sampling of the photothermal signal should be carried out. The duration of this transient regime, which depends on the thermophysical and dimensional properties of the sample studied, increases the total measurement time which is an important parameter, particularly in the case of mapping analysis. To provide freedom from these transient regimes (and possibly slow fluctuations of the source), a procedure enabling automatic compensation for transient effect has been developed.

2. Validation

To validate this method, particularly in the low frequency region (0,1 to 100 Hz), we compared the experimental results obtained with two samples of known properties (one a thick semi-transparent sheet of coloured glass 3,3 millimetres thick, the other a thin

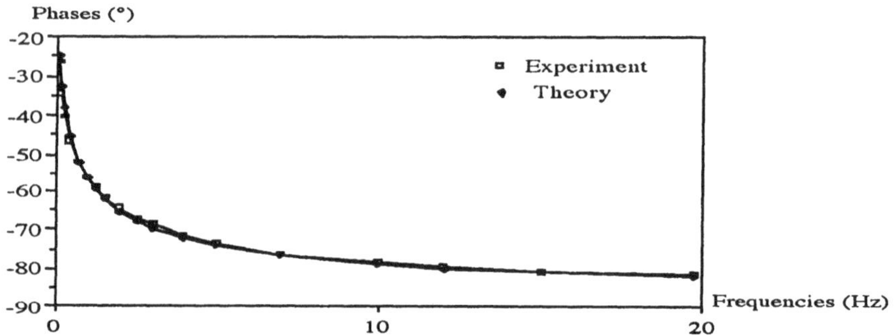

Figure 1. Experimental and theoretical phases for a glass sample.

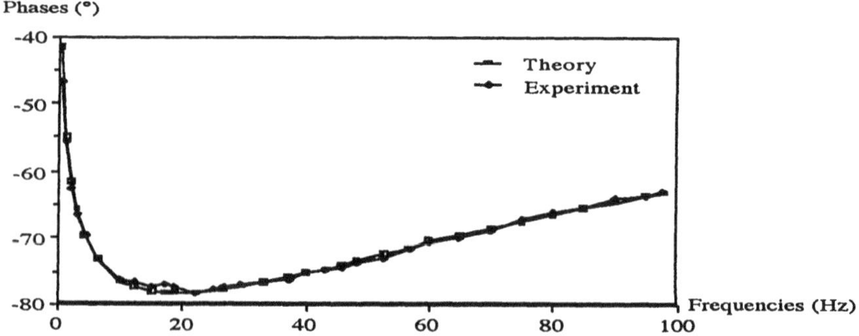

Figure 2. Experimental and theoretical phases for a steel sample.

opaque sheet of steel 100 micrometers thick) with the results of calculations made using a model developed earlier [7].

The thermokinetic model calculates the phase and amplitude of the temperature at the surface of the sample. We make allowance for the Gaussian form of the laser beam. The different layers of the material may be opaque or semi-transparent to the exciting radiation. In both cases, agreement between the theoretical and the experimental results was satisfactory throughout the frequency ranges explored, as shown in Figure 1 (for glass) and Figure 2 (for steel).

3. Application to ceramic coatings

In Figure 3, we show the experimental curves obtained with a two-layer sample. The plasma sprayed coatings on aluminium substrates were prepared in the Ceramic Powder Metallurgy of Renault Research Laboratories. The substrate is metallic and the coating is of zirconia with a thickness of 200 micrometers. There is an adhesion interface caused by the local absence of corundum.

The device which we have developed makes it possible to establish a photothermal map of the sample. We can locate and analyze the defects in the sample by comparing the decrease in temperature (or the transfer function) with that given by a suitable model. Such a defect shows a thermal contrast when examined using a pulse response, or a phase shift if observed with a harmonic response, between the sound zone and the defective zone of the sample.

The results agree well with those calculated using a 1D model developed for semi-transparent and scattering materials [8], in which cases the interface defect was simulated by a contact thermal resistance.

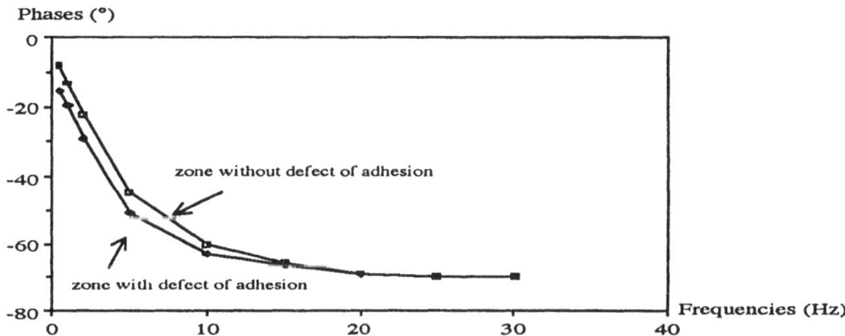

Figure 3. Experimental phases of a ceramic sample presenting an adhesion defect.

References

[1] BEAUDOIN (J.L.), MERIENNE (E.), DANJOUX (R.), EGEE (M.) - Time resolved stimulated IR thermography applied to the subsurface control of materials, Proceedings of the 4th international Topical Meeting on Photoacoustic, Thermal and related Sciences, 1985, Montreal (Canada).

[2]BALAGEAS (D.),BOSCHER (D.),DEOM (A.), FOURNIER (J.) et HENRY (R.) - La thermographie infrarouge: un outil quantitatif à la disposition du thermicien, Rev. Gén. herm., 1988,322, pp. 501-510.
[3] DANJOUX (R.), BEAUDOIN (J.L.), MERIENNE (E.), EGEE (M.), TRETOUT H.) - MECIR: A new apparatus for the control of materials by infrared thermography, Proceedings of the International Meeting: Infrared Technology and Applications, Wembley, London, June 25-28 (1990).
[4] BALAGEAS (D.), KRAPEZ (J.C.) et CIELO (P.) -Pulsed photothermal modeling of layered materials, J. Appl. Phys.1986, 59, 2, pp.348,357.
[5] DARTOIS (R.) - Caractérisation des matériaux bi-couches semi-transparents par radiomètrie photothermique: application aux revêtements de surface et en hématologie, 1986, Reims, Thèse de Doctorat.
[6] MANDELIS (A.) - Time-Delay -Domain and Pseudorandom-Noise Photoacoustic and Photothermal Wave Processes: A Review of the State of the Art, IEEE Transactions on Ultrasonics, Ferroelectrics , and Frequency Control, Vol. UFFC-33, September 1986.
[7] POTIER (F.) - Radiométrie photothermique appiquée à la caractérisation et au contrôle du traitement thermique des métaux, 1989, Reims, Thèse de Doctorat.
[8] HEURET (M.), EGEE (P.), BISSIEUX (C.), KURKA (G.), DANROC (J.) - Characterization of Ceramic Coatings by Photothermal Radiometry. EMRS Congress. May 1990, Strasbourg.

Opto-Thermal Study of Glass Properties

R.E. Imhof[1], *R.M.S. Bindra*[1], *D.J.S. Birch*[1], *P.H. Willson*[1], *J. Locke*[2], *D.G. Sanger*[2], *and N.D. Cowan*[3]

[1]Department of Physics and Applied Physics, Strathclyde University, Glasgow G4 0NG, UK
[2]Central Research and Support Establishment, Home Office Forensic Science Service, Aldermaston RG7 4PN, UK
[3]United Glass Ltd, Porter's Wood, St. Albans AL3 6NY, UK

Abstract. We report experiments that demonstrate applications of opto-thermal transient emission radiometry to non-destructive examination of glasses, including near-surface impurity monitoring, absorption measurement in low loss optical thin films and thickness measurement of protective coatings on glass bottles.

1. Introduction

Opto-thermal transient emission radiometry, or *OTTER* [1], uses pulsed optical excitation and thermal infrared detection to study condensed materials. The non-invasive nature of the technique makes it attractive for industrial monitoring of coating thickness and paint ageing, for example [2], as well as for bio-medical measurements, such as disappearance monitoring of substances on skin *in-vivo* [3]. This article highlights a number of potential application areas in the non-destructive testing of glasses.

Our findings to date indicate that, in general, the opto-thermal impulse response of glass depends strongly on composition, manufacturing process, surface treatment and site probed. The impulse response curves can be analysed to give information about thermo-physical properties within a depth of $<100\mu m$, by least-squares fitting the data to suitable theoretical models. In this work, the model for homogeneous, semi-infinite media [1] is used as a test for near-surface homogeneity.

2. Near-Surface Impurity Distribution

Whilst bulk glass is generally homogeneous, its composition near the surface can be inhomogeneous, if high levels of impurities are present. This is shown in recent observations on float glasses excited with 266nm radiation, as illustrated in figure 1. The bulk glass and float faces emit signals of decaying form, but least-squares analyses suggest a strongly inhomogeneous distribution of impurities in the float face, whereas bulk glass signals show little evidence of inhomogeneity. The response from the non-float face has the form of a delayed thermal wave, whose origin is not yet clear. The impurities in this face may be distributed in such a way that their

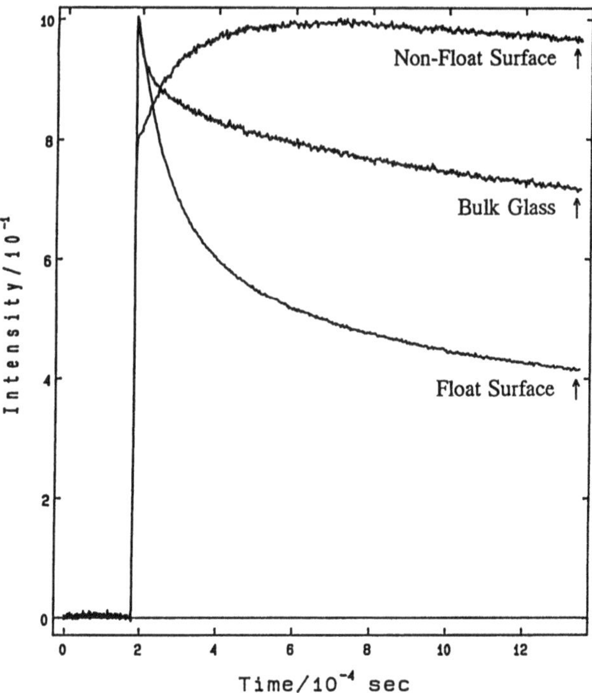

Figure 1.
Opto-thermal signals from a float glass, contrasting the impulse response from the bulk glass with those from the float and non-float faces.

concentrations are low near the surface and increase with depth. Another possibility is that the impurities absorb the 266nm laser radiation much more strongly than the bulk glass, thus forming local hot spots from which thermal waves emanate. In either case, such measurements may be useful for characterising impurity distributions in float glasses, or for identifying glass fragments in forensic investigations.

3. Absorption in Low Loss Films

Weak optical absorption in high quality optical thin films can be measured using opto-thermal techniques. Signals from ZnS/NaF coated interference filters have rounded peaks, indicating their origins from the thin films, an inhomogeneous region within the components. Peak signal strength depends on the heat deposited in the films by the 532nm excitation light, thus giving a direct measure of absorption, with a detection method that is insensitive to reflected or scattered excitation light [4].

4. Protective Coatings on Glass Bottles

Recent experiments show that the thickness of protective coatings applied to glass bottles can be sensed remotely using 266nm excitation radiation and broadband infrared detection. The coatings under investigation consist of a pyrolysed Tin Oxide layer, less than 10nm thick, beneath an organic outer layer of similar thickness. The results are illustrated in figure 2, where the uncoated inside face of a soda lime silicate glass bottle is compared with different coated sites on the outside. Least squares analyses of these data reveal a relatively homogeneous uncoated surface, but the coating clearly causes the response to deviate from this model. Quantitative evaluation awaits the development of a suitable theoretical model, but the opto-thermal data correlate well with independent determinations of inorganic coating thickness, performed using a standard *Hot End Coating Meter*, with the faster opto-thermal decay rates corresponding to greater coating thickness.

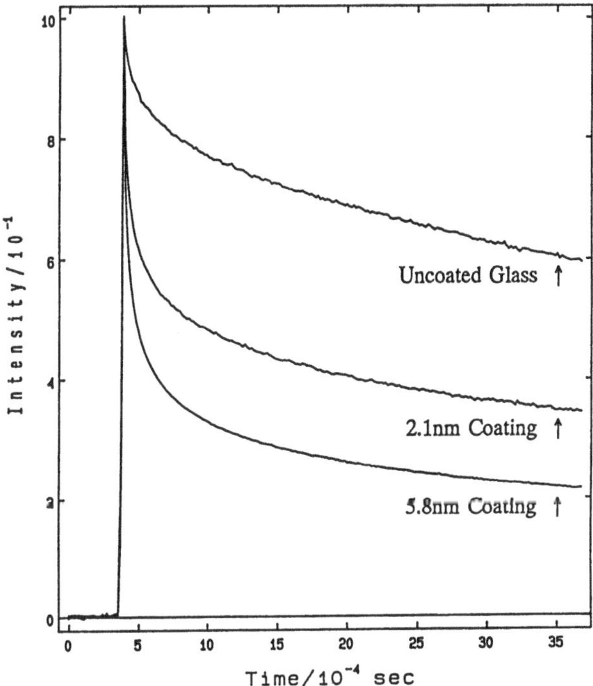

Figure 2.
Opto-thermal impulse response from soda lime silicate bottle glass with and without protective coating.

5. Conclusions

These experiments reveal new thermo-physical effects that may be exploited for remote sensing, non-destructive examination. Further work, especially in theoretical modelling, is required to facilitate the quantitative evaluation of such observations.

Acknowledgments

We thank SERC for financial support.

References

1. R.E. Imhof, D.J.S. Birch, F.R. Thornley, J.R. Gilchrist and T.A. Strivens, J. Phys. E: Sci.Instrum. 17, 521-5, (1984).
2. R.E. Imhof, C.J. Whitters, D.J.S. Birch and T.A. Strivens, Mat. Sci. Eng. B5, 113-7, (1990).
3. R.E. Imhof, C.J. Whitters, D.J.S. Birch, Phys. Med. Biol. 35(1), 95-102, (1990).
4. R.E. Imhof, D.J.S. Birch, M.M. Moksin, J.F. Webb, P.H.Willson and T.A. Strivens, Brit. J. NDT, 33(4), 172-6, (1991).

Non-Invasive Opto-Thermal Measurement of Clear Lacquer Coating Thickness

M.E. Pita de Jesus[1], *R.E. Imhof*[1], *D.J.S. Birch*[1], *J.F. Webb*[1], *P.H. Willson*[1], *and T.A. Strivens*[2]

[1]Department of Physics and Applied Physics, Strathclyde University, Glasgow G4 0NG, UK
[2]ICI Paints Division plc, Wexham Road, Slough SL2 5DS, UK

Abstract. We report experiments that demonstrate the non-invasive measurement of lacquer coating thickness using a thermal wave timing technique.

1. Introduction

Opto-Thermal Transient Emission Radiometry, or *OTTER* [1,2], can be used for measuring clear lacquer coating thickness by exploiting their optical properties to excite thermal waves at the interface with the substrate. The method has promise for on-line monitoring of clearcoat thickness in the production of metallic paint finishes in the manufacture of car bodies.

2. Experimental Details

Experiments were performed with electro-coated steel substrates, coated with clear lacquer of thickness in the range 17-56μm. Two different excitation wavelengths, 532nm and 1.06μm, were used. Both pass through the lacquer coating with minimal attenuation and are strongly absorbed by the substrate. Thermal emissions were studied with both broadband (6-13μm) and selective detection using a range of interference filters. Typical signals for such samples are shown in figure 1. They consist of a superposition of a fast decay, probably associated with thermal emissions from the substrate, and a delayed thermal wave emanating from the substrate and detected as enhanced thermal emission from the coating surface. The figure also shows how the thermal waves are delayed, broadened and attenuated as they traverse increasing thicknesses of lacquer. The measurement of thermal wave peak delay forms the basis of this method of thickness measurement.

Figure 2 shows signal variation with detected wavelength for a 37μm thick coating. The lacquer is effectively opaque at wavelengths less than ~11μm, so that the thermal waves are detected only as they arrive at the surface. It is nearly transparent at wavelengths greater than ~11μm, so these signals are dominated by contributions from the substrate. This ability selectively to examine the surface or the substrate is a useful feature of the technique.

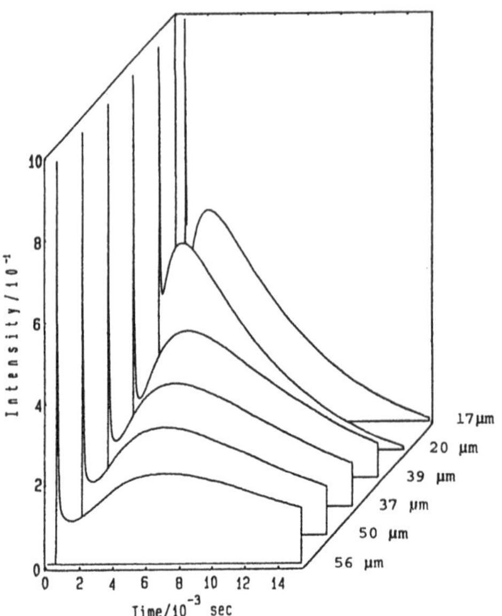

Figure 1.
Lacquer thickness measurement, using 1.06µm excitation and broadband thermal emission. From Back to front, the thickness is 17, 20, 37, 39, 50 and 56µm.

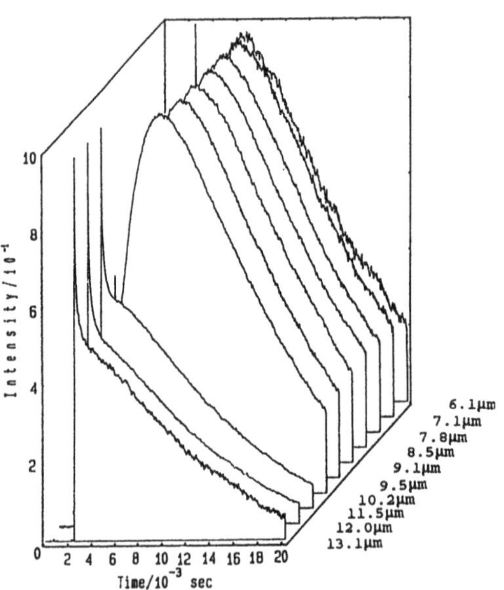

Figure 2.
37µm thick lacquer coating observed with a range of infrared wavelengths. From back to front, the thermal emission wavelengths are 6.1, 7.1, 7.8, 8.5, 9.1, 9.5, 10.2, 11.5, 12.0 and 13.1µm.

3. Results and Discussion

The signals were analysed using a difference of two exponential terms to determine the delay times, Δt, from instant of excitation to the peaks of the humps. Plots of the square root of Δt versus coating thickness produced straight calibration curves, as expected from one-dimensional thermal diffusion theory. The calibrations were independent of excitation wavelength, but not thermal emission wavelength, as illustrated in figure 3 with 532nm data.

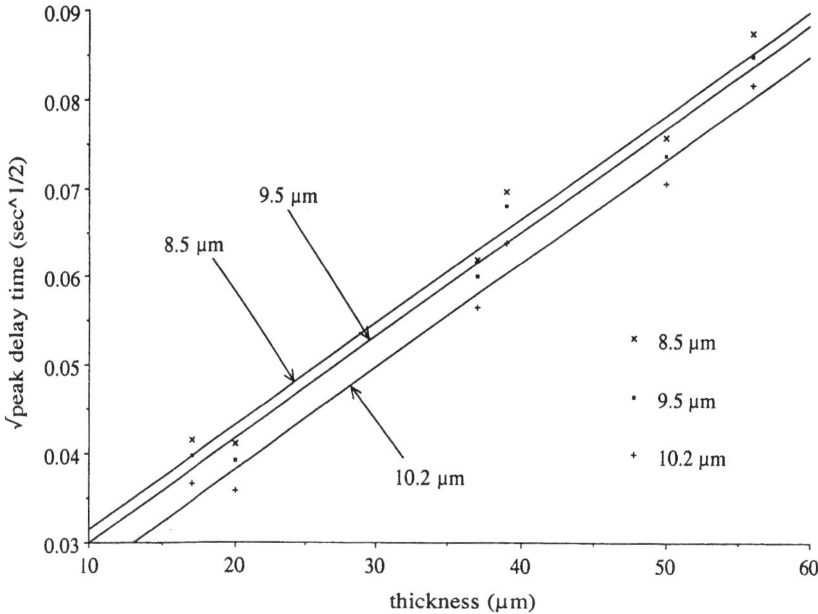

Figure 3.
Thickness calibration lines for 532nm excitation and broadband, 8.5, 9.5 and 10.2µm thermal emission wavelengths.

4. Conclusions

These results show that lacquer coating thickness can be determined with high sensitivity using thermal wave timing measurements. Work on a more realistic theoretical model, that will take the dependence of the signal on thermal emission wavelength into account, is now in progress. The aim is not only to calculate thickness calibration curves, but also to adapt the technique to the measurement of thermal diffusivity.

Acknowledgments

We thank SERC for financial support.

References

1. R.E. Imhof, D.J.S. Birch, F.R. Thornley, J.R. Gilchrist and T.A. Strivens, J Phys E: Sci. Instrum. 17, 521-5, (1984).
2. R.E. Imhof, D.J.S. Birch, M.M. Moksin, J.F. Webb, P.H.Willson and T.A. Strivens, Brit. J. NDT, 33(4), 172-6, (1991).

Thermal Characterization of Low Thermal Diffusivity Glasses

M. Bertolotti[1], *R. Li Voti*[1], *G. Liakhou*[2], *C. Sibilia*[1], *A. Montenero*[3], *and G. Gnappi*[3]

[1]Dipartimento di Energetica, Università di Roma I, Via Scarpa 16, I-00161 Roma, Italy and GNEQP of CNR, Italy
[2]Polytechnical Institute of Kishinev, Kishinev, Moldavia
[3]Istituto di Strutturistica Chimica, Università di Parma, Parma, Italy

The photodeflection technique is a very effective method to measure thermal properties of solid materials. Depending on the material, the technique can be used in its collinear (probe through the transparent sample) or transverse configuration (probe traveling in the air : opaque sample).
In the following we discuss the application of the technique to the case of tranverse configuration to materials with thermal diffusivity lower than the air, because this is the case in which a large error can be introduced in the measure ments.

1- Introduction

Nonlinear thermo-optical properties of doped glasses can be studied with the help of photodeflection method (PD).To increase the accuracy of the measurement it is necessary to evaluate also the thermal parameters such as thermal diffusivity.

We have studied several kinds of glasses, borosilicate glasses doped with semiconductors, sodiodisilicate glasses doped with metals, lead-borosilicate glasses doped with CuCl and Cu_2O, some of them with very low values of thermal diffusivity and the P.D. method has been suitably optimized in both transverse and collinear configuration in the special situation of low thermal diffusivity (with respect to the air).

2- Transverse P.D.

In the transverse configuration one of the most used methods for thermal diffusivity measurements is the "zero crossing" technique [1], in which by moving the probe beam with respect to the pump axis one determines the pump probe offset value at which a zero value of the transverse photodeflection signal occurs. Plotting this offset against the inverse root of the chopper modulationl frequency, the slope of the obtained straight line gives the thermal diffusivity value. However this technique cannot be applied to materials whose thermal diffusivity results lower than that of the surrounding medium, due to the disappearing of the zero crossing.

For low values of the thermal diffusivity parameter also the so called "phase method" [2, 3] falls into error. This method consists into the evaluation of the linear slope of the transverse component of the photodeflection phase signal against the offset y between pump and probe. The linear slope a provides the thermal diffusion length lt , through the relation lt=ay, and therefore the thermal diffusivity is obtained by the expression $lt=(\pi \chi / v)^{1/2}$, where v is the frequency of the chopper pump beam. The accurancy of the measurement is controlled f.e. through a plot of lt against $1/v^{1/2}$, which results a straight line crossing zero. However for low diffusivity materials the experimental data [3,4] show that lt does not go to zero for $v \rightarrow \infty$ and depends on the chopper frequency which is clearly a nonsense.

Looking at the phase signal as a function of the offset y it is important to observe that for offset values larger than the pump beam spot it corresponds the physical situation of point like heat source for which we can write the transverse component of the photodeflection signal as (it is the same for the normal component)

$$\Phi_t = |\Phi_t| \exp[-i\omega t - i(y/lt) + f(z)] \qquad (1)$$

f(z) being a function depending on the vertical offset z. In general the point like approximation is valid for y ›› w, where w is the pump beam laser spot. For materials with low thermal diffusivity values, the function f(z) becomes a slowling varying function of y also [4]. Therefore using the linear slope behavior of the phase signal an error in the measurements of χ is introduced, so that if we plot lt as a function $1/v^{1/2}$ we get a straight line whose intersection with the abscissa axis is not zero, and the obtained thermal diffusivity values are frequency dependent. This does not occur if the vertical offset z is zero.

This last case is the one corresponding to the so called "bounching methods" [5], which can be applied in all cases in which the surface is cleaned. With this method the probe beam is reflected from the surface and the deflection of the reflected beam is studied; this is however not possible if the surface is not polished or in the case of roughness. In this case we need to consider another way to measure the thermal diffusivity.

Very good results can be obtained if we adopt again the "phase methods", but we study the dependence of the thermal diffusivity from the chopper frequency.

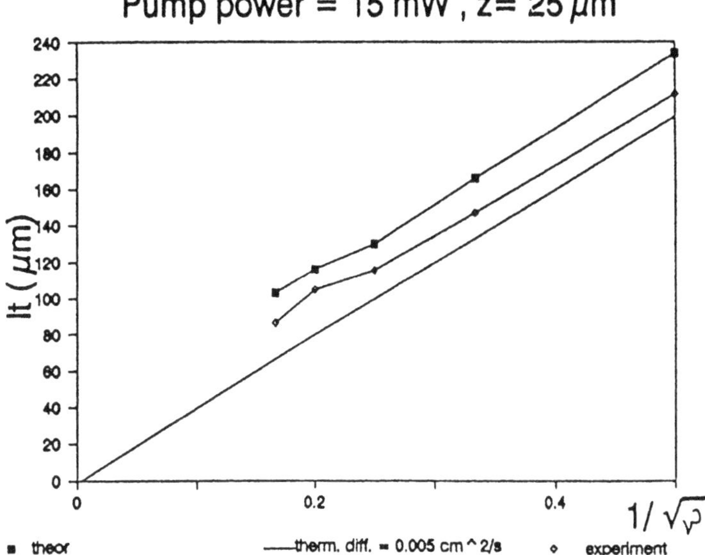

Figure 1 - Thermal diffusion length vs inverse frequency square root

A plot of measured thermal diffusion length as a function of $1/(\nu)^{1/2}$ for different distances z shows that the thermal diffusion length follows the empirical law

$$l_t = A(z) + C/(\nu)^{1/2} \tag{2}$$

where A is the vertical axis intersection and C is a constant which can be estimated taking lt at two different frequencies ν_1 and ν_2 for the same z value

$$C = \frac{l_t(\nu_1) - l_t(\nu_2)}{\frac{1}{\sqrt{(\nu_1)}} - \frac{1}{\sqrt{(\nu_2)}}} \tag{3}$$

Empically agreement with the experiment is obtained by taking $\chi = \pi C^2$ and this result is confirmed by numerical simulations. A(z) than takes the meaning of a remanent term in the diffusieon length for $\nu \rightarrow \infty$ and which depends on z. This "empirical results" have been compared with the theory of PD [4] finding an excellent agreement.

Fig.1 shows the thermal diffusion length plotted as a function of $(1/\nu)^{1/2}$ for different values of the z parameter for a borosilicate glass doped with CdS_xSe_{1-x} . Experimental points and computer symulations are shown in the plot and the agreement is very good.

Work performed with partial support of "Progetto Finalizzato materiali speciali" of CNR, ITALY.

References

1. P.K.Kuo et al - Can J.Phys. 64, 1165 (1986).
2. M.Bertolotti et al., J.Phys. D21, S14, (1988)
3. M:Bertolotti et al., Mat.Science and Engineering B5, 143, (1990)
4. M.Bertolotti et al., Sub. for Publ.
5. A.Salazar et al., J.Appl. Phys.65, 4150, (1989).

Temperature Dependence of Photothermally Modulated FMR Signals: Investigation of Magnetostatic Modes in Yttrium Iron Garnet Films

M. Hoffmann, Th. Orth, O. von Geisau, and J. Pelzl

Institut für Experimentalphysik AG VI, Ruhr-Universität Bochum, Postfach 10 21 48, W-4630 Bochum 1, Fed. Rep. of Germany

Abstract. Photothermally-modulated ferromagnetic resonance (PM-FMR) spectra of magnetostatic modes in yttrium iron garnet have been recorded in the temperature range from 120 K to 580 K. As the Curie temperature is approached, the conventional FMR signal becomes smaller, whereas, as expected from theory, the PM-FMR signal markedly grows. Based on the present result the theoretical description of the PM-FMR signal generation process has been improved.

1 Introduction

Photothermal microwave methods can be used to obtain spatially resolved ferromagnetic resonance (FMR) spectra thus providing information not accessible to conventional FMR measurements. Among the alternative detection schemes (photoacoustic microphone detection (PA-FMR), photothermal laser beam deflection (PD-FMR), photothermally-modulated technique (PM-FMR)) PM-FMR is the most promising one as it has the highest sensitivity, the best spatial resolution and is a real 3-d technique [1,2].

The signal strength of the PM-FMR is proportional to the temperature derivative of the high frequency susceptibility, where the latter quantity is a function of magnetic properties such as spontaneous magnetization or magnetic anisotropy. Temperature dependent measurements provide a suitable tool to separate the different processes contributing to the PM-FMR signal generation.

In this work we have undertaken a detailed study of the temperature dependence of magnetostatic modes (MSMs) [3,1] in yttrium iron garnet (YIG) films by conventional and PM-FMR.

2 Experiments

The investigated sample is a ferrimagnetic YIG film with the dimensions 7.4 mm · 3.8 mm · 18 μm. For both the conventional and the photothermal measurements a conventional X-band EPR-spectrometer with a light-access TE_{102}-cavity with dewar insert (100 K — 700 K) is used. For the PM-FMR measurements an unfocussed 12 mW HeNe-laser periodically illuminates the middle of the sample. The reflected microwave power is detected with a diode and is then fed to a lock-in amplifier. The applied microwave power is 0.17 mW at 9.34 GHz and the modulation frequency for PM-FMR is 1 kHz. The MSMs are investigated in the temperature range from 120 K to 580 K.

3 Results and Discussion

In both conventional FMR and PM-FMR measurements MS surface waves and MS backward volume waves are clearly visible. PM-FMR MSM spectra below 500 K are comparable to those measured at room temperature which are published elsewhere [2]. From 520 K to 560 K the MSM spectra are superimposed by a broad paramagnetic resonance line which for the highest temperatures almost dominates the spectrum while the MSM spectrum becomes invisible (Fig.1). Above the Curie temperature the paramagnetic resonance can only be observed by the conventional technique.

In the left part of Fig.2 theoretical curves for the conventional FMR signal and the PM-FMR signal amplitude are displayed. The calculation of these curves is based on the assumption that at high temperatures in YIG the most prominent contribution to the PM-FMR response arises from the temperature dependence of the saturation magnetization. The FMR signal can be derived from the sum of two Brillouin functions for the saturation magnetization for each of the magnetic sublattices of YIG [4].

The experimental data are displayed in the right part of Fig.2. For temperatures above 360 K the FMR data are in good agreement with theory. When approaching the Curie temperature (560 K) in the real material the transition

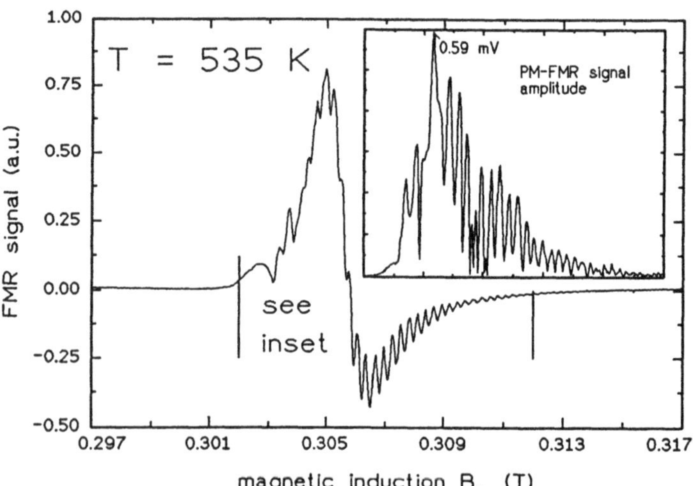

Figure 1: Conventional and PM-FMR signal of magnetostatic modes in a YIG film (7.4 mm · 3.8 mm ·18μm) at 535 K. (microwave frequency = 9.34 GHz, microwave power = 0.14 mW, modulation frequency = 1.0 kHz)

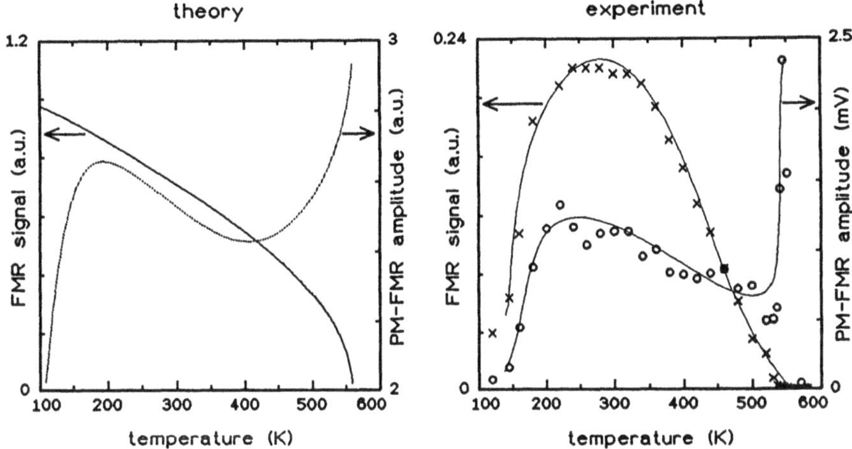

Figure 2: Theoretical and experimental temperature dependence of FMR and PM–FMR signals of magnetostatic modes in YIG. For experimental conditions see Fig. 1.

is smeared out due to short ordering effects not considered in the calculated curves. Below 360 K the slope of the FMR amplitude curve becomes smaller and starts to fall below 260 K. These observations cannot be understood from the theoretical point of view and require some further investigations. The experimental PM–FMR signal amplitudes qualitatively are in good agreement with the calculated curve justifying the assumption that the signal is mainly governed by the temperature coefficient of the spontaneous magnetization.

[1] J.Pelzl, U.Netzelmann, Th.Orth, and R.Kordecki: "Photothermal Imaging Using Microwave Detection", in: J.C.Murphy et al. (eds.): *Photoacoustic and Photothermal Phenomena II*, Springer Series in Optical Sciences vol. 62, p. 2, Heidelberg 1990
[2] M.Hoffmann, O.v.Geisau, S.A.Nikitov, and J.Pelzl, "3D–Imaging of Magnetostatic Modes Using Photothermally Modulated FMR–technique", J. Mag. Magn. Mat. 101 (1991) (in press)
[3] R.W.Damon and J.R.Eshbach, "Magnetostatic Modes of a Ferromagnet Slab", J. Phys. Chem. Solids 19, 308 (1961)
[4] G.Winkler, *Magnetic Garnets*, p. 88, Vieweg, Braunschweig 1981

Analytical Relations Between the Phase of the Photothermal Signal and the Thermal Wavelength

A. Figari

CISE Tecnologie Innovative, Via Reggio Emilia 39,
I-20090 Segrate (Milano), Italy

ABSTRACT Analytical expressions are derived for the phase of the photothermal signal, when the probe beam scans the wavelength λ_{th} of the thermal wave generated by the pump-beam in a sample. An analytical foundation is thus given to a refined experimental technique for measuring λ_{th} and thermal diffusivity α. In particular, the analysis shows that λ_{th} and α can be measured using arbitrary phase shifts relative to the heating spot centre.

The mirage effect is the basis for an absolute and non-contact measurement of the thermal diffusivity α of materials. This application of the photothermal technique origined from heuristic considerations and was supported by numerical calculations and experimental work [1,2]. The esperimental set-up is reported in Ref. 1.

In this communication some meaningful analytical relations are derived, relating the phase ϕ_M of the photothermal signal to the thermal wavelength and the thermal diffusivity. If $\omega=2\pi f$ is the chopper frequency, the thermal wavelength is defined as [1]: $\lambda_{th}=2\pi(2\alpha/\omega)^{½}$. The thermal diffusion length in the medium [3] (or reduced wavelength [4]) is $\mu=\lambda_{th}/2\pi$.

The expression describing the parallel component M of the probe-beam deflection is [1]:

$$M(x,t)=B\ e^{-i\omega t}e^{-i(D^2/2\mu_g^2)}\int_0^\infty k\ dk\ \sin kx\ e^{-k^2d^2/4}F(a)\ \frac{e^{-k_gh}}{k_s} \tag{1}$$

where: B is a constant factor; the integration variable k forms a Fourier variable pair with x; D is the radius of the probe-beam, d the radius of the pump-beam, a the sample thickness, h the probe beam distance from the sample surface, $k_l=(k^2-2i/\mu_l^2)^{½}$ (l=s,g refer to solid, gas, respectively). Finally, F(a) gives the dependence of the signal on backing and can ben omitted for large thickness. In this case, and in the limits $d\to0$ and $h\to0$ the deflection M is proportional to the integral

$$\tilde{M}(x)\equiv\int_0^\infty k(k^2+\beta^2)^{-½}\sin(kx)\ dk \tag{2}$$

where: $\beta^2 \equiv -2i/\mu_s^2$

Now the integral (2) can be solved [5], to obtain:

$$\tilde{M}(x) = \beta K_1(x\beta) \tag{3}$$

with K_1 being the first order Kelvin function. Next one can use a representation of K_1 [5,6], obtaining the useful formula

Springer Series in Optical Sciences, Vol. 69 Photoacoustic and Photothermal Phenomena III
Editor: D. Bićanić

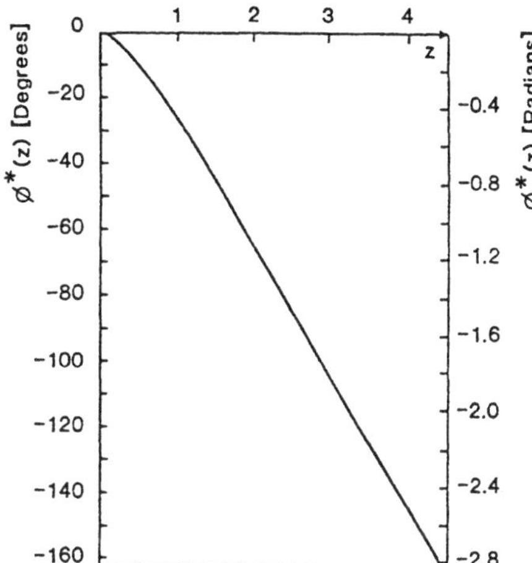

Fig.1: phase of the photothermal signal

$$\tilde{M}(x) = \tilde{\rho}_M(x)\ \exp(i\phi_o) \left[ker_1\left(\sqrt{2}\frac{x}{\mu_s}\right) - i\ kei_1\left(\sqrt{2}\frac{x}{\mu_s}\right)\right] \qquad (4)$$

where $\tilde{\rho}_M(x)$ is an amplitude, $\exp(i\phi_0)$ a constant phase factor. Eq. (4) can be written as

$$\tilde{M}(x) = \rho_M \exp(i\phi_0)\ \exp\left[i\phi_1\left(\sqrt{2}\frac{x}{\mu_s}\right)\right] \qquad (5)$$

where $\rho_M(x)$ is an amplitude, $\phi_1\left(\sqrt{2}\frac{x}{\mu_s}\right)$ a well-known, tabulated function (see e.g. Ref. 6 p. 433).
Eq. (5) gives an analytical foundation to the experimental procedure for measuring λ_{th} via a scanning in the x-variable. Actually one sees that the phase ϕ_M is a function of the ratio $x/\mu_s=(2\pi/\lambda_{th})x$. Putting

$$z = \sqrt{2}\ x/\mu_s \qquad (6)$$

the reference frame can be suitably fixed so that the phase offset of the first order Kelvin function $\phi_1(z)$ vanishes for z=0 [6].
The transformed components of the Kelvin function relative to the new reference frame are indicated with an the asterisk. Fig. 1 shows the phase $\phi^*(z)$ and Fig. 2 the ratio of real to imaginary component of the signal M(x):

$$\frac{Re(M)}{Im(M)} = \frac{ker_1^* z}{kei_1^* z} = cotan[\phi^*(z)] \qquad (7)$$

From the definitions of z (eq. 6), μ_s, $x_o(=2x)$ and taking into account the pump-beam diameter via the constant d one can write:

$$x_o(\phi^*) = A(\phi^*)\ \frac{\lambda_{th}}{2} + d \qquad (8)$$

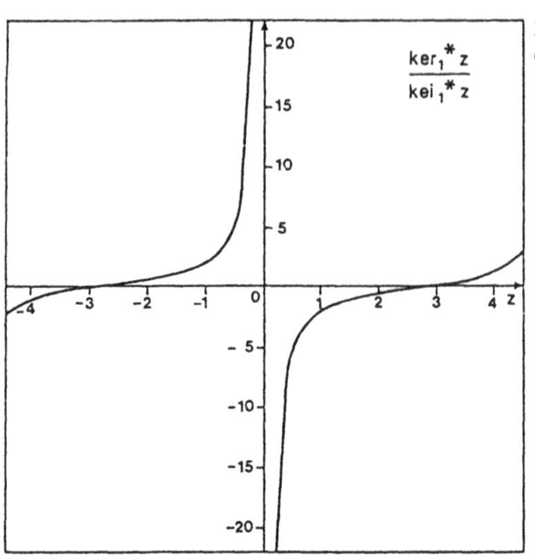

Fig.2: ratio of real to imaginary component of the photothermal signa

with

$$A(\phi^*) = \frac{2}{\sqrt{2\pi}} z \qquad (9)$$

From the definition of λ_{th} one obtains:

$$X_o(\phi^*) = A(\phi^*)\sqrt{\pi\alpha}\; f^{-\frac{1}{2}} + d \qquad (10)$$

Generally, eq. (10) allows one to obtain α by measuring the separation x_o between the points corresponding to an arbitrary phase shift. α can be obtained from the slope $A(\phi^*)\sqrt{\pi\alpha}$ of the x_o versus $1/\sqrt{f}$ plot. The general recipe for obtaining the factor $A(\phi^*)$ is the following: if $\phi^*(z)$ is fixed, eq. (5) gives the corresponding value z (Ref. 6) and eq. (9) gives A. Let us examine two particular cases:

1) the "classic" experiment [1,2] consists in measuring the separation x_o between the points corresponding to a phase shift of $\pm\pi/2$ with respect to the spot centre x=0. If the phase shift $\phi^*(z)=\pi/2$, eq. (5) gives the value $z\simeq 2.666$ (see Ref. [6]). The definition of the factor A (eq. (9)) gives $A\simeq 1.2$, differently from the equations derived in a heuristic way [1];
2) for $\phi^*(z)=\pi/4$ eq. (5) gives the value $z\simeq 1.496$ (see Ref. [6]) and $A\simeq 0.673$.

With this analysis λ_{th} and α can be obtained as mean values calculated from different values $\lambda_{th}(\phi^*)$ and $\alpha(\phi^*)$ corresponding to different phase shifts ϕ^*: this possibility should greatly improve the measurement accuracy.

REFERENCES

1) P.K. Kuo, M.J. Lin, C.B. Reyes, L.D. Favro, R.L. Thomas, D.S. Kim, S.Y. Zhang, L.J. Inglehart, D. Fournier, A.C. Boccara and N. Yacoubi, Can. J. Phys. 64 (9), 1165 (1986)
2) P.K. Kuo, E.D. Sendler, L.D. Favro and R.L. Thomas, Can. J. Phys. 64 (9), 1168 (1986)
3) A. Rosencwaig, Science 218 (4569), 223 (1982)
4) G. Busse and A. Rosencwaig, Appl. Phys. Lett. 36 (10) 815 (1980)
5) A. Erdélyi, W. Magnus, F. Oberhettinger and F.G. Tricomi, Tables of integral transforms, vol. 1, McGraw-Hill Book Company (1954)
6) M. Abramovitz and I. Stegun, Handbook of mathematical functions, Dover Publications, 8th Ed. (1972)

A Highly Sensitive Photothermal Method for Measuring the Optical Losses in the Mirrors of the Gravitational Antenna Virgo

D. Ivanov[1], *N. Man*[1], *J.P. Roger*[2], *and A.C. Boccara*[2]

[1]VIRGO-LAL, Faculté des Sciences, Bât. 208, F-91405 Orsay, France
[2]Laboratoire d'Optique Physique, ESPCI,
10, rue Vauquelin, F-75231 Paris, France

The determination of the optical absorption of low absorbing materials (0.1-1 ppm) requires highly sensitive methods. It was recently found out that the photothermal optic collinear beam deflection method (known as "collinear mirage effect") is the most appropriate one for nondestructive characterization of optical losses of the order of 10^{-7} cm^{-1}. We used this method for measuring the optical losses of the mirrors of the Franco-Italian gravitational interferometric antenna VIRGO. The VIRGO mirror substrates are of high quality fused quartz in form of a cylinder with dimensions : diameter-50 cm, thickness-20 cm and weight-80 Kg. The optical absorption coefficient must be less than 10^{-5} cm^{-1} to avoid all thermal lensing effects.

1 - Introduction

The interferometric antenna VIRGO is a Franco-Italian project for the detection of gravitational signal's amplitudes of the order of 10^{-21} i.e, the interferometer must be able to detect relative length differences of the same order between its two arms[1]. The light power stored in the cavities is of the order of several kWatts and the transmitted light power through the input mirrors is of the order of 10^2 Watts (1.06 µm YAG laser). The resulting thermal effects are unwanted and the antenna s sensitivity depends strongly on the mirrors' quality. The VIRGO mirrors' substrates are of high quality fused quartz in form of cylinder with a diameter of 50 cm, thickness - 20 cm and weight - 80 Kg. To avoid all thermal lensing effects the optical absorption coefficient must be less than 10^{-5} cm^{-1} and therefore the mirrors' substrate must be tested carefully.

2 - Experimental results and discussion

The determination of the optical absorption of low absorbing materials (0.1 - 1 ppm) requires highly sensitive methods. In the past few years an intense research work in photoacoustic spectroscopy and microscopy has been done. These high performance techniques open large possibilities for detection of defects and technological failures with submicron size as well as for nondestructive measuring of optic, elastic and thermal properties of materials and devices. It was found out that the collinear optic beam deflection method (known as "collinear mirage effect") is the most appropriate for nondestructive characterization of optical losses of the order of 10^{-7} cm^{-1} in low absorbing materials and dielectric coatings[2,3]. The well known classical methods for determining the optical losses consists of comparing the measurements of the incident, transmitted and reflected light powers and the best accuracy is of the order of 1%. In the case of laser mirrors, the techniques based on cavity decay time measurements are highly sensitive but these measurements integrate all the optical losses (absorption + scattering + transmission). The transmission can be measured but the absorption and the scattering cannot be separated. By using the mirage techniques the absorption can be measured separately and therefore by joining both techniques, it is possible to obtain the whole necessary information concerning the optical losses in the mirror.

Springer Series in Optical Sciences, Vol. 69 Photoacoustic and Photothermal Phenomena III
Editor: D. Bićanić

OPTICAL BEAM DEFLECTION MONOBLOCK

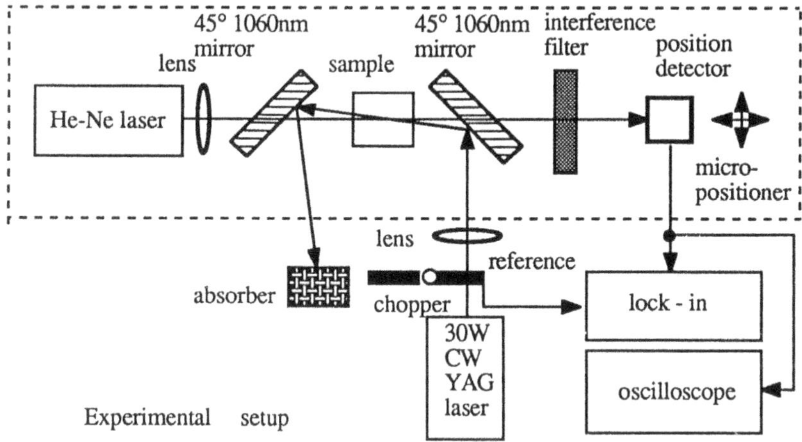

Experimental setup

The absorbed YAG laser power is converted into heat and the resulting change of the refraction coefficient is measured through the deflection of the probe He-Ne laser beam. The photothermal signal is directly created by the absorbed power and therefore it depends only on the incident power and the detector's noise.
When the thermal diffusion length is much smaller that the pump laser beam radius the sample's optical absorption coefficient with respect to a standard sample is given by the formula :

$$\alpha_{sample}[cm]^{-1} = \frac{\alpha_{etalon}\,\mathbf{signal}_{sample}\left(\frac{1}{n}\frac{dn}{dT}\right)_{etalon}(\rho c)_{sample}}{\mathbf{signal}_{etalon}\left(\frac{1}{n}\frac{dn}{dT}\right)_{sample}(\rho c)_{etalon}}$$

The term (1/n).(dn/dT) takes into account the variation of the refraction index with temperature changes. First the setup is adjusted with the help of a standard sample having a large absorption coefficient and afterwards the mirage signal of the silica sample is measured. For most glasses and fused quartz samples the absorption coefficients are of the order of 10^{-3} - 10^{-4} cm^{-1}.
The measurements of such absorption coefficients are performed by using water as standard sample and low pump beam power (2 Watts). The absorption coefficient of water ($\alpha = 0.15 \pm 0.01$ cm^{-1} at 1.06 pm) is determined by traditional transmission measurement. Then, these absorbing silica samples can be used as standard samples in the measurement of weaker absorption coefficients for which high pump beam power are needed. The values of 1/n (dn/dT) and ρc for water and silica are taken in literature.

The German company Heraeus proposed to us two types of high quality silica samples Herasil I Top (natural silica) and Suprasil 310 (synthetic silica) for measuring the optical absorption coefficients. The measurements were made by using a Spectron Laser System 30 Watt CW YAG laser type SL 903-30-CW giving the possibility to increase considerably the sensitivity. The optical absorption coefficients calculated vis-à-vis a standard glass sample having an absorption coefficient of 4.2×10^{-4} cm^{-1} are 4.0×10^{-6} cm^{-1} for Herasil I Top and 2.0×10^{-6} cm^{-1} for Suprasil 310, respectively.

According to the theoretical results developed by computer simulation, these values are just good enough for the VIRGO interferometer, but an improvement by a factor of 10 will be necessary for the next generation of interferometers. Unless it becomes possible to use smaller mirrors, or to obtain large blocks of Suprasil 310, the antenna construction will start with Herasil I Top mirrors. In the future, we may have to use sapphire or diamond as mirrors' substrate, or to find another geometry for the interferometer, which would be less sensitive to thermal problems.

3 - Conclusion

The construction of such high performant systems as VIRGO gravitational wave detector mirrors requires the best of the contemporary advanced materials to be used. However, it is also necessary to develop high accuracy measurement techniques for testing the most important physical parameters. It was found that the collinear optic beam deflection method is a very suitable one for measuring the bulk optical absorption coefficient of the mirrors'substrate as well as for measuring the optical losses in the dielectric coating on the surface.

References

1 - VIRGO, Proposal for the construction of a large interferometric detector of Gravitational waves, 1990

2 - A. C. Boccara, D. Fournier, W. Jackson, N. Amer, Optics Letters, 5,377 (1980)

3 - J. P. Roger, F. Charbonnier, D. Fournier, A. C. Boccara, Proceedings of LASER'88, Lake Tahoe, 1988.

Effect of Photoelasticity on Photothermal Beam Deflection in Transparent Solids

H.J. Seo[1] *and S.I. Yun*[2]

[1]Department of Natural Sciences, Pusan National Institute of Technology, Pusan 608-739, Republic of Korea
[2]Department of Physics and Research Center for Dielectric and Advanced Matter Physics, Pusan National University, Pusan 609-735, Republic of Korea

Abstract. The photothermal beam deflection(PTBD) signals were measured as a function of polarization angle of the probe beam. The PTBD signal within the transparent solids strongly depends on the polarization angle of the probe beam and orientation of the samples. The photoelastic effect in the solid was successfully introduced to analyze the experimental results.

1. Introduction

It is well known that the photothermal beam deflection(PTBD) signal depends on the change of refractive index in the medium adjacent to the periodically heated region/1/. Generally, the refractive index in solids depends on temperature and thermal stress of the medium/2-4/ except for metals and semiconductors in which the refractive index is governed by the behavior of free carriers/5/. However, the effect of thermal stress is negligible for liquids and gases/1/. The photoelastic effect, which is the change of refractive index caused by the thermal stress, has been of great importance in optical materials for high power laser systems/3,4/. In this contribution, PTBD signals of BK7 glass and quartz crystal are measured as a function of probe beam polarization angle, and the experimental results are analyzed with the help of matrix equations for the stress-strain relation, photoelastic effect, and the physical constants found in the literature for these samples.

2. Theory

We can write the change of refractive index for the light polarized with respect to the direction of principal x_i axis as

$$\Delta n_i = \frac{\partial n_i}{\partial T} T(r, t) = \left[\frac{\partial n_{i,T}}{\partial T} + \frac{\partial n_{i,S}}{\partial T} \right] T(r, t). \tag{1}$$

Where $\partial n_i/\partial T$ is called the optical distortion coefficient/3/. Two terms in the parentheses represent the temperature dependence of refractive index at zero stress and the stress-induced photoelastic contribution, respectively. If the probe beam is polarized in the direction of the x_i axis, the deflection angle(ϕ_i) of the probe beam is given by/1/

$$\phi_i = \frac{1}{n_i} \frac{\partial n_i}{\partial T} \int_{path} \nabla_\perp T(r, t) ds. \tag{2}$$

Eq. 2 shows that ϕ_i is proportional to the optical distortion coefficient $\partial n_i/\partial T$. In Eq. 1, $\partial n_{i,S}/\partial T$ can be written by $(\partial n_{i,S}/\partial\sigma_j)(\partial\sigma_j/\partial T)$. The matrix equations for the stress-strain relation and the photoelastic effect are given by $\varepsilon_i = S_{ij}\sigma_j + \alpha_i T(r,t)$/2,6/ and $\Delta B_i = \pi_{ij}\sigma_j$ $(i,j=1,2,\cdots 6)$/2/, respectively. Where the suffix 1,2,3 represent normal components, and 4,5,6 represent shear components. In this expression, ε is the strain, σ is the stress, S is the elastic compliance coefficient, α is the thermal expansion coefficient,

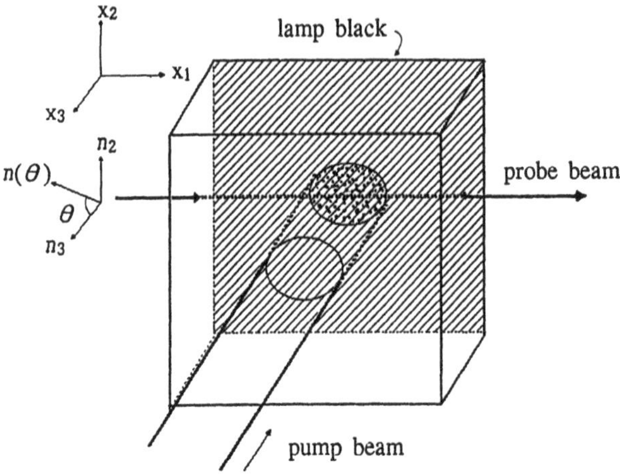

Fig.1. Geometry of the sample and beam propagations for the transverse PTBD measurements.

T(r,t) is the periodic temperature change, B is the dielectric impermeability, and π is the piezo-optic coefficient. The matrix components are determined by crystal symmetry/2/. Fig. 1 shows the geometry of the sample in the transverse PTBD system. The flat x_3 face(the face normal to the x_3 axis and opposite to the illuminated face) of orthorhombic sample is coated with a lamp black. The x_1, x_2, x_3 correspond to the principal axes a,b,c of the quartz crystal. The probe beam is directed parallel to x_1 axis with a polarization angle of θ. If the pump beam diameter is sufficiently larger than that of the probe beam and the sample surface is homogeneously heated by the pump beam, the heat flows only parallel to x_3 axis. These considerations impose the boundary conditions that ε_1, ε_2, σ_3, and all the shear components are zero. Solving $\partial n_{i,S}/\partial T$ for the glass (isotropic) and quartz crystal (trigonal 32) with the above boundary conditions gives

$$\frac{\partial n_{2,S}}{\partial T} = \frac{1}{2} n_1{}^3(\pi_{11}+\pi_{12})\frac{\alpha}{S_{11}+S_{12}}, \quad \frac{\partial n_{3,S}}{\partial T} = \frac{1}{2} n_1{}^3 2\pi_{12}\frac{\alpha}{S_{11}+S_{12}} \quad \text{for glass, and} \tag{3}$$

$$\frac{\partial n_{2,S}}{\partial T} = \frac{1}{2} n_2{}^3(\pi_{11}+\pi_{12})\frac{\alpha_1}{S_{11}+S_{12}}, \quad \frac{\partial n_{3,S}}{\partial T} = \frac{1}{2} n_3{}^3 2\pi_{31}\frac{\alpha_1}{S_{11}+S_{12}} \quad \text{for quartz.} \tag{4}$$

The values of $\partial n_i/\partial T$ in Eq. 1 are obtainable from the published literature/7-10/. The polarization dependent value of the optical distortion coefficient $\partial n(\theta)/\partial T$ at angle θ is obtained by the rotation of the refractive index tensor about x_1 axis/2/; is $(\partial n_2/\partial T)\sin^2\theta$ + $(\partial n_3/\partial T)\cos^2\theta$.

3. Experimental Results and Discussion

For the measurements of the polarization dependent PTBD signal, a polarizer and a rotatable analyzer were used. The azimuth of the polarizer was 45°. The pump and probe beam sources were a 1kW tungsten lamp and a 5mW He-Ne laser, respectively. Fig. 2 displays PTBD signals and $\partial n_i/\partial T$ for BK7 glass and quartz crystal as a function of polarization angle θ. The difference of the PTBD signals for the angle of 0° (x_3-axis) and 90° (x_2-axis) is about 30% for the BK7 glass and 57% for the quartz crystal. These results agree well with the polarization dependent values of $\partial n_i/\partial T$ of the samples. From Eqs. 3 and 4, the values of $\partial n_{2,S}/\partial T$ and $\partial n_{3,S}/\partial T$ are 2.86×10^{-6} ℃$^{-1}$ and 4.92×10^{-6} ℃$^{-1}$ for BK7 glass and 8.03×10^{-6} ℃$^{-1}$ and 12.5×10^{-6} ℃$^{-1}$ for quartz crystal. $\partial n_{i,T}/\partial T$ is 1.43×10^{-6} ℃$^{-1}$ /7/ for BK7 glass, and -5.49×10^{-6} ℃$^{-1}$ (ordinary ray, $\partial n_{2,T}/\partial T$) and -6.53×10^{-6} ℃$^{-1}$ (extraordinary

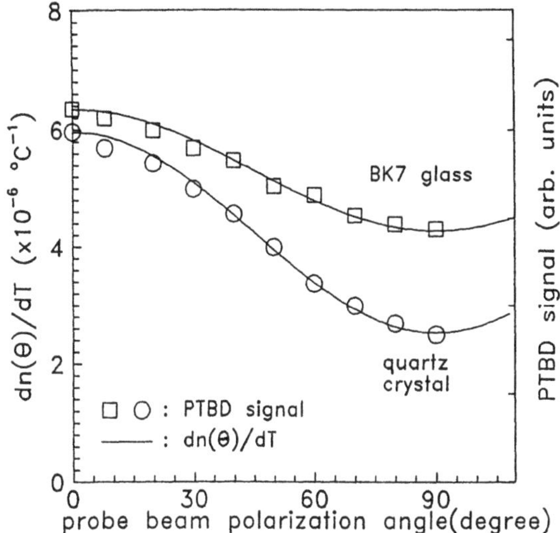

Fig.2. PTBD signal and optical distortion coefficient as a function of polarization angle. The numerical values used for the BK7 glass are $n=1.515(6328Å)$, $\partial n_{3,T}/\partial T=1.43\times10^{-6}$ °C^{-1}, $\alpha=7.40\times10^{-6}$ °C^{-1} /7/, $S_{11}=1.48\times10^{-11}m^2/N$, $S_{12}=3.03\times10^{-12}m^2/N$, $\pi_{11}=3.63\times10^{-13}m^2/N$, and $\pi_{12}=2.25\times10^{-12}m^2/N$ /8/; that for the quartz crystal $n_3=1.543(6328Å)$, $n_2=1.551$ (6328Å), $\partial n_{3,T}/\partial T=-5.49\times10^{-6}$ °C^{-1} (6430Å), $\partial n_{2,T}/\partial T=-6.53\times10^{-6}$ °C^{-1} (6430Å) /9/, $\alpha_1=1.34\times10^{-5}$ °C^{-1}, $S_{11}=1.28\times10^{-11}m^2/N$, $S_{12}=-1.74\times10^{-12}$ m^2/N, $\pi_{11}=1.11\times10^{-12}$ m^2/N, $\pi_{12}=2.50\times10^{-12}m^2/N$, and $\pi_{13}=7.77\times10^{-13}m^2/N$ /10/.

ray, $\partial n_{3,T}/\partial T$)/9/ for quartz crystal. It is clear that the quartz crystal has negative $\partial n_{i,T}/\partial T$ and positive $\partial n_{i,S}/\partial T$, while BK7 glass has all positive values. These are the reasons why we have weaker PTBD signal for the probe beam polarization angle of 90° in the quartz crystal than that in BK7 glass.

In conclusion, we measured the PTBD signals as a function of the probe beam polarization angle. It was possible to explain the dependence of PTBD signal on the polarization angle of the probe beam by the introduction of photoelasticity of the transparent solid medium. We believe that PTBD technique can be used to study the optical distortion in optical components for high power laser systems such as laser windows, optical fibers, and laser rods.

The present studies were supported by 1990 Basic Science Research Institute Program, Contract No. BSRI-90-205, Ministry of Education and by Korea Science and Engineering Foundation through the Science Research Center of Excellence Program, the Republic of Korea.

References

1. W.B. Jackson, N.M. Amer, A.C. Boccara, and D. Fournier, Appl. Opt. 20, 1333(1981).
2. J.F. Nye, *Physical Properties of Crystals* (Oxford U. Press, Oxford, 1957).
3. Claude A. Klein, in *Laser Induced Damage in Optical Materials'80*, NBS Spec. Publ.620, pp.117(1980).
4. J.M. Eggleston, T.J. Kane, K. Kuhn, and J. Unternahrer, and R.L. Byer, IEEE J. Quantum Electron. QE-20, 289(1984).
5. Andrew Skumanich, Daniel Fournier, A. Claude Boccara, and Nabil M. Amer, Appl. Phys. Lett. 47, 402(1985).

6. B.A. Boley and J.H. Weiner, *Theory of Thermal Stresses* (J. Wiley, New York, 1960).
7. T. Izumitani and H. Toratani, J. Non-Crystalline Solids, 40, 611(1980).
8. Luis M. Bernardo and Silverio P. Almeida, Opt. Lett. 7, 442(1982).
9. Walter G. Driscoll(ed.), *Handbook of Optics* (McGraw-Hill, New York, 1978).
10. K.-H. Hellwege(ed.), *Landolt-Bornstein series,* Vol. III/11, (Springer-Verlag, Berlin, 1979).

Contrast Analysis in Thermal Wave Materials Inspection. Part I: Pulsed Excitation

P.M. Patel, S.K. Lau, and D.P. Almond

School of Materials Science, University of Bath,
Claverton Down, Bath, Avon BA2 7AY, UK

A study is presented of the evolution of the thermal contrast at a surface following pulsed thermal excitation. Two parameters are identified for the correlation with the thickness of a coating layer and the thickness of a subsurface air gap. The two parameters are the temperature contrast maximum and the time for this contrast maximum.

Introduction

Pulsed photothermal radiometry (PPTR) and pulsed video thermography (PVT) are two infrared sensing transient thermal wave techniques which are being commercially exploited for quantitative material inspection [1]. Both technique provide the rapid acquisition of the time evolution of the surface temperature at a point (PPTR) or over an area (PVT) on a test sample. A variety of applications have been demonstrated for industrial usage: thickness gauging, defect detection and imaging, thermal property evaluation [2,3]. In this paper a theoretical treatment on the use of temperature contrast analysis is discussed to assess:

a) the depth of a subsurface interface, and
b) the defect depth, thickness and radius.

Theory

The surface temperature contrast, ΔT, between a coated and uncoated sample exposed to plane pulsed surface heating can be expressed as [4]:

$$\Delta T = \frac{I_o}{\sqrt{\pi \rho_1 c_1 k_1 t}} \sum_{n=1}^{\infty} \Gamma^n \exp\left(-\frac{n^2 L_1^2}{\alpha_1 t}\right) \qquad \Gamma = (1-b)/(1+b) \qquad \text{where} \quad b = \sqrt{\frac{\rho_2 c_2 k_2}{\rho_1 c_1 k_1}}$$

where I_o is the optical intensity incident on the sample surface assumed to be fully absorbed at the surface. ρ, c ,k and α are respectively the density , specific heat, thermal conductivity and thermal diffusivity of the sample. L_1 is the interface depth and Γ is the thermal wave reflection coefficient for the interface. An expression can also be derived for time of maximum contrast, t_m:

$$t_m = \frac{2L_1^2}{\alpha_1} \frac{\sum_{n=0}^{\infty} n^2 \Gamma^n \exp(-n^2 L_1^2/\alpha_1 t)}{\sum_{n=0}^{\infty} \Gamma^n \exp(-n^2 L_1^2/\alpha_1 t)}$$

In figure 1 a plot the evolution of the temperature contrast with time for yttria stabilised zirconia (a thermal barrier coating) on stainless steel is presented. For this coating substrate system combination the temperature contrast is negative due to the thermal

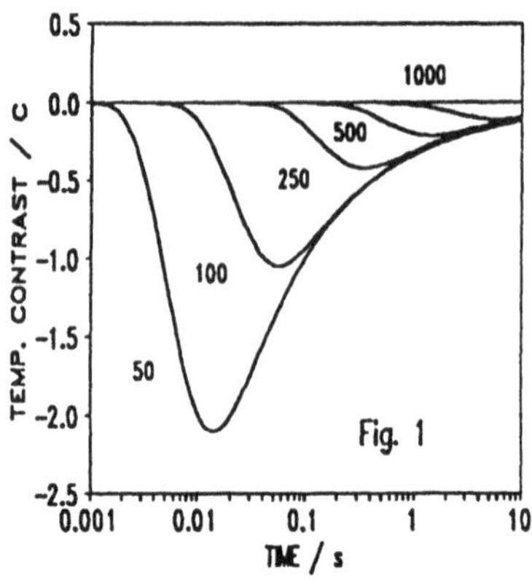

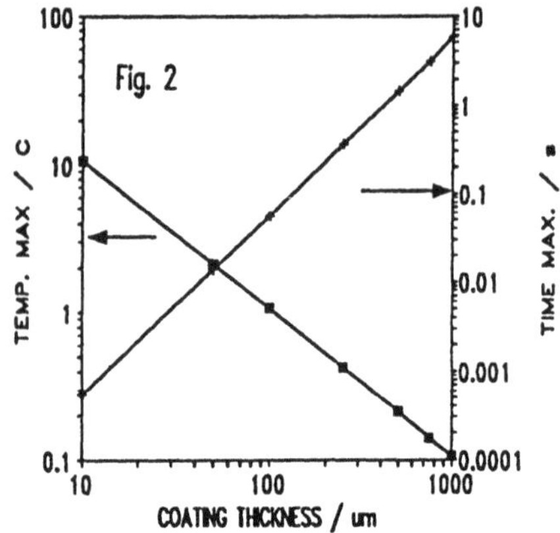

Figure 1. Temperature contrast evolution with time for yttria stabilised zirconia coating on stainless steel. Coating thickness as shown in figure in μm.
Figure 2. A logarithmic plot of maximum temperature contrast and time for maximum contrast with coating thickness.

sinking nature of the conductive substrate. A plot of the magnitude of the maximum temperature contrast ΔT_m, and the time for this maximum contrast, t_m with coating thickness reveals that:- ΔT_m is inversely proportional to the coating thickness and that t_m is directly proportional to the square of the coating thickness, see fig. 2. The t_m values used in fig. 2 were found by solving the equation for t_m graphically. This value was then used to obtain ΔT_m by substituting t_m in the above expression for ΔT.

To obtain the temperature contrast curves for a subsurface defect finite difference numerical techniques were utilised [5]. The relationships summarised in table 1 have been observed for air gap defects in stainless steel with ΔT_m and t_m against the defect depth, defect thickness and defect radius. It is noted that for planar air gap defects of infinite width ΔT_m appears to be inversely proportional to interface depth. This behaviour follows that of the coating/ substrate contrast analysis. With increasing

Table 1: Summary of the relationships of ΔT_m and t_m with defect geometric parameters.

Stainless Steel	Temperature Contrast Maximum	Time for Maximum Contrast
Defect Depth (100 μm air gap)	$\Delta T_m \propto L_1^{-1.18}$	$t_m \propto L_1^{1.38}$
Defect Thickness (10-1000 μm)	Rises to a Maximum	$t_m \propto L_2^{0.61}$
Disc Air Gap Defect	ΔT_m rises to a maximum with increasing radius and defect thickness. $\Delta T_m \propto L_1^n$ with increasing radius	t_m rises to maximum with increasing radius and defect thickness. t_m near linear with defect depth for small defect radii

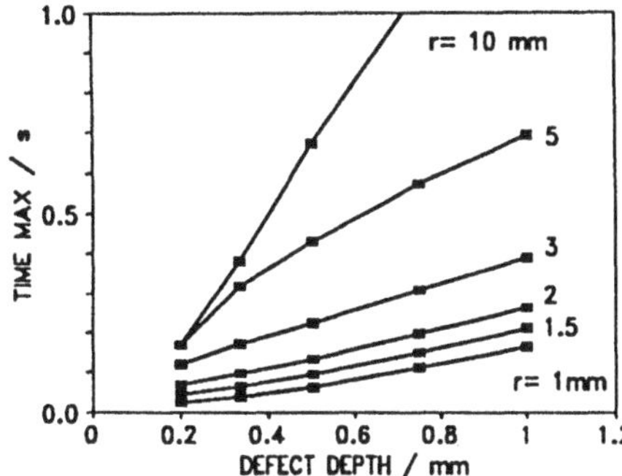

Figure 3. The variation of the time of maximum contrast with defect depth for a range of defect radii. Defect thickness is 0.1 mm.

defect thickness ΔT_m rises to a maximum and exhibits a non-linear relationship with defect thickness. The saturation in contrast occurs because of the loss in the effectiveness of the air gap to conduct heat through the air gap as its thickness increases. It is observed however that t_m shows a near square root dependence with defect thickness.

The contrast response from disc air gap defects is in general complex. There appears to be no simple linear or power relationship with both ΔT_m and t_m to the defect geometric parameters: depth, defect thickness or its radius. In figure 3 a plot of t_m against defect depth is presented for a range of defect radii. For a given defect depth t_m varies with defect radius. This variation increases as the defect depth increases. For small defect radii it appears that t_m is proportional to defect depth for the limited range defect depth considered here: 0.2 to 1mm.

Summary

Two parameters are identified in the analysis of pulsed thermal transients. These are the maximum temperature contrast and time for maximum temperature contrast. Experimentally t_m is usually preferred for quantitative analysis as this parameter is unaffected by spatially varying optical absorption or emissivity on the sample surface. ΔT_m however is strongly affected by these sample surface variations.

Of these two variables the t_m parameter shows some degree of correlation with defect depth and thickness for planar air gaps of infinite width. The use of t_m for correlating with defect depth with data acquired from finite size defects should be used with caution as it is strongly dependent on the defect size.

Acknowledgement

The authors would like to acknowledge the financial support of PowerGen and Science and Engineering Research Council.

Reference

1) P. Potet, Proc. SPIE 1320 p251 (1990)
2) A.C. Tam, Infrared Phys 25(1/2) p305 (1985).
3) W. Reynolds, Brit. J. NDT 26(1) p40 (1984)
4) S.K. Lau, D.P. Almond and P.M. Patel Springer Opt Sci 62 p79 (1990)
5) P.M. Patel and S.K. Lau, Rev. Prog. QNDE 9A p517 (1990)

Contrast Analysis in Thermal Wave Materials Inspection. Part II: Periodic Excitation

P.M. Patel, S.K. Lau, and D.P. Almond

School of Materials Science, University of Bath, Claverton Down, Bath, Avon BA2 7AY, UK

A study is undertaken using phase difference analysis to identify correlations of harmonically varying surface temperature with defect depth, thickness and width. It is found that power relationship exists for planar air gaps of infinite width with: the defect depth, defect thickness and the frequency at which the phase difference between defect and defect free region is a minimum. Phase contrast variation of defects of a finite width is also examined using finite difference numerical analysis.

Introduction

In part I of this paper we examined the temperature contrast behaviour from subsurface defects in homogeneous media with pulsed surface heating [1]. Attention is now turned to periodic surface heating to see if any correlation exists between the measured experimental data, in particular the phase angle, with defect parameters such as its depth, defect thickness or width.

Theory

We assume plane periodic heating of an optically opaque homogeneous medium. The periodic surface temperature variation with frequency, f, of a sample containing a slab like air gap defect lying at a depth L_1 of thickness L_2 is given by the expression [2] :

$$T(f)=\frac{I_o}{2k_1\sigma_1}\left[\frac{1+\Gamma\exp(-2\sigma_1L_1)}{1-\Gamma\exp(-2\sigma_1L_1)}\right] \qquad \Gamma=\frac{\Gamma_1+\Gamma_2\exp(-2\sigma_2L_2)}{1+\Gamma_1\Gamma_2\exp(-2\sigma_2L_2)} \qquad \Gamma_1=\frac{1-b_1}{1+b_1} \qquad \Gamma_2=\frac{1-b_2}{1+b_2}$$

$$b_1=\sqrt{\rho_2c_2k_2/\rho_1c_1k_1} \qquad b_1=\sqrt{\rho_3c_3k_3/\rho_2c_2k_2} \qquad \sigma=(1+j)\sqrt{\pi f/\alpha}$$

where ρ, c, k, and α are respectively the density, specific heat, thermal conductivity and thermal diffusivity. σ is the thermal wavenumber. The indices 1,2 and 3 refer to the coating, air gap and substrate respectively.

Figure 1 illustrates the phase contrast variation with frequency of an air gap in stainless steel at a depth of 0.5 mm for a range of air gap thickness. The characteristic phase variation namely the phase advance to phase lag with increasing frequency is clearly evident.

It is observed that a plot of the frequency at which the phase lag is a minimum, f_m, with defect depth or defect thickness in a logarithmic manner, fig.2a and b, showed a near linear variation . This observation has not been noted previously in frequency domain defect analysis [1,3]. However this kind of relationship has been identified in transient thermal wave defect detection [1,4]. The same behaviour has been found for air gap defects in aluminium and mild steel as shown in figure 2.

To obtain the phase contrast from defects of a finite width a numerical finite difference model was utilised [5]. A grid of equally spaced nodes is used to discretize the sample. In the present analysis a mesh of 50 by 50 is used with a grid spacing of 0.1 thermal

Springer Series in Optical Sciences, Vol. 69 Photoacoustic and Photothermal Phenomena III
Editor: D. Bićanić

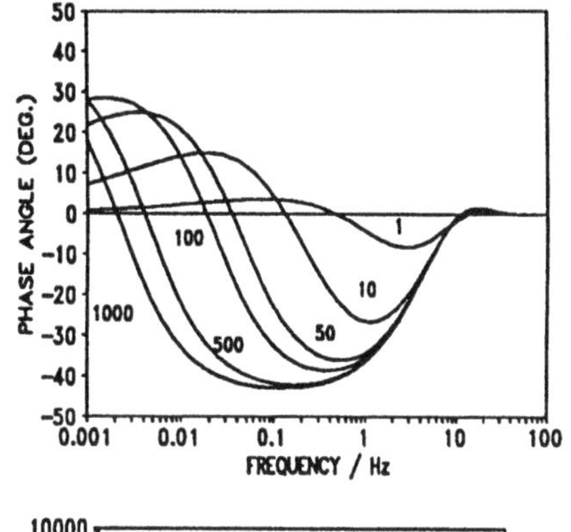

Figure 1: A plot of the phase contrast with frequency for air gap defects at a depth of 500 μm below the sample surface in stainless steel. Defect thickness are 1,10,50,100,500 and 1000 μm.

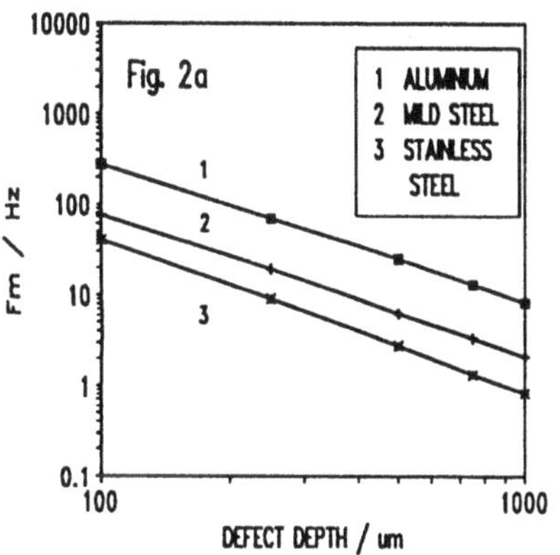

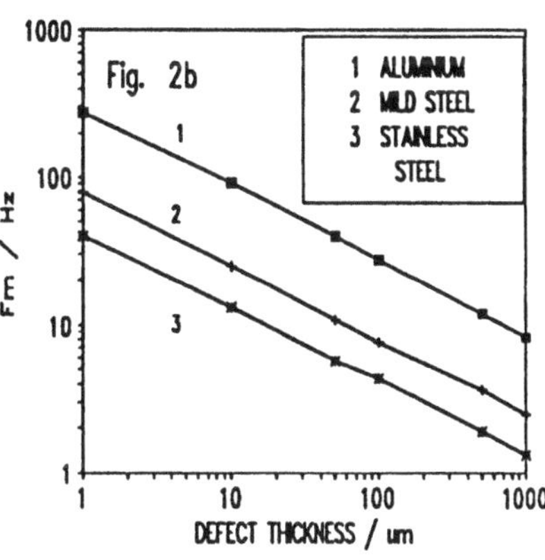

Figure 2a. A plot of f_m against defect depth. L_2 is 0.1 mm
Figure 2b. A plot of f_m against defect thickness. L_1 is 0.5 mm

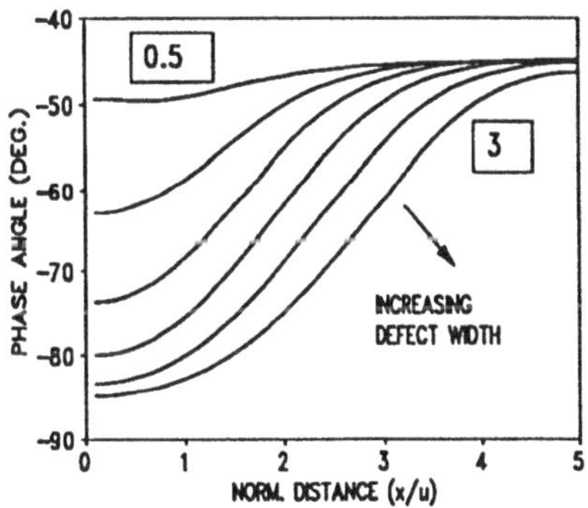

Figure 3: Phase contrast variation with defect width of 0.5, 1, 1.5, 2, 2.5 and 3 thermal diffusion length in stainless steel.

diffusion length. A gaussian elimination algorithm for banded matrices is used to solve complex set of linear equations in double precision arithmetic [6].
Figure 3 shows the phase angle variation of a strip air gap defect at a frequency of 9 Hz for several defect width assuming plane periodic heating.
The depth and thickness of the air gap defect is 0.5 thermal diffusion length. Its width is scaled in thermal diffusion length. The reduction in phase contrast with decreasing defect width is readily apparent from this figure. The reduction in the phase contrast results from the increased thermal diffusion in and around the defect. Thus significantly

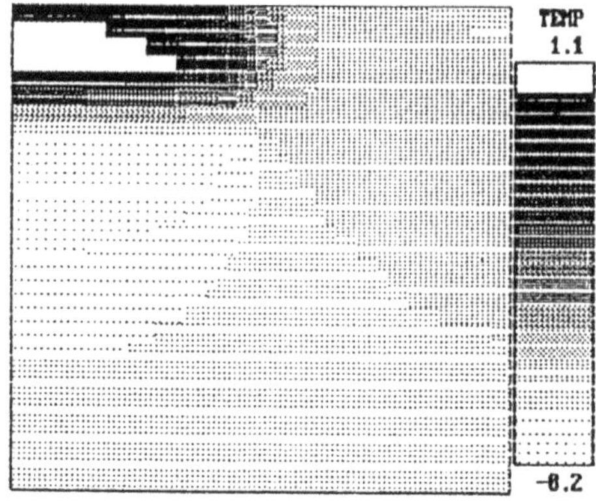

Figure 4. Internal amplitude temperature difference image of a strip air gap defect in stainless steel. Depth and defect thickness 0.5 thermal diffusion length. Defect half width 2.5 thermal diffusion length. Frequency = 9 Hz

weakening the contribution of the reflected thermal waves from the defect to constructively interfere with the surface generated thermal waves. This point is further highlighted in the internal amplitude difference image presented in figure 4. The amplitude difference is constructed from the column subtraction of the temperature values from the defect and defect free region.

Summary

A correlation is found to exist between with defect depth and defect thickness with frequency at which the phase contrast is a minimum for planar air gaps of infinite width. For defect depth $f_m \propto L_1^{-1.5}$ and for defect thickness $f_m \propto$ to $L_2^{-0.46}$. The variation of f_m with defect width appears to be non-linear.

Acknowledgement

The authors are grateful for the financial support of the Science and Engineering Research Council and PowerGen.

References

1) S.K. Lau, P.M. Patel and D.P. Almond, Submitted 7th PAS Conference
2) P.M. Patel and D.P. Almond, Appl. Phys. B43 p9-15 (1987)
3) F.Lepoutre, D. Fouriner and A.C. Boccara J. Appl. Phys 57(4) p1009-1015 (1985)
4) V.P. Vavilov, Proc. SPIE 1313 p178-182 (1990)
5) P.M. Patel and S.K. Lau, Rev. Prog. QNDE 9A 517-524 (1990)
6) R.A. Adey and C.A. Brebbia, "Basic Computational Techniques For Engineers", p61, Pentech Press, London ISBN 0-7273-0204-3 (1983)

Transient Thermal Wave Scattering from Disc Air Gap Defects in Homogeneous Media

P.M. Patel, S.K. Lau, and D.P. Almond

School of Materials Science, University of Bath,
Claverton Down, Bath, Avon BA2 7AY, UK

The alternating direction implicit (ADI) finite difference numerical method is applied to study transient thermal wave scattering from a disc air gap defect lying in a plane parallel to the sample surface in a homogeneous medium.

Introduction

Simple defect geometries which have been numerically or analytically studied in transient thermal inspection include:- the planar air gap interface or contact resistance, the strip air gap defect, and the cylinder and disc air gap defect [1-3]. Finite difference or finite elements methods have been applied to obtain the transient thermal response because of their ease in application to handle heat diffusion in media of varying geometry. In this paper we report on the use of a finite difference method to study transient thermal wave scattering from a disc air gap defect in a homogeneous media.

Numerical Model

The alternating direction implicit (ADI) finite difference algorithm is utilised to solve the heat diffusion equation. The steps involved in the application of this technique are as follows [4]. The sample to be modelled is represented by a set of nodes.

At these node positions the finite difference formulation of the heat diffusion equation is applied in two or three "half" time steps for each of the heat diffusion directions. A set of linear equations are obtained in terms of the present and future half time step. The equations obtained are tri-diagonal and thus can be solved rapidly using the Thomas algorithm. The numerical code developed by the authors is written in Fortran using

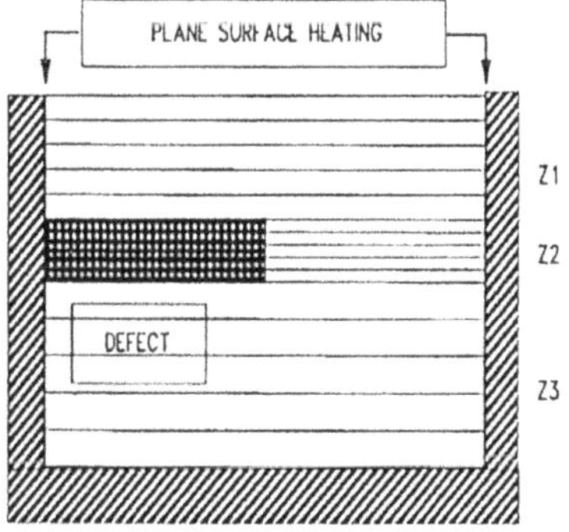

Fig. 1. Finite difference model of disc defect in homogeneous media. Variable spacing in z direction z1,z2 and z3. Hashed region insulated boundaries.

double precision arithmetic and the finite difference grid is over a 100 by 100 node points.

Figure 1 shows the defect geometry and boundary conditions employed in the disc defect modelling. The sample is assumed to be heated by a plane continuous heat source. Only half of the total sample is discretised due to symmetry. Insulating boundary conditions are applied at the symmetry boundary and at boundaries remote from the defect other then the surface.

Discussion

A gray level map of the internal temperature distribution with respect to the semi-infinite variation in a stainless steel sample containing a disc defect at a depth of 0.5mm of radius 5mm at a time of 0.25 second is shown in figure 2. The thermal barrier effect of the defect can be seen as a hot region immediately above the defect and a cold region below it. The lateral heat diffusion effects over the defect radius are also visualised in this figure.

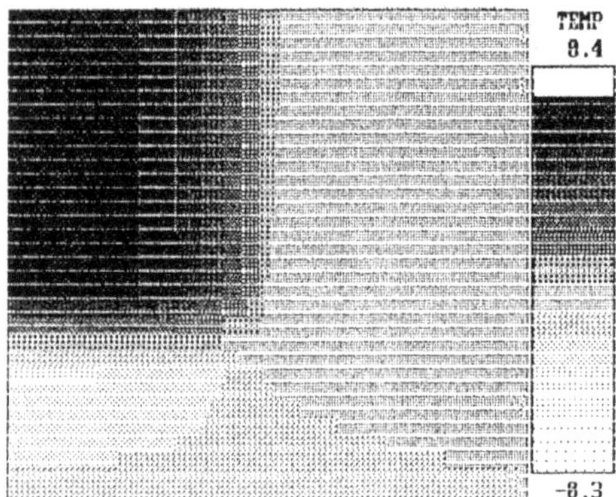

Fig. 2. Internal temperature difference map of a disc defect in stainless steel. Defect radius, depth and thickness are 5, 0.5 and 0.1 mm respectively. Pulsed heating. Time =0.4 s.

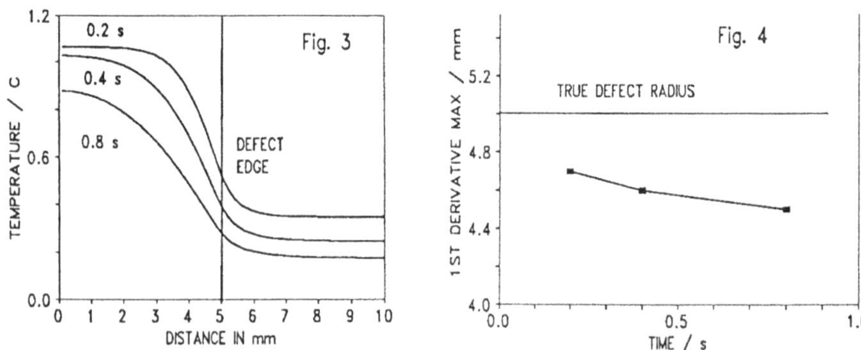

Figure. 3. Surface temperature variation with the radial distance at the times indicated on the figure. Defect parameters identical to those given in fig. 2. (Pulsed surface heating.)

Figure 4. 1st and 2nd spatial derivative of the surface temperature at 0.4 s.

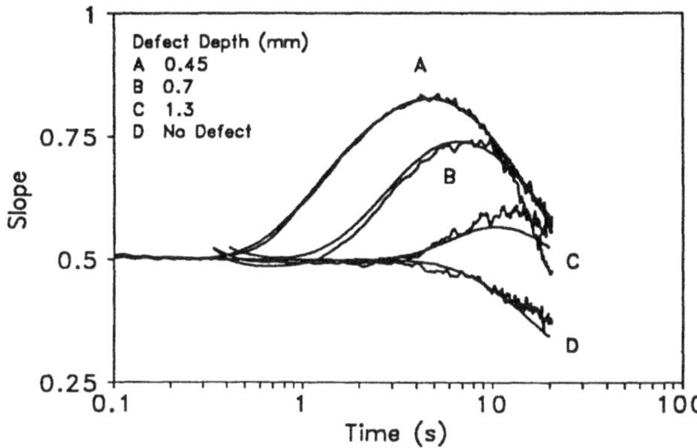

Figure 5. A plot of the step slope response with time for a cylindrical flat top defect in Bakelite. Solid line finite difference fit. Surface flux is modelled as a disc of 6 mm in diameter.

Figure 3 illustrates the surface temperature variation of the above defect sample at three time steps (continuous surface heating). At early times limited heat diffusion has taken place and the surface temperature profile has a flat top. With increasing time increased heat diffusion takes place. Interaction with the defect occurs resulting in a profile that peaks at the defect centre and decays along defect radius.

The spatial temperature derivative is often used in thermographic image analysis to locate defect edges. Figure 4 shows a plot of the variation of the maximum of the 1st derivative at three time steps: 0.2, 0.4 and 0.8 seconds. From this figure it can be seen that the location of this maximum is close to the defect edge at short times. At long times the position of this maximum under estimates the location of the defect edge [5].

Three common methods used to quantify the temporal response from a defect region are: a) the logarithmic temperature - time variation, b) temperature contrast and c) the slope of the Log temperature vs Log time plot. Of these methods, data analysis based on method (c) offers increased sensitivity to subsurface defect parameters. As a test of the application of the above numerical technique figure 5 presents experimental data together with a slope fitting of the step pulse slope verses time response to a defect test sample. The defect is a flat bottomed drilled hole in Bakelite.

Various defect depths have been examined for a single defect diameter of 10 mm and the temperature measurements are taken from the defect centre. The correlation of the numerical analysis with the experimental data is very good. Heat loss effects were of small importance in comparison with the influence of a finite heat source size.

Summary

A brief outline of the application of numerical modelling to study transient thermal wave scattering, image formation and defect detection has been give. Its application in transient thermography in solving the forward inverse problem and thermal tomography has been demonstrated on a simple defect geometry sample.

Acknowledgements

The authors wishes to acknowledges the financial support of PowerGen and the Science and Engineering Research Council.

References

1) P. Cielo
In Proc. 14th Symp. On NDE, San Antoino, (USA) (1983)
2) J. Rantala and J. Hartikainen
Rev. Prog. QNDE. 10A p1051-56 (1991)
3) S.K. Lau, D.P. Almond, P.M. Patel, J. Corbett and M.B.C Quigley
Proc. SPIE 1320 p178-189 (1990)
4) P.M. Patel and S.K. Lau
Rev. Prog. QNDE 9A p519-524 (1990)
5) P.M. Patel, S.K. Lau and D.P. Almond
J. Nondest. Testing and Evaluation (in press)

Dimensionality Viewed from a Thermal Perspective

L.C. Aamodt

Applied Physics Laboratory, Johns Hopkins University,
Johns Hopkins Road, Laurel, MD 20723, USA

Abstract. In an unbounded space the rate that heat flows away from an isolated heated point depends upon the dimensionality of the space. This can be qualitatively related to the number of possible heat paths available for heat to flow. In a bounded space where heat flow is constrained, heat flow at a particular point can be used to define an effective dimensionality of the space at that point.

1. Introduction

The practical world in which we live is three dimensional. Nevertheless, in certain cases, we consider objects in this world to have a lower dimensionality. For example a thin disk or a long rod are often assumed to be two dimensional and one dimensional, respectively, and their behavior seems to justify this assumption. These objects are sometimes viewed as sections or subsets of the three dimensional world in which they exist.

It is possible at times to distinquish individual objects in our three dimensional world and consider them independent of their surroundings. Thus internal motion in a body rigidly clamped at all surfaces is independent of what happens in its external environment if we deal only with mechanical effects. In this case the body can be described as a *bounded* space within the *unbounded* space in which it is located.

In a similar manner if we consider only heat flow in an object, we can speak of an isolated "thermal space" surrounded by regions of zero conductivity isolating it from the surrounding media. Further complications can be introduced by inserting internal volumes, areas, lines, and points of zero conductivity. These spaces can be considered to be *bounded* spaces in contrast to an *unbounded* thermal space where heat can flow in all directions.

2. Thermal Spaces

If an instantaneous heat pulse is applied at a single point in *unbounded* space, the rapidity with which heat diffuses is governed, in part, by the dimensionality of the space. For example, the temperature at the heated point is proportional to $t^{-n/2}$ [1] where n is the dimensionality of the space and t is the elapsed time following the pulse. Solving for n, it is found to be the logarithmic differential, $n = -2\, t/T\, dT/dt$. This temporal dependence provides a means for determining the dimensionality of an *unbounded* space should it not be known, simply place a heat charge [1] in the otherwise unheated space and note how rapidly the temperature at the point of heating decreases.

The rapidity with which heat flows away from the heated point can qualitatively be associated with the number of directions available for heat to flow. As the number of dimensions increase, the number of possible directions for heat to flow also increases, and heat diffuses more rapidly.

The question arises as to whether the rapidity of heat flow in a *bounded* space -- where heat flow is limited or constrained -- can be used to indicate what might be called an "effective" or "thermal dimensionality." To answer this question,we consider dimensionality to be an active agent that helps determine how heat flows and not just a stage (so to speak) on which heat flows.

3. Bounded Thermal Spaces

To procede, we define the "thermal dimensionality" in a *bounded* space to also be n= -2 t/T dT/dt where T is the temperature produced by an instantaneous delta-function heat pulse at the point of application when no other heat source is present. As defined, this dimensionality is localized and can vary from point to point. Under some circumstances, it can also vary with time. This is in contrast to an *unbounded* space where the thermal dimensionality is the same at all points and is independent of time.

To illustrate, consider spaces bounded in various ways. The n-dimensional space shown in Fig. 1a is semi-infinite in one dimension, and infinite in the remaining dimensions. Fig. 1b shows how the thermal dimensionality varies with time at a test point near the boundary. Initially the dimensionality corresponds to that of an *unbounded* space, however, as diffusing heat senses the presence of the boundary, the thermal dimensionality decreases by approximately half a dimension (which is appropriate since one of the two directions in the semi-infinite dimension has been blocked). As heat dissipates, the thermal dimensionality returns to its initial value.

If the space is now bounded on both sides (in one dimension) by insulating regions so that it has a finite width in this dimension, heat is further constrained. The temporal pattern of its thermal dimensionality at the test point shown in Fig. 2a can be seen in Fig. 2b. Here, after heat reaches the spatial boundaries, the thermal dimensionality decreases by one, from n to n -1 and remains at this value. The heating pattern in this space can be simulated by locating an infinite series of image charges equally spaced in the insulating regions. The analytical solution for this case (which is too lengthy to be include in these proceedings) shows that as width of the finite dimension approaches zero there is a smooth transition from an n dimensional to an n -1 dimensional thermal space. Effectively one dimension is "squeezed out."

As indicated, the thermal dimensionality can vary from point to point in a *bounded* thermal space. This is illustrate by the curve $\kappa_0/\kappa = 0$ in Fig. 3 where the test point has been moved closer to one boundary. Here, as heat reaches the nearer boundary, the thermal dimensionality dips to approximately n -.5 (one direction blocked), partially recovers, and then dips to n - 1 (both directions blocked).

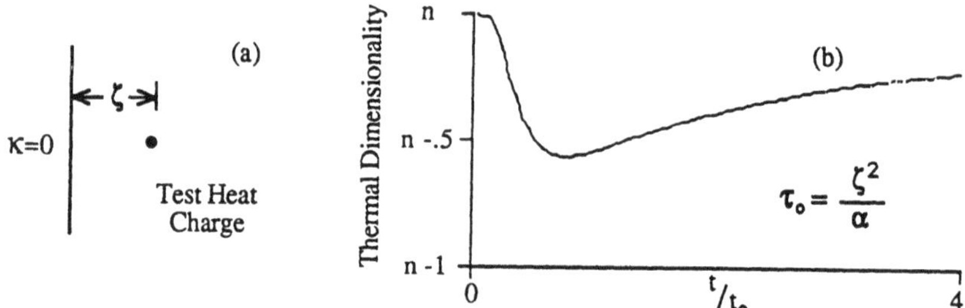

Fig. 1 a) The vertical line represents the edge of an n-dimensional space bounded in one diimension by a region of zero conductivity. b) The thermal dimensionality as a function of time at a test point near the boundary edge.

4. Topographical View of Thermal Spaces

The space shown in Fig.3 can also be interpreted as a *bounded* thermal space with spatially varying thermal properties. These variations in conductivity can be viewed as partially blocking or bounding the space. Topologically the space can be said to have conductivity valleys, hills, and ridges that restrict, or channel, heat flow. These features affect the number of effective paths available for heat to flow away from a point, and thus, point by point, change thermal dimensionality, throughout the space. From this viewpoint, the space in Fig.3a can be interpreted as being a conductive mountain ridge with precipitous sides and the ratio κ_0/κ can be considered to be the conductivity ratio between the mountain peak and the surrounding valley. The thermal dimensionality at a point near one edge of the mountain is shown in

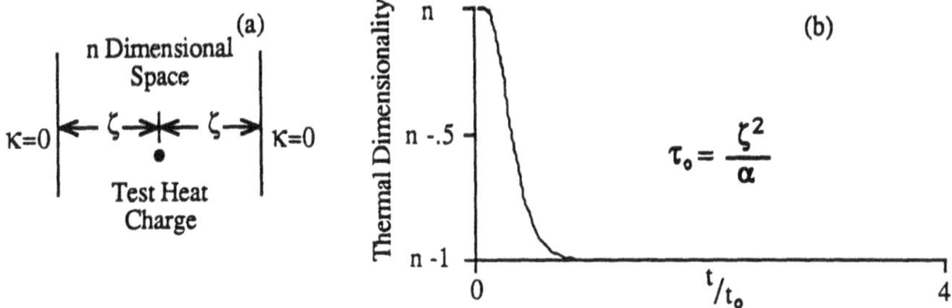

Fig. 2. a) An n-dimensional space bounded on both side in one dimension by regions of zero conductivity. b) The thermal dimensionality as a function of time at a test point symmetrically placed with the bounded dimenson.

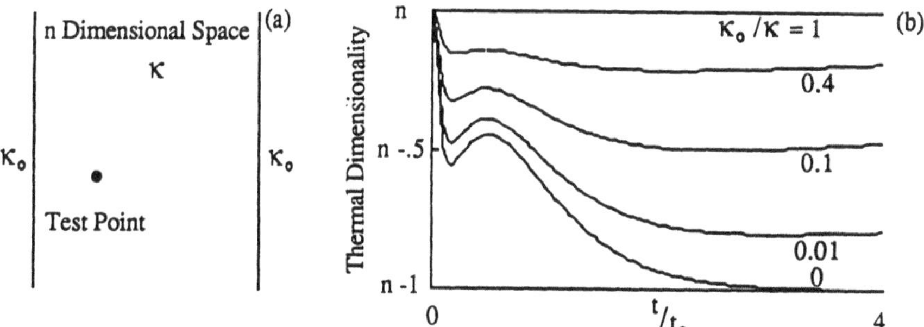

Fig. 3 a) This geometry can be interpreted as an n-dimensional space bounded on both sides by regions with finite conductivity or as an unbounded space with an internal region of conductivity κ surrounded by a region of conductivity, κ_0. b) The thermal dimensionality as a function time at the test point shown in (a).

Fig.3b. As heat reaches the nearer boundary the dimensionality dips (but not to n -.5 as it did in Fig.1), partially recovers, and then dips again (but not to n - 1 as it did in Fig.2). The thermal dimensionality eventually returns to its initial value. This return appears to be associated with the opportunity that heat has to flow to infinity in at least one direction in the bounded or limited dimension. When $\kappa = \kappa_0$, the barrier disappears and the dimensionality remains constant.

Reference

1. H. S. Carslaw, J. C. Jaeger: *Conduction in Solids , 2nd ed.* (Oxford Univ. Press, New York 1959) pp. 255-272

Thickness Determination for Two-Layer Systems of Paint on Polymers with Photothermal Radiometry

W. Karpen[1], *A. Bohnacker*[1], *H.G. Walther*[2], *K. Friedrich*[2], *U. Seidel*[2], *and G. Busse*[1]

[1]Institut für Kunststoffprüfung und Kunststoffkunde, Universität Stuttgart, Pfaffenwaldring 32, W-7000 Stuttgart 80, Fed. Rep. of Germany

[2]Institut für Optik und Quantenelektronik, Universität Jena, Max-Wien-Platz 1, O-6900 Jena, Fed. Rep. of Germany

Abstract. Thickness determination of paint is of interest for industrial applications. While there are conventional methods if the substrate is a metal, only thermal waves can be used if the substrate is a polymer. The purpose of this paper is to extend the work performed on one-layer systems to make it applicable to a two-layer system. Results are shown for systems where either both layers are opaque or where the upper layer is transparent. A method is discussed to determine both thicknesses.

1. Introduction

The application of photothermal radiometry for the inspection of polymer coatings (e.g. paint) on polymer substrates has been described previously /1, 2/. Thickness determination has been performed by comparing the measured phase angle with a curve obtained for step-like cross-sections of known thickness.

However, most coatings consist of more than just one layer. To monitor multilayer systems, one could apply the technique successively for each layer on the production line. Such an approach would not allow to inspect and analyse multilayer systems after they are produced. Therefore we tried to find out which information can be obtained about two-layer coatings deposited on a polymer substrate where the thermal wave reflection coefficient is small at each boundary.

2. The signal surface method

The phase angle of the photothermal signal is a function of both layer thicknesses d_1 and d_2. For calibration one needs samples where each of the two layers shows step-like variations. At a given modulation frequency f_1 this provides a calibration surface φ (f_1, d_1, d_2) where signal phase φ depends on two coordinates d_1 and d_2 representing the two thicknesses. However, if an unknown sample is measured one obtains a certain phase φ_1 which may fit to many combinations of values for d_1 and d_2. These combinations are represented by the intersection of the φ_1-plane with the signal surface φ (f_1, d_1, d_2) and the solid line in the d_1, d_2- plane obtained by vertical projection of this intersection, see fig. 1 ("equal phase-line"). To eliminate this uncertainty one may repeat this process at a different modulation frequency f_2 providing another signal surface φ (f_2, d_1, d_2) from which another intersection and projection is obtained for the observed angle φ_2 (dashed line in d_1, d_2-plane). The point where they cross each other in the d_1, d_2-plane gives the thickness combination which is d_1 = 42 µm and d_2 = 40 µm in the example of fig. 1. The angle between the equal phase-lines may be quite small, also they may be broadened to equal phase areas due to experimental errors in the height of the φ-planes. Thereby the error intervals for d_1 and d_2 may be broadened.

Springer Series in Optical Sciences, Vol. 69 Photoacoustic and Photothermal Phenomena III
Editor: D. Bićanić

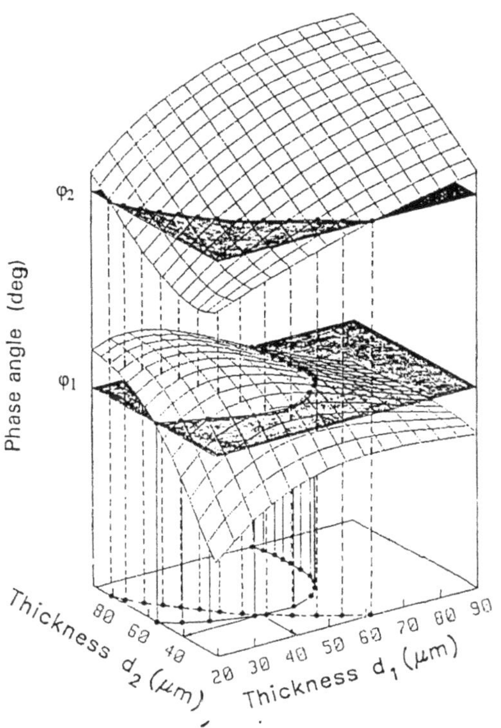

Fig. 1: Signal surfaces φ (f_1, d_1, d_2) and φ (f_2, d_1, d_2) shown together with intersecting planes φ_1 (shaded, bottom) and φ_2 (shaded, top), respectively. Crossing of projected intersection curves (equal phase-lines, solid line for f_1 and dashed line for f_2) provides thicknesses d_1 and d_2.

3. Results

The shape of the signal surfaces depends on the nature of the layers, therefore the phase lines are not always as different as in fig. 1. The result in fig. 2 was found for an unknown thickness combination of two opaque layers. Known thickness values used for calibration (circles) are shown in the d_1, d_2-plane together with three (instead of two) projected equal phase areas obtained at 2.25 Hz, 4.5 Hz, and 18 Hz. The overlapping area of the three areas crossing each other at small angles indicates that d_1

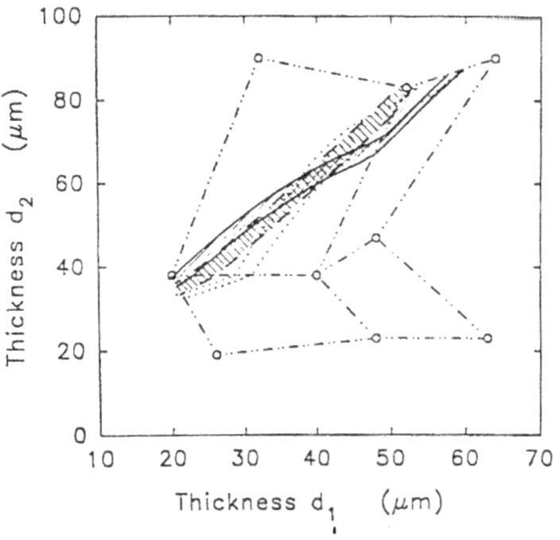

Fig. 2: Intersection of three equal phase-areas (shaded) for two opaque layers in d_1, d_2-plane. Open circles are positions of calibration points.

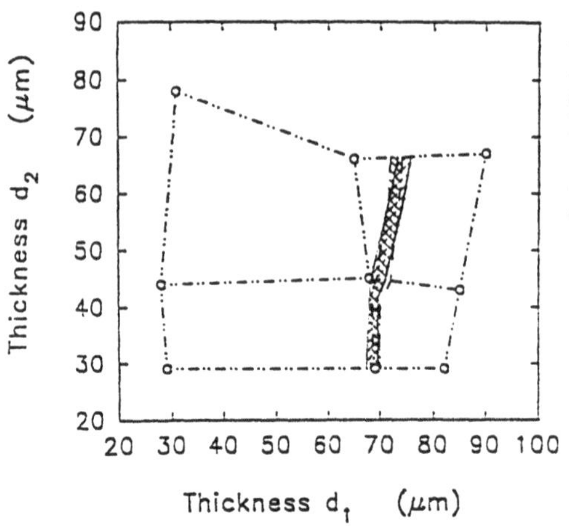

Fig. 3:
Intersection of equal phase-areas for one trans-parent layer on top of an opaque layer (see fig. 2). Open circles are positions of calibration points.

= 40 μm and d_2 = 60 μm are the probable thickness combination. Obviously the error is reduced by using more than only two modulation frequencies or by making the difference between the modulation frequencies larger.

If the upper coating is transparent for the optical radiation the experimental condition is quite different since the thermal wave is generated essentially between the two layers. Therefore the signal at the sample surface is dominated by the directly transmitted part while the reflection at the substrate may be small. Then the signal surfaces have their gradients essentially along one axis giving a projection of equal-phase-lines nearly parallel to d_2 at an extremely small angle. In this case the thickness of the upper layer may be obtained already at one frequency while the second frequency does not help to reduce the uncertainty for the thickness d_2 of the lower layer (fig. 3).

4. Conclusion

Our results show the applicability of the method to opaque two-layer-systems. The expansion to three (or more) layers is possible in principle. However, due to the small reflection coefficients the phase angle changes are generally small giving large errors that will increase with the number of layers.

References

/1/ Busse, G., Vergne, D., Wetsel, B.: In: Photoacoustic and Photothermal Phenomena, ed. by Hess, P.; Pelzl, J. (Springer, Berlin, Heidelberg 1988), pp. 329-332

/2/ Busse, G., Karpen, W.: In: Neue Entwicklungen bei der Lackierung von Kunststoffen, DFO, Düsseldorf 1990), pp. 204-212

Thermal Diffusivity Measurement of Diamond Materials

L. Wei, P.K. Kuo, and R.L. Thomas

Department of Physics and Institute for Manufacturing Research,
Wayne State University, Detroit, MI 48202, USA

Abstract. The thermal wave/mirage effect technique is applied to the determination of thermal properties of polycrystalline diamond slabs and single crystal diamonds of different isotopic compositions. This method is well-suited for these kinds of materials.

1. Introduction

Diamond as a material has many outstanding physical properties. The rareness of large natural single crystals and the high commercial value has prevented large scale applications of many of diamond's usefulness. The only exception is the use as abrasive material of synthetic diamond powder made by high pressure and high temperature processes with a volume exceeding several hundred tons per year. There are two recent developments that are changing the situation. The first is the rapid progress in the technique of making diamond by low temperature, low pressure processes, as witnessed by the increasing number of scientific conferences and publications dedicated to this topic. These processes promise not only high growth rate and large area production of polycrystalline diamonds, but also an end product which has physical properties approaching those of pure diamond crystals. The second is the recent discovery[1] that isotopically enriched diamond crystals are found to have an extraordinarily high thermal conductivity. Single crystal diamonds with 99.9% C^{12} are found to have a thermal conductivity about 50% higher than the best natural diamond crystals which contains 99.1% C^{12}. Since diamond is already the best thermal conductor at room temperature (five times better than copper, a commonly used heat sink material) this fact has obvious scientific and industrial implications.

In order to help optimize the production process, it is important the thermal properties of diamond material be determined accurately, quickly and preferably non-destructively. The thermal wave/mirage effect method developed by the authors in recent years[2,3] has proved to meet this challenge. We report the results of employing this method to measure the thermal diffusivity of thin slab polycrystalline diamonds at room temperature and single crystal diamonds of various isotopic contents over the temperature range 80-300K.

We also introduce a modified Angström's method in which the in-solid mirage effect is used to measure directly the temperature gradient in the interior of a diamond crystal. This method offers better accuracy in low temperature range where the the diffusivity of diamond becomes exceedingly high.

2. The Experimental Methods

The thermal wave/mirage effect method for the measurement of thermal diffusivities of layered materials has been reported in detail.[2,3] Briefly, it uses a modulated laser beam focused on the surface of the sample as an alternating heat source to launch a thermal wave in the solid and the surrounding air. A second laser beam is bounced from the surface to detect the temperature gradient in the heated region. Because this method is non-contact, and it can measure distance and time with great precision, it is very suitable for the determination of thermal diffusivities of thin films or thin slab materials.

For single crystal diamond samples, a thin layer of metal film has to be deposited on the surface to act as heat absorber. The film is necessary for any method which requires optical

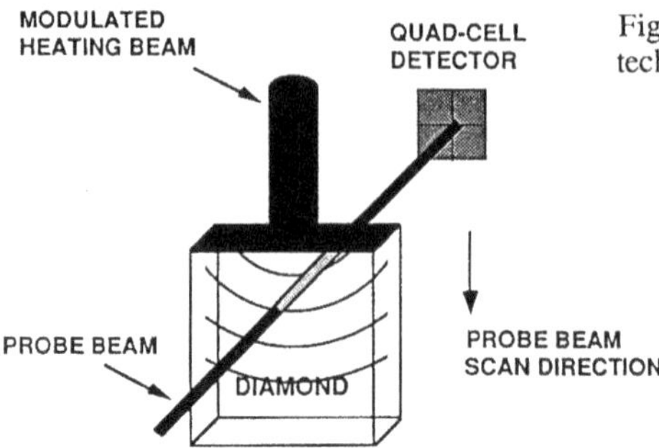

Fig. 1 Schematic diagram of the in-solid-mirage technique in the modified Angström method..

heating or infrared detection. At very low temperature, the diffusivity of diamond is expected to rise about three orders of magnitude over its room temperature value. This makes the presence of a metal film not so negligible. For this reason we have resurrected a method used by Angström more than a hundred years ago[4]. Angström used hot and cold water to launch a thermal wave from one end of a sample and used thermocouples attached to varies locations along the sample to measure the phase and amplitude of the thermal wave as a function of distance of propagation to determine the thermal diffusivity of the sample. In our modified method, the in-solid mirage method is used to replace the thermocouples as the temperature probe. The schematic of the experimental arrangement is shown in Fig. 1. The sample is mounted inside a liquid helium dewar with three optical windows. The He-Ne laser for probe beam and the mirage detector are amounted in a single unit called the Monobloc developed by Charbonnier et al[5]. The use of the Monobloc greatly reduces the difficulty of scanning the probe beam relative to the sample and reduces low frequency noises from mechanical vibrations.

3. Results

We have applied our mirage-in-air technique to a variety of polycrystalline diamond slabs produced by a wide varieties of chemical vapor deposition processes. The resulting thermal diffusivities varied widely with process parameters. The internal consistency of this method

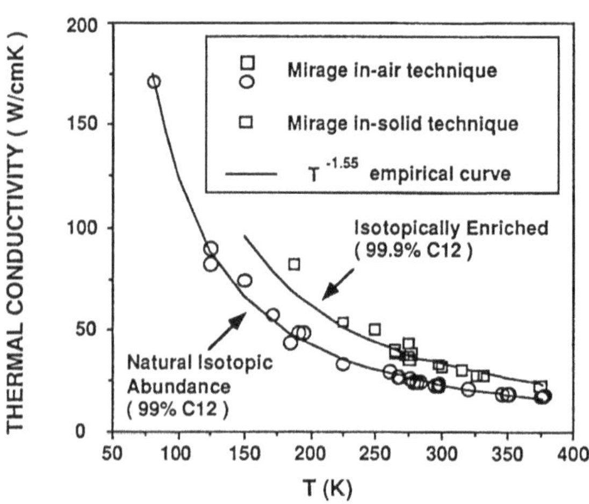

Fig. 2 Results of the mirage in solid and mirage in-air techniques in the measurement of thermalconductivity of single crystal diamonds of different isotopic composition.

has provided manufacturers invaluable information to improve their process control. The comparison with other methods for the determination of thermal properties is currently under way.[6]

We have applied our mirage-in-air technique to single crystal diamonds of different isotopic compositions in the temperature range of 80-300K. The results are summarized in Fig .2.

4. Conclusions

We have demonstrated that the thermal wave/mirage technique is suitable for the determination of thermal properties of diamond materials and it is providing a valuable service to help optimize the production processes.

5. Acknowledgements

This work was partially sponsored by ARO under Contract No. DAAL03-88-K-0089; The General Electric Corporation R&D Center and by the Institute for Manufacturing Research, Wayne State University. We also would like to acknowledge the use of sample supplied by the General Electric Corporation R&D Center and the Double Dee Harris Co.

REFERENCE

1. T.R. Anthony, W.F. Banholzer, J.F. Fleischer, Lanhua Wei, P.K. Kuo, R.L. Thomas, and R.W. Pryor, Phys. Rev. **B42,** 1104 (1990).
2. P.K. Kuo, Lanhua Wei, R.L. Thomas, and R.W. Pryor, Springer Series in Optical Sciences **62**, 124-126, Springer-Verlag, Heidelberg (1990).
3. P.K. Kuo, M.J. Lin, C.B. Reyes, L.D. Favro, R.L. Thomas, D.S. Kim, Shu-yi Zhang, L.J. Inglehart, D. Fournier, A.C. Boccara, and N. Yacoubi, Can. J. Phys., **64**, 1165, (1986); P.K. Kuo, E.D. Sandler, L.D. Favro, and R.L. Thomas, Can. J. Phys., **64**, 1168, (1986).
4. M.A.J. Angström, Philo. Mag. **25**, 181, (1863).
5. F. Charbonnier and D. Fournier, Rev. Sci. Instrum. **57,** 6, (1986).
6. G. Lu, and W.T. Swann, First International Conference on the Applications of Diamond Films and Related Materials, August 17-22, 1991, Auburn, Alabama.

Part IV

Ultrasound and Kinetics, Including Ultrafast Phenomena

Femtosecond Thermoreflectivity

A. Lörincz[1], *R.A. Bakker*[2], *A. Miklós*[1], *and J. Somlói*[1]

[1]Institute of Isotopes, Department of Photophysics,
P.O. Box 77, H-1525 Budapest, Hungary

[2]Delft University of Technology, Department of Physics,
P.O. Box 5, 2600 AA Delft, The Netherlands

Abstract. Approximations treating femtosecond thermoreflectivity are considered. Femtosecond transient reflectivity curves are computed within the hydrodynamic approximation for describing the behaviour of the material in the free-electron domain. Dominant electron-phonon relaxations and dominant electron-impurity scattering are considered in non-equilibrated lattice and electronic subsystems.

1. Introduction

It was about a decade ago when transient thermoreflectivity was first used for the study of transient relaxation of charge carriers in metals [1]. Interest in this type of measurement has been steadily increasing ever since. The advent of lasers and the generation of ultrashort femtosecond laser pulses has been giving the impetus for these studies on the one hand. On the other, the appropriate description of the relaxation process itself is still to be understood.

It is only the pump-and-probe technique that is able to provide information on this extremely short time scale. This technique uses two pulses: one (the pump pulse) to perturb the metal and an other one (the probe pulse) to monitor the results of the perturbation. Monitoring is achieved by chopping the pump pulse sequence, and measuring the intensity of the probe pulses in the case when the pump is "on" and in the case when the pump is "off" and then subtracting the results from each other. Subtraction is usually accomplished by some kind of lock-in techniques [2-4]. Adjusting the pathlengths of the pulses one can modify the delay between the pulses and time resolved data may be collected.

The underlying physical processes to be considered are complex and interlaced, and great care should be taken for concluding. To sketch the problem let us consider what happens after excitation: the photon undergoes scattering in the neighbourhood of the surface of the metal, it is partly reflected and partly absorbed. Photon excitation brings the electronic subsystem to non-equilibrium by leaving energy, momentum and excited species or quasiparticles in the sample. The non-equilibrium electronic subsystem will then relax through electron-electron, electron-phonon collisions and electron-impurity scattering. The first question that comes to mind is whether the concept of temperature is useful for the extremely fast femtosecond processes? Or putting it an other way: under what conditions one may consider the absorbed energy as randomized? This problem, that is the set of approximations needed for allowing the use of the concept of temperature in these considerations is overviewed in the next chapter.

2. Ultrafast photoexcitations in metals

There are two distinct properties of absorption of laser pulses in metals The first is the very small penetration depth. The incoming light will be absorbed in a 10-20 nm surface layer. That may result in large gradients for the transport problem.

The second feature is that conduction electrons and the lattice system are coupled. The absorbed energy is thermalized via three separate processes; that is through electron-electron and electron-phonon collisions and electron-impurity scattering. The electron-electron relaxation may be visualized as follows [5]: the photon energy for visible light is around 2-3 eV. The Fermi energy of good conductors is between 5 eV and 10 eV. The broadening of the Fermi distribution at room temperature is about 0.025 eV. The electron subsystem may be regarded as a gas of quasi-particles having wavenumbers approximately up to the Fermi wavenumber $k_F=(3\pi^2N)^{1/3}$ (N is the electron density). The lifetime of the quasiparticles above the Fermi energy $E_F=\hbar^2k_F^2/2m^*$ (m^* is the effective mass) is inversely related to the square of their excess energy $\Delta E=E-E_F$ [5]. The order of magnitude of collision time is such that the first collision of the excited quasiparticle may occur within the first fs. However, the collision time increases very quickly as the excess energy decreases. The majority of energy randomization happens through this process. However, in case of excess energy of only 0.5 eV the collision time is in the order of 10 fs and the electron-phonon collision will be able to compete. The Drude model gives the rough estimate of 5-10 fs for electron-phonon collision time. In electron-phonon collisions, however, the average energy transferred to the phonon system is approximately kT, where k is the Boltzmann constant, that is approximately 0.025 eV at room temperature, and thus a number of electron-phonon collisions are needed for thermalization. Electron-impurity scattering may contribute to the energy randomization as well. The more disordered the system the larger its role is. This type of scattering may give rise to large angle electron scatterings, that is it may be efficient in thermalizing the electron subsystem.

The energy randomization after the absorption of ultrafast laser pulses in metals may now be described in the following way: the absorbed energy is distributed very quickly in the electronic subsystem, but this process slows down as the excess energy decreases. When the electron-electron collision time becomes approximately equal to the electron-phonon collision time, then the excess energy is relaxed to the phonon system. The question is if temperature may be defined during these relaxations. The correct answer is no. In the following we number the different points to be checked if the relaxation process is to be modelled by coupled thermal diffusion equations (the list is based on the nonequilibrium statistical operator formulation of transient problems [6], details may be found in [7]):

(1) If the very first step of the few fs electron-electron collisions is so fast that this short relaxation makes the initial state of the system irrelevant.

(2) If the number of electrons taking part in this fast relaxation process is so large that all the electrons around the Fermi-surface may be assumed to have taken part in the collisions.

(3) If the electron energy distribution may be regarded as a thermal distribution.

(4) If the electronic motion may be modelled as heat diffusion.

(5) If the electronic heat diffusion connected to the electronic motion may be assumed to be much slower than the electronic screening and the electronic subsystem may be considered as neutral.

(6) If temperature variations are small and the kinetic coefficients depend only slightly on the thermodynamic parameters, for example, on variations of pressure and temperature.

(7) If a term proportional to the product of the temperature gradient and the velocity; that is the $v_i\partial_i T$ term of the Navier-Stokes equation is negligible.

Then the heat diffusion equations read as follows [5]:

$$C_e\frac{\partial T_e}{\partial t}=\kappa\frac{\partial^2 T_e}{\partial z^2} - g(T_e-T_i)+H(z,t) \tag{1}$$

$$C_i\frac{\partial T_i}{\partial t} = g(T_e-T_i) \tag{2}$$

where T_e and T_i are the electronic and lattice temperatures, C_e and C_i are the electronic and lattice heat capacities of unit volumes, respectively, κ is the thermal conductivity, g is the coupling parameter between the subsystems and H(z,t) is the heat source density deposited in the sample.

The point worth to emphasize is that the above model may fit the experimental results well, since the relaxations have diffusion like character anyway. However, if the approximations are not valid then the parameters may be very far from the actual values.

3. The hydrodynamic approach in the collisionless regime

It is a delicate task to determine diffusion processes from optical data. This fact relies in the sophisticated dependence of reflection on material parameters around the penetration depth. Just consider the fact that the pump pulse creates an inhomogeneous layer in the neighbourhood of the surface and it is this inhomogeneity that will change the intensity of the probe beam [8,9].

To unfold the problem we suggest to start in the free-electron domain [10]. This approach has the advantage that it separates the free-electron contribution from that of band-to-band transitions. The two contributions show different behaviour due to their different influence on the optical properties [11].

According to Pines and Nozieres [12] the response of a homogeneous electron gas to time-dependent density perturbations is analogous to that of a fluid in two frequency regions: (a) in the many collision regime where $\omega\tau<<1$ (τ being the collision time), and (b) in the collisionless regime where $\omega\tau>>1$. In case (a) the system has enough time to relax to local equilibrium, while in case (b) strong Coulomb coupling makes the electron gas to react as a jelly. In both regimes the hydrodynamic approach is justified but the parameters of the equations are somewhat different. The hydrodynamic model treats [13,14] all degrees of freedom as collective. That is, there are no individual absorption centres which would undergo scattering and nonradiative processes. Rather the electronic subsystem is treated as a continuum giving rise to currents and reflected waves just as well as a relaxation time and energy dissipation through the flowing currents. That is the dissipated energy of the electric field may be treated classically within the hydrodynamic approximation.

In below we shall discuss two extremes, (i) the case when the energy is dissipated within the electronic subsystem, and (ii) the case when relaxation mostly happens through electron-phonon collisions. Since electron-electron collisions conserve momentum they can not change currents involved in the hydrodynamic approach. The electron impurity scattering shall contribute on this end. Electron-phonon collisions are assumed to give rise to relaxation between the electron and lattice subsystems.

Metal surfaces are known as "spatially dispersive" systems, that is as systems of nonlocal response for excitations with electronic fields. The macroscopic approach based on a hydrodynamic approximation for the material equations gives rise to transversal waves, longitudinal plasma waves and the following boundary conditions: continuity of the tangential component of the electric field $\underline{E}$, of the normal component of $\underline{E}$, of the current density $\underline{j}$ and of the energy current density. The material equation in the hydrodynamic approximation [13,14] may be written as:

$$\frac{\partial \underline{j}}{\partial t} = \frac{\omega_p^2}{4\pi}\underline{E} - \gamma \underline{j} - \frac{3}{5} v_F^2 \, \mathrm{grad}\rho . \tag{3}$$

where $\underline{j}$ is the current, $\underline{E}$ is the electric field, ω_p, v_F and ρ are the plasma frequency, the Fermi velocity and the nonequilibrium charge density, respectively, and γ is the damping parameter, the inverse of the current relaxation time τ. Eq. (3) is an extended Drude model containing the spatial dispersion that leads to plasma waves.

It may be worth noting that τ does not necessarily have a direct physical meaning like the electron-phonon collision time, or the electron-electron relaxation time when the electron-

electron, electron-phonon, phonon-phonon collisions are not well separated in the time domain and τ may have contributions from all the different processes. It is introduced here as a phenomenological parameter of a first order approximation. We assume, that τ does not depend on the frequency of the electric field.

This equation together with Maxwell's equations allow the computation of the travelling, interfering and reflected waves. We assumed a few wavelength long infrared pulse travelling towards the material. Numerical values used in the computations are the spectral width of the envelope function, $k_E = 887.2 cm^{-1}$ and the wavenumber of the centre frequency, $k_c = 8872\ cm^{-1}$.

4. Dissipated heat and heat diffusion according to the hydrodynamic approximation

From Maxwell's equations and with the help the material equation one may derive the energy theorem [13]:

$$-\frac{1}{8\pi}\frac{\partial}{\partial t}\left(\underline{E}^2+\underline{H}^2+\frac{4\pi}{f}\underline{j}^2\right) = \frac{c}{4\pi}div(\underline{E}\times\underline{H})+\frac{\gamma}{f}\underline{j}^2 , \tag{4}$$

where $f=\omega_p^2/4\pi$ within the material.

The rate of energy density change given by Eq. (14) may be summarized as follows: the energy of the radiation field changes according to the radiation energy entering the material and according to the ohmic loss in the material. This last term is proportional to $\underline{j}^2$ and will be the source term in the heat diffusion equation.

In the free electron domain the hydrodynamic approach is consistent from the point of view, that it considers the reflection as a consequence of the collective behaviour of the electronic subsystem and considers dissipated energy as damping of the currents flowing. We shall compute the reflection coefficient for two cases: in the first case we assume, that the coupling constant g of the heat diffusion equations is so small that it may be taken as zero. In this case we assume, that damping of the current of the hydrodynamic equation and thus energy randomization is dominated by electron impurity scattering, and the electron gas shall not relax to the lattice within the time regions considered, that is up to one picosecond. In the other case we assume, that electron-phonon scattering is more efficient, in fact it is so efficient that the relaxation time of the heat diffusion equations $\tau_H=C_e/g$ is as short as 20 fs. We have chosen this value since the low frequency modulated optical reflectivity measurements in the free electron domain show a 20 fs relaxation time for copper at RT, the subject of our computations [15]. This way we assume, that the damping of the current of the hydrodynamic equation is largely due to electron-phonon scattering.

In order to compute the reflectivity curve as a function of time one has to determine the temperature dependence of the material parameters. However, the relative change of ω_p, according to Jones and March [16] is smaller than 10^{-7}/K around room temperature and may be neglected.

In the case of $\tau_H=C_e/g=(20\ fs)^{-1}$ we consider, that the relaxation time of the hydrodynamic equation is dominated by electron-phonon collision. The ω_p coefficient of the hydrodynamic equation depends on the temperature dependent equilibrium electron density n, ($n=n_0(1-\beta T)$), where β is the volume thermal expansion coefficient. Introducing this dependence into the hydrodynamic equation numerical computation shows that the effect is negligibly small.

As far as the current relaxation time, τ is concerned it may depend on the temperature in a strong fashion. For small temperature excursions this dependence should be a linear function of the temperature around room temperature (RT):

$$\tau(z,T) = \tau_{RT}[1+\alpha T(z,t)] \tag{5}$$

The numerical values we have used for α were different for the electron-phonon and electron-

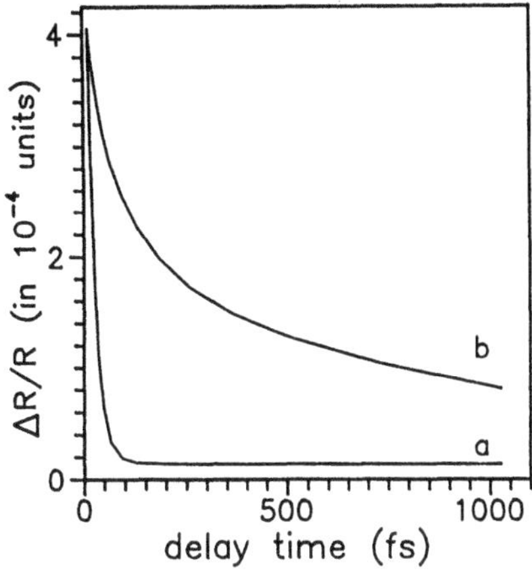

Fig. 1.
Transient reflectance as a function of delay time between the pump and the probe pulses. Curves a and b show the cases of (1) dominant electron-phonon and (2) electron-impurity relaxations, respectively. Temperature coefficients of current relaxation time τ are -0.00421/K and -0.0011/K for curves a and b, respectively.

impurity scatterings. For dominant electron-phonon relaxation the computation is straightforward; τ has the same temperature dependence as the d.c. conductivity σ in accord with the spirit of the hydrodynamic model. The temperature coefficients may be found in handbooks [17].

In the case of dominant electron-impurity scattering the shape of the thermoreflectivity curve may be computed but not the magnitude as we do not know the possibly small temperature coefficient of current relaxation time τ.

The relative change of reflectivity was calculated by computing the reflectivity differences for unheated (pump is "off") and the heated (pump is "on") material. The results are shown in Fig. 1. The upper curve (b) corresponds to the case of dominant electron-impurity scattering, while the lower curve (a) corresponds to dominant electron-phonon scattering. In curve (b) the unknown magnitude of temperature coefficient of the current relaxation time was chosen to fit the curves just after the heating pulse has left.

The decay of dominant electron-phonon relaxation case is much faster. It differs significantly since the electron-lattice equilibration is over on a very short time scale. After the fast relaxation the diffusion process is much slower, the reflectivity curve is almost horizontal on the figure.

At this point it is important to mention that one may try to compute transient reflectivities and transmittivities of thin films within the the hydrodynamic approach, especially since experimental data are available [18-20]. The approach is capable of providing information on the angle dependence of the reflectivity. In this respect it may be worth to emphasize that spatial dispersivity, that is the gradρ term may play a role in this case.

In conclusion, the hydrodynamic approach is simple and flexible enough for using it as a first approximation. The hydrodynamic approach is flexible enough for extending it beyond normal incidence. Comparison with experiments shall indicate points where the classical description fails, and concepts of quantum physics should be introduced.

References

1. G.L. Eesley: Phys. Rev. Lett. **51**, 2140 (1983)
2. P. Bado, S.B. Wilson, and K.R. Wilson: Rev. Sci. Instrum. **53**, 706 (1982)
3. L. Andor, A. Lörincz, J. Siemion, D.D. Smith, and S.A. Rice: Rev. Sci. Instrum. **55**, 64 (1984)
4. G.J. Blanchard and M.J. Wirth: Anal. Chem. **58**, 532 (1986)
5. M.B. Agranat, S.I. Anisimov, and B.I. Makshantev: Appl. Phys. B, **47**, 209 (1988)
6. D.N. Zubarev: *Nonequilibrium Statistical Thermodynamics* (Consultants Bureau, New York, 1974)
7. A. Lörincz and A. Miklós: "Ultrafast Processes" in *Progress in Photothermal and Photoacoustic Phenomena* Ed.: A. Mandelis, to be published by Elsevier
8. A. Miklós and A. Lörincz: J. Appl. Phys. **63**, 2391 (1988)
9. A. Miklós and A. Lörincz: Appl. Phys. B **48**, 261 (1989)
10. J. Somlói, R.A. Bakker and A. Lörincz: Z. Phys., in press
11. A. Lörincz, Z. Bozóki and A. Miklós: J. Appl. Phys. **70**, 941 (1991)
12. D. Pines and P. Nozieres: Theory of Quantum Liqiuds (Benjamin, New York, 1966)
13. F. Forstmann: Z. Physik **32** 385 (1979)
14. K. Kempa and F. Forstmann: Surface Sci. **129**, 516 (1983)
15. R. Rosei, D.W. Lynch: Phys. Rev. B **5** 3883 (1972)
16. W. Jones and N.H. March: Theoretical Solid State Physics (Wiley & Sons, London, 1973)
17. American Institute of Physics Handbook, edited by D.E. Gray (McGraw-Hill, New York, 1982)
18. S.D. Brorson, J.G. Fujimoto and E.P. Ippen: Phys. Rev. Lett. **59**, 1962 (1987)
19. H.E. Elsayed-Ali, T.B. Norris, M.A. Pessot and G.A. Mourou: Phys. Rev. Lett. **58** 1212 (1987)
20. H.E. Elsayed-Ali, T. Juhász, G.O. Smith and W.E. Bron: Phys. Rev. B **43** 4488 (1991)

Photoacoustic Waveforms Generated by Fluid Bodies

G.J. Diebold, T. Sun, and M.I. Khan

Department of Chemistry, Brown University, Providence, RI 02912, USA

The properties of photoacoustic waves generated by absorption of short pulse optical radiation by fluid bodies can be found by solution of the wave equation for pressure. Theorems are given for predicting the time profiles of photoacoustic waves depending on the symmetry of the absorbing body. For long laser pulses, the photoacoustic waveform is proportional to the zero'th derivative in one dimension, the one-half derivative in two dimensions, and the first time derivative of the laser pulse in three dimensions.

The properties of a photoacoustic "phonon maser" constructed as a series of layers of two fluids having different acoustic impedances but with identical transit times across each layer is discussed. The device possesses amplitude resonances in its frequency response, and a spatial directionality dependent on the number of layers.

1. Introduction

Perhaps the central problem in the field of photoacoustics, at least insofar as fluids are concerned, is the determination of the properties of the acoustic wave generated by absorption of optical radiation. It is hardly necessary to explain the importance of a quantitative description of the parameters that determine the amplitude of the acoustic wave since so many applications of the photoacoustic effect are based on direct measurements of the acoustic amplitude. The discussion here however focuses on a second property of the photoacoustic effect–the temporal profile of the photoacoustic wave.[1,2] In particular, several theorems are given that can be used to determine photoacoustic waveforms directly from a knowledge of the spatial profile of the heat deposition by the optical beam. A discussion of the photoacoustic effect produced by layers of absorbing material is given as well.

2. Delta function light pulses

When heat conduction and viscous effects can be ignored, the photoacoustic effect in a fluid is governed by the wave equation for pressure, which can be written in terms of the velocity potential ϕ_v as

$$\left(\nabla^2 - \frac{1}{c^2}\frac{\partial^2}{\partial t^2}\right)\phi_v = \frac{\beta}{\rho_f\, C_P}\, H \tag{1}$$

where the acoustic pressure p is determined from $p = -\rho_f \frac{\partial \phi_v}{\partial t}$, c is the sound speed, ρ_f is the fluid density, β is the isobaric volume expansion coefficient, C_P is the heat capacity per unit mass, t is the time, and H is the heating function, defined as the heat deposited in the fluid per unit volume and time. The solution to the wave equation can be written formally as[30]

$$\phi_v(\mathrm{x}, t) = \frac{-\beta}{4\,\pi\,\rho_f\, C_P}\int_0^t d\,t' \int d\mathrm{x}'\; g(\mathrm{x}, t \mid \mathrm{x}', t')\; H(\mathrm{x}', t') \tag{2}$$

where $g(\mathbf{x}, t \mid \mathbf{x}, t')$ is the Green's function. In one dimension, the Green's function can be written in terms of a Heaviside function, $u(\xi)$ as $g(z,t \mid z',t') = 2\,\pi\, u[(t-t') - \mid z-z' \mid /c]$ where z is the field point. After substitution of the Green's function into Eq. 2, the photoacoustic pressure is found to be

$$p(z,t) = \frac{\beta\, c}{2C_P} \int dz'\, H\left(z', t - \frac{z-z'}{c}\right) . \tag{3}$$

which shows the acoustic wave to be determined as a spatial integral of the heating function evaluated at the retarded time.

Now consider the photoacoustic wave excited by a delta function in time as described by the heating function $H(z,t) = \alpha E_0\, \hat{h}(z)\, \delta(t)$, where α is the optical absorption coefficient, E_0 is the energy fluence in the beam, and $\hat{h}(z)$ is a dimensionless spatial heating function. Substitution of this heating function into Eq. 3 gives an integral that is easily evaluated giving the photoacoustic wave for $t > 0$ as

$$p(z,t) = \frac{\kappa}{2}\, \hat{h}(z - c\,t) \quad \text{or} \quad p(\tau) = \frac{\kappa}{2}\, \hat{h}(-c\,\tau) \tag{4}$$

where $\kappa = \alpha\beta E_0 c^2/C_P$ and where τ is the retarded time from the origin. Equation 4 states simply that *in one dimension the spatial deposition of heat by the light beam is mapped directly into the time profile of the photoacoustic waveform.*

Consider solutions to the wave equation for the velocity potential in three dimensions. For a spherically symmetric deposition of heat, the spatial derivative in Eq 1 reduces to a single radial term. By defining a new variable, $\phi = \phi^\dagger/r$, the wave equation can be reduced to the one-dimensional wave equation,

$$\left(\frac{\partial^2}{\partial\, r^2} - \frac{1}{c^2}\frac{\partial^2}{\partial\, t^2}\right) \phi_v^\dagger(r,t) = \frac{\beta\, r}{\rho_f\, C_P}\, H(r,t)\,, \tag{5}$$

the solution to which can be found using the Green's function method given above. Aside from the modified source term on the right hand side of Eq. 5, an important difference between the one and three dimensional waves is that a spherical wave launched in the direction of the origin undergoes a reflection at the origin and reappears at the field point with an inverted amplitude. The Green's function must therefore include a term that accounts for the reflected wave. Using the procedure given for the one-dimensional case above, the photoacoustic pressure is found to be

$$p(r,t) = \frac{\beta\, c}{2\, C_P\, r} \int dr'\, r' \left[H\left(r', t - \frac{r-r'}{c}\right) - H\left(r', t - \frac{r+r'}{c}\right)\right], \tag{6}$$

For a delta function heating pulse of the form $H(r,t) = \alpha E_0\, \hat{h}(r)\, \delta(t)$ the photoacoustic pressure for $t > 0$ is found from Eq. 6 to be

$$p(r,t) = \frac{\kappa}{2\,r}(r - ct)[\, \hat{h}(r - ct) + \hat{h}(-r + ct)\,] \tag{7}$$

Thus, *for a spherically symmetric deposition of heat, the photoacoustic wave is a mapping of the product of the retarded time from the origin and the symmetrized spatial heating function.*

To obtain a two-dimensional mapping equivalent to Eq. 7, the theorem given in three dimensions can be expressed in cylindrical coordinates (*i.e.* in terms of the z coordinate and ρ, the distance from the z axis) and integrated over the entire z axis giving

$$p(\rho,t) = \kappa \int_{\rho-ct}^{\infty} d\zeta \frac{f'(\zeta) + f'(-\zeta)}{[\,(\zeta + ct)^2 - \rho^2\,]^{\frac{1}{2}}}\,, \tag{8}$$

where $f'(\zeta)$ is the derivative of a source function, the form of which is to be specified. Consider now, a heating function of the form $H(\rho,t) = \alpha E_0 \ \hat{h}(\rho) \ \delta(t)$ It is clear that immediately after the firing of the laser, the spatial profile of the pressure is proportional to the deposition of heat in space and is given by $\kappa \hat{h}(\rho)$, which must equal the pressure given by Eq. 8 at $t = 0$. However, since there are both outgoing and incoming waves, the condition

$$\int_\rho^\infty d\zeta \ \frac{f'(\pm \zeta)}{\sqrt{\zeta^2 - \rho^2}} = \frac{1}{2} \ \hat{h}(\rho) , \tag{9}$$

must obtain (for either the positive or negative sign). It is possible to transform[1] Eq. 9 so that the source function can be obtained from a knowledge of $\hat{h}(\rho)$. The result of the transformation gives the source function as

$$f(\zeta) = -\frac{1}{\pi} \int_\zeta^\infty d\rho \frac{\rho \ \hat{h}(\rho)}{\sqrt{\rho^2 - \zeta^2}} \tag{10}$$

Thus, Eqs. 8 and 10, *constitute a mapping of a cylindrically symmetric heat distribution into the time profile of the photoacoustic waveform.*

The utility of the above three theorems is that photoacoustic waveforms can be found directly from the spatial heating function without having to solve the wave equation for pressure. Good examples of the use of the theorems are provided by the infinite layer, the infinite cylinder, and the sphere. In all three cases, the spatial heating function is a square wave in space, which can be described by $\hat{h}(\hat{\xi}) = \tilde{\Theta}_{0,1} \ (1 - \hat{\xi})$ where $\tilde{\Theta}_{ij}(\hat{\xi})$ is a square wave function that is one for $\hat{\xi}$ between i and j and is zero otherwise, and $\hat{\xi}$ is a generalized coordinate defined as $\hat{\xi} = \xi/\xi_0$, where ξ_0 is the length parameter. Substitution of $\hat{h}(\xi)$ into Eqs. 4, 7, and 10 gives the following expressions for the photoacoustic pressure for $t > 0$ in one, two, and three dimensions, respectively,

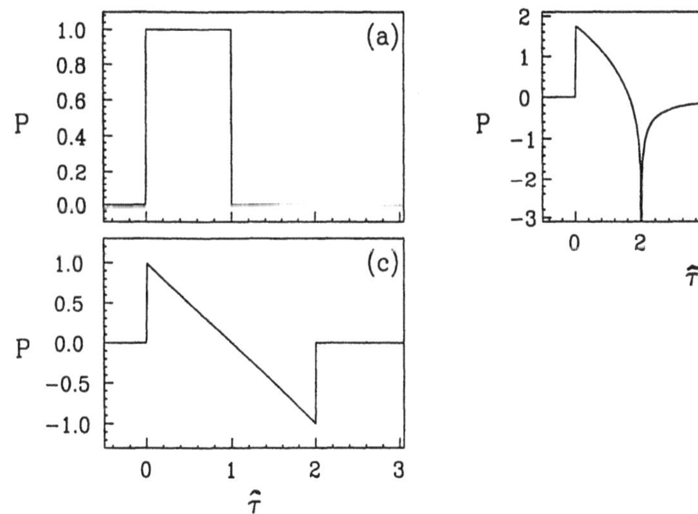

Fig 1 Photoacoustic pressure P in arbitrary units versus the dimensionless time $\hat{\tau}$ for (a) the infinite layer, (b) the infinite cylinder, and (c) the sphere. The width of the square wave in (a), the N-wave in (c), and the section of the wave from $\hat{\tau} = 0$ to $\hat{\tau} = 2$ in (b) corresponds to the transit time of sound across the layer, the droplet, and the cylinder, respectively.

$$p(\hat{\tau}) = \frac{\kappa}{2}\tilde{\Theta}_{0,2}(\hat{\tau}) \tag{11}$$

$$p(\hat{\tau}) = \begin{cases} \frac{\kappa}{\pi\sqrt{2\hat{\rho}}} \int_{-\hat{\tau}}^{0} d\hat{\eta} \frac{(\hat{\eta}+1)}{(\hat{\eta}+\hat{\tau})^{\frac{1}{2}}\,[1-(\hat{\eta}+1)^2]^{\frac{1}{2}}} & 0 < \hat{\tau} < 2 \\ \frac{\kappa}{\pi\sqrt{2\hat{\rho}}} \int_{-2}^{0} d\hat{\eta} \frac{(\hat{\eta}+1)}{(\hat{\eta}+\hat{\tau})^{\frac{1}{2}}\,[1-(\hat{\eta}+1)^2]^{\frac{1}{2}}} & 2 < \hat{\tau} \end{cases} \tag{12}$$

$$p(\hat{\tau}) = \frac{\kappa}{2\,\hat{r}}(1-\hat{\tau})[\tilde{\Theta}_{0,1}(\hat{\tau}) + \tilde{\Theta}_{1,2}(\hat{\tau})\,])\,, \tag{13}$$

where $\hat{\tau}$ is the dimensionless retarded time from the edge of body defined as $\hat{\tau} = (c/\xi_0)[t-(\xi-\xi_0)/c]$, and where the far field limit of the integral is used in Eq. 12. The photoacoustic pressure from Eqs. 11-13 is plotted in Fig. 1. Note that the waveforms in Fig. 1 are identical with those found through a frequency domain solution[2] to the wave equation followed by Fourier transformation of the result.

3. Long light pulses

Consider now the case when the time duration of the exciting laser pulse is far greater than the transit time of sound across the body. In one dimension, a frequency domain acoustic pressure is found by taking a heating function of the form $H(z,t) = \alpha I_0\ \hat{h}(z)\ e^{-i\,\omega\,t}$, where I_0 is the intensity of the optical radiation. Substitution of this heating function into Eq. 3 gives the integral

$$p = \frac{\alpha\ \beta\ I_0 c}{2\ C_P}\ e^{-i\omega\,\tau} \int dz'\ \hat{h}(z')\ e^{-i\,\omega\,z'/c}\,. \tag{14}$$

It can be seen that the argument of the exponential function in Eq. 14 is proportional to the integration variable divided by the wavelength of the radiation. Since this quantity is small by assumption, the exponential function can be set equal to unity, so that the integration over space gives the 0'th moment of the spatial heating function, which can be regarded as a constant here. When Eq. 14 is then Fourier transformed into the time domain, the result is

$$p_l(\hat{\tau}_l) = \frac{\kappa}{2}\delta(\hat{\tau}_l)\,. \tag{15}$$

where $\hat{\tau}_l$ is a dimensionless retarded time from the origin defined by $\hat{\tau}_l = (c/\xi_l)\ [t-\xi/c]$, ξ is a space coordinate, and ξ_l is a length parameter in n dimensions given by

$$\xi_l = \left[\int_0^{\infty} d\xi\ \xi^{n-1}\ \hat{h}(\xi)\right]^{\frac{1}{n}} \tag{16}$$

A similar calculation in three dimensions using Eq. 6 results in the difference of two exponential functions, which can be expressed in terms of a derivative. The three-dimensional equivalent of Eq. 15 is found to be

$$p_l(\hat{\tau}_l) = \frac{\kappa}{\hat{r}_l}\ \frac{d}{d\ \hat{\tau}_l}\ \delta(\hat{\tau}_l)\,. \tag{17}$$

where $r_l = r/\xi_l$.

Now, Eqs. 15 and 17 are useful in determining the long pulse photoacoustic response through the convolution integral

$$P(t) = \frac{1}{2\ \pi}\int_{-\infty}^{\infty} dy\ \tilde{p}_l(y)\ I(t-y)\,, \tag{18}$$

where $\tilde{p}_l$ is a unit response function (formed from Eq. 15 or 17 by dividing by $E_0/2\pi$) and

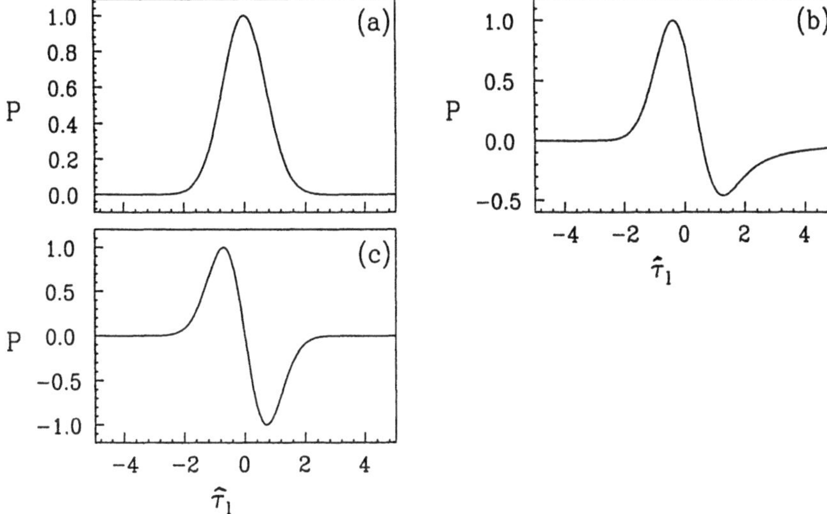

Fig 2 Photoacoustic pressure P in arbitrary units versus the dimensionless time τ_l for a Gaussian light pulse whose duration in time is long compared with the transit time of sound across the irradiated body. The acoustic pressure in one dimension (a), two dimensions (b), and three dimensions (c) is proportional to the 0'th, the (1/2)'th, and the 1'st time derivative of the exciting light pulse, respectively.

where $I(t)$ is the intensity of an arbitrary, long light pulse. Since Eq. 15 is simply a delta function, it is easy to see that the convolution integral requires only that the intensity be evaluated at the retarded time $\hat{\tau}_l$. *Thus, in one dimension the photoacoustic wave is directly proportional to the exciting optical pulse.*

In three dimensions, according to Eq. 17, the convolution integral contains a derivative of a delta function. The convolution integral can be evaluated through integration by parts giving an integral of the product of a delta function with the first derivative of $I(t)$, which is easily evaluated. *Thus, for a spherically symmetric source, the photoacoustic wave is proportional to the first time derivative of the exciting light pulse.*

In two dimensions, the calculation of the long pulse response is complicated, and a multipole expansion of the field must be carried out. The equivalent of Eqs. 15 and 17, however is found to be

$$p_l = \kappa \frac{d}{d\hat{t}_l}\left[\frac{u(\hat{t}_l - \hat{\rho}_l)}{(\hat{t}_l^2 - \hat{\rho}_l^2)^{\frac{1}{2}}}\right], \tag{19}$$

where $\hat{t}_l = ct/\xi_l$ and $\hat{\rho}_l = \rho/\xi_l$. When Eq. 19 is convoluted with an intensity function of the form $I(t) = (E_0/\Theta)\ S(t/\Theta)$, where Θ is a time parameter, the photoacoustic pressure is found to be

$$P(\tilde{\tau}_l) = \frac{\kappa}{\hat{\Theta}^{\frac{3}{2}}\sqrt{2\hat{\rho}}} \int_{-\infty}^{\tilde{\tau}_l} dT\, \frac{d\, S(T)/dT}{\sqrt{\tilde{\tau}_l - T}} \tag{20}$$

(where $\hat{\Theta} = c\Theta/\xi_l$ and $\tilde{\tau}_l = \hat{\tau}_l/\hat{\Theta}_l$) which is of the form of the (1/2)'th derivative of S. *Thus, in two dimensions, the photoacoustic wave is proportional to the (1/2)'th time derivative of the optical exciting pulse.* The long pulse response of bodies in one, two, and three dimensions to Gaussian light beams is shown in Fig. 2.

4. Multiple layers

Consider a device consisting of a N optically thin layers of a material of one thickness placed in contact with transparent layers of a second material with a different acoustic impedance and thickness so that the transit time of sound across each layer is the same. The magnitude of the plane wave emitted by the device can be found by linear superposition of the contributions of each of the N sources. The effect of the n'th layer is to produce a left-going plane wave with amplitude B_n at the left end of the device x_0, and a right-going wave with amplitude A_n at the right end of the device x_{2N-1}, which are determined as solutions to a matrix equation.

Results for $N = 2$, 3, and 5 are shown in Fig. 3 calculated for 2.75 mm thick slabs of Lucite (the absorbing medium), in contact with 1.52 mm layers of water. The pressure amplitude A is plotted versus modulation frequency parameter q. The resonances are seen to sharpen and display more complicated side lobe structure as the number of layers is increased.

A plot of the pressure P versus the retarded time $\hat{\tau}$ is given in Fig. 3 showing the response of the system to short laser pulses. Further properties of the device such as the directionality of the emitted ultrasonic beam are found where the exciting light beam is

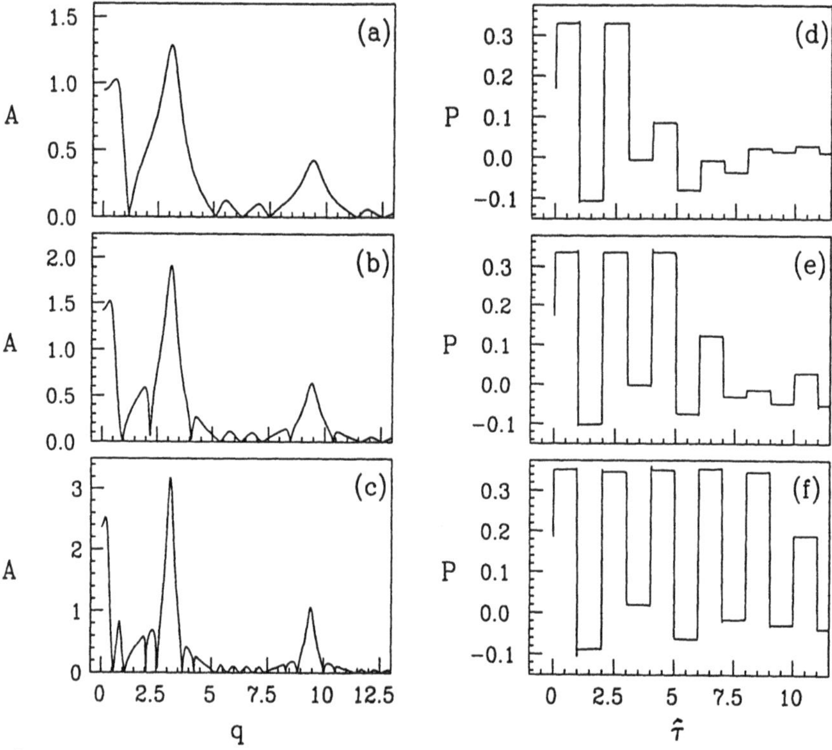

Fig 3 Photoacoustic amplitude A in arbitrary units versus the dimensionless frequency parameter q for (a) two layers, (b) three layers, and (c) five layers excited by sinusoidally modulated optical radiation. Photoacoustic pressure in arbitrary units versus dimensionless time for (d) two layers, (e) three layers, and (f) five layers excited by a single laser pulse whose duration is short compared to the transit time of sound across a single layer.

of a finite spatial extent. Experimental results obtained with a modulated Ar ion laser, and a pulsed Nd:YAG laser confirm the main predictions of the calculations.

The authors are grateful for the support of this research by the U. S. Department of Energy, Office of Basic Energy Studies under Grant ER13235.

REFERENCES

1. A full list of references is too extensive to include here. See G. J. Diebold, T. Sun, and M. I. Khan, "Properties of Photoacoustic Waves in One, Two, and Three Dimensions", to be published.
2. G. J. Diebold, M. I. Khan, and S. M. Park, Science **250**, 101 (1990).
3. P. M. Morse and H. Feschback, *Methods of Theoretical Physics* (McGraw Hill, New York, 1953).

Study of Chemical Relaxation of Dimeric Formic Acid by Photoacoustic Resonance Spectroscopy

A. Winkler, H. Jung, M. Fiedler, and P. Hess

Institute of Physical Chemistry, University of Heidelberg,
Im Neuenheimer Feld 253, W-6900 Heidelberg, Fed. Rep. of Germany

Abstract. Dissociation of dimeric formic acid was studied in the frequency domain by exciting acoustic resonances in a cylindrical cavity. The acoustic resonance profiles of the first radial mode were measured as a function of pressure and temperature. Measurements with buffer gases showed an absorption peak due to chemical relaxation. For temperatures between 283 and 323 K the dissociation rate constants of the dimer were found to vary from 2.2 to 6.9×10^6 1/s and mean $p\tau$ values for the relaxation of the inner degrees of freedom of the monomer/dimer mixture from 2 to 5 nbar s. The activation energy for the dissociation of the dimer was determined to be 23±1 kJ/mol.

1. Introduction

Relaxation effects in gases cause sound dispersion and absorption of acoustic energy [1],[2]. Therefore, a study of the dependence of the sound velocity and absorption on pressure and frequency provides information about the relaxation parameters. The sound wave represents a periodic change of state of the system on a time scale $1/\omega$. The equilibrium processes which are managed by intramolecular collisions follow this change with a specific characteristic time constant called the relaxation time. If the externally applied time scale becomes short compared to the microscopic relaxation time, the collision mechanism can no longer equilibrate all the degrees of freedom. Some degrees of freedom drop out of the energy distribution, leading to a change in the effective heat capacity and therefore to a shift in the sound velocity. If the relaxation processes are very slow or very fast, pressure and density oscillate in phase. However, if external and internal time scales are of the same order of magnitude, a phase lag occurs between pressure and density, leading to a finite value of the integral $\oint p\mathrm{d}V$. Energy is removed from the sound wave but not fed back in phase. This process is responsible for the sound absorption effect connected with relaxation phenomena causing a dispersion of resonance frequency and a broadening of the resonance profiles measured in the present experiment.

2. Experimental

A stainless steel cylinder of 6 cm length and 5 cm diameter was used as the resonant cavity. The standing acoustic waves were excited by an amplitude modulated waveguide carbon dioxide laser. The 9R20 line (9.271 µm) with an average output power of 0.85 watt was used for excitation of the monomers. The laser beam traversed the cavity along the cylinder

Springer Series in Optical Sciences, Vol. 69 Photoacoustic and Photothermal Phenomena III
Editor: D. Bićanić

axis and the transmitted laser power was monitored by a pyroelectric detector. An electrooptic modulator (Pockels cell) was employed to modulate the laser beam. The modulation frequency was generated by a computer controlled frequency synthesizer. For detecting the acoustic signal, an electret microphone mounted at half the cylinder length was used. Its signal was amplified and filtered by a bandpass filter and recorded by a lock-in amplifier. The resonance profiles of the first radial mode of the cylinder were measured for pressures in the range 0.1-10 mbar. The resulting resonance curve was fitted by a Lorentzian profile with the three parameters resonance frequency ν_0, half-width $\Delta\nu$ and signal strength at maximum S_{max}. These parameters contain the information on sound velocity, sound absorption and deposited energy, respectively. The temperature of the cell which was stabilized by a bath was measured by a quartz sensor with a relative precision of 0.03 K. The pressure was recorded with MKS capacitance manometers.

3. Results

The dissociation rate constant and the mean rotational and vibrational relaxation time τ were obtained by fitting a modified version of the theoretical model of the resonator described by Fiedler and Hess [3] to the measured values of frequency dispersion and resonance broadening $\Delta\nu$. Fig. 1 shows a typical behaviour of $\Delta\nu$ as a function of pressure. The absorption peak due to the chemical relaxation is overlayed by the surface losses and the losses due to relaxation of the inner degrees of freedom. These losses increase strongly on decreasing the pressure. Measurements with Ar and He as buffer gases at total pressures in the range of 100-1000 mbar were performed to reduce these high losses. Fig.2 shows the absorption peak due to the chemical relaxation in the presence of 1 bar He. While the dissociation rate constant could be determined very precisely from the measurements of the pure gas, the error in the value for the relaxation time of the inner degrees of freedom is large

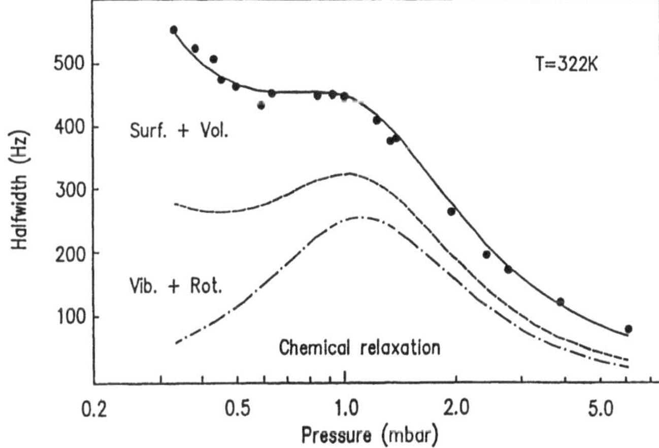

Fig 1: Width of resonance of the first radial mode in the cylindrical cavity as a function of pressure. The different contributions from chemical relaxation, surface and volume losses and vibrational and rotational relaxation are indicated.

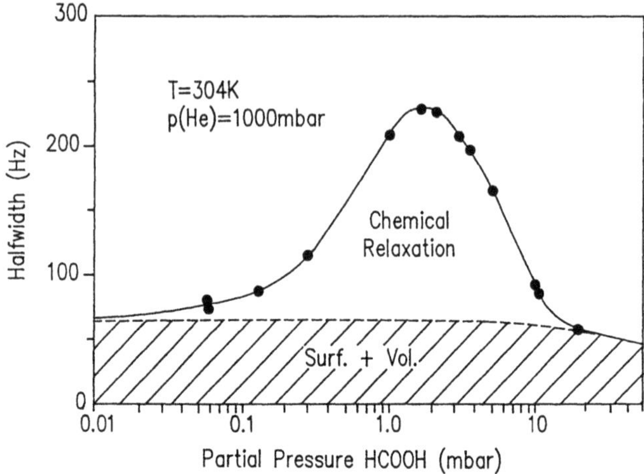

Fig 2: Width of resonance of the first radial mode in the cavity as a function of partial pressure of formic acid in the presence of 1 bar He. The different contributions from surface and volume losses and the chemical relaxation are indicated.

due to uncertainties in the calculation of the surface losses. These losses are proportional to the square root of the thermal conductivity of the gas. A value for this transport coefficient is not available in the literature for formic acid. There is a linear relation between the logarithm of $p\tau$ and $1/T^{1/3}$ as expected from Landau-Teller theory. The value of τ yields a mean value of 5 collisions needed to reach equilibrium. The values of the dissociation rate constants vary between 2.2 and 6.9×10^{6} 1/s for temperatures in the range of 283-323 K. As expected, the logarithm of the measured values is proportional to 1/T. The activation energy can be determined from the Arrhenius plot, yielding $E_A = 23\pm1$ kJ/mol.

References

[1] A.Einstein, Sitzungsber. Dtsch. Akad. Wiss. Berlin, Kl. Math. Phys. Tech. p.380 (1920)

[2] M.Fiedler and P.Hess in *Photoacoustic, Photothermal and Photochemical Processes in Gases*, edited by P.Hess (Springer, Berlin, 1989), pp.85-123

[3] M.Fiedler and P.Hess, J. Chem. Phys., **93**, 8693 (1990)

Characterization of Chemical Reactions by Thermal Lens Spectrometry

M. Franko[1] *and C.D. Tran*[2]

[1]"Jožef Stefan" Institute, Jamova 39, 61111 Ljubljana, Slovenia
[2]Marquette University, Milwaukee, WI 53233, USA

Abstract. Theoretical models describing the formation of the thermal lens during chemical reactions were derived and the effects of changing concentration on the thermal lens strength were evaluated. It was estimated that gas phase reactions on a μs time scale can be followed by this technique. An instrument for simultaneous measurement of absorbance and thermal lens effect was developed and used to verify calculated values. For the S_N2Ar^* photochemical reaction of 1,2-dimethoxy-4-nitrobenzene, the rate constants determined by the thermal lens technique agree well with those by conventional absorption measurements.

1. Introduction

The thermal lens technique has rarely been employed for the characterization of chemical reactions involving weakly absorbing species in spite of its high sensitivity. Furthermore, most of the reported applications were limited to relatively slow reactions. This can be attributed, in part, to the lack of adequate theoretical models which would account for changes in concentration during the thermal lens measurement.

The objective of this work is to examine the possibilities of using the thermal lens spectrometry for kinetic studies of fast chemical reactions, and to assess the type of information which can be obtained using this technique.

2. Theory

Theoretical models describing the formation of the thermal lens during chemical reactions were derived initially. As an example, an equation derived for first order reactions where the measurement is based on the decrease in the concentration of the reactant, is discussed. In such cases the relative change in the probe beam intensity ($\Delta I/I$), obtained from the parabolic approximation [1] for a weak thermal lens is given as:

$$\frac{\Delta I}{I} = \frac{\Theta}{1+t_c/2t} - \Theta\{1-e^{-k_R t} - \frac{k_R t_c e^{-\frac{k_R(t_c+2t)}{2}}}{2}[\ln(1+2t/t_c) + k_R t + \frac{k_R^2 t(t+t_c)}{4} + \frac{k_R^3 t(4t^2+3t_c^2+6tt_c)}{72} + \ldots]\} \quad (1)$$

where Θ is the steady state thermal lens strength for the initial concentration of reactant, t the time after the onset of excitation, t_c the time constant and k_R the reaction rate constant.

The first term in eq. 1 corresponds to the thermal lens strength at constant concentration, while the other terms represent the decrease of the thermal lens strength caused by the decrease of concentration of reactant. The effect of chemical reaction increases when t and t_c increase and disappears when there is no reaction (k_R=0). The equation can be readily adapted for the cases where reaction product is the absorbing species, i.e., thermal lens strength increases as the concentration of reaction product increases.

On the basis of computer simulations it was established that the effect of a first order reaction on the thermal lens signal is negligible when the ratio of the reaction half life time to the time constant is greater than 140. When this ratio is decreased the effect of chemical reaction increases considerably, but, even when $t_{1/2}/t_c$ ratio is as small as 0.6 the effect of chemical reaction is null if the thermal lens signals are measured at times much shorter than t_c ($t_c/t>10$).

It follows, that the measurement of a steady state thermal lens is not recommended in the case of rapid reactions. Instead, several thermal lens signals should be measured during the course of reaction by consecutively exciting the sample for time periods shorter than t_c. But, the initial concentration of the measured species changes for each thermal lens due to chemical reaction. This has to be accounted for in the θ term. Furthermore, the duration of excitation (t), which is kept constant, and the duration of the reaction (t_R) have to be distinguished. When the reaction product is the measured species the time dependence of consecutive thermal lens signal magnitudes $(\Delta I/I)_{t_R}$ during a chemical reaction is given by eq. 2. (θ is a steady state thermal lens strength for the final concentration of reaction product).

$$\left(\frac{\Delta I}{I}\right)_{t_R} = \frac{\Theta}{1+t_c/2t} - \Theta e^{-k_R t_R}\left\{e^{-k_R t} - \frac{1}{1+2t/t_c} + \frac{k_R t_c e^{-\frac{k_R(t_c+2t)}{2}}}{2}\cdot \left[\ln(1+2t/t_c) + k_R t + \frac{k_R^2 t(t+t_c)}{4} + \ldots\right]\right\} \quad (2)$$

3. Experimental

An instrument which is capable of simultaneously measuring the absorbance and the thermal lens effect was developed for kinetic measurements. Its performance is similar to a dual beam (pump/probe) thermal lens instrument [2], except that it is equipped with additional photodiodes for the measurement of absorbance.

The instrument was tested by measuring the rate constant of the nucleophilic aromatic photosubstitution (S_N2Ar^*) reaction of 1,2-dimethoxy-4-nitrobenzene. Up to 150 mW excitation power (457.9 nm) was used to measure the formation of reaction product during the irradiation of the sample with 360 nm UV light.

4. Results and discussion

Kinetic data, simultaneously obtained by the thermal lens (Figure 1) and the absorption techniques, were treated using the modified

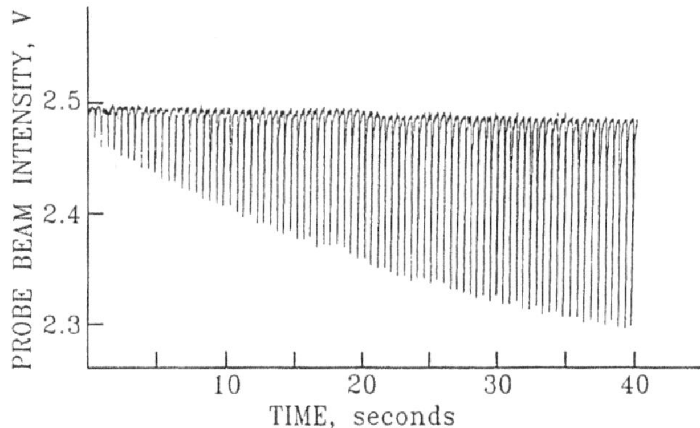

Figure 1.: Thermal lens signals measured during the photochemical reaction of $4.0 \cdot 10^{-6}$ M 1,2-dimethoxy-4-nitrobenzene in 0.5 M KOH.

Gugenheim method [3]. The reaction rate constants of 0.028±0.003 s^{-1} and 0.026±0.001 s^{-1} obtained for 6.0×10^{-5} M 1,2-dimethoxy-4-nitrobenzene by thermal lens and absorption measurements, respectively, indicated very good agreement between the two techniques. More important is the observation that the lowest initial concentration of the reagent needed for the thermal lens measurement (3×10^{-7} M) is 100 times less than that for the absorption method (3×10^{-5} M).

The fastest reaction rate that can be measured using the thermal lens technique is governed by the t_c value which is dependent on the medium of the reaction mixture. Time constants in liquids are usually in the order of ms. Consequently, reactions with half life times shorter than a few ms cannot be characterized by this technique. Conversely, due to their physical properties, gases provide time constants more than 100 times shorter than those of liquids. As a consequence gas phase reactions on a μs time scale can, in principle be followed by this technique.

5. References

1. R. Vyas, R. Gupta: Appl. Opt. **27**, 4701 (1988)
2. H. L. Fang, R. L. Swofford: In "Ultrasensitive Laser Spectroscopy" (Academic Press, New York 1983)
3. P. Zuman, R. Patel: "Techniques in Organic Reaction Kinetics" (Wiley & Sons, New York 1984)

Methane Vibrational Rates from Deconvoluting a Large Data Set with Varying Pressure, Buffer Gases, Temperature, Signal-to-Noise Ratio, and Instrumental Response

J.A. Burt[1], *K. Yang*[1], *J.J. Caldwell*[1,2], *R.N. Halthore*[3,4], and *J.E. Allen*[3]

[1]Department of Physics and Astronomy, York University, 4700 Keele St., North York, Ontario, Canada M3J 1P3

[2]Space Astrophysics Laboratory, Institute for Space and Terrestrial Science, 2700 Steeles Ave. West, Concord, Ontario, Canada L4K 3C8

[3]Laboratory for Extraterrestrial Physics, NASA Goddard Space Flight Center, Code 691, Greenbelt, MD 20771, USA

[4]Applied Research Corporation, Landover, MD 20785, USA

Abstract. Relaxation processes in methane binary mixtures pumped by 3.3 μm laser pulses have been monitored photoacoustically from room temperature down to 77 K. To a first approximation these processes decay by a single dominant exponential which is not at all evident in the digitized microphone recordings. This paper relates the pitfalls, advantages, and algorithms of the involved deconvolution procedure necessary for deriving kinetic the data. Comparison is made with the more common CW method and with the pulsed experiments of earlier workers.

1. Introduction

This paper is not concerned with the numerical values derived for the vibrational relaxation lifetimes in methane but rather with the problems encountered in deconvolving lifetimes from the experimental data. In a future publication we will report the final measured values.

The problem may be simply grasped by viewing fig.1. The light curve is the photoacoustic response at 4.88 Torr, the dark curve is the response when the pressure is halved. The responses have a classical cylinder wave shape since the cell is a cylinder. Due to their vibrational lifetimes, we should see a 1 ms exponential drop in the light curve and a 2 ms drop in the dark one. But if we look at fig.1 we do not immediately see where these exponentials are. Some parts resemble decaying exponentials; which do we pick? Furthermore, the dark curve does not even appear to be much different from the light. Perhaps it is a bit more damped around 2 ms, but the rise times of both curves at 0.5 ms are the same. Very contradictory evidence. Fig.1, then, shows the whole problem: where are the lifetimes? We indicate where in section 3.

What is new in this experiment? New technology, for one. All measurements were made at the Ontario Laser and Lightwave Research Centre using a $ 0.5 M laser which achieved stable, tunable 1 mJ pulses at 3.3 μm, hence providing effectively 1000 times more signal than we could obtain from conventional lab, conventional budget, lasers. Additionally, the photoacoustic measurements cover a very large range of binary mixtures and temperatures through a measurement programme which lasted 2 years. Only a fraction of the data has been deconvolved yet. The data set represents a primary resource for chemical physics. For example, there are only three previous reports [1,2] of sub-ambient methane lifetimes.

The data set is also important in astronomy. Emission from the ν_4 band of methane at 7.8 μm has been observed in the auroral zone on Jupiter and the stratosphere of the giant planets. Atmospheric modelers lack information about the collisional relaxation times for the ν_3 and ν_4 levels in CH_4 over the temperature range of -100 C to -200 C.

Springer Series in Optical Sciences, Vol. 69 Photoacoustic and Photothermal Phenomena III
Editor: D. Bićanić

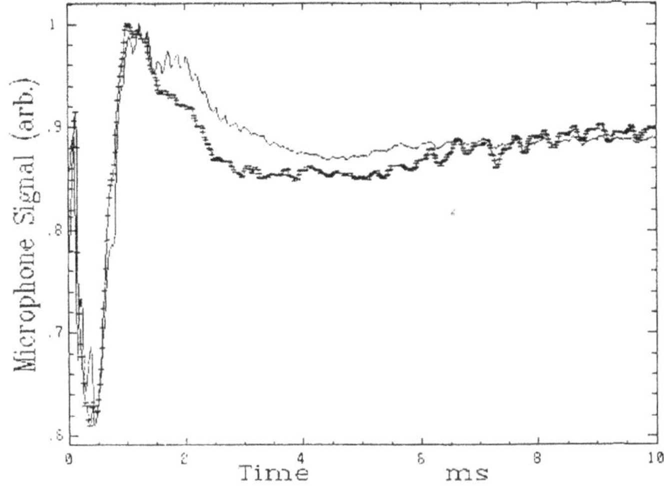

Fig. 1 Photoacoustic cell responses to laser pulse. 4% methane in argon, 23 C.
Light curve: at 4.88 Torr total pressure
Dark curve: at 2.50 Torr total pressure

2. Experimental

The photoacoustic cell used was a 2.5 cm diameter cylinder, 20 cm long, constructed from copper with Brewster's angle windows of ZnSe seated on indium seals for good thermal contact. A stainless steel 1/2 inch microphone (B&K 4134) was mounted flush with the inner wall at the cell centre. A cooling jacket surrounded the microphone/preamp TEE arm which hung down from the cell. The entire cell assembly was contained in a 9 inch QVF glass cross which was evacuated to prevent window condensation.

A pulsed Nd/YAG (Spectra Physics) laser fed a dye laser whose output was heterodyned with a portion of the original beam in a $LiNbO_3$ crystal to produce a 3.3 μm Gaussian beam. The output beam was wavelength tunable and brought to coincidence with the photoacoustic cell axis by a 4 mirror system aligned by a HeNe laser

3. The Deconvolution Procedure

In simple deconvolution one divides the instrumental response, which could be the microphone signal when methane is laser-excited at a pressure of 100 Torr, say, into the response at 1 Torr (all in the Fourier domain). The resulting Fourier-inverted response should then exhibit an exponential decay corresponding to the kinetic processes occurring at 1 Torr. Here we have taken the instrumental response to be simply the 100 Torr response itself, since the relaxation at 100 Torr is 100 times faster than that at 1 Torr hence we have neglected it. Unfortunately, the physics of the acoustic wave generated by a focussed laser pulse changes with pressure. Simply put, at high pressure the molecules in the photoacoustic cell transmit acoustic waves generated by the laser, while at low pressure, the molecules in the vicinity of the initial laser-generated acoustic wavefront can rapidly diffuse away and so may not have occasion to transmit a wave at all.

Few indeed are the photoacoustic relaxation rate papers which discuss deconvolution. All previous photoacoustic methane studies, such as Hess' [3] are CW. The article by Braun [4] has been especially helpful and many of our algorithms are based on his procedures. In fact, it is rare that deconvolution is not required in deriving relaxation rates, however, in some of the early pulsed photoacoustic papers no deconvolution was used nor appeared necessary [5,6].

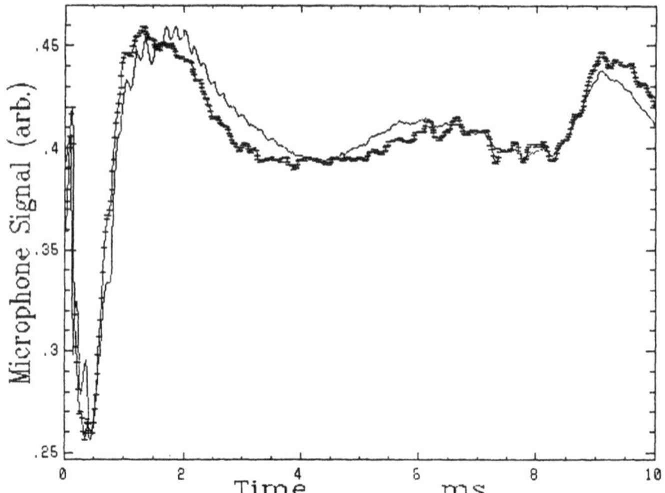

Fig.2 Inverse transforms of the FFTs of the fig. 1 curves divided by the FFT of the cell response at 10.8 Torr

When we applied simple deconvolution we found only noise, just as Braun indicated. Using a modified Wiener filter [4], at times supplemented by a super-Gaussian [7] filter, the noise became signal. The deconvolution procedure, then, requires many adjustments, which find their justification in the theory of inverse problems [8] which embraces deconvolution.

For the fig. 1 curves, we divided their Fourier transforms by the Fourier transform of the cell response at 10.8 Torr. The resulting inverse transforms are shown in fig. 2. The instrumental response is still evident for times < 1 ms. Also, there is a new peak at 9 ms due to the FFT. We took the range from 2.5 to 4.5 ms and fitted exponentials. Data noise is a problem; part of the range for the dark curve shows a slower exponential, as it should since the pressure is halved, and part does not. There is also the problem arising when one tries to fit part of an exponential to a complete one. We repeated the complete deconvolution+fit for a total of 5 responses at different pressures. A good straight line was obtained when the derived rates were plotted against pressure. Other deconvolutions included one where the inherent rate in the 10.8 Torr divisor transform was, itself, first divided out. They all produced similar results.

4. Results and Discussion

For the 23 C methane/argon set, the lifetime-pressure product was found from a plot of rate *vs.* pressure to be 3.5 ms.Torr. Perrin [9] gives a value of 15.4 ms.Torr, but Ponomarev [10] discusses how the laser intensity can reduce the photoacoustically measured lifetime. Applying his correction, the Perrin lifetime would become 9 ms.Torr. The numerical value Ponomarev reports is, however, for CO_2, not CH_4.

We also deconvolved a data set at -28 C, 1% methane in hydrogen, obtaining 1.1 ms.Torr. Moore [11] gives 0.68 ms.Torr at 23 C. Applying the correction established by Madigosky [12] for low temperature, Moore's rate would become 0.98 ms.Torr.

Thus, we have obtained excellent agreement with other published works. Since our deconvolution analysis has only just begun we cannot give a detailed error analysis but note that the data scatter in our rate *vs.* pressure curves is on the order of 20%.

Earlier works in pulsed photoacoustic vibrational relaxation [13, 14, 15, 16] have examined the influence of wall effects and many other effects such as laser excitation spatial and temporal waveforms etc. The deconvolution technique [4] followed in our study accounts for these effects though space does not permit us to go into the details here.

Acknowledgements

The measurements were made at the Ontario Laser and Lightwave Research Centre, Toronto. We wish to express our thanks to Dr. M. Hubert, OLLRC, for cooperation and support. This research was supported by the Space Astrophysics Laboratory of the Institute for Space and Terrestrial Science, Concord, Ont.

References

1 W.M. Madigosky, J. Chem. Phys. **39**, 2704 (1963)
2 M. Rebelo da Silva and F. Lepoutre, "Vibrational Relaxation" in *Physical Acoustics* **XVIII**, edited by W.P. Mason and R.N. Thurston (Academic, San Diego, 1988) p.341, the 2 references are mis-numbered, however.
3 A. Karbach, J. Roeper and P. Hess, Chem. Phys. **82**, 427 (1983)
4 W.Braun, I. Price, B.C. Cadoff, and N.C. Peterson, Intl. J. Chem. Kin. **21**, 1029 (1989)
5 J.M. Weulersse and R. Genier, Appl. Phys. **24**, 363 (1981)
6 E.A. Rohlfing, J. Gelfand, and R.B. Miles, J. Appl. Phys. **53**, 5420 (1982)
7 H.J. Weaver, Theory of Discrete and Continuous Fourier Analysis, Wiley, 1989, New York
8 M. Bertero, P. Brianzi, E.R. Pike, and L. Rebolia, Proc. R. Soc. Lond. A**415**,257(1988)
9 M.Y. Perrin, Chem. Phys. Lett. **94**, 434 (1983)
10 Yu. N. Ponomarev, Infrared Phys. **32**, 377 (1991)
11 J.T. Yardley, M.N. Fertig, and C.B. Moore J. Chem. Phys. **52**, 1450 (1970)
12 W.M. Madigosky, J. Chem. Phys. **39**, 2704 (1963)
13 N.J.G. Smith, C.C. Davis, I.W.M. Smith, J. Chem. Phys. **80**, 6122 (1984)
14 K.M. Beck, A. Ringwelski, R.J. Gordon, Chem. Phys. Lett. **121**, 529 (1985)
15 K.M. Beck, R.J. Gordon, J. Chem. Phys. **87**, 5681 (1987)
16 W. Braun, P. Dagaut, B.C. Cadoff, J. Res, NIST **93**, 643 (1988)

New Aspects of Sound Generation by Light Absorption in Gases

A. Miklós

Institute of Isotopes, Hungarian Academy of Sciences,
P.O. Box 77, H-1525 Budapest, Hungary

Abstract. The problem of acoustic signal generation in gases is investigated in order to understand the results of recent experiments, which contradict to the theory. The experimental results may be explained by assuming that every point along the light beam radiates pressure and velocity waves simultaneously. The physical background of such an assumption is discussed in the paper.

1. Introduction

The physical process of sound generation by light absorption plays an essential role in photoacoustics. It is generally assumed that the photon energy absorbed by a gas molecule is distributed very quickly among the molecules by collisions. This fast relaxation results a small rise of the average energy, that is the temperature of the gas without creating any spatial anisotropy. In case of periodic absorption temperature and pressure waves are generated, but there is no direct momentum generation in the gas. This picture corresponds to the coupled acoustic equations [1] which also contain only one source term:

$$\rho \partial \boldsymbol{v}/\partial t + \nabla p = 0 \tag{1}$$

$$\partial p/\partial t + \rho c^2 \nabla \boldsymbol{v} = (\gamma - 1)\partial w/\partial t \tag{2}$$

where ρ, p, c, $\boldsymbol{v}$, γ and w are the density, pressure, sound velocity, particle velocity, specific heat ratio of the gas, and the volume density of the absorbed light energy, respectively.

A modulated light beam cannot generate longitudinal acoustic wave in a PA cell filled with a weakly absorbing gas [2]. The generated sound pressure in a cell depends only on the acoustic properties of the given experimental arrangement, and on the absorbed power. Therefore all the factors excluding the absorbed power can be included into a numerical constant (cell constant) [3], which plays an important role in the photoacoustic spectroscopy of gases.

2. Experiments

It was found recently [4] at the calibration measurements of an open resonant PA cell, that the cell constant has different values for different types of absorbing gas mixed into pure nitrogen resulting absorber concentrations of the ppm range.

Springer Series in Optical Sciences, Vol. 69 Photoacoustic and Photothermal Phenomena III
Editor: D. Bićanić

In another experiment the modulated beam of a CO_2 laser was sent uniaxially through an open metal tube. The laser, the chopper and the tube were placed into an anechoic room to avoid noise interferences. The sound pressure inside the tube was measured as a function of the chopping frequency. The details of this experiment may be found elsewhere [5], now only the main result is cited: all the longitudinal resonances were observed. This fact contradicts to the present theory, which allows only the excitation of the odd resonances [1].

The experimental results may be fitted to a very simple one dimensional model, which assumes that along the beam point-like monopoles emit pressure and velocity waves simultaneously. In case of infinite medium this waves cancel each other in the longitudinal direction, but in a cell the waves are reflected from the ends, thus the resultant acoustic field can be determined by summing up all the reflected partial waves. The result of such a calculation is presented in [5]. Moreover, recent experimental data suggest [6], that the relative strength of the pressure and velocity generation may also depend on the properties of the excited molecular transition.

Although the assumption of simultaneous pressure and velocity generation seems to be essential for modelling the experiment, it is not clear yet, what kind of physical process leads to this type of velocity generation. The basic ideas of a possible model is given now, the detailed theory will be presented elsewhere [7].

3. Theoretical considerations

The "velocity generation" used in [5] may exist only if the average speed of the thermal motion of the molecules is bigger along the laser beam than in the other directions. Such kind of spatial anisotropy cannot be created by pure thermal processes. But the photon absorption process itself deposits $h\nu/c_0$ momentum and $h/2\pi$ angular momentum (h, ν and c_0 are Planck's constant, light frequency and light velocity, respectively) into every absorbing molecule in an anisotropic way. (As angular momentum is deposited into the medium during illumination, a third phenomenological equation would be needed besides Eqs.1-2. for describing the conservation of angular momentum). The momentum of the photon is parallel with the propagation direction of the laser beam, while the angular momentum is transversal. The direct longitudinal momentum transfer may be neglected, because the photon momentum is very small, but an indirect process of longitudinal momentum production is also present. During the collisions the "inner" angular momentum of the colliding molecule may transform to orbital angular momentum of the other molecule [8], changing also its momentum. The amount of momentum transferred by this kind of indirect process may be estimated as: *transferred angular momentum/characteristic distance of the collision.* The latter quantity can be approximated by the average linear dimension of the molecule. Therefore, the indirect momentum transfer will be much bigger than the direct one, being $c_0/\nu=\lambda$ much bigger than the molecular dimensions.

This indirect momentum-angular momentum transfer is mantained until the gas is irradiated by the light beam, creating the spatially anisotropic momentum (velocity) distribution mentioned above in connection with the term "velocity generation". That is, the average momentum of the gas molecules moving parallel to the laser beam will be bigger, than the momentum of transversal motion, but there is no net momentum generation, because both longitudinal directions are equally probable. If the light is blocked, the spatial isotropy will be quickly restored.

The longitudinal symmetry of the velocity distribution will break down at the boundaries, like tube ends, creating a mean flow, which may be regarded as sound. In other words this kind of sound generation occurs only if reflection is also present.

This velocity source may be easily introduced to the equivalent electric four-pole network of the open tube as two current sources at the in- and output of the network. The amplitude of the acoustic velocity may be approximated as $R\beta w/\rho\omega$, where R is the reflection coefficient of the acoustic particle velocity, $\beta(m^{-1})$ is a constant to be determined by experiments, (its value may be in the order of the inverse of the molecular radius) and ω is the modulation frequency of the light.

These velocity sources cannot be written directly as source terms in the right side of Eq.1., because the described physical process cannot generate vectorial flow in the medium. However, it is possible to modify the Boltzmann equation of the kinetic theory such a way, that the indirect angular momentum-momentum transfer would be included. Based on the modified kinetic model the new phenomenological equations may be determined. Until this enourmous work will be completed, the proposed equivalent network can be used for determining the unknown β quantity from the measurement. On the other hand, knowing the value of β, the velocity contribution can be calculated for any one dimensional PA arrangement.

4. References

1. P.M.Morse, K.U.Ingard:*Theoretical Acoustics*, (McGraw-Hill, New York 1968)
2. A.C.Tam, in *Ultrasensitive Laser Spectroscopy*, ed.by D.S.Kliger, (Academic, New York 1983), Chap.1.
3. C.F.Dewey, Opt. Engineering **13**, 483-488, 1974
4. Gy.Z.Angeli et al., Conf.Digest of 7th Int.Top.Meeting on PA and PT Phen, 96-97, Doorwerth 1991
5. A.Miklos, Z.Bozoki et al., Conf Digest, 411-412, Doorwerth 1991
6. A.Miklos and D.D.Bicanic, submitted to Appl.Phys.**B**.
7. A.Miklos, to be published
8. L.D.Landau, E.M.Lifshitz:*Course of Theoretical Physics*, Vol.10. Kinetic theory (Pergamon, Oxford, 1980)

Photoacoustic Studies of Non-Radiative Energy Transfer in the Flexible Styrylcyanine-Rhodamine B Dye System

M.S.A. Abdel-Mottaleb[1], *S. Negm*[2], *and H. Talaat*[2]

[1]Department of Chemistry, Faculty of Science, Ain Shams University, Abbassia, Cairo, Egypt
[2]Department of Physics, Faculty of Science, Ain Shams University, Abbassia, Cairo, Egypt

Abstract: We have employed (PAS) photoacoustic spectroscopic technique to demonstrate the ability of the styrylcyanine (dye 1) to act as an energy donor via non-radiative transfer of excitational energy process to a suitable acceptor molecule such as rhodamine B (Rh B). Considering the low yield of fluorescence for dye 1 which is characterized by very large Stokes shift (λ_a = 485 nm and λ_f = 613 nm), it follows that the lowest singlet planar charge-transfer state ($S_{1,ct}$) of dye 1 relaxes almost entirely by internal conversion. The non-radiative energy deactivation process can be thought of as having two sources: (i) the Stokes heating (resulting from long range intramolecular geometrical relaxation to yield a weakly fluorescent TICT-like $S_{1,ct}$ state) and (ii) the vibrational relaxation. It is observed that addition of Rh B to the ethanol solution of dye 1 at room temperature completely quenches its PA spectrum. This observation agrees qualitatively with the conclusion that energy transfer process takes place from the planar $S_{1,ct}$ state of dye 1.

1. Introduction:

Photoacoustic (PA) spectroscopy offers a powerful technique to study absorption spectra of highly fluorescent dyes and gives complementary information to fluorescence spectroscopy through detection of non-radiative relaxation process. The data obtained in this way are expected to complement the information obtained by fluorescent spectroscopy. In this work, we have selected two dyes, styrylcyanine (Dye 1) and rhodaminc B (Rh B) where the first is characterized by a lower quantum yield and a larger stokes shift. The PA spectra of the solid state of the two dyes and of pure solutions of the dyes at different concentrations have been recorded. PA studies of excitation transfer in a mixture of the two dyes is reported.

2. Experimental:

Organic dye solutions (styrylcyanine and Rh B) were prepared in ethanol after purification by double distillation. The mixtures of the two dyes were made of equal volumes of donor dye 1 (styrylcyanine) of concentration 10^{-3}M, and different concentrations of the acceptor dye (Rh B). The incident light from

1000 W tungsten halogen lamp was condensed on the entrance slit of a monochromator using two convex lenses. The output light from the exit slit was focussed on the sample inside the cell by using a camera lens. The chopping frequency of the incident light was 11 Hz. The dyes in powder form and in solution were placed in a cylindrical stainless steel cavity of the Princiton Applied Research (PAR) photoacoustic cell. The signal from the PAR cell was lock-in amplified. All the spectra were normalized using a carbon black sample.

3. Results and Discussion:

The PA spectra of different concentrations of styrylcyanine in ethanol as well as the spectra of the dye in solid state is shown in Fig.(1–a), the corresponding PA spectra of Rh B in ethanol is shown in Fig.(1–b). The PA spectra of styrylcyanine and Rh B exhibit peaks at 483 and 550 nm, respectively. The PA signal in the case of solid samples show the regular red shift due to dye ag-

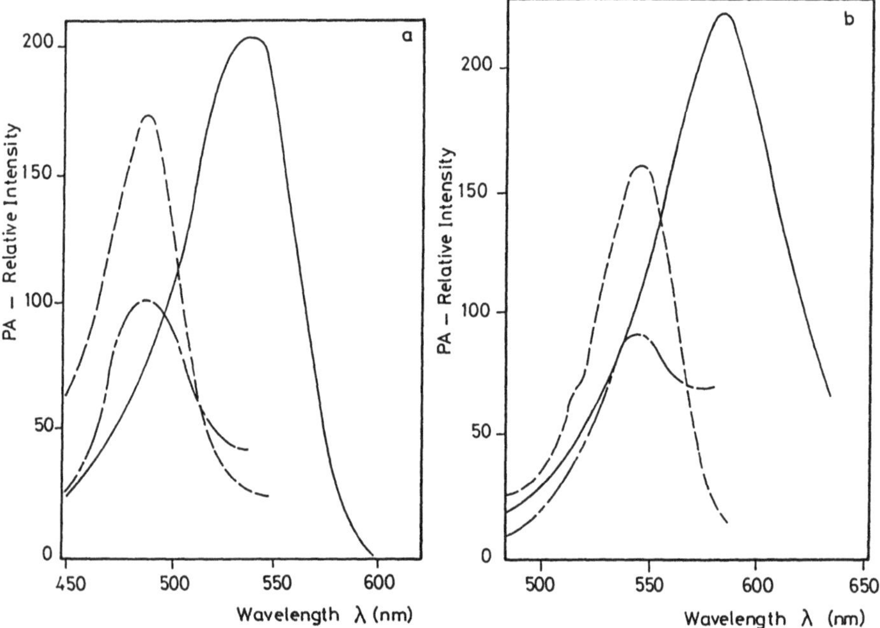

Fig.1. a) Photoacoustic spectra of Styrylcyanine in its solid state (——) and in ethanol 10^{-3} M/L (—-—), 10^{-2} M/L (– – –).

b) Photoacoustic spectra of Rh B in its solid state (——) and in ethanol 10^{-3} M/L (—-—), 10^{-2} M/L (– – –).

The red shift of the signal of the solid samples in the two cases is due to dye aggregate.

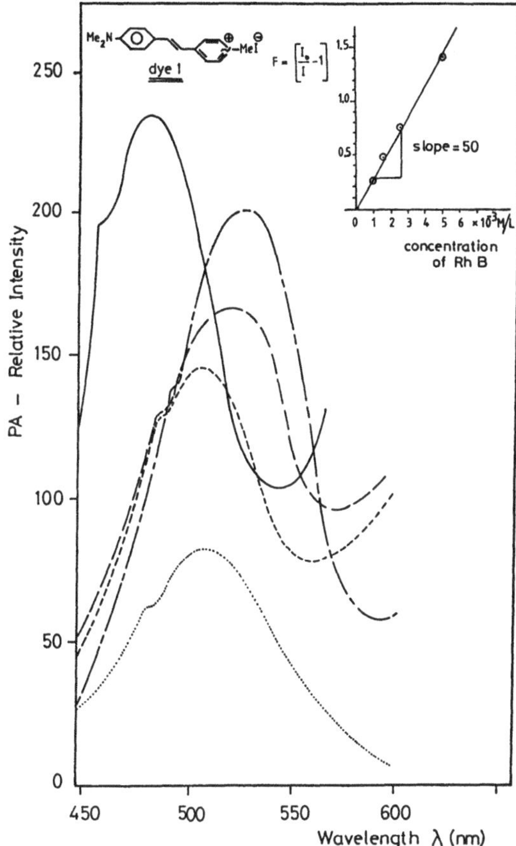

Fig.2. Photoacoustic spectra of Styrylcyanine in ethanol at 10^{-3} M/L. (——)
Photoacoustic spectra of mixture of Styrylcyanine at 10^{-3} M/L with
- 10^{-3} M/L of Rh B. (— – –)
- 1.6×10^{-3} M/L of Rh B. (– – –)
- 2.5×10^{-3} M/L of Rh B. (- - - - -)
- 5×10^{-3} M/L of Rh B. (..........).

gregates. The normalized measured PA spectra for mixtures of styrylcyanine and Rh B are shown in Fig.(2). It is observed that the addition of Rh B to the ethanol solution of dye 1 at room temperature quenches PA spectra for dye 1. This indicates that excitation transfer is taking place in the mixture and that dye 1 acts as an energy donor via non-radiative energy transfer to Rh B through vibrational relaxation and long range intermolecular geometrical relaxation processes. Furthermore, the spectra shows an increase in intensity in going towards larger wavelength λ. This could be attributed to opening of new non-radiative channel through TICT state. It is known from fluorescence studies [1] of the dyes mixture that there are two peaks corresponding to the two dyes. PA spectra, on the other hand, for the mixture show quenching of PA spectra for dye 1 and the peak corresponding RH B by itself increases with increasing concentration between 10^{-3} - 10^{-2} M. This is explained in terms of Rh B acting as the acceptor to the non-radiative energy relaxation of the strained Frank-Condon (FC) state of dye 1 which is immediately reached upon light absorption. The non-radiative relaxation is due to both reorientation relaxation and geometry relaxation of the strained state. Fluorescence studies [1], show a three order of magnitude increase in fluorescence intensity for dye 1 in ethanol glass (77 K) accompanied with a pronounced decrease in Stokes shift due to vibrational relaxation. The fact that the fluorescence spectrum of the mixture of these dyes still exhibits the weak fluorescence band corresponding to the TICT-like $S_{1,ct}$ points to a minor contribution of the ultra fast Stokes relaxation in the fluid ethanol to the overall non-radiative deactivation processes.

The variation of the ratio I_o/I where I_o is the PA signal of the pure donor (styrylcyanine) and I is the PA signal of the donor in the presence of the acceptor dye (Rh B) as a function of concentration of the acceptor shows a straight line with the positive slope indicating that they follow the Stern Volmer relationship

$$\frac{I_o}{I} = 1 + K\tau[A] = 1 + K_{SV}[A]$$

where τ is the lifetime, K is the rate of energy transfer, and [A] is the acceptor concentration. The straight line slope gives the Stern Volmer rate constant $K_{SV} = 50\ LM^{-1}$.

One can conclude from PA and fluorescence results , that energy transfer processes take place from the planar $S_{1,ct}$ state of dye 1 to Rh B. Moreover, the fact that the fluorescence spectrum of the dye mixture still exhibits the weak fluorescence band corresponding to the TICT-like $S_{1,ct}$ state points to a minor contribution of the ultrafast Stokes relaxation in the fluid ethanol to the overall non-radiative deactivation processes. This is reflected in the PA spectrum of dye mixture where a weak long wavelength PA band (above 600 nm) appears demonstrating that the TICT state acts not only as a weak fluorescent state, but also as non-radiative channel.

4. References:

1. M.S.A. Abdel-Mottaleb et al; J. Chem. Soc., Faraday Trans., 2, 1989, 85(11), 1779-1788.

Material Characterization by Interferometric Detection of Acoustically Induced Surface Displacements

P. Weiss and M.W. Sigrist

ETH, Institute for Quantum Electronics, CH-8093 Zürich, Switzerland

Abstract. A fiber-optic interferometer as a new tool for the transient detection of ultrasound is presented. A resolution corresponding to displacement amplitudes of 1 nm at 80 MHz bandwidth is achieved. As a first application the monitoring of the laser-induced spallation of a surface film is demonstrated.

1. Introduction

The investigation of mechanical properties of composite materials is of great interest particularly with respect to the characterization of interface strength and adhesion properties of surface films on solids. Recently, preliminary studies based on a laser spallation technique have been reported where laser-induced stress pulses are applied to spall off the coatings under investigation. In principle, the interface strength could then be deduced from the spallation threshold. However, the experimental precision was insufficient to derive actual interface strength data [1].

We present a different technique that is based on the noncontact detection of the surface displacement of the film, with or without spallation. This displacement is induced by a stress pulse that is generated by the impact of a laser pulse at the opposite (rear) surface of the substrate. A detailed analysis of the transient displacement at and below spallation threshold permits the characterization of adhesion strengths in a nondestructive and noncontact manner.

2. Fiber-optic Interferometer for the Detection of Ultrasound

We generate the acoustic transients by the impact of pulses from a hybrid CO_2 laser system with laser pulses of ca. 180 mJ energy and 150 ns pulsewidth. A thin water layer is applied at the rear surface to enhance the generation of longitudinal waves.

The displacement of the film at the front surface is detected by our new homodyne fiber-optic interferometer [2]. It is a modified version of the interferometer described previously [3,4], adapted to the detection of ultrasound by increasing the bandwidth from 20 kHz to 80 MHz. This instrument offers excellent inherent stability due to the small interferometer cavity size of only a few micrometers and is not very sensitive to the surface quality of the film to be monitored. At present, our system exhibits a noise level of 5 10^{-4} Å/Hz$^{1/2}$, which permits measurements of displacement amplitudes as low as 1 nm at the maximum bandwidth of 80 MHz or of 1 Å at 11 MHz. In the spallation regime, the counting of fringes allows the determination of the speed of the spalled-off piece of film up to 30 m/s over distances of ca. 100 µm.

3. First Results and Discussion

First spallation measurements have been performed on samples consisting of a 0.2 mm thick quartz plate with an evaporated gold film of 300 nm thickness.

Figure 1a shows the CO_2 laser pulse which produced spallation of the gold film. The temporal behaviour of the interferometer signal, recorded in Fig. 1b, is typical for spallation phenomena. First, the film surface is at rest, until the acoustic pulse reaches the front surface, is reflected and produces the tensile stresses which eventually break the substrate-film interface. The displacement of the film during spallation can be seen in the expanded signal, shown in Fig. 1c.

The spalled-off piece of film now moves away from the substrate, clearly monitored in the interferometer signal, which slowly dies away until it vanishes completely when the reflecting gold film has disappeared (Fig. 1b). Fig. 1d shows the interferometer signal produced by the moving piece of film, 4-6 µs after the onset of spallation. The separation between the maxima corresponds to displacements of $\lambda/2$, with λ=633 nm being the interferometer wavelength. The Fourier transform of Fig. 1d, displayed in Fig. 2, yields a distinct frequency of 36.5 MHz corresponding to a film speed of 11.6 m/s. At this point the film has already slowed down from its original speed of 12,3 m/s, derived from an identical analysis of the 1-3 µs section of the signal in Fig. 1b.

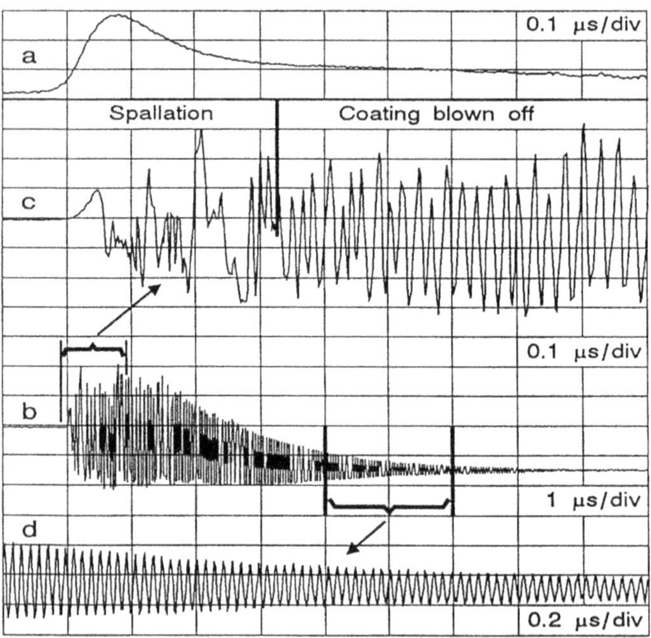

Fig.1. Time dependence of CO_2 laser pulse (a), interferometer signal of induced spallation (b) with expanded sections as indicated (c and d).

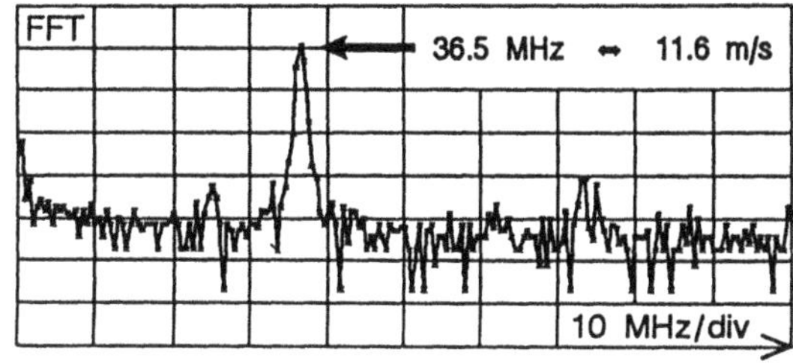

Fig. 2. Frequency spectrum of the signal section displayed in Fig. 1d.

4. Conclusion

A fiber-optic interferometer with high resolution and large bandwidth is proposed for the noncontact characterization of adhesion strengths at substrate-film interfaces. First studies on laser-induced spallation of surface films appear very promising.

This work is supported by the Swiss National Science Foundation, National Research Program 24.

References

1. V. Gupta, A.S. Argon, J.A. Cornie, D.M. Parks: Materials Science and Engineering **A 126**, 105-117 (1990)
2. P. Weiss, Y. Den, M.W. Sigrist: Digest 7th Int. Top. Meeting on Photoacoustic and Photothermal Phenomena, Doorwerth, NL (August 1991), paper O VI/6
3. A.D. Drake, D.C. Leiner: Rev. Sci. Instrum. **55**, 162-165 (1984)
4. D. Rugar, H.J. Mamin, R.Erlandsson, J.E. Stern, B.D. Terris: Rev. Sci. Instrum. **59**, 2337-2340 (1988)

An Optical Calorimeter Based on an Ultrasonic Resonator

O. Oehler, J. Wieland, D. Raillard, and M. Schumacher

ETH, Institute for Quantum Electronics, CH-8093 Zürich, Switzerland

Abstract. The usefulness of a calorimeter based on an ultrasonic resonator is demonstrated by a comparison of the calculated temperature distribution in the resonator cavity with optical absorption measurements. Further bending losses of a light guide were measured calorimetrically.

1. Introduction

The temperature induced detuning of ultrasonic resonators offers possibilities for the construction of calorimeters via the temperature dependence of the velocity of sound in gases. For such purposes the heat source to be measured is positioned in the air gap of the resonator [1, 2, 3]. This leads however to an inhomogeneous temperature distribution which needs to be carefully calibrated to render the calorimeter useful.

It is the purpose of this article to demonstrate the capibility of the device by a comparison of the calculated temperature distribution within the resonator with experimental results.

2. Apparatus layout

An ultrasonic resonator was built, consisting of two identical transducers [4] arranged opposite each other and at a distance of about 6 mm in a thermally controlled metal tube M (Fig. 1a). The transducers (one used as an emitter E, the other as a receiver R) were tuned to an acoustic resonance at a frequency of 220 kHz. One of them was fixed to a piezo-driven translator T. The phase of the signal from the receiver was compared to the one

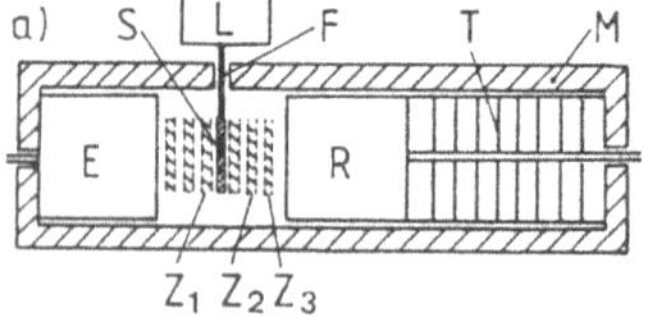

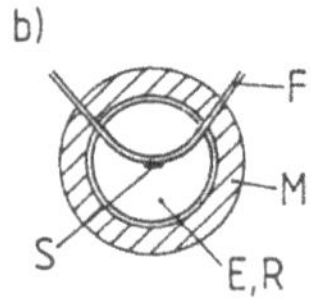

Fig. 1 Ultrasonic calorimeter, composed of two ultrasonic transducers E and R and the Piezo tuning translator T. The periodic heat source is mounted on an optical fiber F and is excited by the laser L.
a) arrangement for measuring the absorption losses in a glass rod S,
b) arrangement for measuring the bending losses in an optical fiber.

Springer Series in Optical Sciences, Vol. 69 Photoacoustic and Photothermal Phenomena III
Editor: D. Bićanić

of the input signal. A signal proportional to the phase difference was used to control the translator via an amplifier. A signal, proportional to the charge at the translator was fed to a digital storage oscilloscope.

The heat source to be measured - a blackend part of a glass fiber F periodically illuminated by a laser L - was positioned at the center of the airgap of the resonator.

3. Experimental

3.1 Resolution of the calorimeter

For calibrating the calorimeter and determining its resolution a calibrated heat source was used. This heat source was realized by the absorption of the intensity-modulated 633 nm radiation of a He-Ne laser which was guided through the fiber to its blackend end.

The spreading of the heat signal was calculated [5] on the basis of a spherical shell model using thermodynamic data of air. At an absorbed light power of 103 μW and a chopping frequency of 0.22 Hz average temperature variations in the resonator cavity of 5.5 mK were found while temperature variations of 6.3 mK were obtained experimentally.

A heat power resolution of 1 μW and a corresponding temperature resolution of 0.05 mK was found with an integration time of 80 sec.

3.2 Optical loss in a wave guide

Laser calorimetry is used for determining extremely low absorption coefficients in solids [6]. Such procedures usually are based on the measurement of temperature changes of a strongly laser irradiated wave guide by a thermocouple. At an optical power of 66.7 W and 5 W a temperature rise of 500 mK [7] and 0.5 mK [8], respectively, were obtained at the sample. It was the subject of our investigations to find out whether the crucial mounting of a thermocouple at the surface of a light guide could be avoided and whether experiments can be carried out at a lower optical power.

For this purpose a colored soft-glass sample S of cylindrical shape and a diameter of 0.6 mm and length of 2 mm was fixed at the tip of a 0.4 mm light guide. The absorption coefficient of the sample was found to be (0.54 ± 0.10) cm^{-1} from cutting measurements. The sample was irradiated by intensity-modulated 633 nm radiation at 0.4 Hz at a power of (840 ± 170) μW, corresponding to an absorbed power in the sample of (86 ± 23) μW. From our calorimetric investigations temperature variations of (18.2 ± 2.6) mK were obtained.

For testing the methods calculations the spreading of the heat signal was based on a thermodynamic model. This model was composed of three cylindrical zones Z_1, Z_2, Z_3 , concentrically surrounding the sample (Fig. 1a). The zones were separated by heat conducting layers. The results of these investigations are given in Table I. It follows that even small temparature changes are detectable. Therefore this method seems to be useful to investigate weakly absorbing samples.

Table I Temperature variations in mK at cylindrical zones Z_1, Z_2, Z_3 surrounding the cylindrical heat source operated at (86 ± 24) μW and 0.4 Hz

sample	111.3 ± 26.3
zone Z_1	52.0 ± 12.3
Z_2	27.1 ± 6.4
Z_3	11.4 ± 2.7
average	30.0 ± 7.1
exp.	18.2 ± 2.6

3.3 Bending losses in a wave guide

It is well known that the light-guiding quality of fibers is reduced in bendings [9] because of a failure of the internal total reflection condition.

In order to measure this effect a bent fiber F of a core diameter of 6 μm and an outer diameter of 125 μm was positioned in the ultrasonic resonator E,R. The exciting radiaton was monitored via its absorption at a black spot S at the surface of the fiber (Fig. 1b). The results of these investigations obtained with 633 nm laser radiation of 58 μW are given in Table II together with the results from transmission attenuation measurements. The quoted error limits correspond to the thermal resolution of the calorimeter. The scatter in the values of the last column is due to the inaccuracy of the definition of the geometric layout.

The method allows the determination of local optical losses of wave guiding systems.

Table II Bending losses of an optical fiber (6 μm, 125 μm) operated at a power of 58 μW

bending radius	bending loss obtaind from transmission attenuation		calorimetric investigation
mm	μW	dB/mm	μW
2.0	10.0	- 1.98	12.0 ± 1.1
2.8	7.2	- 0.91	15.2 ± 1.5
5.1	3.2	- 0.34	4.2 ± 0.9
7.1	2.2	- 0.21	2.0 ± 0.7

4. References

[1] O.Oehler, J.Wieland and S.Friedrich, Helv. Phys. Acta, **61**, 885 (1988).

[2] O.Oehler, S.Friedrich and A. Schäppi, Springer Series in Optical Sciences, **62**, Photoacoustic and Photothermal Phenomena II, 519 (1990).

[3] O.Oehler, P.Rusch and S. Dornbierer, Helv. Phys. Acta, **63**, 533 (1990).

[4] Model E-188/220, Massa Products Corp., Hingham, Mass 02043, USA.

[5] Computer program: Bionik, Dr. H. Weber, Vorstadt 6, CH-5200 Brugg, Switzerland.

[6] J.M.Jewell, G.M.Williams, J.Jaganathan, I.D.Aggarwal and P. Greason, Appl. Phys. Lett. **59**, 1 (1991).

[7] D.A. Pinnow and T.C.Rich, Appl.Opt. **12**, 984 (1973).

[8] M.Hass, J,W.Davisson, H.B.Rosenstock and J.Babiskin, Appl. Opt. **14**, 1128 (1975).

[9] J.A. Arnaud, The Bell System Technical Journal, **53**, 1379 (1974).

Optoacoustic Source as a Tool for the Investigation of Nonlinear Acoustic Effects

T.A. Dunina, S.V. Egerev, and A.T. Skvortsov

N.N. Andreev Acoustics Institute, Academy of Sciences, Shvernika St. 4, 117036 Moscow, Russia

Abstract. The problem of nonlinear acoustic pulse generation due to laser photoablation is examined. The media of interest were those possessing dispersive properties: liquids with bubbles or waveguides. The opportunity of soliton-like acoustic pulses observation is discussed.

Laser pulse generation of short (1-10mcs) acoustic pulses of large magnitude (up to 100 MPa) allows to observe various nonlinear effects in condensed media. In [1] acoustic pulses of thermooptical origin were used for the purpose of nonlinear parameter evaluation. The present paper considers the problem of non-linear acoustic pulse generation due to laser photoablation. The opportunity of so-called "acoustical analogs of non linear optics phenomena" [2] observation is discussed.

The stage of non-linear acoustic pulse generation. The review of previous photoablation studies for water is presented. It appeared to be possible to generalize the experimental results [3-7] and display them in a common plot"peak surface pressure (P_s) vs peak laser intensity (I_0)"(fig.1). Necessary near-field data were obtained with the help of corresponding set of diffraction parameters $D = Rc\tau_0/a^2$, where R is the radius from the source to the observation point, c is the sound velocity, **a** is the characteristic radius of the beam cross section, τ_0 is the laser pulse duration (approximately equal to that of following acoustic pulse). Also, near-field theoretical results [8-10] are displayed. As may be seen, for water with the free surface the following approximation is valid P_s**[MPa]** $\sim$ **0.4**$\cdot I_0$**[MW/sq.cm]**.

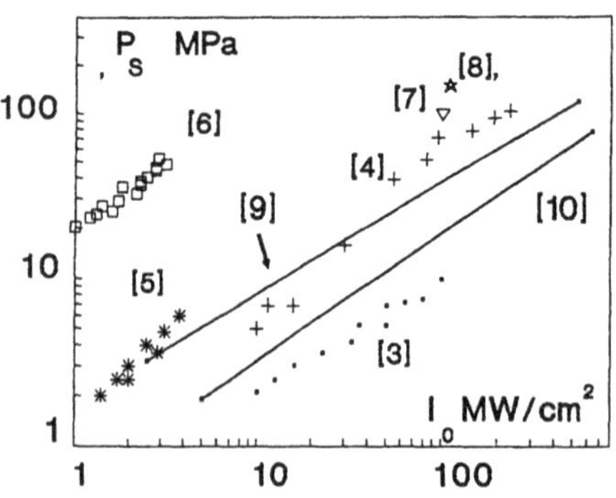

Fig.1. Peak surface pressure vs peak laser intensity. Cumulative plot.

Springer Series in Optical Sciences, Vol. 69 Editor: D. Bićanić Photoacoustic and Photothermal Phenomena III

The results of [6] are worth to be mentioned separately: photoablation in a closed volume gives rise to especially effective optoacoustic (OA) source generation.The second half of the paper deals with propagation of wideband "surface-born" acoustic pulse with magnitude up to 100 MPa and its interaction with non-acoustic motions of condensed media.

The stage of acoustic pulse propagation. The opportunity of observation of non-linear acoustic phenomena stimulated by OA source may be analyzed with the help of the following model using the apparatus of well-known theory of non-linear acoustic beams with narrow angle spectrum ($a/c\tau_0 \ll 1$). We consider medium with weak dispersion ($\omega = ck+\beta k^3$) and weak absorption ($Im(\omega)=\delta k^2$). The equation for acoustic pressure **p** has the form

$$\frac{\partial}{\partial z}\left[\frac{\partial}{\partial \tau} - \alpha p\frac{\partial}{\partial z} - \delta\frac{\partial^2}{\partial z^2} - \beta\frac{\partial^3}{\partial z^3}\right] = -\frac{c}{2}\Delta_\perp p, \quad (1)$$

where z -distance from the surface along the laser beam axis,$\Delta_\perp$- "transverse" Laplacian, $\tau = t - z/c$, $\alpha = \varepsilon / \rho c$, ($\rho$ - mass density). Relative roles of terms of (1) depend both on OA source parameters (P_s, τ_0) and medium parameters ($c,\varepsilon,\beta,\delta$). It is convenient to introduce following characteristic scale distances $L_N= \rho c^3\tau_0/\varepsilon P_s$,$L_D= a /2c\tau_0$,$L_\beta= \tau_0 c^3/\beta^4$, $L_\delta= \sqrt{\delta\tau_0}$,defined respectively by the conditions that pulse displays its nonlinear, diffraction, dispersion and dissipation properties.Various relations between these scales allow to neglect this or that term in (1).So pulse propagation may be analyzed with the help of more simple equations.

In particular, if $L_D \gg L_N, L_\beta, L_\delta$,diffraction term may be neglected ($\Delta_\perp p = 0$), also if $L_\beta \gg L_N, L_\beta, L_\delta$ dispersion term may be neglected ($\beta = 0$) and if $L_\delta \gg L_N,L_D,L_\beta$ dissipation term is small in comparison with the others ($\delta = 0$). As a result, equation (1) obtains forms of Korteveg- de Vries or Burgers equations or their "nearly -2D" analogs (Kadomtsev - Petviashvili equation or Khohlov - Zabolotskaya equation [11]). It is shown that with the help of analytical solutions of the equations mentioned (describing shock waves, solitons, self-focusing, etc.) it is possible to observe corresponding phenomena experimentally. Methods of such OA source and medium parameters matching are also shown.For example, the convenient way of L_N,L_β minimization is to saturate liquid with gas bubbles. Soliton-like acoustic pulse, following laser photoablation of water, was observed as a result [12] (fig.2). It is necessary to note, that due to wideband character of acoustic pulse nonlinear effects may be observed for separate frequency components independently if they obey above mentioned conditions.

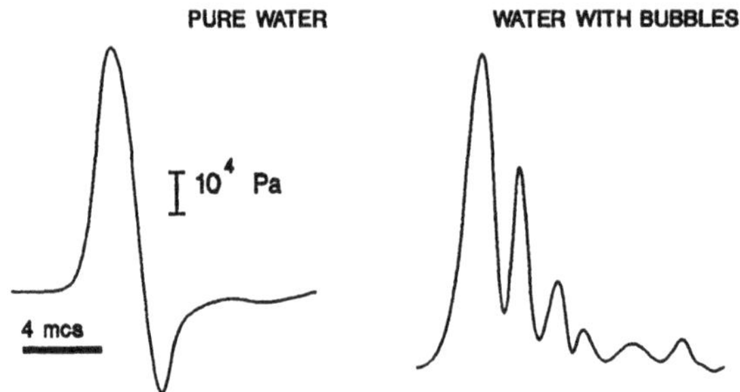

Fig.2. OA-signal profiles

REFERENCES

1.M.W.Sigrist,J.Appl.Phys.,**60**,83(1986).
2.F.V.Bunkin,Yu.A.Kravtsov and G.A.Liakhov, Usp.Phys.Nauk, **149**,391(1986) (in Russian).
3.M.W.Sigrist and F.K.Kneubuhl,J.Acoust.Soc.Am.,**64**,1652(1978).
4.P.Giovaneschi et al.,Appl.Phys.Lett.,**36**,882(1980).
5.A.F.Vitshas et al.,Pis'ma Zh.T.Ph.,**9**,1438(1983)(in Russian).
6.A.F.Vitshas, et al.,Pis'ma Zh.T.Ph.,**14**, I57 (1988) (in Russian).
7.D.Emmony,T.Geerken and N.Klein-Baltink, J.Acoust.Soc.Am., **73**,220(1983).
8.F.D.Feiock and L.K.Goodwin,J.Appl.Phys.,**43**,5061(1980).
9.A.N.Pirri,Phys.of Fluids,**16**,1435,(1973).
10.N.P.Chotiros,In:SPIE v.925 Ocean Optics IX(1988).
11.O.V.Rudenko and S.I.Soluyan, Fundamentals of Nonlinear Acoustics,NY,1977.
12.T.A.Dunina et al.,In: Problems of Nonlinear Acoustics, Nauka, Novosobirsk,1987, Part 1,p.447(in Russian)

Nonlinear Photoeffect at the Boundary of a Semiconductor and Liquid Helium

V.E. Gusev, S.A. Telenkov, and B.V. Zhdanov

International Laser Center, Moscow State University, 119899 Moscow, Russia

Abstract. We investigated longitudinal compression acoustic pulses (propagating in semiconductors) induced by the interband absorption of pulsed-laser radiation near a crystal surface in contact with liquid helium. Theoretical analysis revealed that the thermal flux from the semiconductor to helium initiates heating and boiling of the liquid. It is the pressure of helium vapour that produces the dominant contribution to the process of sound excitation. It was shown that the helium vapour pressure is proportional to the square root of heat flux spreading into the helium.

In the experiments performed at temperature $T_0 = 4.2\ {}^{o}K$ we used irradiation of Ge, Si and GaAs by laser pulses ($\lambda = 0.53\ \mu m$ - light wavelength, $\tau_L = 15$ ns - optical pulse duration, $10^4\ W/cm^2 \leq I_L \leq 10^7\ W/cm^2$-laser intensity). We registered acoustic pulses with $LiNbO_3$, CdS and ZnO piezotransducers. The main result of the investigation was the square root dependence of the amplitudes of compression pulses (launched into the semiconductors) on the laser intensity $A \sim I_L^{1/2}$(see Fig.1).

In accordance with the proposed theoretical model, interband light absorption in a semiconductor initiates heat flow into the helium. The increase of liquid helium temperature leads to its boiling, with acoustic pulses generation followig as a result of helium vapor pressure on the semiconductor boundary. The direct experimental evidence for the proposed mechanism of acoustic pulses excitation was obtained by surrounding the illuminated surface of the semiconductor by vacuum chamber. This led to the significant diminishing of the signal amplitude and appearance of the second (dilatation) phase of the acoustic pulse, typical to optical generation of sound near the free surface of crystal.

Sound wave generation in Ge was considered in detail. It was shown that for $5 \cdot 10^5\ W/cm^2 \leq I_L \leq 10^7\ W/cm^2$ the energy flux

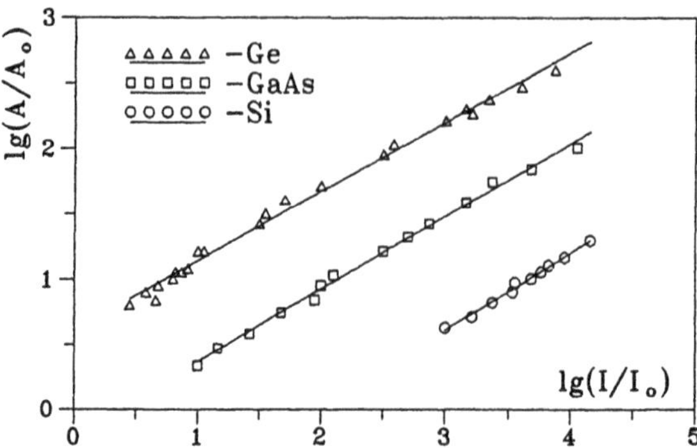

Fig.1. Acoustic signal amplitude as a function of laser intensity. For normalization the values A_o=10mV and $I_o=10^3 W/cm^2$ are used. The slope of the straight lines through the data points is calculated theoretically.

into the helium predominantly caused not boiling of the liquid, but expansion of the already existing vapor. In this situation the vapor pressure depends on the characteristic temperature nonlinearly $P \sim T^{1/2}$. It is also important that the influence of heating on the heat conductivity ($\varkappa \sim 1/T$) and heat capacity ($c_p \sim T$) of the crystal produces a linear relationship between the surface temperature of Ge and the laser pulse intensity ($T \sim I_L$).

The heat flux into the vapor Q_V induces additional vaporization, vapor heating, and expansion with Q_V given by

$$Q_V \sim (\varkappa_V \rho_V C_p^V \tau_L)^{1/2} T \sim \Delta h \rho_V v_B \tau_L + \frac{\gamma_V}{\gamma_V - 1} P v_B \tau_L \qquad (1)$$

Here ρ is the density, τ_L is the laser pulse duration, γ is the adiabatic parameter, P and T are the characteristic pressure and temperature, v_B is the characteristic velocity of the vapor - liquid boundary, Δh is the difference of enthalpies $\Delta h = h_V - h_L$. We used indexes "V", "L", "S" to denote physical parameters of vapor, liquid and the semiconductor. In the case $\rho_V \ll \rho_L$ the velocity v_B is determined by the hydrodynamical motion of the boundary

$$v_B \simeq P/\rho_L c_a^L \qquad (2)$$

where c_a - longitudinal sound velocity. Our estimates showed that the second term at the right hand side of Eq.(1) dominates if $P \geq (4/3)P_o$ (P_o - atmospheric pressure). In this case using the equation of state for ideal gas in the form $P = (\gamma_v-1)\gamma_v^{-1}c_p^v\rho_v T$ and the dependence of $\varkappa_v$ on T ($\varkappa = \varkappa_v^*(T/T^*)^{1/2}$) we obtained from Eq.(1)

$$P \simeq (\rho_L c_a^L)^{2/3}[\varkappa_v^* T^*(\gamma_v-1)\gamma_v^{-1}\tau_L]^{1/3}(T/T^*)^{1/2} \simeq 1.5(T/T^*)^{1/2} \text{ bar} \quad (3)$$

where $T^* \simeq 10\ ^oK$. The characteristic temperature of the semiconductor surface T can be found from the energy balance

$$Q_s \simeq (\varkappa_s \rho_s c_p^s \tau_L)^{1/2} T \simeq \frac{h\nu_L - E_g}{h\nu_L}(1-R)I_L\tau_L \quad (4)$$

Here $Q_s \gg Q_v$ is the heat flux into the semiconductor, $h\nu_L$ and E_g are the energy of light quantum and the forbidden energy gap of the crystal ($h\nu_L > E_g$), and R is the optical reflection coefficient. In the case of Ge for $I \geq 10^4$ W/cm^2 the inequality $T \geq T^*$ is fulfilled and the following approximations $\varkappa_s \simeq \varkappa_s^*(T/T^*)^{-1}$, $c_p^s \simeq c_p^*(T/T^*)$ are valid. Then one can obtain from Eq.(4)

$$T - T^* \simeq 0.2(I_L/10^4 \text{ W/cm}^2)\ ^oK \quad (5)$$

The expressions (3) and (5) describe the square root dependence of acoustic pressure on laser intensity in Ge for $I_L > 5\cdot10^5$ W/cm^2: $P - P_o \simeq 0.2(I_L/10^4 \text{ W/cm}^2)^{1/2}$ bar

This result correlates with our experimental observations (Fig.1).

Probe Beam Detection of IR and UV Laser Polymer Ablation on the Nanosecond Time Scale

L.M. Kukreja[1] *and P. Hess*[2]

[1]Laser and Plasma Technology Division, Bhabha Atomic Research Centre, Bombay-400 085, India

[2]Institute of Physical Chemistry, University of Heidelberg, Im Neuenheimer Feld 253, W-6900 Heidelberg, Fed. Rep. of Germany

Abstract. Time-resolved attenuation of a probe HeNe laser beam by the ablation plume is used to study ablation of polyethyleneterephthalate (PET) and polyimide induced with a KrF excimer laser (248 nm) and a TEA-CO_2 laser (9.2 μm). The delay time of the UV ablation in PET with respect to the laser pulse is found to decrease from about 750 ns to 20 ns when the fluence is increased from the detection threshold of about 8 mJ/cm^2 to 1 J/cm^2. In the case of UV ablation at the cut edges of the polymer sheets a fast signal has been observed that follows the temporal profile of the laser pulse. These data yield important information on the mechanism of the ablation process.

1. Introduction

Time-resolved analysis of laser-induced ablation is important for an understanding of the mechanism of this process [1]. Various techniques have been developed to obtain time-resolved information on laser-induced ablation in polymers [2,3]. Recently Sell et al. have used the acoustic and thermal deflection of a HeNe laser beam to study excimer laser ablation of polymers on a μs-ms time scale [4]. We have used the time-resolved attenuation of a probe HeNe laser beam by the ablation plume to study the laser ablation of polymers with a time resolution of about 2 ns. The important results of these experiments are presented in this paper.

2. Experimental

The experimental setup consisted of a Lambda Physik KrF excimer laser operating at 248 nm or a Lumonics line tunable TEA-CO_2 laser used at 9.2 μm for inducing the ablation in 23μm thick polymer sheets of polyethyleneterephthalate (PET) and 25 μm thick sheets of polyimide. A Uniphase HeNe laser was aligned parallel to the polymer sheets from a touching configuration to a maximum separation of about 3.5 mm to probe the ablation process. A fast PIN photodiode (rise time about 1 ns) was used to detect the attenuation signal from the HeNe laser beam. A second PIN photodiode was used to record the excimer laser pulses, which also triggered the oscilloscope. These photodiodes were positioned in such a way that the transit time of the respective signals from the site of ablation was about the same. The signals were recorded with a 500 MHz Tektronix oscilloscope.

Springer Series in Optical Sciences, Vol. 69 Photoacoustic and Photothermal Phenomena III
Editor: D. Bićanić

3. Results

Figure 1 shows a typical attenuation signal obtained from the ablation of PET with the KrF excimer laser. It was observed that the UV laser-induced ablation signal from a region with cut edges of the polymer sheet was significantly different than that from the normal surface of the polymer. As can be seen in fig. 1, apart from the normal signal obtained from the surface ablation, a much faster signal appeared in the case of edge ablation. When the transit time of the ablation products through the probe laser beam is taken into account, it is found that the fast signal follows the temporal profile of the excimer laser pulse within the time resolution of the detection system. To calculate the transit time, the velocity of the ablation products was measured by a time-of-flight experiment. This was done by determining the slope of the curve between the delay of the ablation signal and the distance of the HeNe laser beam from the polymer surface. The average velocity of the plume from PET in the UV ablation was about $8x10^3$ m/s at the laser fluence of about 1 J/cm^2 and it decreased linearly on decreasing the fluence.

An interesting behaviour of both the IR and UV ablation signals was that their delay with respect to the start of the laser pulse decreased on increasing the fluence. But the fast signal of the edge ablation with the UV laser radiation remained temporally clamped to the laser pulse on varying the fluence. These observations are shown in fig. 2. It can be noticed from this figure that at high fluences the onset time of the normal signal tends to merge with that of the fast signal. The fast signal was absent in the case of IR ablation.

From these results we conclude that the laser-induced ablation is predominantly a thermal process even at 248 nm. The fast signal of the edge ablation with the UV laser radiation appears to orignate from a direct photo-decomposition process.

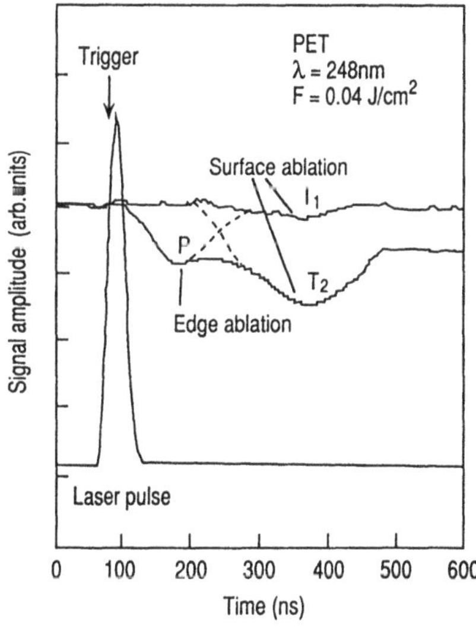

Fig. 1: Time resolved attenuation signals of the ablation in PET with KrF excimer laser radiation at the surface and cut edges of the polymer sheets.

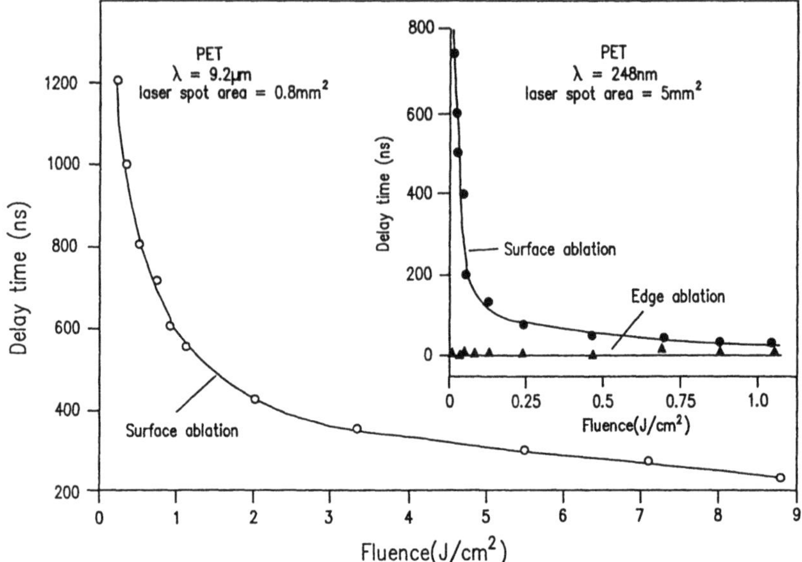

Fig.2: Dependence of the delay between the start of laser pulse and the ablation on the fluence for the TEA-CO_2 laser and the KrF excimer laser. For the TEA- CO_2 laser the delay times of the surface and edge ablations were not resolved.

Acknowledgement: The Authors are thankful to the Deutsche Forschungsgemeinschaft and Fonds der Chemischen Industrie for financial support and to the Alexander von Humboldt Stiftung for a research fellowship to L.M.K.

References

1. E.Sutcliffe and R.Srinivasan: J. Appl. Phys. **60**, 3315 (1986)
2. D.L.Singleton,G.Paraskevopoulos and R.S.Taylor:Appl. Phys. **B 50**, 227(1990)
3. P.E.Dyer and R.Srinivasan: Appl. Phys. Lett. **48**,445(1986)
4. J.A.Sell,D.M.Heffelfinger,P.Ventzek and R.M.Gilgenbach: Appl. Phys. Lett. **55**, 2435 (1989)

Picosecond Photoacoustics in Thin Films of SiO_2 and Si_3N_4

O.B. Wright, T. Hyoguchi, and K. Kawashima

Nippon Steel Corporation, Electronics Research Laboratories,
5-10-1 Fuchinobe, Sagamihara, Kanagawa 229, Japan

Abstract. We present a new optical technique, PATS, or picosecond acoustics for thickness and sound velocity, for the non-destructive testing of thin transparent films on opaque substrates. It is applied to films of silica and to a double layer transparent film of silica on silicon nitride.

Transparent thin films in the thickness range 200 Å to 5 μm have applications in many fields such as in the microelectronics or coatings industries. Conventional ultrasonic techniques at GHz frequencies[1] can be used to probe the acoustic properties of thin films but require contact with the film. We use a non-contact photoacoustic technique with picosecond laser excitation and detection of ultrasonic pulses of frequency ~100 GHz[2] to probe transparent films on opaque substrates[3]. Both the film thickness and the sound velocity can be derived from the data, provided that the film refractive index n is known.

The film reflectance change δR on a picosecond time scale is measured with a pump-probe apparatus incorporating an optical delay line[4]. Laser pulses of duration 3 ps, wavelength $\lambda=630$ nm, pulse energy 0.3 nJ and repetition rate 76 MHz are focused to a spot diameter 10 μm (see Fig. 1a).

The first sample investigated was an a-SiO_2 film of nominal thickness 2000 Å sputtered onto an a-Ge substrate film (4600 Å) on a silica base. Fig. 1b shows the measured $\delta R/R$ variation. In Fig. 1c (upper curve) the background thermoreflectance and electronic contributions are subtracted to highlight the acoustic contribution. Acoustic generation occurs in the substrate, but the unipolar stress pulse is transmitted to the transparent film (see Fig. 1a). The variation is modelled theoretically as the sum of four components: (i) acoustic echoes from the a-SiO_2 film (Fig. 1d), (ii) acoustic echoes from the a-Ge substrate (Fig. 1e), (iii) an interference signal from light reflected by the moving stress pulse in the film (Fig. 1f), and (iv) a signal arising from the ultrafast stress-induced changes in film thickness and from the modulation of light transmitted through the stress pulse (Fig. 1g).

The longitudinal sound velocity for the a-SiO_2 film v can be derived from the period of the interference signal, $\tau_B=\lambda/2nv\cos\theta$, where θ is the probe beam angle in the film ($\approx 34°$) and n is known ($n=1.46$). We obtain $v=5250\pm300$ ms^{-1} ($\tau_B\approx 50$ ps) in reasonable agreement with the

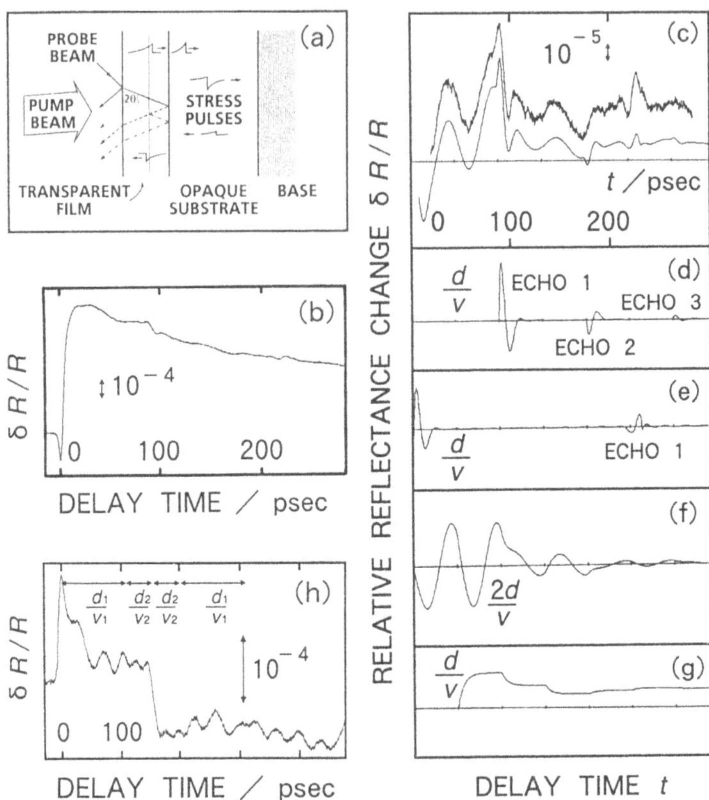

Fig. 1: a) Experimental geometry. The vertical line in the transparent film represents the instantaneous stress pulse position (shown localized for clarity). b) Relative reflectance against delay time for an a-SiO_2/a-Ge film, c) relative reflectance variation with background subtracted (upper curve) and theoretical fit (lower curve), d) echo component from the a-SiO_2 film, e) echo component from the a-Ge substrate film, f) interference component from the stress pulse in the a-SiO_2 film, g) component from film thickness changes and from light transmission through the stress pulse, and h) relative reflectance against delay time for an a-Si_3N_4/a-SiO_2/Cr film. The pump beam angles of incidence for b) and h) are 6° and 20° respectively.

literature[5]. The film thickness d is then obtained using the time d/v determined from the echo component (i), from the discontinuities in the interference signal (iii), or from the signal (iv). We estimate $d = 2300 \pm 200$ Å in fair agreement with the nominal value from the film deposition time. The film-substrate impedance ratio, related to film adhesion, can also be determined from components (i), (iii), or (iv).

The theoretical model is based on thermoelastic generation and non-dispersive, lossless stress propagation. Parameters for the coupling of the optical properties to longitudinal strain η are $dn/d\eta = -0.44$, $dn_1/d\eta = -3$, and $d\kappa_1/d\eta = 1.5$, where $n_1 + i\kappa_1 = 4.2 + 2.3i$ is the substrate index. The value of $dn/d\eta$, derived from the the ratio of components (iii) and (iv), is comparable with the value -0.42 measured at MHz frequencies[6] ($dn/d\eta = -n^3 p_{12}/2$, where p_{12} is the photoelastic tensor coefficient for the s-polarized probe beam). The ratio $dn_1/d\eta : d\kappa_1/d\eta$ is chosen to give the best fit to previous data for a-Ge[4] (cf. the ratio 1.3:2 used in ref. 4). Acoustic parameters are as in ref. 3. Multiple optical reflections are accounted for in the theory (some shown in Fig. 1a). The four components (i) - (iv) are summed and convolved with a Gaussian function (full width at half maximum 3 ps) to simulate the probe pulse width, as shown by the lower curve in Fig. 1c. The observed $\delta R/R$ variation is reproduced well.

A two-layer sputtered transparent film a-Si_3N_4/a-SiO_2 (labelled 2 and 1, respectively) of nominal thickness 5000/5000 Å on a Cr substrate was also tested (with $\theta \approx 0°$). Fig. 1h shows the measured $\delta R/R$ variation. The interference oscillations change in amplitude, phase and frequency at $t = d_1/v_1$ when the stress pulse reaches the bottom of the upper a-Si_3N_4 film. At $t = d_1/v_1 + d_2/v_2$ the sharp drop in signal is caused by the component (iv) [(i) and (ii) are negligible]. The oscillation period for $t < d_1/v_1$ allows v_1 to be derived. For $t > d_1/v_1$ the analysis is complicated by multiple acoustic reflections, but using additional data from an a-SiO_2/a-Si_3N_4 film (inverted order) we obtain $v_1 = 5000$ ms^{-1}, $v_2 = 10200$ ms^{-1}, $d_1 = 5350$ Å, and $d_2 = 4850$ Å, in reasonable agreement with other sources.

In future we will apply the PATS technique to films with practical applications such as passivation layers in microelectronics or anti-corrosion layers in the coatings industry.

1. E. H. Jacobsen, J. Phys. (Paris), 33, C6-25 (1972).
2. C. Thomsen, H. T. Grahn, H. J. Maris and J. Tauc, Phys. Rev. B34, 4129 (1986).
3. O. B. Wright, J. Appl. Phys., 71, (1992) [accepted for publication].
4. O. B. Wright, T. Hyoguchi and K. Kawashima, Jap. J. Appl. Phys., 1B, L131 (1991)
5. C. Thomsen, H. T. Grahn, H. J. Maris and J. Tauc, Optics Comm., 60, 55 (1986).
6. R. W. Dixon, J. Appl. Phys., 38, 5149 (1967).

Nonlinear Regimes of Longitudinal Acoustic Pulse Excitation in Piezocrystals by Ultrashort Laser Action

V.E. Gusev and L.N. Makarova

International Laser Center, Moscow State University, 119899 Moscow, Russia

Abstract. The excitation of longitudinal acoustic pulses as a result of the inverse piezoeffect in electric fields induced by spatial separation of photogenerated electrons and holes was analyzed theoretically. It was shown that in the case of optically-generated electron-hole plasma electroneutrality lack the duration of the excited acoustic pulses decreases with increasing energy of ultrashort laser pulses while their amplitude grows linearly. In the case of sound piezogeneration by Dember's field of quasineutral electron-hole plasma the acoustic pulse amplitude saturates with growth of pulsed-laser intensity while the acoustic pulse broadens.

Nonlinear regime of Rayleigh type surface acoustic wave laser induced piezoexcitation has been recently described /1/. In the present analysis of bulk waves generation we will assume that electrons are the most mobile carriers ($D_e >> D_h$, here $D_{e,h}$ are the diffusion coefficients of free electrons and holes). The process of optoacoustic transformation depends considerably on laser-radiation penetration depth $1/\alpha$ (where α is the light absorption coefficient).

If the region of light absorption is diffusively thin ($1/\alpha << X_s$, where X_s is a scale of spatial separation of photogenerated electrons and holes), then in the case of supersonic diffusion of electrons one can consider that the motion of electrons occurs in the constant electric field of immobile holes $E_h = |q_o| W_{abs}/2\varepsilon h\nu_L$. Here q_o is the electron's charge , W_{abs} is the surface density of absorbed optical energy, ε is the dielectric constant of crystal, $h\nu_L$ is the energy of optical quantum. In this case the nonlinear differential equations governing the nonstationary optically initiated electric field E can be reduced to the Burgers equation

Springer Series in Optical Sciences, Vol. 69 Photoacoustic and Photothermal Phenomena III
Editor: D. Bićanić

$$E_t-\mu_e E E_x - D_e E_{xx}=0$$

subjected to the initial condition $E(x>0,t=0)=0$ and boundary requirements $E(x=0+0,t)=2E_h$, $E(x\Rightarrow\infty,t)\Rightarrow 0$. The characteristic electron and electric-field penetration depth X_s is determined by equality of diffusion and drift electron fluxes: $X_s \sim D_e/\mu_e E_h$ (where μ_e is the electron mobility). The time t_S of stationary electric field formation diminishes with the increase of laser pulse energy: $t_S \sim X_S/\mu_e E_h \sim W_{abs}^{-2}$.

In the case of rapid (supersonic) drift of optically generated electrons and slow (subsonic) motion of holes one can consider that the stationary distribution of nonequilibrium carriers sets up instantaneously (from the acoustic point of view). Then the calculated profile of longitudinal strain U in the acoustic pulse cab be presented in the form:

$$U \sim (q_o E_h/\rho_o C_L^2)(\tau/|\tau|)[1+(C_L|\tau|/X_S)]^{-1}$$

Here ρ_o is equilibrium density of the crystal, C_L is the longitudinal sound velocity and $\tau=t-x/C_L$. The duration τ_a of acoustic pulses is determined by the time of sound propagation across the region of charge separation: $\tau_a \sim X_s/C_L \sim 1/W_{abs}$. In this regime ($X_s \gg 1/\alpha$) the duration of acoustic pulses decreases with increasing energy of laser delta-influence, while the amplitude grows linearly ($U \sim E_h \sim W_{abs}$).

As the spatial separation of positive and negative charges is diminished with rising W_{abs} ($X_s \sim 1/W_{abs}$) then the inequality $1/\alpha < X_s$ is violated for the sufficient energies of optical influence. In the opposite limiting case $1/\alpha \gg X_s$ the electric field (Dember field) caused by diffusive separation of electrons and holes is determined by the expression $E = -(D_e/\mu_e)n_x n^{-1}$ (where $n = n_e = n_h$ is the concentration of quasineutral photogenerated carrier cloud). The analytical calculations were carried out in the laser-δ-pulse regime $n = n_o + (\alpha W_{abs}/2h\nu_L)\exp(-\alpha x)(1 + t/|t|)$ (here n_o is the equilibrium concentration of free charge carriers in piezosemiconductor):

$$U \sim (\alpha|q_o|D_e/2\mu_e\rho_o C_L^2)N(\tau/|\tau|)\exp(-\alpha C_L|\tau|)[1+N\exp(-\alpha C_L|\tau|)]^{-2}$$

Here $N = \alpha W_{abs}/h\nu_L n_o$ is dimensionless energy dependent parameter. The analysis revealed that the amplitude of the

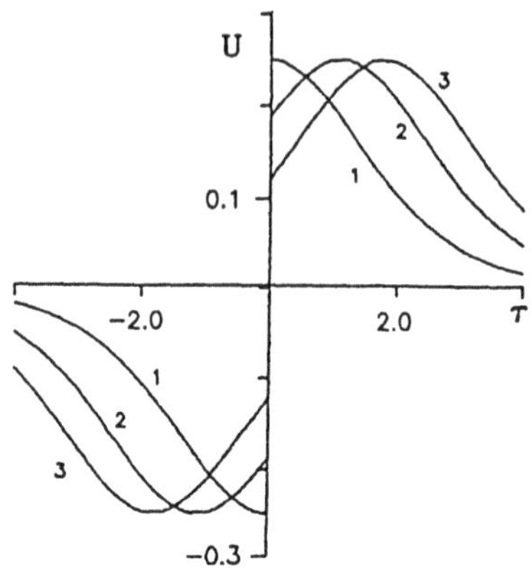

Fig.1. Computed acoustic profiles for increasing energy of laser δ-action in the case of ambipolar diffusion of photoexcited electron-hole plasma. The dimensionless variables

$$\tau = \alpha C_L \tau$$

and

$$U = (\alpha |q_o| D_e / 2\mu_e \rho_o C_L^2)^{-1} U$$

are used.

excited optoacoustic pulses saturates for $N > 1$. The calculated normalized profiles of longitudinal strain U are pictured in Fig.1 for the following values of parameter: N = 1, 3, 6. The growth of the time delay between different-phase peaks of bipolar acoustic pulse (Fig.1) is caused by the displacement of the most effective optoacoustic transformation region into the depth of crystal. This is due to dominant screening of Dember field near the surface, where the carrier concentration reaches its maximum, while the Maxwell's relaxation time reaches its minimum. The optoacoustic pulse duration τ_a is determined by the time of sound propagation across the region of interband light absorption $\tau_a \sim 1/\alpha C_L$ as $N < 1$ and increases as $N > 1$.

Reference

1.V.E.Gusev and L.N.Makarova, J.Phys.D:Appl.Phys.23,1350(1990)

Pulsed Ultrasonic Spectroscopy Based on Laser Thermooptical Generation

A.A. Karabutov, M.P. Matrosov, and N.B. Podymova

Physics Department, Moscow State University, 119899 Moscow, Russia

Abstract. Ultrasonic spectroscopy of some liquid and solid inhomogeneous materials in wide frequency range was carried out by pulse through-transmission technique with digital signal processing. Special media with known opto-acoustic (OA) transfer functions served as OA generators, giving strong wideband pulses on absorbing Nd^{3+}:YAG laser output.

1. Introduction

It is well known that conventional techniques of ultrasonic spectroscopy encounter some difficulties, especially for high-absorptive media. That is, necessity of high energy of signal in wide frequency band, the problem of acoustic bond layer, dynamical errors, etc. Pulsed spectroscopy with digital signal processing [1] is a way to solve this problems, and application of laser source of ultrasound is the next step on this way [2]. Generation of sound directly at the surface of investigated medium has some advantages for nondestructive testing [3], but for the purposes of physical acoustics it is necessary to ensure calibrated sources of sound - opto-acoustic generators (OAG). In this work we propose two of them (see Table). The main requirements to such media are: linearity in great range of laser intensity; high value of thermooptical parameter $\alpha_{opt}\beta c_1/C_p$; weak temperature dependence of expansivity β; wide bandwidth of OA transfer function [4-6].

2. Experimental and discussion

Description of the experimental setup one can find in refs. [4,7]. Nd^{3+}:YAG laser generates 30 mJ / 15 ns pulses that raise up acoustic waves in OAG (mercury or optical filter blue-green glass-22) and partially are reflected by the plate to laser energy meter based on linear OA transformation and make it possible on-line energy control. Acoustic pulses spreading through the investigated medium are received by thin polymer film or thick PZT-19 transducer, digitized and visualized by digital oscilloscope and Fourier-transformed in PC for

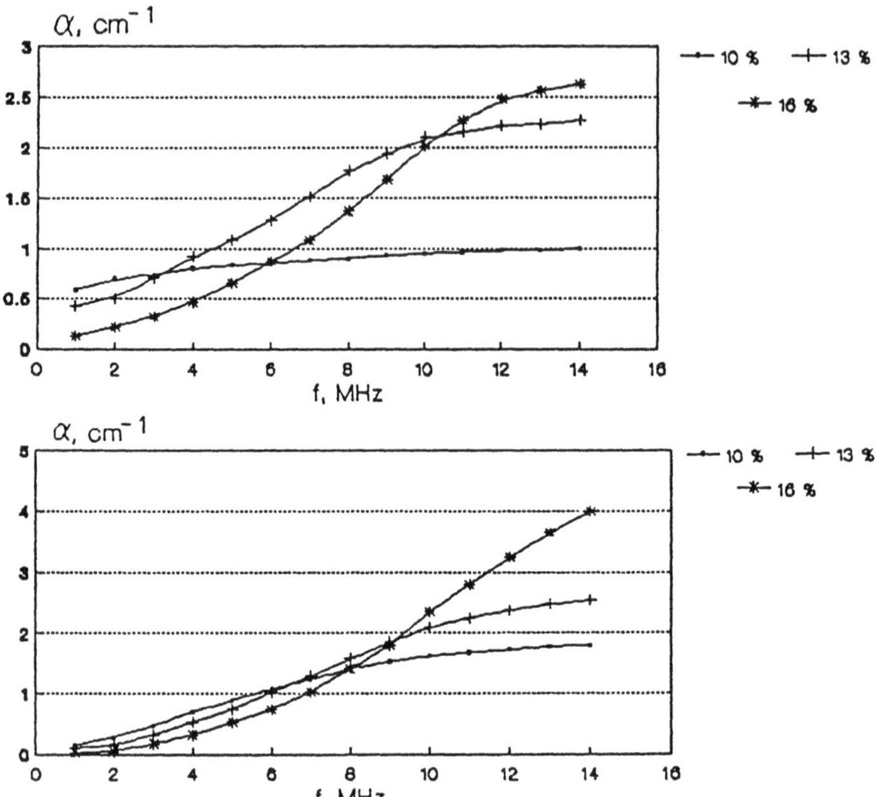

Fig. 1. Attenuation spectra of bentonite clays

comparison with etalon spectrum and extraction of information about ultrasonic attenuation.

One can see on Fig.1(a,b) the spectra of attenuation in two types of the clays: modified by surface-active substance ("A" type) and non-modified ("B" type). Fig.2 exhibits in log scales attenuation spectra of ceramics, and those of graphite-epoxy composite is shown on Fig.3. Three curves correspond to the incident sample, to the sample after 10^4 loading cycles, and to those after 10^5 loading cycles. Working frequency range of any curve corresponds with the maximum error at the edges to be equal 10%. Diffraction correction was conducted in quasi-optical (parabolic) approximation taking into account gauss-like space distribution of energy in optical pulse. Calibration of piezotransducers was carried out by nonlinear transformation technique [4], and spectral analysis exhibits regularity of sensitivity in the working frequency range.

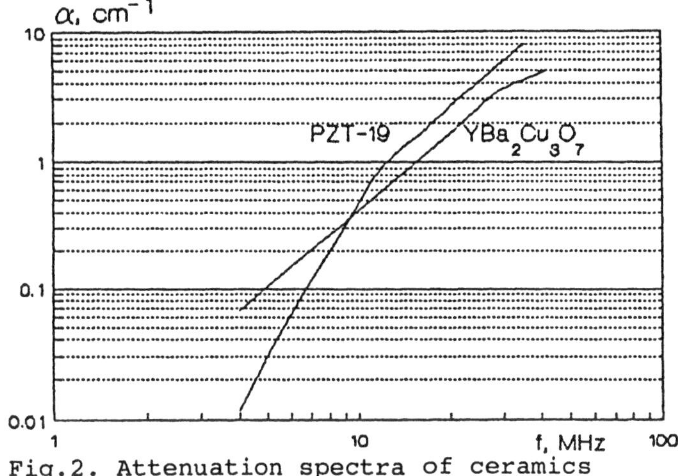

Fig.2. Attenuation spectra of ceramics

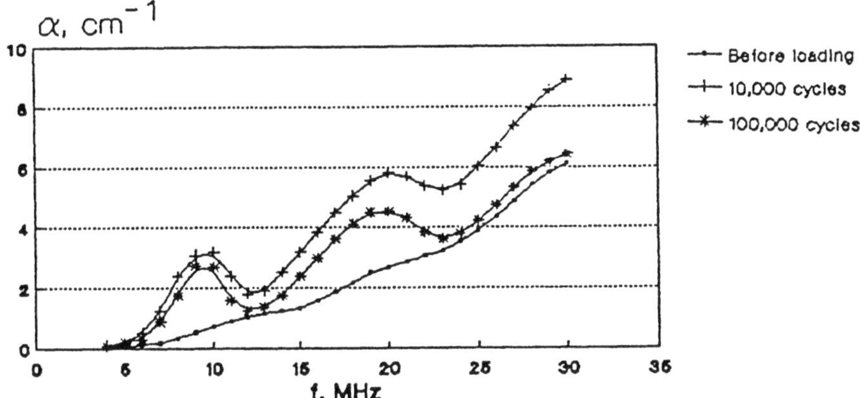

Fig. 3. Attenuation spectra of composites

Further interpretation of obtained attenuation spectra was then conducted using empirical models. Sound propagation in solutions of clays was found to obey Mandelshtam-Leonthovich model with one relaxation time depending on the suspension type and its concentration, the last dependence being exponential with the index of -2.5 for "A" type and -3.0 for "B" type of suspension [4].

Sizes of scatterers in ceramics were calculated using simple scattering model [8]. After plotting attenuation data in the axes $\alpha/f^2(f^2)$ one can differentiate areas of Raleigh, stochastic and diffuse scattering and determine maximum grain size in PZT-19 (50 mkm) and mean size of pores in $YBa_2Cu_3O_7$ (30 mkm). Electron micrographs confirm the calculations [7].

Table. Parameters of OA generators

Subst	α_{opt}, cm^{-1}	ß, K^{-1}	I_{max}, W/cm^2	f_{out}, MHz
BGG-22	140	$2.3*10^{-5}$	$3.0*10^7$	2 - 35
Hg	$>10^4$	$1.8*10^{-4}$	$7.0*10^4$	0.1 - 70

Fatigue changes in the structure of graphite-epoxy composites were detected. Fig. 3 exhibits equidistant arrangement of local maximums that origin after loading the sample. This phenomenon can be caused by the appearance of ring-like cavities (resonators) during the graphite fibres stratification [9]. Mean diameter of the cavities was evaluated (50 mkm). More detailed explanation of obtained results is to be published.

In conclusion, the usage of laser generation of ultrasonic pulses in special media provides high power, stable, wideband and easily adjustable acoustic source, which is useful tool for ultrasonic spectroscopy.

References

1. R.A. Kline: J. Acoust. Soc. Amer. 76, 498 (1984).
2. K.L. Telschow: Rev.Progr.Quant. NDE 7B, 1211 (1988).
3. C.B. Scruby: Ultrasonics 27, 195 (1989).
4. A.A. Karabutov, M.P. Matrosov, N.B. Podymova, and V.A. Pyg: Vestnik Mosc. Univ. Ser. 3 Phys. Astr. 31, 60 (1990).
5. L.V. Burmistrova, A.A. Karabutov, A.I. Portnyagin, O.V. Rudenko, and E.B. Cherepetskaya: Sov.Phys.- Acoust. 24, 655 (1978).
6. S.A. Akhmanov, V.E. Gusev, and A.A. Karabutov: Infrared Phys. 29, 815 (1989).
7. A.A. Karabutov, M.P. Matrosov, and N.B. Podymova: Sov.Phys.- Acoust. 37, 311 (1991).
8. W.N. Reynolds, R.L. Smith: J.Phys.D 17, 109 (1984)
9. Handbook of Composites. Vol.2. Structure and Design. - North-Holland, 1988.

Optical Observation of Laser Induced Lamb Wave Dispersion

R. Hrovatin and J. Možina

Faculty of Mechanical Engineering, University of Ljubljana,
61000 Ljubljana, P.O. Box 394, Slovenia

Abstract. Lamb waves were generated in the ablation regime on a silicon plate by focused Nd:YAG laser pulses with different pulse energies. The surface inclination was detected by deflection of the laser probe beam, which was reflected from the back surface of the sample. Knife-edge method and position sensitive photodiode were used to measure the probe beam deflection. Antisymmetric Lamb wave propagation was observed and presented as signal amplitude propagation. From this data dispersion relations were determined.

Different optoacoustic methods of Lamb wave generation and observation for the purpose of material properties determination have been used by many workers [1,2,3]. Several interferometric methods are widely used in ultrasonic displacement measurements [4,5] as well as total reflection[6] and beam deflection for surface gradient detection[7]. The purpose of the present article is to further examine the use of latter method in the study of laser induced Lamb waves.

In our experimental setup beam deflection method was used. Elastic waves were generated by focused Q-switched Nd:YAG laser pulses with 10ns pulse duration (FWHM) at different pulse energies (Fig.1).

To scan the surface, the mirror and the focusing lens were moved parallel to the specimen surface. The probe 5mW He-Ne laser beam was focused (f = 300 mm) and then reflected from the 500μm thick silicon plate. Its deflection was detected by position sensitive photodiode. Electric

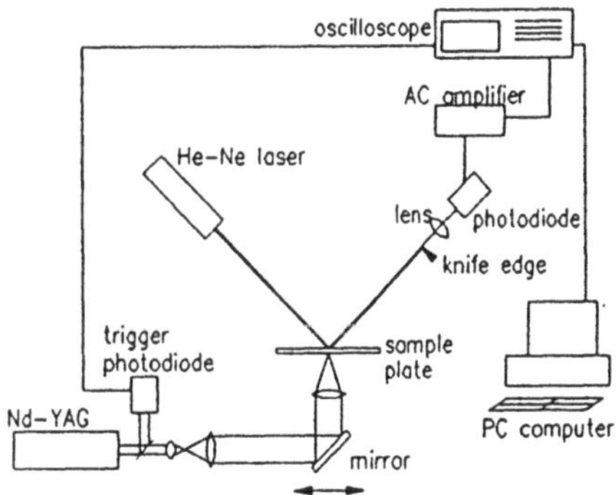

Fig.1. Experimental setup

 Photoacoustic and Photothermal Phenomena III

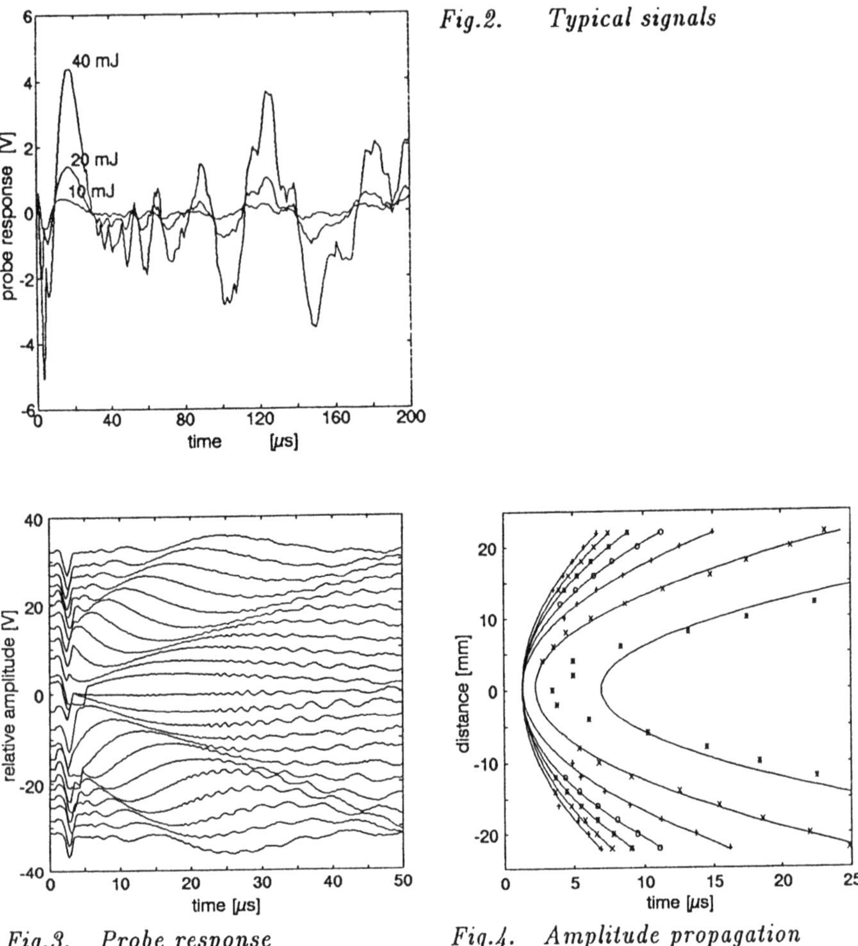

Fig.2. Typical signals

Fig.3. Probe response at various distances

Fig.4. Amplitude propagation

signal was amplified and then recorded digitally with a resolution of 8-bits, using an HP 54502A oscilloscope. It was triggered by Nd:YAG laser pulse, using a fast photodiode. Waveforms were transmitted and stored in a computer via an IEEE-488 interface for analysis. Typical signals at 10mm distance between wave source and probe spot are shown in Fig. 2. Due to the interference effects caused by the reflection from boundary only the first part of the signal is used for further analysis.

Signals recorded in a range from -22 to 20 mm with a step of 2mm (Nd:YAG laser pulse energy: 25 mJ) are presented in Fig.3. Propagation of antisymmetric (a_o) mode of dispersive Lamb wave can be clearly seen. The positions of signal amplitudes were determined by averaging. Results are presented for first seven extremes (Fig.4). In this time range 2nd order polynom least-square fitting is possible.

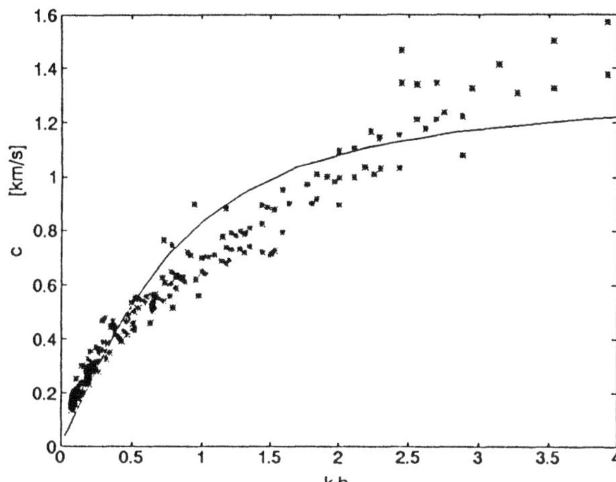

Fig.5. Dispersion relations

The data presented enables us to calculate the corresponding dispersion relations. Calculated values are shown in Fig. 5. Theoretically this relations are relatively complicated. If ω is an angular frequency and k is the corresponding wavenumber then their dependence can be presented in the form of Rayleigh-Lamb equations [8]:

$$\frac{\tanh\, qh}{\tanh\, ph} = \frac{(q^2 + k^2)^2}{4k^2pq} \tag{1}$$

for the antisymmetric modes and for the symmetric modes

$$\frac{\tanh\, qh}{\tanh\, ph} = \frac{4k^2pq}{(q^2 + k^2)^2}\,, \tag{2}$$

where

$$p^2 = k^2 - \frac{\omega^2}{c_L^2}\,, \qquad q^2 = k^2 - \frac{\omega^2}{c_T^2}\,,$$

and c_L, c_T and h are the longitudinal velocity, shear velocity and plate thickness, respectively. Using numerical values c_L=3000 and c_T=1500 m/s reasonable agreement between theoretical data according to Eq.1 and experimental data has been found.

REFERENCES

1. H. Sontag, A. C. Tam, Appl. Phys. Lett., **46**, 725, (1985)
2. J. -D. Aussel, J. -P. Monchlain, Ultrasonics, **27**, 165, (1989)
3. C. B. Scruby, Appl. Phys. Lett., **48**, 100, (1986)
4. D. A. Hutchins, K. Lundgreen, J. Acoust. Soc. Am., **85**, 1441, (1989)
5. C. H. Palmer, R. E. Green, Appl. Opt., **16**, 2333, (1977)
6. M.A.Olmstead et al., Appl.Phys.A, **32**, 141, (1983)
7. H. Sontag, A. C. Tam, Can. J. Phys., **64**, 1330, (1986)
8. I. A. Victorov, Plenum, New York, 1967

Precision Measurement of Laser Induced Surface Acoustic Wave Velocities on Silicon Single Crystals

A. Neubrand[1], *P. Hess*[1], *H. Coufal*[2], *and R.K. Grygier*[2]

[1]Institute of Physical Chemistry, University of Heidelberg, Im Neuenheimer Feld 253, W-6900 Heidelberg, Fed. Rep. of Germany

[2]IBM Research Division, Almaden Research Center, 650 Harry Road, San Jose, CA 95120-6099, USA

Abstract. Surface acoustic waves (SAW) and pseudo surface acoustic waves (PSAW) have been excited by picosecond laser radiation on the (111) and (001) planes of silicon single crystals. Their phase velocities in different directions of propagation have been determined with high precision employing a broadband ferroelectric polymer foil transducer.

1. Introduction

Surface acoustic waves are an important means to investigate the elastic properties and microstructure of the near-surface region of a solid [1]. Narrow-bandwidth techniques employing interdigital transducers on piezoelectric substrates are well developed and are already used in electronic devices and sensor applications. The excitation of broadband acoustic pulses has found increasing interest only recently, when short laser pulses became available. It has been shown that a ringing-free detection of broadband SAW pulses is possible with a polyvinylidene-difluoride (PVDF) foil transducer of about 15MHz bandwidth [2]. In the present experiments, the bandwidth of this type of transducer is extended to about 200 MHz by decreasing its size.

2. Experimental

In the present experiments the transient acoustic pulses are excited on the silicon surface using excimer pumped dye laser pulses of about 10ps duration at a wavelength of 248nm. The laser beam was focussed to a rectangular spot of 10-20μm width and several mm length on the substrate surface by a cylindrical lens. The SAWs were detected by a transducer consisting of a 6μm thick ferroelectric foil of a copolymer of vinylidenedifluoride and trifluoroethylene pressed onto the sample by the edge of a tungsten wedge. Its radius of curvature was about 1μm. The wedge was adjusted parallel to the laser line focus with a precision of a few μm. The signal from this foil was fed into a preamplifier with a bandwidth of 2GHz and 50Ω input impedance. It was averaged and recorded by a Tektronix 7104 transient digitizer with a bandwith of 1GHz. The velocities of SAWs were determined by measuring the propagation time of SAW pulses for different distances between the laser focus and foil transducer for different directions of propagation on the Si(001) and Si(111) surfaces. The silicon used was intrinsic material with a resistivity of 40-200Ωcm. The surfaces were lapped and polished to a surface roughness of less than 2nm rms.

3. Results

For all propagation directions on Si(001) and Si(111), the sensitivity of the foil transducer was high enough to observe SAW or PSAW pulses in the thermoelastic and ablation regime. The halfwidths of the shortest pulses detected were less than 5ns, which corresponds to a bandwidth of about 200MHz. The bandwidth of the observed SAW pulses was not limited by the laser pulse duration or transducer bandwidth, but by the width of the line focus of the laser (the SAW propagation time across the excitation area was several ns for a Si substrate).

For the Si(111) plane the surface wave velocities were measured between the [110] and the [112] directions. The phase velocity increases from 4545m/s in the [110] direction to 4740m/s along [112] (Fig.1). The calculated velocities of the SAW using elastic constants of c_{11}=165.7GPa, c_{12}=63.9GPa, and c_{44}=79.56GPa [3] and a density of ρ=2329kg/m^3 are plotted as a full line in Fig.1. The experimentally determined SAW velocities are slightly higher than the calculated velocities of 4541m/s along [110] and 4736m/s along [112]. Since the elastic constants found in the literature are known only with a precision of about 0.5% and are also slightly temperature dependent, the experimental results agree surprisingly well with the calculated sound velocities. The experimentally determined values of the phase velocity are not exactly symmetric about the [110] direction - this indicates that the orientation of the crystal was inaccurate by 0.6°. The manufacturer gives an accuracy of ±0.5°.

On the Si(001) plane the phase velocity was measured between the [110] direction (0°) and the [010] direction (45°) (Fig.2). The calculated SAW velocities using the elastic constants given above are again about 4m/s lower than the experimentally determined values for propagation directions between 15° and 75° from [110]. Below 17° and above 73° the calculated SAW phase velocity decreases sharply as it approaches the velocity of the slowest transverse bulk wave. Calculations show that this drop in velocity is accompanied by a drastic increase in the penetration depth of the SAW from about 1 SAW wavelength at 18° to more than 10 SAW wavelengths at 12°. This means that the

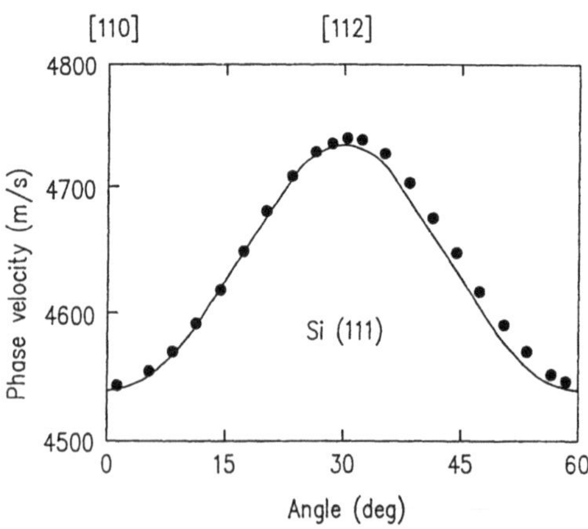

Fig. 1 SAW phase velocities for the (111) plane of silicon as a function of the angle between the propagation direction and the [110] direction. The continuous line represents calculated SAW velocities and the circles are measured SAW velocities.

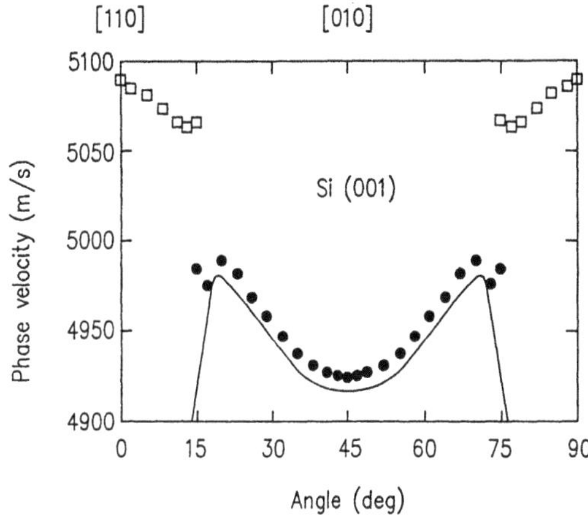

Fig.2 SAW phase velocities for the (001) plane of silicon as a function of the angle between the propagation direction and the [110] direction. The continuous line represents calculated SAW velocities, the circles are measured SAW velocities, and the squares are measured PSAW velocities.

energy of the SAW is less concentrated at the surface, and at angles close to [110] it cannot be observed at the surface even at excitation laser pulse energies above ablation threshold. Instead, another surface wave with a higher velocity and a slightly lower amplitude as compared to the SAW in [010] direction was observed (squares in Fig.2). This wave is a so-called pseudo surface wave with a complex propagation vector and its velocity corresponds to the calculations of *Pratt* and *Lim* [4]. The attenuation of this PSAW was small, and hence it was detected even at large distances from the excitation line for propagation directions up to 15° from [110].

4. References

1. L.Konstantinov, A.Neubrand, and P.Hess: "Surface Acoustic Waves in Solid State Investigations", in *Photoacoustic, Photothermal and Photochemical Processes at Surfaces and in Thin Films*, P.Hess Ed., Top.Curr.Phys., Vol.47, (Springer, Berlin, Heidelberg 1989)
2. A.Neubrand and P.Hess: Mater.Sci. and Eng. **A122**, 33 (1989)
3. B.A.Auld: *Acoustic Fields and Waves in Solids*, Vol.1, (Wiley, New York 1973), p.370
4. R.G.Pratt and T.C.Lim: Appl.Phys.Lett.**15**, 403 (1969)

Photoelastic Law Explains Optically Induced Changes in the Index of Refraction in $Ti^{3+}:Al_2O_3$

H. Eilers [1], *F. Kieseling* [2], *E. Strauss*†, *and W.M. Yen* [1]

[1] Department of Physics, University of Georgia, Athens, GA 30602, USA
[2] Fachbereich Physik, Universität Oldenburg, W-2900 Oldenburg, Fed. Rep. of Germany
†Prof. Dr. Eugen Strauss died on February 17, 1991

Abstract. Photothermal-beam-deflection spectroscopy is used to measure the change in the index of refraction in $Ti^{3+}:Al_2O_3$ after optical excitation. The beam-deflection signal is different depending on the polarization of the probe beam. This anisotropic behaviour can be explained by the photoelastic law based upon the anisotropic volume expansion caused by a strong Jahn-Teller effect of the TiO_6 complex in the excited state.

1.Introduction

Moulton [1] first demonstrated laser action in $Ti^{3+}:Al_2O_3$ in 1982. Since then this material has been the object of increasing interest. Recently, Wall, et al. [2] reported an interferometric investigation of the change in the index of refraction after optical excitation. They explained the nonthermal part of the signal with a harmonic oscillator model, which considers virtual transitions to a charge transfer band. The strong anisotropic behavior of the nonthermal part for different polarizations was attributed to different oscillator strengths.

However, we believe that this anisotropic behavior can be better explained not only by considering the properties of the optically-active ions, but also by considering the elastic properties of the host material. The Jahn-Teller effect (the Jahn-Teller energy is about 3000 cm^{-1}) gives rise to a strong ellipsoidal shaped TiO_6 complex in the excited state. The coupling of the Jahn-Teller mode e_g and the breathing mode a_{1g} gives a different relative length change (strain) in the crystal. This strain can be related to the change in the index of refraction for the different directions via the photoelastic law.

2.Experimental

The beam-deflection measurements were carried out as previously described [3]. The ^2E_g state of the Ti^{3+}-ions was excited using a Q-switched, frequency-doubled Nd:YAG laser, which was operated close to the TEM_{00} mode. The time dependence of the refractive change was probed by a HeNe

Springer Series in Optical Sciences, Vol. 69 Photoacoustic and Photothermal Phenomena III
Editor: D. Bićanić

laser beam propagating collinear ($\angle < 2°$) to the pump beam. The Rayleigh length of both beams were larger than the thickness (3 mm) of the sample. The pump beam was polarized parallel to the c-axis and the probe beam polarization could be adjusted parallel (π) or perpendicular (σ) to the c-axis. The deflection of the probe beam by the sample is proportional to the gradient of the bulk refractive index $grad_r(n)$. For a gaussian pump beam profile the deflection is also proportional to the index change (δn). The deflection was measured using a quadrant photodiode and differential amplifier and recorded with a transient digitizer. The time resolution is limited to about 200 ns by the rise time of the differential amplifier.

3.Results and Discussion

Fig. 1 shows a typical signal for both probe beam polarizations. The data were fit according to

$$f(r,t) = (A+B(1-e^{-\frac{t}{\tau}}))\cdot\frac{2r}{(\sigma^2+4Dt)^2}\cdot e^{-\frac{r^2}{\sigma^2+4Dt}}\cdot N + C\cdot e^{-\frac{t}{\tau}}$$

where A is the prompt heat, B is the delayed heat, C is the nonthermal signal, D is the diffusion constant, N is a normalization factor, and τ is the fluorescence lifetime. The fit results are very good and the ratio for the nonthermal parts of the signal then gives $C_\sigma/C_\pi = 2.75\pm0.6$.

The macroscopic thermal volume change can be described by

$$\Delta V_{th} = \frac{\beta Q}{\rho c_p} = (0.9\pm0.1)\,\text{Å}^3.$$

where β is the thermal volume expansion coefficient, Q is the heat released

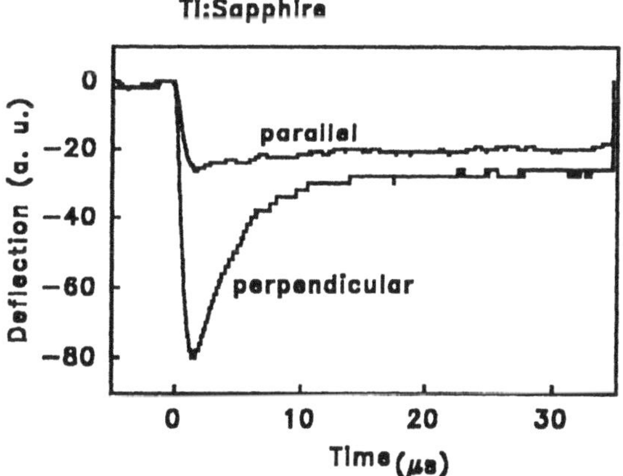

Fig.1: Beam-deflection signal with the probe beam perpendicular and parallel to the c-axis.

per excited Ti^{3+} ion, ρ is the density, and c_p is the specific heat. The thermal signal contains a pure thermal part and a dilatation part [4]. From the ratio of the dilatation part of the thermal signal and the nonthermal signal we get a macroscopic nonthermal volume change of

$$\Delta V_{nth} = (11.8 \pm 0.7)\text{Å}^3.$$

The shape of the Jahn-Teller e_g mode is so that a coupling of this mode to the excited state gives effectively only an expansion perpendicular to the c-axis. The strain for an ellipsoid is $st = \Delta V \cdot N/(2 \cdot V)$, where N is the number of excited Ti^{3+}-ions per unit volume. The photoelastic law [4]

$$\Delta n_i = -n^3/2\ p_{ij} st_j$$

where the p_{ij} are the elements of the strain-optic tensor then gives

$$\Delta n_\sigma / \Delta n_\pi = 3.2 \pm 0.2.$$

If we consider the additional coupling of the totally-symmetric a_{1g} mode with Huang-Rhys parameters of $S_e=9.6$ and $S_a=0.1$ for the Jahn-Teller mode and the totally-symmetric mode, respectively, (determined from the spectroscopic data) we get

$$\Delta n_\sigma / \Delta n_\pi = 2.5 \pm 0.2.$$

This is in reasonable agreement with the experimental value of 2.75±0.6.

4.Acknowledgements This work was supported by DARPA contract # N00014-90-4088.

5.References

[1] P. Moulton, Opt. News, **9**, 1982.
[2] K.F. Wall, R.L. Aggarwal, M.D. Sciacca, H.J. Zeiger, R.E. Fahey, A.J. Strauss, Opt. Lett., **14**, 180, 1989.
[3] E. Strauss, F. Walder, Europhys. Lett., **6**, 713, 1988.
[4] T.S. Narasimhamurty, 'Photoelastic and Electro-Optic Properties of Crystals', (Plenum Press NY, 1981).

On the Possibility of Rayleigh-Type Surface Acoustic Waves Excitation at Frequencies up to 10–100 GHz by Laser Action on a Normally Cut Superlattice

V.E. Gusev

International Laser Center, Moscow State University, 119899 Moscow, Russia

Abstract. The aim of the present publication is to attract attention to the possibility of the use of piezoelectric superlattice (cut perpendicular to its layers) for the excitation of Rayleigh type surface acoustic waves in gigacycle frequency range. The analysis carried out on the basis of the simplest theoretical model has shown that the necessary space modulation of volume sources of acoustic waves can be really achieved in superlattice. This process is contributed by light absorptance and reflectance modulation as well as the confinement of photogenerated free carriers in potential wells realized in composite superlattice.

The process of the periodical surface acoustic wave (SAW) excitation by pulsed-laser action on the 2-component superlattice was investigated theoretically (Fig.1.a). The physical situation $E_{g1}<h\nu_L<E_{g2}$ ($E_{g1,2}$ – forbidden energy gaps of semiconductor layers, $h\nu_L$–energy of optical quantum) was under examination (Fig.1.b). The repetition of absorbing and transparent layers (Fig.1.c, $\alpha_{1,2}$ – coefficients of light absorption) with the spatial period $2.10^2A < d < 2.10^3A$ leads to the spatial modulation of the optoacoustic sources that is necessary for the effective excitation of the Rayleigh SAW at frequency range 10 GHz $< \nu_R <$ 100 GHz. The maximum efficiency is realized at the main frequency of the superlattice with equal widths of the layers : $d_1=d_2=d/2$, $\nu_R=c_R/d$, c_R – Rayleigh SAW velocity. SAW generation as a result of the inverse piezoeffect by nonstationary spatially distributed electric fields (initiated by the diffusive separation of photogenerated electrons and holes) was considered.

We must underline that at the frequencies ν_R>10 GHz the inequality $\omega_R=2\pi\nu_R>>\omega_D^h>>\omega_D^e$ is fulfilled ($\omega_D^{e,h} = c_R^2/D_{e,h}$ – the frequencies at which the wave numbers of electrons (holes) diffusion waves and Rayleigh SAW become equal : $q_D^{e,h} = q_R$,

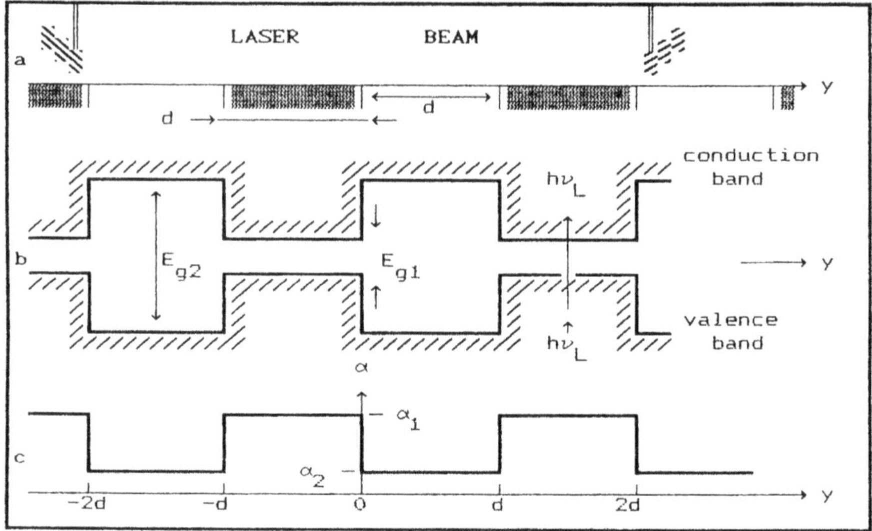

Fig.1 Schematic diagram of laser action on composite superlattice (a), energy bands diagram (b) and spatial variation of optical absorption (c).

$D_{e,h}$ – diffusion coefficients of electrons and holes). At these high frequencies the diffusion motion of nonequilibrium carriers is supersonic. Note that in recent publications /1,2/ on surface acoustic waves laser-induced piezoexcitation only subsonic diffusion had been examined. We had also supposed that interaction between electrons and holes can be neglected : $2\pi\nu_R \gg \omega_{Deb}=(\omega_M^e\omega_D^e)^{1/2}$ ($\omega_M^e = q_0\mu_e n_e/\varepsilon$ – Maxwell relaxation frequency, q_0 – electron's charge, μ_e and n_e – mobility and characteristic concentration of electrons, ε is absolute dielectric constant of the crystal).

The analysis revealed that photogenerated carriers confinement in the potential wells of the considered multilayered structure (Fig.1.a,b) leads to the enhancement of the optoacoustic transformation efficiency:

1) when the light absorption region is diffusively and acoustically thick $0 = \alpha_2 < \alpha_1 \ll q_D^e, q_D^h, q_R$ the efficiency is increased in

$$(\omega^2/\omega_D^e\omega_D^h) \gg 1 \qquad \text{times;}$$

2) when light absorption region is acoustically thick but diffusively thin $q_D^{e,h} \ll \alpha_1 \ll q_R$ the coefficient of the

efficiency enhancement is equal to

$$\omega^{5/2}/(\omega_D^e)^{1/2}\omega_D^h \alpha c_R \gg 1$$

3) when light absorption region is thin both diffusively and acoustically $q_D^{e,h} \ll q_R \ll \alpha_1$ the efficiency of piezoelectric optoacoustic transformation is increased in

$$B^* \omega^{3/2}/(\omega_D^e)^{1/2}\omega_D^h \gg 1$$

times ($B^*=6$, $\omega_D^e \gg \omega_D^h$ in CdS-ZnO superlattice).

This efficiency enhancement is due to the fact that free carriers confinement prevents smoothing of the optoacoustic sources spatial modulation in the process of the supersonic diffusion of the nonequilibrium optically generated electrons and holes.

It is worth noticing that the achieved artificial limitation of free carriers motion along the SAW propagation direction leads to significant deceleration of the decrease of the optoacoustic transformation efficiency with the increase of frequency in all the above mentioned cases: 1) from $\sim |\omega|^{-3}$ to $\sim |\omega|^{-1}$; 2) from $\sim |\omega|^{-3}$ to $\sim |\omega|^{-1/2}$; 3) from $\sim |\omega|^{-2}$ to $\sim |\omega|^{-1/2}$.

Acknowledgements. The author is sincerely indebted to A.R.Baghai-Wadji, V.V.Krylov and B.Djafari-Rouhani for fruitful discussions of the presented ideas.

References

1.V.E.Gusev and L.N.Makarova, J.Phys.D:Appl.Phys.23,1350(1990)
2.V.E.Gusev and L.N.Makarova, Acustica,71,172(1990)

Some Applications of SAW Pulses Photogenerated in Silicon

A.A. Kolomenskii and A.A. Maznev

General Physics Institute, 38 Vavilov St., 117942 Moscow, Russia

Abstract. Surface acoustic waves (SAWs) in silicon wafers have been generated by Nd:YAG or Er:YAG laser irradiation via surface ablation as well as evaporation of a thin water layer. A probe beam deflection technique was employed to record surface acoustic waveforms. High surface accelerations caused by laser-induced SAW pulses were found to lead to removal of micron-size dust particles from the surface. The phonon focusing of the SAW, propagating from a point photoacoustic source was examined using both laser probe measurements and a new visualization technique based on SAW-induced dust removal.

1. Introduction

The SAW generation by pulsed laser irradiation combined with noncontact optical detection schemes has been investigated by various authors and has been demonstrated to be of potential use in applications such as photoacoustic spectroscopy or nondestructive testing [1,2]. The development of the technique for single crystal silicon is of particular interest. In this contribution the generation of high amplitude short duration SAW pulses in a silicon wafer and their application for fine dust particles removal from the surface have been studied. The effect of crystal anisotropy on SAW propagation from the photoacoustic source is also investigated.

2. Experiment and Discussion

To generate SAW pulses two laser sources were used: Q-switched Nd:YAG laser (optical wavelength λ=1.06 μm, pulse duration τ=10 ns, pulse energy E=10 mJ) and Q-switched Er:YAG laser (λ=2.94 μm, τ=250 ns, E=5 mJ). The probe beam deflection technique [3] with a focused beam of He-Ne laser was employed to record the angle of inclination of the surface in the SAW, proportional to the normal surface velocity.

In the first experimental configuration Nd:YAG laser pulses impinged onto the {100} surface of a 0.5-mm-thick silicon wafer. The laser beam was focused to a line of dimensions 10 μm $\times$ 3 mm using a silica cylindrical lens. SAW pulses were generated in the ablation regime. The resulting waveform of the SAW pulse propagating in <110> direction is presented in Fig. 1(a). The waveform obtained is well correlated with a theoretical one assuming the ablation mechanism of SAW generation [2,4].

Springer Series in Optical Sciences, Vol. 69 Photoacoustic and Photothermal Phenomena III
Editor: D. Bićanić

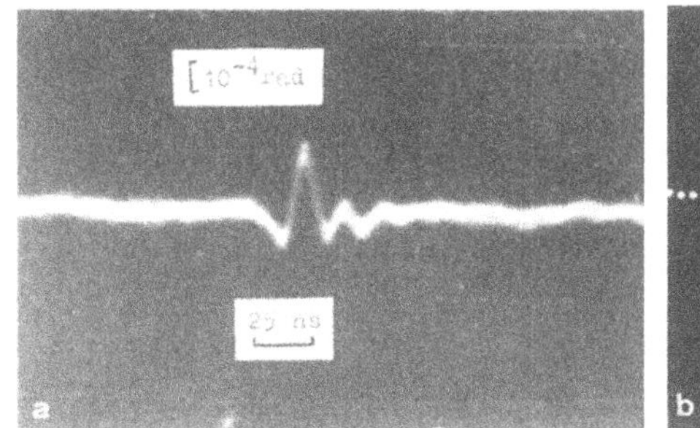

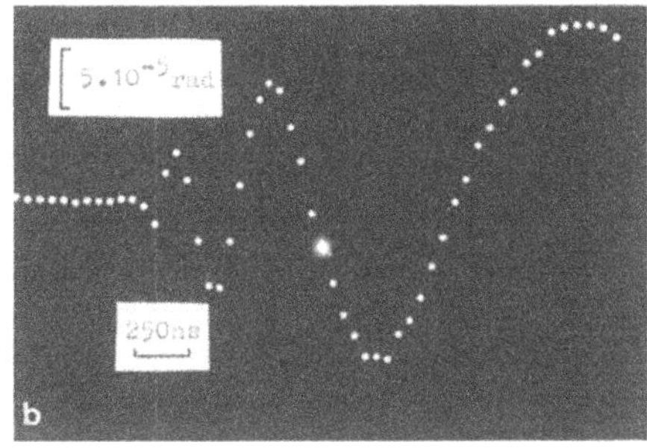

Fig.1 Surface waveforms, generated by (a) Nd:YAG and (b) Er:YAG laser pulses. Source-to-receiver distance 1 cm.

In the above instance the short laser pulse duration caused the acoustic wavelength to be much shorter than the wafer thickness, therefore pure Rayleigh mode was excited. It was not the case, if the erbium laser was employed. The Er:YAG laser pulse was focused into a 30-μm-diam spot on the surface of a wafer, the surface being free or covered with a thin water layer. Extremely high absorbtion coefficient of water (μ=13000 cm^{-1} at λ=2.94 μm) lead to strong heating and evaporation of water and, consequently, to a high amplitude pressure pulse acting on the surface. As a result the powerful acoustic pulse was generated, whose amplitude was two orders of magnitude higher, than of that, generated in ablation regime without water. Because of long duration of the laser pulse the acoustic wavelength was long as compared with the wafer thickness, thereby the lowest antisymmetric Lamb mode was mainly excited. Typical waveform is shown in Fig. 1(b) and it is similar to that observed in [5].

In the experiment with Nd:YAG laser the normal surface acceleration $a=2\times10^8 m/s^2$ produced by SAW pulse was achieved. Such a high acceleration was found to cause the removal of dust particles as small as 1 μm from the silicon surface. Mechanical detachment of the microparticle from the surface occurs if the relation $a > F/m$ is fulfilled, where F is the force of adhesion between the particle and the surface and m is the mass of the particle. Fig. 2(a) shows the photograph of a silicon wafer, which was covered with fine sapphire dust and then was subjected to the photoacoustic treatment. A dark stripe can be seen, which has been cleaned from dust as a result of SAW passage. The size distribution of the dust particles on the surface before and after the photoacoustic cleaning are presented in Fig. 2(b).

Since the dust removal depends on the SAW amplitude a new visualization technique based on the SAW-induced cleaning may be suggested, and we used it to examine SAW propagation anisotropy. Shown in Fig. 3(a) is a dusty {111} surface of silicon wafer after the photoacoustic action, the Nd:YAG laser pulses being focused not into a stripe as in Fig. 2(a), but into a 15-μm-diam Gaussian spot. Clear-cut tracks corresponds, as measurements using the laser probe (Fig. 3(b)) shows, to the sharp maximums of SAW amplitude. Such a distinct anisotropy resulting from the anisotropic elastic crystal properties is referred as a phonon focusing (or channeling). It has been well

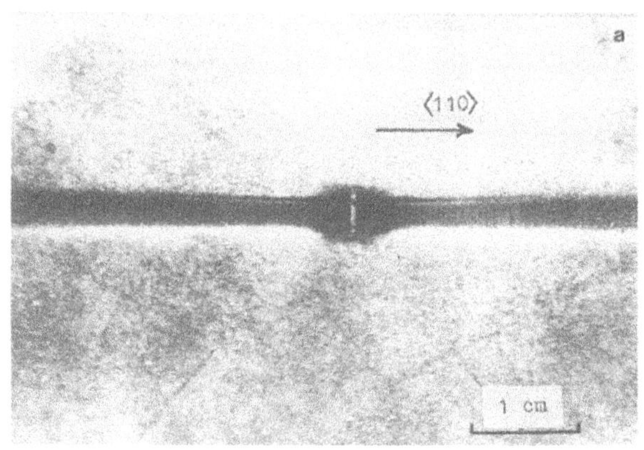

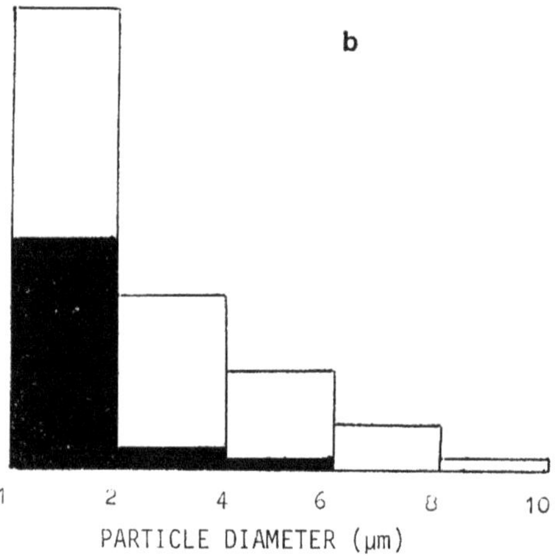

Fig.2 Demonstration of the SAW-induced dust removal. (a) Dusty wafer after the photoacoustic action. (b) Size distributions of the dust particles before (white columns) and after (black columns) photoacoustic action.

investigated for the bulk acoustic waves [6,7], nevertheless any direct observations of this phenomenon for SAW were not conducted to the best of our knowledge. We also observed surface phonon focusing for (100) cut of silicon as well as for (111) cuts of Ge and GaAs. The theoretical analysis of SAW energy flux anisotropy [6] accomplished using available data on angular dependencies of SAW velocities [8] gave a qualitative agreement with the experimental results obtained.

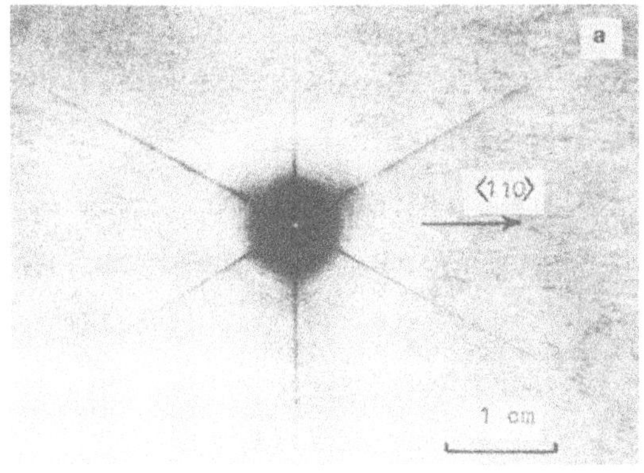

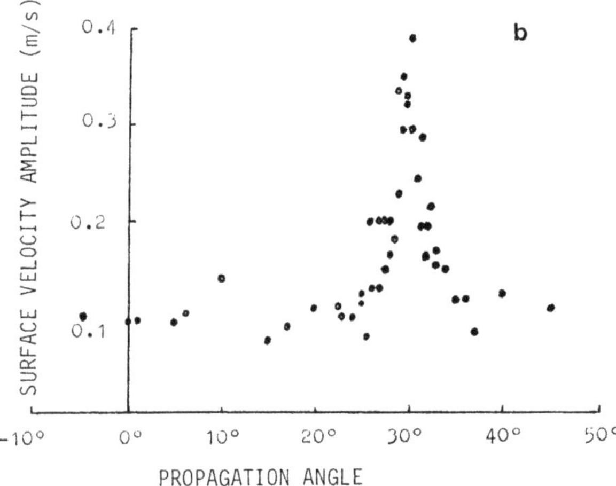

Fig.3 SAW phonon focusing on a silicon {111} surface. (a) Image, arisen as a result of the SAW-induced dust removal. (b) The SAW amplitude dependence on the propagation angle, measured from <110>.

3. Summary

The short duration high amplitude surface elastic pulses were generated in a silicon wafer via both surface ablation regime and evaporation of a thin water layer induced by laser pulsed irradiation. High acceleration achieved in the surface acoustic pulses (up to $2\times10^{8}m/s^{2}$) enables the noncontact removal of micron-size dust particles from the surface. The results obtained allow us to believe, that this new noncontact method of dust removal may be applied to the cleaning of silicon wafers into vacuum chambers as well as to the measurement of the adhesion forces of small particles. A strong anisotropy of SAW amplitude due to phonon focusing effect was observed.

References

1. V.E.Gusev and A.A.Karabutov, Laser optoacoustics, Moscow, Nauka, 1991.
2. D.A.Hutchins and A.C.Tam, IEEE Trans. Ultrason., Ferroelectrics, Freq. Contr. **UFFC-33**, 429 (1986).
3. H.Sontag and A.C.Tam, IEEE Trans. Ultrason., Ferroelectrics, Freq. Contr. **UFFC-33**, 500 (1986).
4. Al.A.Kolomenskii and A.A.Maznev, Akust. Zh. (USSR) **36**, 463, (1990).
5. H.Sontag and A.C.Tam, Appl. Phys. Lett. **46**, 725 (1985).
6. J. Wolfe. Physics Today. December, p.44 (1980).
7. A.G.Every, W.Sache, K.Y.Kym, and M.O.Tompson, Phys. Rev. Lett., **65**, 1446 (1990).
8. Acoustic crystals, edited by M.P.Shaskolskaya, Moscow, Nauka, 1982.

Broad-Band Detection of Laser Excited Surface Acoustic Wave Pulses by a Novel Transducer Employing Ferroelectric Polymers

A. Neubrand[1], *P. Hess*[1], *H. Coufal*[2], *and R.K. Grygier*[2]

[1]Institute of Physical Chemistry, University of Heidelberg, Im Neuenheimer Feld 253, W-6900 Heidelberg, Fed. Rep. of Germany

[2]IBM Research Division, Almaden Research Center, 650 Harry Road, San Jose, CA 95120-6099, USA

Abstract. Broad-band surface acoustic wave transients were excited by ps laser pulses. A novel ferroelectric polymer foil transducer with a bandwidth of 100MHz was employed for their detection. The observed SAW pulse width was limited by materials properties. Phase velocities of surface acoustic waves are given for polycrystalline aluminum, copper, steel and brass samples.

1. Introduction

Surface acoustic waves (SAW) have found increasing interest as a non-destructive method for materials investigations [1]. The excitation of broad-band surface acoustic waves by pulsed lasers allows the analysis of a wide acoustic frequency spectrum [2]. Since SAWs penetrate about one wavelength into the substrate, layers in different depths of the substrate can be probed in one single experiment. Laser-induced SAW transients can be detected with a high bandwidth by optical methods such as interferometers or the deflection of a laser probe [3]. It has been shown recently that a ringing-free detection of broad-band SAW pulses is also possible with a polyvinylidenedifluoride (PVDF) foil transducer of about 15MHz bandwidth [4]. The advantages of such a transducer are its simplicity and ruggedness, and, in contrast to optical methods, it does not require a surface of high optical quality. In the present experiments, the bandwidth of this type of transducer has been extended to about 100 MHz by reduction of the width of the foil detector.

2. Experimental

A typical experimental setup similar to the one described in Ref.4 was employed. Laser pulses of 20ps duration from a Lambda Physik excimer pumped dye laser PSL 4000 operating at a wavelength of 248nm and delivering of the order of 10mJ per pulse are focused on the sample under study. For enhanced focusing and reduction of pulse energy the laser pulses were directed through a slit of 1mm width before passing through a cylindrical lens of 125mm focal length. The focus achieved was about 20μm wide and 20mm long. The pulse energy in the focus was varied between 100μJ and 1mJ depending on the sample. A pulse energy in excess of 1mJ caused ablation in most substrates studied.

The laser-generated SAWs were detected with a detector using a ferroelectric copolymer of vinylidenedifluoride and trifluoroethylene (TrFE) as the active transducer material. A 6μm thick foil made from this material was pressed onto the surface by a tungsten rod with an edge of 6μm radius of curvature. In the direction of SAW propagation the

Springer Series in Optical Sciences, Vol. 69 Photoacoustic and Photothermal Phenomena III
Editor: D. Bićanić

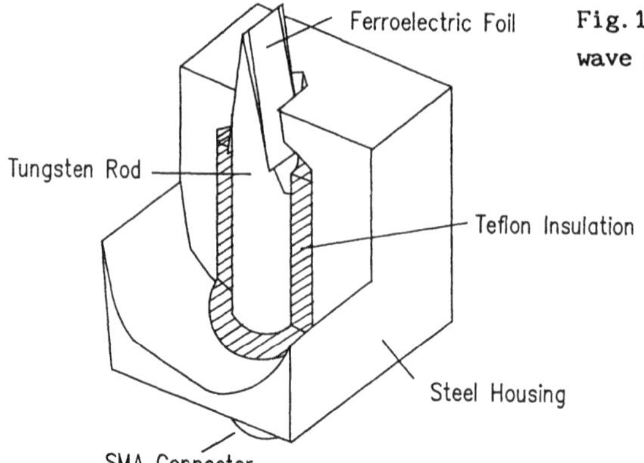

Fig.1 Design of surface acoustic wave foil transducer

contact area extended only several µm, and perpendicular to this direction several mm. The design of this novel detector is shown in Fig.1. The transducer was connected to a low noise voltage preamplifier with 1MΩ input impedance and a bandwidth of 100MHz. The output of the amplifier was averaged and recorded by a Tektronix 7104 transient digitizer with a bandwith of 1GHz. The velocities of SAWs were determined by measuring the propagation time of SAW pulses for different distances between the laser focus and foil transducer.

3. Results and Discussion

The sensitivity of the polymer foil transducer was sufficient to observe SAW pulses excited in polycrystalline samples using laser pulse energies below the ablation threshold. The observed pulse shapes were monopolar for all uncoated metal samples under study (Fig.2). This agrees with measurements using interferometric detection [5] and with theoretical calculations for the normal surface displacements in the near-field of a laser source [2]. This proves that the electrical charges produced by

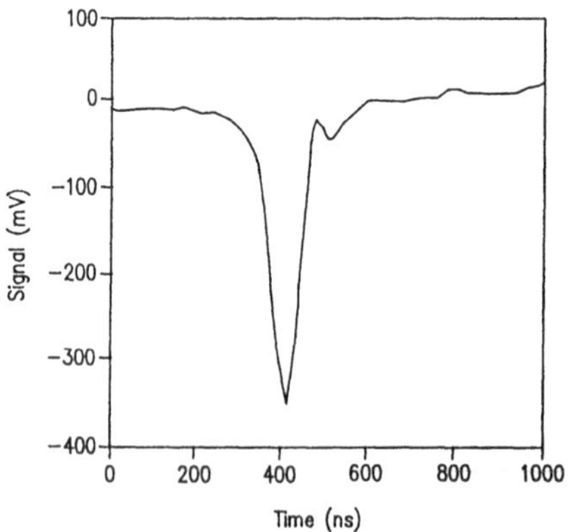

Fig.2 Typical surface acoustic wave signal recorded with foil transducer in the near-field of the laser source. The substrate is stainless steel 304.

Table 1. Sound velocities in polycrystalline metal samples

	v_l (m/s)	v_t (m/s)	v (m/s) measured	v (m/s) calculated
Aluminum	6320 [8]	3130 [8]	2940	2920
Stainless steel 304	5760 [9]	3070 [9]	2890	2848
Copper	4720 [9]	2230 [9]	2130	2087
Brass	4670 [9]	2200 [9]	2000	2059

the foil transducer are proportional to the normal surface displacement and that the transducer is not resonant.

The halfwidth of SAW signals observed on polycrystalline metal samples varied between 60 and 300ns (FWHM). In contrast to this, signals of only about 4ns duration were observed with this type of transducer if single crystals were used as samples [6]. The longer SAW pulse duration in metals is caused by scattering of the SAW at the grain boundaries of the polycrystalline sample. The SAW attenuation due to this so-called Rayleigh scattering increases with the fourth power of sound frequency [7] and thus, even at the shortest propagation distances measured for metal samples, the frequency spectrum of the signal did not reach 100MHz.

The SAW velocities v measured for different materials of technical interest are summarized in Table 1. For comparison, the SAW velocities calculated from longitudinal velocities v_l and shear wave velocities v_t are also given. The agreement is reasonable for all samples. The remaining discrepancies may result from the dependence of the sound velocities on the ratio of grain diameter and acoustic wavelength in polycrystalline samples [8]. This is also supported by the fact that, using a similar experimental arrangement, an excellent agreement between calculated velocities and measured SAW velocities was achieved for silicon single crystals [6]. Thus, the combination of laser generation and polymer foil detection may serve as a simple and precise method to determine SAW velocities in piezoelectric and non-piezoelectric materials.

4. References

1. L.Konstantinov, A.Neubrand,and P.Hess: "Surface Acoustic Waves in Solid State Investigations", in *Photoacoustic, Photothermal and Photochemical Processes at Surfaces and in Thin Films*, P.Hess Ed. Top.Curr.Phys., Vol.47, (Springer, Berlin Heidelberg 1989)
2. A.A.Karabutov: Sov.Phys.Usp. **28**, 1042 (1985)
3. J.-P.Monchalin: IEEE Trans. **UFFC-33**, 485 (1986)
4. A.Neubrand and P.Hess: Mater.Sci. and Eng. **A122**, 33 (1989)
5. A.Neubrand and P.Hess: accepted by J. Appl. Phys. **71** (1992)
6. A.Neubrand, P.Hess, H.Coufal and R.K.Grygier: These proceedings
7. H.Willems and K.Goebbels: Metal Science **15**, 549 (1981)
8. F.E.Stanke and G.S.Kino: J.Acoust.Soc.Am. **75**, 665 (1984)
9. A.C.Tam and H.Coufal: Appl.Phys.Lett.**42**, 33 (1983)

Optoacoustic Investigation of IR Absorption by Diffraction Gratings

A.A. Karabutov, I.A. Kudinov, V.T. Platonenko, and M.A. Sogoyan

Physics Department, Moscow State University, 119899 Moscow, Russia

Abstract. With a help of OA-effect the IR absorption of metal diffraction gratings is considered. The obtained narrow resonances of absorption correspond to maximum efficiency of SEW excitation. The dependence of resonance's width on SEW path, angular and frequency spectrum width is discussed. The deformation of reflected beam is obtained.

1. Introduction

The absorption of radiation by periodic surface differs significantly from that takes place on the plane surface [1]. At the angles of incidence [2]

$$\theta_p=\arcsin|1-l\lambda/d|,\quad \theta_s=\arcsin(1-(l\lambda/d)^2)^{1/2} \qquad (1)$$

the anomalous absorption, connected with the resonant excitation of surface electromagnetic waves (SEW) is possible (λ - wavelength, d - relief period, l - harmonic number).

This anomalies were observed in visible light [3]. For IR these resonances may be vary shape (less then mrad) with high level of absorption [4]. It makes it possible to measure the laser radiation divergence. The aim of our investigations was further optoacoustic measurements of IR absorption by periodic surfaces.

2. Experimental researches

TEA-CO_2-laser, radiated TEM_{00}-mode at λ=10.6 μm was used in our experiments. The acoustic wave, excited in target, was registrated by PZT transducer. The amplitude of oscillations at transducers resonant frequency that is proportional to absorbed energy was measured. The absolute value of coefficient of absorption T was determined optoacoustically.

As an object of investigation the copper grating on a glass substrate was used. The depth of relief was h=0.42 μm, period d=12.09 μm.

The strong absorption - up to 93% - was obtained for S-polarized radiation and resonant excitation of SEW (fig.1). The resonant peak's width (at 1/e level) equaled 20' and was determined by the efficiency of SEW excitation as the angular spectrum of radiation width didn't exceed 7'. The broadening of the angular spectrum by length with f=.3 m gave the broadening of peak of absorption (fig.1) and corresponding decrease of its value.

If the angular spectrum of radiation is wider then the efficiency of SEW excitation, only a part of angular harmonics

Springer Series in Optical Sciences, Vol. 69 Photoacoustic and Photothermal Phenomena III
Editor: D. Bićanić

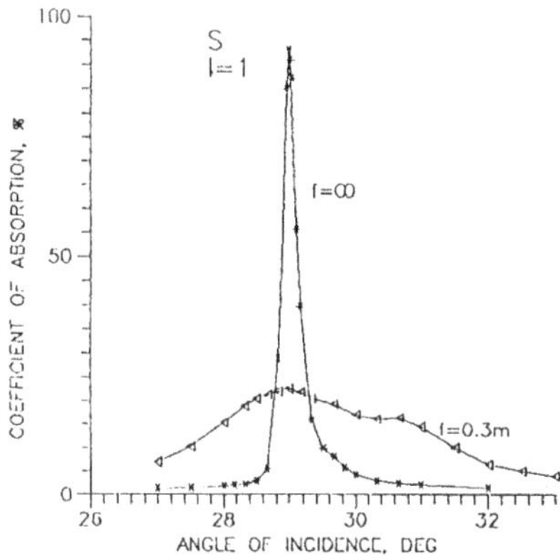

Fig.1 Peak of absorption of gold grating for different spectra of laser radiation

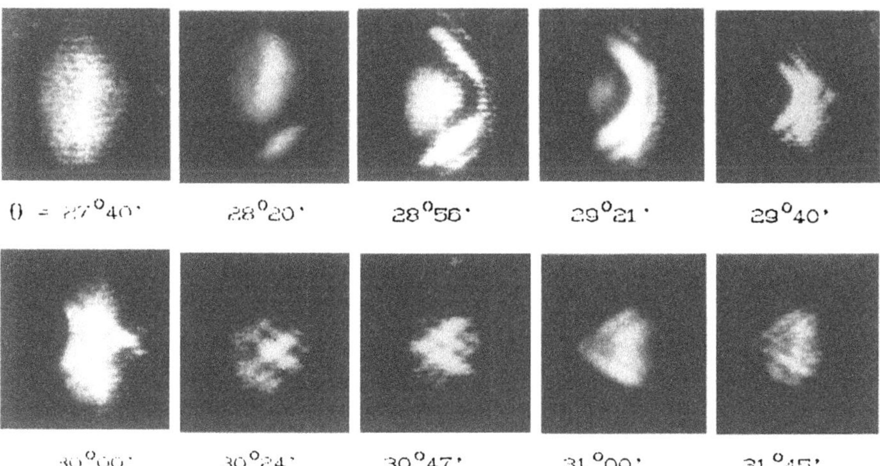

Fig.2 Transformation of reflected beam spot of radiation with broadened angular spectrum at different angles of incidence when SEW excites

is absorbed by grating. It gives a radical transformation of spatial distribution of reflected light. Fig.2 shows the transformation of a reflected light's spot as the angle of incidence is varied in the neighborhood of resonant absorption when the length with f=.3 m was used.

The interference of incident and resonantly absorbed waves leads to arising of two systems of bands in the reflected spot. Their slope are determined by the directions of wave fronts of two resonantly excited SEW (in our case - approximately 30° with $\mathbf{k}$). The interference of this waves give rise to local amplification of light intensity.

3. Conclusions

1. The resonant absorption of IR by resonant excitation of SEW was observed
2. The angular width of laser radiation may be measured with a help of metal grating with shape SEW excitation resonances
3. The transformation of reflected spot and local amplification of light intensity due to resonant excitation of SEW was observed.

4. References

1. Akhmanov S.A., Emelianov V.I., Koroteev N.I., Seminogov V.N. Sov. Phys. Uspehy, 1985, v. 28, p. 1084
2. Emelianov V.I., Seminogov V.N. Quantum electronics (Moscow), 1987, v. 14, p. 47 (in russian)
3. Inagaky T. Motosuga M., Yamamory K, Arakawa E.T. Phys. Rev. B., 1983, v. 28, p. 1740
4. Benditsky A.A., Karabutov A.A., Kudinov I.A., Platonenko V.T., Sogoyan M.A., Chugunov A.V. Moscow University Vestnic. Physics, 1990, v. 31, N 3, p. 43

Part V

Electronic Materials and Semiconductors

Competition Between Thermal and Plasma Waves for the Determination of Electronic Parameters in Semiconductor Samples

D. Fournier

Laboratoire d'Instrumentation, Université P. et M. Curie,
ESPCI, UPRA 0005 CNRS, 10, rue Vauquelin,
F-75005 Paris, France

Abstract : In this paper we will review how the thermal and plasma waves have been revealed to be a good probe for the detection of the dynamics of free carriers and consequently for the determination of electronic parameters.

After a detailed description of the two setups we have used (mirage and photoreflectance), we will explain how to proceed to achieve theoretical calculations to determine both the photocreated carrier density and the temperature rise in either a 1D or a 3D geometry.

We will then illustrate these points with examples describing the evaluation of transport properties for silicon samples with different spatial resolutions.

Finally, we will emphazise that using new experimental conditions (a near IR laser for excitation instead of an argon one and intermediate frequency modulation range) can allow more precise determination of electronic parameters.

1.Introduction

Thermal wave probing has been revealed to be a very interesting way to study various samples in order to solve fundamental or applied physics problems. For semiconductor studies, this technique has demonstrated its power in many cases : spectroscopic investigation of layered or complex samples, thermal diffusivity determination of bulk, layered or heterogeneous samples. Moreover by using the competition between thermal waves and plasma waves, transport properties (minority carriers diffusivity, lifetime, surface recombination velocity...) have been determined with a good precision /1/.

This paper will be divided in three parts. In the first one we recall how to create thermal waves and plasma waves, how these waves can propagate in a semiconductor sample and how they can be detected. In a second part, we will emphasize the ability of a mirage experiment in air to determine the surface recombination velocity of a silicon wafer by following the heat due to the photocreated carriers recombination in the bulk and at the surface of the sample. Moreover by probing the heat and the carrier distributions inside the sample itself, we will show that it is possible to get also precise electronic parameter determination. Finally, in a third section we will explain how a photoreflectance experiment allows the determination of the transport properties and how it is possible to choose convenient experimental parameters (frequency range and excitation wavelength) to facilitate this determination.

2.Photothermal experiments on a semiconductor sample : thermal waves and plasma waves

A photothermal experiment is the "combination" of:
- an illumination of the sample with a pulsed or a modulated pump beam.
- a detection of the surface or volume temperature variation related to the transformation of this light energy into heat.

This detection can be done with different experimental setups depending upon the physical parameters which are under investigation : refractive index, infrared emissivity, acoustic waves or local deformation... In this paper, we will use mainly two kinds of optical detection : the mirage for 1D experiment / 2 / and the photoreflectance for 3D configuration / 3 / setups, which both allow the detection of the surface or the bulk temperature variation with a very good sensitivity.

In a semiconductor sample we also have to take into account the creation of minority carriers when illuminating the sample with a light whose energy is higher than the gap energy of the sample under investigation. The refractive index being also sensitive to the carrier density, the two experimental setups that we are going to describe can follow both the thermal waves and the plasma waves.

a - Experimental setups

Fig.1 describes the mirage set up which allows the recording of the effects induced in the sample. The sample is settled in a "mirage cell"/ 4 / and illuminated with the pump beam (laser beam or monochromator). The heating of the surface leads to a temperature gradient which spreads in the transparent surrounding (air or transparent liquid) and in the sample. The temperature gradient is associated with a refractive index gradient, which can be measured by the deflexion of an HeNe laser probe beam crossing it .The angular deflexion of the probe beam is detected by a silicon position sensor. It is possible to reach the photon noise with this setup, (10^{-11} rad/$\sqrt{}$Hz for the angular deviation which corresponds to 10^{-5} °/$\sqrt{}$Hz for the sample surface temperature variation in air or 10^{-7}°/$\sqrt{}$Hz if the surrounding medium is a transparent liquid such as CCl_4). With this kind of detection it is possible to detect in air thermal waves up to 100 kHz. When probing inside the sample, it is then possible to detect both thermal waves and plasma waves.

Fig. 2 shows a photoreflectance setup scheme. The sample is illuminated with a focussed excitation spot (1 μm^2 to a few hundred μm^2) and the probe beam is reflected at the surface of the sample, in the heated area. In this case, the measurement is done by following the variation of the refractive index of the sample through the reflexion coefficient R which is affected by temperature rise and carrier density changes. It is possible to achieve setups

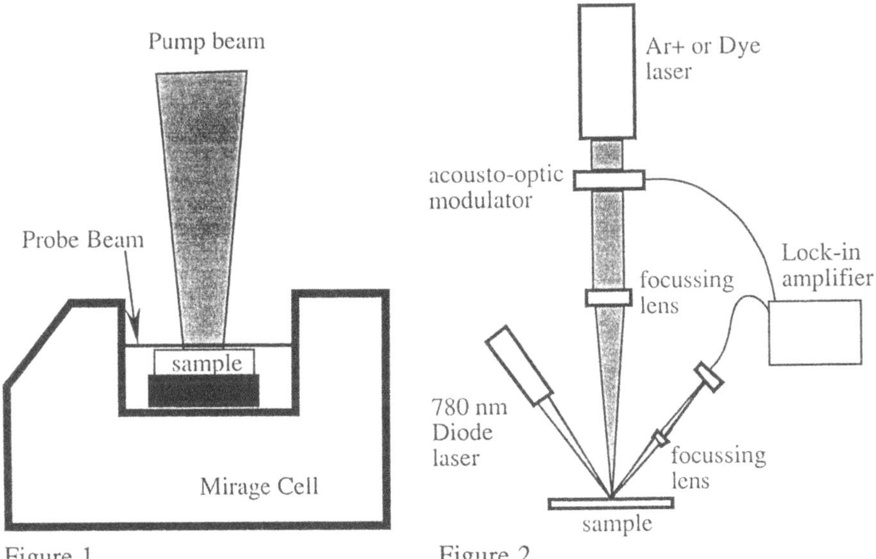

Figure 1

Figure 2

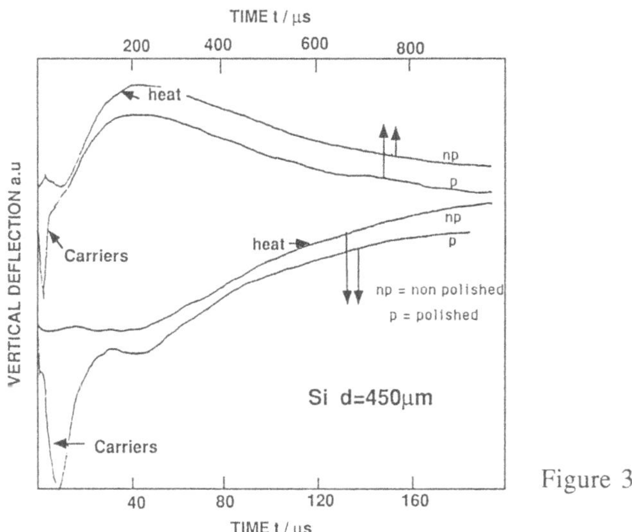

Figure 3

exhibiting a sensitivity of 10^{-6} in $\Delta R/R$ / 3 / or 10^{-7} / 5 / whose frequency range domain spreads up to 20 MHz.

Fig 3 from ref / 6 / shows the result obtained in probing with a "mirage" setup, inside the silicon sample, the two expected waves. We have used a 1.15 μm HeNe laser as probe beam and a 15ns pulsed Nd:Yag laser associated with a doubler as pump beam (0.53 μm). It is clear that one wave is visible at short times, while the other wave appeared at longer times. We will show that the first one is associated with the minority carrier distribution while the other one is the thermal wave. Moreover, the quality of the surface is a dominant factor to observe the plasma wave : when the surface is highly damaged the plasma wave disappears. We have to keep in mind all the results of this experiment in order to choose the better experimental conditions for determining thermal and electronic diffusivities, minority carrier lifetime and surface recombination velocity, with a good precision.

b - Thermal wave

In order to explain what is called a thermal wave, we have to solve the heat diffusion equation in a semi-infinite medium with a plane modulated heat source located at x = 0. The solution is the following :

$$T(x,t) = T_0 \exp(-x/\mu) \exp(j\, [\omega t - x/\mu]) \qquad (1)$$

where : T(x,t) is the temperature at a distance x from the source and T_0 is correlated to the amplitude of the heat source.
μ is the thermal diffusion length which characterizes the heat propagation in the medium and which can be controlled by the modulation frequency $f = \omega / 2\pi$ of the heat source. μ is related to k (thermal conductivity), to ρ (density) and C (specific heat) and f by the following expression :

$$\mu = (k / 2 \pi f \rho C)^{1/2} = (D / \pi f)^{1/2}$$

where D is the thermal diffusivity. The expression (1) could describe the propagation of a scalar optical wave, for which the real part and the imaginary part of the refractive index would be equal. So, although a thermal wave is strongly attenuated, it can reflect, diffract and propagate as a classical one.

Moreover, eq (1) indicates clearly that measuring the phaselag of the complex temperature allows the localization of heat sources in the sample. In reference / 7/it has been clearly demonstrated that the absorbing centers were located on the surface of an amorphous silicon sample by running a high frequency modulation mirage experiment in air.

To know the temperature distribution in the sample and in its surroundings we have to solve the heat diffusion equation in all the media (fluid, sample and backing).If the sample is a complex one (multilayered sample) we have to consider separately each layer of the sample and of the surroundings. At each interface, the incident thermal wave, A_0, is transmitted, A_1, and reflected, B_0, following the thermal properties of each medium. The thermal effusivity $e = (k\sqrt{D})$ is then the analogue to the refractive index :

$$B_0/A_0 = (e_0 - e_1) / (e_0 + e_1)$$
$$A_1/A_0 = 2\, e_0 / (e_0 + e_1)$$

We must underline that a strong thermal mistmatch can be useful for the photothermal investigation of a layered sample. Reference / 8 / provides a good illustration of this characteristic.

Finally, let us underline that as a photothermal experiment is always associated with a theoretical model, it is very important to know with a good accuracy the thermal parameters such as the thermal diffusivity D or the thermal conductivity k of the sample and of its surroundings. We just want to recall briefly that a mirage experiment is able to determine the thermal diffusivity either in the case of a thin layer deposited on a transparent substrate in combining rear and front illumination / 9 /, or in the case of a bulk or layered sample in recording the spatial spreading of the heated spot / 10 /.

c - Plasma wave

When a semiconductor is irradiated with a modulated pump beam whose energy is higher than the band gap, it is well known that free carriers are generated in addition to the heating of the sample. The photocreated carrier motion is also governed by a diffusion equation :

$$D_C \frac{\partial^2 N}{\partial x^2} - \frac{\partial N}{\partial t} - \frac{N}{\tau} + S_N(x,t) = 0$$

where N(x,t) is the density of the photocreated carriers, D_C is their electronic diffusivity and τ their life time. When the source term $S_N(x,t)$ is modulated the solution of this equation can be written as :

$$N(x,t) = N_0 \exp(-\sigma_C x) \exp(j \omega t)$$

where $\sigma_C = 1 / D_C \tau + j \omega / D_C$ is the plasma wave vector.
If $\omega\tau \ll 1$ the plasma wave loses its wave like properties and the process becomes purely diffusive characterized by a diffusion length $\sqrt{D_C \tau}$. In the other case, $\omega\tau \gg 1$ the plasma wave behaviour is analogue to the thermal wave one. In a semiconductor such as silicon and for modulation frequencies in the MHz regime, one can easily have $\omega\tau \gg 1$ and the plasma density behaves then as a critically damped wave.

3.Determination of transport properties in a semiconductor by a mirage experiment

When a semiconductor is irradiated with a modulated pump beam whose energy is higher than the band gap, we have already mentioned that two waves are created in the sample : a thermal wave and a plasma wave which will also act as a distributed heat source. Neglecting the heat transfer to the surrounding fluid, we solve the two following coupled equations in order to know the temperature field T(x,t) and the carrier density distribution N(x,t) / 11 /:

$$D_C \frac{\partial^2 N}{\partial x^2} - \frac{\partial N}{\partial t} - \frac{N}{\tau} + \frac{\partial N}{\partial T}\frac{\partial T}{\partial t} + S_N(x,t) = 0$$

$$k_{th} \left[\frac{\partial^2 T}{\partial x^2} - \frac{1}{Dth}\frac{\partial T}{\partial t}\right] + \frac{Eg}{\tau} Dth\, N + S_T(x,t) = 0$$

The minority carriers are characterized by the diffusivity D_C and their lifetime

τ, while the thermal parameters are D_{th}, the thermal diffusivity and k_{th} the thermal conductivity. $S_N(x,t)$ and $S_T(x,t)$ are respectively the source term for the carrier diffusion equation and for the heat diffusion one. E_g is the semiconductor gap energy / 11 /. In these equations we have neglected the acoustic waves and the thermomechanical and electromechanical coupling contributions.

Since we are in the case of a carrier density smaller than 10^{18} cm^{-3}, it is assumed that the carrier lifetime is independent of the free carrier concentration; and, therefore, Auger recombination will be neglected. Furthermore, we can neglect the $\partial T/\partial t$ $\partial N/\partial T$ term which is related to the carrier creation due to the temperature rise / 12 /. So, we can first solve the carrier diffusion equation and then the heat one,with the carrier density N(x,t) as source term. These resolutions lead to two waves which can be detected separately or not, following the modulation frquency range and the experimental detection.

Let us consider now the photothermal study of a silicon wafer in air, accomplished with a mirage setup, with a 700 mW Ar^+ laser beam to produce a large excitation spot on the surface in order to eliminate the effects of the lateral diffusion and to stay in a 1D diffusion model. The thickness of the n-type wafer was about 360μm, so we have to take into account the recombination processes on both the back and front surfaces. We have used a phenomenological approach by using the surface recombination velocity s to modelize the recombination processes on the surface. In this case, the resolution of the two diffusion equations (carrier distribution and heat distribution) lead to very complicated analytical formulas /13 /.

Let us summarize the main result which is that the surface temperature is due to the contributions of two terms:

- A surface thermal wave correlated to the very fast non-radiative relaxation of the photoexcited free carriers which give the excess of energy of the photon compared to the band gap energy, within a fraction of a picosecond. This is a pure surface wave, in our case, since the excitation is done with an argon laser (200 nm penetration length).

- A volume wave which is due to the non-radiative recombination of the photocreated carriers giving up their remaining energy Eg after a time τ corresponding to their lifetime (on the order of μseconds for silicon).

Depending on the geometry of the experiment (front or back illumination) and on the quality of the surface under investigation (large or small value for the surface recombination velocity), it is possible to reach with a rather good precision the electronic parameters.

We have theoretically studied the four experimental configurations of pump and probe beam to determine the most appropriate one, considering the thickness of the sample and the frequency range for which the signal to noise ratio is sufficient /14 /. We have run calculations for a wafer which exhibits a damaged surface on one side and a polished one on the other side. When excitating a damaged surface (large s), we can probe in the air in front of this surface, but the thermal wave is then dominant throughout the whole frequency range, or behind the polished one. The electronic diffusivity being larger than the thermal diffusivity, we expect the plasma wave to reach the rear surface before the thermal one. When illuminating the polished surface

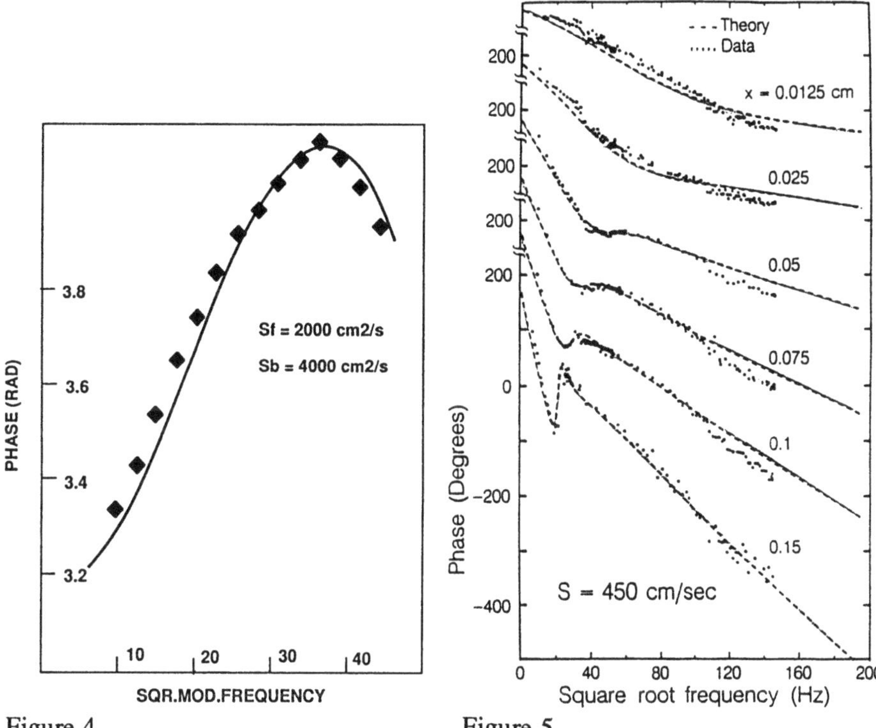

Figure 4

Figure 5

and probing this front surface, the plasma waves are expected even at low frequencies. This can be explained by the random walk of the minority carriers in the bulk of the sample before their recombination at the surface.

To avoid the convolution effect / 15 / due to the finite size of the probe beam we have taken the difference of the results obtained for two configurations (rear and front). Fig.4 from ref / 16 / shows the experimental result and the theoretical fit. We are presently improving our experimental setup (excitation wavelength and modulation frequency) and we hope that it will be possible to determine unambigously the back and front recombination velocities of a silicon wafer with a very simple low frequency mirage experiment.

Finally, figure 5 from ref / 17 / illustrates the propagation of the plasma wave and thermal wave inside the sample when the sample is illuminated by a modulated pump beam. As is clearly demonstrated, the above described model is rather convenient to explain the experimental data. At high frequency the plasma wave is dominant while the thermal wave behaviour is clearly shown at low frequency.

4.Determination of transport properties in a semiconductor with a photoreflectance experiment

In this kind of experiment, the sample is illuminated with a small spot and the measurement is accomplished by following the induced photoreflectance variation of the probe beam reflected on the heated area. We must underline that in the case of a semiconductor sample, the refractive index is affected not only by the temperature variation, but also by the carrier density one: the induced photoreflectance ΔR can be therefore written :

$$\Delta R = (\partial R / \partial T)\, \Delta T + (\partial R / \partial N)\, \Delta N$$

As in the case of the mirage experiment inside the semiconductor, the plasma wave will be detected at short time scale (at large frequencies), while the thermal wave dominates at long time (in the low frequency range). This method of investigation is now currently used in the microelectronic factories. Thermal wave Inc. has indeed developed an instrument using the laser induced modulated reflectance method for monitoring the dose and uniformity of ion implantation in Si and GaAs wafers, for detecting and quantifying the residual damage in silicon wafers that results from chemopolishing and scrubbing processes and for detecting and characterizing radiation damage and contamination in silicon from reactive ion and plasma etch processes / 18 /.

In order to get quantitative information from this experiment, we have to use 3D diffusion equations, to take into account the geometry of the experiment. To obtain the three-dimensional expressions, we have used the integral transformations (Laplace and Hankel) which are convenient for the axial geometry of the problem. By using a numerical calculation of the inverse Hankel transform / 19 /, we are able to calculate the induced photoreflectance signal.

Let us consider now the figure 6 (from ref /5/) in which the amplitude and the phase of the photoreflectance signal of a Σ=25 twinned boundary of a p-type silicon bicrystal have been recorded in the two cases of an impurity free boundary and a decorated one. While the impurity free boundary is well reconstructed and exhibits no electrical activity, decoration with transition metal impurities makes the boundary electrically active and generates large densities of interface traps associated with a strong recombination efficiency / 20 /. This experiment was done with a photothermal microscope built in our laboratory, whose sensitivity in $\Delta R/R$ is 10^{-6}. The photothermal results are in accordance with the electrical ones : only the decorated boundary exhibits a large variation in the photothermal investigation, which can be qualitatively explained as follows:

Far from the boundary the signal is mainly due to the plasma wave while near the boundary the increasing of the signal corresponds to a pure thermal wave associated with a very strong velocity recombination coefficient. The Π phase change corresponds to the reverse sign of $\partial R / \partial T$ and $\partial R / \partial N$. The excitation beam was a 30mW argon laser, the probe beam a 0.1mW HeNe one (1µm spots) and the modulation frequency is in the range of MHz. As expected, it is easy to detect the plasma wave at high frequency.

We want now to underline that it is possible to obtain useful information on the carrier properties by using a 100 kHz modulation frequency and a larger spot than $1 \mu m^2$, this configuration being more convenient in some cases. To enhance the plasma wave with respect to the thermal wave, we can reduce the contribution of the surface thermal wave due to the photon

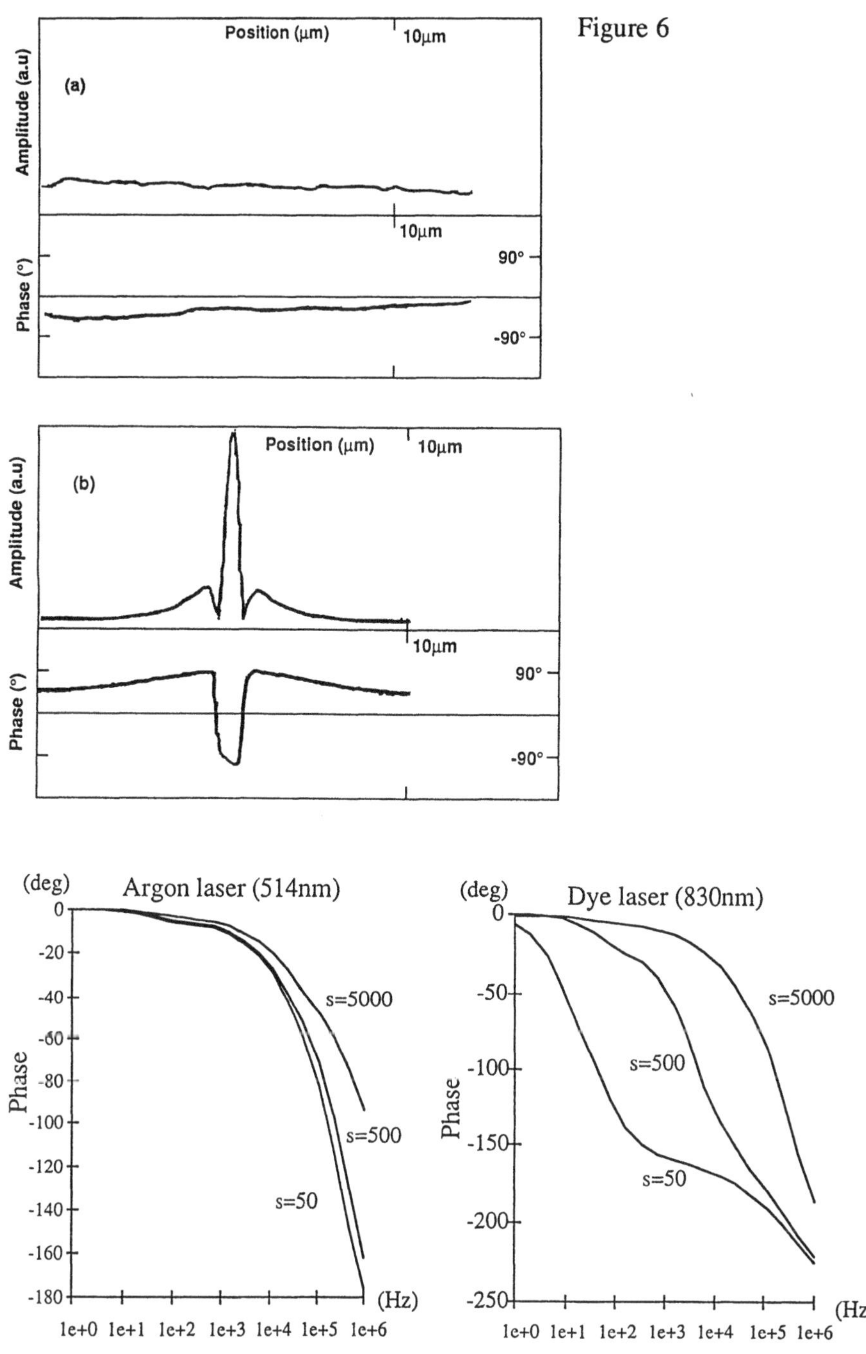

Figure 6

Figure 7

excitation energy in excess of the gap energy by using a dye laser whose energy is smaller than that of an argon laser.

Fig. 7 shows a theoretical calculation, showing that it is then easy to detect the plasma wave even at 100 kHz, by using a 830 nm dye (Styryl 9) laser and a 80 μm diameter spot. Experimental results and their fit to the theoretical model are shown in ref / 21 /.

Conclusion

In conclusion, this paper underlines that thermal waves can be very helpful for the determination of electronic properties in the case of a silicon sample. Photothermal investigation of semiconductors is nowadays well known in the microelectronics world. However, we have demonstrated theoretically that a judicious choice of new experimental conditions for a mirage or a photoreflectance experiment (illumination wavelength, heated spot size) allows one to detect the plasma waves in an intermediate frequency range, which largely facilitates the experiment. Experimental results have confirmed our calculations in the case of n, n^+, p, p^+ type silicon wafers. These recent investigations allow us to believe that this methodology could be extended to other semiconductor samples (GaAs, ...) and more complicated devices (SiMox...).

References

/1/ Mandelis, A., Editor " Photoacoustic and thermal wave phenomena in semiconductors " Elsevier Amsterdam 1987.

Rosencwaig, A., "Photoacoustics and photoacoustic spectroscopy" in Chemical Analysis Elving and Winefordner Editors Vol 57 J.Wiley, New York.1980.

/2/ Boccara, A.C., Fournier, D.and Badoz, J., Appl.Phys.Lett.36 (1979), 130.

/3/ Rosencwaig, A., Opsal,J., Smith,W.L.and Willenborg, D.L., Appl.Phys. Lett. 46 (1985) 1013.

/4/ Charbonnier, F.and Fournier, D., Rev.Sci.Instrum. 57 (1986), 1126.

/5/Inglehart, L.J., Broniatowski, A., Fournier, D., Boccara, A.C.and Lepoutre, F., Appl.Phys.Lett.56 (1990) 18.

/6/ Pelzl, J.,Fournier, D.and Boccara, A.C.,5th Int.Top.Meet.on Photoacoustic and Photothermal Phenomena Heidelberg RFA (1987).

/7/ Chahed, L.Theye, M.L.Fournier, D., Roger, J.P., Boccara, A.C., Li, Y.M., Turner,W.A. and Paul, W., accepted Phys. Rev. (1991).

/8/ Yacoubi, PhD thesis, University of Montpellier (France) (1986).

/9/ Roger, J.P., Lepoutre, F., Fournier, D., and Boccara, A.C., Thin Solid Films, 155, 165-174 (1987).

/10/ Kuo, P.K., Sendler, E.D. Favro, L.D., Thomas, R.L., Can J. Phys. 64 (1986).

/11/ Sablikov, V.A.and Sandormiskii, V.A., Phys.Stat.Sol. B120 (1983), 471.

/12/ Stearns,R.G.and Kino,G.S. in "Photoacoustic and thermal wave phenomena in semiconductors" Mandelis editor, Elsevier Amsterdam 1987.
/13/ Mikoshiba, N., Nakamura, H.and Tsubouchi, K., Proceedings of the IEEE Ultrasonic Symposium San Diego, San Diego Ca (1982)
/14/ Forget, B.C.and Fournier, D.,7th Int.Top.Meet.on Photoacoustic and Photothermal Phenomena Doorweth the Netherlands (1991)
/15/ Legal Lasalle, E.,Lepoutre,F.and Roger, J.P. J.Appl.Phys. 64 (1988),1.
/16/ Forget, B.C.and Fournier, D.,7th Int.Top.Meet.on Photoacoustic and Photothermal Phenomena Doorweth the Netherlands (1991).
/17/ Fournier,D., Boccara, A.C., Skumanich, A.and Amer, N.M. J.Appl.Phys. 59 (1986),787
/18/ Thermal Wave Inc. Fremont, CA 94539 USA.
/19/ Candel, S.M.,Computer physics communication 23 (1981),343.
/20/ Broniatowski,A., Phys.Rev.Lett. 62 (1989), 3074.
/21/ Forget, B.C.and Fournier, D.,7th Int.Top.Meet.on Photoacoustic and Photothermal Phenomena Doorweth the Netherlands (1991).

Photoacoustic Calibration of Silicon Photodiodes

E. Halaška, M. Dienstbier, and P. Sladky

Charles University, Faculty of Mathematics and Physics,
Department of Chemical Physics, Ke Karlovu 3,
CS-121 16 Praha 2, Czechoslovakia

Abstract. We have calibrated silicon photodiodes with P/N and SiO_2/P/N structure. Calibration based on the determination of the internal quantum efficiency of the photodiode is described.

Silicon photodiodes can be used as a precision radiometric standard [1,2,3,4].

Photodiodes used for radiometric measurements are usually operated in the photovoltaic short circuit mode. In this mode, responsivity of the photodiode R(λ) can be expressed:

$$R(\lambda)=(1-\rho(\lambda)).\eta(\lambda,U=0).\lambda/K \qquad K = 1.23985\ \mu m.W/A,$$

where λ is wavelegth, ρ(λ) reflectance of the front surface of the photodiode, U reverse bias voltage, η(λ,U) internal quantum efficiency (ratio of the number of charge carriers measured in an external circuit and the number of photons absorbed within photodiode). Calibration procedure is performed at each wavelength where responsivity data are required. Calibration consists of two measurements - measurement of reflectance and measurement of internal quantum efficiency. Reflectance of the front surface of the photodiode is mainly specular, and its measurement is quite simple. Main experimental problem is the measurement of the internal quantum efficiency of the photodiode.

We employed photacoustics to determine internal quantum efficiency [5,6,7,8]. Experimental setup is shown in fig.1. We measured dependance of PAS (photoacoustic signal) on reverse bias voltage. PAS can be expressed:

$$PAS(\lambda,U)=C.[\Phi_o(\lambda).(1-\rho(\lambda)).h.\nu+U.e.\Phi_o(\lambda)(1-\rho(\lambda)).\eta(\lambda,U)],$$

Springer Series in Optical Sciences, Vol. 69 Photoacoustic and Photothermal Phenomena III
Editor: D. Bićanić

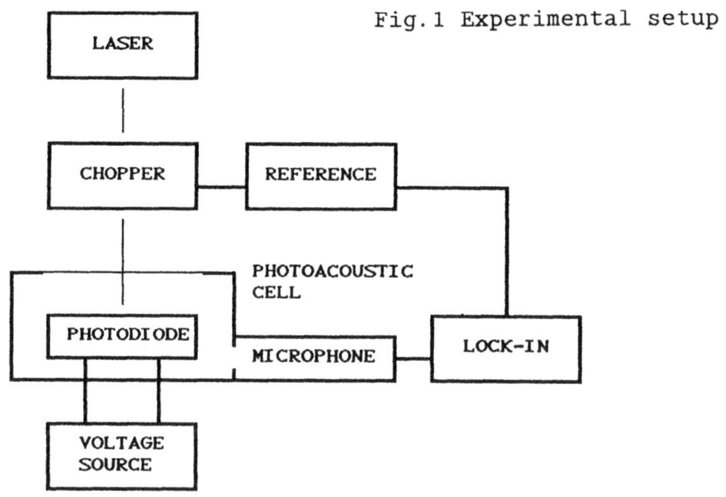

Fig.1 Experimental setup

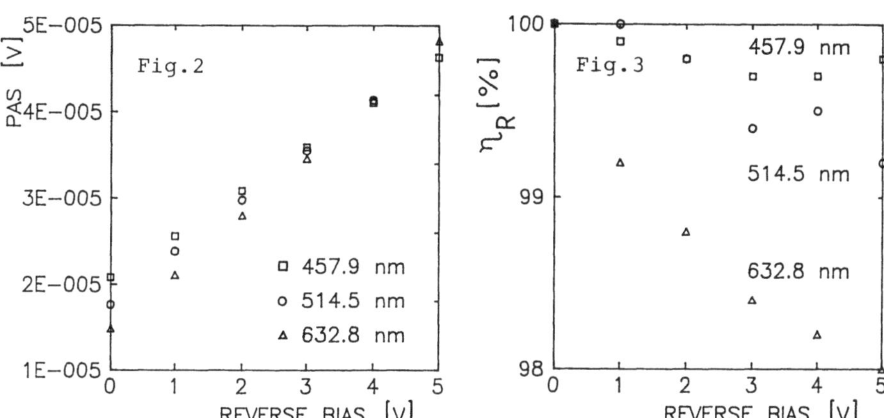

Fig.2 Dependance of PAS on reverse bias voltage (photodiode with P/N structure)

Fig.3 Dependance of η_R on reverse bias voltage (photodiode with P/N structure)

where C is constant, e elementary charge, $\Phi_o(\lambda)$ number of photons incident on photodiode per second, $\rho(\lambda)$ reflectance, λ wavelength, ν frequency of radiation, h Planck's constant, U reverse bias. Results are shown in fig.2.

η depends on λ and U. $\eta(\lambda,U)$ can be expressed:

$$\eta(\lambda,U)=\eta(\lambda,0)/\eta_R(\lambda,U) \qquad (\ \eta_R(\lambda,U)=\eta(\lambda,0)/\eta(\lambda,U)\ ).$$

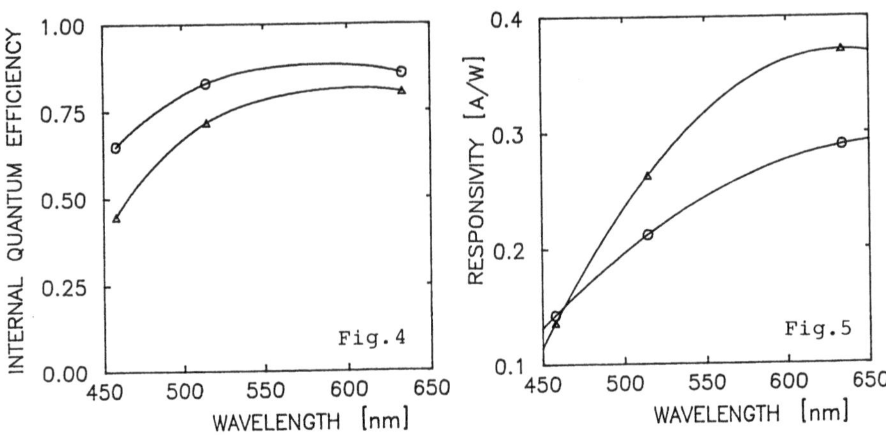

Fig. 4 Spectral dependance of internal quantum efficiency (photodiode with P/N structure - upper curve, photodiode with SiO_2/P/N structure - lower curve; points - measured values)

Fig. 5 Spectral dependance of responsivity (photodiode with SiO_2/P/N structure - upper curve, photodiode with P/N structure - lower curve; points - measured values, curves - quadratic approximation polynoms)

We measured η_R as a function of U (fig. 3), η_R is decreasing as U is increased.

From PAS(λ,U) and $\eta_R(\lambda,U)$ we determined internal quantum efficiency corresponding to zero bias voltage $\eta(\lambda,U=0)$ (fig. 4) and responsivity (fig. 5). We performed these measurements for photodiodes with P/N and SiO_2/P/N structure.

We have estimated uncertainty of the measurement of the internal quantum efficiency to be 5%.

[1] E. F. Zalewski, J. Geist, J. Appl. Opt. 19, 1214 (1980)

[2] J. Geist, E. F. Zalewski, A. R. Schaefer, J. Appl. Opt. 19, 3795 (1980)

[3] R. L. Booker, J. Geist, J. Appl. Opt. 23, 1940 (1984)

[4] E. Halaska, R. Benes, P. Sladky, I. Benc, Czech. J. Phys. A 41, 149 (1991)

[5] D. Cahen, Appl. Phys. Lett. 9, 810 (1978)
[6] W. Thieleman, B. Rheinlander, Solid State Electronics 11, 1111 (1985)
[7] A. Rosencwaig, Photoacoustics an Photoacoustic Spectroscopy, Wiley, New York 1980
[8] E. Halaska, M. Dienstbier, R. Benes, Czech. J. Phys. B 40, 940 (1990)

The Comparative Study of Semiconductor Structures Using Simultaneous Detection of Photoacoustic Signal and Photovoltage

R.M. Burbelo, Yu.B. Utrobin, G.I. Bulakh, A.L. Gulyaev, and I.Ya. Kucherov

Faculty of Physics, Kiev State University, 252601 Kiev, Ukraine

Abstract. In the paper, the results of comparative study of semiconductors by photoacoustic (PA) and photoelectric (PE) microscopy are presented. The latter utilizes the contactless detection of alternating photovoltage as an informative signal. Silicon epitaxial structures and silicon wafers implanted with various dose of ions have been investigated simultaneously.

1. Introduction

It is known, that appreciable electric signal can appear in a semiconducting sample under PA study due to photovoltage generation. For separation of real PA signal in case of piezoelectric detection, the proper electric shielding of a transducer is needed [1,2]. On the other hand, the measurement of alternating photovoltage is an independent method of semiconductor study and provides the realization of PE microscopy of near-surface layers in a sample using the raster scanning by a focused laser beam [3]. The proximity of techniques for PE and PA microscopies and the possibility of contactless capacitive detection of alternating photovoltage allow us to obtain the images of the sample under study by both methods simultaneously, expanding the information obtained.

In this work, the investigations of silicon epitaxial structures and ion-implanted wafers have been performed for the purpose of comparing the techniques mentioned above.

2. Technique and Samples

The detection of PA signal in a sample was carried out by the disk-shaped piezoceramic transducer as in Fig.1. The sample was fixed on the upper electrode of transducer using an acoustic bond. From the lower electrode the signal passed to the lock-in amplifier of the first measuring channel. The piezotransducer was positioned on the movable X-Y stage with a

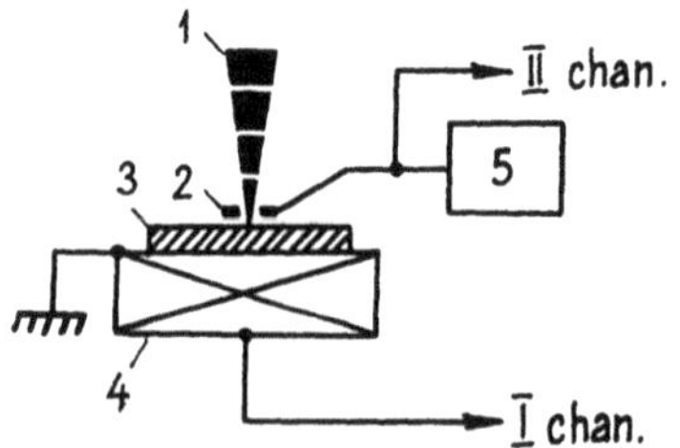

Fig.1. The diagram of measurements. 1- laser beam, 2- capacitive electrode, 3- sample, 4- transducer, 5- alignment system

Springer Series in Optical Sciences, Vol. 69 Photoacoustic and Photothermal Phenomena III
Editor: D. Bićanić

stepping-motor drive. The minimum step of scanning was 10 microns. The capacitive electrode in the shape of a ring was mounted close to sample's surface for photovoltage detection (the second measuring channel). Modulated Ar-ion laser beam (λ = 488 nm) was focused through the electrode on a sample surface into a spot of 10 microns in diameter. Signal measurement, data processing and scanning control were provided by a computer. Images were typed in a half-tone mode with 10 grey levels.

The epitaxial structures were prepared in the following manner. In a p-type silicon wafer the pockets of various topology were made by anisotropic etching to the depth of 40 μm. After that the processes of epitaxial growth of n-type layer up to the thickness more than 40 μm, substrate opening and mirror polishing were carried out respectively. The surface of the samples was optically homogeneous. The contours of pockets could only be seen under suitable lateral illumination.

The second kind of samples were polished wafers of p-type silicon in which the boron ion implantation of 100 keV energy and various doses from 3E11 to 6E13 ions/cm^2 was made through the mask with a sectoral indent. Any annealing of the wafers was not carried out after ion implantation. An optical contrast of the samples surface was absent.

3. Results

In Fig.2a and Fig.2b the results of visualization by means of PA and PE measurements in epitaxial structures are presented respectively. The images were perfomed at the modulation frequency of 80 kHz which corresponded to visualization depth about 20 μm in silicon. The area of visualization equals 4.8x4.8 mm^2.

Analysis of the images shows that the visualization of regions with different types of conductivity takes place in both cases. At the fixed initial phase lag of lock-in detection the mapping of photovoltage gives most distinct images of regions' topology providing for the enhancement of contrast at boundaries. Such a contrast halo around the epitaxial regions appreciably expands their real topology. This is well shown in comparison with PA image, especially at the narrow parts between epitaxial regions. It is seen that a lot of details is present in PA image, pointing to the defectiveness of near-surface layer as a result of improper treatment of the surface. The contrast of p- and n-type regions is worse in general in the case of PA imaging. A "smooth" inhomogeneity of PE signal over the area of image in Fig.2b is not connected with the characteristics of

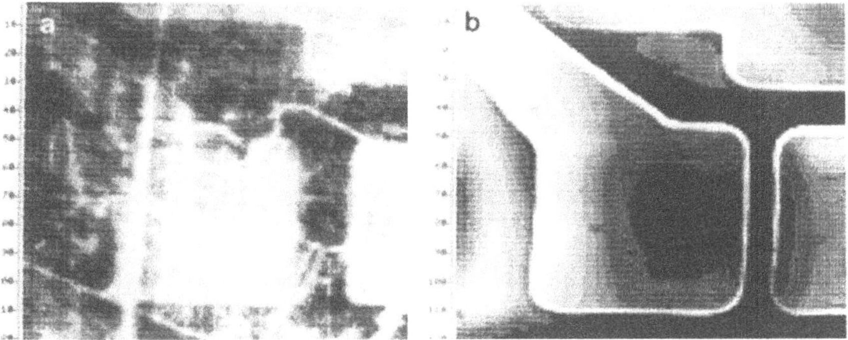

Fig.2. Photoacoustic (a) and photoelectric (b) images of epitaxial pockets in silicon structure.

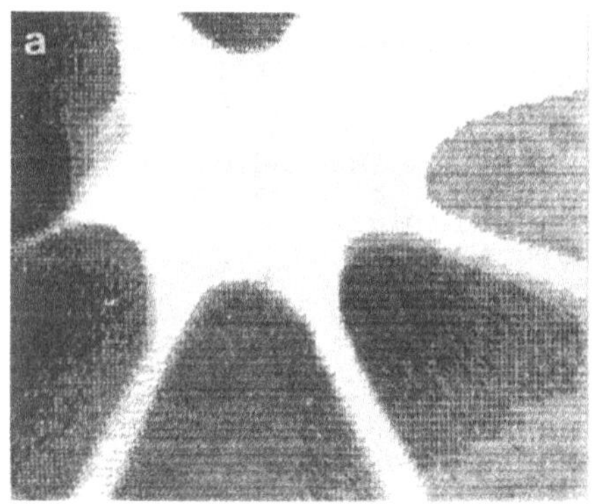

Fig.3. Photoacoustic (a) and photoelectric (b) images of ion-implanted regions in silicon wafer.

the sample. The sample bonding technique can be one of the possible reason of this inhomogeneity.

In the case of ion-implanted silicon samples both PA and PE microscopies provide the visualization of the regions which underwent the implantation with different doses of ions. In Fig.3a the PA image of central part in the ion-implanted wafer with dimension 9.6x9.6 mm^2 is shown. Eight sectoral regions of different dose of implantation are seen. The light central part and one of the sectors have not been affected by ion bombardment. The density of PA signal in implanted sectors corresponds qualitatively to the level of dose. It is also seen that rather high quality of surface finishing has been made. PE image in Fig.3b displays the implanted regions as well, but the correspondence to dose can hardly be observed.

4. Conclusions

The investigations shows a number of differences in images obtained. This indicates that simultaneous sample visualization by both techniques can provide more information obtained in the studies. Such comparative measurements seem to be useful for getting a complementary information about signal formation in complicated semiconductor structures.

References

1. M. Watanabe, in: Photoacoustic and Photothermal Phenomena, Springer Series in Optical Sciences, Vol.58, Heidelberg, 1988,p.254
2. G.I. Bulakh, R.M. Burbelo, O.V. Volchanski, N.K. Zhabitenko, I.Ya. Kucherov, in: Acoustoelectric and Photoacoustic Methods of Material Characterization, Kiev, 1989, p.38 (in Russian)
3. D.I. Bidnic, I.V. Zaginailo, Yu.V. Kovalyov, A.P. Fedchuk, Electronic Industry, No. 6, 80 (1990) (in Russian)

Photoacoustic Characterization of Ion-Implantation Damage in Silicon

T. Matsumori[1], *M. Uchida*[1], *H. Yoshinaga*[1], *J. Kawai*[1], *T. Izumi*[1], *and F. Uehara*[2]

[1]Department of Electronics, Tokai University, Kanagawa 259-12, Japan
[2]Department of Resource and Environmental Science, Tokai University, Kanagawa 259-12, Japan

Abstract. The properties of silicon substrates damaged by self ion-implantation into silicon has been studied by PZT photoacoustic (PA) spectroscopy. The PA spectra in the samples after implantation had a peak near the band-edge and a long decreasing tail towards the shorter wavelength side predominantly due to the wavelength dependence of light transmission depth. As the PA signal at this shorter wavelength range was found to be sensitive to the extent and the structure of damage present in the implanted layers, the PA signal intensity within this range was used for characterizing the implantation-damage. The result indicates that the damage significantly affects the PA signal intensities and it is annealed out through at least four well-defined stages:a largely increasing stage and a largely decreasing stage at 300-450 ℃ and at 450-700 ℃, respectively, and moreover, a slightly increasing stage and a slightly decreasing stage at 750-850 ℃ and at 850-950 ℃, respectively. The first stage suggests the reordering from a highly disordered layer to a crystalline one including a large number of complex defects, the second stage the extinction process of these defects, and the third and fourth stages the generation and extinction of dislocation loops.

1. Introduction

We have studied the furnace annealing behavior of damage produced by self ion-implantation into silicon by observing the photo-acoustic (PA) signals in the damaged layers by the use of the piezoelectric detection technique [1-4].

2. Experimental Results and Discussion

The substrates for implantation were undoped silicon crystals (carrrier concentrations $n \lesssim 2\times10^{12}/cm^3$) grown by the floating zone method. After mechanical polishing and chemical etching, 150 keV Si^+ ions were implanted at room temperature in the (111) surface on the substrates with doses of 10^{12} - 10^{16} ions/cm^2 and with a current density of less than 1 $\mu A/cm^2$ to avoid a temperature rise in the substrates during implantation.

The PA signal measurements were done at room temperature. The light source used was a 150 W halogen lamp and the light-chopping frequency was 80 Hz. The PA signals were detected by a PZT transducer

 Photoacoustic and Photothermal Phenomena III

directly attached to the rear surface of the substrates with a silver paste.

Isochronal furnace annealing of the implanted samples was carried out in the temperature range 100-1000 ℃ for 10 min at each temperature with 50 ℃ steps in an argon atmosphere.

3. Experimental Results and Discussion

Figure 1 shows the near-bandedge PA spectra of silicon samples before and after implantation with a dose range of 10^{12} - 10^{16} ions/cm^2. All of the PA spectra observed in the samples after implantation had a peak at ~1000 nm near the fundamental absorption-edge and a decreasing long tail towards the shorter wavelength side.

This tailing is interpreted to be caused predominantly by the wavelength dependence of the PA signal intensity following the wavelength dependence of the optical transmission depth (which is defined as the reciprocal of absorption coefficient) within the damaged layer where strong optical absorption takes place. As the decreasing tendency of these tails were monotonic and no remarkable structure was observed on the tails, we can assume that the damage layers induced by implantation are optically homogeneous and thus can adopt PA signal intensity at any wavelength within the tails as a measure of implantation induced damage. Consequently, we have employed the PA signal intensity at a wavelength 600 nm as a measure of induced damage and have investigated the annealing behavior of the implanted layers.

Figure 2 shows the isochronal annealing behavior of the PA intensities at a wavelength 600 nm. The result indicates that the annealing curves are composed of at least four well-defined stages: a largely increasing stage S_A, a largely decreasing stage S_B, a slightly increasing stage S_C and a slightly decreasing stage S_D at 300-450 ℃, 450-750 ℃, 750-850 ℃ and 850- 950 ℃, respectively.

The stage S_A suggests the reordering from a highly disordered layer to a crystalline layer including a large number of complex defects, the stage S_B the extinction process of these defects, and the stages S_C and S_D the generation and extinction of dislocation loops, respectively.

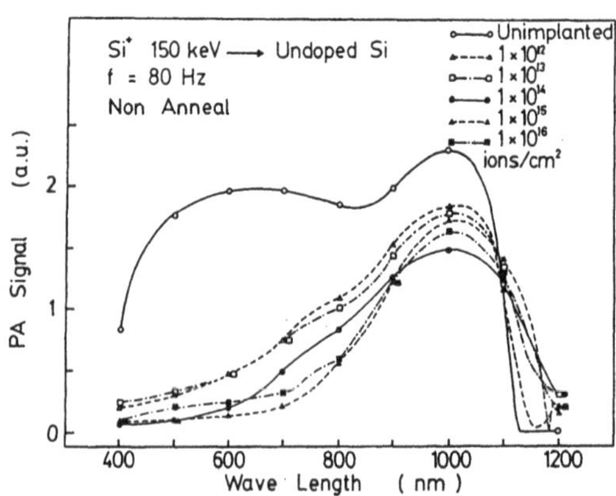

Fig.1. Near-bandedge PA spectra of silicon samples before and after implantation of 150 keV Si^+ ions with a dose range of 10^{12} - 10^{16} ions/cm^2.

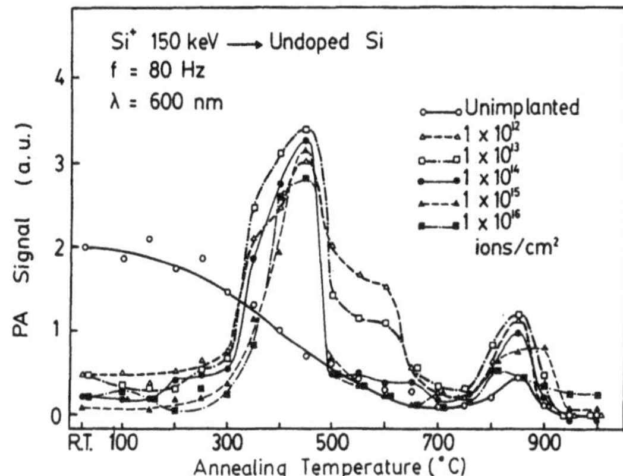

Fig.2. Isochronal annealing curves of the PA signal intensities at a wavelength 600 nm.

References

[1] A.Rosencwaig: J.Appl.Phys. 51 (1980) 2210.
[2] W.Jackson and N.M.Amer: J.Appl.Phys. 51 (1980) 3343.
[3] T.Ikari, S.Shigetomi, and Y.Koga: Photoacoustic and Thermal Wave Phenomena in Semicoductors ed. A.Mandelis (Elsevier, New York, 1987) p.397.
[4] T.Sawada and M.Kasai: Photoacoustic and Thermal Wave Phenomena in Semicoductors ed. A.Mandelis (Elsevier, New York, 1987) p.3.

Photoacoustic Effect in CdS Under Combined Optical Exposure

G.I. Bulakh, R.M. Burbelo, O.V. Volchanski, and I.Ya. Kucherov

Faculty of Physics, Kiev State University, 252601 Kiev, Ukraine

Abstract. This paper describes the study of mechanisms of unmodulated additional illumination (AI) influence on photoacoustic (PA) effect in CdS under piezoelectric detection of PA signal and a role of the alternating photovoltage in the sample during PA measurements. It is shown that in piezosemiconductors the non-thermoelastic component in PA signal is present. It arises due to piezoelectric transformation of photovoltage and depends on the intensity of AI.

1. Introduction

It has been already reported [1-3] on the influence of unmodulated additional illumination on PA signal in CdS crystals. In addition the same behaviour of PA signal in GaAs has been also reported and its non-linear dependence on modulated light intensity has been noted [4]. The data showed the complex character of PA signal formed in such crystals.

This paper describes the experimental study of PA effect in CdS in order to understand the signal formation mechanisms and influence of AI.

2. Samples and Measurements

Monocrystalline CdS samples under study were rectangulars with the dimensions about 5x3x3 mm^3. He-Ne laser (λ=633 nm) or xenon lamp (through a monochromator) were used as PA signal sources. Mechanical chopper was applied for modulation of light beam in the frequency range f = 120 - 2000 Hz. PA signal was detected with either piezoelectric transducer or a gas-microphone cell. A halogen lamp or additional He-Ne laser were used as a source of AI. In the field-effect experiments the electric field could be applied perpendicularly to the exposed sample's surface by the pair of transparent electrodes through dielectric gaps. The photovoltage was detected using capacitive electrodes.

3. Results and Discussion

The dependences of magnitude F and phase of PA signal vs modulated light intensity Im and AI intensity Ia were studied. Spectral and frequency measurements were performed as well. It was established that the influence of AI on PA signal can be observed under piezoelectric detection only. In the samples, which displayed the influence of AI, dependences F(Im) were nonlinear at low values of Ia. Fig.1a shows the curves for F(Im) obtained under the different levels of Ia in the pure CdS

Springer Series in Optical Sciences, Vol. 69 Photoacoustic and Photothermal Phenomena III
Editor: D. Bićanić

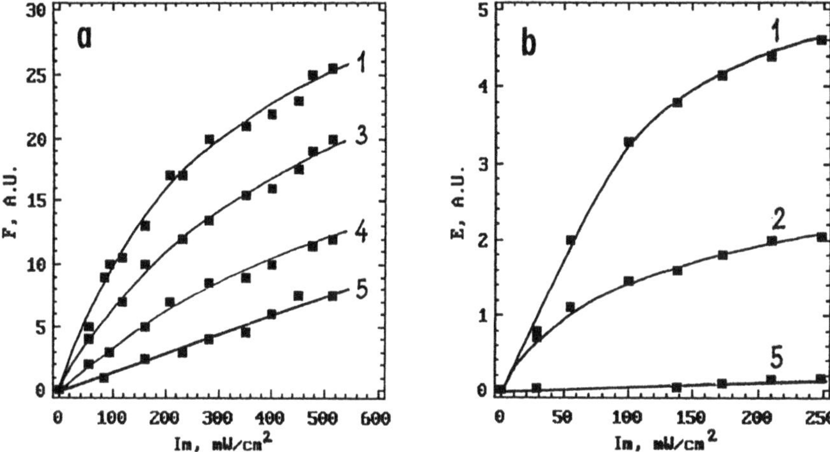

Fig.1. PA signal (a) and alternating photovoltage (b) vs intensity of modulated light at different levels of AI: 1- 0, 2- 30, 3- 80, 4- 150, 5- 250 mW/cm^2

sample. It is seen that the action of AI results in quenching of PA signal. However, in the sample CdS: 0.1%Al the enhancement of PA signal was observed. Under the condition of AI increasing the quenching of non-linearity of the dependences took place in both cases. It is also important to note that in gas-microphone cell the same samples showed the linear dependences.

Basing on these measurements one can conclude that at least two components of PA signal exist in the samples which displayed the influence of AI. One of them is not thermoelastic in nature. Two components of PA signal were separated computationally from experimental dependences F(Im) and F(f) at different values of AI intensity. One of them (F1) had a linear dependence on Im, but its modulation frequency dependence was as 1/f being typical for thermoelastic response. The second one (F2) depended as logIm, and no change in studied frequency range was found.

It should be mentioned that the generation of PA signal in semiconductor can be accompanied by electric signal in a sample itself at the modulation frequency. This signal can reach a considerable value forcing to shield a piezoelectric transducer to prevent its influence [1,5]. In semiconducting monocrystals with piezoelectric properties such photovoltage can directly generate acoustic vibrations. Thus, the capacitive photovoltage was directly measured in the same CdS samples under the same conditions. Its sensitivity to AI has been also found. The spectral and frequency dependences of alternating photovoltage E are in good agreement with those for F2. In Fig.1b the characteristics of E(Im) are shown as an example. The strong influence of external electric field applied to the sample on the value of PA signal and alternating photovoltage has been detected as well.

The dependences of PA signal amplitude and phase on the applied voltage in the field-effect scheme are shown in Fig.2. It is seen, that the influence of external field is rather strong in the case when both components of PA signal are generated (Ia = 0). With the thermoelastic component being dominant (Ia = 250 mW/cm^2) almost no influence of the field was observed. Such behaviour of non-thermoelastic component

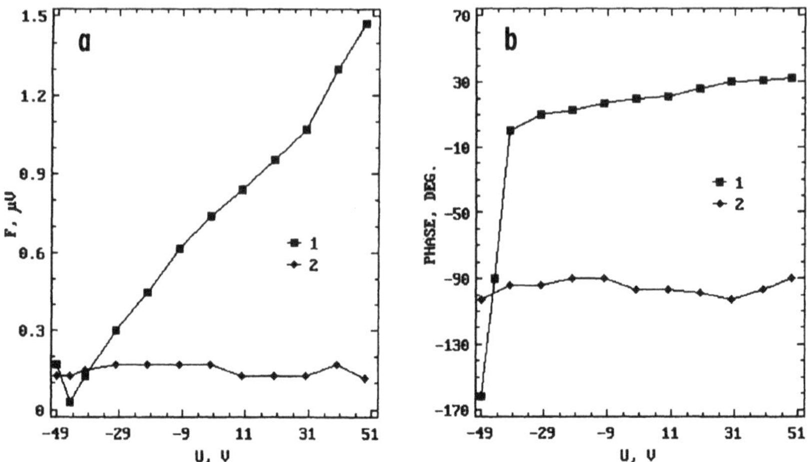

Fig.2. Amplitude (a) and phase (b) of PA signal in pure CdS crystal vs applied electric voltage at two levels of AI (λ = 633 nm): 1- Ia = 0; 2- Ia = 250 mW/cm^2

of PA signal can be explained by the generation of photovoltage at the opposite surfaces of the sample owing to the surface band bending. Only the difference of these two opposite components of photovoltage was detected experimentally. Data obtained allow us to conclude that photovoltage in these experiments is the surface one.

4. Conclusions

Thus, it has been shown in the paper that in CdS crystals PA signal under piezoelectric detection can consist of two components (the first one is thermoelastic in nature, the second is photoelectric) and appears due to direct piezoelectric transformation of surface photovoltage into acoustic vibration. The value of acoustic response of photoelectric nature can considerably exceed that of thermoelastic one.

References

1. G.I. Bulakh, R.M. Burbelo, I.Ya. Kucherov "The investigation of photoacoustic effect in semiconductors", Kiev, UkrNIINTI, No. 1618, (1984)
2. H. Wada, H. Yoshioka, J. Morimoto, T. Miyakawa, Jap.J.Appl. Phys., 24, Suppl. 24-1, 217 (1985)
3. A. Mandelis, E.K.M. Siu, Phys. Rev. B., 34, 7209 (1986)
4. H. Yamamoto, I. Suemune, M. Yamanishi, J.Appl.Phys. 60, 2621 (1986)
5. M. Watanabe, in: Photoacoustic and Photothermal Phenomena. Springer Series in Optical Science. Vol.58, Heidelberg, 1988 p.254

Electron-Induced Ionization and Thermalization in CdS Using Optical Beam Deflection Imaging

G.Y. Chang, R.B. Givens, J.W.M. Spicer, and J.C. Murphy

Applied Physics Laboratory, Johns Hopkins University,
Johns Hopkins Road, Laurel, MD 20723, USA

Abstract. The spatial dependence of electron energy loss in semiconducting cadmium sulphide has been investigated as a function of incident electron beam energy using time resolved internal optical beam deflection techniques. Spatial resolution of approximately 0.5µm has been achieved. Ionization and direct electron injection have been shown to dominate the response for times less than 200 nsec, with local thermalization due to electron relaxation effects dominating for times less than several microseconds. Thermal diffusion dominates the beam deflection response at longer times. From these measurements thermal diffusion coefficients have been measured and primary electron ranges estimated. Time resolved images of the primary electron interaction volume have been obtained.

1. Introduction

This paper presents some further results of an experimental study of subbandgap optical beam deflection (OBD) in CdS using electron beam excitation. Energetic electrons incident on cadmium sulphide penetrate the solid causing internal ionization, and a change in temperature. The depth of penetration, the magnitude of the internal ionization and the relative magnitudes of internal elastic and inelastic electron scattering are beam energy dependent and vary with depth in the specimen. Internal scattering results in a lateral spread of the incident beam whose extent and variation with depth in the solid depends on beam energy. Electron penetration effects may also contribute to the contrast in the SEAM image. A motivation for the current work is to determine electron interaction and loss in a semiconductor to help interpret SEAM images of silicon microcircuits which showed that for beam voltages $E_0 \approx 2$ keV only surface topography was seen while for $15 \leq E_0 \leq 30$ keV subsurface structure was visible at high spatial resolution to depths determined by E_0 [1]. Electron injection, migration and diffusion in semiconducting cadmium sulphide have been studied as a function of beam voltage and position in the sample using time resolved optical beam deflection (OBD).

2. Experimental

A commercial SEM was modified to allow fast blanking of the electron beam with pulse lengths down to 10 nsec, peak currents of approximately 90 µA, and beam energies of $5\text{keV} < E_0 < 30\text{keV}$. The electron beam is incident on a thin edge of an optically clear CdS plate of thickness, d = 35µm. The He-Ne laser probe beam had a beam waist, w ≈ 15µm, inside the sample. It could be positioned with 0.5µm accuracy in depth below the entry surface and in lateral offset from the electron beam center. The response of the probe beam deflection to changes in the electron beam energy and probe beam depth was determined using this system.

3. Results and Discussion

Previous studies have revealed several time-resolved components of the OBD signal associated both with gradients in the electron concentration and with temperature [2]. Figure 1a shows the time development of the normal component of the probe beam deflection at a depth of 11.7 µm

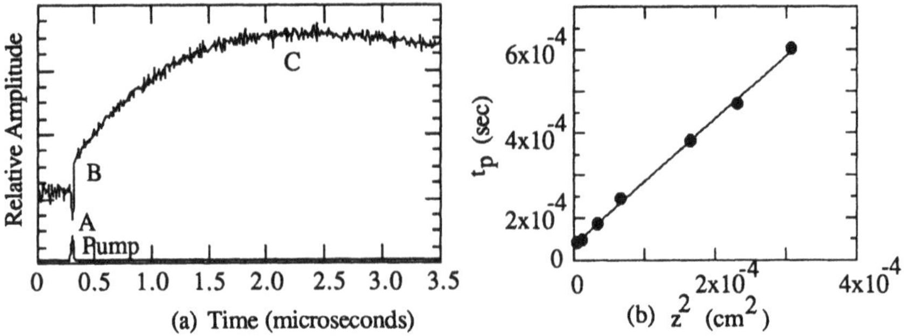

Figure 1 (a) Time-resolved normal component of the optical beam deflection for a probe beam of 11.7 μm at 30 keV. Feature A, the fast negative deflection, followed by B, a prompt positive deflection due to thermalization of diffused electrons and C, a delayed positive deflection due to thermal diffusion. Pulse length is 20 ns. (b) Plot of time of thermal peak (Feature C) as a function of probe beam depth squared.

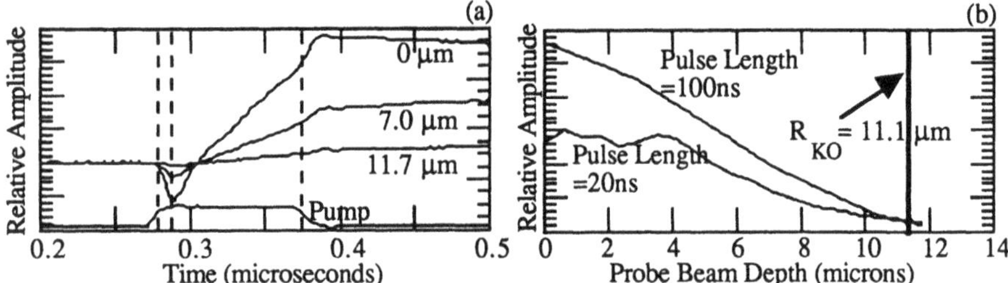

Figure 2 (a) Normal component of the optical beam deflection for 3 probe beam depths showing development of feature A as a function of depth at 30keV. (b) Feature A, direct electron component, is plotted as a function of probe beam depth and is compared with the Kanaya-Okayama Range.

for $E_0 = 30$ keV. At least three signal components, labeled A,B and C, are seen in this figure. The deflection of the first component (A) is negative which is consistent with deflection by a gradient in electron concentration. The other components labeled B and C are positive suggesting their origin is a gradient in temperature. Feature C which indicates the time of the peak signal, t_p, is plotted as a function of z^2 where z is the depth of the probe beam in the sample. The slope (Figure 1b) shows a linear dependence and yields a value for the thermal diffusivity of CdS of 0.099 cm^2/sec in agreement with the literature (0.098-0.12 cm^2/sec). This result also agrees with those obtained by CW methods [3].

Figure 2a shows the signal variation as a function of probe beam depth. At the near surface, z=0 μm, the direct electron interaction dominates the negative going region. Then a prompt thermalization process occurs and the deflection signal begins to increase. Note the inflection point at the end of the pump beam when thermalization processes alone are occurring.

At deeper probe beam depths, z = 11.7 μm, the direct electron component almost vanishes. The expected range of direct electron injection can be calculated using the Kanaya-Okayama Range as follows [4]:

$$R_{KO} = \frac{0.0276\ A\ E_0^{1.67}}{z^{0.889}\ \rho} \quad . \qquad (1)$$

In this expression, R_{KO} denotes the range of electron transport in microns, A the atomic weight (g/mol), E_0 the primary beam energy (keV), Z the atomic number and ρ the density

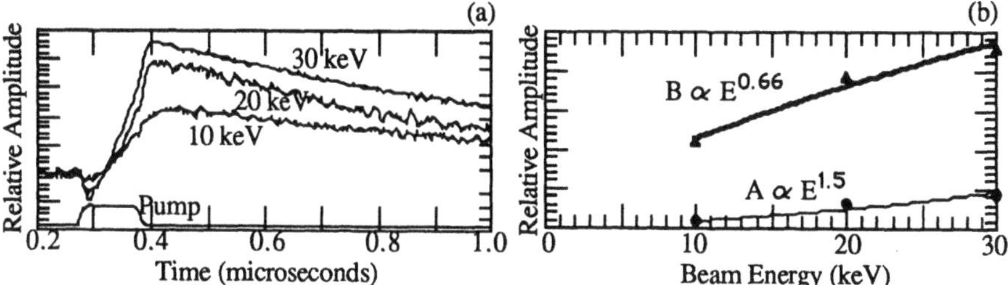

Figure 3 (a) Time-resolved OBD signal showing the behavior of the electronic peak as a function of electron beam energy at z=0 μm. (b) Direct electron interaction A and prompt thermalization component B as a function of electron beam energy.

(g/cm^3). For CdS at 30keV, the calculated value for the electron range is 11.1 μm. Figure 2b shows where the amplitude of direct electron component is plotted as a function of probe beam depth. Note at both 20 ns and 100 ns pulse width, the amplitudes decrease as probe beam depth increases and agrees well with the Kanaya-Okayama range where electron injection terminates. This supports the interpretation that the negative deflection is due to direct electron injection.

The amplitudes of features A and B depend on electron beam energy as shown in Fig. 3a and in the plot of Fig. 3b. The power dependence of the curve (1.5) for component A agrees well with the Kanaya-Okayama Range (1.67) providing further evidence that we are monitoring the direct electron range. Additional work is in progress to investigate and separate out the the recovery rate of electron loss due to electron capture, recombination and thermalization of the initial ionization in region B.

4. Conclusions

Time-resolved measurements of the deflection of a subbandgap probe beam show several components which can be associated with electron injection, thermalization and thermal diffusion. The depth dependence of the thermal diffusion component yields values for the thermal diffusivity agrees well with literature values. The beam energy dependence of the direct electron component shows a similar power dependence as the Kanaya-Okayama Range. Images showing the spatial dependence of the three features in the near surface region have been obtained and compared with Monte Carlo calculations of electron loss. Work is in progress to construct an improved dynamic image of near surface electron injection.

5. References

1. J.C. Murphy, J.W. Maclachan Spicer, L.C. Aamodt, *IEEE Trans.Ultrason., Ferroelec., Freq. Contr.*, UFFC-33, No. 5, 529-541 (1986).
2. G. Chang, R.B. Givens, J.W.M. Spicer, J.C. Murphy, *Review of Progress in Quantitative NDE*, Vol. 10B, 1151-1157 (1991).
3. J.C. Murphy, J.W. Maclachan Spicer, L.C. Aamodt and B.S.H. Royce (Ed.), *Photoacoustic and Photothermal Phenomena II*, Springer-Verlag Berlin, 249-253 (1990).
4. Goldstein, Newbury, Echlin, Joy, Fiori and Lifshin, *Scanning Electron Microscopy and X-Ray Microanalysis*, Plenum Press, New York, 72 (1981).

Topography of Integrated Circuits Detected by Phase Selection in Modulated Optical Reflectance Microscopy

H.S. Li, S.Y. Zhang, J.C. Cheng, Q. Shen, and Y.S. Lu

Laboratory of Photoacoustic Sciences, Institute of Acoustics,
Nanjing University, Nanjing 210008, P.R. of China

The development of semiconductor industry has promoted the improvement of the inspection techniques which can be used for evaluating semiconductor materials and devices. In recent years, the modulated optical reflectance (MOR) technique has attracted great interest because it is a simple, direct and sensitive evaluation method in addition to its noncontact, nondestructive and non-invasive mode.

We have established a microscopic MOR imaging system which can be used to inspect the topography of semiconductor materials and devices, specially for Integrated Circuit (IC), by changing the phase angle setting in the detection of MOR signals. The operation frequency range is (50--300)KHz and a good spatial resolution about 1 μm can be obtained. Some of typical imaging results are shown in Fig.1. From the pictures, we can see that the different structures can be strengthened by different phase angle setting of the reference signal in the lock-in amplifier. In addition, the defects near the surface can also be displayed in the different photographs with different extent, which probably can be used for estimating the depths and areas of the defects.

According to the results of Rosencwaig et al [1,2], the MOR signals are induced by both the plasma wave and the thermal wave. The optical absorption coefficient of the excited beam is very large (about 10^6/m, for silicon and Argon laser 488 nm), a surface absorption approximation is resonable for experimental condition, therefore the plasma wave component of the MOR signal makes no contribution to the depth profile imaging by phase selecting method, and only the thermal wave component must be taken into account in the theoretical investigation.

On the other hand, the thermal wave is generated by three heating sources [3], i.e., instantaneous intraband nonradiative thermalization of the photogenerated carriers (PGC) with photon energy greater than the gap energy, nonradiative surface recombination and nonradive bulk recombination of PGC. Because the first two heating sources appear at the surface of the sample, they also have no contribution to depth profile imaging by phase setting. Therefore, we only consider the last thermal wave source in following theoretical model.

The thermal wave equation can be expressed as [4],

$$\nabla^2 T(r,z) - (i\omega / D_t)\, T(r,z) = -E_g / (\kappa\tau) N(r,z), \tag{1a}$$

with the boundary conditions

$$\partial T(r,z)/\partial z = 0 , \qquad \text{at } z=0 \text{ and } l \tag{1b}$$

where $T(r,z)$ is the thermal wave field, D_t is the thermal diffusivity, κ is the thermal conductivity, l is the thickness of the sample, E_g is the energy gap, and $N(r,z)$ is the plasma wave field. In order to investigate the contribution of the bulk recombination of PGC at point z to the MOR signal, we will find the Green function solution of Eqs.(1a) and (1b), as follow,

Springer Series in Optical Sciences, Vol. 69 Editor: D. Bićanić
Photoacoustic and Photothermal Phenomena III

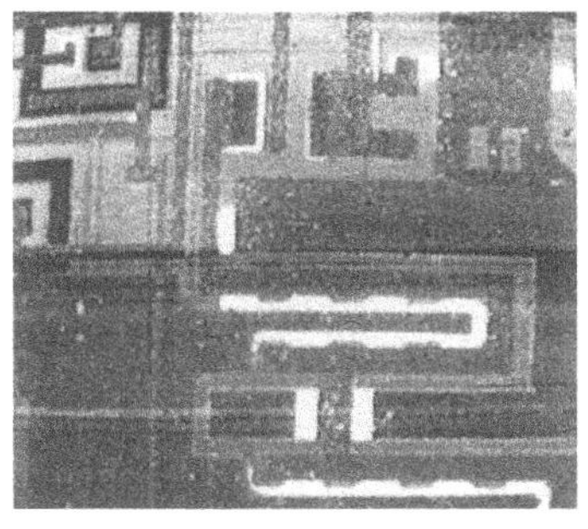

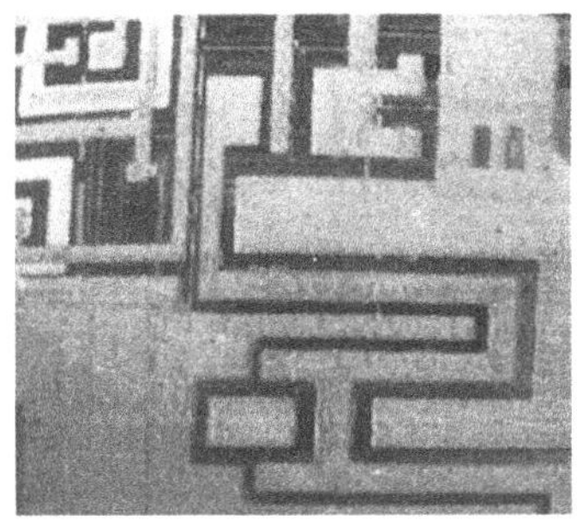

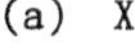

(a) X (b) Y

ϕ=130°

(c) X (d) Y

ϕ=110°

Fig.1. MOR images of IC detected in different phase selecting at 150 KHz

$$T(\delta,z)= - E_g/(\kappa\,\tau)\int_0^L G(\delta,z,z')N(\delta,z')dz' \tag{2a}$$

where $T(\delta,z)$ and $N(\delta,z)$ is the Hankel integrals of $T(r,z)$ and $N(r,z)$, respectively, and $G(\delta,z,z')$ is the Green function, which can be found to be,

$$G(\delta,z,z')=-(2/L)\sum_{n=0}^{\infty}\frac{1}{(n\pi/L)^2+\delta^2+i\omega/D_t}\cos[(n\pi/L)z]\cos[(n\pi/L)z'], \tag{2b}$$

In the practical MOR detection, the MOR signal is mixed with reference signal $V_r(t)$

$$V_r(t)=2\cos(\omega t+\phi) \tag{3}$$

where ϕ is the setted phase.

Therefore, the output V signal of the lock-in amplifier can be expressed as

$$V=\int_0^L V(z')dz' \tag{4a}$$

where $V(z')$ represents the contribution of PGC bulk recombination at point z' to the detected MOR signal,

$$V(z')\sim\int_0^L g(\delta,z')\cos[\phi_0(\delta,z')-\phi]\delta d\delta \tag{4b}$$

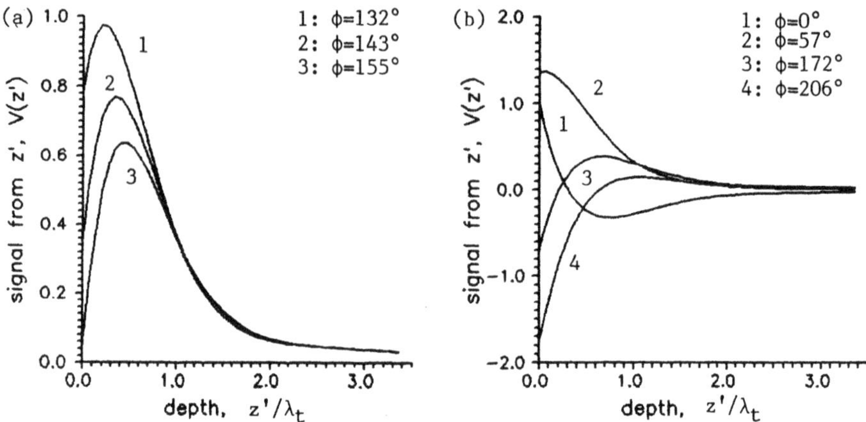

Fig.2. The contribution of bulk nonradiative recombination of PGC from depth z' to MOR signal in different phase setting ϕ (λ_t is the thermal wave length)

where $g(\delta,z')$ can be expressed by,

$$g(\delta,z')=(2/L)\sum_{n=0}^{\infty}\frac{1}{(n\pi/L)^2+\delta^2+i\omega/D_t}\cos[(n\pi/L)z']\,N(\delta,z') \tag{5a}$$

and $\phi_0(\delta,z')$ is the phase angle of $g(\delta,z')$.

We have calculated numerically the distribution $V(z')$ for different setted phases ϕ, as shown in Fig.(2).

From the experimental and theoretical results, we can see that the depth profiling of IC can be obtained by phase selecting in the MOR microcopy. However, the suitable phase angles can only be found in a limited range . For the other phase angle setting, the MOR signals just mainly reflect the surface features of the sample.

References

[1] A. Rosencwaig, J. Opsal, W. L. Smith and D. L. Willenborg, Appl. Phys. Lett., 46, 1013(1985).
[2] A. Rosencwaig, in Photoacoustic and Thermal Wave Phenomena in Semiconductors, edited by A. Mandelis, (Elsevier, New York, 1987) P.97.
[3] Y. S. Lu, S. Y. Zhang, and J. C. Cheng, J. Appl. Phys., 68, 1088(1989).
[4] J. C. Cheng and S. Y. Zhang, to be published in J. Appl. Phys.

Nonlinear Response of Modulated Photoreflectance in Silicon

S.Y. Zhang, Y.S. Lu, and H.S. Li

Laboratory of Photoacoustic Sciences, Institute of Acoustics, Nanjing University, Nanjing 210008, P.R. of China

I. Introduction

In recent years, the modulated photoreflectance (MPR) technique[1] has attracted great attention because of the advantages of nondestructive, noncontact and high spacial resolution. It has been successfully used as a powerful method for monitoring the dose of ion implantation [2] and imaging the subsurface features of integrated circuits [3,4].

So far, the weak pump laser intensity has been used in MPR studies where the dependence of the MPR signal on the pump laser power is linear. However, the nonlinear phenomena in MPR studies of silicon, as the pump laser intensity is sufficiently high, are observed and the MPR signals at the fundamental frequency and the second harmonic frequency are measured in our experiments. In order to explain the experimental results, we establish an one-dimensional model to calculate the MPR signals. The theoretical results are in good agreement with those of the experiments.

II. Experiment

In our experiment, the pump laser beam (Argon ion laser, 488 nm) is modulated and focused on the surface of the silicon wafer with the spot diameter about 2 μm, The pump laser power can be changed continuously between zero and 200 mW. The MPR signals are preamplified by a preamplifier and analyzed by a lock-in amplifier, with which both the fundamental signal and the second harmonic signal are detected separately. In order to diminish the second harmonic distortion of the experimental setup, the modulation function is a square wave with 50% duty cycle and the preamplifier is a wide band one.

The sample is phosphorus-doped silicon <100> wafer with the resistivity about 30 Ω cm.

The dependences of fundamental (ω) and second harmonic (2ω) MPR signals on the pump laser power (I_0) in the modulation frequency 50 kHz are measured as shown in Fig.1. We can see that the fundamental signal increases linearly with I_0 as $I_0 < 20$ mW, but when I_0 is sufficiently high (i.e., $I_0 > 30$ mW), it is approximately quadratic to I_0. On the other hand, the second harmonic signal also increases approximately with I_0 in quadratic relation. But there is a deviation of the second harmonic signal from the quadratic relation at low laser power, which may caused by the equipment noise on the weak 2ω signal.

III. Theory

Three main contributions to the MPR signals in semiconductors [5,6] have been considered, which are: (1) the thermal effect associated with the thermal wave which induces modulation of photoreflectance, (2) the Drude effect associated with the plasma wave which induces modulation of photo-

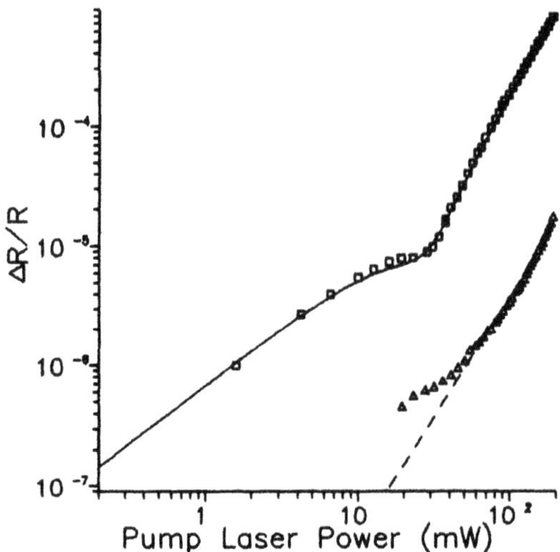

Fig.1. The dependences of MPR signals on pump laser power.

modulation of surface band bending which is induced by the photogenerated carriers (PGC). The third contribution is a second order, nonlinear effect, but the first and the second contributions are linear effects.

Because both the first and the second contributions have been studied by others [4,5], we are mainly interested in the ER effect. Consider the optical nonlinearity [6],

$$\Delta\varepsilon=\varepsilon_0\chi^{(3)}(\Delta E) \ , \tag{1}$$

where $\chi^{(3)}$ is the third order nonlinear optical susceptibility, $\Delta\varepsilon$ is the variation of the dielectric function of the sample, ε_0 is the dielectric function in vacuum, and ΔE is the variation of electric field in the surface space charge region induced by PGC. For a semiconductor sample on open circuit state, ΔE is given by

$$\Delta E=N[ed(\mu_n+\mu_p)]/(\varepsilon\mu_n) \ , \tag{2}$$

where N is the PGC density, e is the electron charge, d is the width of the surface space charge region, μ_n and μ_p are the mobilities of the electrons and holes in the sample respectively, ε is the dielectric function of the sample. So the MPR signal $(\Delta R/R)_e$ generated by the ER effect is:

$$(\Delta R/R)_e=2\varepsilon_0\chi^{(3)}H^2N^2/[n(n^2-1)] \ , \tag{3}$$

where $H=[ed(\mu_n+\mu_p)]/(\varepsilon\mu_n)$, and n is the real part of refractive index of the sample at 633 nm.

With our experimental conditions, we can assume that the Auger recombination does not take place, and the nonlinearities of the plasma waves and the thermal waves can be neglected [7]. The PGC density N and the temperature variation T , which denote the plasma wave and the thermal wave respectively, are linear in the pump laser power I_0. Then we obtain the MPR signal in the sample as follow:

$$\frac{\Delta R}{R}=\mathrm{Re}\left\{\left[AT-\frac{2\lambda^2e^2N_1}{\pi mc^2n(n^2-1)}+\frac{4\varepsilon_0\chi^{(3)}H^2N_0N_1}{n(n^2-1)}\right]e^{i\omega t}+\frac{2\varepsilon_0\chi^{(3)}H^2N_1^2}{n(n^2-1)}e^{2i\omega t}\right\} , \tag{4}$$

where A is the thermal coefficient related to the thermal effect, λ is the probe beam wavelength, m is the effective mass of electron, c is the velocity of light in the sample, N_0 and N_1 are the dc and ac parts of the PGC density respectively,

The first and the second parts in the right side of equation (4) correspond to the first and the second contributions. The third contribution is indicated by the third and the forth parts. We can see from the equation that, there exists a 2ω signal besides the ω signal. The 2ω signal is approximately propertional to I_0^2 , and there is also a quadratic part in the ω signal. When the pump laser intensity is small, the quadratic part is much smaller than the linear parts in the ω signal, and the 2ω signal is very small, so that, the MPR signal of fundamental frequency is linear in the pump laser intensity. However, at sufficiently high pump laser intensity, the MPR signals are dominated by the quadratic parts, and both the ω and the 2ω signals are approximately quadratic in the pump laser intensity.

To interpret the experiment results more precisely, we calculate the dependences of MPR signals in ω and 2ω on the pump laser power numerically. The results are shown in Fig.1. The theoretical results are in good agreement with those of the experiments.

IV. Conclusion

In this paper, nonlinear MPR phenomena are discribed and interpreted quantitatively with numerical calculation. Because the nonlinear MPR signals are the functions of third order nonlinear optical susceptibility which is one of the important parameters of semiconductors in nonlinear optics researches, studying of nonlinear MPR effects will provide a new method for evaluating $\chi^{(3)}$ and also other parameters quantitatively.

References

[1]. A. Rosencwaig, J. Opsal, W.L. Smith, and D.L. Willenborg: Appl. Phys. Lett. 46, p.1013 (1985).

[2]. W.L. Smith, A. Rosencwaig, and D.L. Willenborg: Appl. Phys. Lett. 47, p.584 (1985).

[3]. W.L.Smith, C.G. Welles, and A. Rosencwaig: in "Photoacoustic and Photothermal phenomena II", ed., J.C. Murphy, et al. Springer-Verlag Berlin, Heidelberg, p.146 (1990)

[4]. H.S. Li, S.Y. Zhang, J.C. Cheng, Q. Shen, and Y.S. Lu: in this proceeding.

[5]. A. Rosencwaig: in "photoacoutic and thermal wave Phenomena in semiconductors", ed., A. Mandelis, North Holland, p.97 (1987)

[6]. D.E.Aspnes: in "handbook on semiconductors", Vol 2., ed. M. Balkanski, p.109.(1980)

[7]. J.C. Cheng, S.Y. Zhang, and Y.S. Lu: J. Appl. Phys. 68, p.3865 (1990)

Photomodulated Optical-Reflectance Studies of Germanium: Diagnostic Separation of Temperature and Free-Carrier Effects

R.E. Wagner and A. Mandelis

Photoacoustic and Photothermal Sciences Laboratory, Ontario Laser and Lightwave Reserach Centre, Department of Mechanical Engineering, University of Toronto, Toronto, Ontario, Canada M5S 1A4

Abstract. A method for determining the relative importance of the thermal and Drude (free-carrier) components of the PMOR signal for semiconductors has been demonstrated for Ge. The method depends upon the fitting of PMOR amplitude vs. modulation frequency data to a 3D heat/carrier diffusion model, for different pump beam spot-sizes.

Although the photomodulated optical-reflectance (PMOR) technique has been established as a useful means of determining the ion-implantation dose of an unannealed semiconductor wafer [1], little experimental work has been done with the explicit aim of quantitatively determining the degree to which the temperature and free-carrier (Drude) effects contribute to the PMOR signal for a given sample. One notable exception to this statement is the work of MacDonald *et al.* [2], who examined the variation of the PMOR phase as the pump and probe beams were laterally displaced on a sample of crystalline silicon; they observed a transition from thermal to plasma wave behaviour as the probe separation was increased, in agreement with their theoretical model.

We have developed a different means of estimating the relative magnitudes of the thermal and plasma effects, namely, measuring the PMOR amplitude and phase as a function of the modulation frequency, for different pump beam spot-sizes. This method can yield a good indication as to which effect is dominant, as long as one has some *a priori* knowledge regarding the thermal and electronic transport properties of the material in question. This method is expected to provide dependable results for homogeneous crystalline samples where material properties are well known; the method can also be applied to implanted samples, but the uncertainty in the thermal/electronic properties of the defect layer must be taken into account. Our method is carried out as follows.

The PMOR signal is measured as a function of the modulation frequency, over a typical range of 100 Hz to 5 MHz. These measurements are carried out at several known pump beam spot-sizes, with 1/e radii ranging from 1 to 200 μm. Then the experimental curves are individually fitted to both the temperature and carrier density models [2], with the pump beam spot-size being the free parameter. After the fitting is completed, one can compare the fitted pump beam spot-sizes with the known (measured) values. Now, there are several possible conclusions which can be drawn from this fitting/comparison procedure.

First, if the fitted beam radii for both the temperature and carrier models are very different from the known values, one can conclude either that the assumed thermal/electronic transport parameters are significantly in error, or that some

Springer Series in Optical Sciences, Vol. 69 Photoacoustic and Photothermal Phenomena III
Editor: D. Bićanić

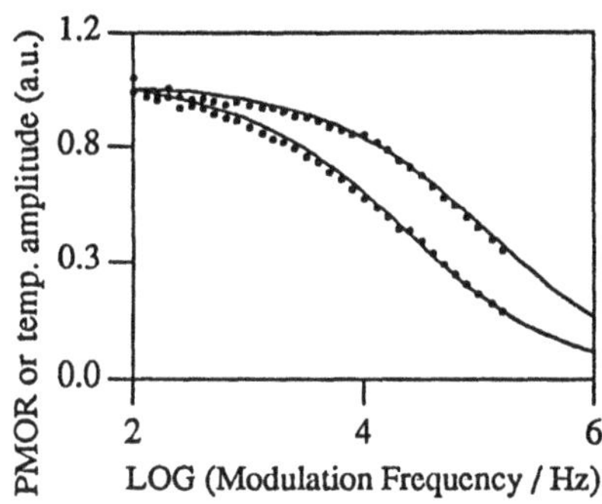

Fig. 1: Fitting of thermal model (solid) to PMOR amplitude data for c-Ge. Fitted 1/e pump beam radii: Upper, 13 μm; Lower, 30 μm.

hybrid of the temperature and free-carrier models must be employed to properly fit the data. (It is assumed that the experimenter has ruled out other effects from contributing to the PMOR signal.) In this case, if the signs of $\partial R/\partial N$ and $\partial R/\partial T$ are known to be different, one can use the measured phase values to provide additional information as to which mechanism is dominant.

On the other hand, if the fitted beam radii for both models are mutually consistent with the known values, then other tests must be performed to independently measure the thermal and plasma effects. Also, as described above, if the signs of the temperature and carrier reflectance coefficients are known to be opposite, phase information can be used to identify the dominant mechanism.

Finally, if only one of the models provides an accurate set of values for the fitted pump beam radii, then we can be quite confident as to which mechanism is dominant, as long as the measured phase values are consistent with the known sign of $\partial R/\partial N$ or $\partial R/\partial T$.

Now, experimental data were obtained on a sample of crystalline germanium (p-type, Ga-doped, (111), ρ = 14 to 16 ohm·cm) at nominal beam radii of 5 and 23 μm, over the frequency range 100 Hz to 160 kHz. Then, using the pump beam radius as an adjustable parameter, the data were fit to the thermal and plasma models (amplitude only) using the following assumed transport parameters: Free carrier lifetime measured by a.c. photoconductivity, 25 μs; carrier diffusion coefficient, 7×10^{-3} m^2/s [3]; surface recombination velocity, 0 m/s; pump beam absorption coefficient, 3.5×10^7 m^{-1} [4]; pump photon energy, 2.1 eV; sample energy gap, 0.66 eV [3]; thermal conductivity, 59 W/(m·K) [3]; specific heat, 310 J/(kg·K) [3]; and density, 5324 kg/m^3 [3].

For the thermal model, the fitted beam radii were 13 and 30 μm respectively, to be compared with the nominal values of 5 and 23 μm; after our measurements were carried out, we found that the cylindrical glass shroud enclosing our samples had defocussed our pump and probe beams to an extent consistent with the fitted beam radii. For the plasma model, the fitted beam radii were much larger than the nominal values. Also, the observed phase values were consistent with the known negative value of $\partial R/\partial T$ at the probe energy of 2.4 eV [5]. Fig. 1 shows the temperature amplitude curves fitted to the experimental PMOR data. Fig. 2 shows the consequential fit of the PMOR phase data to the thermal model, using the beam radii obtained by fitting the PMOR amplitude data. The fit is quite good at lower frequencies, but there is some divergence at higher frequencies (more so for the 30 μm beam); this may indicate that free carrier effects are beginning to play a more

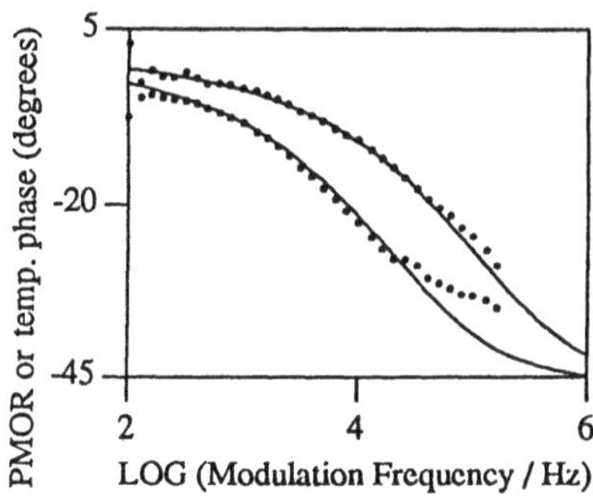

Fig. 2: Consequential fitting of thermal model (solid) to PMOR phase data for c-Ge. Fitted 1/e pump beam radii: Upper, 13 μm; Lower, 30 μm.

important role at higher modulation frequencies, as expected, to which phase values are known to be more sensitive than amplitude values [6].

Overall, our program for independently exploring the temperature and plasma mechanisms of the PMOR effect has yielded considerable success for crystalline germanium. We have clearly studied the PMOR effect in germanium over a range of experimental conditions where the thermal effect is dominant. We are now in the process of creating a hybrid temperature/plasma fitting routine which would yield better fits for both the amplitude and phase data. Also, data are being taken at modulation frequencies up to 3 MHz, on a wider variety of samples.

References

1. Opsal J., Taylor M.W., Smith W.L., and Rosencwaig A. (1987); J. Appl. Phys. **61**, 240.
2. McDonald F.A., Guidotti D., and DelGiudice T.M. (1987) in "Review of Progress in Quantitative Nondestructive Evaluation, Vol. 6B", editted by D.O. Thompson and D.E. Chimenti, (Plenum), pg. 1361.
3. Neuberger M. (1971); "Handbook of Electronic Materials, Vol. 5" (IFI/Plenum).
4. Aspnes D.E. and Studna A.A. (1983); Phys. Rev. B **27**, 985.
5. Batz B. (1966); Solid State Commun. **4**, 241.
6. Opsal J. and Rosencwaig A. (1985); Appl. Phys. Lett. **47**, 498.

Surface Characterization by Modulated Photoreflectance

B.C. Forget and D. Fournier

Laboratoire d'Instrumentation, Université P. et M. Curie,
ESPCI, UPRA 0005 CNRS, 10, rue Vauquelin, F-75005 Paris, France

Abstract - By using the intermediate frequency range (1 kHz to 100 kHz) and selecting a proper excitation wavelength, we demonstrate that it is possible to separate the plasma wave and the thermal wave in a photoreflectance (PR) experiment. We show that this technique permits the determination of surface recombination velocity.

1. Introduction

Photothermal techniques have largely contributed to thermal characterization and to the determination of transport properties in semiconductors /1-3/. The use of intermediate frequency range allows the observation of the passage from a "pure" thermal wave signal, at low frequencies, to a plasma wave signal, or Drude effect, at high frequencies. This passage is characteristic of the electronic transport properties, including surface recombination velocity, of a semiconductor since the heat is transferred to the lattice by the recombination of the photogenerated electron-hole plasma after its diffusion in the sample.

2. Theory: Influence of pump excess energy

The variation of the optical reflectivity is expressed as follows:

$$\Delta R = \frac{\partial R}{\partial N} \Delta N + \frac{\partial R}{\partial T} \Delta T$$

The photogenerated electron-hole plasma density, ΔN, contributes directly to the signal in addition to being a delayed distributed heat source in the bulk of the sample. The thermal contribution can be separated into two terms: one due to the recombination of the electron-hole plasma and one due to the rapid thermalization of the energy in excess of the band gap. This latter term is in no way characteristic of the electronic transport properties of the semiconductor. Therefore, it seems advantageous to minimize this excess energy term by using a laser with a photon energy close to the band gap. Figure 1 shows the increase in contrast of the phase measurement when switching from an Ar^+(514nm) laser to a red laser (625 nm).

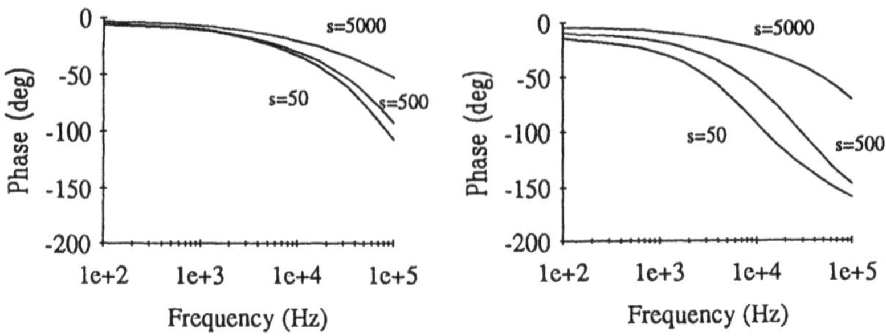

Figure1: The phase of the PR signal for three different values of surface recombination velocities(s = 50, 500, 5000 cm/s). The red excitation (right) gives more contrast than the Ar^+ (left).

3.Experimental Set-up

We have used a classical photoreflectance set-up with a 80µm diameter excitation induced by either a 450 mW Ar+ laser (λ = 514 nm) and a a 400 mW dye laser using Kitton Red dye (λ = 625 nm). This excitation beam is modulated by an acousto-optic modulator from 800 Hz to 100 kHz. The induced variation of the reflected 780nm diode laser is detected by a silicon photodiode, or avalanche photodiode, connected to a lock-in amplifier. The probe beam has a 35 µm diameter and is focussed on the same spot as the pump beam. We have achieved a sensitivity of 10^{-6} in $\Delta R/R$ with this set-up.

4.Results and Discussion

Figure 2 shows the amplitude of the photoreflectance signal for a 300 µm thick, n-type silicon wafer, together with a theoretical calculation /4/ using the numerical fast Hankel transform /5/, for the two excitation beams. The surface of this sample had been polished and, as expected, the surface recombination velocity is very low (s=200 cm/s). We can clearly see that using the 625nm excitation permits us to detect the plasma regime (the rise in amplitude) at lower frequencies than with the 514nm excitation. Experimental data and theoretical calculations for amplitude and phase are in good agreement.

We repeated the experiment on a sample whose surface had been scratched. As expected, the surface recombination velocity is greatly increased by this damage (s=4000cm/s). In this case the difference in signal (Phase is shown in Figure 3) obtained with the two lasers is less apparent. In the case of high surface recombination velocity, the relative importance of the thermal contribution compared to that of the plasma contribution is increased and therefore dominant throughout the frequency range of our experiments.

We are now improving this technique by switching to a Styryl-9 dye which will give us 830 nm excitation. The infrared excitation will reduce even more the

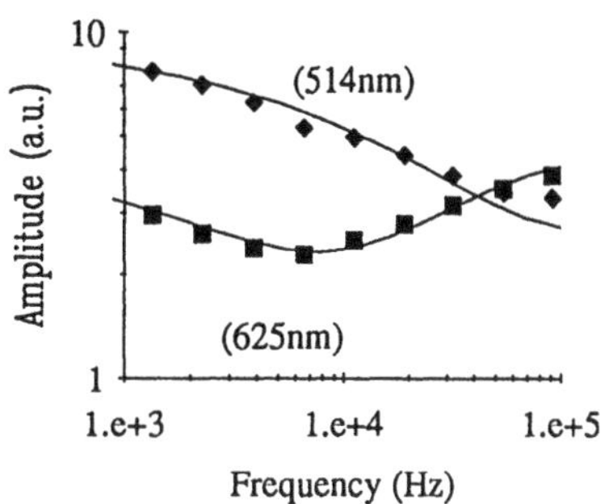

Figure 2: Experimental data and calculation of the amplitude of PR signal for a sample with s=200cm/s for two excitation wavelengths.

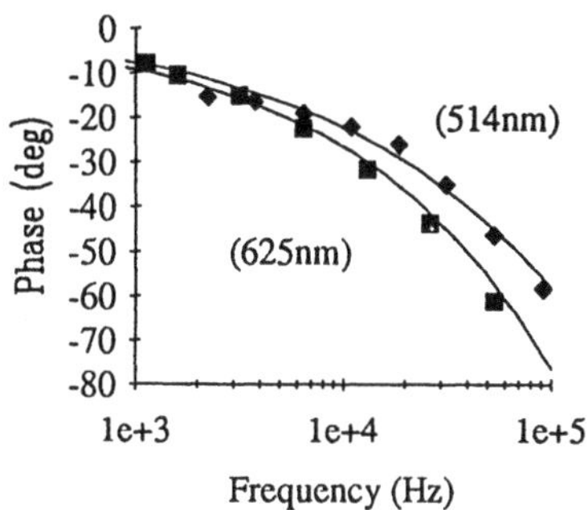

Figure 3: Experimental data and calculation of the phase of PR signal for a sample with s=4000cm/s for two excitation wavelengths.

excess energy contribution, permitting easier determination of surface recombination velocity and other electronic transport properties at lower frequencies as our calculations show.

References

/1/ W.L.Smith, A.Rosencwaig and D.L.Willemborg, Appl.Phys.Lett. **47** (1985), 584.

/2/ A.Pinto Neto, H.Vargas, N.F.Leite and L.C.M.Miranda, Phys.Rev. (1990), 9971.

/3/ F.A.MacDonald, D.Guidotti, and T.M.DelGiudice, from: "Review of Progress in Quantitative Nondestructive Evaluation" Vol. **6B** ed. by D.O. Thompson and D.E. Chimenti , Plenum (1987), 1361.

/4/ N.Mikoshiba, H.Nakamura and K.Tsubouchi, Ultrasonics Symposium Proceedings, IEEE New York (1982), 580.
The model presented here omits the rapid thermalization of energy in excess of the band edge. See /3/.

/5/ S.M.Candel, Computer Phys. Comm. **23** (1981), 343.

Solar Cells Studied with Modulated Reflectance

B. Büchner[1], *R. Osiander*[2], *G. Meyer-Berg*[3], *and P. Korpiun*[3]

[1]Institut für Physikalische Elektronik, Unviversität Stuttgart, W-7000 Stuttgart 80, Fed. Rep. of Germany

[2]Applied Physics Laboratory, Johns Hopkins University, Johns Hopkins Road, Laurel, MD 20723, USA

[3]Physik-Department E13, Technische Universität München, James-Franck-Strasse, W-8046 Garching, Fed. Rep. of Germany

Abstract. Crystalline silicon solar cells and polycrystalline thin film $CuGaSe_2$/(Zn,Cd)S solar cells were examined by two types of experiments based on the measurement of modulated reflectance. We show data collected along the surface of the solar cell in a scan of the pump beam relative to the probe beam. We outline a model for the analysis of the measurement and give preliminary results.

Since a few years modulated reflectance (MR) is a well established method for characterization of semiconductor wafers /1/. Models for other types of photothermal (PT) signals, i.e. PZT–transducer /2/ and Mirage /3/ technique, have been developed for data on doped semiconducting samples including carrier diffusion. A variety of PT measurements on devices, i.e. samples with pn–junction, especially low frequency voltage dependent one, however, could be explained neglecting thermal and carrier diffusion /4/. Recently Mandelis calculated the MR signal of a sample with pn–homojunction /5/. This has been modified taking into account the cylindrical symmetry of remote photoacoustic sensing /6/. Here, we give the corresponding expressions for silicon–solar cells.

The samples studied were devices with different pn–junctions, i.e. homojunctions, in crystalline silicon solar cells or heterojunctions in thin film $Cu_{1-x}Ga_xSe_2$/(Zn, Cd)S solar cells. Front and back contact of the solar cells were shortened, i.e. the applied voltage was zero. It has been shown /4/ that for characterisation of pn–junctions it is preferable to measure at short circuit rather than at open circuit, especially if we want to follow photogenerated carriers. The sample is heated by an acoustooptically modulated Ar^+–laser with 7 mW power and 1.6 μm $1/e^2$ diameter at the sample surface and the MR signal is detected by a 2 mW He–Ne laser, focussed to a diameter of ca. 2.6 μm. A light microscope is used to control the position of pump and probe beam relative to each other and relative to the sample. More details of the experimental set–up can be found in /7/.

A typical behaviour of the MR amplitude versus frequency f between 0.6 and 200 kHz is shown in Fig. 1 for a Si–cell. If only one diffusion process is dominant, there is only one frequency dependent term, $\pi f(a^2+b^2)/4\alpha$, where a and b are the radii of pump and probe beam, respectively, from which the thermal diffusivity α can be determined. Similar analysis is possible, when the pump beam is scanned and the probe beam is at a constant position /8/. The signal is determined as a function of the offset of the pump beam relative to the probe beam center for various modulation frequencies. The amplitude for this type of experiment is plotted as a function of offset in Fig. 2. Here, the sample was a slightly Cu–rich $Cu_{1-x}Ga_xSe_2$/(Zn,Cd)S solar cell and from the difference in widths of the two curves, measured at different frequencies we get a thermal diffusivity of 3.4 mm^2/s for $CuGaSe_2$. This result confirms the estimation of a minority carrier diffusion length of < 0.3 μm for this solar cell /9/, because even at 1.5 MHz the data can be explained by thermal diffusion alone. To check the homogeneity both beams are scanned collinearly (offset zero) along the surface of the solar cell. Long–term (> few minutes) illumination at the same position causes instability in the MR signal of the CGS–based cells.

Springer Series in Optical Sciences, Vol. 69 Photoacoustic and Photothermal Phenomena III
Editor: D. Bićanić

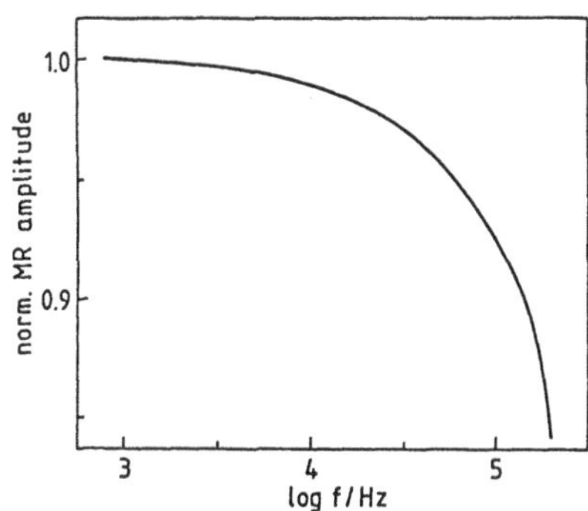

Fig. 1: MR–amplitude of a pnn^+Si solar cell versus frequency.

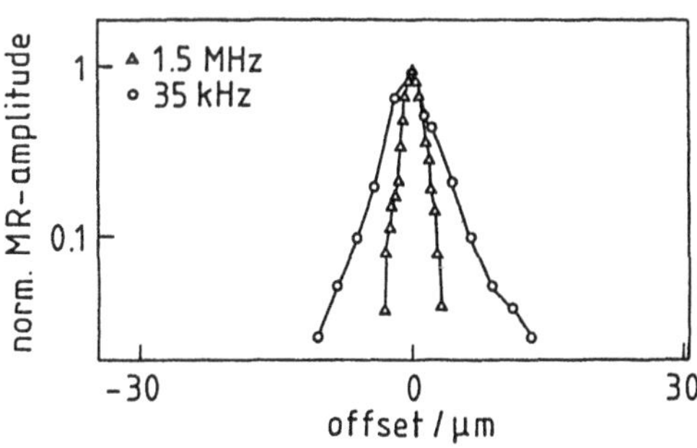

Fig. 2: MR–amplitude for a $Cu_{1-x}Ga_xSe_2$/ Zn,Cd)S solar cell as a function of pump/probe beam offset for two modulation frequencies.

Results for a silicon solar cell are shown in Fig. 3 for modulation frequencies of 800 Hz, 35 kHz, 1.5 MHz. The data can be simulated by a cylinder symmetric extension of the analysis of Mandelis /5/. In the crystalline silicon solar cells with minority carrier diffusion over distances > 250 μm, the signal is related to thermal as well as to carrier diffusion. For a pn–junction, which is d ≈ 0.2 μm below the surface, the MR–signal, S, is given by averaging of the surface temperature along the probe–beam. Because it is Gaussian shaped the averaging can be done analytically resulting in a function of the probe beam/pump beam offset x_o

$$S(x_o) \propto \int_0^\infty \delta d\delta \, J_o(\delta x_o) \exp\left[\frac{-\delta^2(a^2+b^2)}{8}\right] \vartheta(\delta,-d) e^{i\omega t}. \tag{1}$$

Here, J_o is the Bessel function and $\vartheta(\delta,-d)$ the surface temperature. For an approximation of $\beta d << 1$ we have

$$\vartheta(\delta,-d) \propto \frac{\beta}{\sigma(\sigma+\beta)} \left[\Delta E_1 + \frac{(1/L_n)^2 E_g}{(1/L_{\omega n}+\beta)(1/L_{\omega n}+\sigma)}\right], \tag{2}$$

where $\sigma^2 = (\delta^2 + \frac{i\omega}{\alpha})$, L_n minority carrier diffusion length, $(1/L_{\omega n})^2 = [\delta^2+(1+i\omega\tau_n)/L_n^2]$, $\Delta E_1 = h\nu - E_g$, $h\nu$ = energy of incident photons, E_g band gap energy. The difference to the analysis of PA data in /6/ lies in the simpler averaging and the inclusion of lifetime τ_n, caused by high frequencies.

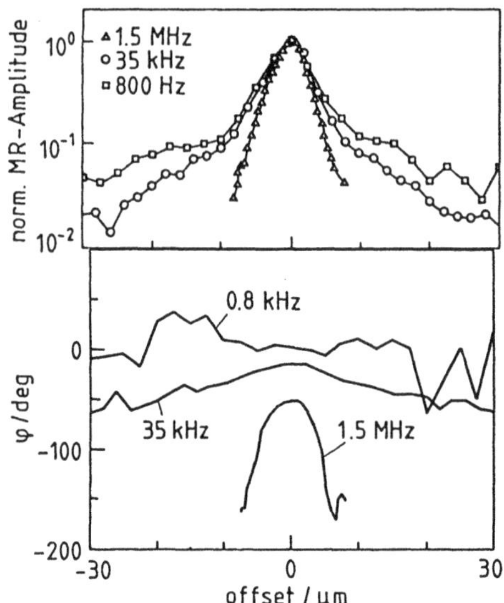

Fig. 3: Amplitude and phase angle of the MR signal for a crystalline silicon solar cell as a function of the pump/probe beam offset for several modulation frequencies.

Conclusion: Our preliminary results show that MR data can be interpreted by thermal diffusion alone in the case of amorphous CGS solar cells. For c–Si–cells there is a qualitative agreement between experiment and calculation taking thermal and carrier diffusion into consideration.

Acknowledgement
We like to thank D. Cahen whose ideas initiated this work and who supplied the c–Si–cells, and R. Klenk for preparation of the CGS cells.

/1/ W.L. Smith, A. Rosencwaig, D. Willenborg, Appl.Phys.Lett. 47, 584, (1985)
/2/ N. Mikoshiba, H. Nakamura, K. Tsubouchi, 1982 IEEE Ultrasonics Symp. Proc. (IEEE, New York, 1982) p.580 and 1983 IEEE Ultrasonics Symp.Proc. (IEEE, New York, 1983)
/3/ D. Fournier, A.C. Boccara, A. Skumanich, N. Amer, J.Appl.Phys. 59, 787 (1986)
/4/ D. Cahen, B. Büchner, F. Decker, M. Wolf, IEEE Transactions on Electronic Devices 37, 498 (1990)
/5/ A. Mandelis, J.Appl.Phys. 66, 5572 (1989)
/6/ B. Büchner, N. Cella, D. Cahen "Low Frequency, PT Measurement of Transport Properties of Crystalline Solar Cells", this issue
/7/ G. Meyer–Berg, R. Osiander, P. Korpiun, "Grain Size..." this issue
/8/ R. Osiander, PhD thesis, TU München, 1991
/9/ R. Klenk, IPE Stuttgart, private communication

Application of the Photothermal Beam Deflection Technique to the Measurement of Si Thermal and Electronic Parameters

A.O. Salnick, A.N. Petrovsky, D.O. Mukhin, and V.V. Zuev

Photoacoustic and Photothermal Methods Laboratory,
Solid State Physics Department, Moscow Physics Engineering Institute,
Kashirskoe shosse 31, 115409 Moscow, Russia

Abstract. The thermal and electronic parameters of wide resistivity range Si samples have been examined by the photothermal beam deflection (PBD) methods with harmonic and pulsed excitation.As a result of many measurements there were exposed the possibilities and deficiencies of PBD methods in such studies.

The application of usually used contact methods to the measurement of semiconductors thermal and electronic characteristics is connected with the problem of good quality contacts. Therefore, the development of new noncontact and nondestructive methods for such studies is very important today.

The PBD technique allows to perform noncontact measurements of several semiconductor transport parameters, as it has been shown earlier by FOURNIER et al. [1]. However, work [1] has been based on obviously inadequate experimental data. The present study is an attempt to supply previous results by new facts concerned with the possibilities and features of the PBD methods in such measurements.

The experimental set up with both harmonic and pulsed excitation has been used. The experimental conditions have been chosen to satisfy the requirements of corresponding theoretical model [2].

The typical results of PBD study of Si samples are presented in Table 1. The values of carrier lifetime (τ_c) obtained by contact method on the same Si samples are also shown in this table.

One can see that the measured Si transport parameters are in good agreement with literature data both for the thermal diffusivity ($\alpha = 0.6 - 0.9$ cm^2/s) and the electronic diffusivity (D=35.1 cm^2/s or 12.5 cm^2/s for the p-type and n-type Si, respectively). Here the error is about 20 %. The accuracy of the carrier lifetime measurements decreases from 30 % for $\tau > 10$ μs down to 50 % for lower τ values.

Our experimental statistic includes the results of measurements of more than 50 different Si samples of both conductivity types. The carrier lifetime was from 0.2 μs up to 1500 μs.

In Table 2 there is presented the comparative analysis of the PBD technique and usually used contact methods, namely the method of conductivity

Table 1. The values of Si transport parameters measured by PBD method and corresponding carrier lifetime values obtained by contact method.

Sample	ρ,Ω	α, cm^2/s	D, cm^2/s	τ, μs	τ_c, μs
n-type	0.05	0.62	-	8	-
p-type	0.38	0.66	32	25	-
n-type	1.	0.75	9	130	150
p-type	1.	0.72	32	65	50
n-type	80.	0.75	10	630	-
n-type	115.	0.66	13	155	150
n-type	600.	0.68	13	135	150
n-type	800.	0.72	10	420	500
n-type	1300.	0.75	9	320	270
p-type	10000.	0.69	-	6	5

modulation in point contact (CMPC) and the photoconductivity drop (PHCD) method.

The efficiency of the PBD method in semiconductor studying is in its' nondestructive and noncontact character and ability of thermal and electronic properties simultaneous analysis. Another one characteristic feature of PBD method is its' ability to perform lifetime measurements on Si samples of extremely low resistivity ($\rho < 1$ Ω). This range of resistivity is unapproachable for contact technique.

However, for some applications such as the measurements on Si samples with low carrier lifetime ($\tau < 1$ μs) the PBD method can't be in competition with special contact technique.

The limitation in τ values could be measured by PBD technique is due to the minimal probe beam depth in experiment. The calculated limit values of τ are about 3 - 5 μs for Si samples [2]. Thus, the data shown in Table 2 are in good agreement with the theoretical predictions.

The deficiencies of PBD methods in semiconductor parameters examination are concerned with the requirement of careful sample treatment including the long time polishing process. Besides, the transverse geometry of the pump and probe beams permits not to perform measurements at the distances more than several carrier diffusion lengths beyond the illuminated sample surface.

Table 2. The characteristics of the PBD method and two contact methods in Si samples study.

Characteristics	CMPC	PHCD	PBD
Transport parameters	τ	τ,D	τ,D,α
Contacts necessity	yes	yes	no
Local measurement ability	no	no	yes
Resistivity range,	> 1.0	> 200.	> 0.05
Limitation in τ, μs	> 3-8	10-10^3	> 4-6
Accuracy of τ measurements, %	25	20	30*

* For $\tau > 10\ \mu$s.

References

1. D.Fournier, A.C.Boccara, A.Skumanich, N.M.Amer, J.Appl.Phys., 59, 787 (1986).

2. A.O.Salnick, Philosophy Doctoral Thesis, Moscow Physics Engineering Institute Publ., Moscow, 1991.

Spatially Resolved Determination of Photogenerated Temperature and Excess Carrier Distribution in GaAsP

G. Li, A. Leyk, and L.J. Balk*

Universität Duisburg, Fachgebiet Werkstoffe der Elektrotechnik, Sonderforschungsbereich 254, Kommandantenstr. 60, W-4100 Duisburg 1, Fed. Rep. of Germany

*On leave from: Shanghai Institute of Ceramics, Academia Sinica, P.O. Box 800-211, Shanghai, P.R. of China

Photo- or electron-acoustic microscopy is an analytical technique which utilizes generation of acoustic waves due to impact of a periodically modulated beam. In case of photo-acoustic microscopy typically lasers are used whereas in electron-acoustic microscopy focused electron beams of primary energies in the medium keV-range are applied. Chopping of the beam intensity can be achieved either mechanically or - for the electron beam - by electrostatic beam blankers. The frequency of this chopping is identical with the frequency of the generated sound waves. Magnitude and phase of the sound wave signal are then correlated with the position of the beam entry point at the specimen surface in a manner as usual for scanning microscopy. Thus, images can be obtained in both magnitude and phase modes. The information gained from these images is related to any kind of material parameter being involved in the conversion of beam energy into the locally generated acoustic wave. This conversion may happen either by direct sound generation or by an intermediate thermal heating of the probed specimen region and its subsequent conversion into acoustic waves due to the thermal expansion coefficient. Therefore, reliable understanding of micrographs can only be obtained by a precise knowledge of the signal generation involved and the according contrast mechanisms. This is especially important for semiconductors, as beam impact may cause simultaneously direct sample heating, generation of excess carriers, and an indirect heating due to non-radiative recombination of these carriers. As a consequence of these different exitations both solely thermo-elastic properties and electronic features may be imaged at the same time /1/.

As the detection of electronic structures is likely correlated with the local distribution of beam generated excess carriers, it is necessary to separate the volumes for direct specimen heating due to beam impact as well as for the excess carrier generation. Methods for this purpose use the internal deflexion of a probe laser beam. In this manner Fournier et al. /2/ were able to do this by photothermal frequency analysis of silicon, whereas Eichler and Massmann /3/ determined local variations of excess carriers for the case of optically induced carrier gratings. For the ternary material $GaAs_xP_{1-x}$ with the mixing parameter x = 0.4 and with an n-doping of 10^{16} - 10^{17} cm^{-3} frequency dependent and spatially resolved experiments are carried out in this work similar to the work by Fournier et al. Due to a band gap of 2.013 eV the use of a He-Ne-laser for measuring the internal mirage effect was possible. The photothermal frequency analysis yielded consistent results to Fournier et al./2/ (fig. 1).

To determine the spatial distributions of temperature rise T and excess carrier concentration N one has to consider that the deflexion of a probe laser within the according volumes is due to a direct heating via the impinging pump beam, an excess carrier related heating, which, however, is directly correlated with the third contribution, the excess carrier concentration itself. For a harmonic excitation the distribution of T and N can be written /4/ (for the example of the x-coordinate) as

$$N(x,t) = A(x) \exp j\,(\omega t + \phi_N) \text{ with } \frac{\partial N}{\partial x} \sim \frac{1}{L_N} \cdot \exp - \frac{x}{L_N}$$

$$T(x,t) = B(x) \exp j\,(\omega t + \phi_T) \text{ with } \frac{\partial T}{\partial x} \sim \frac{1}{L_T} \cdot \exp - \frac{x}{L_T}$$

Springer Series in Optical Sciences, Vol. 69 Photoacoustic and Photothermal Phenomena III
Editor: D. Bićanić

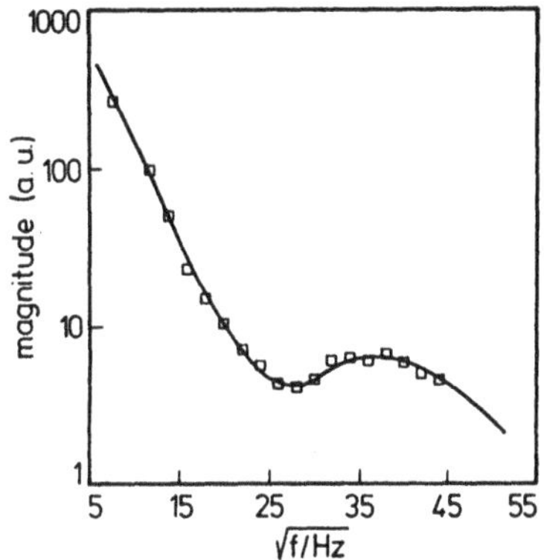

Fig. 1: Frequency dependence of photothermal heating and excess carrier distribution

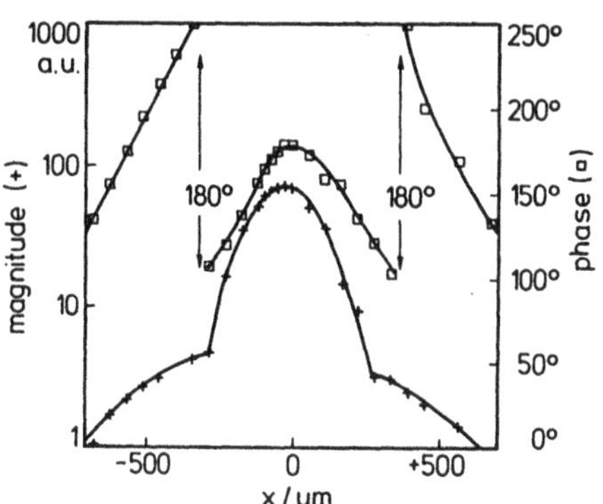

Fig. 2: Spatial dependence of photothermal heating and excess carrier distribution

with L_N and L_T being the diffusion lenghts and ϕ_N and ϕ_T the relative phase shifts. Because of $B(0) > A(0)$, $L_N > L_T$ and $\phi_N \neq \phi_T$ both different spatial behaviour and frequency response are occuring. Especially due to $\phi_N \neq \phi_T$ monitoring of the signal's phase will allow separation of the volume for direct heating from the excess carrier volume. For a fixed frequency a linescan along the specimen surface at a definite penetration depth z will exhibit regions of dominant direct heating and of a dominant contribution of excess carriers (the latter determined undistinguishably for excess carrier related heating and the concentration itself). Fig. 2 is an example of such an experiment taken at a depth of x = 100μm, the region of direct heating being between the two 180°-phase discontinuities.

This principle experiment was extended in this work to an automated two-dimensional measurement. An Nd:YAG laser of 532nm wave length, 2W power, and typically chopped at 3kHz was chosen as pump laser, whereas a 1mW HeNe laser with 632.8nm is used to probe the excited region. Its wavelength is long enough to allow penetration through the examined GaAsP material. Simultaneous monitoring of both magnitudes and phases in x- and y-direction is achieved by a two-dimensional detector diode in connection with two lock-in amplifiers (y is the direction of the probe beam) directly linked to the process computer. Scanning of the probe laser is carried out by means of two computer controlled precision motors capable of a maximum positioning accuracy of 0.1μm. The programs developed for the scanning procedure take into account the possibility that the specimen edge might not be parallel to the initially chosen scan area by fitting the scan automatically to a suitable parallelogram. The signals as delivered by the lock-in outputs are transferred into grey level images with bright for both high magnitude level and large phase angles within the recorded images.

Though having used up to now mechanical chopping only limiting the modulation frequency to a maximum of 3kHz and a 20...30μm diameter for the excitation beam, the deflexion micrographs show a micron sized evaluation of these properties to be feasible. In the magnitude image of fig.3 the axes indicated define the experimental conditions with the x-axis the direction of the excitation beam with (x = 0, y = 0) being its entry point into the sample. The direction of the prober beam is perpendicular to the (x, y)-plane. The magnitude image of fig. 4a is calculated as the total of both components (in x- and in y-direction, respectively), whereas for the phase image of fig. 4b only the x-value is monitored.

The results gained up to now show a resolution of about 1μm. Furthermore the images of fig. 4 reveal an apparent diffusion length for the temperature - for the chosen modulation frequency of 3kHz - of 80μm and for the excess carrier concentration of 200μm, which is in good agreement with the theoretical values. The region of direct temperature heating is clearly visible, for instance, as the dark area in the phase images.

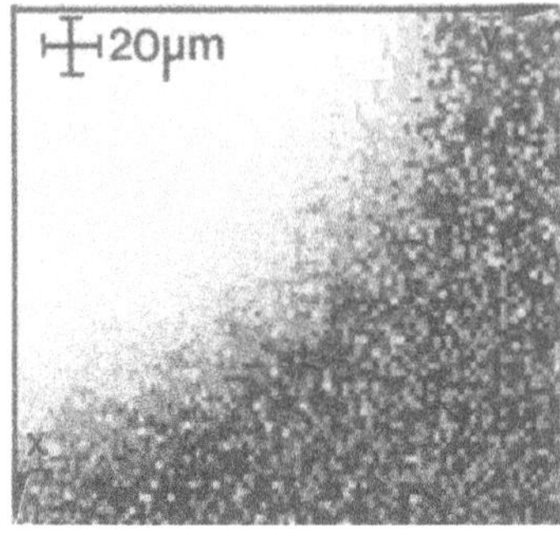

Fig. 3: Micrograph of deflexion magnitude with definition of beam direction and sample orientation

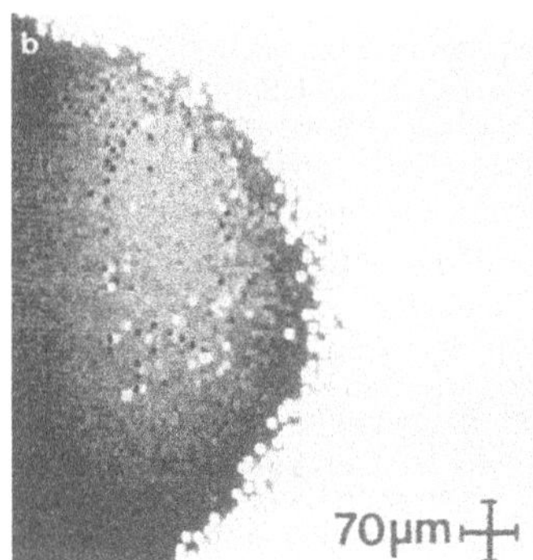

Fig. 4:

a) total amount of magnitude b) phase change in x-direction

Deflexion micrographs for GaAsP

Furthermore, a dark notch ring comes up in the magnitude images which correlates with the notch visible in the frequency scan of fig.1.

/1/ L.J. Balk, Advances in Electronics and Electron Physics. **71**, 1-73 (1988)

/2/ D. Fournier, C. Boccara, A. Skumanich, and M. Amer, J.Appl.Phys. **59** (3), 787-795 (1985)

/3/ H.J. Eicheler and F. Massmann, J.Appl.Phys. **53** (4) 3237-3242 (1982)

/4/ L.W. Casperson, Appl.Opt. **12** (10), 2434 (1973)

Subgap Photothermal Deflection Spectroscopy of Silicon Wafers

G. Amato, G. Benedetto, L. Boarino, and R. Spagnolo

Istituto Elettrotecnico Nazionale "G. Ferraris",
Strada delle Cacce 91, I-10135 Torino, Italy

Abstract. Results are reported of subgap absorption spectra of doped crystalline silicon wafers. The spectra are determined by photothermal deflection technique. The exponential tail slope and the optical gaps are evaluated. The dependence of tail slopes and of gap energies on the doping level is shown to agree with theory. Moreover, the free-carrier absorption is shown to be an effective mechanism also in lightly-doped silicon.

1. Introduction

Photothermal methods have achieved great success in recent years for the determination of subgap optical absorption coefficient of amorphous semiconductors, which is particularly interesting because it is directly connected to the gap-state profile. For heavily doped crystalline semiconductors, subgap absorption coefficient is essentially due to impurity level-to-band absorption and free carrier absorption [1-2]. Also for this kind of semiconductors PDS can be a powerful tool of characterization, owing to its high sensitivity (absorptance values of the order of 10^{-5}), which allows measurements also in the case of low dopant concentrations. The considered samples are single-crystal Floating Zone (FZ) n-type and p-type silicon wafers with dopant concentrations N_d in the range $2 \cdot 10^{12}$ - $2 \cdot 10^{18}$ cm^{-3} and different surface finish, lapped, etched and polished.

2. Experimental results and discussion

The measurements are performed in a conventional PDS set-up, with CCl_4 as the deflecting medium. The light source is a 1000 W Xe arc lamp, mechanically chopped at 12 Hz. The probe beam is a 2 mW He-Ne laser. A lateral position sensor is used to determine the beam deflection. The considered light wavelength range is 900 - 1600 nm. The signal is referred to the lamp spectrum as it is obtained on a carbon black sample inserted in a photoacoustic cell at each measurement wavelength. The absolute values of the absorption coefficient are obtained by normalizing the signal to values corresponding to the spectral zone where the sample is opaque and taking into account the sample thickness.

Fig. 1 shows the optical absorption spectra of four n-doped chemically etched samples. In the experimental spectra three different behaviours can be detected. For energies from 0.9 to 1.1 eV an exponential rise of the optical absorption coefficient is evident; at energies just below the exponential tail, an intersubband absorption is sometimes present; at lower energies a monotonic increase is evident in heavily doped samples, which can be ascribed to free carrier absorption. An interpretative hypothesis for this behaviour based on the contribution of different absorption mechanism is suggested in [3].

The characteristics of the samples are reported in Table 1 together with the energy gap E_g and the slope of the exponential tail E_e. These are determined as follows. The energy gap values are obtained by the extrapolation at higher energies of the exponential tails of the spectra: the energy at which the absorption coefficient is 10^1 cm^{-1} is considered representative of the optical gap and significant in the comparison of samples with different dopant concentration. The experimental values are well correlated with the dopant concentration.

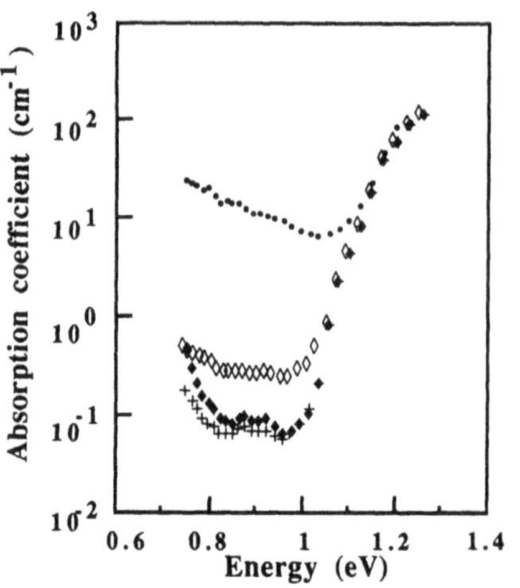

Fig. 1. Absorption spectra of four n-doped chemically etched samples; +, $2 \cdot 10^{12}$ cm^{-3}; ♦, $8.5 \cdot 10^{13}$ cm^{-3}; ◊, $7.5 \cdot 10^{15}$ cm^{-3}; •, $2 \cdot 10^{18}$ cm^{-3};

Fig.2. Comparison between the absorption spectra of a n-doped (- - -) and a p-doped (—) sample with approximately the same dopant concentration.

Tab. 1- Sample characteristics and parameters

wafer type	dopant concentration [atoms cm^{-3}]		thickness [μm]	Eg [eV]	Ee [meV]
A	10^{18}	n	380	1.116	41
B	$2 \cdot 10^{18}$	p	380	1.077	56
C	$7.5 \cdot 10^{15}$	n	300	1.127	33
D	$8.5 \cdot 10^{13}$	n	350	1.131	23
E	$2 \cdot 10^{12}$	n	350	1.131	28

The parameter E_e is calculated fitting the exponential part of the spectra by means of the relationship:

$$\alpha = \alpha_0 \exp (h\nu / E_e) \qquad (1)$$

where hν is the energy of the incident light. The values of these slopes range from 28 meV at $N_d=10^{12}$ cm^{-3} to 56 meV at $N_d=10^{18}$ cm^{-3}. In Fig. 2 the spectra of two samples are shown, respectively n-doped and p-doped, with approximately the same dopant concentration, $N_d=10^{18}$ cm^{-3}. As it can be observed in the figure, the absorption coefficient of p-doped sample at low energies is considerably higher than that of n-doped sample. This is in good agreement with Drude theory [2] which correlates the free carrier absorption to their conductivity effective mass.

Fig. 3 shows the influence of surface finish. The figure reports the spectra of two samples with the same dopant concentration, $N_d=7.5 \cdot 10^{15}$ cm^{-3}, one lapped and the other chemically etched. The spectrum of a corresponding mechanically polished sample, which is not reported in the figure, overlaps that of the etched one. For the lapped sample a loss of definition in the fundamental absorption edge can be observed. This can be interpreted as an effect of the surface roughness, which is particularly important in the intermediate spectrum region corresponding to tail-to-tail transitions, where the sample is semitransparent.

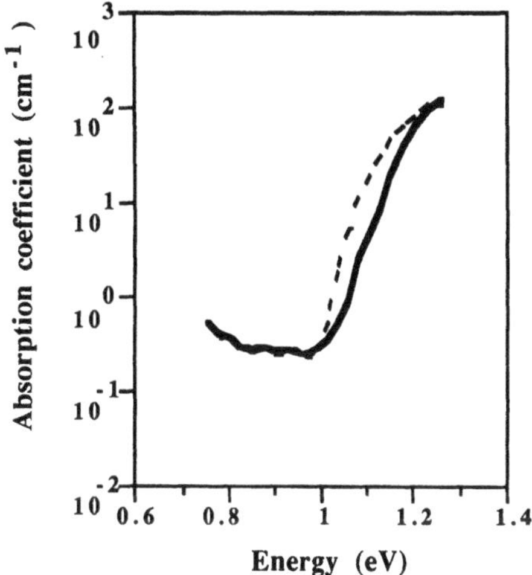

Fig. 3. Comparison between the absorption spectra of two samples with different surface finishing: etched (—) and lapped (- - -)

3. Conclusions

The paper reports one of the first application of PDS to the characterization of subgap absorption coefficient of doped crystalline silicon wafers. Also in this case, as for amorphous semiconductors, PDS has shown to be a simple and powerful technique. Information on the effect of surface finishing processes and on the contribution of different mechanisms, as free carrier and intersubband absorption, can be easily obtained also in the case of low dopant concentration.

References

1. D.K. Schroeder, R.N. Thomas and J.C. Swartz, IEEE Trans. Elect. Devices, ED-25 (1978), 254.
2. Y. Mishima, M. Hirose and Y. Hosaka, J. Appl. Phys. 51 (1980) 1157.
3. G. Amato, G. Benedetto, L. Boarino, R. Spagnolo, to appear on Materials Letters

Theory of Pulsed Photothermal Beam Deflection Signal Formation in Semiconductors

A.N. Petrovsky, A.O. Salnick, V.V. Zuev, V.V. Grigoryev, and M.M. Mekhtiev

Photoacoustic and Photothermal Methods Laboratory,
Solid State Physics Department, Moscow Physics Engineering Institute,
Kashirskoe shosse 31, 115409 Moscow, Russia

Abstract. The complete theory of pulsed PBD signal formation in semiconductors has been developed to be correct for any excitation irradiation and samples parameters. The simplified expressions for the photoinduced carrier concentration and sample temperature distribution are presented.

The photothermal beam deflection (PBD) technique is a prospective new method for semiconductor study. The theory of PBD signal formation in semiconductor with harmonic excitation has been described earlier by FOURNIER et al. [1] but for pulsed excitation the complete theory has not been developed. PELZL et al. [2] have described recently a simple theoretical model of the pulsed PBD method. However, in this theory some important physical processes that can affect significantly the PBD signal have been ignored.

In the present report we have described a complete theory of pulsed PBD signal formation that includes all the main physical processes in semiconductors following the absorption of pulsed pump laser irradiation.

Let us consider that the photoinduced carrier concentration $N(x,t)$ appears due to an external pulsed source with photon flux $I_o(cm^2/s)$ with pulse duration t_p. The carrier generation rate is $I_o\beta\exp(-\beta x)\theta(t)\theta(t_p-t)$ where β is the absorption coefficient (cm^{-1}), $\theta(t)$ is the unit function and x is the depth below sample surface.

Having ignored the internal electric field as is usual in such models we can write the carrier diffusion equation as follows

$$\frac{dN}{dt}=D\frac{d^2N}{dx^2}-\frac{N}{\tau}+I_o\beta\exp(-\beta x)\theta(t)\theta(t_p-t) \qquad (1)$$

with boundary conditions

$$D\frac{dN}{dx}\bigg|_{x=0} = SN(0,t);\ N(x,0)=0 \qquad (2)$$

where D is the minority carrier diffusivity (cm^2/s), τ is the minority carrier lifetime (s) and S is the surface recombination velocity (cm^2/s). We suppose a semiinfinite sample, so at $x \to \infty$, $N(x,t) \to 0$.

Springer Series in Optical Sciences, Vol. 69 Photoacoustic and Photothermal Phenomena III
Editor: D. Bićanić

The thermal diffusion equation can be written as

$$\frac{dT}{dt}=æ\frac{d^2T}{dx^2}+\frac{E_g æN}{k\tau}+\frac{(h\nu-E_g)æ}{k}I_o\beta\exp(-\beta x)\theta(t)\theta(t_p-t) \tag{3}$$

where æ is the thermal diffusivity and k is the thermal conductivity. The boundary conditions are

$$-k\frac{dT}{dx}\bigg|_{x=0}=E_g SN(0,t);\ T(x,0)=0 \tag{4}$$

Using the corresponding Green's function the solution of (1) may be expressed as

$$N(x,t)=\frac{1}{2}I_o\beta\exp(-t/\tau)\int_0^t\theta(\tau')\theta(t_p-\tau')\exp(\tau'/\tau)\Bigg\{\Big[\exp(-\beta x)\mathrm{erfc}(-)+$$

$$+\exp(\beta x)\mathrm{erfc}(+)C_s\Big]-\frac{2S\exp\left[\frac{S}{D}x+\frac{S^2}{D}(t-\tau')\right]}{(\beta D-S)}\mathrm{erfc}(x\&S)\Bigg\}\,d\tau' \tag{5}$$

where $\mathrm{erfc}(\pm)=\mathrm{erfc}\left[\beta\sqrt{D(t-\tau')}\pm\frac{x}{\sqrt{4D(t-\tau')}}\right];\ C_s=1+\frac{2S}{\beta D-S}$

$$\mathrm{erfc}(x\&S)=\mathrm{erfc}\left[\frac{x}{\sqrt{4D(t-\tau')}}+S\sqrt{(t-\tau')/D}\right]$$

This equation is correct for all values of β, τ, and t. One can see that equation (5) is too complicated for analytical treatment. However, in real experimental situation it can be simplified significantly using the following evident conditions

$$\beta^{-1}<< D/S;\ 2D(t-\tau')/x;\ \sqrt{D(t-\tau')} \tag{6}$$

Then the expression for N(x,t) can be reduced to

$$N(x,t)=I_o\exp(-t/\tau)\int_0^t\theta(\tau')\theta(t_p-\tau')\exp(\tau'/\tau)\Bigg\{\frac{\exp(-x^2/4D(t-\tau'))}{\sqrt{\pi D(t-\tau')}}-$$

$$-\frac{S}{D}\exp\left[\frac{S}{D}x+\frac{S^2}{D}(t-\tau')\right]\mathrm{erfc}(x\&S)\Bigg\}\,d\tau' \tag{7}$$

Conditions (6) are satisfied if $(t-\tau') >> \max(x/2\beta D;1/\beta^2D)$. Let's evaluate the values in right part for n-Si sample. If the pump laser wavelength is λ=0.63 µm ($h\nu$=1.96 eV), β= 4 10^3 cm^{-1}, D =10 cm^2/s and x=0.05 cm the values in brackets are 6.25 10^{-7}s and 6.25 10^{-9} s, respectively. Thus, a limitation on t - τ' of 1 µs gives a reasonable value for the experimental applications

The same treatment of the thermal diffusion equation (3) gives the following form for the sample temperature T(x,t)

$$T(x,t)=\frac{1}{2}E_g I_o t_p \frac{\text{æ}\exp\left[\frac{\text{æ}t}{(D-\text{æ})\tau}\right]}{\sqrt{(D-\text{æ})\tau}}\left\{\exp(+)\left[\mathrm{erf}(D+)-\mathrm{erf}(\text{æ}+)\right]+\right.$$

$$\left.+\exp(-)\left[\mathrm{erf}(D-)-\mathrm{erf}(\text{æ}-)\right]\right\}+\frac{(h\nu-E_g)I_o t_p}{\sqrt{\pi k}}\sqrt{\frac{\text{æ}}{t}}\exp\left[-\frac{x^2}{4\text{æ}}\right] \quad (8)$$

where $\exp(\pm)=\exp\left[\frac{\pm x}{\sqrt{(D-\text{æ})\tau}}\right]$; $\mathrm{erf}(D^{\pm})=\mathrm{erf}\left[\sqrt{\frac{Dt}{(D-\text{æ})\tau}}\pm\frac{x}{2\sqrt{Dt}}\right]$

$$\mathrm{erf}(\text{æ}^{\pm})=\mathrm{erf}\left[\sqrt{\frac{\text{æ}t}{(D-\text{æ})\tau}}\pm\frac{x}{2\sqrt{\text{æ}t}}\right]$$

We have considered here simplifying conditions similar to those described above. We have also supposed $S \ll \sqrt{D/\tau}$. As in the case of carrier concentration distribution all the conditions used can be shown to be correct in practice.

Finally, we have obtained the equations for the carrier concentration (7) and the temperature distribution (8) suitable for total PBD signal analysis in semiconductors.

References

1. Fournier D., Boccara A.C., Skumanich A., Amer N.M., J.Appl. Phys., 59, 787 (1986).

2. Pelzl J., Fournier D., Boccara A.C., Springer Series in Optical Sciences, v.58, 241, 1987.

Photothermal Deflection Spectroscopy Study of the Effect of Ion Dose on the Subgap Absorption of Defects in Ion Implanted Layers of GaAs and Si

U. Zammit, F. Gasparrini, M. Marinelli, R. Pizzoferrato, A. Agostini, F. Mercuri, F. Scudieri, and S. Martellucci

Dipartimento Ingegneria Meccanica, II Università di Roma "Tor Vergata", Via Emanuele Carnevale, I-00173 Roma, Italy

Abstract Subgap optical absorption investigations have been used to study the influence of increasing ion dose on the properties of ion implanted layers of compound (GaAs) and elemental (Si) semiconductors. A common trend has been observed for ion implanted Si and GaAs in the behaviour of the subgap absorption spectra as a function of the implantation dose, and it depended on the presence or not of amorphous material in the implanted layer.

In this work we report on subgap optical absorption studies which have been carried out by Photothermal Deflection Spectroscopy on room temperature implanted layers of GaAs and Si, as a function of the implantation dose, in order to compare the features which are observed in ion implanted compound and elemental semiconductors. The measurements were performed at 340 Hz using CCl_4 as coupling fluid.

The samples consisted of 200 nm implanted layer on top of 0.5 mm crystalline substrate and a two layers model was used to deconvolve the absorption due to the layer alone. Fig. 1 shows the absorption spectra for unimplanted single crystalline semi insulating (SI) GaAs sample as well as for GaAs layers implanted with increasing doses of 100 keV Si ions. The unimplanted crystalline sample spectrum is characterized by a steep decrease of the absorption coefficient in the fundamental band edge region and a much slower decrease at lower energies. In spite of the great absorption values of the layers found in the band edge region there is no saturation of the signal due to the small thickness of the layer and to the small absorption of the crystalline substrate

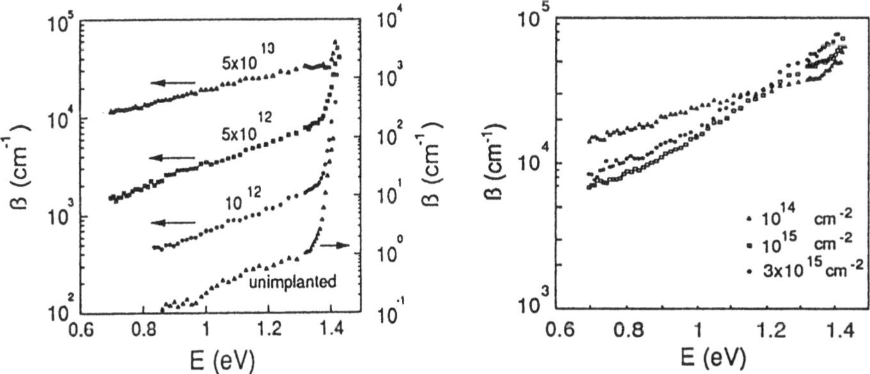

Fig. 1 Subgap absorption spectra for GaAs layer implanted with increasing doses of 100 keV Si ions

Springer Series in Optical Sciences, Vol. 69 Photoacoustic and Photothermal Phenomena III
Editor: D. Bićanić

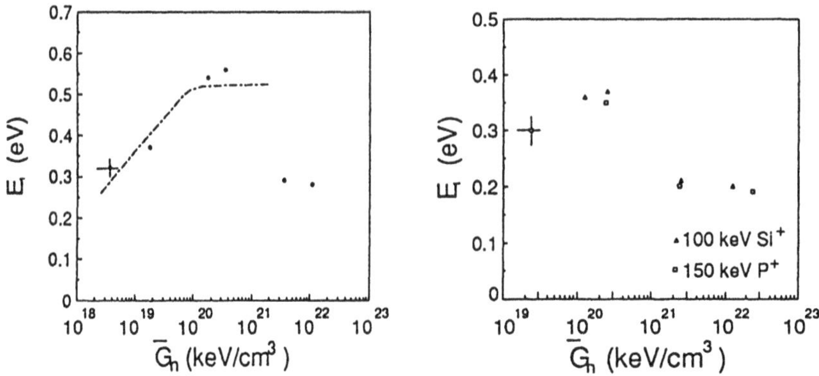

Fig. 2 Inverse logarithmic slope E_1 of ion implanted GaAs (left) and Si (right) layers absorption spectra as a function of $\overline{G}_n$, the mean energy density deposited by the implanted ions.

in the subgap region. In the case of the implanted layers, there is a second region, at lower energies, also exhibiting a practically linear behaviour in the semilog plot. In such a region there is a progressive increase in the absorption values for increasing implantation doses. Moreover, the slope of the data curves decreases up to doses values of 10^{14} cm^{-2}, while it increases and saturates for larger doses. Fig. 2 shows the inverse logarithmic slope values, E_1, relative to the lower energy region as a function of $\overline{G}_n$, the mean energy per unit volume deposited in the target by the implanted ions, which increases for increasing ion dose and was calculated by a Monte Carlo computer code. The results show that, for implantation conditions such that the implanted layer consists of damaged crystalline material, E_1 ranges between 0.3 and 0.52 eV for increasing $\overline{G}_n$ values. For larger values of the implantation dose, for which there is formation of amorphous material, the E_1 drops to 0.3 eV and saturates.

The features which have been observed in the case of ion implanted GaAs, as a function of the implantation dose, are also found in the case of implanted Si as also shown in fig 2. The results are shown to be independent from the implanted species.

The trends which have been observed as a function of the implantation dose in GaAs and Si suggest a common mechanism in the evolution of localized gap states energy distribution responsible for the observed features. In particular, an increase of E_1 (decrease in the absorption vs energy slope) requires a mechanism, involving localized gap states, through which the absorption due to electronic transitions at lower energies gets enhanced more than the one due to transitions at larger energies. By assuming a peaked distribution of localized gap states associated with the defects, as already.considered for hydrogenated amorphous Si [1], we have calculated the evolution of their distribution, in terms of its width and of the energy position of its peak, by fitting the experimental data with the convolution integral of the initial and final density of states

$$\alpha(E)E = K\int \varepsilon^{1/2}\left\{N_v \exp\left[\frac{-\varepsilon - E_g + E}{E_o}\right] + N_r \exp-\left[\frac{(\varepsilon + E_r - E)^2}{2E_w^2}\right]\right\}d\varepsilon$$

Table I Gap states distribution parameters as a function of ion dose for implanted GaAs

Ion dose (cm^{-2})	E_r (eV)	E_w(eV)	N_r/N_v
10^{12}	1.26±0.10	0.37±0.05	(2.7±0.8)×10^{-3}
5×10^{12}	1.25±0.10	0.42±0.05	(4.6±0.8)×10^{-3}
5×10^{13}	1.19±0.10	0.49±0.05	(2.5±0.8)×10^{-2}
10^{14}	1.10±0.10	0.47±0.05	(4.3±0.8)×10^{-2}
10^{15}	1.39±0.10	0.41±0.05	(7.7±0.8)×10^{-2}
3×10^{15}	1.39±0.10	0.45±0.05	(9.1±0.8)×10^{-2}

We assumed only transitions between localized states above the valence band edge, consisting of valence band tail states and a Gaussian distribution of defect density of states, and a parabolic distribution of extended conduction band states [1]. N_v and N_r represent the density of states, respectively, of the band tail distribution at the valence band edge and of the defect states at the peak of their distribution. E_r is the distance of the distribution peak from the conduction band edge E_c and E_g is the gap energy. The results for GaAs are reported in Table I.

It is shown that it is mostly the distribution peak position rather than its width which is affected by the dose increase. E_r initially shifts towards E_r and this corresponds to an increase in E_1. When the implanted material becomes amorphous, E_r moves towards the valence band edge corresponding to the decrease observed in the E_1 values. The progressive increase in the ratio N_r/N_v with increasing ion dose is reflected in the increase of the subgap absorption as pointed out earlier on, and tends to saturate as the material becomes amorphous. Similar results have been obtained in the case of the implanted Si.

REFERENCES

1. A. Triska, J. Kocka, M. Vanecek, in "Disordered Semiconductors", Eds. M. A. Kastner, G. A. Thomas and S. R. Ovshinsky, (Plenum Press, New York and London 1987) p. 459 and references therein.

Influence of Carrier Recombination on the Pulsed Photothermal Beam Deflection Signal in Semiconductors

A.N. Petrovsky[1], *A.O. Salnick*[1], *V.V. Zuev*[1], *D.O. Mukhin*[1], *M.M. Mekhtiev*[1], *J. Pelzl*[2], *A.C. Boccara*[3], and *D. Fournier*[3]

[1]Photoacoustic and Photothermal Methods Laboratory, Solid State Physics Department, Moscow Physics Engineering Institute, Kashirskoe shosse 31, 115409 Moscow, Russia

[2]Institut für Experimentalphysik AG VI, Ruhr-Universität Bochum, W-4630 Bochum 1, Fed. Rep. of Germany

[3]Laboratoire d'Optique Physique, ESPCI, 10, rue Vauquelin, F-75231 Paris Cedex 05, France

Abstract. The carrier recombination effect on the thermal response of pulsed photothermal beam deflection (PBD) signal in Si has been analyzed theoretically and experimentally. This effect displays as a small depression in the PBD-signal time dependence at the probe beam depths close to the carrier diffusion length.

The photothermal beam deflection (PBD) technique has been proved to be a very sensitive method to probe semiconductors local transport properties. It relies on laser probe beam deflection by the local change of the refraction gradient due to an optical pump beam.

Both, harmonic [1] and pulsed [2] pump beams have been used for the carriers population and thermal wave generation in Si. Although the pulsed technique provides an immediate response of the whole frequency range its analytical treatment presents several difficulties compared to that of a frequency domain experiment. It is particularly difficult to treat the carrier recombination effect on the thermal response. This has been investigated in the present study.

The absorption of a pump pulse (with quantum energy $h\nu$, light flux I_0 and pump pulse duration t_p) by semiconductor leads to sample temperature $T(x,t)$ changes. The thermal energy sources occur due to both fast carrier thermalization and carrier recombination. The first of these sources has the form $(h\nu-E_g)I_0\beta\exp(-\beta x)\Theta(t)\Theta(t_p-t)$, where β is the sample absorption coefficient, E_g is the semiconductor bandgap and $\Theta(t)$ is the unit function. The second source is equal to $E_gN(x,t)/\tau$, where $N(x,t)$ is the photoinduced carrier concentration and τ is the minority carrier lifetime. Thus, the thermal diffusion equation in one-dimensional model is:

$$\frac{dT}{dt} = \alpha \frac{d^2T}{dx^2} + \frac{E_g\alpha N}{k\tau} + \frac{(h\nu - E_g)\alpha}{\kappa} I_0\beta\exp(-\beta x)\Theta(t)\Theta(t_p-t) \tag{1}$$

Springer Series in Optical Sciences, Vol. 69 Photoacoustic and Photothermal Phenomena III
Editor: D. Bićanić

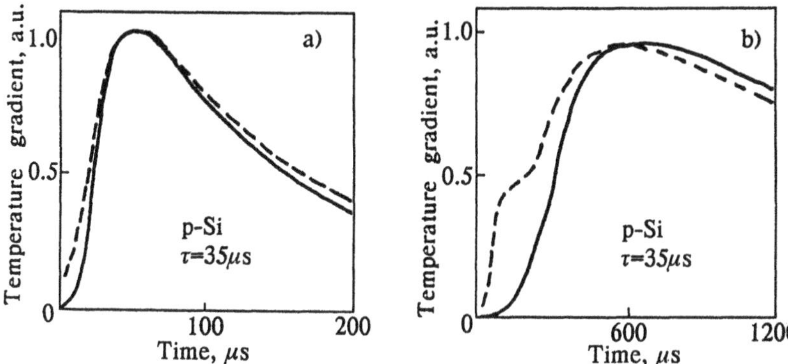

Fig.1 Calculated temperature gradient time dependencies for the "pure" thermal response (solid lines) and the thermal response affected by carrier recombination (dashed lines) at different depths (x) below p-Si sample surface: a) x=150μm, b) x=550μm.

where α and k are the sample thermal diffusivity and thermal conductivity, respectively.

The boundary conditions are

$$-k \left.\frac{dT}{dx}\right|_{x=0} = E_g SN(0,t), \quad T(x,0) = 0 \tag{2}$$

where S is the sample surface recombination velocity.

The resulting solution of (1) with boundary conditions (2) is too complicated for analytical treatment. To simplify we suppose the following evident stipulations to be fulfilled: i) $t_p \ll \tau$, ii) $\beta^{-1} \ll D/S$ and iii) $S \ll (D/\tau)^{1/2}$, where D is the carrier diffusivity. Then the expression for the temperature gradient can be reduced to a form suitable for analysis.

The normalized temperature gradient time dependencies calculated according to the theoretical model discussed above are presented in Fig.1. The effect of carrier recombination appears as a small depression in the intermediate time range between the "pure" carrier and the "pure" heat response.

One can see this effect to be most pronounced for the depths close to the carrier diffusion length L_d (L_d = 400μm for Si sample in Fig.1). It's due to great non-uniformity in carrier concentration distribution at these depths. The influence of carrier recombination appears to occur both in the characteristic depression of thermal gradient time dependence shape and total PBD-signal increasing.

The experimental set up used has been the same as described earlier [2]. It consisted of a Rb pulsed excitation laser (λ=0.69μm, t_p=40μs), He-Ne probe laser (λ=1.15μm), silicon position sensor and programmable digital memory oscilloscope. The n-type and p-type Si samples with carrier lifetime from 35μs up to 630μs have been examined to detect the carrier recombination effect. Special cares have been maintained to avoid the influence of accessory processes on the PBD-signal.

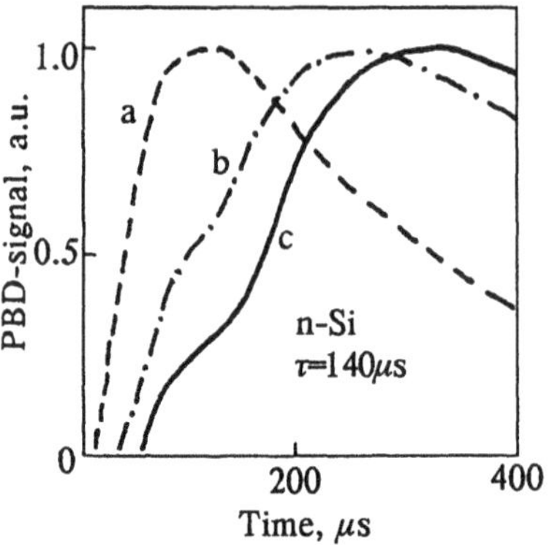

Fig.2 Experimental PBD-signal time dependencies at different probe beam depths (x) for n-Si sample:
a) x=200μm, b) x=250μm, c) x=350μm

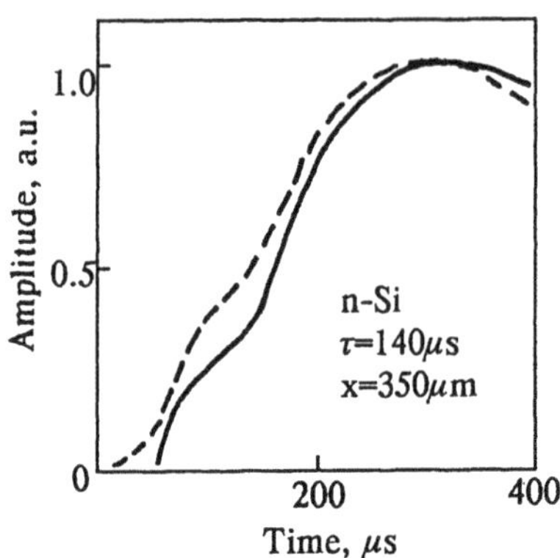

Fig.3 Experimental PBD-signal (solid line) and calculated temperature gradient (dashed line) amplitude time dependencies for n-Si sample.

Fig.2 shows the typical PBD-signal time dependencies observed in the present study. For small probe beam depths (compared to L_d) variations in the signal time dependence shape have not been found. To the contrary at the depths close to L_d the depression in the time dependence becomes visible. These same effects were observed for all the samples used but of different intensities.

Experimentally observed depressions in the time position is found to correspond to the carrier lifetime values, as predicted theoretically. The values of τ evaluated by this way appear to be in reasonable agreement with those calculated using the procedure described previously [2].

For comparison, the experimental PBD-signal time dependence together with theoretically calculated thermal response for this sample are presented in Fig.3. One can observe the good agreement in theory and experiment. Here some discrepancy in time dependence shape is due to the carrier response which has not been included to the theoretical curve.

References

1. D.Fournier, A.C.Boccara, A.Skumanich, N.M.Amer, J.Appl.Phys., 59, 787 (1986).
2. J.Pelzl, D.Fournier, A.C.Boccara, *Springer Series in Optical Sciences*, v.58, 241 (1987).

Thermal Diffusivity Measurements in Semiconductors – Influence of the Carrier Diffusion and Recombination

S. Surnev and D. Ivanov

Faculty of Physics, Sofia University, Sofia 1126, Bulgaria

Abstract. The theory developed by Kuo for determination of the thermal diffusivity by means of OBD technique was extended to include the effects of diffusion, bulk and surface recombination of photoexcited minority carriers in semiconductors. Numerical calculations of the transverse deflection have shown that when the exciting wavelength is approximately equal to the semiconductor band gap deviation from the Kuo's theory should be expected. This results in an error in the thermal diffusivity calculated from the slope of zero crossing-inverse root frequency curves.

1. Introduction

The photothermal deflection method (OBD or "mirage" technique) for determination of the thermal diffusivity was first introduced by Kuo and coworkers [1]. It consists in measurement of the transverse probe beam deflection generated by a highly focused light source. This method is valid only for optically and thermally thick samples. Recently it has been generalized by Salazar et al. [2] to cover all types of samples. However the correct application of this method for semiconductors is not clear. For these materials the incident light generates photoexcited carriers which can further influence the sample temperature distribution by their diffusion, bulk and surface recombination. This effect has already been considered within the framework of 1 dimensional OBD detection scheme by Fournier et al. [3].

Here the effect of the carrier diffusion and recombination on the thermal diffusivity of semiconductors as measured by the Kuo's method was studied theoretically.

2. Theory

In order to calculate the resulting transverse deflection we have to solve the coupled set of carrier and temperature diffusion equations in 3 dimensions. This will be accomplished in the case of thermally and optically thick sample. Three heat sources are considered: (i) instantaneous intraband nonradiative thermalization, (ii) interband nonradiative bulk recombination and (iii) nonradiative surface recombination. The solutions for the carrier and temperature distribution are given by:

$$N(r,t) = \frac{1}{2} \int_0^\infty \lambda \, J_o(\lambda r) \, N(\lambda) \, \exp(-\zeta z) \, \exp(j\omega t) \, d\lambda \qquad (1)$$

$$T(r,t) = \frac{1}{2} \int_0^\infty \lambda J_o(\lambda r) [T_1(\lambda) \exp(-\zeta z) + T_2(\lambda) \exp(-\beta_s z)] \exp(j\omega t) d\lambda \quad (2)$$

where the functions $N(\lambda)$, $T_1(\lambda)$ and $T_2(\lambda)$ are determined from the boundary conditions. The transverse mirage deflection is given by:

$$\varphi_t = -\frac{1}{n} \frac{dn}{dT} \int_0^\infty \lambda \sin(\lambda y) E(\lambda) \exp(-\beta_g z) \exp(j\omega t) d\lambda \quad (3)$$

where $E(\lambda)$ is the Hankel transform of the surface temperature distribution. Using this model we obtain:

$$E(\lambda) = \frac{P_o \exp(-\lambda^2 a^2/4)}{2\pi k_s \beta_s} \left[(1-\varepsilon) + \frac{\varepsilon}{D\tau(\zeta+s/D)(\beta_s+\zeta)} + \frac{s\varepsilon}{D(\zeta+s/D)} \right] \quad (4)$$

Here $\zeta^2=\gamma^2+\lambda^2$, $\gamma^2=(1+j\omega\tau)/(D\tau)$, $\beta_i^2=j\omega/\alpha_i+\lambda^2$, $i=g,s$ and $\varepsilon=E_g/h\nu$. D, τ and s are the minority carrier diffusion, lifetime and surface recombination velocity, respectively. The other parameters have their usual meaning. The first term within the square brackets describes the effect of the heating due to the instantaneous thermalization of carriers. The other two terms describe this effect caused by the nonradiative bulk and surface carrier recombination.

3. Results and Discussion

The results of the present theory will be illustrated for the case of a Si crystal. The material constants for Si, used in the calculations, are listed in Table 1 [3,4]. The other parameters are as follows: α_g(thermal diffusivity of air)=0.2 $cm^2 s^{-1}$, a(pump beam radius)=100 μm, z(probe beam normal offset)=100 μm and P_o(pump beam power)=500 mW. Two regimes can be considered:

(1) ε is approximately equal to zero, i.e. when the exciting energy is approximately equal to the semiconductor band gap. In this case source (i) is the only present and eq.(4) coincides with the expression of $E(\lambda)$ for an arbitrary thermally and optically thick solid. Figure 1 shows the phase of the transverse deflection vs. the transverse offset, y, calculated at modulation frequencies of 100, 500 and 1000 Hz. The arrows indicate the positions of zero crossings. As seen from the inset of fig.1 the plot of zero crossing vs. the inverse root frequency is a straight line, the slope of which according to [1] is equal to $\sqrt{\pi\alpha_s}$. In this case the method of Kuo for determination of the thermal diffusivity is valid.

(2) ε is approximately equal to unity when the source term (i) is absent. It is evident from fig.2 that the phase-transverse offset curves in this case are broader than the ones in the first case (fig.1). This is due to the carrier diffusion and

Table 1: Material constants of Si used in the calculations.

Thermal conductivity	$k(W\ cm^{-1}K^{-1})$	1.57
Thermal diffusivity	$\alpha_s(cm^2s^{-1})$	0.9
Diffusion constant	$D(cm^2s^{-1})$	30.0
Surface recombination velocity	$s(cm\ s^{-1})$	450.0
Carrier lifetime	$\tau(\mu s)$	45.0

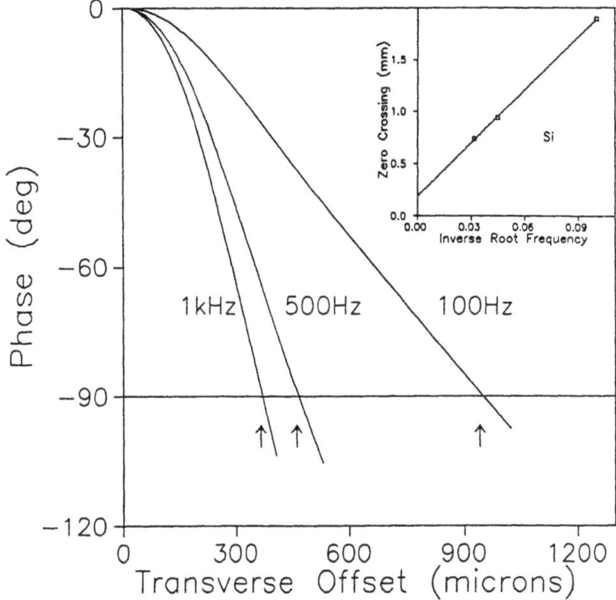

Figure 1: Plots of the phase of the transverse deflection as a function of the transverse offset at different frequencies for the case (1). Inset: Zero crossing vs. the inverse root frequency.

recombination processes. At higher frequencies a dip is observed which is explained by the interference of plasma and thermal waves [3]. It is obvious that the method of Kuo can not be applied in this case. A correction can be made by fitting the experimental OBD curves to the theoretical ones.

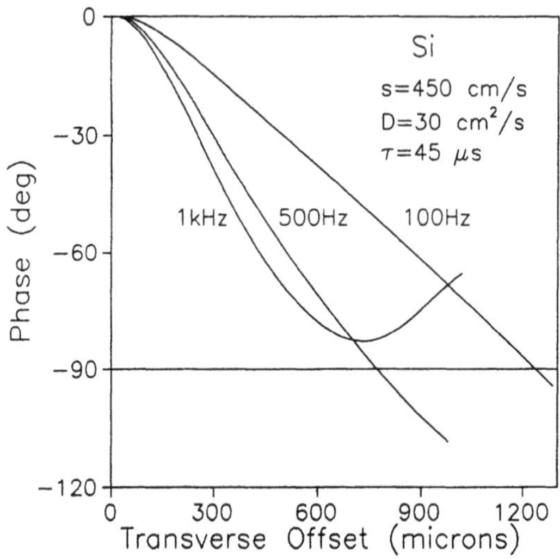

Figure 2: Plots of the phase of the transverse deflection as a function of the transverse offset at different frequencies for the case (2).

References

1. P.K.Kuo, E.D. Sendler and R.L. Thomas, Can. J. Phys. 64 (1986) 1168.

2. A. Salazar, A. Sanchez-Lavega and J. Fernandez, J. Appl. Phys. 65 (1989) 4150.

3. D. Fournier, C. Boccara, A. Skumanich and N. Amer, J. Appl. Phys. 59 (1986) 787.

4. N. Mikoshiba, H. Nakamura and K. Tsubouchi, Ultrasonics Symp. Proc. (IEEE, New York 1982) p.580.

Low Frequency, Photothermal Measurement of Transport Properties of Crystalline Solar Cells

B. Büchner, N. Cella+, and D. Cahen*

Department of Structural Chemistry, The Weizmann Institute of Science, 76100 Rehovot, Israel
Present addresses: *Institut für Physikalische Elektronik, Universität Stuttgart, W-7000 Stuttgart 80, Fed. Rep. of Germany
+Inst. Politécnico do Rio de Janeiro, Nova Friburgo, RJ, Brazil

Abstract. A photoacoustic (PA) method that had been used for analyzing energy fluxes in photovoltaic pn-junction devices /1,2/ has been refined so that data on single crystal Si solar cells show already at such low modulation frequencies as 200Hz the effects of electronic carrier diffusion in the base. Thus photothermal methods are able to examine actual transport properties, in addition to their use for device characterization. This is done by remote sensing of the PA-signal by varying the distance between the area of photo-excitation and detector area.

Separation of the contributions of photogenerated and injected carriers to power losses is possible due to their different dependency on voltage and, furthermore, the **location** of photogeneration and injection is different. In a partially illuminated solar cell light generates carriers only inside the illuminated part of the sample while injection occurs over the whole cell. The contribution of the latter can be eliminated by measuring at reverse bias or short circuit ($V\leq 0$).

Photogenerated carriers give rise to two types of heat sources:

1.) In the illuminated area part of the absorbed energy is converted into heat instantaneously (on the time scale of the experiment) by relaxation of hot carriers to band edges. It corresponds to $h\nu - E_g$, if monochromatic light of energy $h\nu$ is used and E_g is the bandgap energy.
2.) Over an area, determined by minority carrier diffusion length all of the remaining photon energy that is not converted into electrical energy, is released. The photogenerated carriers release E_g -qV, upon recombination when a voltage, V, is applied. This occurs at an average distance of a minority carrier diffusion lenght from their generation or upon recombination at the surface.

The spatial distribution of the heat sources due to fast relaxation and those due to recombination after diffusion can be measured in a photothermal experiment via the oscillating surface temperature. Our novel approach lies in measuring the surface temperature as a function of the **lateral distance** from the illuminated area. To that end we fix a small photoacoustic (PA) chamber ($2r_c$ = 2.4mm) on a single crystal solar cell with a total area of a few cm^2. The distance, X_0, between a slightly focussed modulated Ar^+ laser beam (2a = 0.9-1mm) and the PA chamber is varied along the solar cell surface up to 3-4mm, while the applied voltage is kept constant controlled by a potentiostat. In a conventional photoacoustic experiment on solar cells with illumination inside the measuring chamber the signal is proportional to the dissipated power. The latter can be described in terms of measured currents and voltages /5/. The PA-signal as a function of X_0 for our remote sensing is given by averaging over the cylindrical shape of the PA-chamber, c.f. Fig.1. The origin is in the center of illumination and after integration over z-direction we have

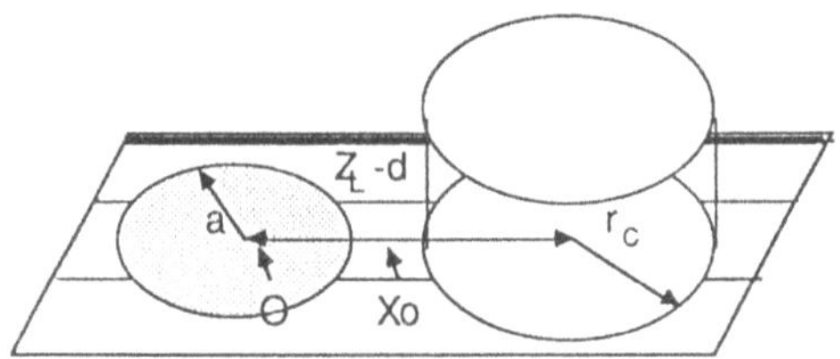

Fig.1: Model of arrangement of illumination, PA chamber on solar cell.

$$S_{PA}(X_0) \propto \int_0^{\varphi_a} \int_{r_-}^{r_+} \int_0^{\infty} \delta\, d\delta\, r\, dr\, d\varphi \frac{1}{\sigma_g} \vartheta(\delta, -d)\, J_0(\delta r)\, e^{i\omega t} \qquad (1)$$

where $\varphi_a = \operatorname{asin}(r_C / X_0)$; $r\pm = X_0 \cos\varphi \pm (r_C^2 - X_0^2 \sin^2\varphi)^{1/2}$; $\sigma_g = \sqrt{\delta^2 + \frac{i\omega}{\alpha_g}}$

f=modulation frequency, α_g = thermal diffusivity of filling gas, J_0 is the Bessel function and $\vartheta(\delta, -d)$ is the temperature at the sample surface (z= -d). The signal is described by eq.(1) only, if the radial flux out of the chamber can be neglected, i.e. for frequencies =100Hz. $\vartheta(\delta,-d)$ has to be calculated from the thermal diffusion equation including minority carrier diffusion in the p- and n-type region of the sample. A 1-dim. model was published recently by Mandelis /4/. We generalize the solution including finite sample thickness and cylinder symmetry in a similar way Mikoshiba et al. /5/ did.

The inhomogeneous thermal diffusion equation is

$$\frac{1}{\alpha}\frac{\partial\vartheta}{\partial t} = \Delta\vartheta + \frac{\Delta E_2}{\tau_n \kappa} n_p + \frac{\beta N_0 \Delta E_1}{\kappa} e^{-\beta(z+d)+i\omega t} \frac{2}{\pi a^2} e^{-2\left(\frac{r}{a}\right)^2} \qquad (2)$$

where n_p is the density of electrons in the basis, β the optical absorption coefficient, N_0 the incident photon flux, κ the thermal conductivity, $\Delta E_1 = h\nu - E_g$; $\Delta E_2 = E_g - qV$; τ_n = lifetime.

With appropriate boundary conditions /4/ we first get an expression strongly depending on properties of the hole diffusion in the emitter, but in the approximation of a small junction depth, d=ca.0.2μm, all these terms cancel and we are left only with electronic diffusion in the p-type basis, characterized by L_n, the diffusion length. We end up with

$$\vartheta(\delta, -d) \propto \frac{\beta}{\sigma(\sigma+\beta)} \left\{ \Delta E_1 + \frac{(1/D_n\tau_n)\Delta E_2}{(1/L_n+\beta)(1/L_n+\sigma)} \right\} e^{-(\delta a)^2/8} \qquad (3)$$

α= thermal diffusivity of the sample;

$$\sigma = \sqrt{\delta^2 + \frac{i\omega}{\alpha}}; \quad \frac{1}{L_n} = \sqrt{\delta^2 + \frac{1}{D_n\tau_n}}.$$

Some curves are calculated in Fig.2. The results are in agreement to the measured curves shown in Fig.3.

Advantages of our arrangement are very simple sample preparation and use of low frequencies. Only the surface has to be accessible and there is no need for cutting of cells like for EBIC or beam deflection /6/ measurements

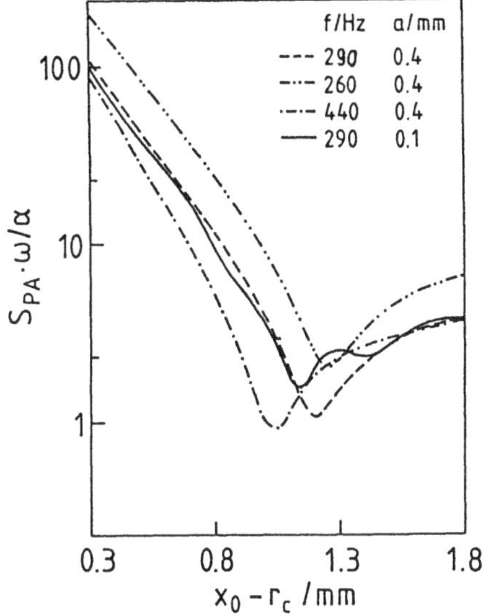

Fig.2: Calculation of distance dependent signal amplitudes for various frequencies and laser radii.

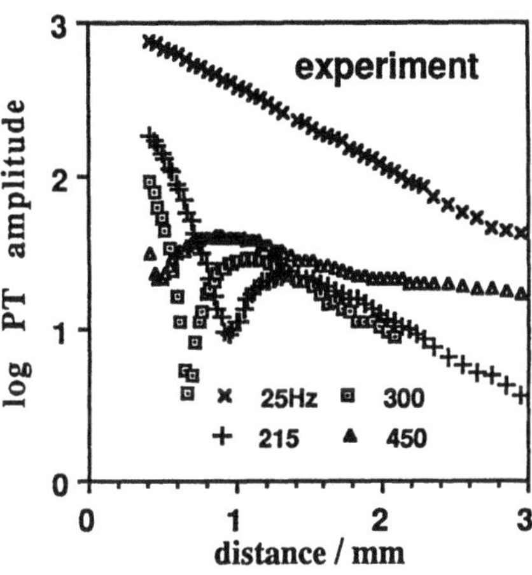

Fig.3: Experimental results for various frequencies. Distance is zero at $X_0=a+r_c$.

inside cells. Also the experimental set-up is much simpler than that for MHz requiring modulated reflectance /7,8/.

/1/ D.Cahen, B.Büchner, F.Decker, M.Wolf, IEEE Transactions on Electron Devices **37**, 498 (1990)

/2/ D.Cahen, B.Büchner, H.Flaisher, M.Wolf, in Photoacoustic and Photothermal Phenomena II, J.C. Murphy et al. eds. Springer Series in Optical Sciences, vol. 62 p.397

/3/ H.Flaisher, D.Cahen, M.Wolf, J.Appl.Phys. **66**, 1832 (1989)

/4/ A.Mandelis, J.Appl.Phys. **66**, 5572 (1989)

/5/ N.Mikoshiba, H.Nakamura, K. Tsubouchi, 1982 IEEE Ultrasonics Symp. Proc. (IEEE, New York, 1982) p.580; and 1983 IEEE Ultrasonics Symp. Proc. (IEEE, New York, 1983)

/6/ D.Fournier, A.C.Boccara, A.Skumanich, N.M.Amer, J.Appl.Phys. **59**, 787 (1986)

/7/ A.Rosencwaig, W.L.Smith, D.L.Willenborg, Appl.Phys.Lett. **46**, 1013 (1985)

/8/ B.Büchner, et al., Solar cells studied by Moldulated Reflectance, this issue

Photothermal Characterization of Materials Suitable for Thin Film Solar Cells

B. Büchner and H.W. Schock

Institut für Physikalische Elektronik, Universität Stuttgart, W-7000 Stuttgart 80, Fed. Rep. of Germany

Abstract. Chalcopyrite structured Cu(In,Ga)Se_2 are studied for their utility as absorber material for low cost, but rather efficient solar cells. The pn-junctions in such solar cells are heterojunctions, consisting of absorber (mainly p-type) and window (often n-type). In many preparation processes there is an additional buffer layer between absorber and window to overcome mismatches. Absorber and window material are up to 3 µm thick. Concerning the um thickness and the strength of the optical absorption these semiconductors have quite similar properties as thin layers of a-Si. Photothermal beam deflection spectroscopy (PDS) is a well established method to study the optical absorption, especially the subbandgap absorption, of such films /1/.

1. Absorber layer

Many parameters influence the fabrication of high efficiency solar cells out of these polycrystalline absorbers and need optimization; i.e. composition, size of crystallites, optical absorption length, minority carrier diffusion length. Surface treatments are known to improve the efficiency of $Cu_{1-x}Ga_xSe_2$-based solar cells /2/. We studied their influence on the subbandgap optical absorption. High subbandgap absorption is not desirable and is correlated to bad electrical behaviour of the solar cells. In Fig.1 we show the influence of KCN-etch on the absorption at 1.1eV, an energy smaller than the bandgap (1.6eV). The Copper content increases to the right side of the plot, so does the subbandgap absorption. The reduction of the absorption after etching behaves generally in accordance to the one of electrical resistivity or Hall-mobility /2/, but the order of magnitude of the change is smaller than for these measurements.

2. Window layer

The window layer in our case is the n-type region of the pn-junction and should have a small electrical resistivity. To that end the window material is highly Ga- or Al-doped in the case of (Zn,Cd)S and ZnO, respectively. A limitation for the doping level is that the window has to be transparent over the solar spectrum. As an example Fig.2 shows the large subbandgap absorption in Ga-doped compared to the undoped (Zn,Cd)S. The energy dependence of the subbandgap absorption can be described by a power law with an exponent of 3.7. This value is close to the one attributed to free carriers scattering on ionized impurities /3/.

3. Buffer layer

Good solar cells with ZnO as window layer can only be obtained, if a very thin (Zn,Cd)S layer is introduced between the absorber and the window lay-

Springer Series in Optical Sciences, Vol. 69 Photoacoustic and Photothermal Phenomena III
Editor: D. Bićanić

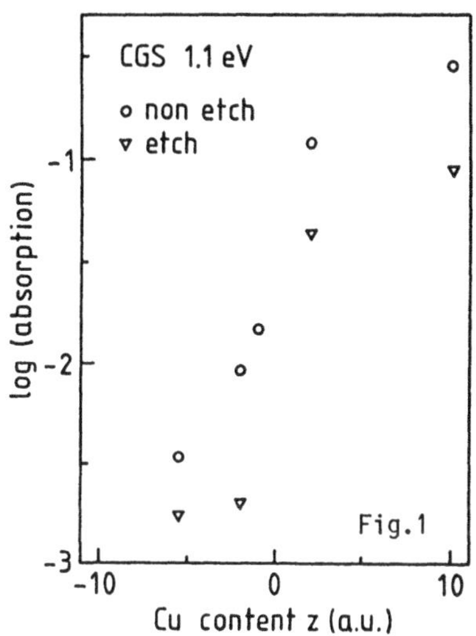

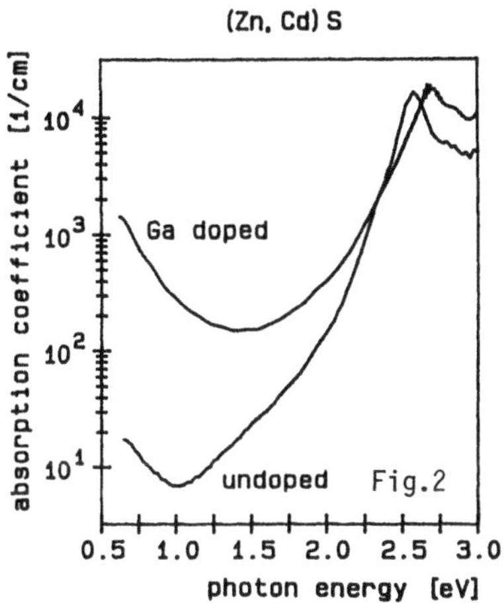

<u>Figure 1:</u> Subbandgap absorption at 1.1 eV of $CuGaSe_2$ before and after surface treatment. The Cu-content increases with the x-axis.

<u>Figure 2:</u> Optical absorption coefficient versus photon energy for undoped and Ga-doped (Zn,Cd)S, 2.2 μm thickness on C7059 glass.

er. One possibility for their fabrication is by chemical bath deposition. Such films can be less than 1nm thick and their characterization is not straight forward. We show that it is possible to determine thicknesses of various directly on glass deposited films relatively to each other and even absolutely from PDS data by knowing one thickness and by assuming the optical absorption to be equal for all films. For known optical absorption coefficient over a wide range of energy, PDS data, that are proportional to the absorption, can be calculated for various film thicknesses l. The onset of signal saturation, i.e. $\exp(-\beta l) \ll 1$, where β is the optical absorption coefficient, is shifted towards lower energies for thicker samples and is thus a measure of sample thickness.

Conclusion

PDS-measurements provide useful information on the subbandgap absorption of the single layers of thin film solar cells.

Acknowledgement: B.B. thanks R.Klenk, R.Klinger, M.Ruckh and R.Schäffler for providing samples and E.Lotter for the possibility to measure on his PDS set-up.

/1/ N.M.Amer, W.B.Jackson in <u>Semiconductors and Semimetals,</u> Vol.**21B**, J.I. Pankove ed., Academic Press, New York, 1984, ch.3

/2/ R.Klenk et al., Proc. of 9th PV Solar Energy Conference, Kluwer Academic Press, p.469, 1989

/3/ J.I. Pankove, <u>Optical Processes in Semiconductors</u> (Dover Publ., New York, 1975) ch.3H

Photothermal Deflection Spectroscopy and Electron Spin Resonance in Amorphous Silicon Based Alloys

G. Amato, G. Benedetto, L. Boarino, and R. Spagnolo

Istituto Elettrotecnico Nazionale "G. Ferraris",
Strada delle Cacce 91, I-10135 Torino, Italy

Abstract. Results are reported of combined Photothermal Deflection Spectroscopy and Electron Spin Resonance in binary amorphous hydrogenated silicon-based alloys. The problem of the different sensitivity of the two techniques to charged and neutral defects is highlighted, and the differences with respect to the case of amorphous silicon explained.

1. Introduction

Photothermal Deflection Spectroscopy (PDS) and Electron Spin Resonance (ESR) are versatile experimental tools for the investigation of localized defects [1].The first one (being sensitive to unpaired spins in the material) gives information on spin dependent properties and therefore on paramagnetic defects. PDS, on the other hand, is sensitive to all optical absorption processes that produce heat in the material, that is all optical transitions where the photoexcited carriers recombine nonradiatively. The important optical (electronic) transitions that contribute to PDS signal are denoted by T_i (i=1..6) and shown in fig.1. E_v, E_i, E_F and E_c are respectively the valence band, defect states, Fermi level and conduction band energy. ΔE_{corr} is the correlation energy. Fig. 2 reports a typical PDS absorption curve for a-SiC:H alloy. The contributions of the different transitions T_i are schematically reported.

In recent years, a method largely used for the determination of the number of defects is that suggested by Jackson and Amer [1] (see the insert in fig.2). In this approach a spectrum of defect related absorption coefficient is obtained by subtracting the exponential component from

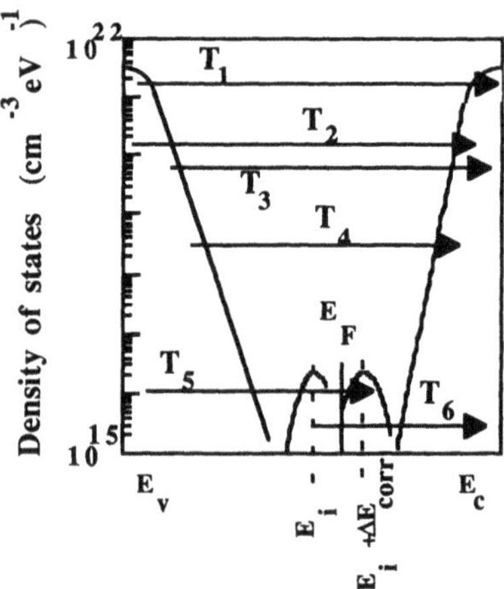

Fig.1 Main optical transitions in amorphous semiconductors

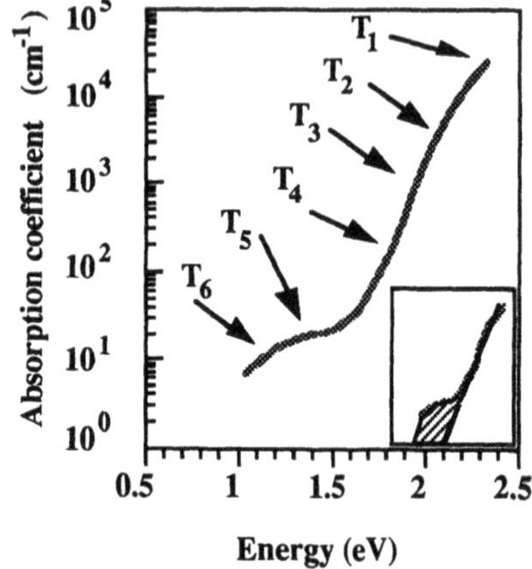

Fig.2 Absorption coefficient for an a-SiC:H sample. The contributions of different transitions are displayed. Insert: the dashed area reprents the integral proportional to the density of states, according to ref. [1].

Springer Series in Optical Sciences, Vol. 69 Photoacoustic and Photothermal Phenomena III
Editor: D. Bićanić

the experimental absorption values. The residual absorption coefficient is then integrated with respect to photon energy to the low energy limit of the spectrum and multiplied by a constant which, in the case of a-Si:H is taken to be 7.9 10^{15} cm^{-2} eV^{-1}. A good agreement between the number of defects detrmined by this method and the number of spins determined by ESR has been found for a-Si:H [1,2]. The validity of this integration rule has been recently questioned [3], the most serious problem being the incorrect application of a sum formula originally derived to evaluate Si-H bond density. If the integration rule can be taken as an empirical method in the case in which a single type of significant defect - in a-Si:H generally presumed to be the dangling bond - mainly contributes to the absorption, the situation is much more complex in silicon based alloys which have multiple types of defects. Aim of this work is to compare results obtained by ESR and PDS in a-SiC:H and a-SiGe:H alloys.

2. Results and discussion

Four a-SiC:H samples were produced by glow discharge at different substrate temperatures between 100 and 250 °C. Three a-SiGe:H samples were produced by sputter assisted plasma chemical vapor deposition, with different values of the flow of SiH_4.

Tab. 1 - Normalized elemental compositions of the samples: $Si_xGe_yH_1$ for SG samples; $Si_xC_1H_z$ for SC samples

	Si	Ge	C	H
SG5	1.3	4.3	-	1
SG7	2.1	3.5	-	1
SG10	4	1.6	-	1
SC100	5.7	-	1	3.7
SC150	5.7	-	1	2.7
SC200	5.7	-	1	2.7
SC250	5.7	-	1	2.4

In tab. 1 are reported the elemental compositions of the samples. The a-SiGe:H samples are labelled SG followed by the SiH_4 flow in SCCM; the a-SiC:H samples by the substrate temperatures. In Fig. 3 the ESR spin densities as a function of the ratio $R=N_s/I$ is reported for the different samples. N_s represents the density of spins while I is the integrated low energy portion of the absorption spectra.

The non-linearity between the two quantities is evident, even if a correlation is present, and seems to increase if the number of defects increases. The results are in agreement with those obtained by other authors [4]. Furthermore, if the two curves are extrapolated down to $N_s=10^{16}$ cm^{-3} (which is the density of spins commonly found in a-Si:H), we find $R \approx 10^{16}$ cm^{-2} eV (close to the suggested value [1] for a-Si:H).

This result could suggest that alloying Si with another element increases the ratio $R=N_s/I$, say, the number of unpaired spins increases more rapidly than the integrated absorption.

Furthermore, the rate of this increase seems to depend on the element to whom Si is alloyed. We explain this effect as a consequence of creating multiple types of defects in the semiconductor gap. These new defects, with their own associated matrix elements cannot be simply treated as dangling bonds in a-Si:H. For this reason, the empirical integration rule which holds in the case of a-Si:H could have no significance in the case of Si-based alloys, since it treates all defects in the same way. As an example, defects associated with alloying processes, like microvoids, could give rise to a subgap contribution to absorption which cannot be related to simple electronic transitions between extended and localized states.

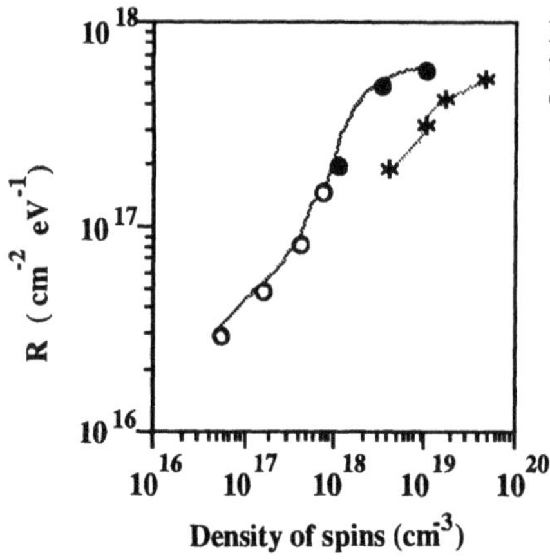

Fig. 3 ESR spin densities as a function of the ratio $R=N_S/I$: ∗, a-SiC:H; •, a-SiGe:H; o, a-SiGe:H (from ref. [4])

3. Conclusions

It has been shown that PDS is a sensitive tool for the characterization of amorphous silicon-based alloys.

The mismatch between PDS and ESR results has been attributed to the different energy ranges involved in the two methods and mainly to the poor deconvolution methods that have been applied until now on PDS spectra.

The situation is even more complicated in alloys like a-SiC:H and a-SiGe:H which have multiple types of defects with their associated optical matrix element.

References

1. N.M. Amer, W.B. Jackson in :"Semiconductors and Semimetals" J.I. Pankowe ed. 21 part B, (Academic Press, London), 1984.
2. S. Yamasaki, Phil. Mag. B 56, 79, 1987.
3. Y. M. Li, W. Paul Mat. Res. Soc. Symp. Proc. 219 (MRS, Pittsburgh, 1991).
4 M. Stutzmann, C. C. Tsai, R. A. Street, Jour non-Cryst. Solids 97/98, 1011 (1987)

Single Frequency Thermal Diffusivity Measurements in Semiconductors by Photothermal Deflection Spectroscopy

F. Gasparrini, U. Zammit, M. Marinelli, R. Pizzoferrato, and A. Agostini

Dipartimento Ingegneria Meccanica, II Università di Roma
"Tor Vergata", Via Emanuele Carnevale, I-00173 Roma, Italy

Abstract Absolute measurements of the thermal diffusivity of several semiconductors have been performed by the analysis of the Photothermal Deflection Spectroscopy phase spectra obtained at a single value of the modulation frequency. A simple procedure is presented which enables to work under the sample thermally thick condition thus improving the sensitivity of the measurements.

The knowledge of the thermal diffusivity of semiconductors is important due to its relation to the power dissipation problem in microelectronic and optoelectronic devices which limits their performance. In this paper we report on the absolute determination of the thermal diffusivity, D, in semiconductors by PDS measurements carried out at a single value of the modulation frequency. The method is based on the analysis of phase subgap spectra of the Photothermal Deflection Spectroscopy signal, similarly to what previously reported [1], which, however, has a limited sensitivity due to the restrictions of operations to relatively low frequencies. In this work we show that a very simple procedure enables to work even at large modulation frequencies adopted and to increase considerably the sensitivity of the measurements.

Fig 1 shows the PDS signal relative phase spectra, normalized with respect to the values obtained in the spectral region where the sample is optically opaque, for a 500 μm thick GaAs sample, and for the same sample with a carbon black layer on its rear surface, obtained at 290 Hz. The behaviour of the phase can be schematically explained as follows. Though in an absorbing sample the light is absorbed in part or in the entire sample volume, one may think of an effective "centre" of the heating source whose position depends on the sample optical absorption depth, $\mu_\beta = 1/\beta$, (β is the optical absorption coefficient) and thermal diffusion length, $\mu = (D/\pi f)^{1/2}$ (f is the modulation frequency). The further is the position of this "centre" from the sample front surface, the greater will be the phase lag (the smaller the phase value) with respect to a reference signal. In the region above the semiconductor band gap the sample is optically opaque and the "centre" is located near the sample front surface. As the sample becomes progressively more transparent over the band edge region, the "centre" position moves away from the sample front surface, due to the progressive increase of μ_β, and the phase value correspondingly decreases. As μ_β becomes much larger than μ the "centre" cannot change its position any longer and the phase will reach a constant value. The phase values are therefore do not depend on the actual optical absorption coefficient values in the spectral regions where the sample is optically opaque and where it is optically thin. The difference between the phase values obtained in such spectral

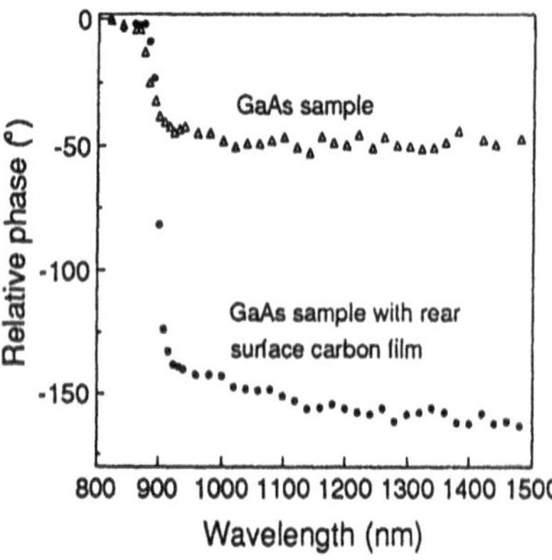

Fig. 1 PDS signal relative phase for an uncoated GaAs sample and for rear surface carbon film coated GaAs sample.

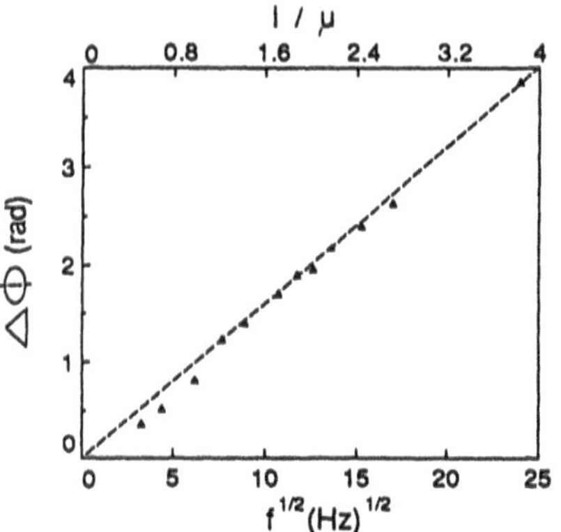

Fig. 2 ΔΦ as a function of the square root of the modulation frequency for a rear surface carbon coated GaAs sample.

regions, ΔΦ, is therefore only a function of the sample thermal parameters and of the modulation frequency. The value of ΔΦ increases as the modulation frequency is increased until it saturates at 45° when the sample becomes thermally thick, for samples where only bulk absorption is involved. In such a case, which corresponds to the results shown by the triangles in fig. 1, and to some of the spectra reported in ref. [1], ΔΦ becomes practically independent from the sample thermal parameters. The frequency range where ΔΦ depends on the sample thermal parameter is therefore restricted to the one corresponding to the sample not being thermally thick, where relatively small values of ΔΦ may be achieved, thus limiting the sensitivity of the measurements.

The sensitivity may considerably be increased by depositing a thin carbon layer on the sample rear surface. Since the carbon film is optically opaque, in the spectral region where the sample is optically transparent, the presence of such a layer shifts the heating "centre" considerably towards the sample rear surface. The thermal wave has then to go through the entire semiconductor sample thickness before it is detected, at the sample front surface, by the probe laser beam . The value of ΔΦ increases substantially, as shown in fig. 1. In such a case, $\Delta\Phi = l/\mu$ where l is the sample thickness, as can be obtained form the expression of the PDS signal for a two absorbing layers sample [2]. ΔΦ increases with frequency and the sensitivity of the measurements from the sample thermal diffusivity may correspondingly be increased. Fig. 2 shows ΔΦ vs $f^{1/2}$ behaviour. The thermally thick condition applies when l/μ is equal or greater than about 1.3 corresponding to ΔΦ values greater or equal to about 1.3 radians (~70°). Therefore, whatever is the investigated sample or operation frequency, provided the ΔΦ value which is obtained at a single frequency exceeds 70°, the sample thermal diffusivity may readily be obtained from the above shown simple expression of ΔΦ and with a single frequency measurement. When the sample thermally thick condition cannot be achieved, the thermal diffusivity measurements may still be performed using, however, the full expression for the PDS signal for the evaluation of ΔΦ. The measured values of thermal diffusivity are reported in the table below for the GaAs

Table I. Measured and literature values of thermal diffusivity of semiconductor samples together with respective l/μ values (see text).

Sample	l/μ	Measured value (cm^2s^{-1})	Literature value (cm^2s^{-1})
GaAs	≥ 1.3	0.29±0.03	0.20-0.34 [3]
Si	1.7	0.95±0.03	0.85-0.95 [3]
Ge	1.7	0.31±0.03	0.27-0.33 [3]
InP	1.5	0.50±0.03	0.46 [4]

sample, a 1.2 mm thick Si sample, a 1.2 mm thick Ge sample and a 400 μm thick InP sample, together with the respective values of l/μ at which measurements have been performed and the diffusivity values reported in literature. The measured values are in good agreement with the ones reported in literature.

REFERENCES

1. N. Yacoubi and M. Fathallah, Proceed. of " 5th International Topical Meeting on Photoacoustic and Photothernal Phenomena ", Heidelberg 1987, Eds. P. Hess, J. Pezl
2. (Springer-Verlag Berlin, Heidelberg 1988) p. 341.
   A. Mandelis, Y.C. Teng, B.S.C. Royce, J. Appl. Phys, **50**, 718 (1979).
3. Y. S. Toulukian, R.W. Powell, C.Y. Ho and M. C. Nicolaou, "Thermal Diffusivity" (IFVPlenum, New York, Washington, 1970), Vol 10.
4. O. Pessoa Jr., C. L. Cesar, N. A. Patel, H. Vargas, C. C. Ghizoni and L. C. M. Miranda, J. Appl. Phys., **59** ,1316 (1986).

Reflection and Transmission of Thermal Waves and Plasma Waves in Silicon Wafers

B.C. Forget and D. Fournier

Laboratoire d'Instrumentation, Université P. et M. Curie,
ESPCI, UPRA 0005 CNRS, 10, rue Vauquelin, F-75005 Paris, France

Abstract - Using a combination of different configurations for the excitation and the detection, we explore the possibilities of low frequency Mirage experiments to determine simultaneously recombination velocities (SRV) of both surfaces of a silicon wafer. The experimental results presented are in good accordance with the theoretical model.

1. Introduction

In recent years, photothermal techniques have largely contributed to the characterization of surfaces and to the determination transport properties of semiconductor materials /1-3/. In this paper, we explore the possibilities of low frequency Mirage detection experiment to determine concurrently the surface recombination velocities of both surfaces of a semiconductor wafer. This determination is possible by using a combination of different configurations for the excitation and the detection. We present here theoretical study of different configurations and experimental fit.

2. Experimental Setup

We have used the Mirage cell developed in our laboratory /4/ with a 700 mW Ar^+ laser excitation. The beam was not focussed on the surface in order to eliminate the effects of lateral diffusion. The samples under investigation were 300 μm thick n-type silicon wafers (~1 Ωcm). One of the surfaces was polished and the other deliberately damaged to ensure a significant difference in the surface recombination velocities. The modulation frequencies used range from 100 Hz to 2500 Hz.

3. Theoretical study

We have studied the four experimental configurations of pump and probe beam position in order to determine the most appropriate ones considering the thickness of the sample and the frequency range for which the signal over noise is sufficient.

When exciting the damaged surface, we can probe in the air either in front of this surface or behind the polished one. In the first case the thermal wave is dominant through out the whole frequency range and it is impossible to discriminate the recombination velocity of the rear surface (See Figure 1, dashed line). In the second case (see Figure 2) we expect the plasma wave to reach the rear surface before the

Springer Series in Optical Sciences, Vol. 69 Photoacoustic and Photothermal Phenomena III
Editor: D. Bićanić

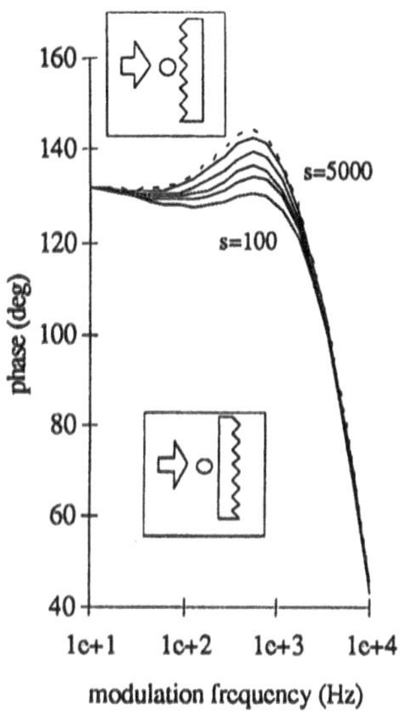

Figure 1: Theoretical calculations for one SRV=10000 cm/s and the other varying from 100 to 5000 cm/s

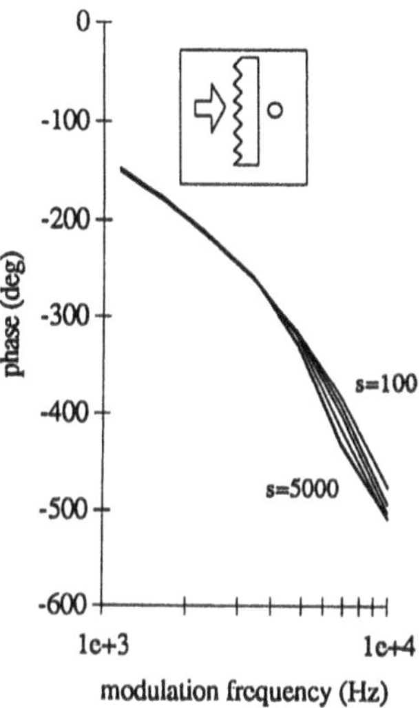

Figure 2: Theoretical calculations for front SRV=10000 cm/s and back SRV varying from 100 to 5000 cm/s

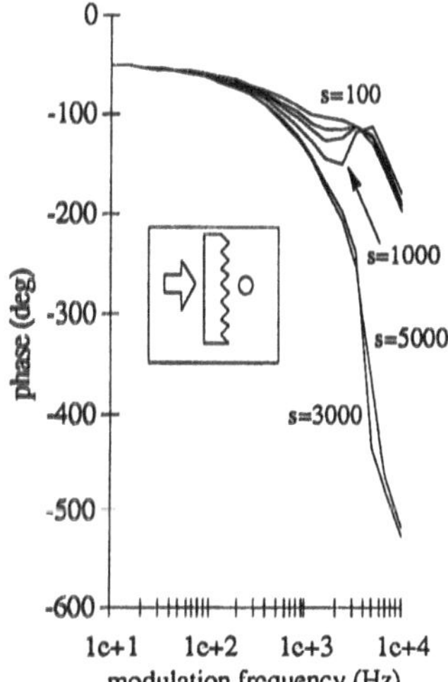

Figure 3: Theoretical calculations for back SRV=10000 cm/s and front SRV varying from 100 to 5000 cm/s

thermal one, the electronic diffusivity of photogenerated carriers being larger than the thermal one, and therefore to observe an increased sensitivity to the rear surface at high frequencies. Furthermore, as expected, the phase lag is larger for "pure" thermal signals (high surface recombination velocities).

When illuminating the polished surface and probing this front surface, the plasma wave are clearly present at low frequencies as seen in Figure 1 (solid lines). This can be explained by the recombination at the front surface of plasma waves reflected by the back surface. In the last configuration, shown in Figure 3, the information of the front surface recombination velocity is carried by the thermal wave generated by recombination at this surface and thus is detected at low frequencies, corresponding to long propagation time.

4. Results and Discussion

Direct measurement of the phase is complicated by the convolution effect due to the finite size of the probe beam. The phase difference between a pair of configurations does not depend on this finite size /5/. Therefore, we use this difference to fit our experimental results to theoretical calculations. To choose the appropriate pair of configurations to use, we must not only consider the sensitivity in phase, but also take into account the relative amplitude of the two configurations. We took the phase difference between front and rear detection while illuminating the polished surface. To fit our results we have used the heat and plasma diffusion model given in ref.6. Experimental results, presented in figure 4, are in very good accordance with the model and permit us to determine the value of SRV for both surfaces: 700 cm/s for the polished surface and 4000 cm/s for the damaged one.

We are presently improving our experimental setup, in particular by the use of a variable wavelength excitation, to make better use of all four pump and probe beam configurations in order to obtain non ambiguous and simultaneous determination of the surface recombination velocities of both surfaces of a semiconductor wafer.

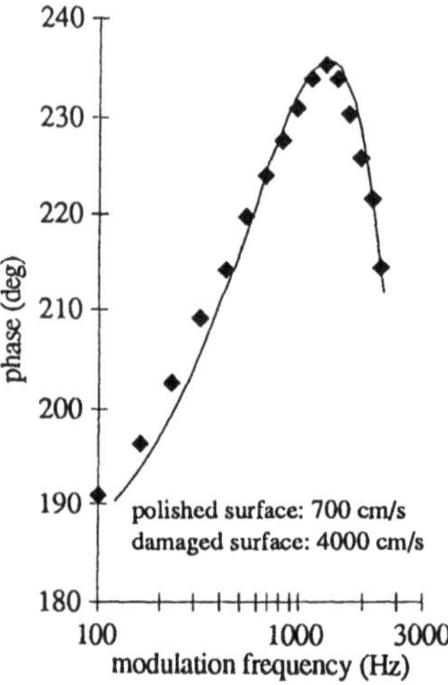

Figure 4: Experimental data (♦), theoretical calulation (----) and fitted values of SRV for both surfaces

References

/1/ D.Fournier, A.C.Boccara, A.Skumanich, N.M.Amer,J.App.Phys.**59**,787 (1985).

/2/ W.L.Smith, A.Rosencwaig, D.L.Willenborg, Appl.Phys.Lett.**47**, 584 (1985).

/3/ A.Pinto Neto,H.Vargas,N.F.Leite,L.C.M.Miranda, Phys.Rev., **41**, 9971 (1990).

/4/ F.Charbonnier and D.Fournier, Rev.Sci.Instrum., **57**, 1126, (1986).

/5/ J.P.Roger,F.Lepoutre,D.Fournier,A.C.Boccara,Thin Sol.Films,**155**,165 (1987).

/6/ V.A.Sablikov and V.B.Sandomirskii, Phys.Stat.Solid.**B120**, 471(1983).

Out-of-Phase Photopyroelectric (PPE) Spectroscopy on Amorphous Semiconductors

C. Christofides and A. Mandelis

Photoacoustic and Photothermal Sciences Laboratory,
Department of Mechanical Engineering, and
Center for Hydrogen and Electrochemical Studies,
University of Toronto, Toronto, Ontario, Canada M5S 1A4

Abstract: A PPE spectrometer with self normalization capability has been used for both conventional transmission and thermal wave spectroscopic measurements of amorphous *Si* thin films, deposited on *c–Si* substrates. Optical-absorption-coefficient spectra were obtained from these measurements and the superior dynamic range of the out-of-phase photopyroelectric signal was established due to its zero-background compensation capability. Instrumental sensitivity limits of $\beta t \approx 5 \times 10^{-3}$ were demonstrated.

Recently, Mandelis *et al.* [1] first performed PPE spectroscopic measurements of an *a–Si:H* thin film on quartz. Christofides *et al.* [2] have extended the use of *PPES* to spectroscopic measurements of crystalline semiconducting wafers.

In this work we present the first application of *PPES* on very thin absorbing amorphous layers deposited on absorbing crystalline *Si* substrates. For this study a few *a–Si* samples of various thicknesses, *t*, deposited on *Si* were used. The samples were put on the sample-cell detector in order to have thermal contact with the black painted *PVDF* in the case of configuration "A". In case of configuration "B" the sample was put on the ring spacer over the *PVDF* pyroelectric film and at a distance of 200 μm, away from the detector surface, so that there would be no thermal wave communication to the pyroelectric.

Using the configuration "B", measurements were made on two pure crystal silicon samples obtained from different wafers. At wavelength $\lambda = 1000$ *nm*, no signal was produced due to the opacity of the substrate. Consequently, we started the experiment at 1150*nm*. The chosen modulation frequency, f, was 45*Hz*, in order to satisfy the thermally thick condition for the 52μm *PVDF*. Complete substrate transparency was considered to be at $\lambda > 1200nm$, in which region no quadrature signal could be registered from a *c–Si* sample. As was expected, it has been found that the *PPE* transmitted signal S_O increases as a function of wavelength. At a given wavelength, it was observed during the same experiment that the *PPE* signal due to transmission decreases when the thickness of the amorphous layer increases. Finally, the phase of all the different pure *c–Si* samples was found to be constant as a function of the wavelength; *i.e.* there was no quadrature contribution from delayed synchronous thermal signals S_T. The measurement of the optical absorption coefficient as a function of λ cannot be performed

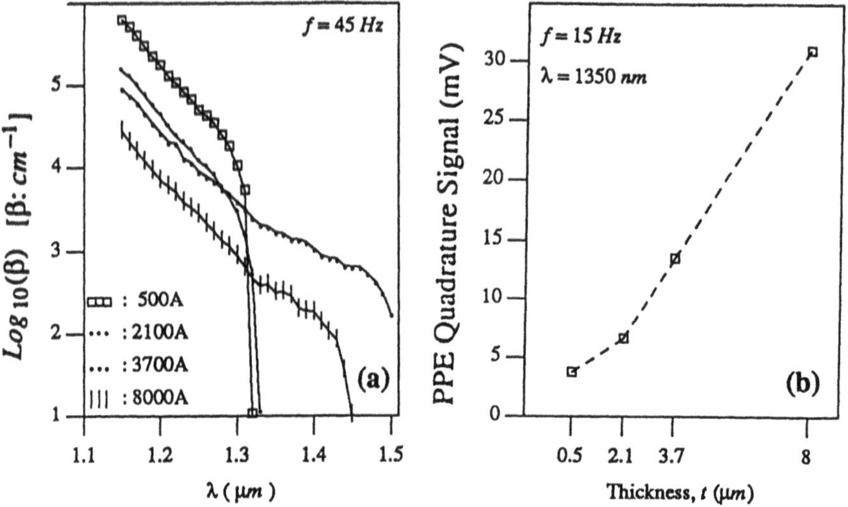

Fig. 1 (a)Variation of β as a function of the wavelength for various amorphous layer thicknesses. (b) *PPE* quadrature signal dependence on deposited layer thickness.

from the transmission measurements alone, because of the unknown reflectance coefficients between the interfaces: film-air and film-substrate.

Several PPE spectra have been obtained at $15 Hz$ using configuration "A". As was expected, the normalized signal, S_T, increased and phase difference, Φ, decreased with increasing wavelength (note that Φ varies from negative values to zero). It has been proven that in the case of thin film the optical absorption coefficient can be given by the relation [3]:

$$\beta(\lambda) = -\frac{1}{t} Ln\{ 1 + \frac{S_O(\lambda)\ \tan[\Phi(\lambda)]}{(1 - R_1)\exp(-qL)\sin(qL)} \} \tag{1}$$

where L is the thickness of the substrate and $q = \sqrt{\pi f / \alpha}$ with α the thermal diffusivity of the substrate. R_1 is the reflectivity of film-air. By combining thermal-wave and transmission mode data and by taking into account the calculated value for the thermal diffusivity α of silicon, the optical absorption coefficient of the amorphous layer sample was determined from Eq.(1). In Fig. 1(a) we present the optical absorption spectrum of the amorphous layers obtained at 15 Hz. These spectra were obtained in a self-normalizing fashion by using the in-phase signal channel in the role of the normalizing spectrum, instead of the signal from the reference sensor. These two signals were found to be similar, as the thin surface layer absorption, S_T, was essentially a negligible perturbation to the much stronger direct transmission, S_O.

As was shown, it is practically impossible to obtain the quantitative optical absorption coefficient spectrum from transmission spectra since it is not easy to determine very low absorptances, as well as the reflectivity between film and substrate. On the other hand, proper elimination of the large background by out-of-phase PPE detection shows absorption spectra of the deposited thin-film amorphous *Si* closely resembling each other in Fig. 1(a). The noticeable trend to somewhat increased $\beta(\lambda)$ with decreasing layer thickness is most pronounced with the 500 A thick film and may due to an increase in the importance of extraneous absorption at the interface between *a–Si* and substrate.

Figure 1(b) shows the change of *PPES* signal with *a–Si* thickness for our samples and demonstrates the ability of the present technique to monitor layers even thinner than our minimum 500 A. An advantage of *PPES* over *PDS*, in this context, lies in the avoidance of a liquid interface between the electronically sensitive microelectronic device structure and a fluid of high refractive index gradient, $\partial n/\partial T$. A further advantage of the present out-of-phase *PPE* determination of $\beta(\lambda)$ in the weak absorption regime, over earlier investigations [3,4] is the elimination of the requirement for a reference photon counting sensor for quadrature detection with a transparent substrate: spectral self-normalization can be efficiently performed in the transparent regime by utilizing the in-phase signal as the normalizing reference.

The sensitivity of the *PPE* quadrature technique is seen to be better than the minimum thickness detection in these measurements, for which $\beta(\lambda)]\,(t_{min}) \approx 5 \times 10^{-3}$. According to Tanaka [4] with the photothermal beam deflection (PDS) technique one can obtain sensitivities of $\beta\, t \approx 10^{-4}$ by immersing the sample in CCl_4.

[1] A. Mandelis, P.E. Wagner, K Ghandi, R. Baltman, and P. Dao, Phys. Rev. B **39,** 5254 (1989).

[2] C. Christofides, A. Mandelis, and K. Ghandi, Rev. Sci. Instrum., **61,** 2360 (1990).

[3] C. Christofides, A. Mandelis, A. Engel, M. Bisson, and G. Harling, Can. J. Phys., **69,** 317 (1991).

[4] K. Tanaka, in *Photoacoustic and Thermal Wave Phenomena in Semiconductors*, Edited by A. Mandelis (North-Holland, New York 1987) p. 441.

Electron Acoustic Microscopy of Defects in Plastically Deformed Si Crystals

V.G. Eremenko

Institute of Microelectronics Technology and
High Purity Materials, Russian Academy of Sciences,
142432 Chernogolovka, Moscow District, Russia

Abstract. The first results of the SEAM imaging of quasi-two-dimensional defects associated with dislocation slip planes and platelets of Si hexagonal phase are presented. It is suggested, that SEAM contrast has mainly thermal nature.

1. Introduction

It is well known that mobility and electrical activity of dislocations essentially depend on conditions of thermal treatment during plastic deformation or subsequent annealing. However, so far details of this process are not known yet.

As a result of such interaction new quasi-two-dimensional defects were observed to be associated with slip planes swept by moving dislocations during plastic deformation [1]. The EBIC observations revealed dark contrast and the width of image is about 15-20 μm [2]. The last our investigations of electrical activity of the slip planes [3] have demonstrated, that near by slip planes extensive n-p-n structure forms, which can't be explained on the basis only of the EBIC data. Nevertheless, it was confirmed, that in the regions adjacent to the slip plane thin band with high enough impurity concentration exists. The present paper is the first application of the SEAM technique directed to understanding of these phenomena.

2. Experimental Conditions

The investigations were carried out on Czochralski grown n--type Si samples with phosphorus and oxygen content respectively of 10^{14} cm^{-3} and 10^{18} cm^{-3}. The samples were plastically deformed by bending and indentation as described in [1,4]. A basic elements of the SEAM system and conditions for recording of the images have been reported earlier [5]. Only an amplitude images in the modulation frequency range from 50 to 200 kHz were registered. All SEAM images were recorded in a vicinity of antiresonance frequencies.

3. Results

The typical SEAM image of a slip plane traces after moving about 40 dislocations and corresponding SE image are shown in Fig.1a,b. It is seen first of all, that the contrast of a slip plane traces is presented by narrow dark bands and their

Springer Series in Optical Sciences, Vol. 69 Photoacoustic and Photothermal Phenomena III
Editor: D. Bićanić

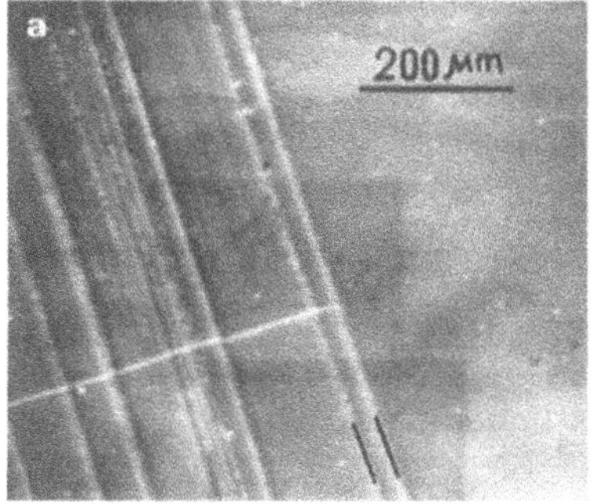

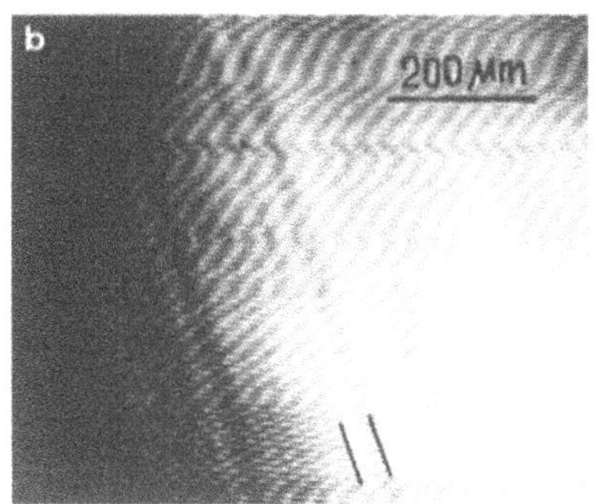

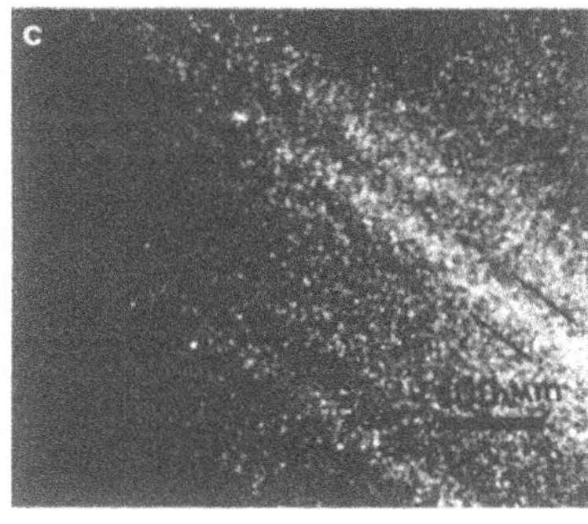

Fig.1 a. SE image of traces on (100) surface
b. SEAM image. Frequency 96.3 kHz; signal level 56 μV
c. SEAM image of the same region after annealing; 95.9 kHz; signal level 5 μV

position concides with position in the SE image. The EBIC imaging of a similar traces gives the image width about 15 - -20 μm. However in our case the image width of dark band is two times or more. Becides, the SEAM images reveal additional peculiarities of the contrast, namely, the diffuse white regions arise at both sides of dark bands. It is seen more distinctly in the same specimen annealed during 30 minutes at 700°C [Fig.1c].
Therefore, the total SEAM image of a slip plane traces has width about 100 - 150 μm, that is much more than EBIC one. It is interesting to compare the SEAM images with results of Photoresponce measuring when focused laser beam is scanned across isolated quasi-two-dimensional defect [3]. It was revealed that n-p-n structure exists near by slip planes with anomally wide p - regions up to 100 - 200 μm. The fact, that the width of SEAM image and photoresponce signal are practically equal allows us to assume, that the structure of electrically active centers and their distribution in the volume adjacent to slip plane are more complicated. From this point of view probably only part of impurity complexes take part in the EBIC contrast formation and a complete distribution of impurities and complexes is reflected by the SEAM images. An estimation show that concentration of defects in dark band of the image is about 10^{15} cm^{-3}. The bright bands correspond apparently more pure regions of a crystal, because of gettering effects by moving dislocations.

Thus, it is suggested that the nature of the SEAM contrast of slip planes is local changing of a thermal conductivity. Similar contrast was observed by Balk et al. [6] in denuded zones near by grain boundary in poly-silicon. Of course, the

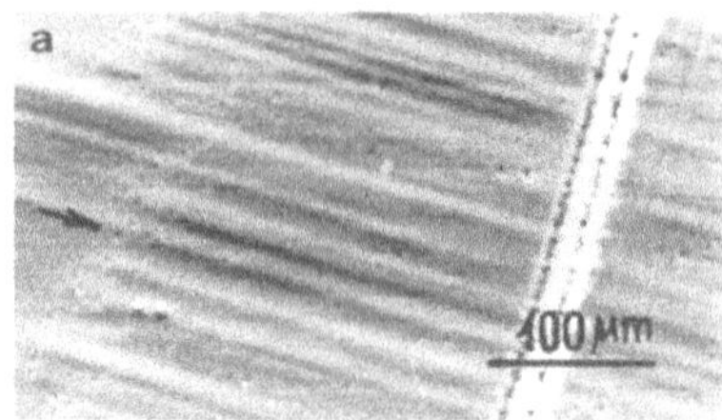

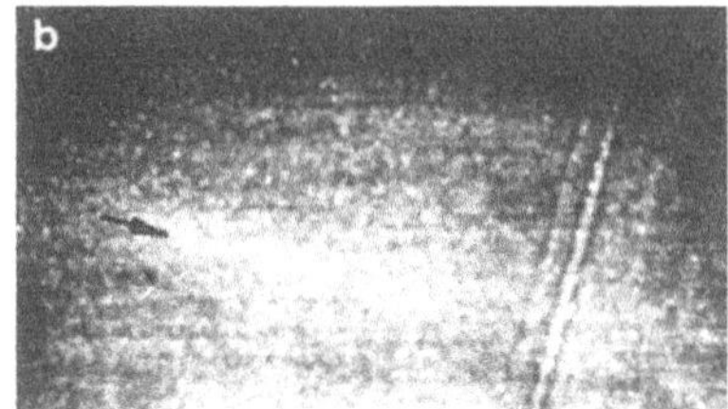

Fig.2 a. SE image; b. SEAM image, 101.4 kHz; 17.7 µV.

recombination source of heat exists, but has low intensity. This follows also from Modulated Optical Reflectance (MOR) investigations of dense dislocation rows [7] and as well as from results of Bivas et al.[8]. Similar results are obtained in the SEAM imaging of traces behind individual 60°--dislocations. However, only a part of traces gives the contrast [Fig.2] and futher investigations, combined with EBIC and MOR imaging are necessary. The first SEAM imaging of Hexagonal Si phase revealed weak bright contrast less than one per cent. Weak signal is very lakely stipulated for a small volume of phase investigated.

4. Acknowledgements

Author thanks Dr. Yakimov E.B. and Dr.Zaitsev S.I. for useful discussions of results and Mrs. Dubinina M.G. for specimen preparation.

5. References

1. Bondarenko I.E., Eremenko V.G.,Farber B.Ya.,Nikitenko V.I., Yakimov E.B., Phys. Stat. Sol. (a) 68, p.53 (1981)
2. Bondarenko I.E., Blumtritt H.,Heydenreich J.,Kazmiruk V.V., Yakimov E.B., Phys. Stat. Sol. (a) 95, p.173 (1986)
3. Eremenko V.G., Trofimov O.V., Semicond. Sci. and Technol. (to be published)
4. Eremenko V.G., Nikitenko V.I., Phys. Stat. Sol.(a) 14, p.317 (1972)
5. Aristov V.V., Gurtovoi V.L., Eremenko V.G., Microscopy of Semicond. Mater., Inst. Phys. Conf., Ser.87, p.685 (1987)
6. Balk L.J., Richard G., Kultcher N., Inst. Phys. Conf., Ser.76, p.343 (1985)
7. Bailey J., Weber E.R., Opsal J., J. of Cryst. Growth,103, p.217 (1990)
8. Bivas A., Smith W.L., Kola R.R., Radzimski Z.J., Rozgonyi G., J. of Cryst. Growth, 103, p.200 (1990)

Nonlinear Photoacoustic Effect in Semiconductors: Second Harmonic Wave Behavior

J.C. Cheng, S.Y. Zhang, Y.S. Lu, and Z.Q. Wang

Laboratory of Photoacoustic Sciences, Institute of Acoustics, Nanjing University, Nanjing 210008, P.R. of China

The nonlinear photoacoustic (PA) and photothermal (PT) phenomena have attracted considerable attention in recent years. Balk et al [1] reported the nonlinear effects observed in Scanning Electron Acoustic Microscopy (SEAM), and pointed out that comparing with the fundamental wave imaging, the second harmonic imaging displayed different structures of materials. Wetsel et al [2] investigated the nonlinear phenomena in Photothermal-Optical-Beam-Deflection (POBD) imaging, and presented a nonlinear thermal resistance model. We have studied [3] the nonlinear piezoelectric PA effect of silicon wafers in an applied dc electric field. The dependences of the fundamental frequency PA amplitude and phase on laser power have been measured and explained by considering the nonlinear surface and bulk Auger recombinations of the photogenerated carriers (PGC). Now, both experimental and theoretical investigations on the second harmonic wave behavior of the nonlinear piezoelectric PA effect in semiconductor silicon wafer are reported. The "subsurface approximation" method presented by Cheng et al [3] is extended to calculate a second harmonic component. A good agreement between the experiment and the theory is obtained.

I. Experiments

The sample is composed of silicon(111) wafer doped with n-type antimony in a rectangular region in which the dopant concentration is about $10^{18}-10^{19}/cm^3$. Two aluminum electrodes are placed on the two sides of the doped region surface and have Ohmic contact with the wafer. An external dc electric field is applied to the sample parallel to the surface. An argon ion laser beam modulated by an acousto-optic modulator illuminates the silicon wafer. The PZT transducer used to detect the PA signal provides the electric signal to a lock-in amplifier via a preamplifier. The experimental results of the dependence of the second harmonic wave PA amplitude on the laser power in different applied dc electric fields are shown in Fig.1.

II. Theory

According to Ref[3], the nonlinear equation expressing the PGC diffusion in the n-type semicondutor samples can be written as,

$$\partial P(x,t)/\partial t-D_e\partial^2P(x,t)/\partial x^2=I_0/(2h\nu)(1+\cos\omega t)\exp(-\beta x)$$
$$-P(x,t)[1+P(x,t)/N_0]^2/\tau_0, \qquad (1)$$

with the boundary conditions

$$D_e\partial P(x,t)/\partial x=S_0[1+P(x,t)/N_0]P(x,t), \qquad x=0 \qquad (2a)$$

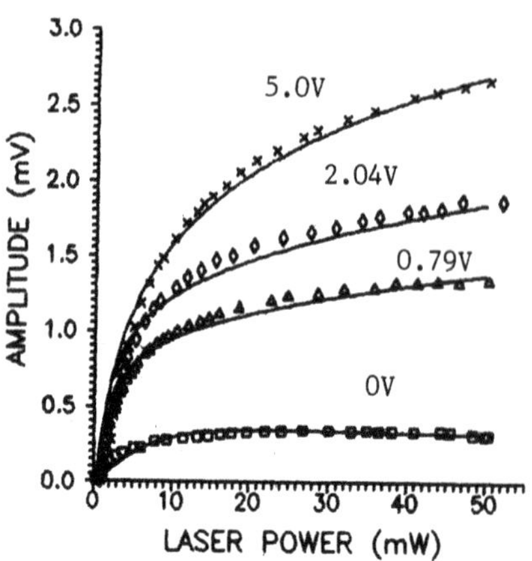

Fig.1 Dependences of second harmonic wave amplitude on laser power at different applied voltages.

$$P(x,t)=0 \,. \qquad x\to\infty \qquad (2b)$$

where $P(x,t)$ is the excess PGC density, D_e is the ambipolar diffusion coefficient, I_0 is the intensity of incident laser beam, ω is the modulation frequency of laser beam, β is the optical absorption coefficient of the sample, $h\nu$ is the energy of photon, τ_0 is the Auger recombination lifetime of PGC at equilibrium, S_0 is the surface recombination rate of PGC, and N_0 is the majority-carrier density at equilbrium.

The thermal wave equation is as follow,

$$\partial T(x,t)/\partial t-D_t\partial^2T(x,t)/\partial x^2=E_gD_t[1+P(x,t)/N_0]^2P(x,t)/(\kappa\tau_0)$$
$$+eE^2D_t(\mu_n+\mu_p)P(x,t)/\kappa$$
$$+I_0(h\nu-E_g)D_t/(2h\nu)(1+\cos\omega t)\exp(-\beta x)/\kappa\,, \qquad (3)$$

with boundary conditions

$$\partial T(x,t)/\partial x=-S_0E_g[1+P(x,t)/N_0]P(x,t), \qquad x=0 \qquad (4a)$$

$$T(x,t)=0 \qquad x\to\infty \qquad (4b)$$

where $T(x,t)$ is the temperature variation, D_t is the thermal diffusion coefficient, κ is the thermal conductivity, μ_n and μ_p are the drift mobilities of electron and hole respectively, E_g is the energy bandgap, E is the applied dc electric field, and e is electron charge.

In Eq.(3), the terms in the right side represent the thermal wave sources due to interband nonradiative bulk Auger recombination of PGC, instantaneous Joule thermalization of PGC in the applied electric field, and instantaneous intraband nonradiative thermalization respectively. Meanwhile, the right side of Eq.(4a) is due to interband nonradiative surface Auger recombination of PGC.

The elastic wave equations in semiconductor sample and transducer are,

$$\partial^2U_s(x,t)/\partial x^2-1/C_s^2\partial^2U_s(x,t)/\partial t^2=\beta_t\partial T(x,t)/\partial x+\beta_e\partial P(x,t)/\partial x, \qquad 0<x<d \qquad (5a)$$

$$\partial^2U_t(x,t)/\partial x^2-1/C_t^2\partial^2U_t(x,t)/\partial t^2=0, \qquad d<x<L \qquad (5b)$$

respectively, with boundary conditions

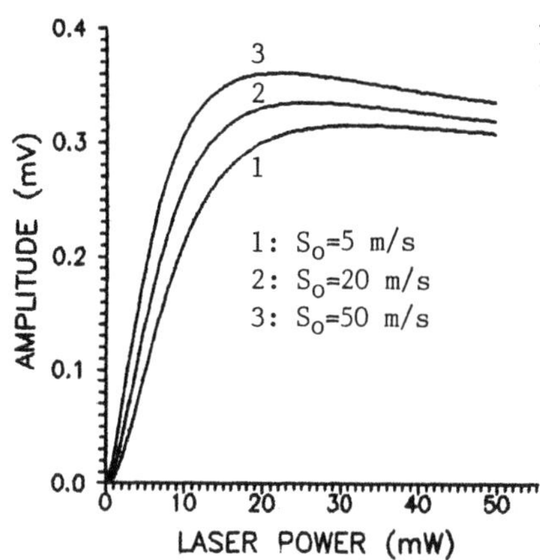

Fig.2 Influences of surface recombination velocity on second harmonic wave amplitude.

$$\partial U_s(x,t)/\partial x = \beta_t T(x,t) + \beta_e P(x,t), \quad x=0 \qquad (6a)$$

$$\partial U_s(x,t)/\partial x = \eta\,\partial U_t(x,t)/\partial x, \quad x=d \qquad (6b)$$

$$U_s(x,t) = U_t(x,t), \quad x=d \qquad (6c)$$

$$\partial U_t(x,t)/\partial x = 0. \quad x=L \qquad (6d)$$

where $U_j(x,t)$ is the particle displacement, j = s, t represent the sample and transducer respectively, and

$$\beta_t=(3\lambda_s+2\mu_s)/(\lambda_s+2\mu_s)\alpha_t,\ \beta_e=(3\lambda_s+2\mu_s)/(\lambda_s+2\mu_s)\alpha_e,\ \eta=(\lambda_s+2\mu_s)/(\lambda_t+2\mu_t)$$

(λ_j,μ_j) are Lame constants, C_j is the acoustic velocity, α_t is thermal expansion coefficient of sample, α_e is electronic-strain coefficient, d and L-d are the thicknesses of silicon wafer and transducer respectively.

Employing the "subsurface approximation" presented in Ref[3], we have calculated numerically the dependence of the second harmonic wave amplitude on the laser intensity, also as shown in Fig.1. The results show that: (1) when the laser power is not large enough, the amplitude of the second harmonic wave increases quadratically; (2) when the laser power is very large, the amplitude approaches a saturation value; (3) as the applied electric field increases, this saturation value increases. In addition, we have also calculated the inflence of the surface recombination velocity on the PA amplitude curves, as shown in Fig.2.

III.Conclusion

We have presented the theoretical and experimental studies for the second harmonic wave behavior of nonlinear photoacoustic effect in semiconductors. The "subsurface approximation" has been extended to calculate second harmonic wave component. The results are significant for the study of the generation mechanism of the nonlinear photoacoustic signal and for the characterization of the nonlinear processes in semiconductors.

References

[1] L. J. Balk , IEEE Trans. Magnetics 20 (1984) 1466.
[2] G. C. Wetsel and J. B. Spicer, Can. J. Phys. 64 (1986) 1269.
[3] J.C. Cheng, S.Y. Zhang, and Y.S.Lu, J.Appl.Phys. 68 (1990) 3865.

Part VI

New Topics, Experimental Techniques and Instrumentation

Some Theoretical and Experimental Aspects of Photothermal Imaging

H.G. Walther[1], *K. Friedrich*[1], *R. Göring*[2], *K. Haupt*[1], *and U. Seidel*[1]

[1]Institute of Optics and Quantum Electronics, University of Jena, Max-Wien-Platz 1, O-6900 Jena, Fed. Rep. of Germany

[2]Institute of Applied Physics, University of Jena, Max-Wien-Platz 1, O-6900 Jena, Fed. Rep. of Germany

Abstract. A phase interference technique is presented which has a higher SNR than the commonly used Mirage technique and is suitable for being integrated into a laser scanning microscope. For better interpretation of photothermal images the photothermal point spread function is introduced. By means of it the expedient conditions of getting high contrast images can be deduced.

1. Introduction

Recently photothermal measuring techniques have been used successfully to detect hidden patterns in the interior of opaque or strongly scattering samples. The field of this socalled photothermal imaging includes the nondestructive evaluation of materials and components with pattern size typically in the mm range /1/ as well as the photothermal microscopy with sub-μm resolution aspired /2/. Objects of considerable practical interest are for instance ceramics, laminates, compounds, overcoated surfaces and doped and patterned semiconductors. To enlarge the field of application of photothermal imaging two essential problems have to be solved:

1) The increasing of the signal-noise-ratio (SNR) of the photothermal detection
2) The "definition" of the imaging has to be known, this means the degree to which a photothermal image faithfully represents the shape and size of the subsurface object /3/.

2. Improvement of the SNR

2.1. General approach

The detectability of small and weak thermal patterns is limited by the SNR of the used photothermal technique. Many effects are aimed at the increase of it. First of all the experimenter has to select such a detection scheme which is appropriate for the sample under study. Besides by performing a favourable time regime of sample illumination and/or signal detection the amount of noise can be efficiently suppressed. Finally the application of compensation methods and interference techniques leads to a considerable increase of the SNR. Such techniques which probe the thermally induced sample surface deformation are already reported in the literature /4-7/. Mostly Michelson-type-interferometers are used, whereby the vibrating sample surface acts

Photoacoustic and Photothermal Phenomena III

as the one mirror. To become independent on the optical quality of the specimen surface we followed a suggestion published in /8, 9/, to probe interferometrically the temperature induced refractive index modulation within the air above the locally heated solid state sample. This technique we call Mirage interferometry.

2.2. Photothermal Mirage Interferometer

To perform photothermal microscopy the interferometric set-up was placed onto the specimen stage of a laser scanning microscope, as shown in Fig. 1. A He-Ne-laser probe beam (1) running parallelly with the specimen surface is divided by an adjustable phase grating (G1) into three rays one of them penetrating the thermal lens in air. The grating G1 is imaged by the lens L1 at the plane of a second grating G2 which is appropriately aligned to mix the different diffraction orders and to provide an additional phase shift of 90°. At the image plane P of the lens L2 placed behind G2 we can observe an interference pattern of oscillating points of intensity. As shown in /10/ the difference between the intensity of the plus and minus first order is proportional to the overall phase modulation in the thermal lens. Signal contributions of unmodulated probe light cancel out.

As a heating beam an Ar^+-laser was coupled into the tube of the microscope. By delaying this beam along the focal line of the zeroth order probe beam we realised a line scan over the field of view of the microscope objective (O). The line feed was carried out by a computer driven translation stage. The sensitivity of this photothermal microscope can be illustrated by the fact, that we could receive signals from Si-wafers at 10 kHz with a SNR better than 50 corresponding to a minimum detectable optical path length difference of $2 \cdot 10^{-13}$ m/$\sqrt{}$Hz. Above that the signal is rather insensitive to mechanical vibrations of the detector and to spot fluctuations of the probe laser.

The incorporation of a photothermal option into an existing microscope requires a further reduction of the Mirage interferometer. Integrated optical components seem to be promising to meet this necessity. As a first step in Fig.2 we show a photothermal interferometer whose heart is a double layered Y-branch (1)which is connected by monomode fibres with a well stabilised laser diode (2) and a photodiode (3) also. The double light channel exit of the Y-branch is imaged by the lens (4) onto an adjustable mirror (5) and back onto the Y-branch. The double

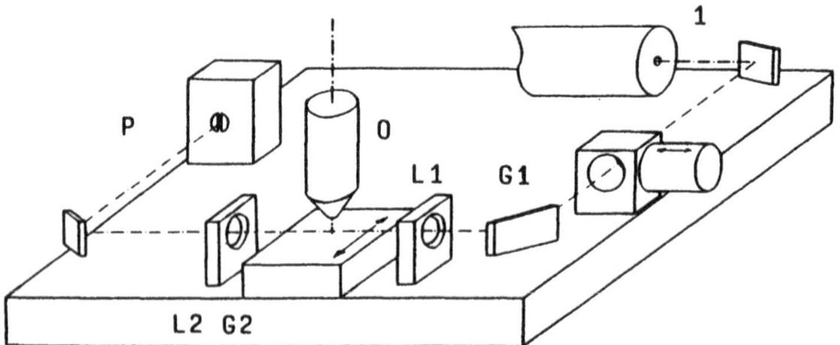

Fig.1 Scheme of the photothermal Mirage interferometer mounted onto the specimen stage of a laser scanning microscope

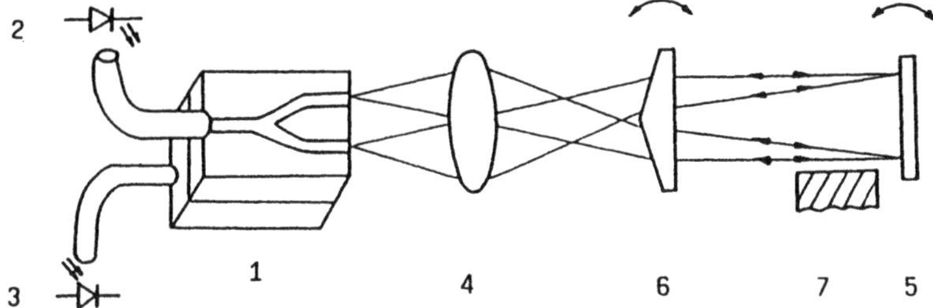

Fig.2 Mirage interferometer with a double layered Y-branch

prism (6) makes the partial beams parallel. The sample under study (7) is placed near position (5) and therefore the thermal lens is penetrated twice by the probe beam. At exit (3) probe beam interferences due to thermal lens action are observable.

3. Photothermal Point Spread Function

The quality of the photothermal imaging can be evaluated by its definition, resolving power and contrast /2, 3/. When studying these problems it is advantageous to use the formalism of point spread function (PSF). This function descended from optical microscopy /11/ describes the blurring of the image of a point object as the result of the physical mechanism of image formation.

In the case of photothermics the scattering of heavily damped thermal waves from thermal inhomogeneities makes it possible to image such structures. We consider the sample being illuminated by a localized and modulated heating beam which forces a periodic variation of the sample surface temperature. Our discussion is restricted to such photothermal techniques which yield a signal proportional to the averaged surface temperature oscillation. This condition is met when detectors with sufficiently large aperture will be used.

To calculate the PSF we have to solve the heat diffusion equation with spatially dependent coefficients. If the deviation of them from the thermal mean values are small enough we succeed in performing a perturbation theoretical approach /12, 13/. With the thermal defect strengthes $A=\Delta k/k_0$, $B=\Delta(\varrho\cdot c)/(\varrho\cdot c)_0$ we finally get the PSF of a surface absorber.

$$G(r) = \frac{\pi}{2}\sigma^2 \frac{1}{\sqrt{r^2+z_0^2}} e^{-\sigma(z_0+\sqrt{r^2+z_0^2})} \left[\frac{A z_0}{\sqrt{r^2+z_0^2}} \left(1+\frac{1}{\sigma\sqrt{r^2+z_0^2}}\right) + B \right] \quad \textbf{(1)}$$

with $\sigma^2 = 2i/\mu^2$, μ is the thermal diffusion length of the homogeneous material, r and z_0 are lateral and vertical offset between point defect and heating beam. When we scan an extended inhomogeneity of shape F(x,y) localized at depth z_0 below the surface by heating laser profile W(x,y), then the signal deviation ΔS from the unperturbated signal S_0 is given by the multiple convolution integral

$$\Delta S = S_0\,[W{**}F{**}G] \quad \textbf{(2)}$$

4. Contrast Function

It is appropriate to introduce contrast functions K_φ and K_S of phase and amplitude which immediately can be derived from measurements. With the complex valued signals $S_0=|S_0|\exp(i\varphi_0)$, $S=|S|\exp(i\varphi)$ we define

$$K_\varphi = arg(S) - arg(S_0) = \Im m\, G$$
$$K_S = \frac{(S-S_0)}{S_0} = \Re e\, G \tag{3}$$

In Fig.3 and 5 essential properties of the contrast functions of a hidden point defect are shown, taking into account only deviations of thermal conductivity. The calculated contrast functions just above the defect show extrema and zero crossings at certain modulation frequencies and - as expected - they vanish at high

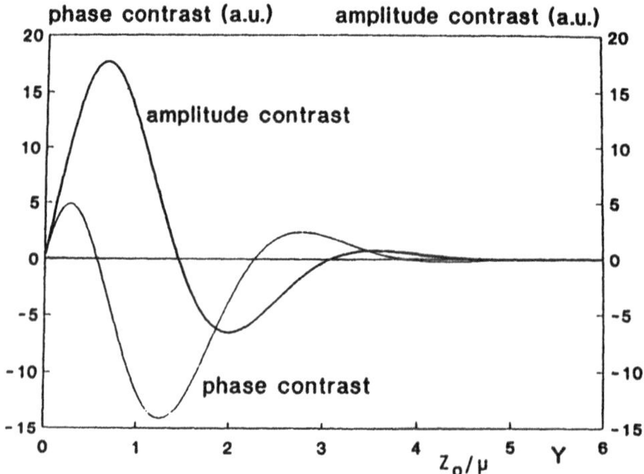

Fig.3 Calculated contrast functions K_φ and K_S above the defect locus in dependence on modulation frequency

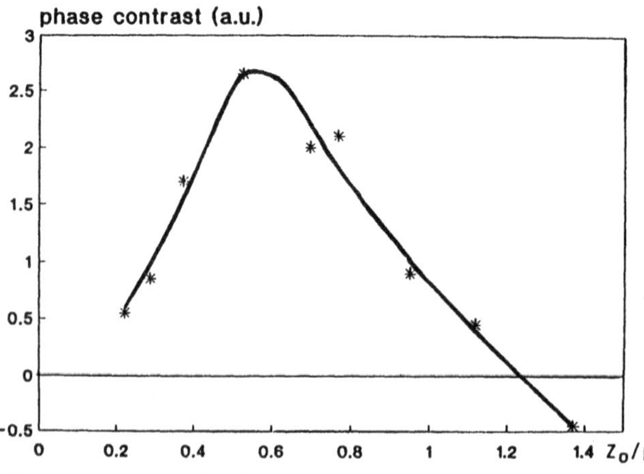

Fig.4 Measured phase contrast of tin defects in brass

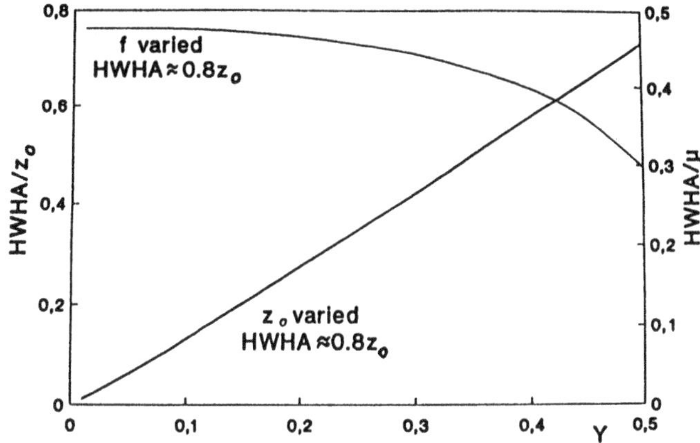

Fig.5 Calculated half width at half amplitude of the phase contrast function vs. z_0/μ

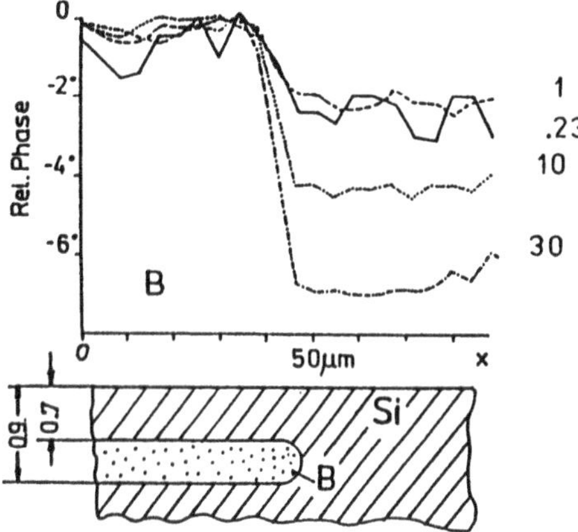

Fig.6 Photothermal line scans over the border between boron doped and undoped silicon at various modulation frequencies

modulation frequencies or large defect depthes. This behavior we could experimentally confirm, as shown in Fig.4 by measuring of the phase contrast of tin inhomogeneities in brass. There exists a distinct maxima at a modulation frequency of about 40 Hz. The measurement has been performed by Mirage interferometry. Photothermal radiometric measurements provide the same result. The calculated half width at half amplitude HWHA of the phase contrast is shown in Fig.5. We conclude that for $z_0<\mu$ the HWHA is independent on modulation frequency and is approximately $0.8\cdot z_0$. As an illustration of this fact we scanned photothermally a silicon wafer which was doped with boron ions along a limited stripe. Because of the implantation conditions this stripe was buried beneath the wafer surface about 700 nm. Measured phase scans across the border of the doped region showed that the width of the phase jump between the doped an undoped material is independent of modulation frequency though it was varied by more than two orders of magnitude(Fig.6).

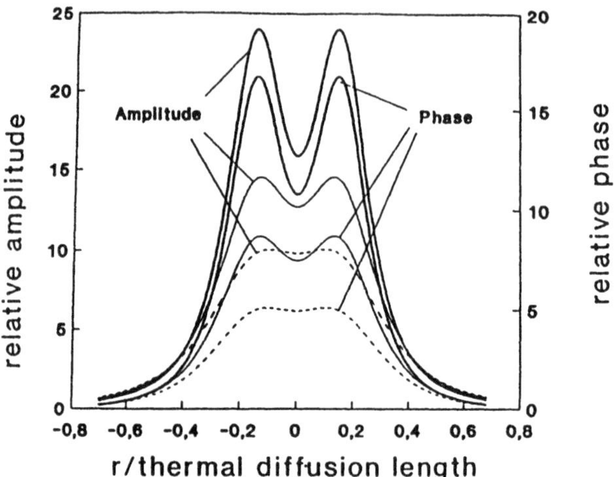

Fig.7 Resulting contrast of 2 hidden point defects at different depthes z_0/μ=0.1, 0.2, 0.3

The HWHA limits the possible lateral resolution. Calculated contrast functions of 2 neighbouring point defects are presented in Fig.7 showing the well known decrease in resolution with increasing defect depth.

5. Acknowledgement

The authors are indebted to G. Busse, B.K. Bein and K.L. Muratikov for brainstorming discussions and useful advices. Experimental work was supported by PHOTOTHERM Dr.Petry GmbH,Saarbrüken.

References

/1/ G.Busse, H.G.Walther, in "Progress in Photothermal and Photoacoustic Sciences and Technology", Vol.1 ed. by A.Mandelis, Elsevier Appl. Science, London and NY.(1991)

/2/ M.Munidasa, A.Mandelis, see /1/

/3/ G.C. Wetsel, F.A. McDonald, J.Appl.Phys. 56, 3081, (1984)

/4/ Z.Sodnik, H.J.Tiziani, Opt.Commun. 58, 295, (1986)

/5/ B.L.Yordanov, J.I.Burov, in "Photoacoustic and Photothermal Phenomena II", Springer Ser.Opt., Vol.62, (1990)

/6/ M.Paul, B.Betz, W.Arnold, Appl.Phys.Lett. 50, 1569, (1987)

/7/ H.Takamutsu, Y.Nishimoto, Y.Nakai, Jap.J.Appl.Phys. 29, 2847, (1990)

/8/ L.Chen, S.Y.Zhang, Appl.Phys.Lett. 50, 1340, (1987)

/9/ A.L.Glasov, K.L.Muratikov, Int.Journ.Optoelectr. 4, 589, (1989)

/10/ H.G. Walther, K. Friedrich, K. Haupt, A.L. Glasov, K.L. Muratikov, Appl.Phys.Lett. 57, 1600, (1990)

/11/ M.Born, E.Wolf, Principles of Optics, Pergamon Press Oxford, 1989

/12/ V.Gusev, Ts.Velinov, K.Bransalov, Semicond.Sci.Techn.4, 20, (1989)

/13/ A.L.Glasov, K.L.Muratikov, J.Techn.Phys.(in russ.) 57, 2184, (1987)

Photoacoustic and Photothermal Investigations of Thin Films

E. Matthias, H. Grönbeck, E. Hunger, J. Jauregui, H. Pietsch, M. Reichling, S. Petzoldt, E. Welsch[1], and Z.L. Wu[2]

Fachbereich Physik, Freie Universität Berlin,
Arnimallee 14, W-1000 Berlin 33, Fed. Rep. of Germany
On leave from [1]Institute of Optics and Quantum Electronics,
Friedrich-Schiller-University Jena, Max-Wien Platz 1,
O-6900 Jena, Fed. Rep. of Germany
[2]Shangai Institute of Optics and Fine Mechanics,
Academia Sinica, P.O. Box 800-211, Shanghai, P.R. of China

Abstract. Three different techniques - mirage effect, photothermal displacement, and thermoreflectance - have been installed for investigating ablation thresholds and thermal properties of thin films. By monitoring the shock wave signal the mirage effect was used to measure ablation thresholds of optical coatings. For multilayer systems of type polymer-metal SiO_2 changes in the shock wave signal allowed to distinguish between the ablation of different layers. The photothermal displacement method was applied to measure periodicity and decay of transient thermal gratings on surfaces of thin films. This technique yields information about lateral and vertical heat diffusion and is, therefore, sensitive to anisotropic heat transport in thin films. The potential of this technique for precise measurements of thermal diffusivities is illustrated for thin films of gold and amorphous diamond.

1. Introduction

During the past decade thermal properties of thin films have received increased attention [1,2]. They are inherently connected with the quality of thin films and are, therefore, most appropriate for their characterization. Precise measurements of thermal diffusivities provide information about finite size effects, structural anisotropies, interface absorption, defect and impurity absorption, and incubation effects. Also, it has been suggested by several workers in the field [3,4] that optical damage thresholds are strongly correlated to thermal properties of thin films, which means that these also play an important role for laser processing of thin films. Hence, the interest in thermal data ranges from optical, protective, and polymer coatings to semiconducting and metallic films as well as to high-T_c superconducting films.

These prospects led us to start an activity where several techniques like mirage effect, photothermal displacement and thermoreflectance are employed to yield complementary information about the same or similar film material. We also find it important to work both in the time and frequency

Springer Series in Optical Sciences, Vol. 69 Photoacoustic and Photothermal Phenomena III
Editor: D. Bićanić

domain in order to have the possibility for cross-checking the results. One ultimate goal of this project is to quantitatively correlate ablation thresholds to thermal properties of thin films. This contribution summarizes the present status of the work on thin films in our laboratory by presenting informative examples for each technique.

2. Damage thresholds measured by the mirage effect

The mirage effect caused by the transient change of refractive index [5-7] measures both acoustic and thermal waves emerging from a surface when irradiating it with laser pulses of sufficiently large fluence. They can be distinguished by their range as well as by their speed. This technique is particularly well suited for studying ablation thresholds and the development of laser-induced plasmas, in which case the thermal wave is substituted by the plasma wave. A typical signal obtained from a Kapton foil with a fluence above the ablation threshold is displayed in the lower part of Fig. 1. It shows a fast signal which we attribute to the refractive index gradient caused by the shock or blast wave. The huge later signal reflects the slowly expanding plasma. Deflection signals from PMMA recorded at 20 times higher fluences are very similar in shape, hence, we consider the one in Fig. 1 as being generally representative for different types of material. In fact, it very closely resembles the pictures taken by an ultrafast imaging technique at above threshold fluences with polyimide [8] and PMMA [9]. Notice, however, that the "clean" signal in Fig. 1 can, for certain materials, be badly distorted by probe laser scattering from large size fragments leaving the surface.

When applying the mirage technique to the investigation of ablation thresholds both the shock and plasma signals in Fig. 1 yield identical results, for example for polyimide. Because of its superior reproducibility, however, we prefer to use the fast shock wave signal for measuring optical damage thresholds. An example is shown in Fig. 2, where the deflection signal is plotted as a function of fluence for both pure

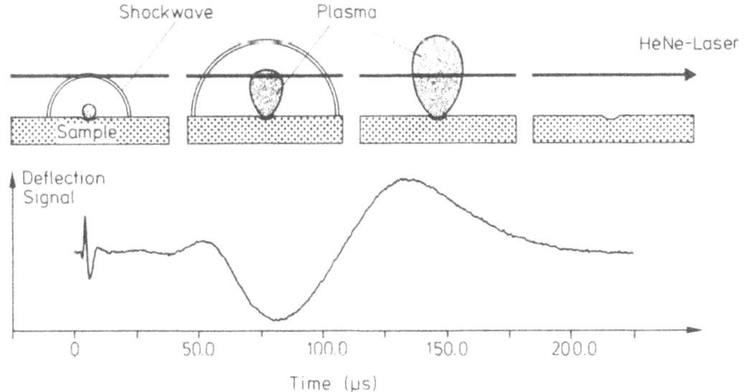

Fig. 1: Mirage effect signal from polyimide at a fluence of 0.12 J/cm^2 (lower part). The upper part sketches the origin of the two different refractive index changes.

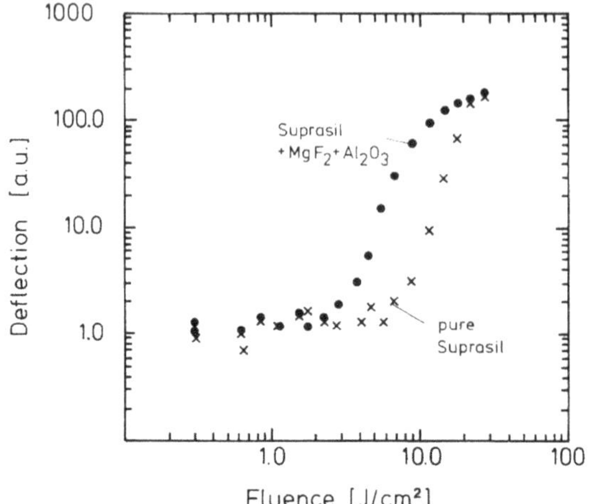

Fig. 2: Reduction of optical damage threshold by an optical coating on suprasil, measured by the shock wave mirage deflection signal at a pump laser wavelength of 248 nm [10].

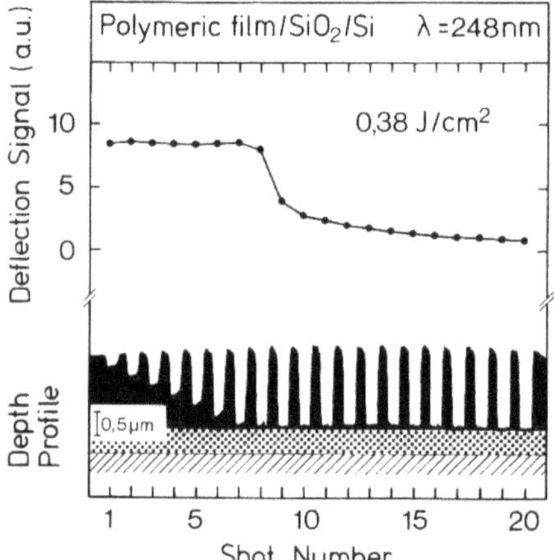

Fig. 3: Change of the shock wave mirage deflection signal at the transition from a photoresist film to a SiO_2 layer [12]. The lower part of the figure indicates the thickness scale and shows the depth profile of the photoresist layer.

suprasil and suprasil antireflex-coated with MgF_2 and Al_2O_3. Obviously, the coating causes a substantial lowering of the damage threshold. More detailed studies show that the reduction is due to the Al_2O_3-layer, including its interface while the MgF_2 coating alone causes no decrease. The underlying mechanism for this behavior remains to be investigated and calls for a measurement of the thermal diffusivities of these coatings.

For a given laser fluence the amplitudes of both signals in Fig. 1 depend on the absorbance of the material. Consequently the shock wave signal can also be utilized to distinguish between the ablation of layers of different materials [11]. This is illustrated in Fig. 3, where we can see that about 8 laser shots of 0.38 J/cm^2 at 248 nm are needed to remove a 1.5 μm thick film of photoresist. The deflection signal undergoes a step-like change when the laser light reaches the SiO_2 layer which has a much higher ablation

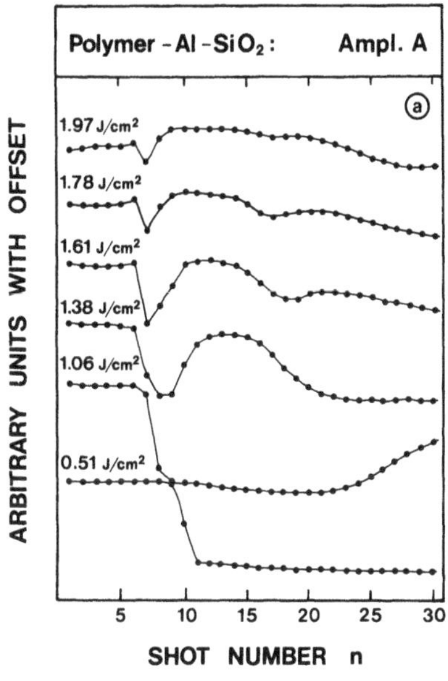

Fig. 4: Shock wave mirage deflection signals for a multilayer system [12] exposed to various fluences at 248 nm as a function of laser shots. The first step indicates penetration of the polymer layer, the following increase and maximum reflects ablation of the Al layer.

threshold and consequently no shock wave is generated. The small and decreasing signal above 8 shots is due to cleaning of residual photoresist and etching of the walls. Measurements of the etch depth profile were carried out to verify this interpretation and are shown in the lower part of Fig. 3.

This technique still works for multilayer systems provided the damage thresholds of the materials involved differ significantly. To demonstrate this point the film sequence polymer-metal-SiO_2 was irradiated with 248 nm light at various fluences [13]. The result for a sample with an Al-layer is shown in Fig. 4. Again, in the fluence range used the ablation of the photoresist starts at the first shot. The fluence of 0.51 J/cm^2 is too low to damage the Al-film with 30 shots. For 1.06 J/cm^2 we notice an increase of the deflection signal after about 25 shots, indicating the onset of metal ablation. At 1.38 J/cm^2 this onset is shifted to 10 shots, and starts still earlier at higher fluences. One can even notice an indication of SiO_2 breakup around 20 shots at 1.61 J/cm^2. Clearly, the contrast in the deflection signal between different layers is lost at higher fluences (> 1.7 J/cm^2), but within a certain fluence range it is possible to monitor the ablation of individual layers in multilayer systems [14].

3. Thermal diffusivities measured by displacement detection of transient thermal gratings

The pulsed photothermal displacement technique [15] is in principle an ideal tool for measuring the thermal diffusivity κ of thin films since the thermal diffusion length, $L_{th} = 2(\kappa\tau_0)^{1/2}$, is of the order of micrometer or less for a laser pulse length, τ_0, of nanoseconds or shorter. To our knowledge, however, only few attempts have been made to apply this

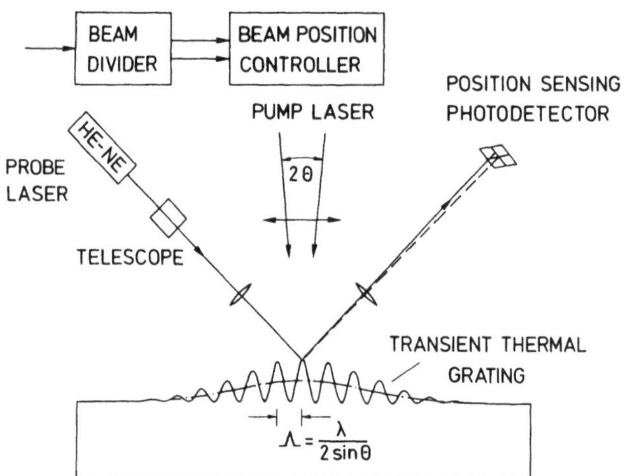

Fig. 5: Principle experimental scheme for photothermal displacement measurements on transient thermal gratings.

technique to thin films [16]. The reason may be that the intensities required for a good signal-to-noise ratio often come close to the damage thresholds of thin films. In the following we want to demonstrate that this technique can be considerably refined by combining it with the excitation of transient thermal gratings (TTG's) [17] on the film surface.

The use of TTG's instead of a circular beam spot brings several advantages: (1) It defines a direction in the surface plane and thereby becomes sensitive to anisotropic lateral heat diffusion. (2) Under certain conditions [18] theory predicts the following simple relation between the effective decay rate τ_{eff}^{-1} and the periodicity Λ of the grating:

$$\tau_{eff}^{-1} = \left(\frac{2}{a^2} + \frac{4\pi^2}{\Lambda^2}\right) \kappa + \tau_z^{-1} \quad , \tag{1}$$

where a is the extension of the grating perpendicular to its periodicity, and τ_z^{-1} takes into account the vertical heat diffusion. Hence a measurement of the exponential decay time, the periodicity, and the width of the grating leads to values for the thermal diffusivity and the vertical diffusion rate. (3) As long as the theoretical approximations are valid this technique leads to a good accuracy for the thermal diffusivity since τ_{eff}, Λ, and a can be measured rather precisely. Notice, however, that strictly speaking the technique determines the diffusivity constant for heat transport in the plane of the film.

The experimental set-up for probing the surface displacement of TTG's by reflecting a HeNe-laser beam is sketched in Fig. 5. The change in the reflection angle is recorded by a quadrant diode. The HeNe-laser is always focussed to the same surface spot (beam diameter about 50 μm) while the interference pattern generated by the two pump beams (wavelength 543 nm, pulsewidth 4 ns) is moved across this

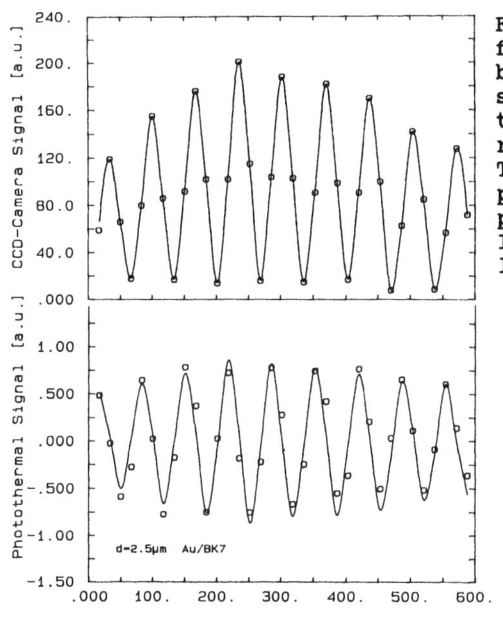

Fig. 6: TTG on a 2.5 μm gold film (lower part) generated by the interference pattern shown in the upper part of the figure. The latter was recorded by a CCD camera. The solid line in the upper part is a fit to the data points yielding Λ = 67.3 μm. Its derivative is the solid line in the lower part.

spot. Before a grating is measured the interference pattern is recorded by a CCD-camera to control its general shape and to measure its width a. Scans of the CCD-camera recording and of the corresponding TTG on the surface of a 2.5 μm gold film on BK7 glass are compared in Fig. 6. The phase shift between both patterns is $\pi/2$ since the displacement technique measures the derivative of the grating. We notice the excellent agreement between the two different sets of data. It also is obvious that a fit to the grating in the lower part of Fig. 6 will lead to an accurate value of the periodicity Λ.

The second important quantity needed for Eq.(1) is the effective decay rate. In Fig. 7 we show typical decays of TTG's on gold films of different thicknesses on a BK7 glass substrate [19]. Notice that in each case the decay is well described by a single exponential function indicated by solid lines. The striking difference between the decay constants for films of 100 nm and 2.5 μm reflects the influence of finite size effects, the increase of structural defects with decreasing thickness [20], and an increase of vertical heat flow with increasing thickness [19]. The latter, however, is not as important as one might expect, since the straight lines in Fig. 8, fitted to the data according to Eq.(1), intersect the ordinate at roughly the same value. The strong thickness dependence of the slopes corresponds to the different decay times in Fig. 7, and shows the drastic reduction of the thermal diffusivity with film thickness.

The fact that the straight lines in Fig. 8 fit the data well gives faith in the theoretical model and the validity of Eq.(1). However, as discussed in Ref. [19] our results for the thermal diffusivities are significantly smaller compared to the values obtained by other techniques. For example, photothermal reflectance measurements with MHz modulated

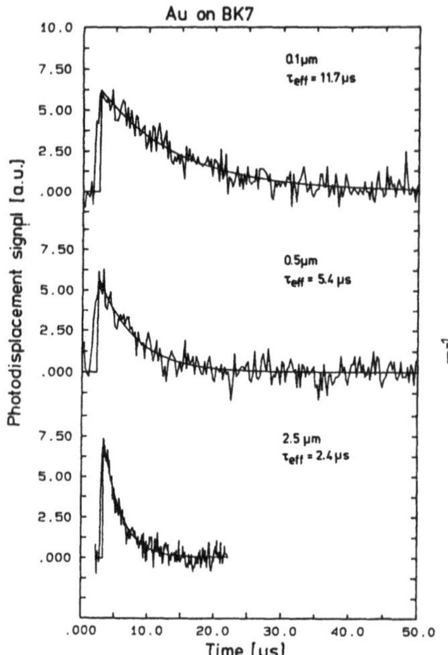

Fig. 7: Decay of TTG's on gold films of different thicknesses. Solid lines are fits to the data resulting in the decay constants shown.

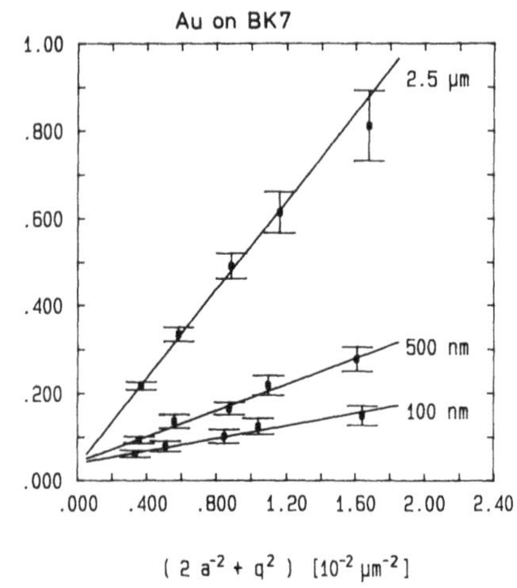

Fig. 8. Averaged effective decay rates for gold films of different thicknesses plotted according to Eq.(1). The slope of each line yields the thermal diffusivity. The results are in cm^2s^{-1}: 0.07, 0.15, 0.50 for the thicknesses 100, 500, and 2500 nm, respectively.

Ar^+-laser light carried out at our laboratory [21] reproduced for the 2.5 µm film the bulk value of 1.25 cm^2s^{-1}, in contrast to 0.50 cm^2s^{-1} obtained from the fit in Fig. 8. For a 1 µm film the thermoreflectance measurement yields 1.0 cm^2s^{-1} while the TTG displacement technique gives 0.4 cm^2s^{-1}. The latter technique does not even replicate the bulk value when applied to a 210 µm gold plate, instead we obtain the 25 % lower value of 0.94 cm^2s^{-1}. The deviations are systematically toward lower values and much larger than the experimental uncertainty of about 5 %. We propose that these are caused by surface texture, island structure and other defects [22], and that they reflect the sensitivity of the TTG displacement technique to lateral heat diffusion. Further evidence for this is obtained by studying the influence of surface morphology on the thermal diffusivity. Fig. 9a shows a microscopy picture of the gold plate surface for which the measurement resulted in a 25 % reduced value of κ. The random texture apparently diminishes the lateral heat conduction. Such is not the case for the polished Ni surface (Fig. 9b) for which we measure κ(Ni) = 0.23±0.01 cm^2s^{-1}, in perfect agreement with the literature bulk value of 0.22 cm^2s^{-1} [18]. The surface of a Ni-foil shown in Fig. 9c provides an example for a strongly anisotropic lateral heat diffusion. If we align the wave vector, $q = 2\pi/\Lambda$, of the TTG parallel to the grooves we find $\kappa_{\parallel}$ = 0.19±0.03 cm^2s^{-1}, while the thermal diffusivity perpen-

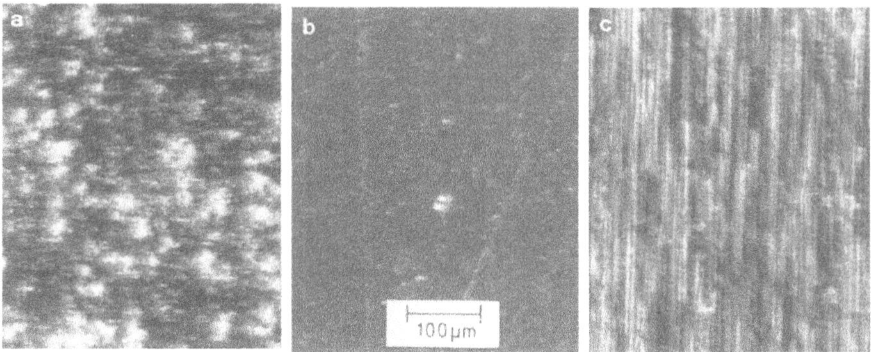

Fig. 9: Dependence of thermal diffusivities on surface morphology. (a) surface structure of 210 μm Au-plate (purity 99.8. %); (b) polished surface of Ni (99.98 %); (c) surface structure of 20 μm Ni-foil (99.98 %).

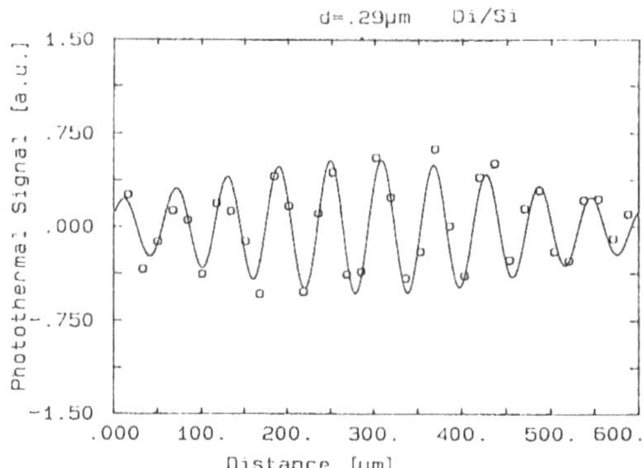

Fig. 10: TTG pattern on a 0.29 μm amorphic diamond film on silicon [24]. The solid line represents a fit to the data, yielding Λ = 59 μm.

dicular to the grooves is reduced to $\kappa_{\perp} = 0.05 \pm 0.02$ cm^{2}s^{-1}. These examples confirm that the TTG displacment technique measures predominantly the lateral heat flow in a rather shallow layer defined by the grating amplitude, and that q^2 is the leading term in Eq.(1). Hence this technique is ideally suited for studying the structural quality of thin films.

Due to their small optical absorption depth and the poor thermal conductivity of the substrate, metal films on glass represent a favorable case for the TTG displacement technique. The question is whether Eq.(1) will still be applicable when the optical absorption depth is comparable or even larger than the film thickness. To a first approximation one expects that the technique is useful as long as a TTG can be detected. For very thin films, however, it must be checked that the TTG is not formed at the substrate surface. A most interesting case close to the limit of validity are hydrogen-free "amorphic diamond" films with a typical optical absorption length of 0.2 μm [23]. A well developed TTG on a 0.29 μm thick film of this kind is shown in Fig. 10, and a

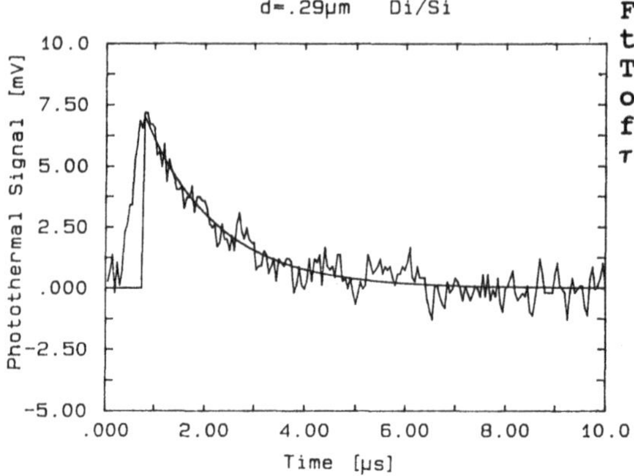

Fig. 11: Typical decay of the TTG shown in Fig. 10. The solid line is the fit of a single exponential function to the data with $\tau_{eff} = 1,45\ \mu s$.

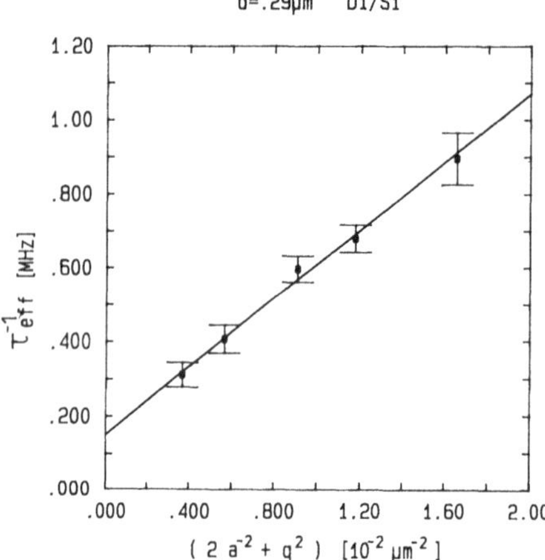

Fig. 12: Averaged effective decay rates for five different TTG's on a 0.29 µm amorphic diamond film [24] plotted according to Eq.(1). The solid line represents a fit to the data.

typical decay curve can be seen in Fig. 11. Many of those decay curves were measured and the results averaged to give the effective decay rates plotted in Fig. 12 for five different gratings in the form dictated by Eq.(1). The straight line is a fit to the data with the result $\kappa = (0.45 \pm .03)\mathrm{cm}^2\mathrm{s}^{-1}$ and $\tau_z = 6.7\ \mu s$. Again the data are perfectly well described by a straight line, which makes us confident that thermal diffusivities of diamond films can be measured by this technique down to thicknesses of about 0.2 µm, depending on their absorption coefficient. This offers the possibility to utilize this type of measurement for a quality control of thin diamond films.

4. Conclusions

It was shown that the mirage deflection signal, caused by the shock wave emerging from the surface, is an ideal tool to measure optical damage thresholds of coatings. It can also be successfully used to monitor the selective ablation of thin films whenever the damage thresholds of film and substrate differ. Within a certain fluence range, determined by the ablation thresholds of the materials involved, this is still true for multilayer systems. Therefore the shock wave detection has great potential for in-situ controlling laser structuring of thin films.

In order to obtain thermal diffusivities with good accuracy the displacement detection of TTG's was introduced. The power of the technique was demonstrated for gold films on glass and amorphous diamond films on Si. It was shown that the TTG displacement technique is most sensitive for lateral heat diffusion. Therefore the results may be affected by the surface texture. Since the TTG introduces a direction, this technique is well suited for studying anisotropic lateral heat transport.

This work was supported by the Deutsche Forschungsgemeinschaft, Sfb 337, and by the Bundesministerium für Forschung und Technologie, contract No. 13N5627. E.W. and Z.L.W. acknowledge fellowships from the Alexander von Humboldt-Stiftung. J.J. was supported by the Deutscher Akademischer Austauschdienst (DAAD).

References

1. P. Hess (editor): Photoacoustic, Photothermal and Photochemical Processes at Surfaces and in Thin Films, Topics in Current Physics Vol. 47, Springer Verlag Berlin-Heidelberg 1989
2. J.C. Murphy, J.W. MacLachlan Spicer, L.C. Aamodt, B.S.H. Royce (editors): Photoacoustic and Photothermal Phenomena II, Part III, Springer Ser. Opt. Sci. 62 (1990)
3. E. Welsch, H.G. Walther, D. Schäfer, R. Wolf: Thin Solid Films 152, 433 (1987)
4. J.C. Lambropoulos, M.R. Jolly, C.A. Amsden, S.E. Gilman, M.J. Sinicropi, D. Diakomihalis, S.D. Jacobs: J. Appl. Phys. 66, 4230 (1989)
5. W.B. Jackson, N.M. Amer, A.C. Boccara, D. Fournier: Appl. Opt. 20, 1333 (1981)
6. H.M. Lai, K. Young: J. Acoust. Soc. Am. 72, 2000 (1982)
7. S. Petzoldt, A.P. Elg, M. Reichling, J. Reif, E. Matthias: Appl. Phys. Lett. 53, 2005 (1988)
8. R. Srinivasan, K.G. Casey, B. Braren, M. Yeh: J. Appl. Phys. 67, 1604 (1990)
9. R. Srinivasan, B. Braren, K.G. Casey: J. Appl. Phys. 68, 1842 (1990)
10. Samples were kindly provided by Wild-Leitz AG
11. E. Hunger, S. Petzoldt, H. Pietsch, J. Reif, and E. Matthias: SPIE Vol. 1441, 283 (1990)
12. The sample was kindly provided by Dr. E. Neske, Fraunhofer Institut für physikalische Meßtechnik, Freiburg
13. E. Hunger, H. Pietsch, S. Petzoldt, E. Matthias: Appl. Surf. Sci, in print
14. E. Hunger, H. Pietsch, S. Petzoldt, E. Matthias: SPIE Vol. 1598, to be published

15. C. Karner, A. Mandel, F. Träger: Appl. Phys. A38, 19 (1985)
16. Q. Kong, J. Fischer, F. Träger: p. 179 in Photoacoustic and Photothermal Phenomena, editors: P. Hess, J. Pelzl, Springer Ser. Opt. Sci. 58 (1988)
17. H. Eichler, P. Günter, D.W. Pohl: Laser-Induced Dynamic Gratings, Springer Ser. Opt. Sci. 50 (1986)
18. J. Jauregui, E. Matthias: to be published, and contributed paper to this conference
19. J. Jauregui, Z.L. Wu, D. Schäfer, E. Matthias: contributed paper to this conference
20. P. Nath, K.L. Chopra: Thin Solid Films 20, 53 (1974)
21. M. Reichling, Thesis, Freie Universität Berlin 1991
22. I.J. Hodgkinson, P.W. Wilson: CCR Critical Reviews in Solid State and Material Sciences 15, 27 (1988)
23. F. Davanloo, E.M. Juengerman, D.R. Jander, T.J. Lee, C.B. Collins: J. Appl. Phys. 67, 2081 (1990)
24. The sample was kindly provided by professor C.B. Collins, University of Texas at Dallas

Signal Enhancement and Noise Suppression Considerations in Photothermal Spectroscopy

A.C. Tam

IBM Research Division, Almaden Research Center,
650 Harry Road, San Jose, CA 95120-6099, USA

Abstract. This paper provides physical discussions of factors controlling sensitivity in photothermal spectroscopy, in particular, photoacoustic spectroscopy and probe-beam deflection spectroscopy. We consider the physical basis for signal generation and enhancement methods for pulsed and continuous-modulated excitations as well as the various sources of noises which become significant when the absorption approaches the part-per-million level or below. Proper signal enhancement and noise suppression are essential to make photothermal spectroscopy a sensitive tool.

1. Introduction

Photothermal (PT) spectroscopy detects certain transient thermal effects in, on, or near a sample irradiated by a modulated excitation beam. It relies on the existence of a "thermal de-excitation" branch (see Fig. 1), which occurs for most materials. As this beam is scanned in wavelength, the PT effect is correspondingly monitored either in magnitude or in shape so that an "excitation spectrum" of the sample is obtained. This technique of PT spectroscopy is in principle "zero-background," which means that if there is no optical absorption, there is no PT signal. This is in contrast to conventional "subtractive" techniques involving the measurement of transmitted or reflected beams from the sample.

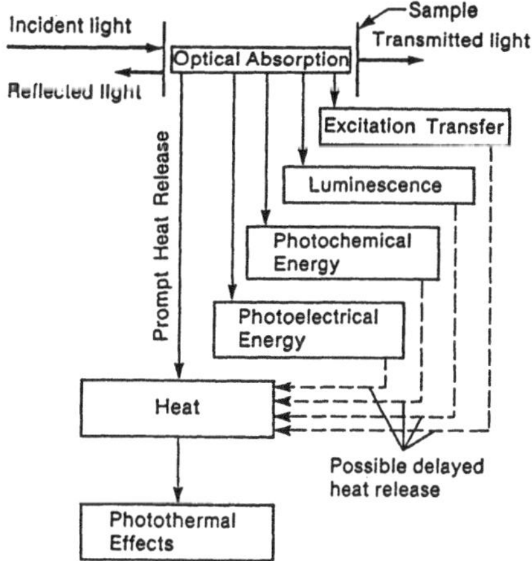

Figure 1. Schematics of optical absorption and de-excitation channels that can lead to "prompt" and "delayed" photothermal effects.

Table 1. Comparison of various techniques for weak optical absorption measurement

Technique	Examples	Conditions or Restrictions	Typical sensitivity for absorption (cm^{-1})
Gross subtraction (subtract exit intensity from incident intensity to get absorption)	Transmission; Reflection	Transmitted or reflected beam available for measurement	$10^{-2} - 10^{-4}$
Net macroscopic effect (measure macroscopic effects produced by the absorbed energy)	Photothermal calorimetry; Photoacoustics; Probe-beam refraction; Probe-beam reflection; Photothermal radiometry; Photoconductivity; Optogalvanic effect	Sample should produce heat or other macroscopic effects with sufficient efficiency. Intermolecular collisions or interactions usually needed.	$10^{-5} - 10^{-10}$
Net microscopic effect (measure quantum effects after exciting a molecule)	Fluorescence; Photoionization; Quantum-state change	Sample is placed in special environments (e.g., vacuum) so that the detected quantum effects are not quenched.	$< 10^{-10}$

Table 2. The heating of a sample due to optical absorption can result in various photothermal effects and provide the corresponding detection method

Photothermal effects	Detection methods (applied to sample S or to adjacent coupling fluid F)
Temperature rise	Photothermal calorimetry (S or F)
Pressure change	Direct photoacoustic detection (S) Indirect photoacoustic detection (F)
Refractive-index change (Thermal or acoustic)	Probe-beam refraction (S or F) Probe-beam diffraction (S or F) Probe-beam interference (S or F)
Surface deformation (Thermal or acoustic)	Probe-beam deflection (S) Probe-beam interference (S)
Thermal emission change	Photothermal radiometry (S)
Reflectivity/absorptivity change	Transient thermal reflectance (S) Transient piezo-reflectance (S) Optical transmission monitoring (S or F)

Hence, it is true that PT spectroscopy, if properly optimized in signal detection and noise suppression, should generally provide improved absorption sensitivity compared to traditional subtractive methods. On the other hand, since most if not all PT detection schemes involve macroscopic heat generation (i.e., significant amount of energy), they could hardly compete with truly "ultra-sensitive" spectroscopic tools which detect microscopic quantum effects. Thus, PT spectroscopy can be generally made "sensitive" but not "ultra-sensitive." The situation is summarized in Table 1. It is noted, however, that although the PT detection is less sensitive compared to detections of quantum microscopic effects, it does offer the advantages of simplicity, convenience, and general applicability. It is simple because only a relatively easy and inexpensive experimental arrangement is required. It is convenient, since little sample preparation is needed. Furthermore, it is broadly applicable to diverse types of samples (including "difficult" samples like opaque materials, aerosols, powders, flames, and so on), since nonradiative thermal relaxations occur very generally (even though they may only be partial in certain highly fluorescent or chemically active systems), while ion attachment, diffusion, recombinations, and quenching phenomena frequently impose serious limitations on ultra-sensitive detection schemes involving ionization or luminescence.

Most PT effects can be detected *directly* on or inside the irradiated sample, or *indirectly* in some coupling fluid adjacent to the sample due to thermal (or acoustic) coupling. A listing of these direct and indirect PT effects is given in Table 2. Generally, direct PT effects offer higher detectivity for weak absorption compared to indirect PT effects. On the other hand, indirect PT detection may be advantageous for cases when the sample is opaque, highly light-scattering, inhomogeneous, or layered. The purpose of this paper is to review signal enhancements and noise reduction for several common PT detection schemes, and consider how high sensitivity can be achieved.

2. Photoacoustic Detection

PA generation due to thermal expansion mechanisms can be classified as direct or indirect as defined earlier. PA generation can also be classified according to the two excitation modes: the continuous-wave (CW) modulation mode, whereby the excitation beam is modulated at near 50% duty cycle, and the pulsed mode, whereby the excitation beam is of very low duty cycle but high peak power. In the CW case, the signal is typically analyzed in the frequency domain; amplitude and phase of one or several Fourier components are measured and narrow-band filters can be used to suppress noise. In the pulsed technique, however, the signal is acquired and analyzed in the time domain, making simple gating techniques for noise suppression possible. The features of the two PA techniques are compared in Table 3. In pulsed PA measurements, the excitation pulse is typically short ($\lesssim \mu$sec) and the acoustic propagation distance during the excitation pulse is typically much smaller than the dimension of the sample; hence, in most cases, the PA pulse shape is independent of boundary reflections, and the sample can often be treated as infinite in extent. In CW modulated PA measurements, the modulation

Table 3. Comparison of CW modulated and pulsed PA techniques

	CW modulated	Pulsed
Feature of the modulation	high duty cycle, low peak power	low duty cycle, high peak power
Acoustic detection method	low frequency transducer with lock-in amplifier	high frequency transducer with boxcar or transient recorder
PA generation efficiency	low	high
Signal integration time	long	short
Thermal diffusion effects	may be important	usually negligible
Acoustic boundary conditions	important	unimportant
Unmodulated background heating	usually substantial	usually small

frequency is typically below several kHz, and the acoustic propagation distance during a period is typically much larger than the sample cell. In such cases, we must first determine the acoustic eigenmodes of the sample cell, and the excitation beam will excite an eigenmode to an amplitude depending on the overlap integral of the photothermal source with the eigenmode, and the Q-factor of the mode (see Tam [1]).

We consider in the following the parameters affecting signal and noise for a typical case of direct PA detection, and of CW-modulated indirect PA detection.

2.1 Direct PA Detection

Good detection sensitivity has been demonstrated using direct PA detection technique both in the pulsed excitation mode (pulse duration in the range of 10^{-8} to 10^{-6} sec) and in the CW-modulated excitation mode (excitation frequency in the range of $10^2 - 10^5$ Hz, usually at an acoustic resonance frequency). Here, we consider the factors that would optimize the signal for pulsed excitation. These considerations are also similar for the case of CW modulated excitation.

We consider here signal generation using a semi-quantitative theory for weak absorption measurement, as schematically shown in Fig. 2. The laser excitation pulse width τ_L is assumed short enough so that thermal diffusion is negligible (this usually means that the excitation pulse is much shorter than one millisecond). The initial expansion ΔR of a certain "PA source radius" R (to be defined later) immediately after the laser pulse is given by

$$\pi(R+\Delta R)^2\ell - \pi R^2\ell = \beta V_1 \Delta T\,, \tag{1}$$

with the temperature rise just after the laser pulse being

$$\Delta T = \frac{E\alpha\ell}{\rho V_1 C_p}\,, \tag{2}$$

where ℓ is the length of the PA source (assumed long), β is the expansion

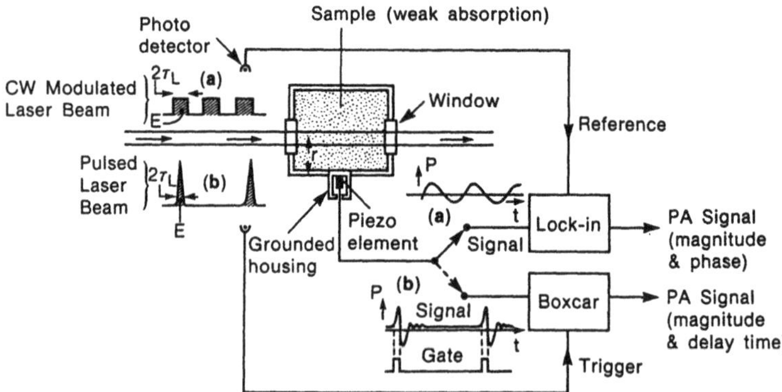

Figure 2. Direct PA detection in a weakly absorbing sample using either (a) the CW modulated excitation, or (b) the pulsed excitation.

coefficient, $V_1 = \pi R^2 \ell$ is the source volume, E is the laser pulse energy, α is the absorption coefficient (assuming $\alpha\ell << 1$), ρ is the density, and C_p is the specific heat at constant pressure. Combining Eqs. (1)-(2) and assuming $\Delta R << R$ (true in all cases we are considering), we have

$$\Delta R = \frac{\beta E \alpha}{2\pi\rho C_p R} . \tag{3}$$

The peak displacement U(r) at the observation point, at distance r from the PA source (for $r << \ell$), scales as $r^{-1/2}$ because of conservation of acoustic energy, as described by Landau and Lifshitz [2] for a cylindrical acoustic wave:

$$U(r) = \Delta R (R/r)^{1/2} = \frac{\beta E \alpha}{2\pi\rho C_p R^{1/2} r^{1/2}} . \tag{4}$$

The ratio of the peak acoustic pressure P(r) at position r to the acoustic displacement rate $U(r)/\tau$ is the acoustic impedance [3] $v\rho$

$$P(r) \approx v\rho U(r)/\tau , \tag{5}$$

where v is the acoustic velocity and τ is a characteristic width of the PA pulse (to be defined later).

Substituting Eq. (4) into (5), we derive the following expression for the peak PA pressure at the observation location r:

$$P(r) \approx \frac{\beta E \alpha v}{2\pi C_p \tau R^{1/2} r^{1/2}} \tag{6}$$

$$= \left[\frac{\beta E \alpha}{2\pi\rho C_p R} \right] \left[\frac{\rho v}{\tau} \right] \left[\frac{R}{r} \right]^{1/2} . \tag{6'}$$

The form of Eq. (6′) clearly indicates the various factors for the PA pressure generation: the three brackets indicate the thermo-elastic source displacement factor, the pressure/displacement ratio, and the cylindrical propagation factor.

Now we need to examine what are the meanings of the "characteristic width" τ and the "PA source radius" R. These depend on whether the laser beam radius R_0 is much smaller or much larger than the acoustic propagation distance $v\tau_L$ during the laser pulse. These cases are called "small-radius" limit and "large-radius" limit, respectively. The values to be assigned to τ and R for these two cases are given in Table 4. More general expressions [4] (true also when neither R_0 nor $v\tau_L$ is much larger than the other) are:

$$R = [(v\tau_L)^2 + R_0^2]^{1/2} , \tag{7}$$

$$\tau = [\tau_L^2 + (R_0/v)^2]^{1/2} , \tag{7'}$$

which are related by

$$R = \tau v . \tag{7''}$$

Table 4. Assigned values for R and τ for the small-radius limit ($R_0 << v\tau_L$) and for the large-radius limit ($R_0 >> v\tau_L$)

	Small laser radius for the PA source	Large laser radius for the PA source
"PA source radius" R	$v\tau_L$	R_0
"PA pulse width" τ	τ_L	R_0/v

Using Eq. (7″), we can re-write Eq. (6) into the following more compact form:

$$P(r) \approx \frac{\beta E \alpha v^{1/2}}{2\pi C_p \tau^{3/2} r^{1/2}} , \tag{8}$$

or equivalently,

$$P(r) \approx \frac{\beta E \alpha v^2}{2\pi C_p R^{3/2} r^{1/2}} . \tag{8'}$$

Limiting approximation for small PA source radius (i.e., $\tau \to \tau_L$) and for large PA source radius (i.e., $R \to R_0$) are easily obtained from Eqs. (8) and (8′), providing:

$$P(r) \xrightarrow[\text{source radius}]{\text{small PA}} P_s(r) \approx \frac{\beta E \alpha v^{1/2}}{2\pi C_p \tau_L^{3/2} r^{1/2}} \tag{9}$$

$$P(r) \xrightarrow[\text{source radius}]{\text{large PA}} P_\ell(r) \approx \frac{\beta E \alpha v^2}{2\pi C_p R_0^{3/2} r^{1/2}} . \tag{9'}$$

Comparing Eqs. (9) and (9′), we conclude [4] that

$$\frac{P_\ell(r)}{P_s(r)} = \left(\frac{v\tau_L}{R_0} \right)^{3/2} < 1 , \tag{10}$$

which shows that the PA pulse amplitude weakens quickly ($\sim R_0^{-3/2}$) as R_0 increases from the value $v\tau_L$ (when signal is largest). This is intuitively appealing, since for the large-radius case ($R_0 > v\tau_L$), the contributions from different radial positions in the source do not simply add up coherently because of the long acoustic transit time across the diameter.

The acoustic amplitudes $P_s(r)$ and $P_\ell(r)$ given in Eqs. (9) and (9′) resemble those given by Nelson and Patel [5] (1981) and by Naugol'nykh [6] (1977) in that all pressure amplitudes go as $\beta E\alpha/C_p$. While the different authors indicate different dependences on v, τ_L, R_0, and r, we believe that the forms given in Eqs. (9) and (9′) give the correct functional dependence on the various thermoelastic, ultrasonic and geometrical parameters. They are consistent with rigorous theoretical treatments [4] They can also be used approximately for the CW modulated excitation case (assuming heat diffusion is negligible) with $4\tau_L$ being changed to the modulation period and $E/2\tau_L$ being changed to the intensity. They also provide the following guidance towards signal enhancement for measuring weak absorption:

- Increase expansion coefficient β (11a)

- Increase laser energy E (11b)

- Increase acoustic velocity v (11c)

- Decrease specific heat C_p (11d)

- Decrease laser pulse duration τ_L (11e)

- Decrease laser beam radius R_0 (11f)

- Decrease observation distance r (11g)

However, there are certain limitations to the above recipe for "signal enhancements." The first limitation is due to non-linear absorption and breakdown of the sample, thus limiting the peak laser intensity of $E/(\pi R_0^2 \tau_L)$. The second limitation is due to the "cross-over" between $(v\tau_L)$ and R_0 (Eqs. 7 and 7') which means that, for example, reducing an initially large R_0 is no more useful once it passes below $v\tau_L$ if τ_L is not reducible. The third limitation is due to the requirement of high detection bandwidth when both τ_L and (R_0/v) are small, i.e., τ is small according to Eq. (7'); since the PA pulse is really "bi-polar," i.e., composed of a compression "peak" followed by a rarefaction "trough" approximately 2τ later [4], a detection bandwidth of at least $1/(4\tau)$ is required, or otherwise the compression part would partially cancel the rarefaction part of the detected signal.

2.2 CW-modulated indirect PA Detection

Indirect PA detection (schematically shown in Fig. 3) using a coupling fluid (usually a gas) to monitor the PT absorption by the sample is usually done in conjunction with CW-modulated excitation (rather than pulsed excitation) because microphones of limited bandwidth are usually employed. Consider the case of an excitation beam of radius r, modulated at frequency f, incident on the sample of thickness ℓ, thermal conductivity k_s and diffusivity D_s in a cylindrical cell of radius R and coupling gas thickness ℓ_g. Let the sample optical attenuation coefficient be α at the excitation wavelength, and the optical penetration length be $\mu_\alpha = 1/\alpha$. The sample is assumed to be mounted on a *long transparent* backing material of thermal conductivity k_b and diffusivity D_b. Let the laser beam intensity just below the sample surface be represented by

$$I(t) = I_0(1 + \sin 2\pi f t) . \tag{12}$$

The corresponding thermal diffusion lengths μ_s and μ_b in the sample and backing materials are given by

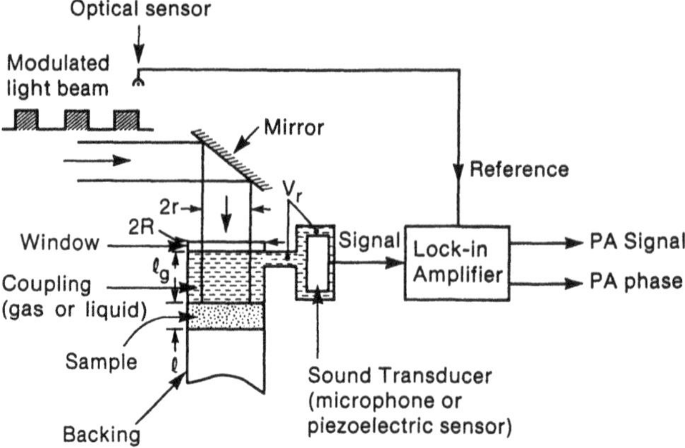

Figure 3. Indirect PA detection using a transparent coupling fluid in contact with the sample.

$$\mu_s = (D_s/\pi f)^{1/2} \tag{13}$$

$$\mu_b = (D_b/\pi f)^{1/2}\,. \tag{13'}$$

The smallest of the three lengths μ_s, ℓ, and μ_α defines the fraction of the modulated laser intensity I_0 that effectively produces a modulated surface temperature of amplitude θ_0. If μ_s is the shortest (thermally thick limit), this fraction is $\alpha\mu_s$. If ℓ is the shortest (thin sample limit), this fraction is $\alpha\ell$. If μ_α is the shortest (optically thick limit), this fraction is $\alpha\mu_\alpha = 1$. Hence, in general, the effective modulated heat source is

$$H_m \approx I_0\{\alpha\mu_s, \alpha\ell, 1\}_{min} \tag{14}$$

where the notation $\{\ \}_{min}$ indicated the selection of the smallest quantity inside the braces.

Now, the modulated heat source H_m produces a modulated sample surface temperature θ_0 which drives the heat conduction either into the sample or into the backing, depending on whether $\mu_s < \ell$ or $\ell < \mu_s$ respectively:

$$H_m \approx \theta_0 \frac{k_{s,b}}{\mu_{s,b}} \tag{15}$$

where the subscripts for k and μ are to be s if $\mu_s < \ell$, and to be b if $\ell < \mu_s$. Combining Eqs. (14) and (15), we have

$$\theta_0 = I_0 \frac{\mu_{s,b}}{k_{s,b}} \{\alpha\mu_s, \alpha\ell, 1\}_{min}\,. \tag{16}$$

This sample surface temperature modulation θ_0 is thermally coupled to an active volume V_{act} of the gas, given by

$$V_{act} \approx \pi r^2 \{\mu_g, \ell_g\}_{min} \tag{17}$$

where μ_g is the gas thermal diffusion length and l_g is the gas column thickness. Using the ideal-gas law, the amplitude δV of the volume change of V_{act} is

$$\delta V = V_{act}\theta_0/T\,, \tag{18}$$

where T is the absolute temperature. Now the volume fluctuation δV causes a pressure fluctuation δP at the microphone. Assuming the adiabatic pressure-volume relation, we have

$$\delta P = \gamma P \delta V/V_0\,, \tag{19}$$

where γ is the ratio of the specific heats and V_0 is the total cell volume, given by

$$V_0 = \pi R^2 \ell_g + V_r\,, \tag{20}$$

where V_r is the residual volume in the PA cell (Tam and Wong [7]; Korpiun et al. [8,9]). Combining Eqs. (17)-(20), we obtain the PA amplitude δP as

$$\delta P \approx \frac{\gamma P r^2 \{\mu_g, \ell_g\}_{min}}{T(R^2 \ell_g + V_r/\pi)} \theta_0 \tag{21}$$

and using Eq. (16) for θ_0, we have

$$\delta P \approx \frac{\gamma P r^2 I_0 \{\mu_g, \ell_g\}_{min} \{\alpha\mu_s, \alpha\ell, 1\}_{min}}{T(R^2 \ell_g + V_r/\pi)} \frac{\mu_{s,b}}{k_{s,b}}\,. \tag{22}$$

Equation (22), which is true for all the different cases of thermally thick limit, thin sample limit, or optically thick limit, is consistent with the early mathematical derivation of Rosencwaig and Gersho [10] who only considered the conditions of $\ell_g >> \mu_g$ and $V_r << \pi R^2 \ell_g$. Equation (22) provides the following recipe for signal enhancement for a fixed sample absorption coefficient α:

- Increase the ratio of specific heats of the coupling gas γ (22a)
- Increase pressure P (22b)
- Increase absorbed power ($\pi r^2 I_0$) (22c)
- Increase diffusion lengths μ_g, μ_s or μ_b (22d)
- Increase sample thickness ℓ (22e)

- Decrease temperature T (22f)

- Decrease cell radius R (22g)

- Decrease gas column thickness ℓ_g (22h)

- Decrease residual cell volume V_r (22i)

- Decrease thermal conductivity k_s or k_b (22j)

Again, some limitations on the above signal-enhancement scheme are worth considering. First, the thermal diffusion lengths μ_s or μ_b are best increased (22d) by reducing the corresponding density or specific heat, or the chopping frequency f (which also increases μ_g) rather than by increasing the thermal conductivity since this would contradict recommendation (22j). Secondly, increasing sample thickness ℓ is useful only as long as it is smaller than μ_s and $1/\alpha$. Thirdly, when ℓ_g is decreased (say, from a large initial value), this results in a signal enhancement until $\ell_g \approx \mu_g$ (more accurately [7], when $\ell_g = 1.8\,\mu_g$); further reduction in ℓ_g has the opposite effect of degrading the signal because of heat conduction at the front window.

2.3 Other Factors for Signal Enhancement

We have considered in the above the factors that would enhance thermo-elastic generation of the PA signal, either directly or indirectly. There are other tricks that can be used besides the optimization of thermoelastic, acoustic or geometrical parameters. Some of these are listed below.

First, the useful optical intensity can be increased. For example, for a highly reflective sample, its reflectivity could be reduced (e.g., by refractive index matching at the incidence location). For a weakly absorbing sample, it could be put in a "multi-pass" optical-beam configuration [11] or even "intra-cavity," i.e., inside the excitation laser cavity [12,13].

Secondly, acoustic detection efficiency can be enhanced by using a detector with high sensitivity and good acoustic impedance matching to the medium in which it is located. Matched detectors used in high-sensitivity PA detection of both pulsed and CW modulated types include microphones and probe-beam deflection in gaseous samples (which have low acoustic impedance), piezoelectric transducers in condensed samples (which have high acoustic impedance), and surface interdigital transducer for surface absorbance [14] (with interdigital spacing matching the half-wavelength of the surface wave). Furthermore, proper choice of geometry and excitation frequency can provide acoustic wave "focusing" effects [15] or acoustic gas-resonance effects [13,16-19] — these can frequently increase the signal amplitude by more than an order of magnitude. Acoustic resonances of different vibrational or "drum" modes of the sample (if suitably mounted) can also greatly increase the signal.

Thirdly, the samples can be chosen or specially treated to provide so-called "PA amplification." For example, if the sample is photo-chemically active with an associated volume change or heat release, this can greatly enhance [20,21] the direct or indirect PA signal amplitude. Also, if the sample contains an adsorbed volatile species, its indirect PA signal strength can be greatly enhanced [22,23] due to gas desorption. Furthermore, PA amplification (or demagnification) could also occur when the sample is on the borderline of some phase transition by proper choice of the ambient conditions (e.g., temperature, pressure, magnetic field, etc.).

2.4 Noise Consideration

The improvement of detection sensitivity always hinges upon two factors besides signal averaging; increasing the signal and decreasing the noise. We already discuss how to increase the signal. Here, we summarize various common types of noise encountered in pulsed and CW modulated PA detection, as well as in many other PT detection schemes, as listed in Table 5 together with some schemes for their reduction. The noise sources are listed in order of decreasing magnitude typically. Many experimental papers have described how to reduce some of the larger noise sources. Fundamental noise limits, which include Brownian motion noise, and statistical fluctuations of the excitation and deexcitation processes, are seldom approached.

For many cases, sensitivity is limited by "unwanted absorption" at "unwanted places" producing an "unwanted signal" which is called "noise." These unwanted places are usually the transducer surface, windows, cell walls, or sample substrates or holders, irradiated by direct, reflected, or scattered excitation light. Effects due to scattered light on transducer surface or cell walls can be reduced by making these surfaces highly reflecting and thermally conductive [24,25] and by proper time-gating for pulsed excitation [24-26]. Effects due to absorption at windows which are frequently exposed to the full excitation beam can be reduced by Brewster window geometry [13], muffler [19], window elimination [27], time-gating for pulsed excitation [24-26],

Table 5. Possible noise sources and corresponding noise reduction schemes

Origin of noise	Noise reduction	Noise more serious for either pulsed or CW modulated excitation
1. Undesired absorption (transducer, window, cell wall, or substrate)	Proper choice of transducer location and shielding, window location or no window, cell wall material, bandpass or time-gate, and modulation technique for background subtraction	CW-modulated
2. Laser intensity or energy fluctuation	Normalization	Pulsed
3. Vibration and acoustic disturbance	Vibration isolation; avoid ambient noise peaks; go to higher frequency	CW modulated
4. rf interference (i.e., electromagnetic pick-up)	Shield transducer and electronics; minimize transducer cable length; avoid ground loops and line power	Pulsed
5. Amplifier noise	Optimize amplifier	Both
6. Electrostriction	Avoid excessive intensity; signal shape is different from thermo-elastic case	Pulsed
7. Laser pointing noise	Laser stabilization; minimize turbulence	Both
8. Johnson noise	Reduce temperature and bandwidth	Both
9. Fundamental noises (e.g., Brownian motion, deexcitation probabilities, statistical fluctuations)	Not avoidable—represent the ultimate limits	Both

differential excitation of a sample cell and reference cell [18], and various modulation schemes, e.g., Stark-modulated excitation [28]. Similar techniques are possible to reduce substrate-absorption effects; in particular, schemes to modulate the irradiation location, excitation wavelength, or polarization [29].

From the above discussions, we can see that whether the pulsed type or the CW modulated type provides a higher detection sensitivity depends on the details of the experimental arrangement and the origins of the noise; both types have been used to demonstrate high sensitivities. The main advantages of CW modulated detection are the high duty cycle for excitation and the sensitivity of lock-in detection, and so under circumstances when random noise dominates, CW modulated detection is frequently preferable. On the other hand, when systematic noise dominates (like noise due to window, PA cell wall, or substrate absorption), pulsed PA detection is usually preferred, since it provides strong noise suppression by time-gating of the PA signal; the trade-off here is that the time gate is typically short, and the pulse repetition rate is generally low, so that the actual "signal-gathering time" is typically much shorter than the CW modulated case.

3. PT Probe-Beam Refraction

There are several varieties [30] of what we call here PT probe-beam refraction (PBR) schemes, which rely on the use of a probe laser beam (usually CW) to monitor the refractive index gradient generated inside or outside a sample irradiated by an excitation beam; these are named "direct" or "indirect" PT refraction signals respectively. If the probe beam is collinear with the excitation beam, the technique is generally called "thermal lensing" measurement. In general, the probe beam can be parallel to but displaced from the excitation beam (to be called parallel PBR) or at any angle with respect to the excitation beam. For the purpose of considering parameters to enhance the signal, we consider here only the parallel PBR case. These considerations are similar for other angles and various extent of overlap between the probe and excitation beam. probe-beam-refraction configuration.

3.1 Signal

We derive here theoretically the parallel PBR deflection angle ϕ (Fig. 4), assuming short pulsed excitation and fast thermal de-excitation, and a narrow probe beam at a radial distance r from the axis of the excitation beam with a Gaussian profile. The energy density E(r) at r is written as:

$$E(r) = \frac{2E_0}{\pi a^2} \exp(-2r^2/a^2) \,, \tag{23}$$

where E_0 is the incident pulse energy of the excitation beam, a is the excitation beam $(1/e^2)$-radius. The corresponding temperature gradient at r immediately after the short pulsed excitation is

$$\frac{dT}{dr} = \frac{-8E_0\alpha}{\rho C_p \pi a^2} \frac{r \exp(-2r^2/a^2)}{a^2} \,, \tag{24}$$

where ρ = density, C_p = specific heat, and α = absorption coefficient

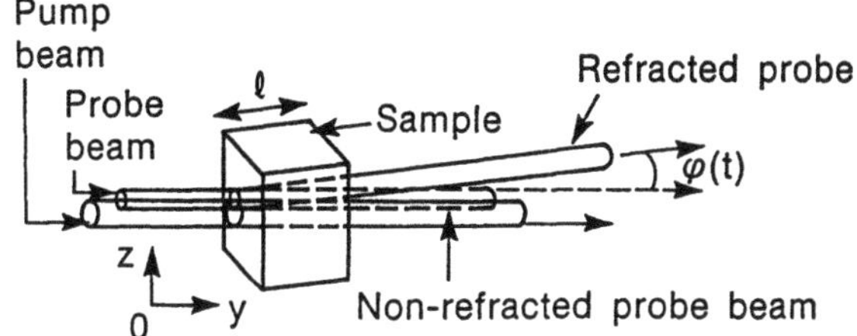

Figure 4. Typical parallel probe-beam-refraction configuration.

(assumed small). For small probe deflection ϕ, geometrical optics gives

$$\frac{\phi n_0}{\ell} = \frac{dn}{dr} = \frac{dn}{dT}\frac{dT}{dr} , \tag{25}$$

where ℓ is the optical path length, and n is the sample refractive index of ambient value n_0. Combining Eqs. (24) and (25), we have

$$\phi(r) = \frac{\ell}{n_0}\frac{dn}{dT}\frac{(-8E_0\alpha)}{\rho C_p \pi a^2}\frac{r\exp(-2r^2/a^2)}{a^2} . \tag{26}$$

Note that the parallel PBR angle $\phi(r)$ given in Eq. (26) for the probe located at displacement r from the pump beam axis is derived assuming the following conditions: (a) weak absorption, (b) short excitation duration, (c) narrow probe beam, (d) sample thermal de-excitation time τ is small, (e) no thermal diffusion (i.e., observation shortly after excitation) and (f) no flow of the sample. Equation (26) is basically the same as given by Boccara et al. [31] or by Rose, Vyas, and Gupta [32]. Equation (26) suggests the following signal enhancement schemes for detecting weak absorption α:

- Increase interaction length ℓ (27a)
- Increase the temperature coefficient of the refractive index dn/dT (27b)
- Increase excitation pulse energy E_0 (27c)
- Decrease density ρ (27d)
- Decrease specific heat C_p (27e)
- Decrease excitation beam radius a (27f)
- Locate the probe beam at $r = a/2$ (27g)

The effect of thermal diffusion is not included in the above discussion. It has the effect of making the excitation beam radius a larger:

$$a^2 \rightarrow a^2 + 8Dt\,. \tag{28}$$

The signal $\phi(r,t)$ at time t is obtainable from the expression $\phi(r)$ in Eq. (26) for time 0 by the substitution in Eq. (28).

While much of the earlier PBR work is done in air or in a gas [30] and good sensitivity and useful applications have been demonstrated [33], the PBR signal can be much enhanced in a liquid coupling medium [34,35] (27b). Also, while angular deflection can be detected by using a knife-edge, aperture or split-detector, novel schemes like an etalon analyzer [36] can be useful to enhance the sensitivity.

3.2 Noise

Various noise sources are possible for PBR monitoring of PT refractive-index gradients. Some are similar to the PA case: undesired absorption, excitation laser intensity or pulse fluctuations, vibration, rf interference, amplifier noise, electrostriction, probe- and excitation-laser pointing stability, probe beam "shot noise" and other fundamental noise terms. Usually, the main source of noise encountered in PBR (as in the PA case) is "unwanted absorptions" by windows or by background fluids. The effects of background absorption can again be reduced by modulation techniques when the "sample molecules" are sensitive to the modulation while the "background molecules" are insensitive. Such modulations can be, e.g., wavelength modulation [37] or polarization modulation [38], as for the PA detection case. On top of these noises, there is also the effect of probe beam wandering or modulations due to any vibrations, turbulence, convection currents, gas flow, or plasma oscillation inside the probe laser. These "probe-beam-related" noises are frequently also dominating noise contributions; thus sensitive PBR detection really requires the use of a highly stable probe beam and good control of any unwanted beam-wandering. Vibration effects can be reduced by vibrational isolation, e.g., using alternating stacks of heavy metal plates and air-bags [39].

Conclusion

This review paper provides physical insights into the question of signal generation and hence signal enhancement for several typical PT detection schemes, namely, direct and indirect PA detection, and probe-beam refraction detection. These semi-quantitative yet analytic and intuitively appealing results agree with more rigorous earlier analysis [4,30]. The noise sources for these measurement configurations are discussed. Examples are selected from the literature to illustrate some of the concepts. Sensitivities achieved for PT spectroscopy of bulk materials are typically in the range of 10^{-5} to $10^{-10}\mathrm{cm}^{-1}$, with higher detectability typically possible when some signal/noise enhancement techniques as discussed here are utilized. Surface absorption as small as 10^{-9} has also been demonstrated [14].

Recent examples of practical applications of PT trace detections are given by Meyer and Sigrist [40] for atmospheric pollutants, Oehler and Blum [41] for plant gas-exchange, and Bicanic and co-workers [42-45] for innovative use in agriculture. Good sensitivity for PT detection is useful not only for chemical analysis of weak absorptions and trace constituents, but also for detections of small objects like thin films [29], aerosols [19] or even single particles [46-48].

References

1. A. C. Tam, in *Ultrasensitive Laser Spectroscopy* Chapter 1, edited by D. Kliger (Academic Press, New York, 1983).
2. L. D. Landau and E. M. Lifshitz, *Fluid Mechanics* Chapter VIII, translated by J. B. Sykes and W. H. Reid (Pergamon, Oxford, 1959).
3. J. Blitz, *Fundamentals of Ultrasonics*, Chapter I, (Butterworths, London, 1963).
4. A. C. Tam, *Rev. Mod. Phys.* **58**, 381, 1986.
5. E. T. Nelson and C. K. N. Patel, *Opt. Lett.***6**, 354, 1981.
6. K. A. Naugol'nykh, *Sov. Phys. Acoust.* **23**, 98, 1977.
7. A. C. Tam and Y. H. Wong, *Appl. Phys. Lett.* **36**, 471, 1980.
8. P. Korpiun, B. Büchner, A. C. Tam and Y. H. Wong, *Appl. Phys. Lett.* **46**, 1039, 1985.
9. P. Korpiun, B. Büchner, A. C. Tam and Y. H. Wong, *J. Appl. Phys.* **59**, 2944, 1986.
10. A. Rosencwaig and A. Gersho, *J. Appl. Phys.* **47**, 64, 1976.
11. K. P. Koch and W. Lahmann, *Appl. Phys. Lett.* **32**, 289, 1978.
12. R. G. Bray and M.J. Berry, *J. Chem. Phys.* **71**, 4909, 1979.
13. F. J. M. Harren, F. G. C. Bijnen, T. Reuss, L. A. C. J. Voesenek and C. W. P. M. Blom, *Appl. Phys.* **B-50**, 137, 1990.
14. M. Y. A. Raja, D. W. Reicher, S. R. J. Brueck, J. R. McNeil and D. E. Oates, *Opt. Lett.* **15**, 66, 1990.
15. H. Sontag and A. C. Tam, *Appl. Phys. Lett.* **46**, 725, 1985.
16. P. Hess, in *Topics in Current Chemistry*, edited by F. L. Boschke, Vol. 111 (Springer, Berlin, 1983).
17. M. W. Sigrist, S. Bernegger and P. L. Meyer, *Infrared Phys.* **29**, 805, 1989.
18. S. Bernegger and M. W. Sigrist, *Infrared Phys.* **30**, 375, 1990.
19. M. M. Malley and D. J. Golich, *Bull. Am. Phys. Soc.* **35**, 1526, July/August 1990.
20. R. R. Chance and M. L. Shand, *J. Chem. Phys.* **72**, 948, 1980.
21. G. J. Diebold and J. S. Hayden, *Chem. Phys.* **49**, 429, 1980.
22. P. Ganguly and T. Somasundaram, *Appl. Phys. Lett.* **43**, 160, 1983.
23. J.-P. Monchalin, L. Bertrand, G. Rousset and F. Lepoutre, *J. Appl. Phys.* **56**, 190, 1984.
24. C. K. N. Patel and A. C. Tam, *Rev. Mod. Phys.* **53**, 517, 1981.
25. J. V. Beitz, M. M. Doxtader, V. A. Maroni, S. Okajima and D. T. Reed, *Rev. Sci. Instr.* **61**, 1395, 1990.
26. P. V. Cvijin, D. A. Gilmore and G. H. Atkinson, *Appl. Spectrosc.* **42**, 770, 1988.
27. M. W. Sigrist, *J. Appl. Phys.* **60**, R83, 1986.
28. H. Sauren, D. Bicanic, H. Jalink, and J. Reuss, *J. Appl. Phys.* **66**, 5085, 1989.
29. H. Coufal, F. Träger, T. J. Chuang and A. C. Tam, *Surf. Sci.* **145**, L504, 1984.
30. A. C. Tam, in *Photothermal Investigations of Solids and Fluids* Chapter 1, edited by J. A. Sell, (Academic Press, New York, 1988).
31. A. C. Boccara, D. Fournier, W. Jackson and N. M. Amer, *Opt. Lett.* **5**, 377, 1980.
32. A. Rose, R. Vyas and R. Gupta, *Appl. Opt.* **24**, 4626, 1986.
33. J. A. Sell, D. M. Heffelfinger, P. Ventzek and R. M. Gilgenbach, *Appl. Phys. Lett.* **55**, 2435, 1989.
34. S. Nonomura, H. Hayashi and S. Nitta, *Rev. Sci. Instr.* **60**, 657, 1989.

35. A. Skumanich and J. C. Scott, Mol. Cryst. Liq. Cryst. **183**, 365, 1990.
36. S. E. Bialkowski and Z. F. He, *Anal. Chem.* **60**, 2674, 1988.
37. C. D. Tran and M. Franko, *J. Phys. E. Sci. Instrum.* **22**, 586, 1989.
38. C. D. Tran and M. Xu, *Rev. Sci. Instr.* **60**, 3207, 1989.
39. A. C. Tam and H. Schroeder, *IEEE Trans. Mag.* **25**, 2629, 1989.
40. P. Meyer and M. W. Sigrist, *Rev. Scien. Instrum.* **61**, 1779 (1990).
41. O. Oehler and H. Blum, *Springer Series in Optical Sciences* **62**, (Springer Verlag Berlin, 1990), p. 369.
42. D. Bicanic, B. Zuidberg, H. Jalink, A. Miklos, K. Hartmans and A. van Es., *Appl. Spectr.* **44**, 263 (1990).
43. F. Harren, J. Reuss, E. Woltering and D. Bicanic *Appl. Spectr.* **44**, 1360 (1990).
44. H. Sauren, D. Bicanic, W. Hillen, H. Jalink, K. van Asselt, J. Quist and J. Reuss, *Appl. Optics* **29**, 2679 (1990).
45. E. Strauss, J. P. Favier, D. Bicanic, K. van Asselt and M. Lubbers, *Analyst* **117**, 77 (1991).
46. H. B. Lin and A. J. Campillo, *Appl. Opt.* **24**, 422, 1985.
47. H. Sontag and A. C. Tam, *Can. J. Phys.* **64**, 1121, 1986.
48. M. A. Taubenblatt, *Appl. Phys. Lett.* **52**, 951, 1988.

Analytical Thermal Lens Spectrometry: Past, Present and Future Prospects

C.D. Tran

Department of Chemistry, Marquette University, Milwaukee, WI 53233, USA

1. Introduction

Thermal lens techniques have been demonstrated to be sensitive methods for low absorbance measurements. They can be used for the determination of absorptivities as low as 10^{-7}. Although an increasing number of applications of thermal lens techniques to chemical analysis have been reported during the past few years, the methods are still relatively underutilized compared to the other spectral methods. This is surprising inasmuch as the techniques offer substantial advantages over conventional spectral methods with respect to noncontacting, in-situ capability, precision, sensitivity, expanded dynamic range and minimum sample required.

The limited use of the thermal lens techniques is probably due to a combination of factors including the low selectivity and the lack of efficient and systematic methods to fully exploit the unique features of the techniques. In this overview, initially, an instrumental approach will be described in which the development of novel instruments which have higher selectivity, namely the multiwavelength thermal lens spectrophotometers and the thermal lens-circular dichroism spectropolarimeters, will be will be delineated. Subsequently, the exploitation of organized media such as reversed micelles, crown ethers to improve the thermal physical properties of the environment around the solute will be reported. It will be demonstrated that the improvement in the thermal physical properties induced by these organized media will provide substantial enhancement in the sensitivity as well as the selectivity of the thermal lens techniques. Finally, the future of these techniques will be forecasted. Elaboration will be specifically focused on the anticipated synergistic use of these newly developed instruments and methods for the direct and in realtime determination of trace chemical species of small volume (μl), multicomponent samples whose concentration can be as low as 10^{-11} M; and the exploitation of their unique features for the determination of data which are currently unobtainable using existing techniques.

2. Instrumentation Development

As mentioned earlier, in spite of their advantages, the thermal lens techniques are not as widely used as other spectrochemical methods. This is mainly due to the lack of selectivity. In fact, the majority of published work on the thermal lens techniques are based on the use of instruments which have only one excitation wavelength. The use of only a single wavelength severly hinders the use of these promising techniques for the identification of unknown samples as well as for the analysis of mixtures of compounds [1-9]. It is known that one of the most powerful means by which to increase the selectivity is to measure absorption at two different wavelengths simultaneously. This is because the ratio of the absorption at two wavelengths will provide fingerprints and identification of a substance in a mixture, thereby introducing selectivity into the technique. Based on such considerations we have recently developed the first dual wavelength pump/probe configuration thermal lens spectrometer which is capable of simultaneously measuring the thermal lens signal at two different wavelengths [10-13]. As shown in Figure 1, in this instrument, the two excitation beams were derived from the same argon ion laser operating in the multiline mode. The sample was excited by these two wavelengths alternatively and the corresponding thermal

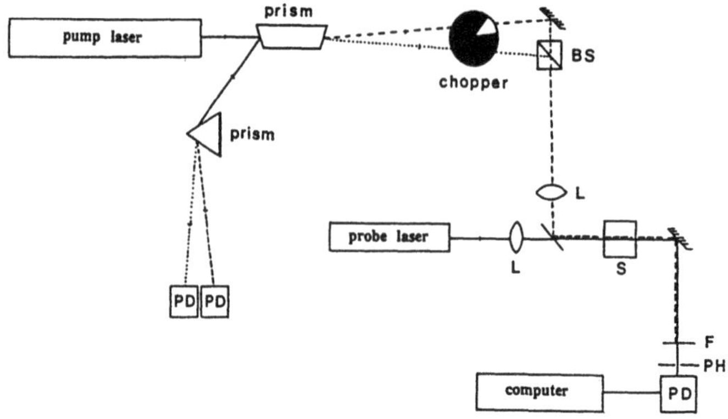

Figure 1. Schematic diagram of the double beam, dual wavelength thermal lens spectrometer: B.S., beamsplitter; L, lens; S, sample; F. filter; pH, pinhole; PD, photodiode.

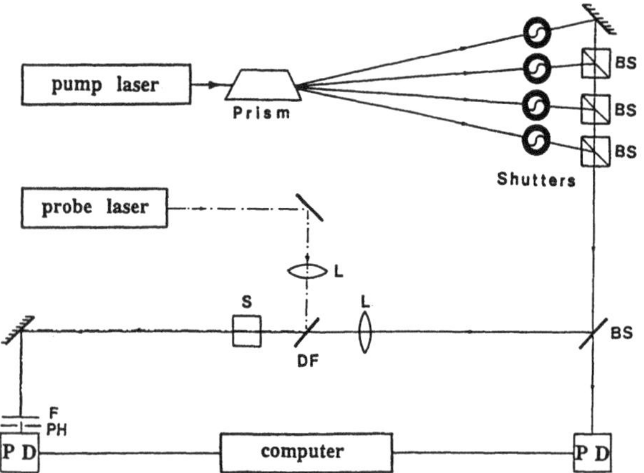

Figure 2. Schematic diagram of the four-color thermal lens spectrophotometer: B.S., beamsplitter; L, lens; D.F., dichroic filter; S, sample; F, filter; PD, photodiode.

lens signals were monitored by a He-Ne probe laser. Compared to single wavelength techniques, the advantage of this dual wavelength apparatus includes its improved selectivity. Furthermore, it enables the determination of trace chemical species in the presence of interfering species present at much higher concentrations. Using this apparatus, the detection limit for praseodymium ion in the presence of 1.0×10^{-2} M concentration of nickel glycinate complexes (25 mW excitation beams modulated at 1 Hz) is estimated to be 4.7×10^{-5} M[7] [10]. The dual wavelength capability also enables this spectrometer to be used for the detection of two component mixtures as well as to determine the pHs of solutions at very low indicator concentrations. In fact, metal ions such as iron(II) bipyridyl and ferroin as well as lanthanide ions in two component mixtures were determined at concentration as low as 10^{-9} M. The solution pHs, ranging from 3 to 10, were also determined accurately by this apparatus utilizing such indicators as phenol red, methyl orange and thymol blue whose concentrations were in order of 10^{-9} M [13].

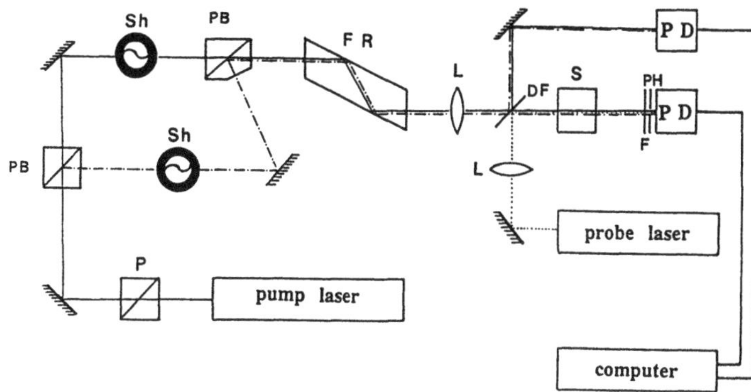

Figure 3. Schematic diagram of the thermal lens circular dichroism spectropolarimeter: P, polarizer; PB, polarizing beamsplitter; Sh, electronic shutter; DF, dichroic filter; FR, Fresnel rhomb; L, lens; S, sample; PD, photodiode; PC, microcomputer.

As can be gleaned from these results, the use of multiwavelength excitation techniques introduces much needed selectivity into the system. The induced selectivity is expected to be directly proportional to the number of excitation wavelengths. We have, in fact, demonstrated recently that samples which have up to four different components can be quantitatively determined in one measurement without any pretreatment, by use of the four wavelength thermal lens spectrophotometer which was recently developed in our laboratory (Figure 2) [14].

Selectivity enhancement can also be achieved by varying the polarization of the excitation beam. The use of left circularly polarized laser light (LCPL) and right circularly polarized laser light (RCPL) for excitation is of particular interest since it enables the measurement of circular dichroism (CD). As shown in Figure 3, in the thermal lens-circular dichroism spectropolarimeter (TL-CD) which we developed recently, the two excitation beams, LCPL and RCPL, were derived from the same argon ion laser [15,16]. The chiral sample was sequentially excited by these two beams, and the corresponding thermal lens signals were monitored by a He-Ne probe laser collinearly overlapping or perpendicularly intersecting with the two pump beams inside the sample. The apparatus is about three orders of magnitude more sensitive than the conventional circular dichroism spectropolarimeter. In addition, it can be used to detect chiral samples having very small volumes. A detection limits of 5 ng of optically active $[Co(en)_3]^{3+}$ complexes whose volume as small as 8 μl has been estimated for this apparatus using 37 mW excitation power modulated at 1.1 Hz.

The small sample capability of the TL-CD spectropolarimeter enables it to be successfully used for the chiral detection of high performance liquid chromatography (HPLC) [17]. In this TL-CD-HPLC chiral detector which is shown in Figure 4, the chromatographic effluent was sequentially excited by LCPL and RCPL; both of these excitation beams were derived from the same argon ion laser whose linearly polarized output was transformed into CPL by means of a Pockels cell. The heat generated as a consequence of the sample absorption of the LCPL and RCPL was measured by the He-Ne probe laser beam collinearly overlapping with the two excitation beams. A lock-in amplifier was used to measure the TL-CD signal which corresponds to the difference in the thermal lens signals produced by the LCPL and RCPL excitation beams. In addition to its high sensitivity, the advantages of this TL-CD chiral detector include its ability to provide, for the first time, directly and in real time, information on the chirality (i.e., circular dichroism) and optical purity of chiral samples. A detection limit of 7.2 ng was achieved for (-)-tris(ethylenediamine)cobalt(III) (k' = 0.45) as well as for the (+)-tris-(ethylenediamine)cobalt(III) (k' = 1.40) when these two enantiomers were chromatographically separated from the corresponding racemic mixture through the use of bis(μ-d-tartrato)diantimonate(III) ion pair reversed phase chromatography. This detection limit

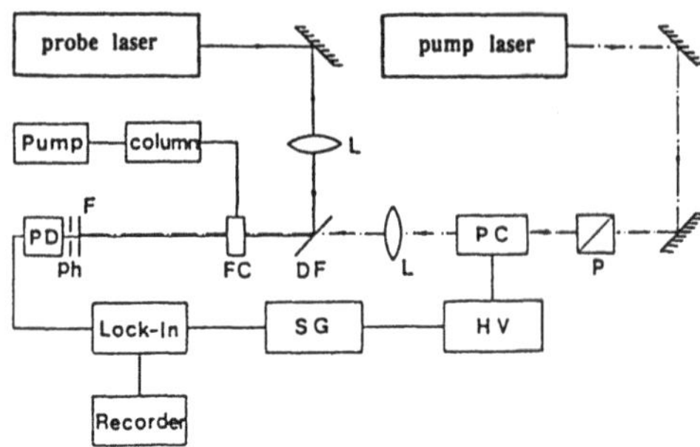

Figure 4. Schematic diagram of the thermal lens-circular dichroism detector for HPLC: P, prism polarizer; PC, Pockels cell; HV, high voltage power supply; SG, signal generator; L, lens; DF, dichroic filter; FC, microflow cell; F, interference filter; PH, pinhole; PD, photodiode.

was found by using a 10 μl flow cell (5 mm path length) and 6 mW excitation laser beam (λ = 514.5 nm) modulated at 2 Hz [17].

3. Chemically Induced Enhancement Methods

For weak absorbing species, the thermal lens signal, which is measured as the relative change in the laser beam center intensity in the far field, $\Delta I_{bc}/I_{bc}$, is related to the excitation laser power and sample absorbance A by

$$\frac{\Delta I_{bc}}{I_{bc}} = \frac{1.206\,P(dn/dT)A}{\lambda k} \tag{1}$$

where λ is the wavelength, and dn/dT and k are the temperature coefficient of the refractive index and thermal conductivity of the solvent, respectively. It is thus clear that in addition to the sample absorbance and laser excitation power, the thermal lens signal intensity is also dependent on the thermal physical properties of the solvent, namely dn/dT and k values of the solvent. As shown in eq. 1, for the same analyte concentration, higher thermal lens signal intensity can be achieved by choosing a solvent with high dn/dT and low k values. Aqueous solutions are the worst medium for thermal lens measurements owing to water's low dn/dT and high k values. Generally, non polar solvents such as CCl_4 and hydrocarbons are good thermo-optical solvents because they have high dn/dT and low k values [18-21]. In fact, it has been calculated and experimentally verified by us that at the same laser intensity, thermal lens measurements in CCl_4 and n-pentane are 38 and 40 times higher than those in water [4,5]. The dependency of the thermal lens signal on the thermal physical properties of solvent can, in principle, be exploited to enhance the sensitivity and selectivity of the technique. One such possibility is based on the used of organized assemblies such as micelles, reversed micelles and crown ethers to modify the thermal physical properties, namely dn/dT and k values, of the environmental around the analyte in order to enhance its thermal lens signal intensity.

Reversed Micelles: Amphiphilic surfactant molecules form micelles in aqueous solution and reversed micelles in nonpolar solvent. In the reversed micelle, the surfactant's polar headgroups outline a water soluble polar core of the micelle shielded by the nonpolar hydrocarbon tails [22-27]. Consequently, the reversed micelle contains a small water pool in which a water soluble analyte can be dissolved. Because the size of the water pool

Table I. Relative Thermal Lens Signal Intensity of Tb^{3+} in Different Media

Surfactant[a]	Solvent	
	water	1
	dioxane	22
AOT	dioxane	25
AOT	n-nonane	15
AOT	n-heptane	21
AOT	n-pentane	23
AOT	cyclohexane	20
AOT	CCl_4	38

[a]Concentration of surfactant used were 0.20 M.

containing the dissolved analyte is relatively small, the thermo-optical properties of the micellar solution are expected to be similar to those of the bulk nonpolar solvent. As a consequence, the solubilization of such analytes in reversed micelle is expected to provide an enhancement in the thermal lens compared to that possible in bulk water alone.

Table I shows the effects of solvents and reversed micelles on the thermal lens signal intensity of terbium chloride (4). The lanthanide ion was chosen because of its high hydrophilicity: it is very soluble in water, sparingly soluble in polar solvents and insoluble in nonpolar solvents [4]. As expected, the thermal lens signal amplitude of Tb^{3+} in methanol is 10 times higher than that in water. The enhancement is further increased to 15 times when Tb^{3+} is in 1-butanol. Substantial enhancement was found, however, when Tb^{3+} is solubilized in AOT reversed micelles. Sodium bis (2-ethyl hexyl) sulfosuccinate or Aerosol-OT (AOT) was used to investigate the effect of reversed micelles because this anionic surfactant can form reversed micelles in a variety of nonpolar solvents and can solubilized large amounts of water [22,27].

It is interesting to note that in spite of the fact that the Tb^{3+} ions are soluble in the water pool of the reversed micelles, the thermal lens signal intensities of the Tb^{3+} are much higher than that in pure water. The enhancements were found to be dependent on the particular nonpolar solvent in which the reversed micelles are formed. Under identical conditions, Tb^{3+} exhibits 38-fold enhancement when it is solubilized in AOT/CCl_4 reversed micelles as compared to that in pure water. The enhancement factor decreases to 23 in AOT/n-pentane, 20 in AOT/cyclohexane, 21 in AOT/n-heptane and 15 in AOT/nonane.

Taking advantage of the fact that AOT forms reversed micelles in dioxane [22,27] and terbium chloride is soluble in this solvent, thermal lens measurements were also performed for Tb^{3+} in pure dioxane and for Tb^{3+} solubilized in the water pool of AOT/dioxane reversed micelles. The thermal lens of Tb^{3+} in pure dioxane is 22 times more than that in pure water and is the same, within experimental error, with that of Tb^{3+} in AOT/dioxane (25X; Table I). These results clearly demonstrate that the surfactant molecules do not produce observable effect on the dn/dT and k values of the dioxane solution. Additional information can be obtained from the relative thermal lens signal intensity of Tb^{3+} in AOT/CCl_4, AOT/hydrocarbons and AOT/cyclohexane reversed micellar systems compared to that in pure water. It is known that AOT forms its largest reversed micelle in cyclohexane as the aggregation number, N, in this solvent is between 45 and 65 [22,27]. Medium size micelles can be found in hydrocarbons (N = 25 - 31) with the CCl_4 provides the smallest (N = 17) [22-27]. Any micellar effect, if existent, is expected to provide a larger enhancement for the medium leading to large size micelles whereas smaller enhancements are expected for the smaller size micelles. However, the observed enhancement values are in the reversed order: AOT/CCl_4 > AOT/hydrocarbons > AOT/cyclohexane.

Information of the micellar effect on the thermal conductivity, k, is relatively scarce and inconclusive. It seems that in the absence of aggregation such as the case of AOT in CCl_4 without any solubilized water, the surfactant has no observable effect on k because the k values were found to be the same with and without the surfactant [4]. Adding water to the solution leads to the formation of AOT reversed micelles in CCl_4 [22-27]. The thermal conductivity of the mixture was found to dependent on its chemical composition [4]. The information is still inconclusive at the moment but it seems to suggest that at low solubilized water concentration, the k value for the mixture is higher than that for pure CCl_4 solution [4]. This, according to eq. 3, would result in lowering the thermal lens signal intensity. Because the observed thermal lens signals were in fact, enhanced by AOT/CCl_4 reversed micelles, the reversed micellar effect on the k value of the solution is very small at low solubilized water concentration.

Collectively, these results seem to suggest that the effect of the reversed micelles on the dn/dT and k values of the bulk nonpolar solvents is rather small. The thermal lens signal magnitude of Tb^{3+} is AOT reversed micellar systems is governed mainly by the thermo-optical properties of the nonpolar solvent even though the metal ions are solubilized in the water pool. This deduction can be further explained in terms of the relative size of the water pool and the laser beam spot size. The concentrations of the solubilized water in the reversed micellar systems used in this work were kept very low ($\omega = [H_2O]/[AOT] < 1$) so that the diameter of those water pools is estimated to be less than 20Å [22-27]. This value is smaller than the spot size of the laser beam at the sample which is estimated to be 0.10 mm. Therefore, the measured thermal lens is strongly effected by the thermo-optical properties of the bulk nonpolar solvent of the reverse micelle.

Crown Ethers: Crown ethers are group of macrocyclic ligands whose chemistry is of particular interest because of their ability to form complexes with a variety of metal ions [22-25, 27-36]. This complex formation is of extreme importance in analytical chemistry, particularly separation science because it enables the selective extraction of metal ions from aqueous solution into organic phase [27-36]. It is thus possible to enhance not only the sensitivity but also the selectivity of the thermal lens detection of such metal ions as lanthanide ions. This anticipation is based on the fact that crown ether (host) can selectively form an inclusion complex with a metal ion (guest) and this complex formation will introduce the hydrophobicity into the analyte (i.e., metal ion). As a consequence, the complex will be extracted from the thermo-optically poor water into a nonpolar solvent that has relatively better thermo-optical properties. The thermal lens signal intensity of the extracted inclusion complexes is expectedly higher than that in pure water. Selectivity enhancement in this case would depend on the size selectivity of the macrocyclic host, i.e., the formation of inclusion complexes between the crown ethers such as 18-crown-6, 15-crown-5 and 12-crown-4 and the lanthanide ion can only be achieved when the size of the metal ion is comparable to the cavity size of the crown ether.

Recently, we were able to exploit this possibility to develop a novel method to enhance the sensitivity and selectivity of the thermal lens detection of lanthanide ions [37,38]. In this method, the rare earth ions were selectively extracted from water to organic solvent with the use of crown ether, e.g., 18-crown-6, 15-crown-5 as synergistic extractant and lithium benzoate as the counter ion. For instance, by use of 15-crown-5, neodymium ion (Nd^{3+}) was extracted from aqueous solution into organic phase such as ethyl acetate, heptanol, octanol, 1,2-dichroroethane, methylene chloride and chloroform [37]. Compared to that in water, the thermal lens signal intensity of Nd^{3+} was substantially enhanced when it is extracted into the organic phase: up to 24-fold enhancement was observed in dichloroethane, 20X in $CHCl_3$, 18X in CH_2Cl_{2X}, 14X, 13X and 11X in octanol, heptanol and ethyl acetate, respectively [37].

The induced selectivity is demonstrated by experiments in which a series of lanthanide ions ranging from Pr^{3+} to Er^{3+} were extracted from the aqueous phase into the organic phase using different ethers, namely 12-crown-4 (12-C-4), 15-crown-5 (15-C-5) and 18-crown-6 (18-C-6). As listed in the Table II, only small amounts of lanthanide ions (between 8% and 12%) were extracted into the chloroform phase when 12-C-4 was used as the extracting agent. The low extraction yields can be explained in terms of the mismatching in sizes of the crown ether and the lanthanide ions; i.e., the cavity of 12-C-4, which is between 1.2 and 1.5 Å, is

Table II. Size Effect on the Extractions of Lanthanide Ions by Crown Ethers

	Extraction yield, %		
metal ion	18-crown-6	15-crown-5	12-crown-4
Pr^{3+}	28	10	6
Nd^{3+}	29	14	6
Sm^{3+}	30	20	8
Eu^{3+}	31	25	a
Tb^{3+}	35	30	a
Dy^{3+}	37	33	a
Ho^{3+}	38	31	11
Er^{3+}	41	39	12

[a]Signals were too small to be measured accurately.

too small to accommodate the lanthanide ions whose ionic diameters are between 1.76 and 2.02 Å [39]. Substantial improvement in the extraction yield can be accomplished when 15-C-5 or 18-C-6 was used as the extracting agent. For instance, up to 39% and 41% of Er^{3+} ions can be extracted from the aqueous phase into the chloroform phase with the use of 15-C-5 and 18-C-6, respectively. The enhancement in the extraction yield is probably due to the fact that the cavities of these crown ethers are sufficiently large (the cavities of 15-C-5 and 18-C-6 are 1.7-2-2 and 2.6-32 Å, respectively [40]) to accommodate the lanthanide ions and thus enable them to be extracted into the organic phase. As expected, these two crown ethers also offer noticeable selectivity in the extraction processes: at the same extraction conditions as those for the Er^{3+} ions, only 10% and 28% of the Pr^{3+} ions were extracted into the organic phase by 15-C-5 and 18-C-6, respectively. The size selection provided by the 15-C-5 is relatively better than that by the 18-C-6 even though, for the same ion, the extraction yields obtained by using the latter are higher than the former. This is probably because the cavity of the 18-C-6 is larger and hence more flexible than that of 15-C-5. It can, therefore, adopt a variety of conformations to accommodate the lanthanide ions having different sizes. Consequently, a rather low selectivity is obtained. The larger size of 18-C-6 also makes it more lipophilic than 15-C-5 and hence enables it to extract more metal ion from the aqueous phase into the organic phase.

4. Exploitation of Unique Characteristics of Thermal Lens Technique

Up to now, the dependency of the thermal lens signal on the thermal physical properties of solvent has only been exploited to enhance the sensitivity and selectivity of the technique. Conceivably, one can also take advantage of this dependency in the reverse way, i.e., to use the thermal lens technique as a precise and sensitive method for the determination of the thermal physical properties of solvents, namely their dn/dT and k values. These two thermal physical properties can be individually determined by measuring the transient of the thermal lens. This is because the thermal lens, when recorded as the time dependent change in the far field beam center intensity ($I_{bc}(t)$) after the onset of excitation laser illumination ($I_{bc}(0)$), is given by:

$$I_{bc}(t) = I_{bc}(0)\ [1 + (\frac{\theta}{1 + t_c/t}) + 1/2\ (\frac{\theta}{1 + t_c/t})^2]^{-1} \tag{2}$$

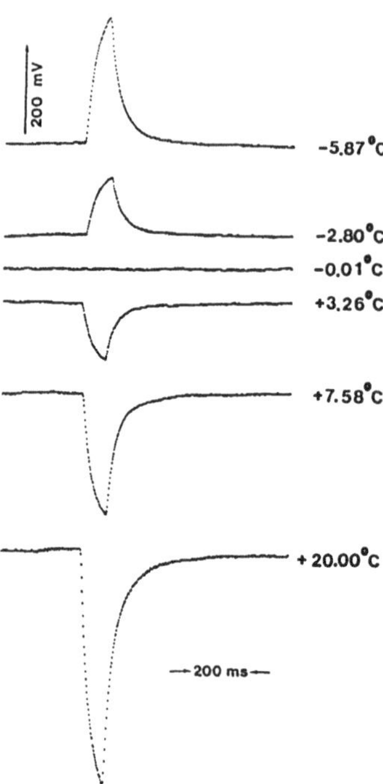

Figure 5. Thermal lens signals of aqueous solution at different temperatures.

where θ is the strength of the thermal lens and is given by eq 1. The thermal time constant t_c which depends on the spot size, ω, of the excitation beam in the sample, the density, ρ, specific heat capacity C_p and thermal conductivity, k, of the solvent, is given by:

$$t_c = (\omega^2 \rho C_p)/4k \tag{3}$$

Thus, θ and t_c values can be obtained by nonlinear curve fitting of the thermal lens transient to eq. 2. Substitution of the t_c value into eq. 3 will yield either the thermal conductivity or the heat capacity of the micellar system. The dn/dT value can then be calculated from eq. 1 using the θ and k values. The accuracy of the method can be evaluated by performing the thermal lens measurement on a solvent whose thermal conductivity, density and specific heat capacity values are known and comparing the t_c values obtained by thermal lens method with the literature values.

The dn/dT and k values of various solutions including aqueous and organic solvents such as methanol, carbon tetrachloride, chloroform, DMF, DMSO and heptane were accurately determined by use of this method [41-43]. In addition to their scientific and technological importance, the obtained dn/dT and k values can provide insight into the structure of solutions. Of particular interest is the information on the effect of the temperature, electrolytes and surfactants on the structure of water.

From the study of the temperature effect on thermal lens measurements in water, it was found that the magnitude and sign of the thermal lens signal intensity are strongly dependent on the temperature of the aqueous solution [41,42]. As shown in Figure 5, dependent on whether the measurements are performed at temperatures lower or higher than -0.01°C, the photoinduced thermal lens can have either a positive (converging) or negative (diverging) focal length. At precisely -0.01 ± 0.04°C, no thermal lens signal could be observed. This is because the dn/dT values of water are positive at T < -0.01°C, negative at T > -0.01°C, and equal to zero (i.e., maximum refractive index) at T = -0.01°C [41,42].

Using the method described above, the k and dn/dT values of water at different temperatures were calculated from the corresponding thermal lens transient signals. The

Table III. Thermal Conductivity and dn/dT Values of Water at Different Temperatures, Measured by Thermal Lens Technique.

Temperature	thermal conductivity mW cm^{-1} K^{-1}	dn/dT x 10^{-5} K^{-1}
+17.98	6.1 ± 0.1	-8.6 ± 0.2
+16.13	6.1 ± 0.1	-8.2 ± 0.2
+14.31	6.1 ± 0.1	-7.4 ± 0.2
+12.52	6.1 ± 0.1	-6.7 ± 0.2
+10.68	5.9 ± 0.1	-5.8 ± 0.2
+8.86	5.8 ± 0.1	-4.8 ± 0.1
+7.13	5.9 ± 0.2	-4.0 ± 0.2
+5.15	5.7 ± 0.1	-2.8 ± 0.1
+3.34	6.0 ± 0.2	-1.7 ± 0.1
+1.26	5.3 ± 0.2	-0.6 ± 0.1
-0.76	7 ± 2	+0.4 ± 0.1
-2.60	5 ± 1	+1.3 ± 0.3
-4.47	5.3 ± 0.4	+2.2 ± 0.2
-6.78	5.4 ± 0.2	+3.6 ± 0.1
-8.44	5.6 ± 0.1	+5.1 ± 0.2

results are shown in Table III [43]. As listed, within experimental error, the k value remains constant at different temperatures. The dn/dT value, however, undergoes significant change as the temperature varies. Increasing temperature leads to an increase in the dn/dT value. For instance, increase the temperature by 16 degrees (from 1.26°C to 17.98°C) leads to an 14 fold increase in the dn/dT value [43]. This observation can be explained in terms of the structure of water.

The structure of water is known to be greatly affected by the temperature of the solution. Increasing the temperature of the aqueous solution will have the "hydrogen bond breaking effect" [44-53]. As a consequence, the water has less hydrogen bonds, becomes less structure (or less order) and hence can have large changes in the density and in the refractive index with temperature (i.e., higher dρ/dT and dn/dT values).

Effects of electrolytes and surfactants on the thermal lens signal have also been investigated [43]. The study was undertaken because it has been reported, based on the IR, Raman and NMR studies, that adding certain electrolytes or surfactants into water has the same effect as heating up the solution [44-53]. Therefore, it may be possible to enhance the sensitivity of the thermal lens technique by adding electrolytes to the aqueous solution. Since there are many types of electrolytes and surfactants, namely, the so-called "water structure former" and "structure breaker", it is important to systematically investigate this possibility and whether it is possible to elucidate the enhancement mechanism [44-53]. It was found that electrolytes did, indeed provide up to a 2-fold enhancement in the thermal lens signal [43]. Higher enhancement (up to 8-fold) can also achieve when surfactants were added to the solution [43]. In both cased, the enhancement was found to be due to the effect of the electrolyte not on the thermal conductivity but rather on the dn/dT of the solution. Interestingly, the effect of electrolytes and surfactants seems to be different from the temperature effect [41,42]. This conclusion was derived when the change in dn/dT upon the addition of electrolytes or surfactants was deconvoluted into two parts. The first, which results from the difference in the specific refractivity and dn/dT of the electrolyte or surfactant solution as compared to that of pure water, has a stronger effect on the enhancement than the second, which is caused by the alteration of the hydrogen-bonding network of water [43].

Recently, another novel application of the thermal lens techniques has been reported. The new application is based on the use of the thermal lens technique for kinetic determinations of fast chemical reactions [54-56]. It has been demonstrated that in addition to advantages such as high accuracy and precision, the technique can be used to determine kinetics with reagents whose concentrations are about 100 times lower than those required by conventional methods. As a consequence, this thermal lens method is particularly valuable for kinetic determination under stopped flow condition where due to its high sensitivity, reactants can be used at much lower concentrations. Thus, it will make possible, for the first time, the determination of bimolecular reactions which to date are too fast to be measured (i.e, few μs) [54-56].

Collectively the recent development in the instrumentation and methodology not only make the thermal lens technique to be one of the most sensitive and selective spectrochemical methods for trace chemical analysis but also enable it to be successfully used for measurements of important data which are currently unobtainable using existing techniques. Such data include: (1) the direct (i.e., without pretreatment) determination of (chemical and stereochemical) structures, concentrations of multicomponent chemical and biological samples having small volume and low concentrations, the insight into the structure of liquids in the bulk, adsorbed and interfacial states and (3) the kinetic determination of fast chemical reactions in normal and confined volume (e.g., water pools of reversed micelles, interfacial water).

ACKNOWLEDGMENT The author is grateful to his former coworkers, namely Drs. Mladen Franko, Minren Xu, Weifeng (Kerry) Zhang for their dedications and hard working. Financial support for this work provided by the National Institutes of Health is gratefully acknowledged.

References

1. J. M. Harris in "Analytical Applications of Lasers", edited by E. H. Piepmeier (John Wiley, New York, 1986) Chap. 13.
2. N. J. Dovichi, CRC Crit. Rev. Anal. Chem. 1987, 17, 357.
3. N. Teramae and J. D. Winefordner, Appl. Spectrosc. 41, 164 (1987).
4. C. D. Tran, Anal. Chem. 60, 182 (1988).
5. C. D. Tran and T. A. Van Fleet, Anal. Chem. 60, 2478 (1988).
6. C. D. Tran, Analyst, 112, 1417 (1987).
7. C. D. Tran, Anal. Chem. 58, 1714 (1986).
8. C. D. Tran, Appl. Spectrosc. 40, 1108 (1986).
9. C. D. Tran, Appl. Spectrosc., 41, 512 (1987).
10. M. Franko and C. D. Tran, Anal. Chem. 60, 1925 (1988).
11. C. D. Tran and M. Franko, J. Phys. E: Sci. Instrum. 22, 586 (1989).
12. C. D. Tran and M. Xu, Appl. Spectrosc. 43, 1056 (1989).
13. M. Franko and C. D. Tran, Appl. Spectrosc. 43, 661 (1989).
14. M. Xu and C. D. Tran, Anal. Chim. Acta 235, 445 (1990).
15. C. D. Tran and M. Xu, Rev. Sci. Instrum. 60, 3207 (1989).
16. M. Xu and C. D. Tran, Appl. Spectrosc. 44, 962 (1990).
17. M. Xu and C. D. Tran, Anal. Chem. 62, 2467 (1990).
18. Y. S. Toulukian, P. E. Liley and S. K. Saxena in "Thermophysical Properties of Matter, T.P.R.C. Data Series", (IFI-Plenum Press, New York, 1970).
19. R. C. Weast, "CRC Handbook of Chemistry and Physics" (CRC Press Inc., Boca Raton, Florida, 1985).
20. N. V. Tederberg, "Thermal Conductivity of Gases and Liquids" (MIT Press: Cambridge, 1965).
21. J. Timmermans, "Physics-Chemical Constants of Pure Organic Compounds" (Elsevier: Amsterdam, 1965, Vol. 1 and 2).
22. J. H. Fendler, "Membrane Mimetic Chemistry" (Wiley: New York, 1982).
23. D. W. Armstrong, Sep. Purif. Methods, 14, 212 (1985).
24. W. L. Hinze and D. W. Armstrong "Ordered Media in Chemical Separations" (American Chemical Society, Washington, D.C., 1987).

25. W. L. Hinze, Sep. Purif. Methods 10, 159 (1981).
26. D. W. Armstrong, Anal. Chem. 59, 84A (1987).
27. J. H. Fendler and E. J. Fendler, "Catalysis in Micellar and Macromolecular Systems", (Academic Press, New York, 1975).
28. B. S. Mohite and S. M. Khopkar, Anal. Chem. 59, 1200 (1987).
29. L. M. Tsay, J. S. Shih and S. C. Wu, Analyst, 108, 1108 (1983).
30. W. Wenji, C. Bozhong, J. Zhong-Kao and W. J. Ailing, Radioanal. Chem. 76, 49 (1983).
31. V. K. Manchanda and C. A. Chang, Anal. Chem. 59, 813 (1987).
32. Y. Hasegawa and S. Haruna, S. Solv. Ext. Ion Exch. 2, 451 (1984).
33. Y. Hasegawa, M. Masuda, K. Hirose and Y. Fukuhara, Solv. Ext. Ion Exch. 5, 255 (1987).
34. H. F. Aly, S. M. Khalifa, J. D. Navratil and M. T. Saba, Solv. Ext. Ion Exch. 3, 623 (1985).
35. D. D. Ensor, G. R. McDonald and C. G. Pippin, Anal. Chem. 58, 1814 (1986).
36. J. Tang and and C. M. Wai, Anal. Chem. 58, 3235 (1986).
37. C. D. Tran and W. Zhang, Anal. Chem. 62, 830 (1990).
38. C. D. Tran and W. Zhang, Anal. Chem. 62, 835 (1990).
39. J. -C. Bunzli in "Handbook on the Physics and Chemistry of Rare Earths", edited by K.
A. Gschneidner and L. Eyring, (North-Holland, Amsterdam, 1987) Vol. 9, Chap. 60.
40. K. L. Cheng, K. Ueno and T. Imamura, "Hankbook of Organic Analytical Reagents", (CRC Press, Boca Raton, Fl, 1982).
41. M. Franko and C. D. Tran, Anal. Chem. 61, 1660 (1989).
42. M. Franko and C. D. Tran, Chem. Phys. Lett 158, 31 (1989).
43. M. Franko and C. D. Tran, J. Phys. Chem. 95, 6688 (1991).
44. K. A. Hartman, Jr.; J. Phys. Chem. 70, 270 (1966).
45. J. Stangret and J. Kostrowicki, J. Sol. Chem. 17, 165 (1988).
46. G. E. Walrafen, M. S. Hokmabadi and W. H. Yang, J. Chem. Phys. 85, 6964 (1986).
47. F. Rull and J. A. de Saja, J. Raman Spectrosc. 17, 167 (1986).
48. M. M. Marciacq-Rauselot and M. Lucas, J. Phys. Chem. 77, 1056 (1973).
49. R. Bhanumathi and S. K. Vijayalakshamma, J. Phys. Chem. 90, 4666 (1986).
50. R. E. Verrall in "Water - a comprehensive treatise", edited by F. Franks (Plenum Pres, New York, 1974) Vol. 3.
51. F. Frank and D. S. Reid in "Water a Comprehensive Treatise", edited by F. Franks (Plenum Press, New York, 1973) Vol. 2, Chapter 5.
52. M. Klose and J. U. Naberuchin, "Water - struktur und dynamik", (Akademie Verlag, Berlin, 1986).
53. D. Eisenberg and W. Kauzmann, "The Structure and Properties of Water" (Oxford University Press, Oxford, U.K., 1969).
54. M. Franko and C. D. Tran, Rev. Sci. Instrum. 62, 2430 (1991).
55. M. Franko and C. D. Tran, Rev. Sci. Instrum. 62, 2438 (1991).
56. M. Franko and C. D. Tran, this volume.

Stimulated Spectroscopic Ellipsometry and Imaging in the Space, Time and Frequency Domain: A New Photothermal Probe

J. Gang[1], *J.P. Roger*[1], *A.C. Boccara*[1], and *J.L. Stehlé*[2]

[1]Laboratoire d'Optique Physique, ESPCI, UPR A0005 du CNRS, 10, rue Vauquelin, F-75005 Paris, France

[2]SOPRA, 68, rue P. Joigneaux, F-92270 Bois-Colombes, France

Abstract. Using a commercial spectroscopic ellipsometer, we have been able to follow in time (or in frequency) the variation of the ellipsometric signal when the sample is irradiated by a pulsed (or a modulated) excitation. The theory leads to the quantitative measurement of the static and dynamical optical parameters. Imaging with a 100 μm resolution has been performed on semiconductor samples.

Modulation spectroscopy /1/ has been found very powerful to evaluate the electronic properties of solid materials submitted to light, heat, electric field...

In such experiments, most of the time, a probe beam is reflected at the sample surface and monitors the variation of the sample reflectivity. The associated intensity variation carries only *one* single piece of information /2/. In an ellipsometric measurement *two* parameters are available and there have been some preliminary experiments to study the evolutions of these parameters under external excitation /3/, /4/, /5/, /6/.

We present here stimulated ellipsometry under external excitation using a *commercially available ellipsometer.* Fig. 1 shows the experimental set-up that we have built with the SOPRA ellipsometer in order to perform the detection under pulsed (a) or modulated (b) excitation.

In traditional ellipsometry /7/, the determined quantity is:

$$\rho = r_p / r_s = tg\psi \cdot e^{i\Delta} \tag{1}$$

where r_p and r_s are the reflection coefficients for the electric vector components parallel and perpendicular to the plane of incidence respectively.

We have developed a theoretical approach and sampling method which allow us to determine the static and dynamic (light induced) ellipsometric parameters from our measurements. Under an external excitation, the intensity detected by the detector can be expressed:

$$I_c(A) = I_o \cdot (1 + \alpha\cdot\cos 2A + \beta\cdot\sin 2A) + \delta I_o + (I_o\cdot\delta\alpha + \alpha\cdot\delta I_o)\cdot\cos 2A + (I_o\cdot\delta\beta + \beta\cdot\delta I_o)\cdot\sin 2A \tag{2}$$

in which, the first term is for traditional ellipsometry, A is the azimuth of analyzer.

For a position of the polarizer, we have measured the signals for 4 successive azimuths of the analyzer. Then, the parameters and their variations are deduced. The variations are:

$$\delta tg\psi = tgA\cdot\delta\alpha / (1 - \alpha)\cdot(1 - \alpha^2)^{1/2} \tag{3a}$$

$$\delta\cos\Delta = \delta\beta / (1 - \alpha^2)^{1/2} + \alpha\cdot\beta\cdot\delta\alpha / (1 - \alpha^2)^{3/2} \tag{3b}$$

Springer Series in Optical Sciences, Vol. 69 Photoacoustic and Photothermal Phenomena III
Editor: D. Bićanić

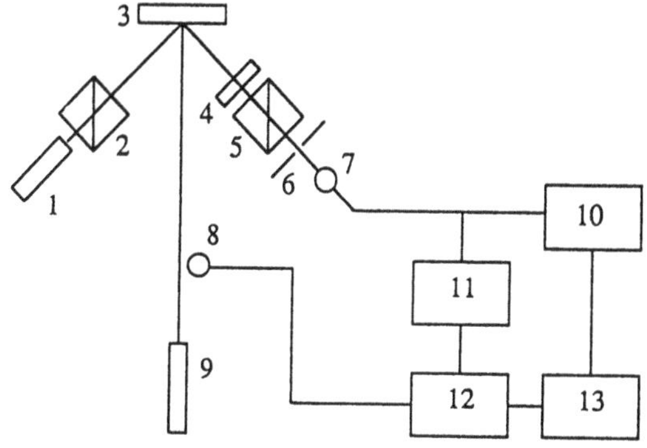

Fig.1: Experimental set-up: a) for pulsed excitation, 1.probe beam, 2.polarizer, 3.sample, 4.optical filter, 5.analyzer, 6.aperture, 7.8.photodetector, 9.pulsed excitation, 10.voltmeter, 11.pre-amplifier, 12.digital oscilloscope, 13.computer, b) for modulated excitation, 8.optical chopper, 9.continuous light source, 12.lock-in amplifier

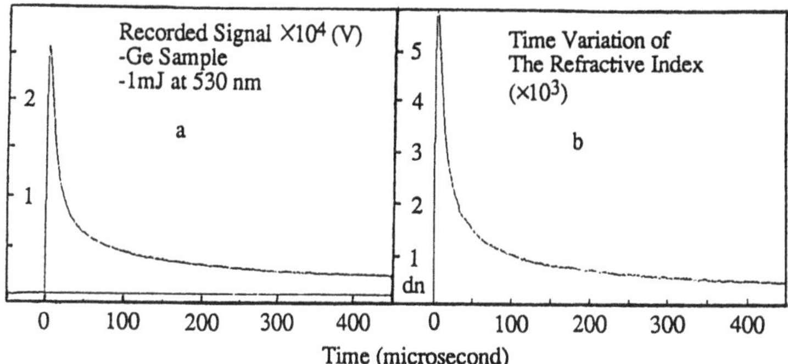

Fig.2: Time dependence of the signal (a) and of the refractive index (b) for a Ge sample

For a bulk sample, the variation δN of the complex refractive index N(=n+ik) can be deduced:

$$\delta N = 2 \cdot \sin^2\phi_0 \cdot tg^2\phi_0 \cdot (\rho - 1) \cdot \delta\rho / N \cdot (\rho + 1)^3 \qquad (4)$$

ϕ_0 is the incident angle.

As an illustration, Fig. 2a shows the signal of a Ge sample obtained for one (over 4 successive) position of the analyzer under the conditions of a pulsed laser as excitation and He-Ne laser as probe beam. The signal to noise is rather good despite of the low level of the excitation ($\approx$ 1 mJ/pulse). The minimum variation in the real part n of the refractive index in such conditions is of the order of $\delta n \approx 10^{-5}$.

Fig. 2b shows the time variation of n. At long time scale the diffusion in the bulk exhibits a $t^{-1/2}$ behaviour, whereas below 10 µs carriers modify the signal.

In the case of a modulated excitation, we have been able to use a broadband probe source and then to perform a spectroscopic analysis of the photoinduced

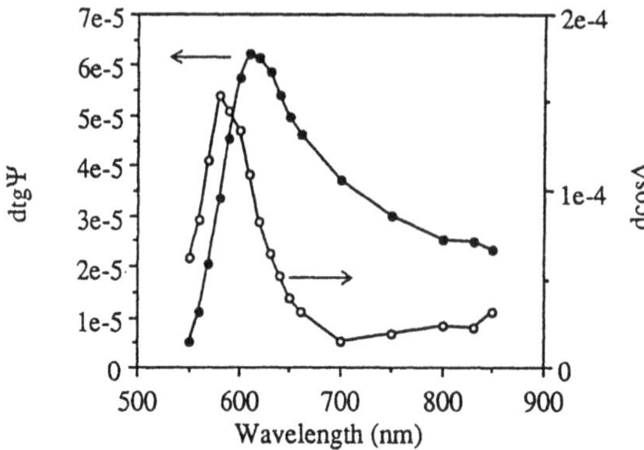

Fig.3: Spectra of the photoinduced variation of ellipsometric parameters for a Ge sample. External excitation: 2W, 515nm, Ar^+ laser modulated at 1500Hz.

variation of the ellipsometric parameters as shown in Fig. 3. The strong anomaly in the 550 - 600 nm wavelength range can be mainly related to a thermoreflectance effect as observed by Batz /1/. The same set-up has been successfully used for studying implanted semiconductors, Simox structures and polymer films deposited on semiconductor.

At last, we would like to underline the *imaging* capabilities of our set-up. The above described operations have been made using a 100 µm diameter probe beam which allows mapping of our sample at an intermediate scale.

We would like to thank MRT for financial support and Madame Monfrey for her encouragements.

The first author gratefully acknowledges the support of K. C. Wong Education Foundation, Hong Kong.

References:

/1/ M. Cardona, *Modulation Spectroscopy,* Solid State Physics, n°11 (Academic Press, New York, 1969).

/2/ J. Opsal, M.W. Taylor, W. Lee Smith and A. Rosencwaig, J. Appl. Phys. **61**, 240 (1987).

/3/ D.H. Auston and C.V. Shank, Phys. Rev. Lett. **32**, 1120 (1974).

/4/ G.E. Jellison, Jr., and D.H. Lowndes, Appl. Opt. **24**, 2948 (1985).

/5/ J.Th. Zettler, Th. Dittrich and L. Schrottke, Phys. Stat. Sol. (a) **119**, K91 (1990).

/6/ JIN Gang, J.P. Roger and A.C. Boccara, International Workshop of Spectroscopic Ellipsometry and User's Training, Paris, April 23-26, 1990.

/7/ R.M.A. Azzam and N.M. Bashara, *Ellipsometry and polarized light* (North Holland Publishing Company, 1977).

Photothermal Induced Frequency Conversion of a Double Modulated Laser Beam in Layer Structures

H.D. Geiler[1], *M. Wagner*[1], *P. Kowalski*[2], *and R. Bleiß*[2]

[1]Friedrich-Schiller-Universität Jena, Institut für Festkörperphysik, Helmholtzweg 5, O-6900 Jena, Fed. Rep. of Germany

[2]Jenoptik Carl Zeiss Jena GmbH, Göschwitzer Str. 33, O-6905 Jena-Göschwitz, Fed. Rep. of Germany

Abstract: The double modulation of a laser beam created by two laser diodes is used to generate the differential frequency in the sample by photothermal modulation of the refractive index. That part of the reflected intensity representing the differential frequency depends on the amplitude and phase of the photothermal response and is applied to measure the accumulation of radiation defects in ion implanted layers.

1.Introduction

The technique of optical detection of laser excited charge carrier waves and thermowaves by measuring the response modulated refractive index was successful used for analyzing near-surface properties of materials [1,2,3]. Most of these methods use two distinct laser beams with different wave lengths: a modulated one for excitation (pump beam) and an unmodulated one for the detection by measuring the reflectivity (probe beam) [3,4]. Some optical arrangements have been published in which only one laser beam is used for both excitation and detection [5,6]. This paper presents a single beam technique basing on the intensity modulation of a laser beam with two frequencies supplied by two modulated laser diodes. The twofold photothermal modulation of the refractive index causes a frequency conversion to higher harmonics and the sum and difference frequencies.

2. Modulation and frequency conversion

The intensities I_1, I_2 of two laser diodes are modulated with two frequencies Ω_1, Ω_2 and the modulation indices m_1, m_2 ($m_1, m_2 \approx 1$), respectively. The excitation generated in the sample by a beam of the intensity $I = I_1 + I_2$ leads to the optical response R(t) (reflection or transmission):

$$R(t) = R_o\{1 + m_{R1} I_1 m_1 \cos(\Omega_1 t + \phi_1) + m_{R2} I_2 m_2 \cos(\Omega_2 t + \phi_2)\} \tag{1}$$

where R_0 is the mean response signal (mean transmission or reflexion coefficient) and m_{R1} and m_{R2} are the modulation indices of the optical response determined by the carrier wave and the thermowave [8,10]. The phases ϕ_1 and ϕ_2 describe the phase shifts of the response with respect to the exciting modulation $\cos\Omega_1 t$ and $\cos\Omega_2 t$, respectively.

The transmitted or reflected intensity $I_{R\alpha}(t)$ of the partial intensity I_α follows the equation

$$I_{R\alpha}(t) = I_\alpha[1 + m_\alpha \cos(\Omega_\alpha t)] \times R(t) \qquad ; \alpha = 1,2 \tag{2}$$

Thus, besides the original frequencies Ω_1 and Ω_2 the measured intensity contains the intermodulation products $2\Omega_1$, $2\Omega_2$, $\Omega_1 + \Omega_2$, $\Omega_1 - \Omega_2$. Thus, for determining the modulation index $m_{R\alpha}$ of the sample governed by the photothermal effects it is rather promising to detect the differential frequency $\Omega_1 - \Omega_2 = \Omega_-$ [13] because of the stable detection frequency in the case of a synchronous frequency sweep. Assuming a small dispersion of the photothermal response (this is a good approximation for small differences between Ω_1 and

Ω_2) one gets $m_{R1} \approx m_{R2} = m_R$ and $\Phi_1 \approx \Phi_2 = \phi$. For the partial intensity $I_{R\alpha}(t)$ its content can be derived from equation (2):

$$I^-_{R1}(t) = m_R R_o M \cos(\Omega_- t - \phi)/2$$
$$I^-_{R2}(t) = m_R R_o M \cos(\Omega_- t + \phi)/2 \qquad (3)$$

with the modulation factor $M = I_1 I_2 m_1 m_2$.
In the case of detecting the intensity $I^-_R = I^-_{R1} + I^-_{R2}$ one obtains from equ.(3)

$$I^-_R(t) = m_R R_o M \cos(\Omega_- t) \cos\Phi \qquad (4)$$

This means, when detecting the differential frequency Ω_- the phase information Φ will be lost, i.e. it cannot be detected independently of the amplitude by phase sensitive techniques but it will additionally influence the signal amplitude. For $\Phi = 90^\circ$ the signal will drop to zero. When detecting the partial beams separately a pure phase dependent signal can be gained.

3. Experimental setup

The experimental setup is shown in fig.1. Two laser diodes ($\lambda_1 = 820$nm, $\lambda_2 = 817$nm) are modulated with the frequencies 1.04308 MHz and 1.00000 MHz, respectively. The detection system was tuned to the differential frequency of $\Omega_- = 43.08$ kHz. The linearly polarized beams were taken as partial beams ($I_1 = I_2$) to be recombined by means of a 50% beam splitter. After passing a polarizing splitter and a quarterwave plate the laser beam was focused with a microscope objective to a spot diameter of 1.5 µm on the sample. The reflected part was selected by the polarizing splitter and guided to the response photodiode. A portion of the incoming beam was gained from a beam splitter and was taken as reference beam after having passed a neutral density wedge to realize a balance detection principle. The minimum detectable differential frequency modulation produced by the sample response depends on three factors:
(i) the actual differential frequency content of the detector signal for zero sample response (due to cross talk and intermodulation, which are minimized using the methods summarized in [13]),
(ii) the noise produced by the laser beam detection (suppression of excess noise by the balance detection with the reference detector, shot noise),
(iii) the pure detector/amplifier noise.
For best detector balance (exact adjustment of the neutral density wedge for zero mean differential photocurrent) the differential frequency interference at zero response corresponded to a modulation index of 10^{-7}. For a typical value of the photocurrent of 10^{-4} A

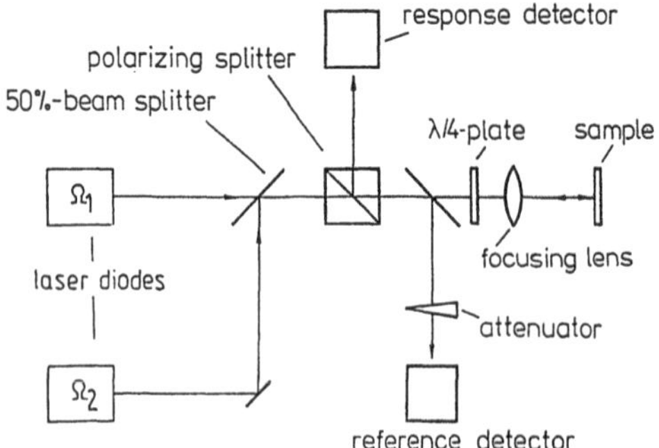

Fig.1: Experimental set-up applied for frequency conversion by photothermal modulation of the refractive index

the noise was determined to be $6\times10^{-7}(Hz)^{-1/2}$ which is only slightly above the shot noise limit of $8x10^{-8}(Hz)^{-1/2}$.
At the detector output the signal was measured using a lock in amplifier which provided the further limitation of the bandwidth as well as the correct sign of the signal.

4. Results

Some typical results gained from ion-implanted GaAs and Si samples illustrate the efficiency of the measuring technique. To study defect accumulation Cr-compensated GaAs wafers implanted with 300 keV Se^+-ions at room temperature were investigated. In Fig. 2 the dependence of the response signal on the dose of Se^+ -implantation with different current densities is presented. The amorphization threshold determined by Rutherford backscattering spectrometry is found for an implantation dose of 5×10^{13} cm^{-2} and 3×10^{13} cm^{-2} in the case of the dose rate of 10 $nAcm^{-2}$ and 30 $nAcm^{-2}$, respectively. This values correspond with the first maximum of the response signal versus the dose. The result agrees with the dose rate effect published in [9]. Because of the very small magnitude of the product of the modulation frequency with the carrier life time governing in compensated GaAs mainly the phase of the charge carrier response [11], the total phase factor in equ. (4) can be expected to be nearly constant for implantation doses lower than the amorphization threshold. Hence, the measured signal is determined mainly by the amplitude of the photothermal response and can be directly compared with results gained by classical modulated reflectivity [7] introducing the conversion factor $K_R=I^-_R/(MR_0)$. In the range of high dose implantation above the amorphization threshold the oscillations of the signal can be explained by the growing thickness of the amorphous layer (the first minimum corresponds to 200nm and the second maximume to 280nm, both thicknesses determined by RBS). Using the the complex optical refractive index of crystalline and amorphous (Se-implanted) GaAs $n_c=3.79+i0.22$ and $n_a=4.22+i0.99$ [9,10] and applying the matrix formalismus given in [11], reflectivity and energy deposition in the amorphous layer were calculated. The oscillation cannot be explained by a one layer optical interference effect [7] alone, because for a thickness of the amorphous layer above 200nm the energy deposition is nearly constant. Hence, the charge carrier wave (diffusion length $l_c\approx3\times10^{-5}$cm) should be responsible for the second oscillation of the signal.

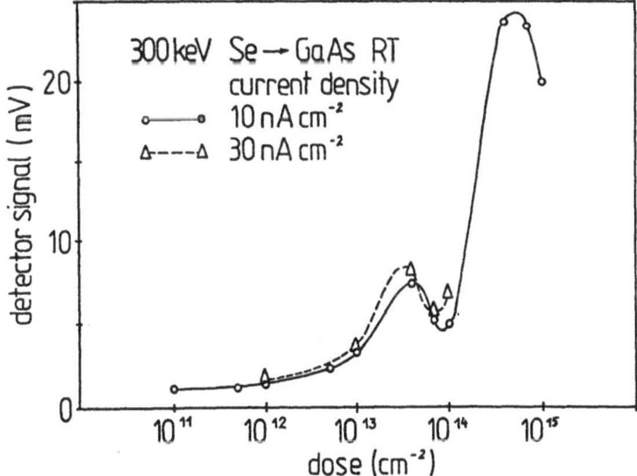

Fig.2 Detector output signal versus the implantation dose for compensated GaAs implanted with 300 keV-Se^+ ions at room temperature and for different dose rates (laser power of the photothermal measurement on the sample 10 mW).

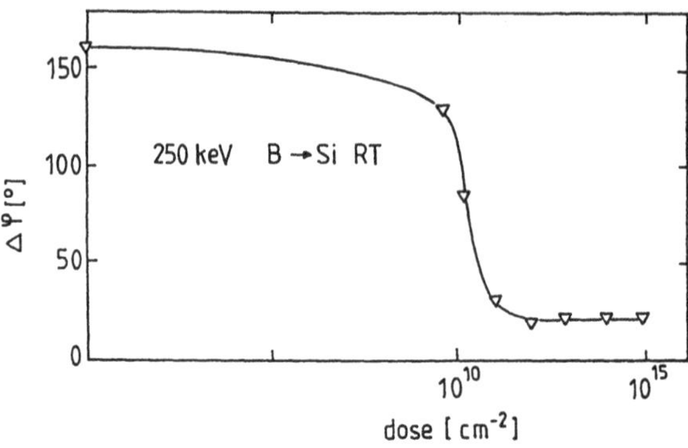

Fig.3 Phase dependence on the dose in the case of low dose implantation of B into Si wafer of low conductivity (1KΩcm)

To demonstrate the influence of the phase Si crystals of high purity were implanted with B^+ ions at very low doses to create point defects and change the carrier life time in the range of 10^{-7}s. the results of the phase determination shows fig.3. The drastic change of the phase can be explained by the phase of the carrier wave if the product of carrier life time and modulation frequency reaches nearly 1. Because the response phase crossing $\pi/2$ causes the signal to change the sign (see in equ.(4)) an precise measurement of the carrier life time is possible sweeping the modulation frequency.

5. Conclusions

The dose dependence can be measured with the same accuracy as gained by two beam methods [3,4,7]. High sensitivity is realized also in the low dose region by the phase dependence of the signal. In addition the system provides the following technical advantages: continous frequency sweeping in the range from 100kHz to 10 MHz by stable low frequency detection, possibility of integration of the small module in optical measuring systems for in-situ or on-line measurement or inspection.

References

[1] A.Mandelis (editor),"Photoacoustic and Thermal Wave Phenomena in Semicon ductors"; North-Holland Pub., NewYork,1987

[2] B.J.Kirby, L.A.Larson, Ru-Yu Liang; Nucl.Inst.Meth. B21(1987)550

[3] A.Rosencwaig in ref [1] p.97

[4] C.Christofides, I.A.Vitkin, A.Mandelis; J.Appl.Phys.67(1990)2815,2822

[5] A.Lörincz; Appl.Phys. B47(1988)35

[6] L.Chen, K.H.Yang, S.Y.Zhang; Appl.Phys.Lett. 50(1984)1349

[7] J.Opsal, A.Rosencwaig in "Photoaccoustic and Photothermal Phenomena" ; Ed.by P.Hess, J.Pelzl, Springer Series in optical Sci., vol.58, Berlin 1988, p.224

[8] J.Opsal; in"Photoacoustic and Photothermal Phenomena II" ed. by J.C.Murphy, J.W.Maclachlan Spicer, L.C.Aamodt, B.S.H.Royce; Springer series in Optical Sci.Vol.62, Springer Verl., Heidelberg 1990, p.140

[9] E.Wendler, W.Wesch, G.Götz; Nucl.Instr.Meth.B52(1990)57

[10] W.Wesch, E.Wendler, G.Götz, N.P.Kikelidse; J.Appl.Phys.65(1989)519
[11] H.D.Geiler,K.Hehl, D.Stock; phys.stat.sol.(a)78(1983)193
[12] M.Wagner, N.Winkler, H.D.Geiler; Appl.Surf.Sci.50(1991)376
[13] M.Wagner, H.D.Geiler; J.Meas.Sci.Technol. (1991) in press.

Cavity-Enhanced Photothermal Spectroscopy: A Tool for High Sensitivity Experiments

H.A. Schuessler, J. Tang, and E.C. Benck

Department of Physics, Texas A&M University,
College Station, Texas 77843-4242, USA

Abstract. The absorption of thin films, multilayer coatings, solid and liquid surfaces was studied with cavity-enhanced photothermal spectroscopy.

1. Introduction

The novel technique of cavity-enhanced photothermal spectroscopy (CEPTS) [1] is ultrasensitive and can determine surface-specific optical absorbances $\alpha(\omega) \approx 10^{-6}$ at a laser power density of 10^4 W/cm^2 and with a resolution in the few μm range. Like standard PTS [2] it is based on the small amount of heat generated by a weakly absorbing surface during laser irradiation, and the simultaneously occurring changes in the index of refraction of the adjacent transparent medium. However, the cavity-enhanced variant of PTS achieves increased sensitivity since the sample is placed inside an external resonant cavity. In essence the standard photothermal signal gets amplified by the quality factor of the high finesse optical cavity. The method is quite different from intracavity PTS [3] where gain changes of the active laser medium produce the signal. In the present pump laser-probe laser geometry using an external cavity the sample is decoupled from the laser gain and optical feedback is avoided. The system is easy to align and samples can be readily exchanged without realignment.

2. Experiment

Fig. 1 depicts the experimental setup which is made rather compact by employing optical fibers. A tunable cw dye laser or an argon ion laser beam is focused onto the sample located inside a high-finesse confocal spherical etalon which serves as the optical cavity. The photothermal effect is monitored by a weak HeNe-laser probe beam. This single-mode HeNe-laser is first split into two beams using a fiber optical beam splitter and is then mode matched with the output lens to the optical cavity. One subbeam is used as the probe beam and grazes the sample surface inside the enclosing medium. The other subbeam is used to stabilize the etalon by locking it to the side of a transmission fringe, and, thereby, to the HeNe-laser frequency. In the present experiment, the samples were placed in air, but in principle any transparent medium will work and a medium with a high thermal gradient of the index of refraction can be used to advantage.

Springer Series in Optical Sciences, Vol. 69 Photoacoustic and Photothermal Phenomena III
Editor: D. Bićanić

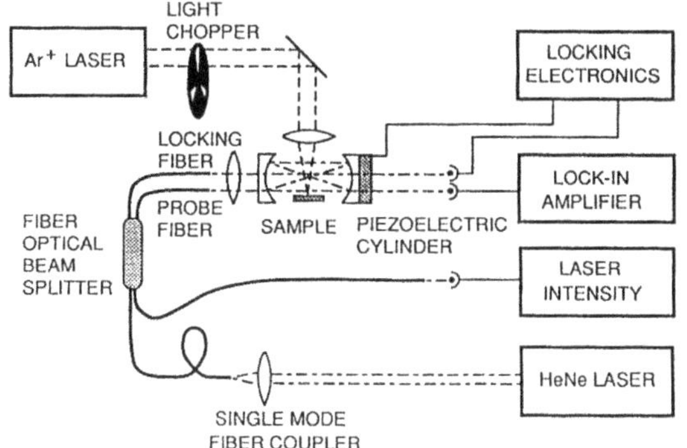

Fig. 1. Schematic diagram of the experimental setup.

3. Theory

Thermal modelling calculations [4] to determine the temperature distribution $U_i((r,z,t)$ in the medium above the sample (i = 1), in the sample (i = 2), and in the supporting substrate (i = 3) have been carried out by solving the appropriate heat diffusion equation in each medium using also the proper boundary conditions. They show that in addition to the optical absorbance $\alpha(\omega)$ also the thermal diffusivity $a_i^2 = \kappa_i/\rho_i C_i$ (κ_i = heat conductivity, ρ_i = density and C_i = specific heat) of the thin film sample can be determined. The discussion was carried out for both cw and modulated pump laser irradiation and is in quantitative agreement with the experimental results.

4. Results

A large variety of different thin film samples was studied. On the one hand they reach from high reflectivity laser gyro mirrors ($\alpha \sim 10^{-6}$) consisting of dielectric multilayer coatings, to thin metal films. On the other hand, transparent media containing small amounts of absorbers or dyes were also investigated. In both cases the weak absorption occurred simultaneously with the strong optical reflection or transmission of the pump laser beam which approached unity. Under these conditions, minute surface specific absorptions are in other methods not observable due to the strong level of the pump laser light. However, since PTS is based on detecting the effects of the heat produced by the absorbing sample, the measurement is not influenced by the directly reflected, scattered or transmitted pump laser light.

Fig. 2 displays as an example the CEPT signal for a high temperature superconducting film and another film of identical metal composition, which is oxygen deficient and therefore nonsuperconducting. The PT signal obtained at room temperature is plotted versus the modulation frequency of the pump beam. The different frequency dependences of the two signals reflect the different thermal conductivities of the two films. Comparatively for unmodulated cw pump laser irradiation, the

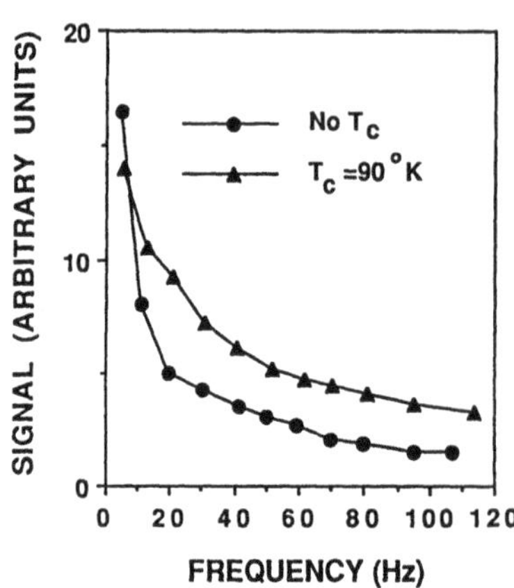

Fig. 2. CEPT signal versus the modulation frequency of the pump beam.

absorbance ($\alpha(\omega) = 0.18(2)$) was the same for both films and independent of laser frequency in the blue-green spectral range studied.

The spatial resolution of the present setup was investigated using a Ronchi grating as the sample. When the pump beam is scanned over the ruling, the optical absorption in the opaque parts will quickly set up a periodic temperature distribution above the grating, yielding a periodic photothermal signal. The measurements [5] demonstrated a resolution in the few μm range.

This work was supported by the Texas Advanced Research Program, NSF, and the Center for Energy and Mineral Resources at Texas A&M University.

References

[1] H. A. Schuessler, S. H. Chen, Z. Rong, and Z. C. Tang, Optics Lett. **16**, 608 (1991).

[2] D. Fournier, A. C. Boccara, N. M. Amer, and R. Gerlach, Appl. Phys. Lett. **37**, 519 (1980).

[3] K. H. Fung and H. B. Lin, Appl. Optics **23**, 1238 (1984).

[4] S. H. Chen, Z. C. Tang, Z. Rong, and H. A. Schuessler, SPIE Proc. **1208**, 149 (1990).

[5] H. A. Schuessler, S. H. Chen, Z. C. Tang, and Z. Rong, Appl. Surface Science **48/49**, 254 (1991).

A New Differential Probe System for the Detection of Thermal and Plasma Waves

M. Liu, M.B. Suddendorf, and M.G. Somekh

Department of Electrical and Electronic Engineering,
University of Nottingham, Nottingham, UK

This paper discusses a new non-contact, non-destructive evaluation (NDE) system for examining materials such as semiconductors. It is based on two parallel, heterodyne interferometers, which probe the specimen to detect pump beam induced changes in sample reflectivity and surface displacement. A frequency doubled Nd-YAG laser emitting 45mW at 532nm is intensity modulated at frequencies up to 4MHz and used as the pump source. The two probe beams are focused onto the sample together with the pump beam, thus enabling one to monitor the propagation of thermal and plasma waves in a very flexible manner. The system configuration and some preliminary results, demonstrating the versatility of the system, are presented.

Introduction

Thermal wave characterisation techniques are becoming increasingly important for both, non-destructive testing and semiconductor characterisation. Various techniques are presently being used which rely on laser induced changes in the optical reflectivity. The most commonly used of these is the Therma-probe system [1] which has proved successful for the characterisation of of semiconductors. The use of a heterodyne interferometer to detect the thermally induced sample displacement and changes in the phase of the refractive index has been developed by Martin et al. [2]. This system has also been successfully exploited for the measurement of variations in the properties of semiconductors particularly those effecting carrier lifetime [3]. A problem arises however when one whishes to make quantitative measurements of lifetime with high spatial resolution - since it is often not possible to distinguish between recombination of carriers and diffusion out of the area probed [4]. We assert that the use of two asymmetrical probe beams will enable us to perform these measurements without compromising spatial resolution. The system presented here uses two parallel interferometers, which give two independent measures of the time varying changes in amplitude and phase of the sample reflectivity as well as the photodisplacement of the sample, induced by a pump beam.

Optical system and electronics

The optical configuration is shown in fig.1. The 532nm beam emitted from the frequency doubled Nd-YAG Laser is intensity modulated at frequencies up to 4 MHz using an acousto-optic modulator. The beam is reflected off a dichroic mirror and focused onto the sample surface having a time-averaged power on the sample of approximately 10mW. The 0.8mW beam of a HeNe laser (633nm) and a Bragg-cell are employed to generate the two probe beams. The Bragg-cell is driven by two frequencies ω_1 and ω_2, resulting in two first order beams which are frequency shifted with respect to the common reference and angular separated by an amount proportional to the difference in the drive frequencies ω_2-ω_1. A projection system consisting of two achromatic doublets and an infinity corrected microscope objective is employed to focus the probe beams perpendicular and adjacent to the pump beam onto the sample. The back-reflected beams follow their original paths, recombine by undergoing another frequency shift in the Bragg-cell, and illuminate the photodetector. The reference beam is reflected off a mirror and is then incident onto the photodetector where it interferes with the probe beams. The probe beams are low-index amplitude (AM) and phase modulated (PM) due to the periodic change of the complex refractive index and the thermal expansion of the sample. The electronic system uses heterodyne detection at an

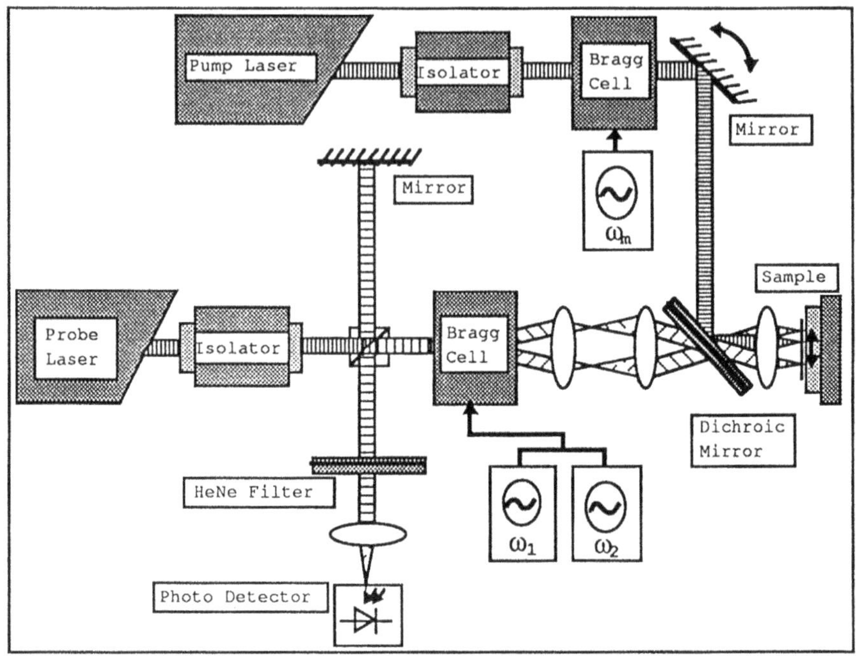

Fig.1: The optical configuration

intermediate frequency of 10.7MHz, which can detect either the amplitude or phase modulation imposed on the interference signals. Microphonics will be common to the probe beams and will therefore be removed in the phase comparison.

Experimental Details and Results

First measurements have been made on a glass substrate coated with a thin layer of Aluminium (coating thickness less than 1/10 of a thermal diff. length). On this sample the measurements have been made by keeping the probe beam separation fixed and scanning the pump beam along a line across the probe beams. The drive frequencies for the Bragg-cell were 62 and 92 MHz, resulting in a probe beam separation of about 6μm, and the pump beam was scanned over a range of about 40μm.

Figure 2 shows a set of results obtained from the coated glass sample. The pump beam was modulated at frequencies between 10 and 100kHz and scanned in the manner described above. The rate at which the PM amplitude signals decay from their peak values, increases with modulation frequency as does the rate of phase change. The ω dependence of the amplitude decay and maximal phase gradient predicted by the one-dimensional model for the thermal expansion ($1/\omega$ and $1/\sqrt{\omega}$ respectively) is not satisfied by our results. This is not surprising since in the frequency range used, the thermal diffusion length in glass is still greater than the spot size of pump and probe beams (1μm). This necessitates a proper 3D treatment which will be the subject of a separate paper.

To demonstrate the ability to detect amplitude and phase modulated signals i.e. different contrast mechanisms, a layered silicon structure with a surface profile shown in Fig. 3b,c) has been used in a second measurement. In this case, though, only one probe beam has been used with the pump beam superimposed on it and the sample was scanned through 250μm. Figure 3a shows that the AM signal is maximum on the layered structure whereas the PM signal is greatest on bulk silicon.

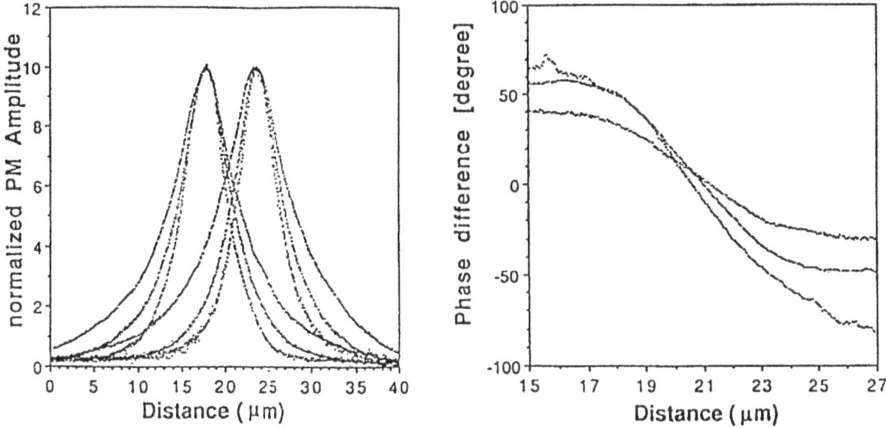

Fig. 2: Amplitude and Phase response (two beam system) on the Al-coating. Plots shown for modulation frequencies of 10, 40 and 100kHz. The width of the amplitude response decreases for higher modulation frequency whereas the phase gradient increases.

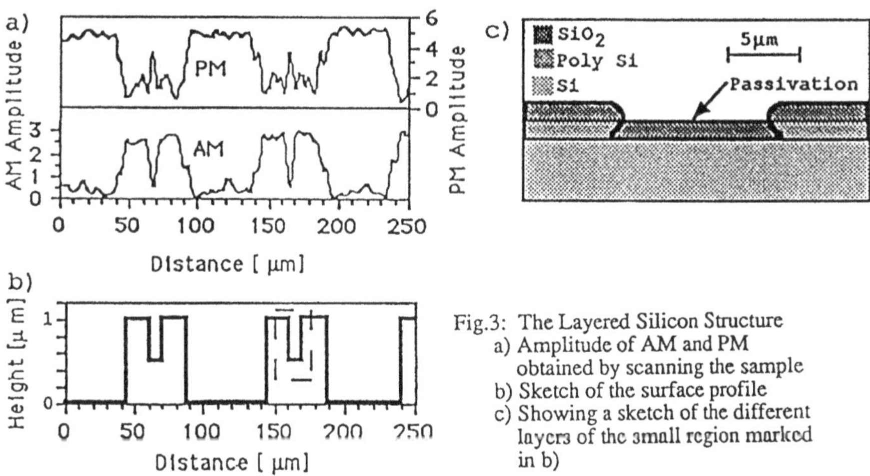

Fig.3: The Layered Silicon Structure
a) Amplitude of AM and PM obtained by scanning the sample
b) Sketch of the surface profile
c) Showing a sketch of the different layers of the small region marked in b)

Without going into details of the reasons for this behaviour, it shows clearly the ability of the instrument to expose different contrast mechanisms.

Conclusion

Although the prototype implementation of the system is only just completed, there are a wide range of applications of the technique. These rely on the flexibility achieved by using two probe beams and on the great variety of output that can be produced as well as on the ability to vary the lateral position of all beams electronically. The inclusion of a differential phase profilometer based on the same optical configuration, which has been developed in our group [5], will allow the comparison between topographical and compositional variations on the sample and hence, further contribute to the versatility of the instrument. Immediate applications include semiconductor characterisation as described in [3], however, the extra probe beam will give access to both carrier transport and lifetime properties of the material. The ability to separate the probe beams will make the instrument effective for the imaging of lateral discontinuities such as cracks and steps.

Acknowledgement

This work is supported by the Science and Engineering Research Council.

References

[1] A Rosencwaig, Ch.5 in "Photoacoustic and Thermal Wave Phenomena in Semiconductors" (Ed. A Mandelis) North Holland, 1987.
[2] Y Martin, E A Ash, Phil Trans Roy. Soc. Lond. A320, pp257-269, 1986.
[3] G M Crean, M G Somekh, C Jeynes, R P Webb, ESSDERC, Berlin, pp. 929-933, 1989.
[4] S J Sheard, M G Somekh, T Hiller, Proc. Symp. Acoustic Thermal Wave and Optical Characterisation of Materials (Ed G M Crean et al.), North Holland, p155-159, 1990.
[5] M S Valera, M G Somekh, R K Appel, Electronics Letters, Vol. 27, No.9, p719, 1991.

Pulsed Photoacoustic Technique for the Measurement of Laser Damage Threshold in Bulk Polymers

A.V.R. Kumar, G. Padmaja, P. Radhakrishnan, V.P.N. Nampoori, and C.P.G. Vallabhan

Laser Division, Department of Physics, Cochin University of Science and Technology, Kochi 682 022, Kerala, India

Abstract. Pulsed photoacoustic technique is used to determine the laser induced damage threshold of bulk polymer sample of polytetrafluroethylene (Teflon). The experimental method for determination of damage threshold of samples under the irradiation from the 1.06 μm output of a Q-switched Nd:YAG laser using pulsed photoacoustic (PA) technique which employs a piezoelectric transducer as a detector has been described. It is observed that an abrupt increase in the PA signal amplitude occurs at the threshold energy density causing surface damage. There also exist distinct values for the threshold corresponding to different mechanisms operative in causing surface damage.

1. Introduction

Absorption of pulsed radiation gives rise to transient thermal variations due to heat released through non-radiative de-excitation channels, thereby causing pressure waves to be generated in the sample. By using a suitable acoustic transducer, it is possible to study the photoacoustic signals which arise from the non-radiative processes occurring in the sample. For a solid sample, there exists a threshold laser energy density above which ablation of the sample surface occurs. During ablation, many of the sample parameters influencing the thermal balance of the system are altered. Compared to the various methods like surface morphological studies, photothermal deflection technique etc., for the determination of the laser damage threshold of materials, photoacoustic technique is simple and can be used in-situ to determine this quantity in bulk and thin film samples. The present paper describes the use of pulsed photoacoustics to determine the laser damage threshold of bulk polytetrafluroethylene (Teflon) sample.

2. Experimental Technique

Figure 1 shows the schematic diagram of the experimental set up for determination of the damage threshold of bulk polymers. The laser is a Q-switched Nd:YAG laser (Quanta Ray, DCR-11) giving 1.06 μm wavelength having a pulse width of 10 nsec operating in the single shot mode. The sample, in the form of a disk of diameter 2.5 cms and thickness 5 mm is pressure contacted on to the surface of a piezo-electric transducer (PZT) mounted inside a steel enclosure. The surface of the sample is polished using 400 grade silicon carbide powder. The laser beam is focussed on to the sample surface to a diameter of ~1 mm. An on-line laser energy meter (Delta Developments) monitors the laser energy for each pulse. Care is taken to see that the laser does not irradiate the transducer casing and to expose fresh area of sample surface to the laser for each observation. The transducer and the sample are placed in an irradiation chamber.

Springer Series in Optical Sciences, Vol. 69 Photoacoustic and Photothermal Phenomena III
Editor: D. Bićanić

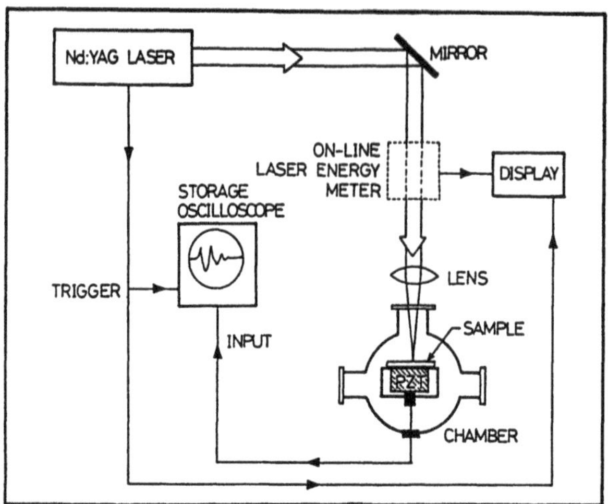

Fig.1. Schematic diagram of the experimental setup.

3. Results and Discussion

The absorption of the laser pulse by the sample produces the PA signal from the transducer which is monitored on a 100 MHz storage oscilloscope (Tektronix model 466) using a 50 ohm termination. As the laser energy is gradually increased, the amplitude of the PA signal pulse is monitored. A typical CRT trace of the PA signal pulse (laser energy density ~1.8 J/cm^2) is shown in Fig.2. The plot of the variation of the PA signal amplitude and the laser energy density is shown in Fig.3. From the figure, it is seen that there are clear changes of slope for the plot at two different regions **A** and **B**. The variation at **A** results from an abrupt increase in the PA signal amplitude. This is in accordance with the observations of previous workers [1,2,3]. The threshold at **A** (energy density ~1.78 J/cm^2) can be attributed to the surface morphology of the sample, which characterizes the damage due to impurities, inclusion and surface

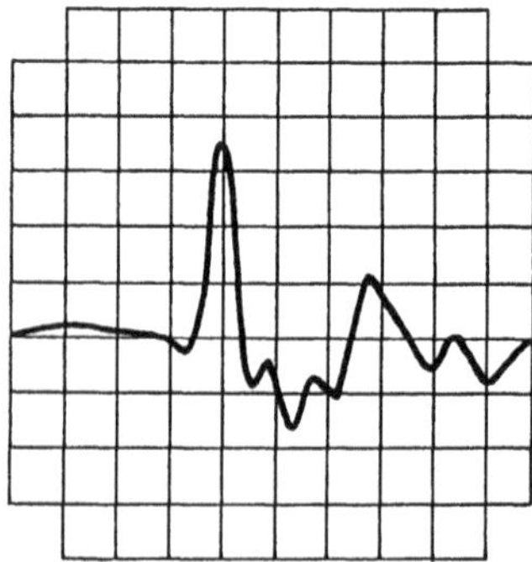

Fig.2. Typical CRT trace of the PA signal pulse from the transducer. (Horizontal : 2 µsec/div., Vertical : 50 mV/div.)

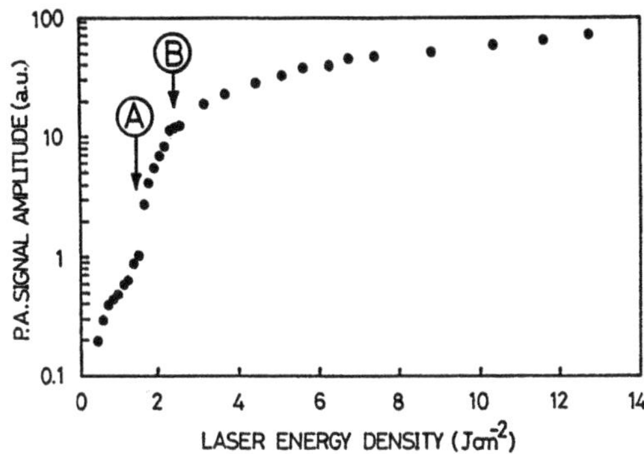

Fig.3. The variation of PA signal with laser energy density for bulk Teflon sample.

inhomogeneity. This value of threshold energy density agrees well with the results obtained for similar samples by other optical methods [4]. At higher laser energy density, there is yet another variation in the PA signal amplitude at B (laser energy density $\sim$2.85 J/cm^2). This can be attributed to the damage initiated by bond breaking and subsequent plasma formation and melting of the sample. At still higher laser energies, the plasma intensity increases and so does the PA signal, but it tends to saturate due to the fact that at greater laser energies, the plasma density is sufficient to cause absorption of the laser beam, thus saturating the PA signal.

4. Conclusion

The laser ablation threshold for bulk Teflon sample has been determined using pulsed photoacoustic technique using a piezo-electric transducer. The results show two distinct values for the ablation threshold corresponding to damages initiated by surface characteristics as well as melting and ablative processes in the sample surface.

5. Acknowledgements

The authors wish to thank the Council of Scientific and Industrial Research, New Delhi, the Department of Science and Technology and the Ministry of Human Resource Development (Government of India) for financial assistances.

6. References

1. Rosencwaig, A. and Willis, J.B., Appl.Phys.Lett., **36** (1980), p.667.
2. Harada, Y. Kanemitsu, Y., Tanaka, Y., Nakano, N., Kuroda, N and Yamanaka, K., J.Phys:D Appl.Phys., **22** (1989), p.569.
3. Dienstbier, M., Benes, R., Rejfir, P. and Sladky, P., Appl.Phys.B, **51** (1990), p.137.
4. Milam, D., Appl.Opt., **16** (1977), p.1204.

Depth Profiling of Polymer Laminates by Step-Scan Fourier Transform Infrared Photoacoustic Spectroscopy

R.M. Dittmar[1], *J.L. Chao*[2], *and A. Palmer*[1]

[1]Department of Chemistry, Duke University, Durham, NC 27706, USA
[2]IBM Corporation, Research Triangle Park, NC 27709, USA

Abstract. The use of step–scan Fourier transform infrared (FT–IR) spectroscopy combined with photoacoustic (PA) detection is used to depth profile laminated polymeric materials. Depth profiling can be performed using a combination of the modulation frequency dependence and phase lag of the PA signal.

1. Introduction

Fourier transform infrared (FT–IR) spectroscopy combined with photoacoustic (PA) detection is used for the analysis of a large variety of materials. While most commercial FT–IR spectrometers operate with continuous–scan interferometers, there is a growing interest in step–scan interferometers due to their intrinsically more facile application to dynamic spectroscopy, including condensed–phase PAS. PAS can be considered a dynamic (time–dependent) process due to the slow diffusion of the thermal wave as it travels from the point of absorption to the surface of the sample. The main advantage of step–scan interferometry for PAS is that a constant modulation frequency is applied over the entire spectral range to generate the PA signal with a thermal diffusion depth independent of wavelength. As a result, the sampling depth (in the absence of saturation) is constant across the spectrum for each frequency used. In contrast, with continuous–scan FT–IR each wavelength, λ, is modulated at its own frequency, $F=2v/\lambda$, where v is the mirror velocity. In the step–scan mode, the single modulation frequency allows a two–phase lock–in amplifier to be used to extract phase information with respect to the modulation and to extract the in–phase (I) and quadrature (Q) components of the PA signal. The extraction of PA phase data expands the capability of FT–IR PAS for the spectral depth profiling of heterogeneous materials.[1] This capability has not been fully taken advantage of until recently due to the difficulty in separating the instrumental phase from the sample phase in continuous–scan FT–IR. However, in step–scan FT–IR the easily accessible sample phase is completely separated from the instrumental phase. In addition, the instrumental phase for the I and Q interferograms is identical. The application of PA spectral depth profiling to polymer laminates is described in this paper.

The use of the PA phase for depth profiling was first illustrated by Kirkbright and co–workers in 1976.[2] More recently Mongeau *et al.*, have described a method for the separation of surface and bulk absorption in photothermal spectroscopy using the photothermal beam deflection (PBD) phase.[3] The PA phase relationships

Springer Series in Optical Sciences, Vol. 69 Photoacoustic and Photothermal Phenomena III
Editor: D. Bićanić

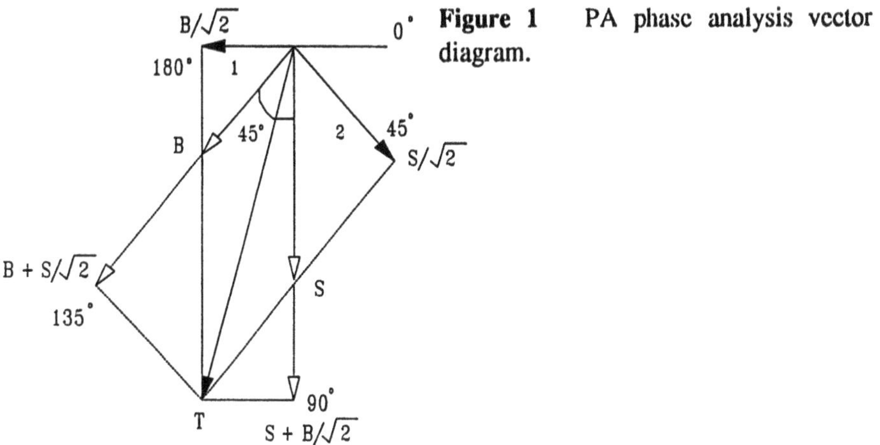

Figure 1 PA phase analysis vector diagram.

for the surface, bulk, and total signal in a two-layered sample consisting of a thermally thin, weakly absorbing surface layer on a thermally thick, weakly absorbing substrate are illustrated in Fig. 1.

To eliminate instrumental phase contributions, it is necessary to establish a phase reference to determine the correct angle of detection for a surface absorption. One way to do this is to use a completely saturated strong absorber as a reference sample, since it will have all of its absorption localized near the surface. The lock-in amplifier phase setting required to maximize the signal from this saturated absorber in one channel of a two-phase lock-in amplifier will be the same phase angle as that required to maximize the signal from a thermally thin surface absorber in the same lock-in amplifier channel. When the reference is replaced by the sample, the bulk signal multiplied by cos 45° (vector 1, Fig. 1) should be in the channel orthogonal to the reference phase. The surface signal free of contribution from the bulk may be detected by advancing the phase by 45° (if the substrate is weakly absorbing), to give a signal equal to the surface signal multiplied by cos 45° (vector 2, Fig. 1).

The technique outlined above is a method of depth-profiling which, strictly speaking, works only for thermally thin, weakly absorbing films on thermally thick, weakly absorbing substrates. However, in practice it can be extended qualitatively to samples which do not meet these strict criteria. When such deviations from ideal conditions occur, empirical adjustment of the detection phase may be necessary, and different phase separations may be required for different regions of the spectrum. Rather than collecting data at numerous phase angles, the phase of one set of interferograms (I and Q) can be rotated as necessary by use of a computer program based on simple vector mathematics in order to achieve maximum spectral distinction.

A more traditional way of obtaining depth profiling information by FT-IR PAS is to change the thermal diffusion length (by changing the modulation frequency) and to monitor the relative intensities of peaks that originate from different depths in the sample. In continuous-scan FT-IR this type of depth profiling is complicated by the range of depths sampled over the spectral range due to the wavelength dependent modulation frequency. However, in step-scan FT-IR each spectrum corresponds to a constant sampling depth (in the absence of saturation). Therefore, the analysis of these step-scan FT-IR spectra is greatly simplified. The combined use of the phase lag and the modulation frequency dependence of the PA signal allows for a greater range of depths and resolution in depth profiling. For example, in a three layer sample, the spectra of the top two layers could be phase separated at a high modulation frequency while the bottom bulk layer could be separated from the top two layers by going to a lower modulation frequency. One limitation of using the modulation frequency dependance for depth profiling is an interferometer's limited range of modulation frequencies. For example using the step-scan FT-IR spectrometer designed in our lab, the phase modulation frequency is limited to a range of approximately 100 -1000 Hz. However, this frequency range is large enough for many useful and interesting applications. More sophisticated step-scan FT-IR spectrometers will now have this frequency limitation.

2. Results and Discussion

The step-scan FT-IR PAS spectra of a variety of laminar polymeric materials have been collected using a modified Nicolet IR42 FT-IR spectrometer that is capable of collecting data in both the step-scan and the continuous-scan modes.[4] As an example of the depth profiling results, data are illustrated here for a laminate consisting of a 12 μm ethylene-vinyl acetate (EVA) copolymer film cast onto a 60 μm thick polypropylene (PP) substrate.

Spectra B and D of Fig. 2 show two vector components of the spectrum of the EVA/PP laminate separated by 135° of a single 100 Hz scan. Calculating the square root of the sum of the squares of these two vectors gives the 100 Hz magnitude spectrum (spectrum E). Note that the magnitude spectrum is a composite of the EVA and PP spectra. According to the theory illustrated in Fig. 1, the two spectra B and D in Fig. 2 should represent the phase-separated surface and bulk spectra, respectively. Spectra of pure PP and pure EVA are also shown in Fig. 2 A and C, respectively, for comparison.

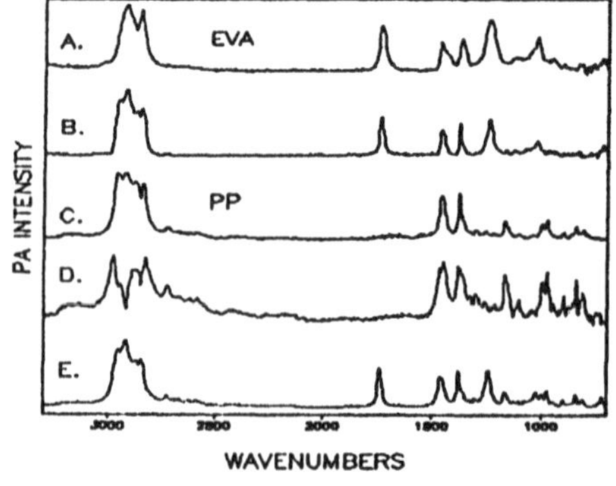

Figure 2 Step-scan FT-IR PA spectra of EVA/PP. A. Pure EVA, B. 45° (rel.) spectrum of laminate, C. Pure PP, D. 180° (rel.) spectrum of laminate, and E. $(B^2+D^2)^{1/2}$.

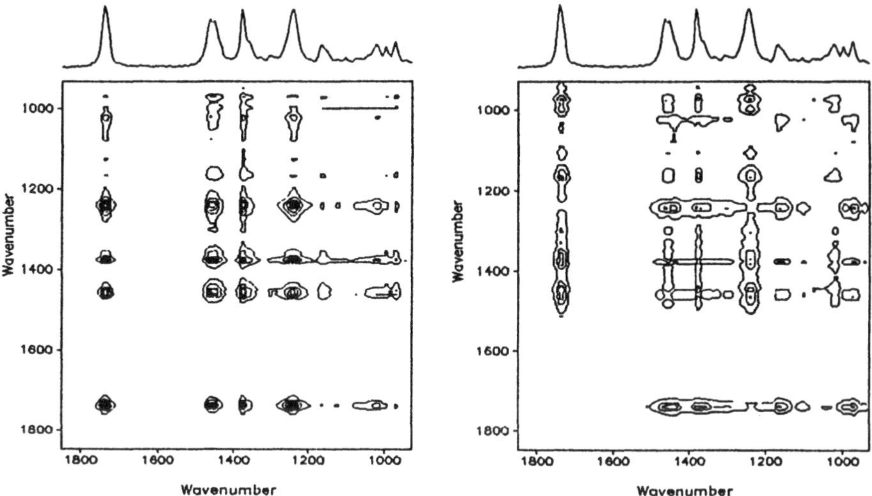

Figure 3 Synchronous 2D Map of EVA/PP.

Figure 4 Asynchronous 2D Map of EVA/PP.

The discrepancies in some regions between the phase–separated surface and bulk spectra of the laminate (B and D) and the true spectra of EVA and PP (A and C) can be explained by consideration of the differences in absorption strength of the various bands in the surface and bulk layers. Strong absorption in the surface masks and attenuates absorption by the bulk at those wavelengths (eg., 3000 – 2850 cm^{-1}), whereas strong absorption in the bulk shifts the phase of that signal toward the phase of the surface signal.

Another method to enhance the depth information from these PA spectra is the use of 2D FT–IR correlation maps.[5] In the 2D FT–IR analysis, the response to an external perturbation is measured by recording the I and Q components of the IR absorption, emission, or reflection. The I and Q spectra are then used to calculate the 2D frequency correlation maps.[6] In 2D FT–IR PAS the external perturbation is the PT effect itself and the time–dependent process is the thermal diffusion. The 2D correlation maps then become alternative representations of the depth profile information in the PA spectra of heterogeneous samples. Off–diagonal peaks in the synchronous correlation 2D FT–IR PA map connects frequencies of bands in components at the same depth. On the other hand, off–diagonal peaks in the asynchronous correlation map connects peaks from components at different depths. Illustrated in Figs. 3 and 4 are the synchronous and asynchronous spectra, respectively, of the same EVA on PP sample illustrated in Fig. 2.

Spectral depth profiling by phase analysis works most efficiently and simply when neither the surface layer nor the bulk is strongly absorbing and when the surface layer is thermally thin. However, modifications of this technique have allowed for the distinction of non–ideal surface and bulk layers as well. These modifications include spectral separations only in wavelength regions where the ideal conditions are met, the combined use of the phase lag and modulation frequency dependence of the PA signal, and the use of 2D correlation maps that can potentially extract more information than the vector components of the signal. General application of these techniques to heterogeneous samples with depth dependent absorptions is the subject of ongoing work.

1. R.M. Dittmar, J.L. Chao, and R.A. Palmer, Appl. Spectrosc. **45**, 1104 (1991).
2. M.J. Adams, B.C. Breadle, A.A. King, and G.F. Kirkbright, Analyst **101**, 553 (1976).
3. B. Mongeau, G. Rousset, and L. Bertrand, Can. J. Phys. **64**, 1056 (1986).
4. C.J. Manning, R.A. Palmer, and J.L. Chao, Rev. Sci. Instrum. **62**, 1219 (1991).
5. I. Noda, J. Am. Chem. Soc. **111**, 8116 (1989).
6. R.A. Palmer, C.J. Manning, J.L. Chao, I. Noda, A.E. Dowrey, and C. Marcott, Appl. Spectrosc. **45**, 12 (1991).

Methodical Studies of the Alignment Problem and Data Filtering in Photothermal Beam Deflection Experiments

B.K. Bein[1], *M. Woelker*[1], *J. Pelzl*[1], *and H.G. Walther*[2]

[1]Institut für Experimentalphysik VI, Ruhr-Universität, Postfach 102148, W-4630 Bochum, Fed. Rep. of Germany

[2]Institut für Optik and Quantenelektronik, Friedrich-Schiller-Universität Jena, Max-Wien-Platz 1, O-6900 Jena, Fed. Rep. of Germany

Abstract: In this methodological study of the photothermal beam deflection experiment with two laser beams, two aspects have been treated, which are essential for the reproducibility and quantitative interpretation: The smoothing of the measured raw data by convolution filtering and the alignment of the probe beam relative to the pump beam and sample surface.

1. Convolution Filtering of Measured Data

The performance of beam deflection experiments in measuring localized low-amplitude thermal waves is relatively bad. Acceptable results are only achieved by using the lock-in technique with narrow bandwidths leading to long data sampling times. Under such conditions, however, the complete system of a beam deflection experiment consisting of the sample, the thermal wave excitation and detection devices can be affected by long-time drifting. Under long-time heating, for example, the thermal expansion of the sample can disturb the adjustment between the probe beam and the solid surface, making comparative and reproducible measurements impossible. To improve the reliability of the experiment, even if the integration time of the lock-in amplifier is reduced, a convolution filter procedure [1] has been applied to the measured raw data. Figure 1 shows the transverse deflection signals of a lateral scan through the temperature mountain, found in the gas layer infront of a thermally thick Cu sample. The deflection signals are presented in the form quadrature versus in-phase contributions. At constant modulation frequency (14 Hz) and constant distance between the probe beam and sample surface (0.7 mm), the raw signals (connected in Fig.1 by thin straight lines) were registered as function of the coordinate along the surface (25 µm step width). Using a lock-in integration time constant of 1.25 s, twelve measurements were accumulated. Although the raw data points roughly follow the characteristic double loop predicted by theory [2], a quantitative interpretation of the signals seems to be impossible due to statistical fluctuations and external perturbations. This is only possible for the smoothed data (full line in Fig. 1), obtained after numerical filtering. The fluctuations of low ampli-

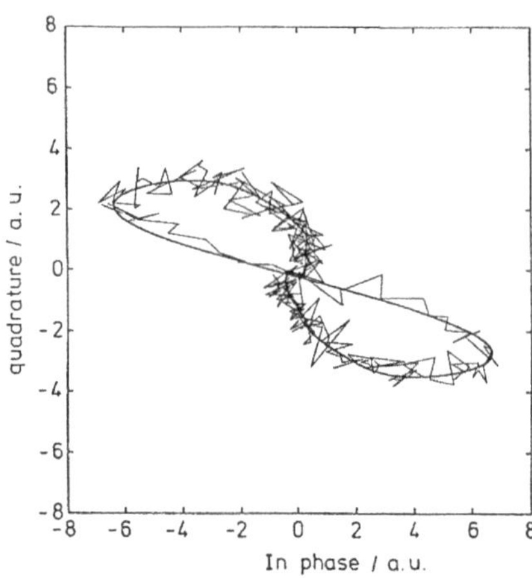

Fig.1: Quadrature versus in-phase contributions for the transverse deflection signal of a lateral scan: Comparison of the measured and filtered data.

tude have been eliminated by the filter procedure that relies on the spatial spectral distribution of a complete lateral scan across the temperature mountain. Using the low-pass Butterworth filter [3], the spectral distribution is preferentially smoothed at the short length scale disturbances. The effect of filtering can be described by an equivalent time constant leading to an enhancement of the signal to noise ratio by a factor of about 15.

2. Alignment of the Probe Beam Relative to Pump Beam and Sample

To establish the conditions for auto-alignment, which means that the positioning of the probe beam relative to the center of the heated spot can be determined by the beam deflection experiment itself, a scan system consisting of four prisms on two movable stages (Fig.2) was developed, which allows computer-controlled lateral and vertical scan measurements, while the other involved components (pump and probe laser, sample, and detector) remain at fixed positions. The alignment problem has been studied by systematic lateral and vertical scans across the temperature mountain in the gas region, whereby the geometrical parameters required for the quantitative signal interpretation have been determined.

Once the position of the probe beam relative to the center of the heating spot is known, and after applying the numerical filter procedure, the transverse deflection signals can be interpreted, e.g. with respect to the first zeros located to the right and the left of the center of the heating spot. Figure 3 shows the results of a frequency dependent study, where the zeros measured for the in-phase contributions are compared with the values calculated from an analytical formula [4],

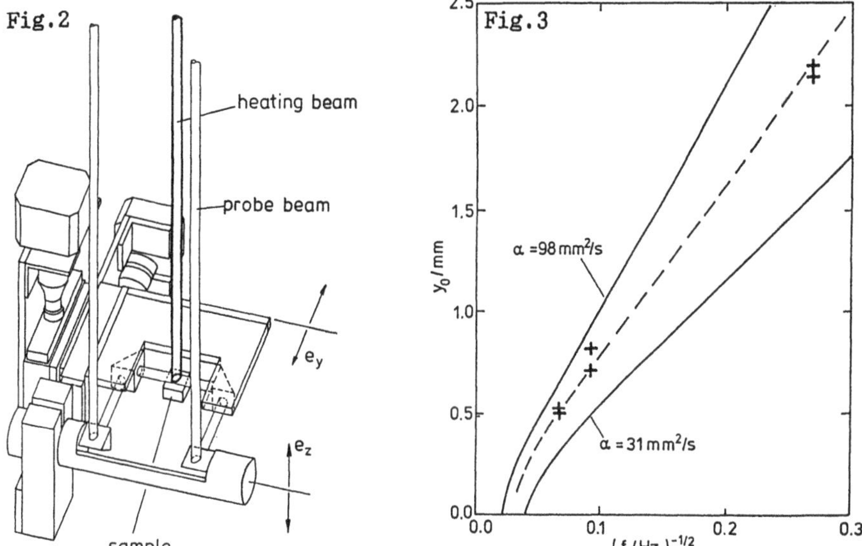

Fig.2: Scan unit consisting of four prisms on two movable computer-controlled stages.

Fig.3: Zeros of the transverse deflection in-phase contributions, measured for an Al sample as function of the modulation frequency f at a constant distance between the probe beam and the sample surface: Comparison of measurement (+) and theory (full and broken lines).

$$2\ y_0 = 1.2013\ (\lambda_s/2)\ [1 - 2.7719\ z_0{}^2/(\lambda_s/2)^2]^{1/2}$$

where $\lambda_s = \sqrt{4\pi\alpha_s/f}$ is the thermal wavelength of the solid sample and z_0 is the distance between the probe beam and the solid surface. The measured values (+) are approximated in Fig.3 by a curve calculated with a thermal diffusivity value $\alpha_s = 61 \cdot 10^{-6}$ m²/s, which is between the limits for Aluminium and air, the values of the solid sample and the surrounding gas, respectively.

3. Conclusions

For good thermal conductors of smooth surface the alignment of the probe beam relative to the heated spot has been determined by the beam deflection experiment itself. This position control is required for reproducible measurements and the quantitative signal interpretation and can be used for the auto-alignment in imaging applications. The extrapolation to smaller characteristic lengths, shorter thermal diffusion lengths and smaller pump and probe beam diameters seems to be possible. With the filter procedure in the Fourier space the necessary measurement time can be reduced and experimental details become available that otherwise would be hidden by external perturbations or statistical fluctuations.

References:

1 A.V. Oppenheim and R.W. Schafer, Digital Signal Processing (Prentice Hall, 1975).
2 L.J. Inglehart, Optical Beam Deflection Detection of Thermal Waves in Opaque Solids, Ph.D. thesis, Wayne State University, 1984.
3 S.A. Azizi, Entwurf und Realisierung Digitaler Filter (Oldenbourg, 1988).
4 M. Woelker, B.K. Bein, J. Pelzl, H.G. Walther, J.Appl.Phys. 70 (1991), 603-610.

Nondestructive Testing of Electric Contacts by Time-Resolved Infrared Radiometry

U. Netzelmann

Fraunhofer Institute for Nondestructive Testing, The University, Bldg. 37, W-6600 Saarbrücken, Fed. Rep. of Germany

Abstract. The bonding of small electric contacts consisting of small platelets of silver alloys welded to thin copper sheets has been tested by time-resolved photothermal radiometry with step-function surface heating. The contact platelet can be regarded as a heat sink coupled to the substrate by the thermal resistance of the welding joint. Identification of good and poor bonding is possible within 50 ms by analysing the normalized signal of the temperature response.

1. Introduction

Electric contacts consisting of small platelets of silver alloys arc-welded to thin copper sheets are produced in rates of several hundreds per minute. A 100% quality control during production requires fast and contact-free techniques for characterization of the bonding quality. A thermal testing technique has the advantage of making it possible to determine one of the most important operational parameters of the bond: The ability to conduct the heat away which is released in the contact during the switching process.

2. Experiments

We have studied silver alloy contact platelets with a size of about 2 mm x 3 mm and 0.7 mm thickness welded to copper sheets acting as a spring (thickness $d = 0.2$ mm). The mechanical and thermal contact between the platelet and the copper spring is only achieved by two linear welding stripes of about 1.5 mm x 0.3 mm size (Fig. 1). As sample geometry and its thermal parameters would require low modulation frequencies and long integration times in a narrowband photothermal experiment, we have used time-resolved photothermal radiometry with a step-function surface heating [1]. Due to the high reflectivity and the low emissivity of the contact surface, the heat input was realized by broad laser light illumination of the copper substrate from the side opposite to the platelet.

In Fig. 2 typical thermal response curves corresponding to different contacts are shown in a square-root/linear plot. After a transient occuring at very short times due to a thin graphite layer (< 10 μm) on the copper surface, a linear increase of temperature is observed in all cases, which corresponds to the thermal transit time of $0.4\, d^2/\alpha$ (about 0.2 ms) of the copper spring (α: thermal diffusivity). The curves in Fig. 2 have been normalized in this region. For the case of a spring without a platelet, after this time the slope increases due to the absence of a heat flow through the backside of the spring until finally transverse heat diffusion dominates. For the well bonded case, the platelet acts as a heat sink coupled by the thermal resistance of the welding joints. Therefore, the temperature increases at a significantly slower rate and with a different curvature, than in the uncoupled case.

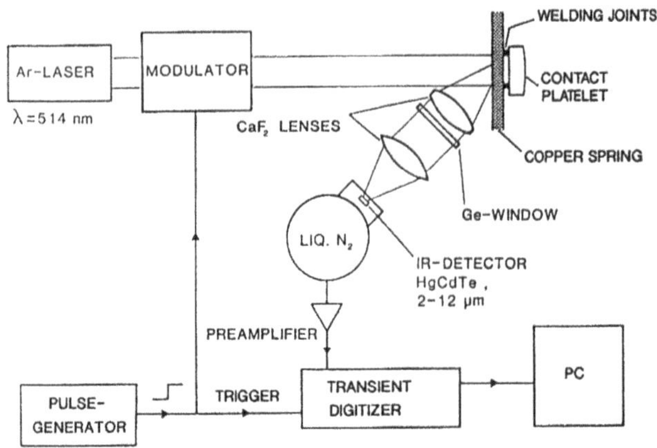

Fig. 1: Block diagram of the experiment and the geometry of the sample

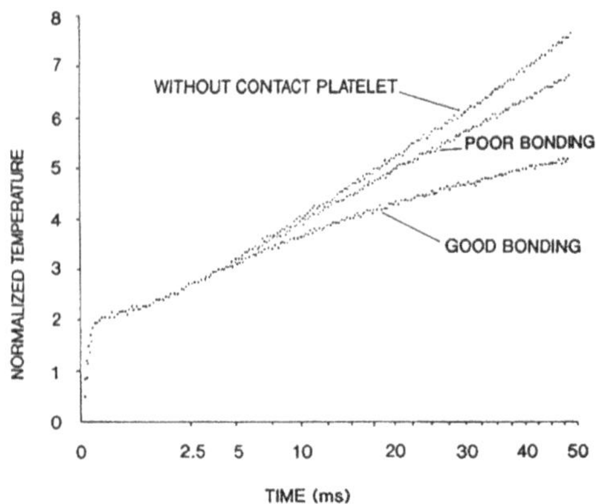

Fig. 2: Thermal response curves after step-function heating for different kinds of bonding

Detailed theoretical description of the temperature curve is complicated due to the geometries involved. A more global description should be possible by concepts used previously for analysing thermal coupling of particles to substrates [2]. The platelet is heated with a time constant of $\rho cV/G$, where G is the thermal conductance of the bond (ρ: density, c: heat capacity, V: volume).

3. Comparison with High-Frequency Ultrasound Measurements

The results have been compared with high-frequency ultrasound C-scans measured with narrowband excitation at 40 MHz by using a focussing polymer film transducer. The C-scan shown in Fig. 3 has been obtained from the same side as in the thermal experiment with a time gate set to the bonding

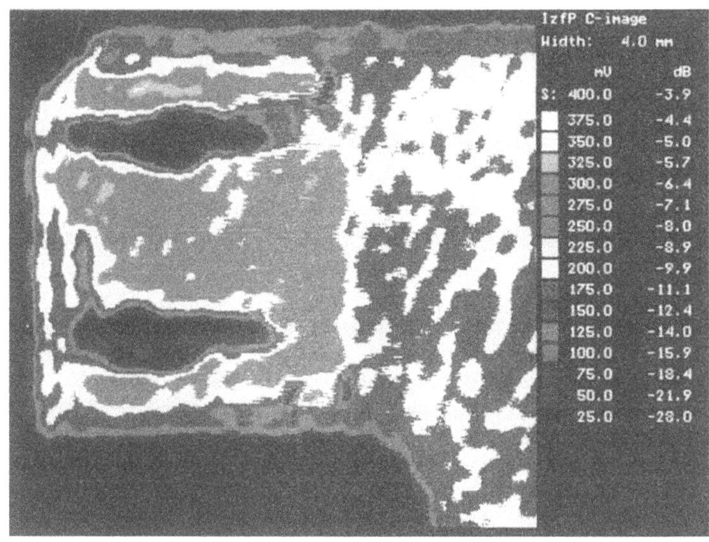

Fig. 3: High-frequency ultrasound (40 MHz) C-scan of an electric contact obtained from the back side

plane. The two welding stripes are visible as areas of low reflection due to the absence of the backwall reflection. However, surface roughness at poorly welded joints influences the ultrasound reflection in a similar way.

4. Conclusion

Provided that no thick layers have to be passed thermally, time-resolved radiometric techniques appear to be attractive testing tools for a fast characterization of welding joints even with complex geometries involved.

References:

[1] J. W. Maclachlan Spicer, W. D. Kerns, L. C. Aamodt, and J. C. Murphy, Rev. Progr. Quant. NDE **9B**, 1169 (1988)

[2] S. Utterback, F. H. Dacol, H. Ermert, and R. L. Melcher, Appl. Phys. Lett. **46**, 1054 (1985)

Radiometric Detection of Small-Amplitude Thermal Waves Between 500 K and 1000 K

R. Hüttner[1], *B.K. Bein*[1], *J.H. Gu*[1], *J. Pelzl*[1], *D.L. Balageas*[2], *and A.A. Déom*[2]

[1]Institut für Experimentalphysik VI, Ruhr-Universität, Postfach 102148, W-4630 Bochum, Fed. Rep. of Germany
[2]Division de Thermophysique, ONERA, B.P. 72, F-92322 Châtillon, France

Abstract: A thermal wave experiment based on IR detection has been developed, by which temperature oscillations of less than 100 mK are detected at average sample temperatures of 500 - 1000 K. The complete experimental setup includes a heating furnace to heat the sample in inert gas or vacuum. The thermal diffusivity and thermal depth profiles of graphite have been measured, thus demonstrating the potentiality of this method for remote sensing and in-situ measurements in non-accessible hostile environments.

1. Experimental Device

Thermal waves based on photoacoustic detection were proved to be valuable tools for depth-dependent studies of plasma surface modifications [1] and for measurements of thermal properties between 300 K and 1000 K [2]. In this work, the photoacoustic detection has been replaced by the radiometric one [3], to study the possibilities of remote in-situ analysis of surface modifications, e.g. in non accessible plasma devices for nuclear fusion. The complete experiment (Fig.1) to test the radiometric detection and to measure thermal properties at high temperatures, consists of three main components: The thermal wave excitation system, the vacuum cell to keep the sample at the desired temperature and protect it against oxidation and corrosion, and the IR detection system. Thermal waves are excited in the sample by absorption of a light beam, modulated by an electro-optic modulator or mechanical chopper C1. The sample S can be illuminated from the left and the right hand side, to allow the detection of both transmitted and reflected thermal waves. The corresponding radiometric signals, which are small fluctuations of the high radiation level related to the average sample temperature, can be filtered from the detector response by a vector lock-in. Chopper C2 is used when the total IR emission at the average sample temperature is measured. This second signal yields information about the surface emissivity and can be used for the calibration of the thermal wave signals.

In view of applications in large plasma devices, the scaling of thermal wave detection with larger distances between the sample and the first IR lens has been studied by using two configurations where the distance between the sample and the first lens was equal to, respectively twice the focus length. The radiation emitted by a sample surface area of about 7 mm² was collected within a solid angle of 0.11 sr, respectively 0.03 sr. To study the influence of

Springer Series in Optical Sciences, Vol. 69 Photoacoustic and Photothermal Phenomena III
Editor: D. Bićanić

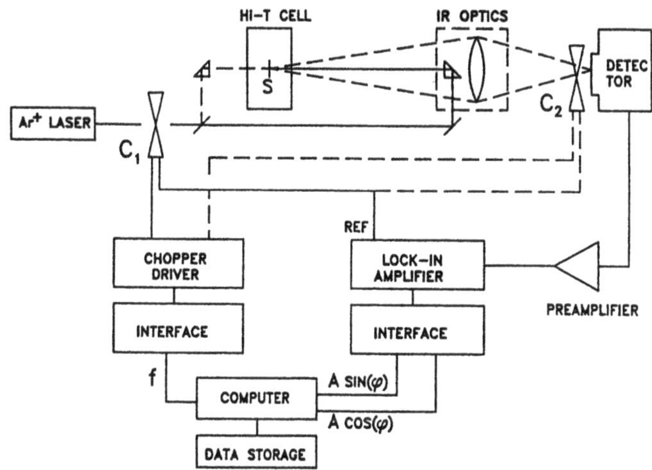

Fig.1: Schematic of the experiment, allowing to measure the IR signals of reflected and transmitted thermal waves, as well as the total IR emission at the average sample temperature.

the limitation of the detection wavelength interval, various IR filters and detectors have been used: an InSb- and an InAs detector and a broad band pyroelectric one.

2. Theory of IR Signal Generation

The radiometric signal related to the thermal waves and the signal of the total IR emission at the average sample temperature,

$$\delta M(T_s) = C' \cdot \epsilon(T_s, \lambda_1, \lambda_2) \int_{\lambda_1}^{\lambda_2} d\lambda \ \partial w(\lambda, T_s)/\partial T_s \ D^*(\lambda) \cdot \delta T_s \ , \quad (1)$$

respectively

$$M(T_s) = C' \cdot \epsilon(T_s, \lambda_1, \lambda_2) \int_{\lambda_1}^{\lambda_2} d\lambda \ w(\lambda, T_s) \ D^*(\lambda) \ , \quad (2)$$

depend on the temperature-dependent spectral distribution of the blackbody radiation $w(\lambda, T_s)$, the detector characteristics $D^*(\lambda)$, and the detection wavelength interval $\lambda_1 < \lambda < \lambda_2$. The quantity $\epsilon(T_s, \lambda_1, \lambda_2)$, introduced in analogy to the wavelength-averaged total emissivity ϵ, is an effective emissivity

$$\epsilon(T_s, \lambda_1, \lambda_2) = \frac{\int_{\lambda_1}^{\lambda_2} d\lambda \ \epsilon(\lambda, T_s) \ w(\lambda, T_s) \ D^*(\lambda)}{\int_{\lambda_1}^{\lambda_2} d\lambda \ w(\lambda, T_s) \ D^*(\lambda)} \ , \quad (3)$$

which characterizes the sample surface in the detected wavelength interval. Based on measurements of the signal $M(T_s)$ of the total IR emission at the average sample temperature, the combined quantity $C' \cdot \epsilon(T_s, \lambda_1, \lambda_2)$ can be eliminated, and thus the thermal wave δT_s can be measured directly, independent of the spectral emissivity $\epsilon(\lambda, T_s)$ which in plasma surface interactions may change as a

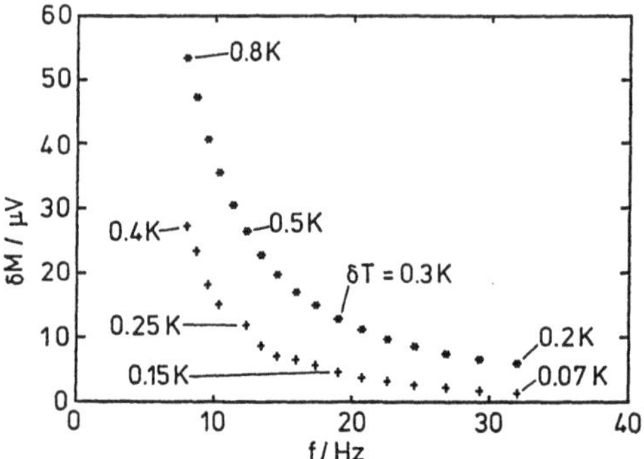

Fig.2: Radiometric signals of thermal waves, measured as function of the heating modulation frequency.

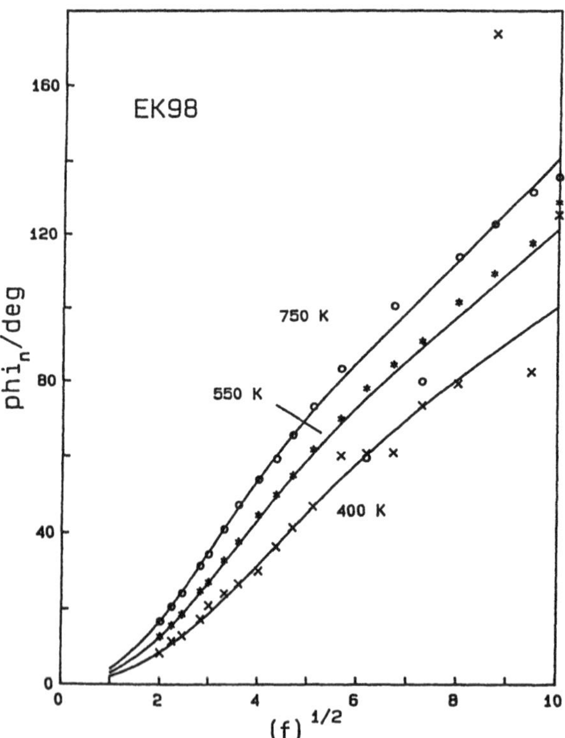

Fig.3: Normalized phases of transmitted thermal waves, measured for two graphite samples of different thickness.

function of time. The thermal waves $\delta T_s(f,T_s)$ can be registered as functions of the modulation frequency and average sample temperature and can be interpreted with respect to amplitude and phase.

3. Experimental Results

To demonstrate the resolution of small temperature oscillations at an average sample temperature background of 775 K, Figure 2 shows the radiometric signal $\delta M(f)$ measured with the InSb detector for thermal waves in transmission. The temperature amplitudes, measured at a modulation frequency of 32 Hz at the rear surface of Ta samples of 0.5 and 1 mm thickness, were 200 mK and 70 mK. Figure 3 shows normalized phases measured for two graphite samples at different temperatures. The deduced values of the thermal diffusivity support the results obtained by high-temperature photoacoustics [2]. Figure 4 presents measurements of reflected thermal waves, recorded from different graphite surfaces. In principle, the signals obtained by IR detection (O) at 700 K show the same characteristic behavior measured by photoacoustic detection on different graphite samples before and after plasma contact [4].

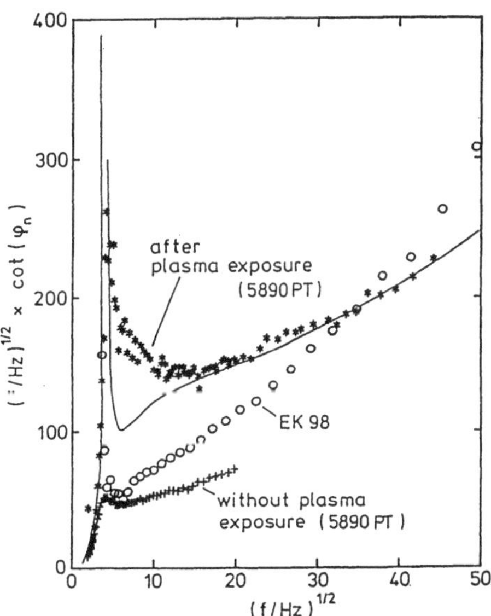

Fig.4: Phase shifts of reflected thermal waves, measured for rough graphite surfaces: Comparison of IR detection (O) with photoacoustic measurements.

4. Conclusions and Outlook

Thermal waves have been measured on metal and graphite samples, where at sample background temperatures of 500 K - 1000 K temperature amplitudes of less than 100 mK have been resolved. In measurements of reflected thermal waves, it has been shown that the sensitivity of the radiometric detection to surface structures or modifications is comparable to that of photoacoustic detection.

Scaling considerations based on measurements at different heating powers for thermal wave excitation and at different configurations of the IR system have lead to the conclusion, that the solid angle of the collected radiation can further be reduced. This corresponds to larger distances between the sample and the first IR lens. Distances of 3 m - 6 m, required for in-situ studies in large plasma devices, are realizable if a higher laser beam power (4.5 - 15 W) is available, if numerical data filtering is used in addition to the lock-in technique, and if the detection wavelength interval is chosen according to the average sample temperature.

Acknowledgements: The development of this experiment was supported by Max-Planck-Inst. für Plasmaphysik, Garching. Part of the work has been done in the frame of the French-German program PROCOPE.

References:

1 B.K. Bein, M. Wojczak, J. Pelzl, Materials Science and Engineering A122 (1989), 93-99.

2 B.K. Bein, H.W. Schmidt, D. Krüger, J. Gibkes, J. Pelzl, Materials Science and Engineering A122, 107-112 (1989).

3 P.E. Nordal, S.O. Kanstad, Infrared Physics 25, 295-304 (1985).

4 B.K. Bein and J. Pelzl, In: PLASMA DIAGNOSTICS, Vol.2, Surface Analysis and Interactions, (Eds. O. Auciello, D.L. Flamm), Academic Press, 1989, 211-326.

Photothermal Imaging of Electric Current Distributions in Thin Metallic Films

M. Reichling[1], *E. Welsch*[1,2], *C. Göbel*[1], *Z.L. Wu*[1,3], *D. Schäfer*[4], *and E. Matthias*[1]

[1]Institut für Atom- und Festkörperphysik, Freie Universität Berlin, Arnimallee 14, W-1000 Berlin 33, Fed. Rep. of Germany
[2]Institut für Optik und Quantenelektronik, Friedrich-Schiller-Universität Jena, Max-Wien-Platz 1, O-6900 Jena, Fed. Rep. of Germany
[3]Shanghai Institute of Optics and Fine Mechanics, Academia Sinica, P.O. Box 800-211, Shanghai, P.R. of China
[4]Zentralinstitut für Optik und Spektroskopie, Rudower Chaussee 6, O-1199 Berlin, Fed. Rep. of Germany

Abstract. The thermoreflectance technique is used to monitor periodic surface temperature variations induced by ac currents in thin gold films. It is shown that the thermoreflectance pattern directly images the current density distribution in the film provided thermal broadening is avoided by a proper choice of the modulation frequency. The method is applied for the absolute calibration of photothermal reflectivity and displacement measurements.

1. Introduction

Several techniques have been applied to measure temperature variations induced by ac currents in metals and semiconductors. Fanton and Kino /1/ used an interferometric technique to detect the surface displacement of a tungsten film, Heinrich et al. /2/ demonstrated the capability of the modulated reflectance technique to monitor fast electronic processes in semiconductor devices and Mandelis et al. /3/ performed imaging experiments on semiconductors applying combined optical and electric current excitation. In this paper we investigate the temperature field in a thin gold film on a glass substrate resulting from an ac current in the sample. The modulation frequency is chosen high enough to avoid thermal broadening i.e. the thermal diffusion length is smaller than the size of typical structures of the current density distribution. Since the film is homogeneous across the thickness an image of the surface temperature distribution is a direct probe for the squared current density pattern in the film. With a good degree of accuracy the temperature rise T in any point of the sample can be assumed to be proportional to the power density $p=\sigma j(x,y)^2$ generated in that point. The constant of proportionality can be calculated by a simple one-dimensional heat flow model considering heat losses to the substrate. We found experimental evidence that also the surface displacement is a direct measure for j^2 and can be calculated using a simple linear expansion model neglecting any lateral effects. Due to this simplicity in a precise description of surface temperature and displacement with a simple model such an experiment is ideally suited for the absolute calibration of photothermal measurements.

2. Experimental

A schematic of the sample and the experimental setup is shown in Fig. 1. The sample is a 2.5 μm thick rectangular (10mm x 30mm) gold film on a quartz substrate connected to the ac current source by two small feeding stripes (width 2mm). The current density of

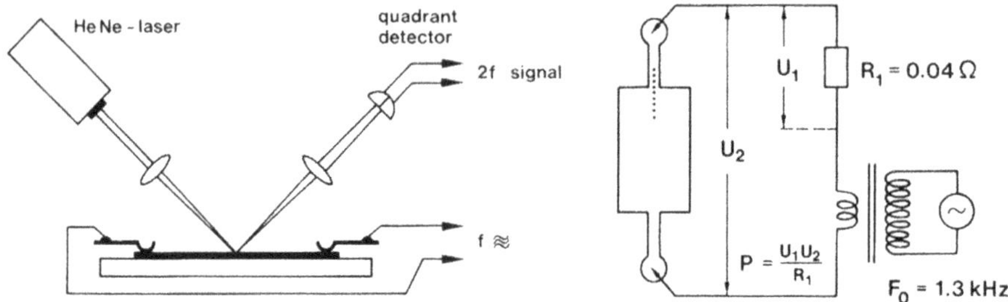

Fig. 1 Experimental setup for thermoreflectance measurements. Dotted line marks region for one-dimensional scan (see Fig. 3).

$j_{ss} \approx 10^5 A/cm^2$ in the stripe causes a dc temperature rise of approximately 80 ^{0}C. The surface reflectivity change and displacement are detected by a HeNe-probe laser beam focussed onto the sample. The difference or sum signal of a quadrant detector sampling the reflected HeNe laser light is processed by a Lock In amplifier operating at twice the current modulation frequency of 1.3 kHz. The sample is mounted on a computer controlled translation stage that allows two-dimensional scanning of the surface. The obtainable lateral resolution of the images is limited by the focal diameter of the probe beam that is approximately 5 μm.

3. Results

Fig. 2 shows images of amplitude (a) and phase (b) of the thermoreflectance signal for a two-dimensional scan over 9 mm^2 in the contact region of the sample. A constant signal in the feeding stripe is followed by a sharp decrease in amplitude in the transition region to the main rectangular part. The fact that the phase is almost constant over the whole sample can be regarded as experimental evidence for the assumption that lateral heat transport does not influence the signal significantly and the thermoreflectance pattern, in fact, is a probe for the local heat generation.
This finding is also supported by a comparison of the experimental data to a theoretical calculation of the current density distribution in the sample. Fig. 2c shows a graphical representation of the square of the current density calculated under the assumption of a

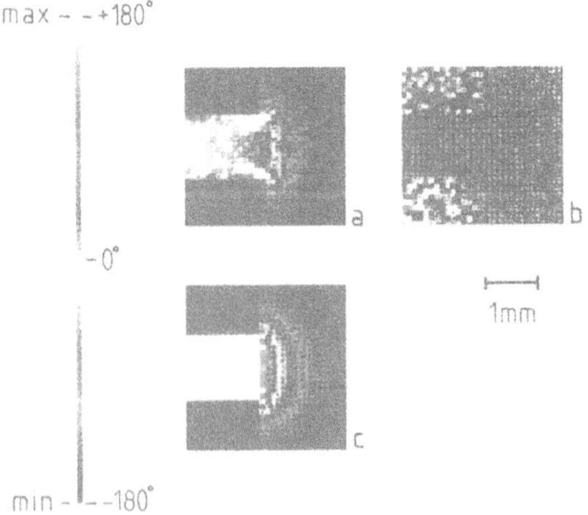

Fig. 2
Amplitude (a) and phase (b) of the thermoreflectance image at 2.6 kHz in the contact region of the 2.5 μm Au-film on quartz.

For comparison: square of the current density distribution (c) calculated from a theoretical model assuming a constant current density in the feeding stripe.

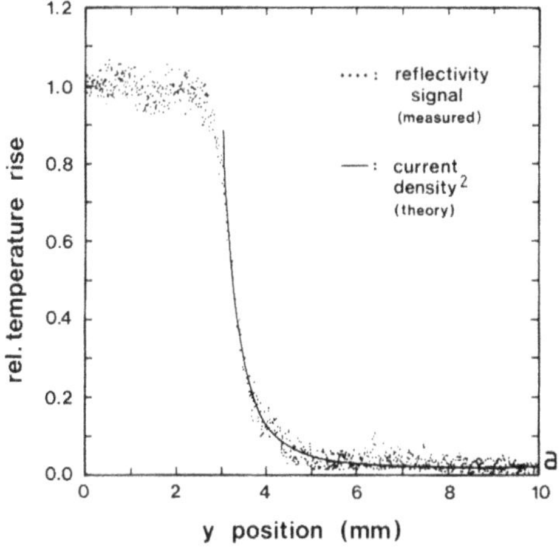

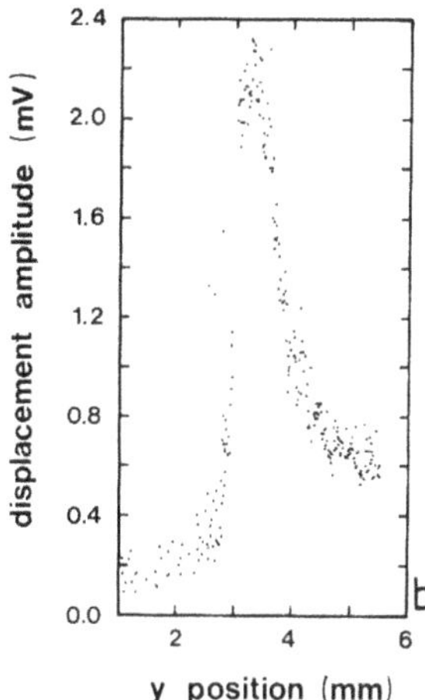

Fig. 3 Reflectivity (a) and displacement (b) signal for a one-dimensional scan across the contact region (see Fig. 1). The solid line represents a theoretical calculation of the squared current density.

constant current density in the whole contact stripe. Generally there is a good agreement between theory and experiment. The discrepancy appearing at the end of the feeding stripe is due to the incomplete theoretical model that neglects the distortion of the homogeneous field at the end of the stripe a phenomenon that has to be taken into account in a physically more realistic model. This effect can be studied more quantitatively in Fig. 3a where the reflectivity signal and the calculated temperature riseR along a straight line in the center of the stripe are compared.

Due to the large temperature gradient in the transition region this scan also yields a strong displacement signal represented in Fig. 3b. As predicted by the simple linear expansion model for the film displacement the line integral of this curve is almost identical with the temperature profile.

The measurements from Fig. 3 have been used for the absolute calibration of the experimental setup. We found calibration factors of $\Gamma_T=(3.67\pm0.10)$mV/K for thermoreflectivity and $\Gamma_D=(3.9\pm0.40)10^{-8}$ mV for the surface slope, i.e. an output voltage of $U_{pp}=3.67$ mV results from a periodic surface temperature rise of 1K and a voltage of $U_{pp}=3.92$ mV results from a periodic change of the surface slope by 10^{-8}.

This work was supported by the Deutsche Forschungsgemeinschaft, Sfb 337.
E. Welsch and Z.L. Wu acknowledge fellowships from the A.v. Humboldt foundation.

4. References

1. J.T. Fanton, G.S. Kino, Appl.Phys.Lett. **51**(2), 66 (1987)
2. H.K. Heinrich, B.R. Hemenway, K.A. McGroddy, D.M. Bloom, Electronics Letters **22**(12), 651 (1986)
3. A. Mandelis, A. Williams, E.K.M. Sin, J. Appl. Phys. **63**(1), 92 (1988)

Induction Heating as a Selective Heat Source in Fast Thermal Non-Destructive Evaluation

R. Lehtiniemi, J. Hartikainen, J. Varis, and M. Luukkala

Department of Physics, University of Helsinki,
Siltavuorenpenger 20D, SF-00170 Helsinki, Finland

Abstract. The application of radio-frequency induction heating as a selective heat source in fast thermal non-destructive testing system is described. The sample is heated through a scanning induction coil and the increased infrared emission from the surface of the sample is monitored by a line scanner. Because only conductive materials warm up in the induction field, dielectrics and conductors can be clearly distinguished in the thermal images. Both the spatial resolution and the selectivity of the system are demonstrated with test samples.

1. Introduction

Induction heating has formerly been applied to thermal non-destructive testing [1, 2], even in some industrial-scale systems [3], but generally laser heating has been much more popular. Because a focussed laser beam is an extremely localized heating method, it enables high spatial resolution. However, induction heating allows selective heating of conductive objects.

2. The Measurement System

The measurement system consists of a microcomputer-controlled scanning radio-frequency (RF) induction heating equipment combined with a multi-use fast photothermal evaluation system based on a simple and inexpensive infrared (IR) line scanner [4 - 6]. A schematic diagram of the entire system is shown in Fig. 1.

The RF signal is obtained from an ordinary signal generator and amplified by a wideband 200 W amplifier. The high power RF signal is then fed through an impedance matching circuitry to the induction coil designed to produce a linear heating pattern. Because of the sufficiently high 1 MHz – 30 MHz induction frequency, the skin depth remains small and accordingly only a very thin surface layer of the conductive material is heated.

The line scanner consists of a 8 – 12 μm range HgCdTe detector, a fast 100 Hz deflection mirror and a germanium IR lens, and it is focussed on a line parallel to the heating coil. Thermal image of the sample is formed by moving the sample perpendicularly to the heating line.

Entire system is controlled by a microcomputer with a 12-bit multifunction interface board. The I/O board also digitizes the data obtained from the IR detector. The data is displayed as pseudocolor images on the computer screen and the images

Springer Series in Optical Sciences, Vol. 69 Photoacoustic and Photothermal Phenomena III
Editor: D. Bićanić

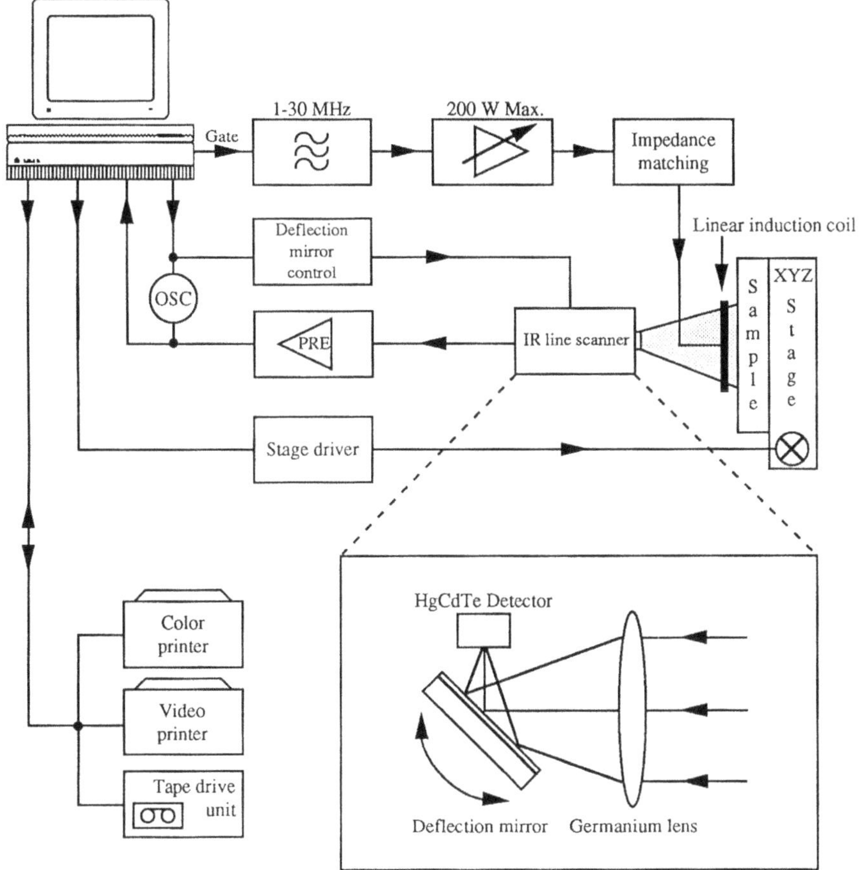

Figure 1. A block diagram of the measurement system.

can be enhanced by means of digital image processing with either the measurement software or commercial programs. The data can be stored on the computer hard disk and/or on the tape drive unit and hardcopies can be obtained with either an inexpensive ink-jet colour printer or a 256 gray shades video printer.

3. Result Examples

Fig. 2 shows a thermal image of a stainless steel ruler recorded using the system. Because the contrast variations in the picture are due to the variations in emissivity of the sample surface, the image demonstrates the spatial resolution of the system.

Fig. 3 illustrates the selective heating of the conducting objects by induction. Fiber composite test sample consists of multiple layers of glass and carbon fibers.

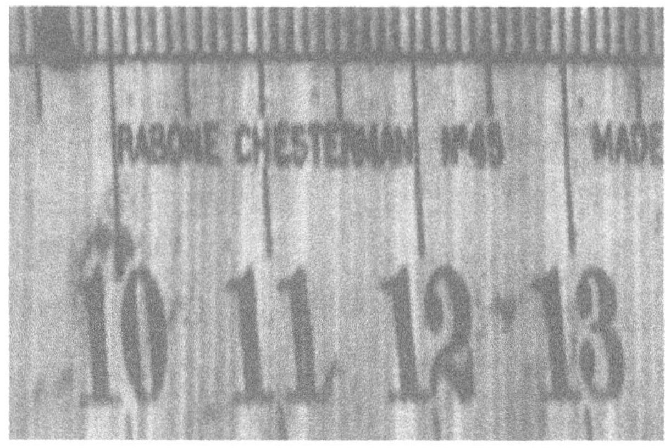

Figure 2. A thermal image of a stainless steel ruler heated through induction. The scale of the ruler is in centimeters.

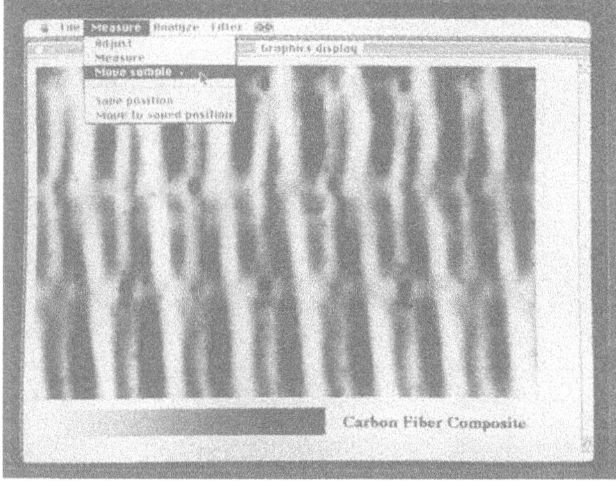

Figure 3. A thermal image of a carbon fiber composite sample. The size of the imaged area is 7 cm x 4 cm.

Carbon fibers have much better electrical conductivity than glass fibers and thus in practise only the carbon fiber bundles warm up and are easily detectable in thermal images. The conductivity of the carbon fibers is also highly anisotropic, for which reason the fibers parallel to the induction coil are heated most. Fig. 3 also demonstrates the graphical user interface of the system.

4. Conclusions

The induction heating is proved to be a considerable and practical alternative for the traditional laser heating in thermal non-destructive evaluation, especially when the sample consists of materials whose electrical conductivities differ clearly from each other and the defects are situated near the surface of the sample.

References

1. J. Saniie, M. Luukkala, A. Lehto, and R. Rajala, Electr. Lett. **18**, 651 (1982)
2. D. M. Heath and W. P. Winfree, in *Review of Progress in Quantitative Nondestructive Evaluation Vol. 9A*, edited by D. O. Thompson and D. E. Chimenti (Plenum Press, New York, 1990), p. 577
3. K.-J. Kremer, in *Proceedings of 3rd European Conference on Nondestructive Testing*, edited by C.M. Rosa (Italian Society for NDT, Brescia, 1984), Vol. 4, p. 171
4. J. Hartikainen, Appl. Phys. Lett. **55**, 1188 (1989)
5. J. Hartikainen, Rev. Sci. Instr. **60**, 1334 (1989)
6. J. Hartikainen, R. Lehtiniemi, and M. Luukkala, in *Review of Progress in Quantitative Nondestructive Evaluation*, edited by D. O. Thompson and D. E. Chimenti (Plenum Press, New York, 1991)

Competition Between Normal and Transverse Heat Flow in Time-Resolved Infrared Radiometry (TRIR)

J.W.M. Spicer, L.C. Aamodt, W.D. Kerns, and J.C. Murphy

Applied Physics Laboratory, Johns Hopkins University,
Johns Hopkins Road, Laurel, MD 20723, USA

Abstract. The time-resolved infrared radiometry (TRIR) technique involves monitoring the surface temperature of a specimen during the application of a step-function heating pulse. In the present work we address the competition between normal and transverse heat flow through structured media in determining the TRIR temperature-time$^{1/2}$ profile. Understanding this competition is critical if TRIR data are to be interpreted correctly to provide quantitative information about specimen properties.

1. Introduction

The TRIR technique [1] differs from other pulsed IR techniques [2,3] in that we monitor the development of surface temperature during the application of a step-function heating pulse to the specimen rather than monitoring the temperature decay after an instantaneous pulse. Evaluation of the temperature-time response as a function of position on the specimen allows information about local specimen properties and structure to be obtained. For example, the depth of a defect and the thermal resistance of the defect can be determined simultaneously in the same measurement. TRIR images of the specimen are constructed by displaying the temperature as a function of position for set time delays after the onset of the heating pulse.

Characterization of defect structures which lie parallel to the sample surface such as disbonds in coatings and delaminations in composites are well addressed by a one-dimensional, multilayer model [4] assuming area heating such as shown in Fig. 1(a). Here the thermal transit time is determined by the depth of the defect and the through-thickness thermal diffusivity. Detection of vertical defects which lie normal to the sample surface is not possible in this case and the use of a localized heating source is required. This situation is illustrated in Fig. 1(b) and indicates a second thermal transit time which is dependent on the diameter of the heating beam. In the following sections we give examples of TRIR characterization of structured specimens by monitoring normal heat flow and transverse heat flow.

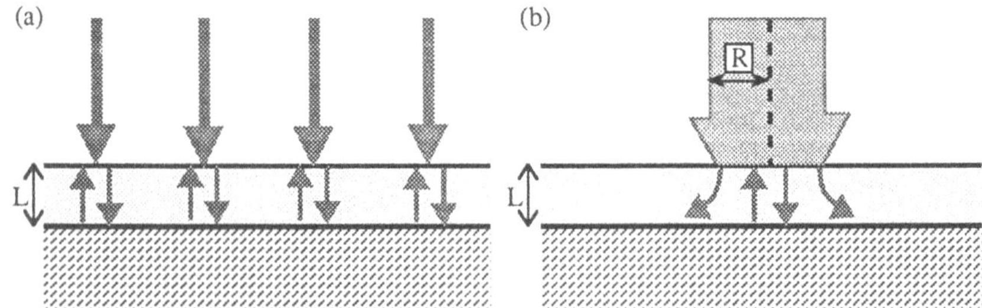

Fig. 1 Schematic representation of normal (a) and transverse (b) heat flow patterns in a layered structure. The normal thermal transit time, τ_N, is proportional to L^2/α_N where L is the coating thickness and α_N is the through-thickness thermal diffusivity of the top layer. The transverse transit time, τ_T, is proportional to R^2/α_T where R is the heating beam radius and α_T is the transverse thermal diffusivity.

Springer Series in Optical Sciences, Vol. 69 Photoacoustic and Photothermal Phenomena III
Editor: D. Bićanić

2. Normal Heat Flow in Multilayered Systems

Figure 2(a) shows a series of TRIR temperature vs time$^{1/2}$ curves for a zirconia thermal barrier coating on a superalloy substrate which exhibited a region of spalled coating. These data are plotted as a function of time$^{1/2}$ since this linearizes the response expected during heating of a thermally-thick medium and provides a convenient reference curve. For all areas on the coating, the temperature rises as time$^{1/2}$ until about 0.3 sec$^{1/2}$ which is the thermal transit time for the coating. After this time the slope of the curve decreases for regions well-removed from the spall due to the flow of heat into the more thermally conductive metal substrate. For regions adjacent to the spall, however, the slope of the curve increases due to the thermally insulating nature of the air layer between the disbonded coating and the substrate. Of most significance is the range of TRIR responses for the intermediate positions between these two extreme cases, indicating the potential for quantitatively evaluating the severity of the coating disbond.

Figure 2(b) shows the theoretical temperature-time curves from a one-dimensional multilayer theory [4] for a model system consisting of the coating layer, an intervening air layer, and a thermally-thick substrate (see inset) which represents the disbonded coating system for a series of air layer thicknesses. Note that the range of responses mirrors those found in Fig. 2(a). Destructive analysis of the specimen after the TRIR measurements were performed allowed a comparison of the amount of coating-substrate separation with the TRIR response as a function of position on the coating. Good general agreement was obtained although further work is in progress to refine the multilayer model to incorporate the effects of local points of attachment of the coating and the presence of oxidation products at the interface.

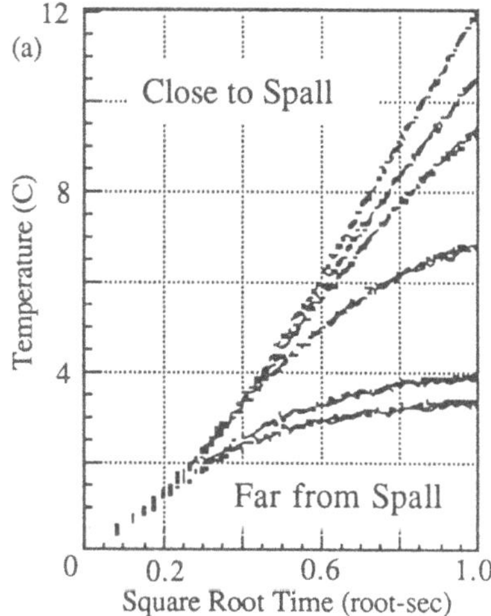

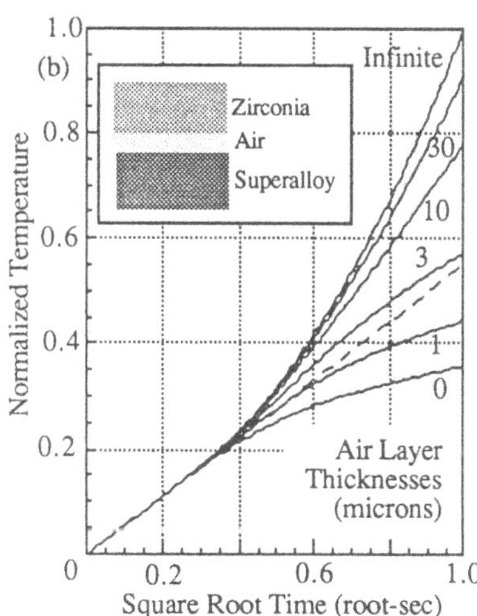

Fig. 2 (a) Experimental TRIR temperature-time$^{1/2}$ responses for a series of locations on a disbonded zirconia coating on a superalloy substrate. (b) Theoretical temperature-time$^{1/2}$ responses for a series of air layer thicknesses between a zirconia coating and a superalloy substrate.

3. Transverse Heat Flow for Detection of Vertical Defects

While the results above show the dependence of the TRIR response on the structure of a multilayer system due to variations in normal heat flow, it is important to understand the effects of transverse heat flow on the TRIR response since this can produce effects which may be incorrectly interpreted as representing well-bonded regions of coating. This is demonstrated in Fig. 3(a) which shows the experimental and theoretical TRIR responses for a thermally-thick

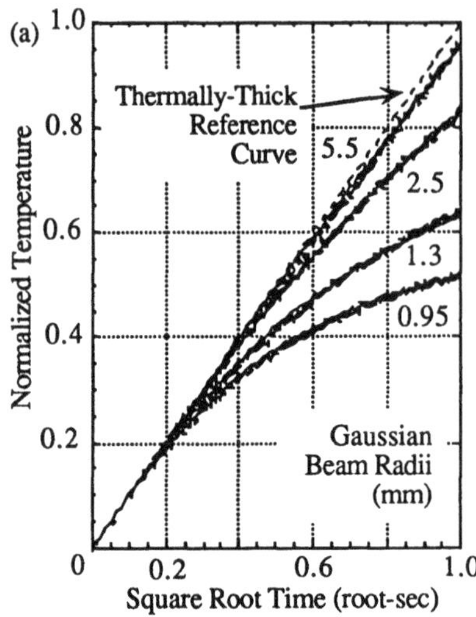

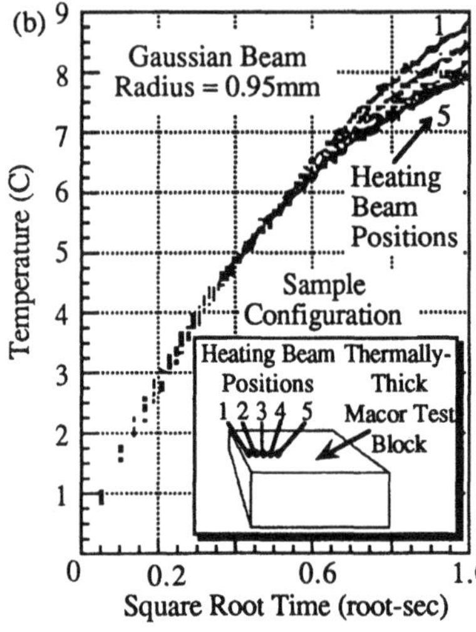

Fig. 3 (a) Experimental and theoretical TRIR responses as a function of beam radius in a thermally-thick ceramic block. (b) Experimental TRIR responses for a 0.95 mm Gaussian beam radius as a function of offset from the edge of a thermally-thick ceramic block.

ceramic specimen for a series of different Gaussian beam radii. Note that as the beam radius decreases, the time of deviation from the time$^{1/2}$ dependence decreases due to the decrease in the transverse transit time shown in Fig. 1(b).

The use of transverse heat flow for detecting vertical defects is demonstrated in Fig. 3(b) which shows the TRIR response for a small Gaussian radius as a function of offset from the edge of a thermally-thick ceramic sample. For positions well-removed from the edge of the sample (e.g. 4 and 5) the TRIR response is similar to that shown in Fig. 3(a) for the 0.95 mm beam. As the heating beam nears the sample edge, the TRIR response shows a slope increase due to the presence of the vertical boundary. The use of this effect in an imaging context has been applied in TRIR X-Y scans of composite samples subjected to impact loading. While overall regions of damage can be identified when a broad area heating source is used, there is a marked increase in image resolution when a line heating source is used. While the origin of this contrast is still under investigation, we feel it is due to the detection of vertical defect structures in the composite which can be imaged by transverse heat flow with a line heating source but not with the purely normal heat flow induced by an area source.

4. References

1. J.W.M. Spicer, W.D. Kerns, L.C. Aamodt and J.C. Murphy, *J. Nondestruct. Eval.* **8**(2), 107-120 (1989).
2. P. Cielo, *J. Appl. Phys.* **56**(1), 230-234 (1984).
3. D.L. Balageas, J.C. Krapez and P. Cielo, *J. Appl. Phys.* **59**(2), 348-357 (1986).
4. L.C. Aamodt, J.W.M. Spicer and J.C. Murphy, *J. Appl. Phys.* **68**(12), 6087-6098 (1990).

Real Time Thermal Wave Tomography

L.D. Favro, H.J. Jin, P.K. Kuo, R.L. Thomas, and Y.X. Wang

Department of Physics and Institute for Manufacturing Research, Wayne State University, Detroit, MI 48202, USA

Abstract. We report a thermal wave tomographic method that utilizes real-time techniques. A reference cooling curve, corresponding to a surface region beneath which there are no subsurface scatterers, is subtracted from the time dependence of every pixel of the image as it is generated. This is followed by a pixel-by-pixel determination of the peak contrast times of these subtracted curves and the peak times are stored as a 512 x 480 tomogram, all of which is done in real-time. We also store a peak-contrast image in which each individual pixel is displayed as it appeared when its contrast with the background was maximum. Currently, we are able to carry out this tomographic imaging at sampling rates up to 10 Hz. The process has been implemented with two different systems, both using an Inframetrics IR-600 Camera for data collection. In one of the systems the real-time processor is a series of DataCube boards on a VME bus controlled by a Sun 3/160 workstation. The other system is comprised of a Perceptics NuVision processor under the control of a Macintosh II microcomputer.

1. Introduction

Vavilov et al [1] have recently described a technique for making tomographic thermal wave images. Their method involves recording a succession of thermal wave images after a flash-heating pulse, followed by a numerical search for the temporal peak of the reflected thermal wave from subsurface features on a pixel-by-pixel basis. Since the peak time is related to the depth of the scatterer, this information enables one to separate time (or depth) slices of the total sequence of images. The result is a thermal wave tomogram. Since their process involves post-processing and a search through a large number of stored images, it is memory-intensive and cannot be done in real time. In the present paper, we report a real-time thermal wave tomographic method which accomplishes the same result, but does so with real-time techniques which avoid the storage of a large number of images.

2. Principle of Operation

In order to describe the concept of thermal wave tomography, we first describe the detailed time dependence of the surface temperature of the sample following flash heating, both at featureless regions and at regions above subsurface defects. When the flash occurs, the surface temperature in all regions rises very rapidly to a peak which is essentially independent of spatial position. Following the peak, the temperature falls more or less exponentially, with the rate of decline varying both in time and from point to point on the surface as the result of thermal wave scattering from subsurface defects. A typical cooling curve corresponding to some point on the surface can be thought of as being comprised of two components. One component is simply that which would occur if the material underneath that region contained no thermal wave scatterers. The other component results from the thermal wave reflected from any such scatterer. This reflected thermal wave slows the rate of cooling. To illustrate this principle, in Fig. 1 we show experimentally obtained heating and cooling curves corresponding to several subsurface flat bottom holes at different depths in a polymer test specimen. It can be seen from Fig 1 that there are varying cooling rates over different regions, as expected on the basis of the description above. In Fig. 1, the curve with the fastest cooling rate corresponds to a surface region beneath which there are no subsurface scatterers. In our tomographic method we make use of this curve as a reference and subtract this reference from the time dependence of every

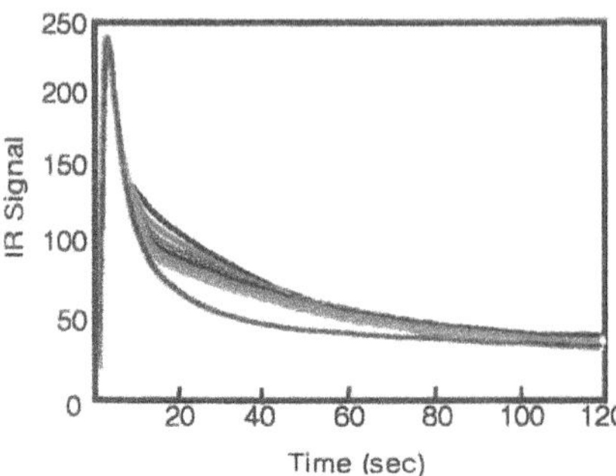

Fig. 1 Heating and cooling curves over six subsurface flat bottom holes at different depths in a plastic sample. The reference (lowest) curve corresponds to a region away from the holes.

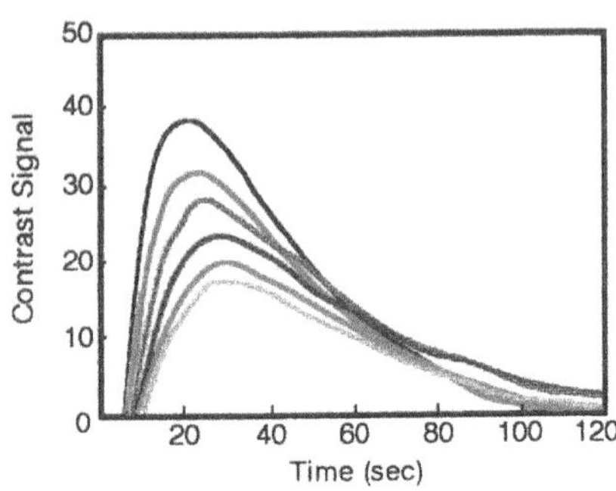

Fig. 2 Subtracted curves (expanded in scale) for the reference curve subtracted from the six curves above the subsurface holes of Fig. 1.

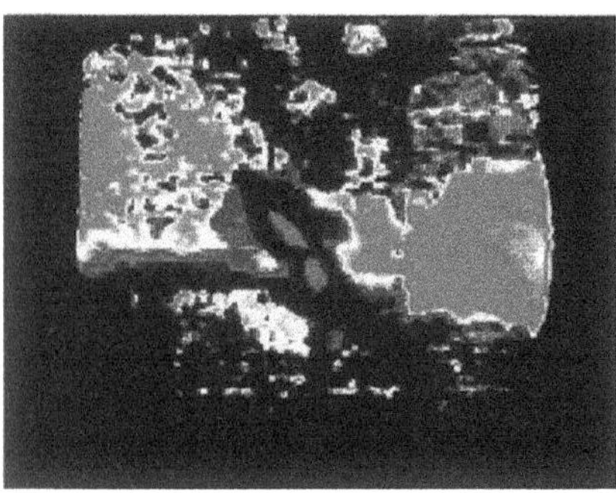

Fig. 3 Thermal wave tomogram of a sample of graphite-epoxy material with impact damage. Red indicates longer time (greater depth), blue shorter time (lesser depth).

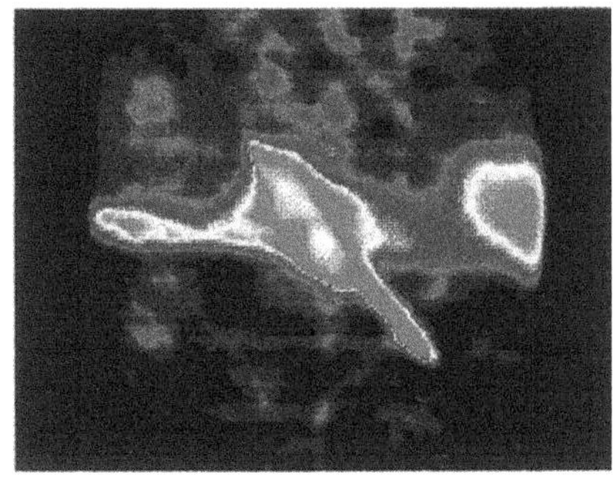

Fig. 4 Peak contrast image of the impact damage sample shown in Fig. 3. Red and white indicate greater contrast, blue lesser contrast.

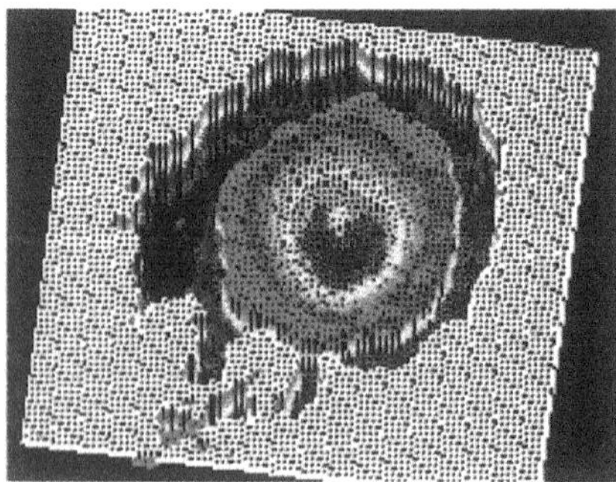

Fig. 5 A tomogram of a second graphite-epoxy impact damage specimen, shown as a perspective plot. Increasing time (depth) is plotted downward.

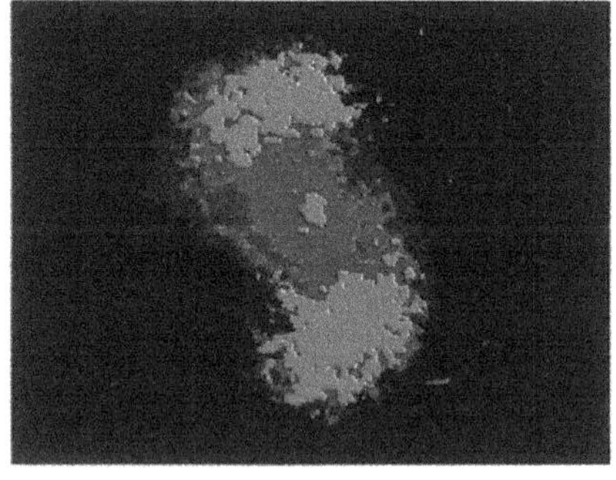

Fig. 6 A tomogram of a third graphite-epoxy sample showing impact damage which spirals to the right as the depth increases.

pixel of the image as it is generated. Examples of such subtracted curves (expanded in scale) are shown in Fig. 2 for the six selected regions corresponding to the cooling curves of Fig. 1. Because these curves have had the background curve subtracted, they correspond to plots of the reflected thermal waves from the subsurface holes. Although the peaks are broadened because of the diffusive nature of the heat propagation, the peaks of the subtracted curves (see Fig. 2) are seen to vary both in height and in temporal position, with the peaks corresponding to the deeper holes being lower and later. In our real-time method, we carry out a pixel-by-pixel determination of the peaks of these subtracted curves. This is accomplished by smoothing the curves to avoid spurious peaks caused by noise in the signal. The resulting 512 x 480 collection of peak times is stored as a pseudo-image in a buffer of the real-time processor. Thus, when the heating and cooling cycle is completed, a real-time tomogram is available in the buffer for viewing. In addition, a second buffer contains a "peak contrast" image. This second image displays each individual pixel as it appeared at the time when its contrast with the background of the image was maximum. Currently, we have the ability to produce both tomographic and peak contrast images simultaneously with a single heating flash, and with post-heating sampling rates up to 10 Hz.

3. Results

In Fig. 3 we show a tomogram of a graphite-epoxy sample which had been subjected to an impact which caused damage which extends from directly under the point of impact out to the edges of the sample in the horizontal direction. The sample was viewed from the side opposite the impact. The tomogram shows that the damage near the edges is further from the surface (white & red color) than the damage under the impact site (light blue color). In addition, there are other areas of the sample which indicate reflections from intermediate depths (yellow color) which may indicate porosity or incomplete bonding of the plies. Fig. 4 is a peak contrast image of the sample shown in Fig. 3. This image shows each pixel of the image as it appeared when its contrast was maximum. The greatest contrast (white color) corresponds to the shallowest defects as expected for heat diffusion. In Fig. 5 we show another sample of impact damage in a graphite-epoxy specimen. This sample is viewed from the impact side and the data are presented as a color perspective plot with greater time plotted downward to give the subjective impression of depth in the image. This impact caused shallow damage directly under and immediately surrounding the point of impact, with a wider ring of much deeper damage caused when the shock wave from the impact neared the rear surface. A final impact damage specimen is shown in Fig. 6. The damage in this specimen tends to spiral to the right as it goes deeper. This presumably results from the rotation of the fiber direction in the deeper plies.

The different depths in the tomograms shown here have been displayed through the use of pseudocolor because of space limitations in this publication. It is clear that it is possible to split the data into many images in such a way that each image shows a finite thickness "slice" of the sample, thus justifying the use of the word "tomogram" for these images. In a typical application, eight to ten such slices would be used to display the internal structure of a sample.

4. Acknowledgements

This work was sponsored by ARO under Contract No. DAAL03-88-K-0089; by the Center for Advanced Nondestructive Evaluation, operated by the Ames Laboratory, USDOE, for the Air Force Wright Aeronautical Laboratories under Contract No. W-7405-ENG-82 with Iowa State University; and by the Institute for Manufacturing Research, Wayne State University.

Reference

1. V. Vavilov, T. Ahmed, H.J. Jin, R.L. Thomas and L.D. Favro, Sov. J. NDT 12 (1990).

A Quantitative Analysis of Pulsed Video Thermographic Imaging of Subsurface Defects

S.K. Lau, D.P. Almond, and P.M. Patel

School of Materials Science, University of Bath,
Claverton Down, Bath BA2 7AY, UK

Abstract. Pulsed video thermography is quantitatively analysed using the images of a set of well characterised defects in a steel test-piece. An analytical finite defect model is presented which provides an understanding of the imaging process. The analysis is compared with the simulations of a numerical 3-d finite difference model.

1. Introduction

Pulsed video thermography (PVT)[1] is a non-destructive thermal wave technique which has recently evolved through the commercial development of video compatible infrared cameras and fast computational power. In its basic form, an infrared camera is utilised to record at fast video frame rate a large surface area's transient thermal response, photothermally heated by an intense flash of light. In this paper, the ability of PVT to image a series of well characterised subsurface defects is quantitatively assessed.

2. Theory

The 1-d thermal response of a semi-infinite homogeneous solid following the surface excitation of a Dirac delta heat pulse of intensity Q is given by [2]:

$$T(z,t)=\frac{Q}{\sqrt{\pi\rho c k t}}\exp\left(-\frac{z^2}{4\alpha t}\right) \tag{1}$$

in which T is the temperature at depth z beneath the surface at time t and $\alpha=k/\rho c$ is the thermal diffusivity of the solid (ρ,c,k are the density, specific heat and thermal conductivity resp.).

Consider the influence of a large air gap defect lying a distance L beneath the surface (z=0) and parallel to it. This air gap will be very much less conductive than the solid and it will "reflect" much of the incident thermal energy, coming from the pulse excitation, back towards the surface, enhancing the build-up of surface temperature in region containing the defect as "hot spot". The resulting surface temperature can be shown to be given by superimposing the solution of eqn.(1)

$$T(z=0,t)=\frac{Q}{\sqrt{\pi\rho c k t}}\left[1+2\sum_{n=1}^{\infty}\Gamma^n\exp\left(-\frac{n^2L^2}{\alpha t}\right)\right] \tag{2}$$

in which Γ is the thermal reflection coefficient of the air gap interface. The summation appearing in the second term is included to account for the effective multiple heat reflections between the defect and the surface.

For a finite air gap defect, however, only a portion of the initial uniform thermal wavefront that is incident on the defect surface will be reflected back towards the surface. An assumption at the outset is that the defect is also uniformly heated across the whole of its surface. The image formation of the defect on the surface is then

Springer Series in Optical Sciences, Vol. 69 Photoacoustic and Photothermal Phenomena III
Editor: D. Bićanić

basically a Huygen's superposition of thermal waves re-emitted from the secondary point heat sources of the defect surface. An assumption is made here that there is no interaction between neighbouring point heat sources and only the first total reflection ($\Gamma=1$) by the defect is considered. The transient solution is obtained from first solving the problem in the frequency domain followed by a Laplace inversion. Using spherical polar coordinates (r,θ,φ), the surface temperature at any point on the surface, distance x from the central axis of a circular defect of radius a, heated by a periodic source of angular frequency, ω, is given by:

$$\overline{T}(x,\omega)=\frac{Q}{k\sigma}\left[1+\frac{2\sigma}{\pi}\exp(-\sigma L)\int_0^a\int_0^\pi \exp\frac{(-\sigma R)}{R}\,r\,d\theta\,dr\right] \qquad R=\sqrt{z^2+r^2+x^2-2rx\cos\theta} \tag{3}$$

where $\sigma=(1+j)/\mu$; $j=\sqrt{-1}$ and $\mu=\sqrt{2\alpha/\omega}$ is the thermal diffusion length. The basic one-reflection time dependence of surface temperature contrast over the center of the defect is given by:

$$T(x=0,t)=\frac{Q}{\sqrt{\pi\rho ckt}}\left[\exp\left(-\frac{L^2}{\alpha t}\right)-\exp\left(-\frac{(L+\sqrt{a^2+L^2})^2}{4\alpha t}\right)\right] \tag{4}$$

3. Experimental Results

The mild steel defect test piece studied contained a series of 0.1mm deep circular air gaps of various diameters all at a depth of about 0.8mm beneath the surface [3]. The experimental PVT system has been described before [1] and the results were stored on video tape. An image analyser was used to subtract background (no defect region) from defect centre signals to extract the video thermal contrast data. The characteristic rise of temperature contrast to a peak followed by a decay at long times is a feature of finite defect response. Figs.1 to 3 show the curve fits to the extracted PVT contrast data of 2,2.8 and 4mm diameter defects. Curves B and C are the 1-d contrast response predicted for an infinitely wide defect (eqn.2: setting $\Gamma=1$, $n=\infty$ and 1 respectively). The larger defect in fig.3 follows, as expected, more closer to the 1-d response. Curve A in each figure is the corresponding 3-d numerical finite difference simulation for 1mm thick disc air gap defects of appropriate diameters. Curve 1 is the single reflection finite defect model (eqn.4).

4. Discussion and Conclusions

Deviations from the response predicted by 1-d thermal diffusion theory have been found in PVT data obtained from real finite defects. A first order finite defect theory is presented and compared with the calculations obtained from a dedicated 3-d numerical finite difference defect model, both show to a good order of agreement with the real defect data at short times. Both experimental and theoretical studies show the most useful data dominated by short wavelength thermal waves at short times can be used to assess defect depth reliably when image degradation by diffraction is at a minimum. On comparing curves 1 and A in each figure, it appears that the finite defect thermal wave model provides a good approximation to the characteristic responses of all the small defects using only a single defect reflection. The discrepancies of temperature contrast at long times are believed to be contributed from an overestimation of the reflection strength of the assumed uniformly heated defect surface without considering defect edge effect. However, neither the model nor the numerical simulation account for the experimental response at long times. For increasing defect size, multiple reflections of long wavelength components have been discussed in [3].

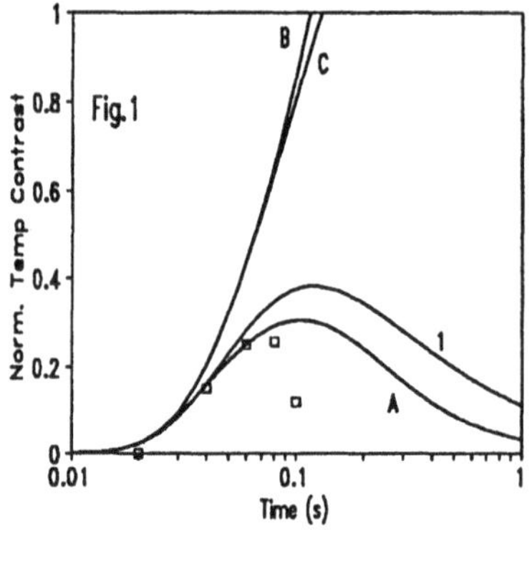

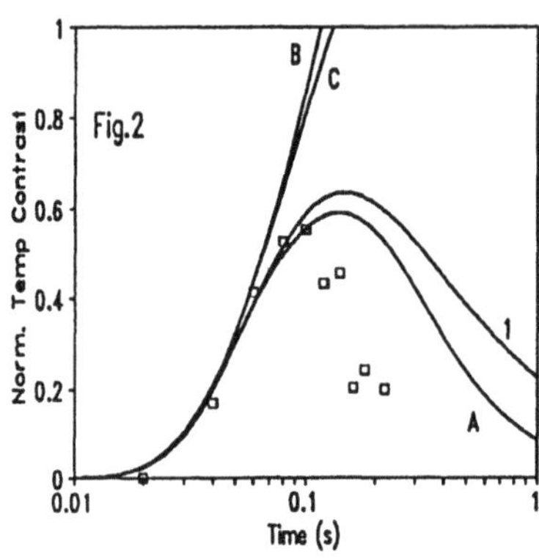

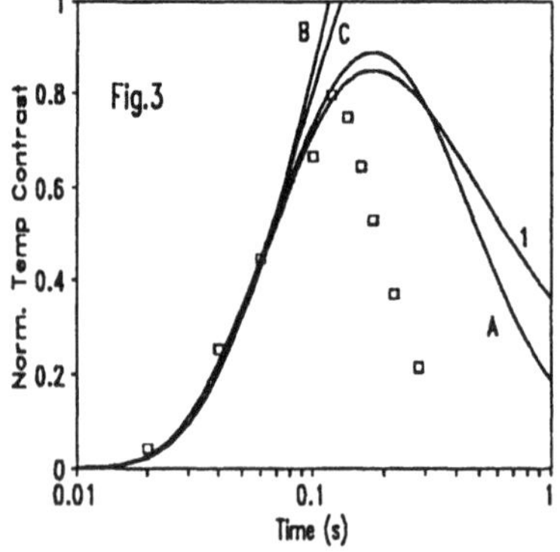

Curve fits to normalised PVT temperature contrast data of fig.1)2mm, fig.2)2.8mm and fig.3)4mm dia. defects. 1)single reflection finite defect response, A)3-d numerical finite difference simulation of 1mm thick disc air gap defect, B) and C) 1-d infinite air gap defect response [eqn.2: setting $\Gamma=1$, $n=\infty$ and 1 resp.]

5. Acknowledgements

The authors wish to acknowledge the financial support of the PowerGen plc (U.K.).

6. References

1. J.M.Milne and W.N.Reynolds. SPIE **520** (1984) 119-122.
2. H.S.Carslaw and J.C.Jaeger. Conduction of Heat in Solids. Claredon Press (1959) 256
3. S.K.Lau, D.P.Almond, P.M.Patel, J.Corbett and M.B.C.Quigley. SPIE **1320** (1990) 178-189.

Photodeflection and Interferometric Methods in Thermal Wave Spectroscopy and Microscopy of Solids

A.L. Glazov and K.L. Muratikov

A.F. Ioffe Physical-Technical Institute, Academy of Sciences, 194021 St. Petersburg, Russia

Abstract. Photodeflection and interferometric techniques for thermal wave detection are compared. Conditions are determined under which the interferometric detection is preferable to the photodeflection one.

At present, the photodeflection (PD) method is widely used in thermal wave spectroscopy and microscopy of solids. It is based on measurements of probe laser beam deviation resulting from interaction with the thermal lens (TL). At the same time, an application of the interferometric technique has been demonstrated for studying similar problems [1-3]. For this reason, it is important to compare the potentialities of the PD and interferometric methods and also to determine the optimum conditions for their application.

PD experiments are often carried out over a wide range of modulation frequencies of the pump radiation. In this case the TL sizes become comparable with the width of the probe laser beam and it is necessary to use the results of wave optics for analyzing the behavior of the PD and interferometric signals [4, 5]. These papers contain expressions for the PD signals obtained under these conditions. Similarly, the following expression can be obtained for the interferometric signal in the "mirage" effect geometry :

$$S_i(\omega)=\frac{mKI}{\lambda}\cdot\frac{dn}{dT}\exp\left[\frac{i\omega r^2}{4\varkappa}\right]\int_{-\infty}^{\infty}\exp(-\gamma z_o)\operatorname{erfc}\left[\frac{\gamma r}{2}-\frac{z_o}{r}\right]\hat{T}_s(0,k_y,\omega)dk_y, \quad (1)$$

where K is the proportionality coefficient of the light intensity on the photodetector and the electric signal; m is the coefficient characterizing intensity losses due to the probe beam splitting in an interferometric scheme; $\omega=2\pi f$ is the modulation frequency of the pump laser intensity; I, λ and r are the probe beam intensity, wavelength and radius respectively; dn/dT is the temperature coefficient of the refractive index; $\varkappa$ is the thermal diffusivity of the air; $\gamma=(k_y^2+i\omega/\varkappa)^{1/2}$; z_o is the height of the probe beam axis over an object surface; $\hat{T}_s(k_x,k_y,\omega)$ is the Fourier-transform of the surface temperature.

Let us compare the behavior of the PD and interferometric signals for a relatively simple case when the length of temperature waves forming the TL is greater than the radius of the probe beam and much less than the pump beam radius. Then the expressions from papers [4,5] and expression (1) can be consi-

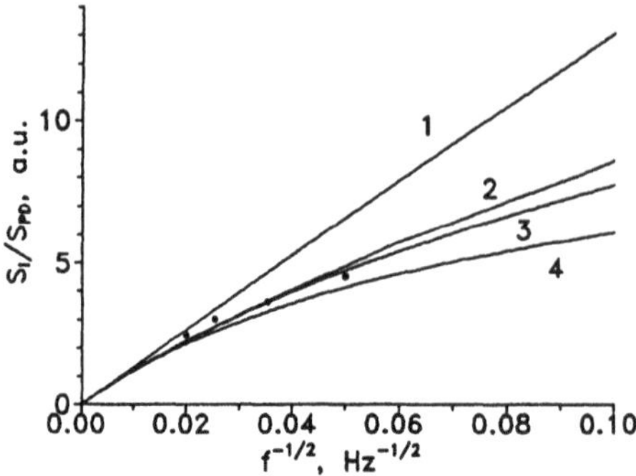

Fig.1. The dependence of the magnitude ratio of the interferometric and PD signals on the modulation frequency. Curve 1 corresponds to the case of 1D thermal waves. Curves 2, 3 and 4 correspond to absorption coefficient α=1,100 and $10^4 cm^{-1}$ respectively. For all curves r = 24μ; for curves 2-4, z_o=50μ and the pump beam radius equals 20μ. Points represent experimental results for GaAs at λ=1.06μ.

derably simplified. The signal amplitude ratio has the following form:

$$\frac{S_i(\omega)}{S_p(\omega)} = m\left(\frac{\pi}{2}\right)^{1/2} \cdot \frac{l(\omega)}{r}, \qquad (2)$$

where $S_p(\omega)$ is the PD signal; $l(\omega)=(2\varkappa/\omega)^{1/2}$.

Therefore, the interferometric signal will be greater than the PD one at modulation frequencies $\omega<\omega_o= \pi m^2\varkappa/r^2$. Under common conditions of the PD experiments, ω_- usually ranges from several to dozens kHz and the interferometric technique can provide a greater signal compared to the PD one. It should be noted that the ratio (2) is independent of an absorption coefficient of the object. However, this is valid only in the 1D approximation of thermal diffusion processes. A detailed analysis of more general 3D case may be made only numerically.

The behavior of the interferometric and PD signals from highly absorbing objects has been analyzed in [6]. Therefore, in this paper we shall deal with a case of weakly absorbing objects. Fig.1 represents the dependence of the interferometric and PD signal ratio on the modulation frequency. Theoretical and experimental results obtained for GaAs wafer show that the interferometric signal is still greater than PD one at not very high frequencies. Moreover, one can conclude from the comparison with cited results [6], that the signal ratio in the case of weak absorption is somewhat larger compared to strong absorption case and the advantage increases with decreasing the frequency .

To compare sensitivities of the techniques a noise analysis is required. First, there are several noise kinds which are the same for the both methods, namely, noises caused by probe laser intensity fluctuations and electronic noises of a recording system. On the other hand, there is an additional "pointing" noise for the PD technique which play no role in the interferometric detection. This means that the total noise level for the interferometric method does not exceed that for PD one. The possibility of obtaining a signal of a greater magnitude by means of the interferometric method results in its greater sensitivity.

The corresponding expressions for the PD and interferometric signals in a geometry with parallel probe and pump beams are given in [4]. According to these results, the signal ratio is independent of the sample absorptivity and the interferometric signal always exceeds the PD one. Hence, in the parallel geometry the interferometric method is able to provide higher sensitivity up to very high modulation frequencies.

This paper does not consider a realization of the interferometric method advantages.An optical system with a large acoustic noise immunity has been proposed and studied in [2,3] that is capable to realize the above mentioned advantages.

To conclude, the results presented in the paper demonstrate that in many cases the interferometric method can be more effective than the PD one. In this connection, it is interesting to investigate more profoundly its possible applications in thermal wave microscopy and spectroscopy.

References

1. L.Chen,S.Y.Zhang, Appl.Phys.Lett. **50**,1340(1987).
2. A.L.Glazov, K.L.Muratikov, Int. J. of Optoelectronics **4**, 589 (1989).
3. H.G.Walther,K.Friedrich,K.Haupt,K.Muratikov,A.Glazov,Appl.Phys. Lett. **57**, 1600 (1990).
4. A.L.Glazov,K.L.Muratikov, Optics Comm. **84**, 283(1991).
5. A.L.Glazov,K.L.Muratikov,Zh.Tech.Fiz.(in Russian) **61**, N11,187 (1991)
6. A.L.Glazov,K.L.Muratikov,Pis'ma Zh.Tech.Fiz.(in Russian) **16**,81 (1990).

Photodeflection Signal Formation from Solids Within the Framework of Wave Optics

A.L. Glazov and K.L. Muratikov

A.F. Ioffe Physical-Technical Institute, Academy of Sciences, 194021 St. Petersburg, Russia

Abstract. An influence of wave optics effect on parameters of the photodeflection signal is studied theoretically and experimentally. It is shown that these effects play an important role at a high modulation frequency of pump laser radiation.

At present, the photodeflecton (PD) method is widely used in thermal wave study of solids. It is necessary to correctly interpret various experimental results in order to determine sample parameters quantitatively. Nowadays, the parameter determination by the PD technique is based on a PD signal calculation in terms of geometrical optics. An account of a finite probe laser beam radius has been shown [1,2] to be very important. However, in this approach, the effects related to the interference of individual laser rays in a plane of photosensitive elements are neglected. For this reason, it is necessary to consider the PD signal formation within the framework of wave optics.

To determine the PD signal, a probe beam intensity distribution in the photodetector plane after interaction with a thermal lens (TL) have to be found. Fig.1 shows an arrangement for a sample, laser beams and photoreceiver in the "mirage" geometry experiment. Under common PD experimental conditions, the calculation may be made using the Fresnel-Kirhhoff integral in an approximate form

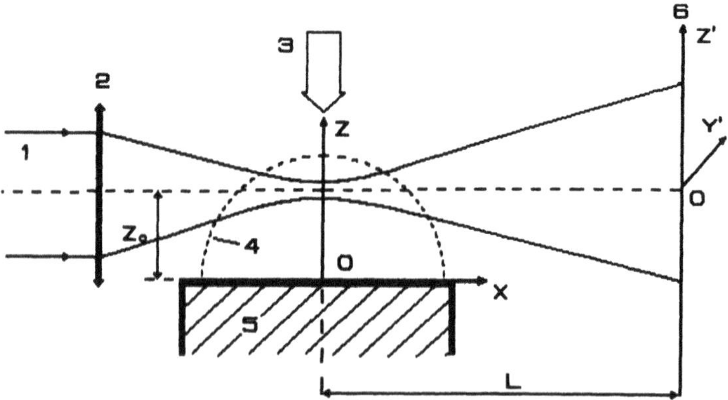

Fig.1. Schematic diagram of the PD set-up in the "mirage" effect geometry. 1)The probe laser beam; 2)the objective; 3) pump radiation; 4)the thermal lens; 5) the sample; 6) the photodetector plane.

Springer Series in Optical Sciences, Vol. 69 Photoacoustic and Photothermal Phenomena III
Editor: D. Bićanić

$$U(y',z',t)=\frac{i}{\lambda L}\int dy\int dz U(y,z,t)\exp\left[-i\pi/(\lambda L)((y-y')^2+(z-z')^2)\right], \quad (1)$$

where λ is the probe beam wavelength; L is the distance between the TL and the photodetector plane; $U(y,z,t)$ is the complex amplitude in the probe laser beam just after its interaction with the TL; y' and z' are the coordinates in the photodetector plane.

If $U(y',z',t)$ is known the PD signal can be expressed as

$$S(t)=K\int_{-\infty}^{\infty}dy\int_{0}^{\infty}dz\left[U_o^*(y',z')\delta U(y',z',t)+U_o(y',z')\delta U^*(y',z',t)\right], \quad (2)$$

where K is the proportionality coefficient of the light intensity on the photodetector and the electric signal; $U_o(y',z')$ is the probe beam amplitude in the photodetector plane when there is no TL; $\delta U(y',z',t)$ is the amplitude deviation tue to the TL.

For brevity, we omit the long calculations and present only the finite result for periodical heating with frequency $\omega=2\pi f$:

$$S(\omega)=\frac{KI}{\sqrt{\pi}\lambda r}\cdot\frac{dn}{dT}\int_{0}^{\infty}dz\,\mathrm{Im}\left[\mathrm{erf}\left((z-z_o)\left\{i\left(\frac{\pi}{\lambda L}-\frac{1}{r^2}\right)\right\}^{1/2}\right)\right]\int_{-\infty}^{\infty}dk_y\left[\exp\left(-\gamma z-k_y^2r^2/4-2(z-z_o)^2r^{-2}\right)\hat{T}_s(0,k_y,\omega)\right]dk_y \quad (3)$$

where I and r are the probe beam intensity and radius respectively; dn/dT is the temperature coefficient of the refractive index; $\varkappa$ is the thermal diffusivity of the air; $\gamma=(k_y^2+i\omega/\varkappa)^{1/2}$; z_o is the height of the probe beam over the surface; $\hat{T}_s(k_x,k_y,\omega)$ is the Fourier-transform of an object surface temperature.

It should be noted that (1) takes the form of a common expression for the PD signal derived from the geometrical optics at $r=0$. Otherwise, these expressions are different at $r>0$, and the differences have to be considered more profoundly.

Let us analyze specific features in the behavior of the PD signals calculated in terms of geometrical and wave optics and study amplitude and phase dependencies on the modulation frequency. The geometrical optics data have been computed using results from papers [1,2]. Fig.2 represents both the calculated and experimental results obtained for a stainless steel specimen. The figure shows that there is considerable differences between the results obtained in terms of geometrical and wave optics. Moreover, the differences take place in spite of the attempts to take into account a finite probe beam size in the framework of geometrical optics. It is seen that the deviations of the signal phase component are greater than those of the amplitude. In the both cases the deviations increase with the modulation frequency of the pump beam intensity.

To conclude, the presented theoretical and experimental results show that wave optics effect should be taken into account to increase the accuracy of sample parameter determination by the PD technique.

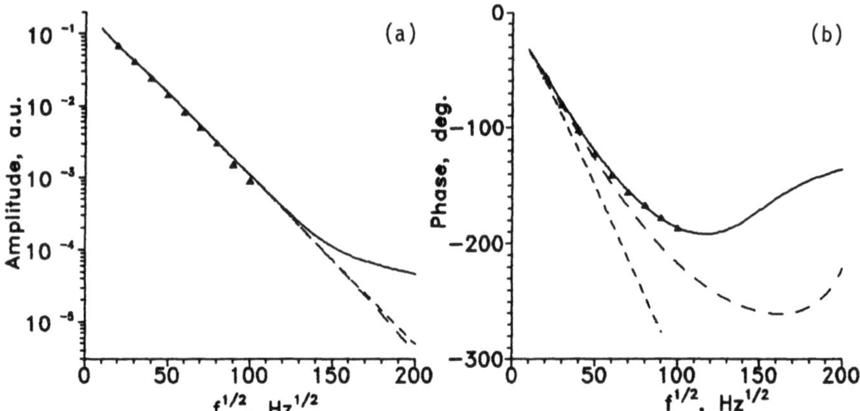

Fig.2. PD signal amplitude (a) and phase (b) dependence on the modulation frequency. ——— :Wave optics at r=49μ; — — : geometrical optics at r=49μ; - - -:geometrical optics at r=0. The normal offset z_0 equals 150μ.

References

1.F.A.McDonald,G.C.Wetsel,G.E.Jamieson, Can. J. Phys. **64**, 1265 (1986).

2.E.Legal La Salle,F.Lepoutre,J.P.Roger,J.Appl.Phys.**64**,1(1988).

3.A.L.Glazov,K.L.Muratikov,Optics Comm. (to be published).

4.A.L.Glazov,K.L.Muratikov,Sov.Phys.Tech.Phys. (to be published).

Pulsed Laser-Induced Temperature Changes Measured by the Photothermal Deformation Technique

J. Jauregui, Z.L. Wu, and E. Matthias*

Fachbereich Physik, Freie Universität Berlin,
Arnimallee 14, W-1000 Berlin 33, Fed. Rep. of Germany
*On leave from: Shanghai Institute of Optics and Fine Mechanics, Academia Sinica, P.O. Box 800-211, Shanghai, P.R. of China

For laser-interactions with materials one of the key parameters is the temperature rise at the surface during irradiation. This problem has been treated in the past by many authors, both for modulated cw [1] and pulsed [2-4] laser light. Nevertheless, the experimental verification of the absolute temperature change is still in an unsatisfactory state, although several investigations along these lines have been reported in recent years [5-11]. Among the experimental methods, the measurement of photothermal deformation [7,9] appears to be most attractive since this technique is able to monitor transient thermal processes at the surface in a non-contact way and with good time resolution and high sensitivity [12,13].

In this contribution the time-resolved photothermal deformation technique was applied as a non-contacting thermometer for calibrating temperature rises due to short pulse laser irradiation on polished polycrystalline Al and Cu surfaces. A pulsed dye laser (4 ns FWHM, 543 nm) was used as a pump. Typical deformation signals are displayed in Fig. 1. By recording a two-dimensional deformation profile we made sure that the traces in Fig. 1 cut through the maximum. A typical time-dependence of the deformation signal for copper near maximum slope is shown in Fig. 2.

The dependence on intensity of the signal at the maximum slope r_m (arrow in Fig. 1), integrated over 1 μs around the maximum in time t_m (arrow in Fig. 2) is shown in Fig. 3 for Cu, together with the variation of the reflectivity. We notice that the deformation signal depends linearly on intensity up to the melting point where an abrupt breakdown of the bulge

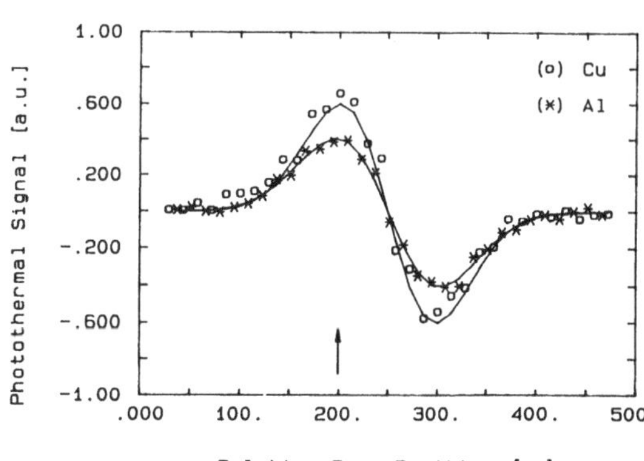

Fig. 1.: Photothermal deformation signals of Cu and Al surfaces as a function of relative pump beam position. The pump laser intensity was in both cases 10 MW/cm^2. The time integration interval was 1 μs.

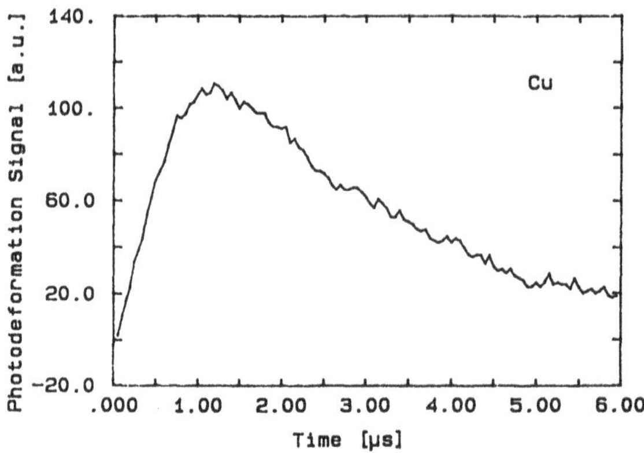

Fig. 2: Time-resolved photothermal deformation signal of Cu, measured near the maximum slope of the bulge.

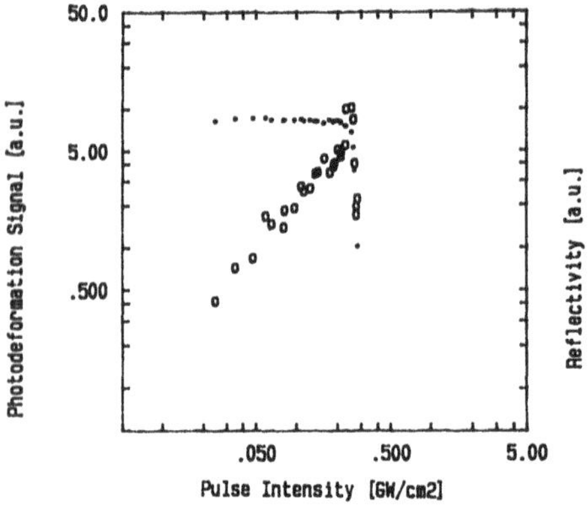

Fig. 3: Dependence on intensity of reflectivity (dots) and deformation (circles), integrated over 1 μs around the maximum for a polycrystalline Cu surface.

occurs. This behavior is verified by the reflectivity results. Our experiments differ from previously reported ones [7,9] in that the earlier work needed a combination of both techniques.

The maximum deformation signal can be compared to theoretical predictions. It can be shown [2,14] that for a given laser pulse length τ_0 the peak temperature T_m and therefore the maximum of the deformation signal is independent of time and given by

$$\frac{dU_z}{dr}(r_m,t_m) \propto \beta_{th}T_m = \beta_{th}S_F I_0(1-R)(\tau_0\kappa)^{1/2}\ K^{-1} \quad .$$

Here, I_0 is the laser intensity and R the reflectivity of the surface. β_{th} is the linear expansion coefficient, and κ, K are the thermal diffusivity and conductivity, respectively. The dependence of T_m on pulse shape is described by the factor S_F (for a gaussian pulse $S_F \sim 1$). This enables us to calculate the peak temperature rise as a function of laser intensity. Neglecting the temperature dependencies of R, κ, K, and β_{th}, the result is the solid line in Fig. 4. There is good

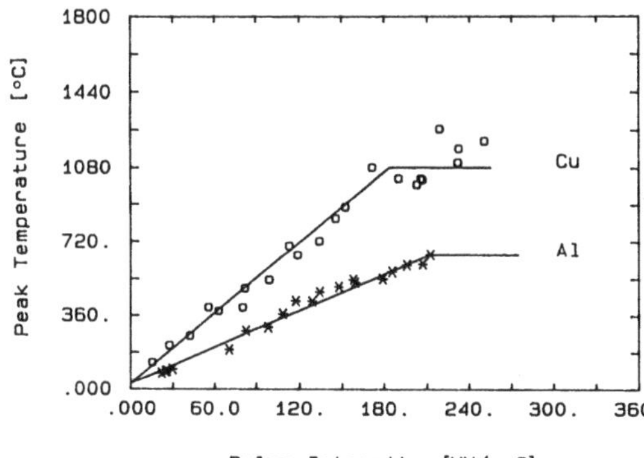

Fig. 4: Experimental data and theoretical predictions (solid lines) for the maximum temperature rise at polished polycrystalline Cu and Al surfaces as a function of laser intensity. Melting points are indicated by horizontal lines. (λ = 543 nm, 4 ns FWHM)

agreement between experimental and theoretical data, proving that the pulsed photodisplacement technique is useful for calibrating the temperature rise caused by laser irradiation. It must be emphasized, however, that the approximate linearity between laser intensity and peak temperature rise is valid only for small time intervals around the maximum deformation signal. As soon as lateral heat diffusion becomes effective the linearity between deformation amplitude and temperature will no longer exist [7,14].

This work was supported by the Deutsche Forschungsgemeinschaft, Sfb 337. Z.L.W. acknowledges a fellowship from the AvH-Foundation and J.J. one from the Deutsche Akademische Austauschdienst (DAAD).

References

1. W. Jackson and N.M. Amer: J. Appl. Phys. 51, 3343 (1980)
2. J. H. Bechtel: J. Appl. Phys. 46, 1585 (1975)
3. F.A. Mc Donald et al.: Proc. 1986, IEEE Ultrasonic Symposium (IEEE, New York 1986), p. 403
4. B.C. Li: J. Appl. Phys. 68, 2 (1990)
5. R. Zenobi et al.: Chem. Phys. Lett. 150, 361 (1988)
6. P.J. Timans et al.: Appl. Phys. Lett. 53, 1844 (1988)
7. Q. Kong et al.: in Photoacoustic and Photothermal Phenomena, P. Hess and J. Pelzl (Eds.), Springer Series in Optical Sciences 58, 179 (1988)
8. J.M. Hicks et al.: Phys. Rev. Lett. 61, 2588 (1988)
9. J. Fischer et al.: in Photoacoustic and Photothermal Phenomena II, J.C. Murphy et al. (Eds.), Springer Series in Optical Sciences 62, 539 (1990)
10. J. Loarel et al.: Appl. Opt. 29, 979 (1990)
11. H.E. Elsayer-Ali and J.W. Herman: Appl. Phys. Lett. 57, 1508 (1990)
12. M.A. Olmstead et al.: Appl. Phys. A32, 141 (1983)
13. C. Karner et al.: Appl. Phys. A38, 19 (1985)
14. J. Jauregui, Z.L. Wu, E. Matthias, to be published

Measurement of Thermal Diffusivities Using Transient Thermal Gratings

J. Jauregui and E. Matthias

Fachbereich Physik, Freie Universität Berlin,
Arnimallee 14, W-1000 Berlin 33, Fed. Rep. of Germany

Laser-induced transient thermal gratings (TTG's) have been generated on polished polycrystalline nickel and copper surfaces and detected by the time-resolved photothermal displacement technique. The purpose was to test the potential of this technique for measuring thermal diffusivities.

Up to now TTG's have only been observed by diffraction, and few attempts were made to utilize them for determining thermal diffusivities [1,2]. However, no satisfactory agreement with bulk values was obtained. For example, for colored glass and silicon 35 % lower values were found [2]. We suspect that these deviations were caused by data analysis using the 1-dimensional solution of the heat conduction equation. In this contribution we show that a simplified 3-dimensional treatment of the data leads to excellent agreement between measured and literature [3] values for the thermal diffusivities of polycrystalline nickel and copper.

Experimentally, TTG's were generated on metal surfaces by interference of two pump beams from a pulsed dye laser (543 nm, 4 ns FWHM) [4]. The grating period was altered by varying the angle between the two pump beams. The gratings were scanned by reflecting a He-Ne probe laser beam with ~ 60 μm focal diameter. A typical grating pattern on a nickel surface is shown in Fig. 1. The solid line represents the derivative of the interference pattern recorded with a CCD camera. The decay of this grating can be seen in Fig. 2 together with the exponential fit. TTG's and their time decay recorded on copper surfaces were measured with similar quality.

The 1-dimensional solution of the heat conduction equation leads to the relation [1], $\kappa = \Lambda^2(4\pi^2\tau_y)^{-1}$, between thermal diffusivity κ, grating periodicity Λ, and decay time

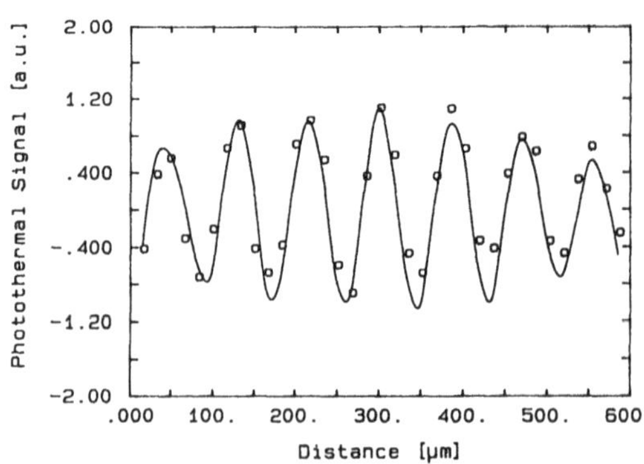

Fig. 1: Transient thermal grating on polycrystalline nickel surface. The solid line corresponds to Λ = 86 μm.

Springer Series in Optical Sciences, Vol. 69 Photoacoustic and Photothermal Phenomena III
Editor: D. Bićanić

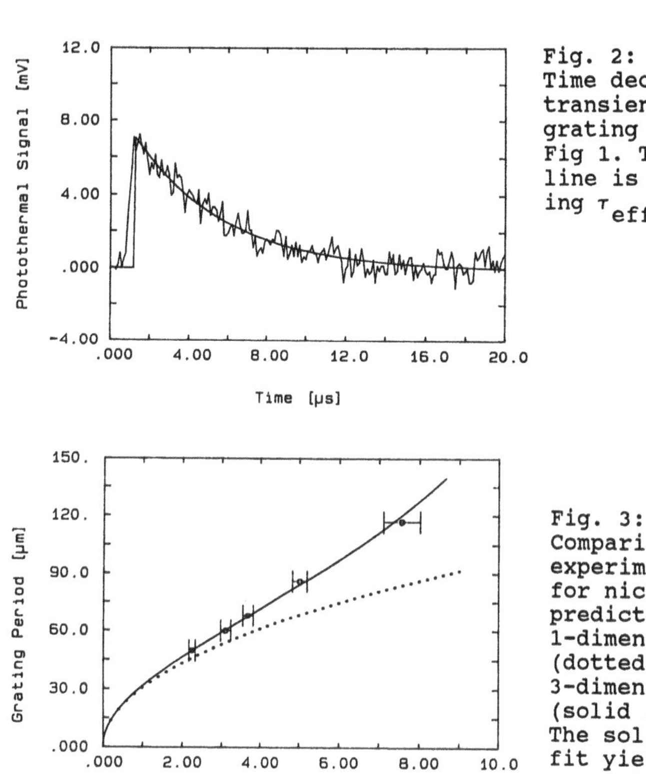

Fig. 2: Time decay of the transient thermal grating shown in Fig 1. The solid line is a fit yielding $\tau_{eff} = 4.8\ \mu s$.

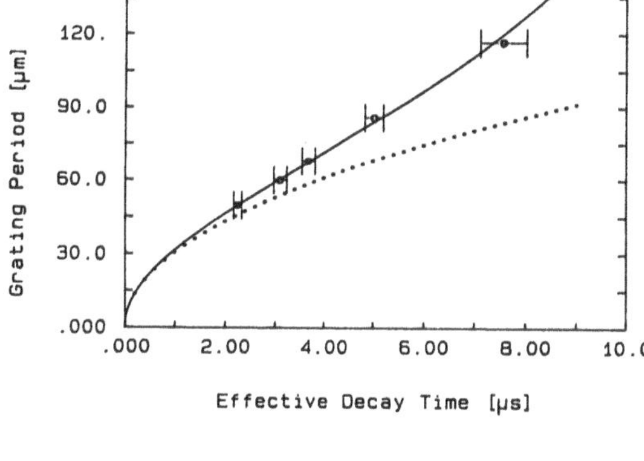

Fig. 3: Comparison of experimental data for nickel with predictions of the 1-dimensional (dotted line) and 3-dimensional (solid line) model. The solid line is a fit yielding $\kappa = 0.23\ cm^2\ s^{-1}$, τ_z=17 μs with a=73 μm.

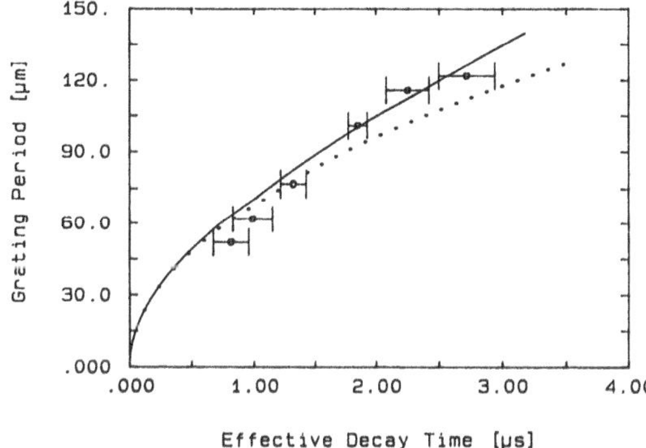

Fig. 4: Comparison of experimental data for copper with predictions of the 1-dimensional (dotted line) and 3-dimensional (solid line) model. The solid line was calculated using $\kappa = 1.17\ cm^2\ s^{-1}$, a = 85μm and τ_z=15 μs.

τ_y resulting from a single exponential decay. Here it is assumed that the grating extends along the y-direction with the wave vector $q = 2\pi/\Lambda$. This simple relationship was previously used to measured κ [1,2]. However, when plotted for nickel and copper (dotted lines in Figs. 3 and 4), using $\kappa(Ni) = 0.23\ cm^2\ s^{-1}$ and $\kappa(Cu) = 1.17\ cm^2\ s^{-1}$, respectively, this prediction does not fit the experimental data for copper and even less than for nickel. Much better agreement (c.f. solid lines in Figs. 3 and 4) is obtained with the same values of κ when using a model where we write the effective heat diffusion rate at the surface (z=0) and in the center of the grating (x=0) as a sum of the decoupled diffusion rates along the three axes [4]:

$$\tau_{eff}^{-1} = \frac{1}{2}\tau_x^{-1} + \tau_y^{-1} + \tau_z^{-1} \quad .$$

The grating width a along the x-axis is contained in $\tau_x = a^2/4\kappa$. The first two terms can be obtained with certain approximations from the solution of the thermoelastic wave equation for the vertical component U_z of the displacement [4]. The vertical diffusion rate τ_z^{-1} is added intuitively but turns out to be of minor importance.

In conclusion we propose a model which yields the thermal diffusivity from the periodicity and the single exponential decay constant of the grating. Analysis of the data by this model leads to good agreement between the resulting values for the thermal diffusivities and those quoted in the literature. In addition it shows that TTG's decay predominantly by lateral heat diffusion.

This work was supported by the Deutsche Forschungsgemeinschaft, Sfb 337. One of us (J.J.) acknowledges a fellowship from the Deutsche Akademische Austauschdienst (DAAD).

References

1. H. Eichler, G. Salje, H. Stahl: J.Appl.Phys, 44, 5383 (1973)
2. A. Harata, H. Nishimura, T. Sawada: Appl.Phys.Lett. 57, 132 (1990)
3. M. Liezers and R.M. Miller: Photoacoustic and Photothermal Phenomena, editors: P. Hess, J. Pelzl, Springer Ser. Opt. Sci. 58, 344 (1988)
4. J. Jauregui and E. Matthias, to be published

Measurement of the Thermal Radial Diffusivity of Anisotropic Materials by the Converging Thermal Wave Technique – Application to NDE: First Results in the Detection of Normal Cracks

F. Enguehard, A. Déom, and D. Balageas

ONERA, Division de Thermophysique, B.P. 72,
F-92322 Chatillon Cedex, France

Abstract. The converging thermal wave (CTW) technique is used for measuring the thermal radial diffusivity of orthotropic materials. In this technique, the surface of the sample is heated on an annular pattern, and the temperature rise at the center of the annulus is recorded. From this temperature-time history, the radial diffusivity is identified using a novel procedure. Application to a carbon-epoxy composite material is presented.

The ability of the CTW technique to detect cracks normal to the heated surface of the sample is also demonstrated.

1. Principle of the CTW technique

The principle of the CTW technique was first proposed by Cielo et al. [1]: the surface of the sample is impinged by a pulsed heat flux on an annular pattern, and the temperature rise at the center of the annulus is recorded. From the experimental temperature-time history $\theta_{exp}(t)$ of this point, one can plot an effective radial diffusivity versus time curve $\kappa_r(t)$ using a novel procedure which is presented in the next section.

The experimental apparatus is presented in detail in ref. [2].

2. Theory for a homogeneous orthotropic sample

An analytical study of the experiment was made [2], taking into account the finite thickness e and the thermal anisotropy ($\lambda_r \neq \lambda_z$) of the sample, and convective heat losses (coefficient h) on its front side. The thermogram $\theta_{th}(t)$ at the center of the heating annulus (of radius R) was found to be a function of three non-dimensional variables: $Fo_r = \alpha_r t/R^2$, $A = (e/\sqrt{\alpha_z})/(R/\sqrt{\alpha_r})$ and $Bi = hR/\sqrt{\lambda_r \lambda_z}$; its general shape is shown in fig. 1.

Analytical calculations showed that in the usual case of thick enough samples ($e \approx 1$ mm), convective heat losses on the front side have a very poor effect on the precision of the determination of α_r; therefore these heat losses can rationally be neglected. The thermogram $\theta_{th}(t)$ has then the following expression:

$$\theta_{th}(t) = \frac{K}{Fo_r} \exp\left(\frac{-1}{4Fo_r}\right) g\left(\pi^2 Fo_z\right), \tag{1}$$

K being a multiplicative constant, $Fo_z = \alpha_z t/e^2 = Fo_r/A^2$, and g being the function:

$$g(x) = 1 + 2 \sum_{n=1}^{\infty} \exp(-n^2 x). \tag{2}$$

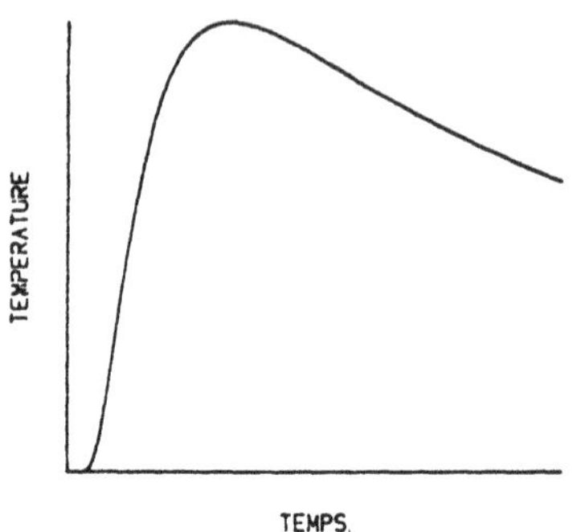

Fig. 1: general shape of the theoretical thermogram $\theta_{th}(t)$.

One notices that:

$$\tilde{\theta}_{th}(t) = \frac{\theta_{th}(t)}{g\left(\pi^2 Fo_z\right)} = \frac{K}{Fo_r} \exp\left(\frac{-1}{4Fo_r}\right) \tag{3}$$

is a function of Fo_r only (which has the same shape as the curve of fig. 1); thus, provided the thermal longitudinal diffusivity α_z of the sample is known, one gets rid of the longitudinal diffusion effect by dividing the experimental thermogram $\theta_{exp}(t)$ by $g\left(\pi^2 Fo_z\right)$. An effective radial diffusivity $\kappa_r(t)$ can then be calculated at each time t of the thermogram, either by numerical inversion of the thermogram $\tilde{\theta}_{exp}(t)$ using the mathematical expression of the theoretical thermogram $\tilde{\theta}_{th}(t)$, or by a method using its logarithmic derivative with respect to time (for more details see [2]); if the adequation of the model to the experiment is good, $\kappa_r(t)$ is a constant equal to the radial diffusivity α_r.

3. Radial diffusivity identification: theory, validation

The experimental apparatus was tested on a 0.8 mm thick carbon-epoxy sample; the obtained experimental thermogram is shown in fig. 2a. Its signal to noise ratio is quite satisfactory, due to the important thermal anisotropy, the low thickness and the high emissivity of the sample. On the other hand, its shape at short times is not the one predicted by the theory; the peak that one observes is probably due to light diffusion of the laser beam inside the converging lens and the axicon, which superposes a light spread heating of the whole front side of the sample upon the main annular heating. Up to now we haven't succeeded in getting rid of this peak in the experimental thermogram; however it is possible to subtract it as follows.

Consider a sample of finite thickness e, of longitudinal diffusivity α_z, subject to a uniform pulsed heat flux, with negligible heat losses; the surface temperature has the following expression:

$$\theta_{ld}(t) = K' \, g\left(\pi^2 Fo_z\right), \tag{4}$$

K' being a multiplicative constant, and g and Fo_z having the expressions given in the preceding section. Consequently, in the case of our experiment, the thermogram $\theta_{th}(t)$ at the center of the heating annulus has two components: a radial diffusion one $\theta_{rd}(t)$ (given by expression (1)), and a longitudinal diffusion one $\theta_{ld}(t)$ (given by expression (4)):

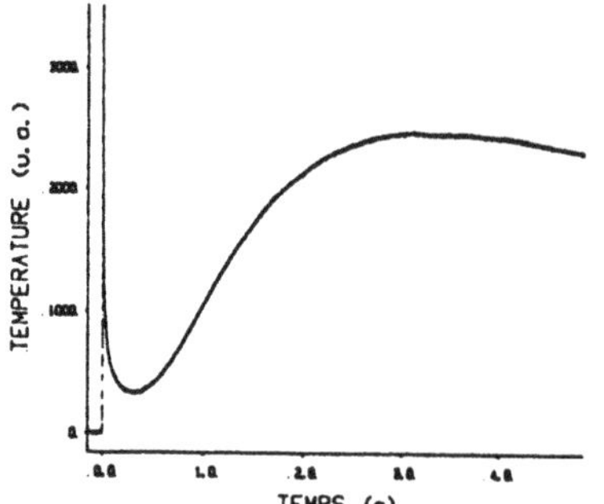

Fig. 2a: experimental thermogram $\theta_{exp}(t)$ obtained on a carbon-epoxy sample.

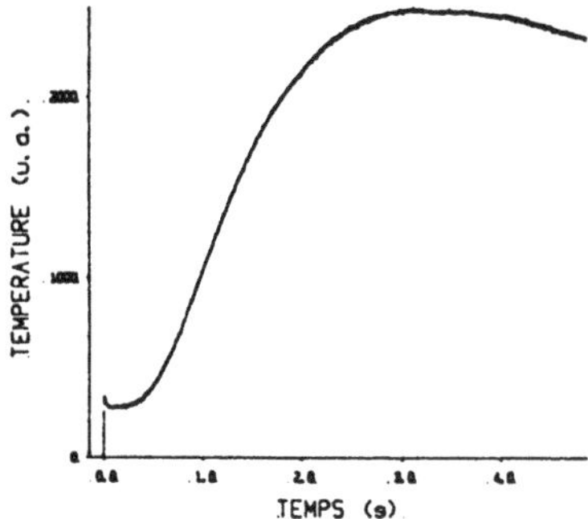

Fig. 2b: thermogram $\tilde{\theta}_{exp}(t)$ of the carbon-epoxy sample.

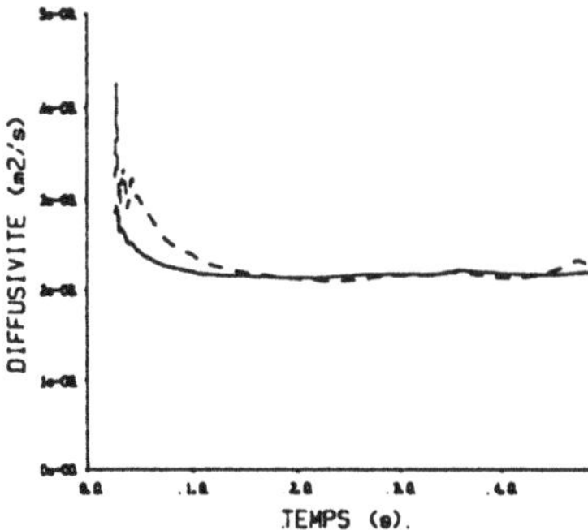

Fig. 2c: effective radial diffusivity curves $\kappa_r(t)$ obtained (—) by the numerical inversion method (- -) by the logarithmic derivative method.

$$\theta_{th}(t) = \theta_{rd}(t) + \theta_{ld}(t) = \frac{K}{Fo_r} \exp\left(\frac{-1}{4Fo_r}\right) g\left(\pi^2 Fo_z\right) + K' g\left(\pi^2 Fo_z\right). \tag{5}$$

Once again, one notices that:

$$\tilde{\theta}_{th}(t) = \frac{\theta_{th}(t)}{g\left(\pi^2 Fo_z\right)} = \frac{K}{Fo_r} \exp\left(\frac{-1}{4Fo_r}\right) + K'; \tag{6}$$

so, provided α_z is known, if one divides the experimental thermogram $\theta_{exp}(t)$ by $g\left(\pi^2 Fo_z\right)$, one obtains a new thermogram $\tilde{\theta}_{exp}(t)$ given by expression (6), in which the constant K' can easily be evaluated since it is the value of $\tilde{\theta}_{exp}(t)$ at short times. After subtraction of K', the thermogram has the expression given by equation (4), and the numerical inversion and logarithmic derivative methods can be used.

The thermogram $\tilde{\theta}_{exp}(t)$ of the carbon-epoxy sample is shown in fig. 2b (the longitudinal diffusivity α_z was measured using a classical flash technique; we found α_z = 6.3 10^{-7} m^2/s): the peak has disappeared, it has been replaced by an offset K' and a horizontal slope. After subtraction of K' and numerical filtering of the curve, one can plot two effective radial diffusivity curves $\kappa_r(t)$ using the numerical inversion and logarithmic derivative methods. Those two curves are shown in fig. 2c; they yield a weighed average value of the radial diffusivity $<\kappa_r>$ = 2.19 10^{-6} m^2/s and a weighed average relative error ε = 1.8 % (the weighing function used results from a sensitivity study [2]).

4. Application to NDE: first results in the detection of normal cracks

The experimental apparatus can also be used for the detection of normal cracks located in the circular area delimited by the heating annulus: the maximum temperature increase of the radial diffusion component is then expected to be lower and more delayed in time in the case of a fractured zone than in the case of a healthy one.

A 6 µm wide normal crack was performed in a 1 mm thick ferrite ceramic sample, and this sample was submitted to the annular heating experiment, first on a healthy zone, then with the crack inside the heating annulus. The radial diffusion components of the obtained experimental thermograms are shown in fig. 3a: they are quite consistent with what was expected. The radial diffusivity curves $\kappa_r(t)$ obtained by the numerical inversion method are shown in fig. 3b: as expected, the values $\kappa_r(t)$ are lower in the case of the fractured zone ($\langle\kappa_r\rangle = 2.32\ 10^{-6}\ m^2/s$) than in the case of the healthy one ($\langle\kappa_r\rangle = 2.52\ 10^{-6}\ m^2/s$); on the other hand, the weighed average relative error ε is greater in the case of the fractured zone ($\varepsilon = 3.5$ %) than in the case of the healthy one ($\varepsilon = 1.4$ %).

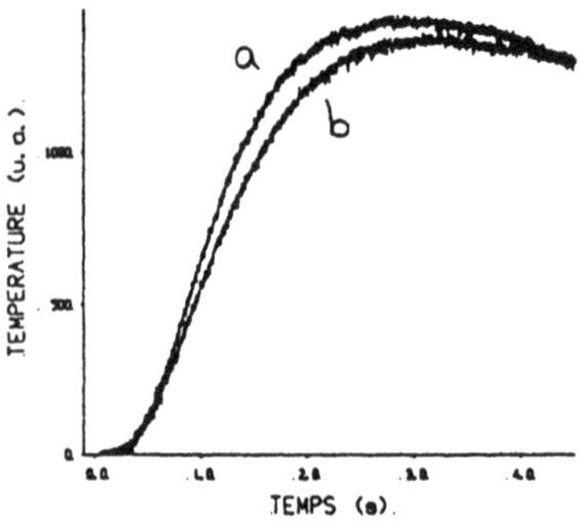

Fig. 3a: radial diffusion components of the experimental thermograms obtained on a ferrite ceramic sample (a) on a healthy zone (b) on a fractured one.

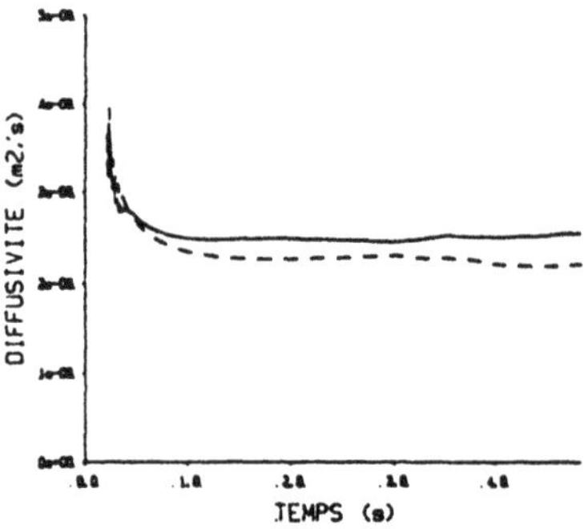

Fig. 3b: radial diffusivity curves $\kappa_r(t)$ obtained by the numerical inversion method on the ferrite ceramic sample (—) on the healthy zone (- -) on the fractured one.

5. Conclusions

The complete identification procedure for thermal radial diffusivity measurement was presented and validated by an experiment on a carbon-epoxy sample.

The ability of NDE of normal cracks using this method was also demonstrated.

References

[1]: P. Cielo, L.A. Utracki, M. Lamontagne: "Thermal Diffusivity Measurements by the Converging Thermal Wave Technique", Can. J. Phys. 64, 1172 (1986).

[2]: F. Enguehard, D. Boscher, A. Déom, D. Balageas: "Measurement of the Thermal Radial Diffusivity of Anisotropic Materials by the Converging Thermal Wave Technique", in "Acoustic, Thermal Wave and Optical Characterization of Materials", Ed. G.M. Crean, M. Locatelli and J. McGilp, North-Holland, Amsterdam, p. 181, 1990.

Optimization of a Mirage Setup for Thermal Diffusivity Measurements

Ph. Forge, F. Lepoutre, and M.Z. Silva*

Laboratoire d'Optique Physique, UPRA0005 CNRS, ESPCI,
10, rue Vauquelin, F-75005 Paris, France
*Present address: ONERA, Division DES/SB; B.P. 72,
F-92320 Chatillon, France

Abstract: We discuss the sensitivity of the mirage deflection (parallel Φ_t and perpendicular Φ_n to the sample surface) to the experimental parameters (beams radii, probe-surface distance, probe orientation with respect to the surface) in order to optimize an experimental setup devoted to thermal diffusivity measurements.

1. Introduction

The mirage effect has been shown by many authors to be an efficient tool for thermal diffusivity measurements. Even more the method is probably unique for measurements on coated materials especially when the layer thickness range from 10 to 100 microns.

Nevertheless and though industrial interest for thermal data is increasing, up to now the Mirage method still remain an experimental system of research laboratories. One of the reasons of its poor development is the real difficulty encountered by the users to interpret precisely their experimental results. We propose in this paper theorerical and experimental analysis of this problem in order to enhance the reliability and the precision of the thermal diffusivity measurements.

Our contribution is a sensitivity study and a few examples of measurements.

2. Sensitivity study

A sketch of the detection with the definitions of the deflections Φ_n and Φ_t is given in figure 1. Let us first recall that we will discuss only the phase of Φ_n or Φ_t since it has been shown that they don't depend on the optical characteristics of the sample which is the case for the amplitude. We have studied the conditions in which Φ_n or

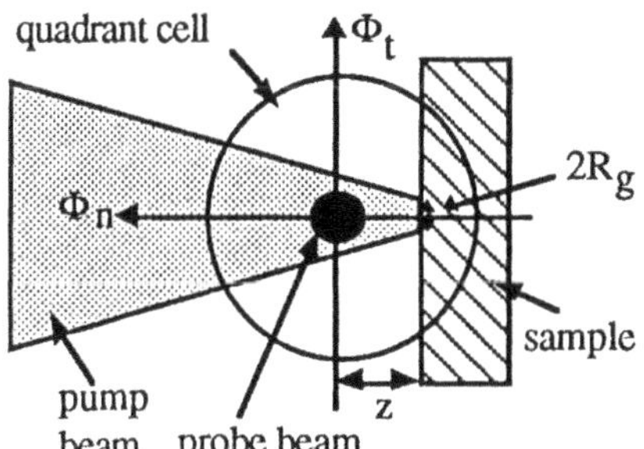

Fig. 1: geometric arrangement of the mirage detection

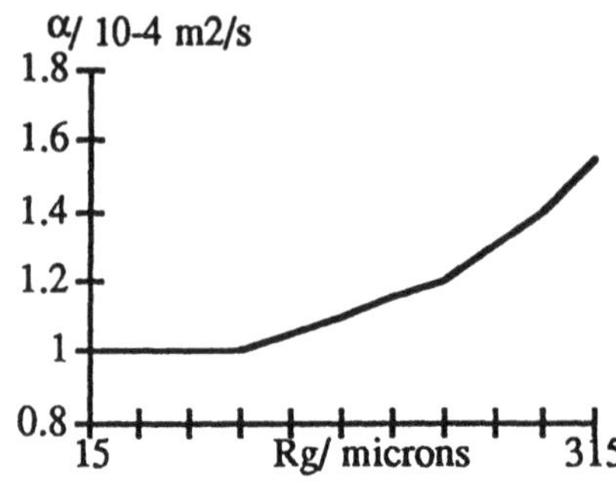

Fig 2: Variation of the thermal diffusivity deduced from a Φ_t inversion as function of the value of R_g used for the theorical model; $\alpha=10^{-4}$ m^2 s^{-1}; Z=100μm

Φ_t are preferable for diffusivity measurements. This study is summarized as follows:

-Φ_t is totally insensitive to the thermal diffusivity of transparent coatings. In this case Φ_n must be used exclusively.

-Φ_t and Φ_n are equivalent in all the other cases but their sensitivities to the experimental parameters are very different, following the value of α. Roughly one can say that Φ_n is generally preferable for the determination of the low diffusivity ($\alpha<2.10^{-5}$ m^2s^{-1}) while the measurements using Φ_t can be inversed very easily for the large thermal diffusivity ($\alpha>2.10^{-5}m^2s^{-1}$) provided that Rg is kept smaller than 100 μm. In particular we found that two experimental parameters must be known precisely:

- the pump beam radius R_g on the sample surface
- the distance Z between the probe and the sample surface which, in addition, must be kept constant if the distance R between pump and probe is varied during the experiment.

Caution must be also taken about the parallelism between the sample surface and the Φ_t axis of the quadrant cell (see figure 1). The sensitivity to the other parameters (probe beam radius, pump profile, modulation frequency) is less important. As a typical example of this result of the sensitivity study, figure 2 shows the effect of R_g on the thermal diffusivity determination of a bulk sample using Φ_t mesearuments.

3.Measurements

The experimental setup used was described elsewhere /1/. The experimental procedure was automated and mechanical adjustments allow the knowledge of Rg with a precision of 2 μm and z with a precision of 5 μm.

The first example is a Φ_n measurement of the thermal diffusivity α of a ZrO_2 (9 μm thick) coating on Mo substrate (see figure 3). The precision on α is about ±10%. In such case a very high precision on Z (about 1 μm) would be necessary to increase the precision on α.

The second example , figure 4, shows experimental points obtained on two varieties of pipe exchangers (1:Inconel 600, 2:Inconel 690). The two varieties are very well separated ($\alpha1=4.10^{-6}$ m^2s^{-1};$\alpha2=3.10^{-6}$ m^2s^{-1}) and the precision is better than ±5%. It must be noticed that an increase of the frequency would have decreased the precision of the measurement (for instance $\Delta\alpha/\alpha=30\%$ at 225Hz).

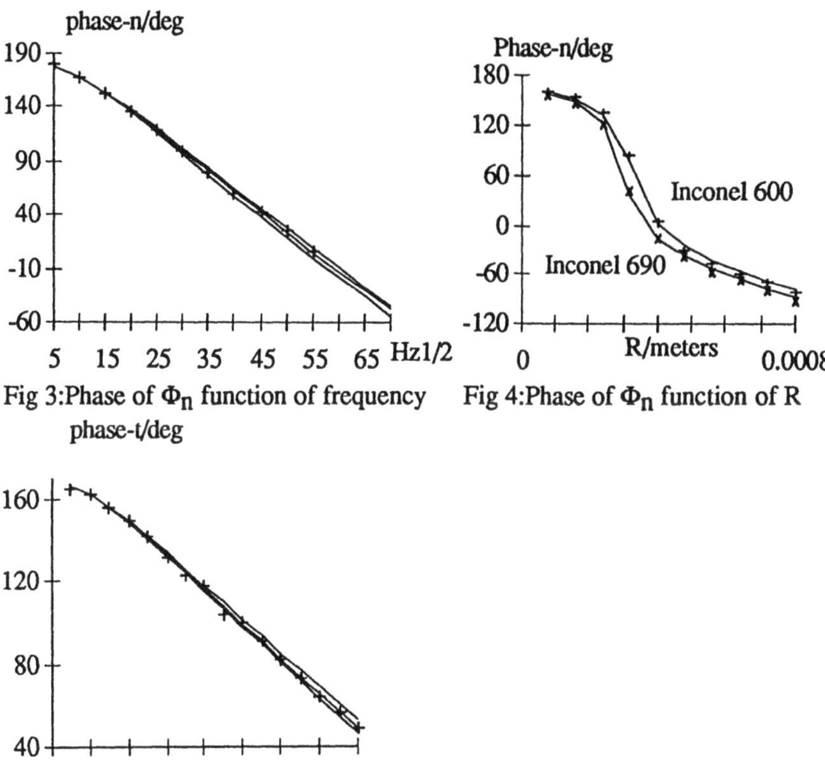

Fig 3:Phase of Φ_n function of frequency

Fig 4:Phase of Φ_n function of R

Fig 5:Phase of Φ_t function of R

The last example is the Φ_t determination of the diffusivity of a substrate of AlN covered with 25 µm thick Mo coating (see figure 5). The method of inversion cannot be used so that Φ_n measurements are also possible. We have found a precision of about 10% with the two measurements.

4. Conclusions

The mirage effect can be used for thermal diffusivities measurements provided that the experimental set up allows a precise (typically ±1µm) knowledge of Rg (pump beam radius) and z (probe - surface distance). The accuracy of the diffusivity determination is then of about 5%, but care must be taken in the choice of Φ_n, Φ_t and the modulation frequency. In the case of Φ_t, when the thermal diffusion length is smaller than the thickness of the material of unknown diffusivity, a fast inversion procedure can be used.

References

/1/ F.Charbonnier and D.Fournier, Rev.Sci.Instrum., **57**, 1126, (1986).

Inversion of Pulsed Thermal Wave Images for Recovery of the Shape of the Object

D.J. Crowther, L.D. Favro, P.K. Kuo, and R.L. Thomas

Department of Physics and Institute for Manufacturing Research, Wayne State University, Detroit, MI 48202, USA

Abstract. An outstanding problem of thermal wave imaging has been the blurring of the images because of the lateral diffusion of heat in the sample. We demonstrate here, both theoretically and experimentally, that this blurring is not due to any inherent loss of information in the diffusion process, but that it can, in fact, be removed by appropriate processing of the image. The result is a very significant improvement of the resolution of the thermal wave imaging process.

1. Introduction

Thermal wave images of subsurface features tend to be blurred by the lateral diffusion of the heat scattered from those features. In this paper we illustrate a technique for removing this blurring in images of planar subsurface features such as delaminations. This technique is based on a model which allows us to develop an algorithm which performs a mathematical inversion of the scattering process when it is applied to experimental data. The idea behind the algorithm is first to manipulate the equations of scattering theory, as applied to the heat equation, in such a way that the surface temperature of a sample is expressed as a convolution of a "heat spread" function with a function which describes the shape of the scatterer. The Fourier transform of the surface temperature can then expressed as a simple product of the Fourier transforms of the heat spread function and the shape function of the scatterer. The algorithm then consists of performing an FFT on the experimental image of an unknown scatterer, dividing the resulting Fourier transform by the transform of the (known) theoretical heat spread function, and finally doing an inverse FFT to obtain the shape of the scatterer. A practical difficulty arises from the finite accuracy of the scattering data, which is typically digitized at an 8-bit level. We demonstrate that this difficulty can be overcome by applying a low-pass filter to the Fourier transform of the experimental image.

2. The Model

Because of space limitations in this publication, the model presented here will be limited to planar scatterers which are parallel to the sample surface, and which have an insulating boundary condition on their surface. Furthermore, we will ignore multiple scattering effects in the model presented here. Our model then is equivalent to the first-order Born approximation. With these restrictions, the theoretical temperature contrast (the temperature resulting from the scattering without the background temperature decay) resulting from a uniform surface heat pulse applied at $t=0$ can be expressed as

$$\Delta T(x,y,t) = -\frac{A}{2\pi}\left(\frac{1}{4\pi\alpha t}\right)^{1/2}\frac{\partial}{\partial \ell}\iint dx'\,dy'\,\frac{e^{-\frac{\left[\left[(x-x')^2+(y-y')^2+\ell^2\right]^{1/2}+\ell\right]^2}{4\alpha t}}}{\left[(x-x')^2+(y-y')^2+\ell^2\right]^{1/2}}\,f(x',y')\,. \quad (1)$$

Here ℓ is the depth of the scatterer, α is the thermal diffusivity of the sample, and $f(x,y)$ represents the shape of the scatterer. The overall constant A is determined by the intensity of

Springer Series in Optical Sciences, Vol. 69 Photoacoustic and Photothermal Phenomena III
Editor: D. Bićanić

Fig. 1 Recovered shape of a rectangular object from simulated scattering data, demonstrating that the algorithm is capable of complete recovery of the shape.

Fig. 2 Result of an attempted shape recovery when the scattering data used in in Fig. 1 are truncated to 8-bits to simulate the digitization of experimental data.

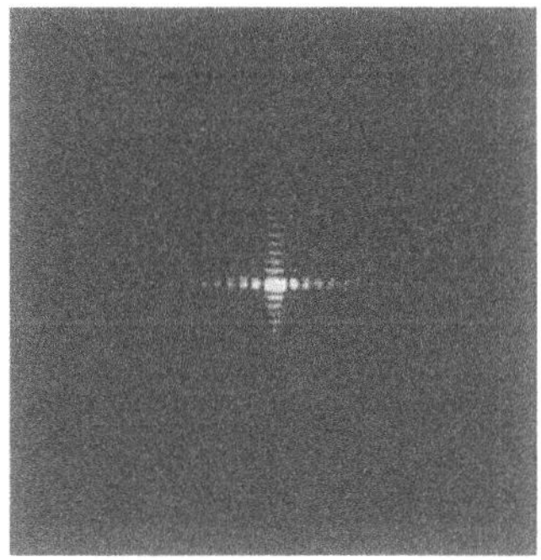

Fig. 3 Fourier transform of the simulated data which led to Fig. 1.

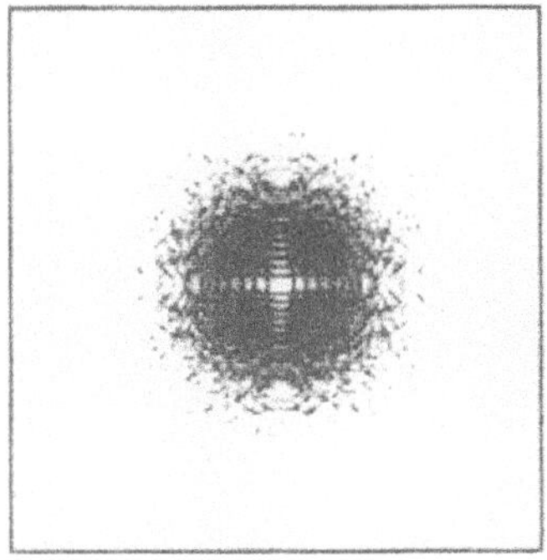

Fig. 4 Fourier transform of the 8-bit truncated data which led to Fig. 2.

the original heat pulse. A detailed derivation of this formula and a formula for a more general model will be published elsewhere.

Since this expression is a convolution of a "heat spread" function with the scatterer's shape function f(x,y), the shape function can in principle be obtained by dividing the Fourier transform of the contrast image by the Fourier transform of the heat spread function. We have demonstrated that this deconvolution procedure works as expected by calculating simulated images of subsurface features of a variety of shapes, and then applying the procedure to the simulated image. The result is a shape function which is indistinguishable from the original. However, experimental data do not have unlimited accuracy. In particular, many infrared cameras which are used to obtain thermal wave images produce data with only 8-bit accuracy. In order to investigate the effect of this lack of accuracy, we have truncated the data in the simulated images to simulate the accuracy the digitization of the experimental data. The result is that, even with 12-bit truncation, the original shape appears to be completely lost after the deconvolution. An example of this effect, shown with 8-bit truncation to simulate typical experimental accuracy, is shown in Figs. 1 and 2. Fig. 1 shows the original shape, and Fig. 2 shows the result of simulation of a thermal wave image by application of Eq. 1, truncation of the resulting data to 8-bits, and deconvolution by the procedure described above. This result would seem to preclude the use of our algorithm to perform useful inverse scattering calculations on experimental data. However, an examination of the Fourier transforms of our simulated images reveals the origin of this problem and suggests a solution to it. Figs. 3 and 4, display the Fourier transforms of the images in Figs. 1 and 2. The transform of the truncated image data, Fig. 4, contains a large amount of high frequency "digitization noise". Only the central, low frequency, portion of the transform remains unchanged. This suggests that low pass filtering of the transform before application of the inversion process might make it possible to recover some features of the shape of the scatterer. This assumption proves to be correct. After filtering, the rectangular shape of the simulated scatterer can be almost completely recovered. Space does not allow displaying them here, but the only difference between the original shape and the recovered one is a slight rounding of the corners due to the loss of high frequency information.

3. Results

We have applied our algorithm to a variety of 8-bit experimental images, both with and without low pass filtering of their Fourier transforms. Without filtering, the results are as expected; the recovered images show no hint of the scatterer's shape. After filtering, the results are spectacularly different. We obtain almost complete recovery of the objects' shapes, again with just a slight rounding of the corners. Space does not permit the display of many such images here, so we will exhibit just one. It is an image of a cross-shaped flat bottom hole in the rear surface of a stainless steel plate. The raw thermal wave image is shown in Fig. 5, and the recovered image is shown in Fig. 6. It should be noted that, in the recovered image, the left arm of the cross appears to be slightly thicker than the right arm. This detail is faithful to the original shape of the cross which was only crudely milled into the sample.

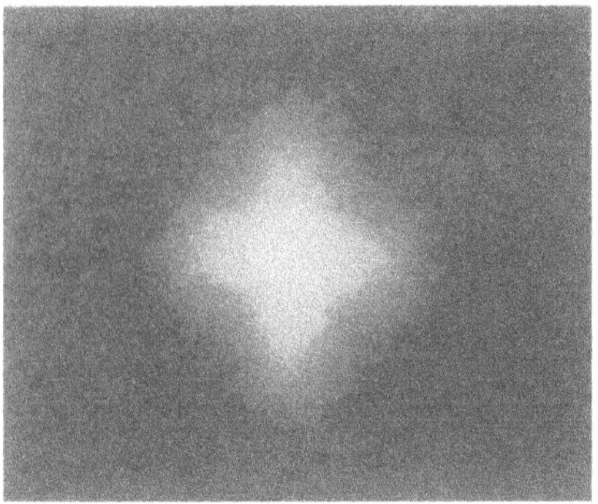

Fig. 5 Raw thermal wave image of a flat bottom hole in the shape of a cross cut into the rear surface of a stainless steel plate.

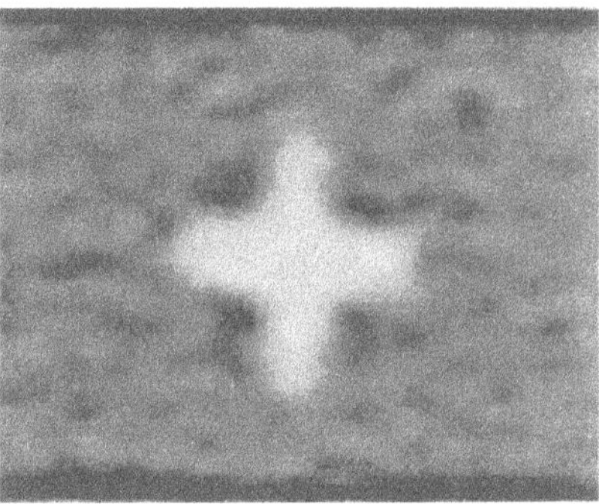

Fig. 6 Recovered cross shape from the image of Fig. 5.

4. Conclusions

We have demonstrated that it is possible to remove the blurring of experimental thermal wave images by inverting them with the inverse Born approximation. While some loss of high frequency information occurs as a result of the removal of digitization noise with a low pass filter, the recovered shapes accurately reproduce the shapes and sizes of the scatterers. More accurate digitization of the experimental data should result in even more accurate recovery of the object's shape. The implications for the use of thermal wave imaging for defect detection and sizing are clear.

5. Acknowledgements

This work was sponsored by the Center for Advanced Nondestructive Evaluation, operated by the Ames Laboratory, USDOE, for the Air Force Wright Aeronautical Laboratories under Contract No. W-7405-ENG-82 with Iowa State University; by the FAA Center for Aviation Systems Reliability, and by the Institute for Manufacturing Research, Wayne State University.

Energy Dependent Characteristics of Optoacoustic Pulses Measured by Optical Probe

J. Diaci and J. Možina

University of Ljubljana, Faculty of Mechanical Engineering,
P.O. Box 394, 61000 Ljubljana, Slovenia

Abstract. Waveform characteristics of optoacoustic pulses detected by optical probe during laser induced material transformation are examined. A set of optical probe responses was recorded by varying excitation energy and distance between the irradiated spot and the probe beam. Analysing the detected signals we found the dependence of waveform duration on energy consistent with the predictions of the point explosion model. On this basis it is possible to determine optical to acoustical energy conversion efficiency using relatively simple and inexpensive instrumentation.

Optoacoustic transients are important source of information about laser-solid interaction processes [1,...,6]. Recent studies [5,6,7] have shown that shock wavefront propagation of strong acoustic waves produced in gas during laser induced transformation of solid surfaces can be explained in terms of the point explosion model [9]. We have demonstrated lately [8] that also the waveforms detected by optical probe closely follow the theoretical predictions of the model. The purpose of this paper is to examine further waveform characteristics of optical probe responses.

Our experimental set-up is shown in Fig. 1. A Q-switched Nd:YAG laser excitation beam (total energy: 65 mJ at 1.06 μm, FWHM duration: 16

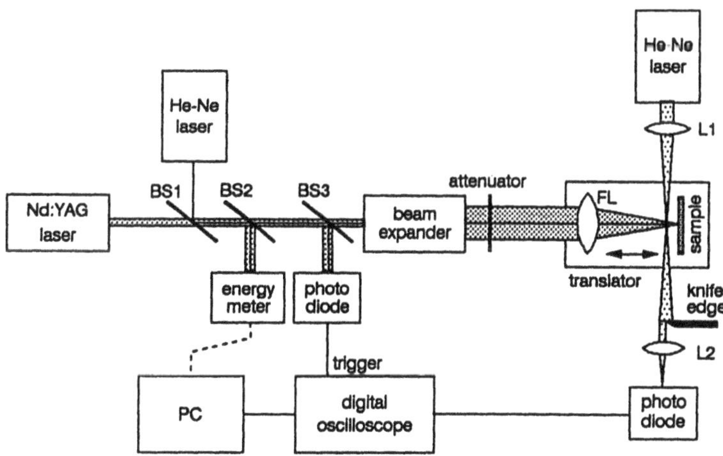

Fig. 1. Schematic of the experimental set-up

ns) is passed through a beam expander and focused by a short focal length to a spot diameter of 0.5 mm. The focusing lens and the sample assembly are mounted together on a mechanical translator driven by a stepping motor, allowing precise setting of the distance between the probe beam and the sample while maintaining constant focusation of the excitation beam. The probe beam from a 5-mW He-Ne laser is lying in a plane passing through the center of the irradiated spot. Deflection of the probe beam is detected by a knife edge and a fast photodiode.

Bare steel samples were irradiated by sequences of laser pulses. A set of optical probe responses was acquired by varying excitation energy E_e at each selected distance r_s between the irradiated spot and the probe beam. The excitation energy was changed by inserting neutral density filters into the excitation beam.

We found the following waveform characteristics by analysis of the detected signals, illustrated in Fig. 2:

- either decrease of the excitation energy or increase of the distance r_s or both results in decrease of the compression and rarefaction amplitude (the decrease of the former is larger than the decrease of the latter) and in increase of the transit time;
- decrease of the excitation energy at constant distance r_s leads to decrease of the total waveform duration t_w, with the decrease of the rarefaction duration which is larger than the decrease of the compression duration;

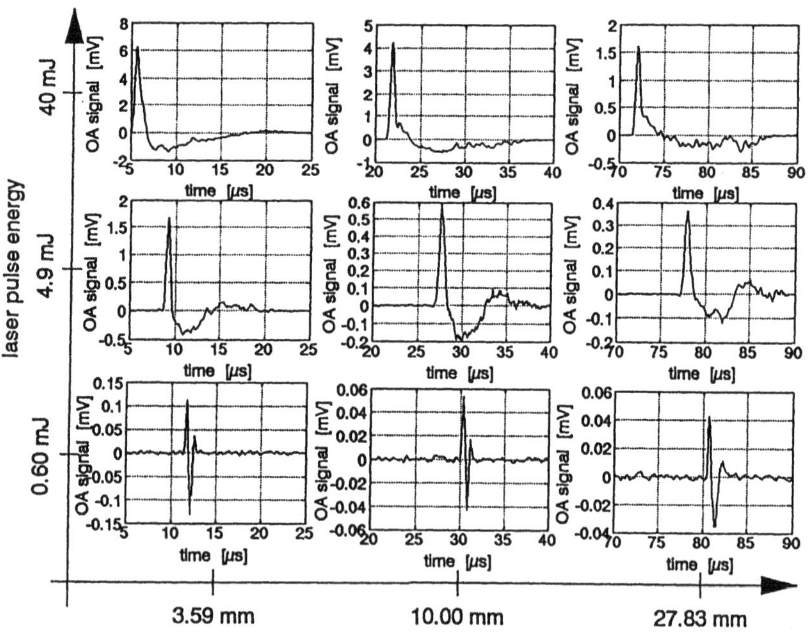

Fig.2. Examples of optoacoustic pulses measured by optical probe

- increase of the distance r_s at constant excitation energy leads to very weak increase of the compression duration and the total waveform duration;

This observations are consistent with the predictions of the point explosion model. The solution, given in terms of dimensionless space $\xi = r/r_c$ and time $\tau = t/t_c$ coordinates [9], has two scaling parameters:

$$r_c = (E/p_0)^{1/3} \text{ and } t_c = r_c/c_0 \tag{1}$$

that depend on blast energy E, ambient pressure p_0 and sound velocity c_0 of the undisturbed gas. Using the theoretical dependence of the dimensionless waveform duration τ_w on the dimensionless space coordinate ξ_s of the probe [10] it is possible to determine blast energy:

$$E = p_0 (c_0 t_w/\tau_w)^3 \tag{2}$$

and thereby optical to acoustical energy conversion efficiency. In conclusion, it is possible to determine optical to acoustical energy conversion efficiency using relatively simple instrumentation.

References

1. A. Rosencwaig and J. B. Willis, Appl Phys Lett **36**, 667 (1980).
2. J. Diaci and J. Možina, in: "Ultrasonics Int. 87 Conf. Proc.", Butterworth, Guildford, 96 (1987).
3. S. Petzoldt, et al., Appl Phys Lett **52**, 2005 (1988).
4. G. Koren, Appl Phys Lett **51**, 569 (1987).
5. P.E. Dyer and J. Sidhu, J Appl Phys **64**, 4657 (1988).
6. T. Zyung, et al., J Appl Phys **65**, 4548 (1989).
7. J. Diaci and J. Možina, in: "Ultrasonics Int. 89 Conf. Proc.", Butterworth, Guildford, 724 (1989).
8. J. Diaci and J. Možina, in: "Physical Acoustics: Fundamentals and Applications", O. Leroy and M.A. Breazeale (Eds.), Plenum, 291 (1991).
9. L. I. Sedov, "Metody podobya i razmernosti v mekhanike", Nauka, Moskva (1972).
10. H.L. Brode, J Appl Phys **26**, 766 (1954).

Photopyroelectric Thermal Wave Spatial and Depth Resolved Imaging with Ray Optic Tomographic Reconstruction

M. Munidasa and A. Mandelis

Photothermal and Optoelectronic Diagnostics Laboratory,
Department of Mechanical Engineering, University of Toronto,
5 King's College Rd., Toronto, Ontario, Canada M5S 1A4

Abstract: We report a capacitive photopyroelectric tomographic detection technique to obtain spatial and depth resolved photothermal scanning imaging of solid samples. Unlike the two-dimensional "projection" images obtained by conventional photothermal detection methods, data are collected in a tomographic scanning sequence for a given cross-section of the sample. The possibility of using tomographic reconstruction for depth-resolved imaging of subsurface defects is demonstrated assuming ray-like propagation of thermal waves. Limitations of this reconstruction method are discussed.

Recently, a pyroelectric instrumentation[1] using polyvinylidene fluoride (PVDF) films electroded on one side only, and a remote metal tip capable of sampling the local thermal wave field value in the back of the sample has been developed. This permits spatially resolved scans of this field. Images obtained from this and other photothermal methods so far are two-dimensional "projections" of sub-surface features obtained in a two dimensional raster scan without regard to the actual depth position of scatterers. Although work has been done in obtaining depth-resolved information on sub-surface features using equivalent experimental techniques[2] (and Mirage effect scanning[3]), and depth profiling of layered samples[4], no work has been reported in obtaining tomographic images using thermal waves.

In this paper we report the preliminary work performed towards reconstructing a thermal wave tomographic image using the pyroelectric detection instrumentation[1] reported earlier. Ray-like propagation of thermal waves has been assumed[5] in order to demonstrate the possibility of thermal wave tomography. Since both modulated heating source and detector pin are localized, and can be scanned independently, it is possible to perform a limited angle tomographic scan using this instrument. Aluminum samples of thickness 2 mm with different sub-surface defects have been used in this experiment.

Scans across cross-sections to be imaged were performed as shown in Fig.1(a). For each laser position, data were collected at several pin positions along a straight line which defined the cross-sectional plane (PQRS of Fig.1(a)) to be imaged. Phase data (laser/pin-position vs. the

Springer Series in Optical Sciences, Vol. 69 Photoacoustic and Photothermal Phenomena III
Editor: D. Bićanić

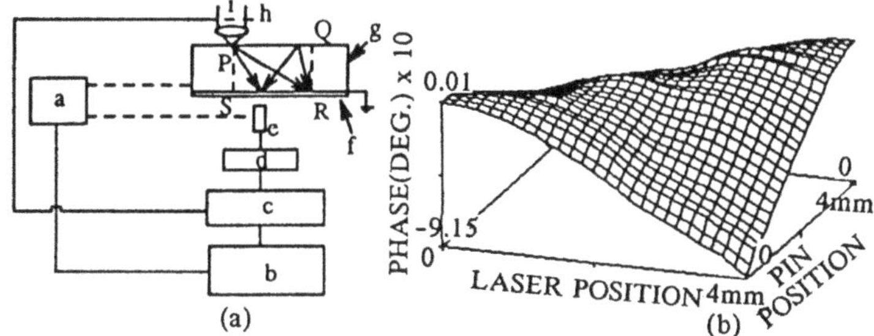

Fig.1 (a) Experimental setup. a:stepping motor scanner, b:microcomputer, c:lock-in amplifier, d:pre-amp, e:metal pin, f:PVDF film, g:sample, h:chopper, i:laser beam. (b) Scan data for a cross-section PQRS (Fig.1(a)) with two holes of diameter 0.5 mm at 20 Hz.

phase of the signal) obtained from such a tomographic scan across a cross-section with two holes of diameter 0.5 mm at depths of 0.55 and 1.45 mm (hole centers are horizontaly separated by 0.8 mm) taken at 20 Hz are shown in Fig.1(b) (e.g. laser-position P and pin-position S in Fig.1(a) refers to the coordinate (0mm,0mm) in Fig.1(b)).

Based on ray-like propagation of thermal waves in one-dimension:

$$T = T_0 \exp(-x/\mu) \cos(\omega t - x/\mu),$$

the average value of the thermal diffusivity, α, given by

$$\alpha = \pi f \mu^2$$

along the lines joining the heated spot and the pin can be calculated. By averaging over each line, values of α for each individual pixel are calculated. The value of α (used as the imaged quantity) for each pixel is weighed by the distance a ray travels across that particular pixel.

Figure 2(a) shows the reconstructed tomogram of the sub-surface holes obtained by using the data shown in Fig.1(b). Here, the actual position of the holes are indicated by circles. Comparing tomograms of a 0.5 mm hole with a 1 mm hole cross-section have shown that the smaller feature is more accurately reproduced. All the reconstructed images show an elongation of the hole cross-section along the vertical direction. It has also been seen that increasing frequency tends to move the position of the sub-surface defects towards the front surface from its actual position. This has been partially eliminated by averaging the two reconstructed images obtained from the data collected from both sides of the sample. These distortions strongly suggest that a proper field inversion based on the recently proven Laplace Diffraction Theorem[6] is likely necessary in order to enhance the fidelity of the image reconstruction.

Figure 2(b) shows the phase-reconstructed tomogram of a fatigue crack which starts at the surface and propagates into the material. It can be

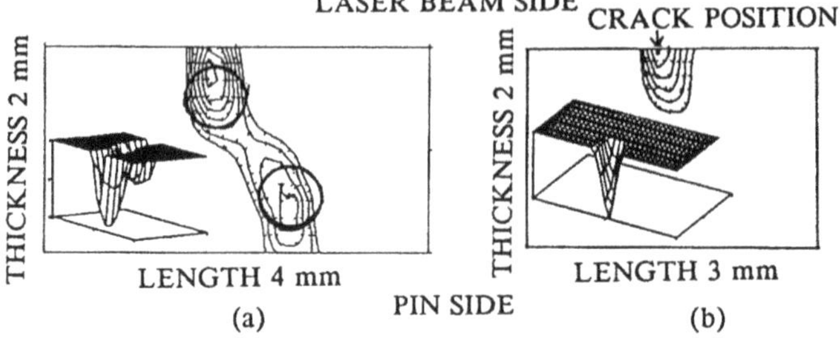

Fig.2 Reconstructed images (contour maps) (a) from Fig.1(b) data, and (b) of a fatigue crack. Corresponding inserts show the same images in 3-dimensional perspective plots. Background has been eliminated to enhance the defect.

seen that the crude reconstruction correctly images the decreasing importance of the crack with depth toward determining the value of the thermal diffusivity. This may be interpreted as a decreasing extent of the crack at greater depths. From Fig.2(b) it appears that the effective crack depth is about 500μm below the surface.

Although a simple ray-optic reconstruction method has been used, the results are encouraging and suggestive of the potential and limitations of the method, as well as the importance of improved image resolution and the need for reconstruction using the diffractive nature of thermal waves.

1. Mieszkowski,M., and Mandelis,A., "Photoacoustic and Photothermal Phenomena II", J.C.Murphy, J.W.Maclachlan Spicer, L.C.Aamodt, and B.S.H.Royce, Eds., Springer-Verlag, Berlin, 507-510 (1990).
2. Busse,G., and Renk,K.F., Appl.Phys.Lett. **42**, 366-368 (1983).
3. Fournier,D., Lepoutre,F., and Boccara,A.C., J.Phys.(Paris) **C6-44**, supl.10, 479-482 (1983).
4. Uejima,A., Itoga,F., and Sugitani,Y., Anal.Sci. **2**, 113-117, (1986).
5. Burt.,J.A., J.Phys.(Paris) **C6-44**, supl.10, 453-457, (1983).
6. Mandelis,A., J.Phys.A: Math. Gen. **24**, 2485 (1991).

Evaluation of Magnetic Disk Heads by Photoacoustic Microscope

T. Hoshimiya, M. Ohta, and M. Kojima

Department of Applied Physics, Tohoku-Gakuin University, Tagajyo, Miyagi 985, Japan

Photoacoustic images of magnetic disk heads were obtained and a qualititative evaluation for a gap or an isolation area of the head was performed and evaluated with the aid of amplitude, phase and "hologram" images for the first time.

1. Introduction

Recently, photoacoustic microscope(PAM) has been applied to measure and evaluate samples with small inner dimension and structures. As current development of the information science has stimulated high-density magnetic recording techniques, miniaturization of every parts of recording media motivates a need to measure the finished size of a built-in magnetic head and to evaluate its inner cracks nondestructively. In a point of view descrived above, we have developed an application of PAM to magnetic recording. In this study, photoacoustic images of magnetic disk heads were obtained, and a qualititative evaluation for a gap or an isolation area of the head was performed and evaluated with the aid of amplitude, phase and "hologram" images for the first time.

2. Apparatus and samples

A Photoacoustic imaging device used is basically the same as that used for a paper chromatography[1]. Furthermore, a lock-in amplifier(NF Circuit Block model LI-574 or 5610) and a microscope objective of a long working diatance typo with a magnification of 40 is used. Magnetic disk heads used as samples were that used in floppy-disk(FD) and that used in hard-disk(HD), both of which are made of calcium titanate($CaTiO_3$) ceramics, furthermore, ferrite FD head was also measured.

3. Definition of the "hologram"

In a PAM, a phase image gives a more detailed information on a sample than a thermally-difused amplitude image , while an amplitude image gives an intuitive information on whole of the sample. To amalgamate both advantages, one of the author(T.H) introduced the concept of "hologram" in the field of photoacoustics[2].

To define a "hologram" of a photoacoustric image of a sample, there are two possibilities to choose the "hologram" using an amplitude image $A(x,y)$ and a phase image $\phi(x,y)$ as $A(x,y)*[1+\cos(\phi(x,y))]$ or $A(x,y)^2*[1+\cos(\phi(x,y))]$. However, in this paper, a "hologram" of a sample $H(x,y)$ is defined by

$$H(x,y)=A(x,y)^2*[1+\cos(\phi(x,y))]. \quad (1)$$

And we would like to let a discussion of which definition is better pending here.

4. Results and discussion

Fig. 1 (a) and (b) represents an amplitude image and a phase image of a gap area of FD disk head, respectively. Phase image shows a better resolution, and the achieved resolution of the fabricated PAM with the microscope objective was up to about 2 to 3 micron meter for a 256 times 256 dot image such that the gap with 2 micron meter distance was recognizable. In addition, an obtained thermal image of the gap area showed a remarkable edge effect (confinement of the thermal wave) for a gap or an isolation area of the head. This is interpreted as the thermal wave generated in a $CaTiO_3$ region is reflected at an interface between a $CaTiO_3$ and a silicate glass, thermal properties of which are different.

Furthermore, we observed a surface crack in a ferrite region of FD disk head with PAM. And a phase image of a glass used as a filler for a magnetic circuit of HD disk head, obtained at the reverse side, clearly suggested an existence of a bubble or an undersurface structure, shown in Figures 2. In Figures (a) to (c), amplitude, phase and "hologram" images are shown, respectively. Width of the glass region was about 200 micron meter. All data were taken at the power level of 260mW and the modulation frequency of 90 Hz, the corresponding thermal diffusion length calculated being about 40 micron meter.

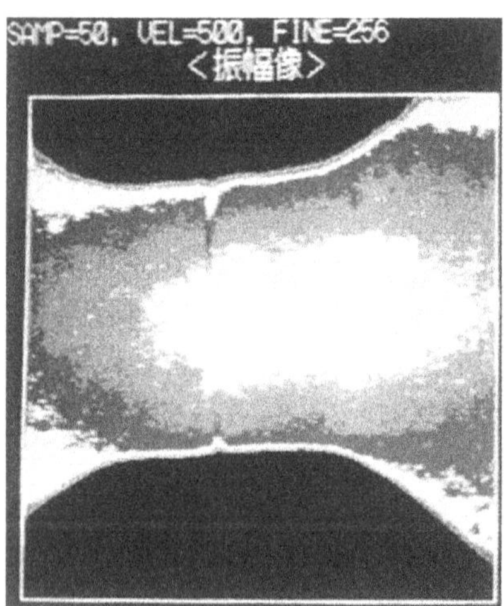

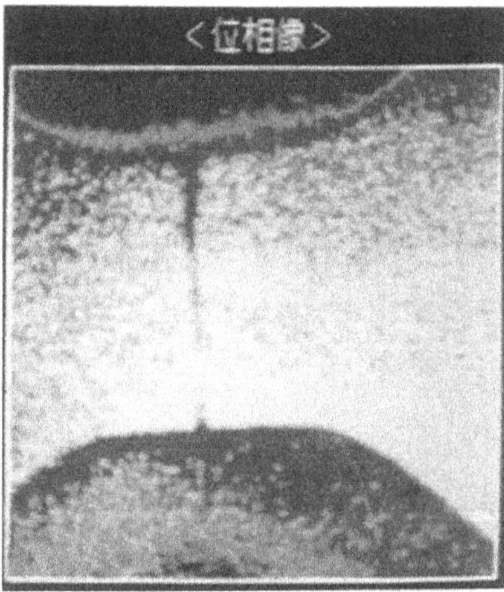

Fig. 1 An amplitude image (a) and a phase image(b) of a gap area of FD disk head.

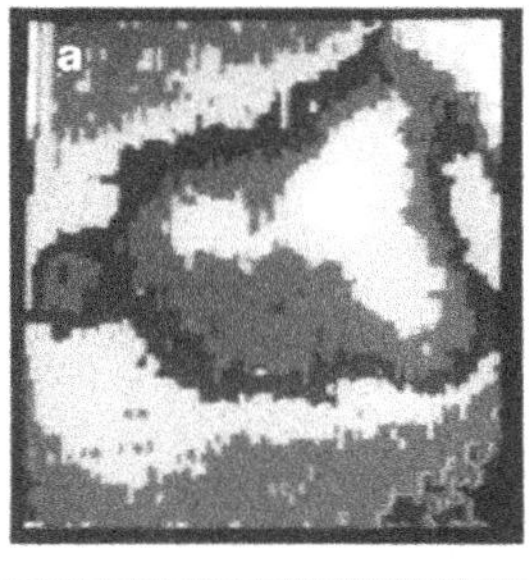

(a)

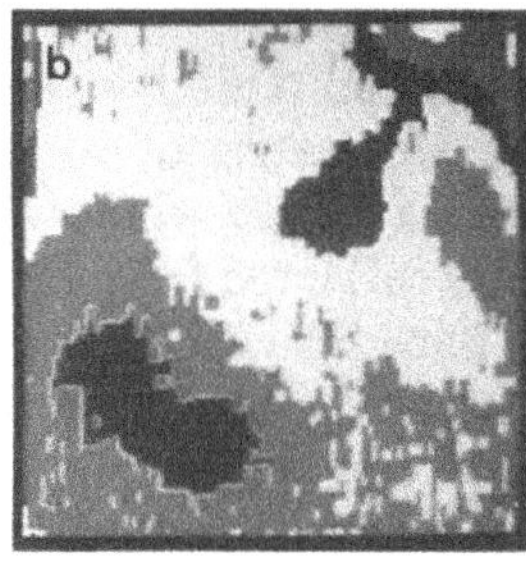

(b)

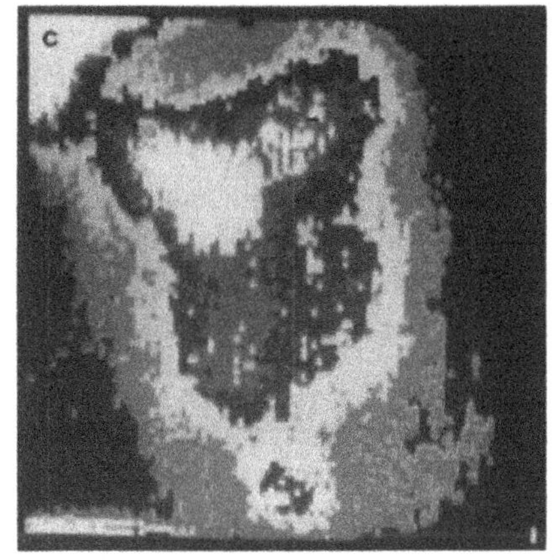

(c)

Fig. 2 (a)Amplitude, (b)phase and (c) "hologram" images of a glass which suggest a bubble.

5. Conclusion

In conclusion, a nondestructive evaluation of magnetic disk heads using PAM was performed qualititatively in this paper. A current requirement for a high-density magnetic recording will make a quantitative and precise measurement of inner dimension of magnetic disk heads by PAM with periodical or random modulation essential in the near future.

References

1. T. Hoshimiya: "Rapid-scan and High-fidelity Dual-mode Photoacoustic Microscope and its Applications to Paper Chromatography." in "Photoacoustic and photothermal phenomena II", Ed. by J. Murphy(Springer Verlag, 1990).
2. T. Hoshimiya: "Analysis and simulation of images by gas-microphone photoacoustics and optical-probe method.", Abstract of the 36th Canadian Spectroscopic Conference(St. Catharine, Aug., 1990).

Rate-Window Transient Thermomodulation Spectrometry: Technique and Measurements in Semiconductors

Z. Chen and A. Mandelis

Photothermal and Optoelectronic Diagnostics Laboratory,
Ontario Laser and Lightwave Research Centre, Department of Mechanical Engineering, University of Toronto,
Toronto, Ontario, Canada M5S 1A4

Abstract: A new photothermal measurement technique suitable for the characterization of thermal and electronic transport parameters of wide classes of materials is presented in this paper. By setting a "rate window" to the temporal evolution of the photothermal response reflected from the sample surface, this method provides a non-contact, non-destructive way to diagnose microelectronic and optoelectronic junction devices, where electron-hole plasma de-excitation time constants are long comparing to lock-in limited rate windows. The basic theory regarding this technique and some experimental results are shown in this paper.

In its present, first realization, the photothermal reflectance (PTR) detection scheme was adopted[1,2] due to its ability to access high modulation frequencies at acceptable signal-to-noise ratios (SNR). The experimental set-up is shown in Fig.1. An electrical pulse mixed with an AC ripple in the 2 - 100 KHz frequency range, described by

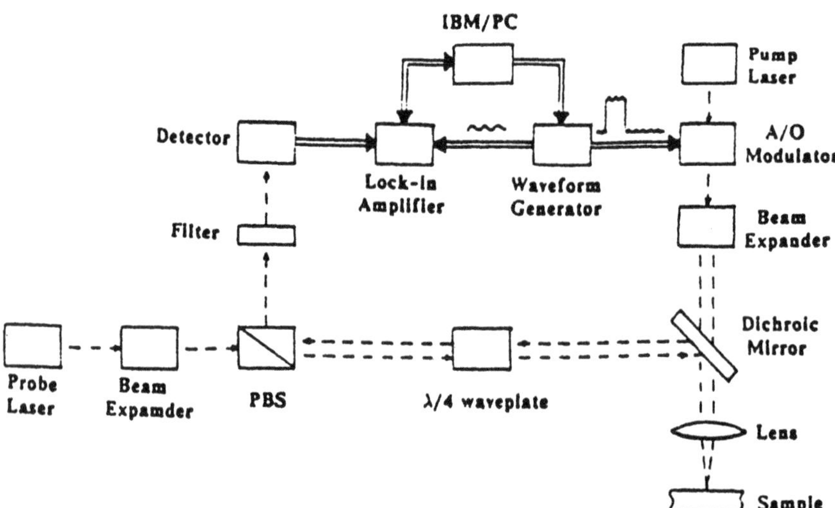

Fig.1 The Experimental set-up. A modulated laser pulse is used as a pump beam and a two channel lock-in amplifier with a fast time constant is employed to track the temporal evolution at the AC ripple frequency.

Springer Series in Optical Sciences, Vol. 69 Photoacoustic and Photothermal Phenomena III
Editor: D. Bićanić

$$F(t) = \begin{cases} 1 + me^{i\omega_0 t}; & 0<t<\tau \\ 0; & t>\tau \end{cases} \qquad (1)$$

with m << 1, is used as the input to the acousto-optic modulator (AOM). The 514 nm line of an Ar^+ ion laser is thus an intensity-modulated pulse of duration τ, typically 100 ms. Both pump and He-Ne 632.8 nm probe beam are focused collinearly on the sample surface. The reflected He-Ne laser beam is directed to a Si photodetector in the conventional manner[1,2]. The output from the detector is connected to a phase-sensitive, two-channel lock-in amplifier. Here, the lock-in amplifier can be used to track the temporal evolution of the photothermal reflectance response of the sample after the end of the optical pulse with a short time constant, or to obtain the Fourier components with frequency of reference signal with a long time constant.

The experimental system is controlled by an IBM PS/2 computer and a dual-gate integrator on the transient signals is performed by suitable software manipulations. Using this "rate-window" method[3], the transient in-phase or quadrature magnitude is sampled at two different time t_1 and t_2 after the pulse. The difference between the two signal levels at t_1 and t_2 is the output of ΔS_I. For a given rate window, signal ΔS_I scanning as a function of modulation frequency f_0 produces a maximum at a characteristic frequency, ω_e, which depends on the thermal diffusivity and the pump beam spot size. This can be shown by direct analysis of the Fourier content of the thermal-wave problem with a laser source temporal behaviour given by Eq.(1). Consideration of the three-dimensional thermal-wave problem gives[4]

$$\omega_e = const. \times \alpha_s / W^2 \qquad (2)$$

where α_s is the thermal diffusivity and W^2 is the beam spotsize, assuming a Gaussian profile:

$$A(r) = \frac{\eta P_0}{\pi W^2 k_s} e^{-2r^2/W^2} \qquad (3)$$

In Eq.(3), η is a nonradiative energy conversion efficiency, P_0 is the incident laser intensity, and K_s is the material thermal conductivity. In the case of good thermal conductors, such as crystalline semiconductors, the frequency transient decay time constant is limited by the lock-in response time constant (1 ms); therefore, the time-gated signal produces information about the thermal diffusivity only through the magnitude of the transient. Results are shown in Figs.2 and 3. Figure 2 shows in-phase experimental results from c-Si with the laser beam focused to different spotsizes. The frequency $f_e = \omega_e / 2\pi$, which corresponds to the maximum ΔS_I signal, is 50 KHz when the pump beam is focused through a microscope objective to 1 μm diameter spot. The peak shifts down to 32

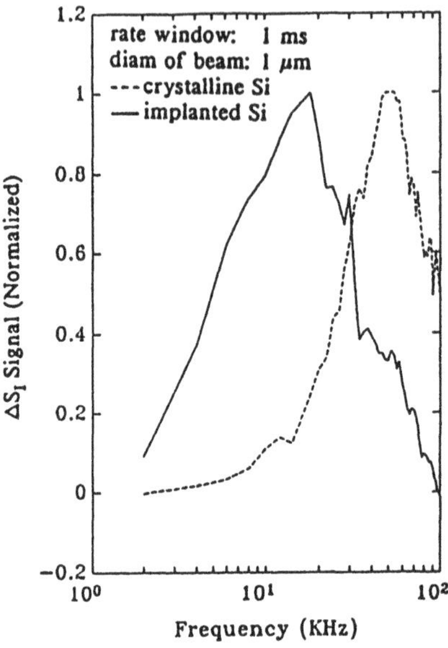

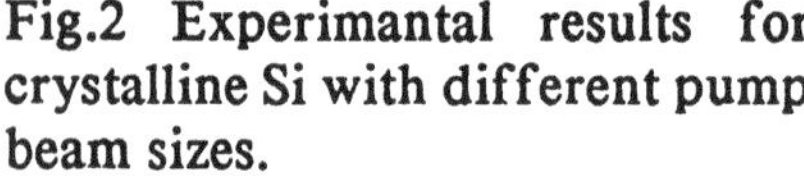
Fig.2 Experimantal results for crystalline Si with different pump beam sizes.

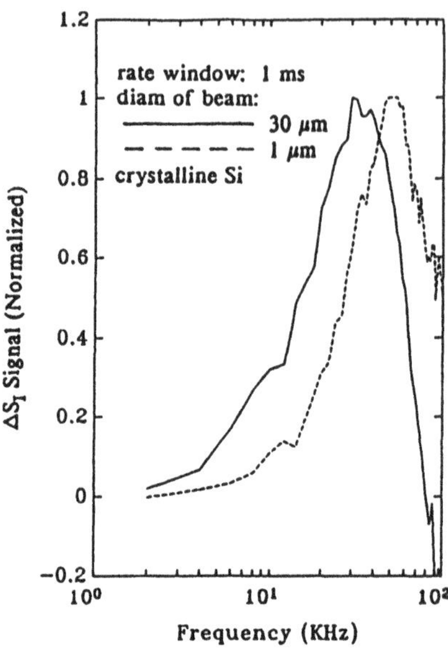

Fig3. Experimental results for differemt materials with same pump beam sizes.

KHz when lens focusing is used to 30 μm beam spot size, in agreement with Eq.(2). Figure 3 shows similar in-phase results with the pump beam fixed at a spot diameter of 1μm. The solid curve represents results from a P^+ implanted Si sample with a dose of 10^{16} cm^{-2} at 150 KeV ion energy. The frequency f_e (16 KHz) is seen to be approximately one-third of that obtained with crystalline Si (50 KHz). In view of Eq.(2) these data can be interpreted that the thermal diffusivity α_s for our implanted Si sample is about one-third of that for c-Si.

The merits of this new method lie i) in its superior dynamic range and SNR to conventional laser-pulse based transient photothermal techniques; ii) in its ability to highlight characteristic thermal transport-related time constants via the rate-window concept; and iii) in its promise for direct adaptation to non-contact, non-destructive microelectronic and optoelectronic junction device diagnostics, as shown by the related (but contacting) technique of Deep Level Transient Spectroscopy (DLTS)[3].

1. W.L.Smith, A.Rosencwaig, and D.Willenborg, Appl.Phys.Lett. **47**, 584 (1985).
2. I.A.Vitkin, C.Christofides, and A.Mandelis, Appl.Phys.Lett. **54**, 2392 (1989).
3. D.V. Lang, J.Appl.Phys. **45**, 3023(1974).
4. A.Mandelis and Z.H.Chen, unpublished.

Open Photoacoustic Cell Spectroscopy of Transparent Samples

N.F. Leite and L.C.M. Miranda

Laboratório Associado de Sensores e Materiais,
Instituto Nacional de Pesquisas Espaciais,
Caixa Postal 515, 12201 São José dos Campos, SP, Brazil

Abstract. In this paper we describe the use of the so - called open - photoacoustic cell (OPC) for both spectroscopic and thermal characterization of transparent samples. The samples used to demonstrate the usefulness of the OPC configuration are 0.6% Cobalt - doped soda - lime glass and tourmaline. It is shown that absolute values of non - radiative relaxation time and the optical absorption coefficient at given wavelengths are readly obtained. It is also shown that a complete thermal properties (thermal diffusivity and conductivity) characterization can be straightforwardly performed.

The open - photoacoustic cell (OPC) configuration consists of mounting the sample directly onto a circular electret microphone [1]. It is an open - cell detection in the sense that the sample is placed on top of the detection system itself, as in the case of piezo- and pyro- electric detections. This minimal - volume PA detection has been applied in the last two years to the characterization of both thermal and transport properties of several samples. These applications include the monitoring of adhesive curing [2], the measurement of thermal diffusivity of doped polymers [3], two - layer systems [4] and semiconductors [5]. In this paper, we explore the enhanced signal-to-noise ratio of the OPC detection to the complete optical and thermal characterization of transparent samples.

For the important spectroscopic case of optically transparent and thermally thick samples, the microphone output signal is given by:

$$V = V_0 \frac{j\omega\tau_E}{(1+j\omega\tau_E)} G \frac{I_0}{T_0 \ell_g \sigma_g k_s \sigma_s} F_s \, e^{j\omega t} \quad , \tag{1}$$

where T_0 is the ambient temperature, I_0 is the incident light intensiy at a given wavelength, $V_0 = (\ell_b \ell_m \sigma_0)/(\ell_b \epsilon + \ell_m \epsilon_0)$, with σ_0 being the charge density per cm^2, ϵ the dielectric constant, ℓ_b the static back plate air gap thickness and ℓ_m the electret membrane thickness. Furthermore, $\tau_E = RC$ is the electronic time constant, $G = (cosh(\ell_g\sigma_g) - 1)/sinh(\ell_g\sigma_g)$ is the gas chambre response function and F_s is given by:

$$F_s = \left[\frac{1}{1+j\omega\tau} \frac{1}{1-r} + \beta' \frac{k_s}{k_m} \left(\frac{\alpha_m}{\alpha_s} \right)^{1/2} F_m \right] e^{-\beta \ell_s} \tag{2}$$

with

$$F_m = cosh(\ell_m \sigma_m)/sinh(\ell_m \sigma_m) \quad , \quad r = \beta/\sigma_s. \tag{3}$$

In the above equations ℓ_i, k_i and α_i are the length, thermal conductivity and diffusivity of material i. Here the subscript i denotes the sample (s) and gas (g) respectively, $\sigma_i = (1+j)a_i$, with $a_i = (\pi f/\alpha_i)^{1/2}$ being the thermal diffusion coefficient of material i, β the optical coefficient absorption and $\beta' = 1 - r$. Finally, τ denotes the sample non - radiative relaxation time at the incident light wavelength.

Eq(2) tells us that for thermally thick samples the OPC signal is essecially proportional to the sample transmitance,$exp(-\beta \ell_s)$ and is dominated by the second term of F_s, i.e.,

$$F_s \simeq \beta' \frac{k_s}{k_m} \left(\frac{\alpha_m}{\alpha_s} \right)^{1/2} F_m e^{-\beta \ell_s} \ . \tag{4}$$

By an appropriate normalization procedure consisting of dividing the sample signal V by the microphone direct incidence signal V_d (in which case the sample is replaced by a transparent optical glass window like BK-7) the optical absorption spectrum is readly obtained from

$$\frac{V_d - V}{V_d} = 1 - e^{-\beta \ell_s} \ . \tag{5}$$

The complete optical and thermal characterization at given wavelength is obtained from a data fitting procedure of the OPC signal and phase data, as a function of the modulation frequency, to Eqs.(2) and (3). The values obtained for the thermal diffusivity and conductivity were 0.0045 cm^2/s and 27.7 mW/cmK for the Co - doped soda - lime glass and 0.0099 cm^2/s and 9.95 mW/cmK for the tourmaline sample. In the Table 1. we summarize the wavelength dependent parameters we got for 0.6% Co - doped soda - lime glass and tourmaline.

Table 1. Optical and thermal properties obtained using the open cell method.

	Co-Doped Soda-Lime			Tourmaline
$\lambda(nm)$	540	600	650	720
$\tau(ms)$	0.29	0.21	0.27	26.5
$\beta(cm^{-1})$	9.25	11.80	10.36	15.5

References

1. L. F. Perondi and L. C. M. Miranda: J. Appl. Phys. **62**, 2955 (1987).
2. A. Torres-Filho, L. F. Perondi and L. C. M. Miranda: J. Appl. Polym. Sci. **35**, 103 (1988).
3. N. F. Leite, A. Torres-Filho, R. Stempniak and L. C. M. Miranda: J. Appl. Phys. **66**, 407 (1989).
4. A. M. Mansanares, A. C. Bento, H. Vargas, N. F. Leite and L. C. M. Miranda: Phys. Rev.(B) **42**, 4477 (1990).
5. A. Pinto-Neto, H. Vargas, N. F. Leite and L. C. M. Miranda: Phys. Rev. (B) **40**, 3294 (1989).

A Reliable Optothermal Sensor

P. Helander

AB Varilab, Fullerstavägen 32, S-14144 Huddinge, Sweden

Abstract. Temperature variations in a sample are measured by means of a thermistor. The sensor is easy to use, has a rugged construction and is not sensitive to vibrations. A typical applications is analysis of liquid samples at low modulation frequencies.

1. Introduction

A common feature of most photoacoustic and photothermal techniques is that optical absorption close to the sample surface generates the measured signal.

A common way to perform this kind of experiment is to use a closed chamber. The sample in the chamber is illuminated with intensity modulated radiation. The subsequent pressure variations are detected by a microphone [1]. Other means to measure the temperature variations at the sample surface are either by an infrared detector [2] or by an transparent conductive film [3]. It is also possible to use a thin solid plate in contact with the illuminated surface. Periodic thermal expansions of the plate can be measured by a piezoelectric crystal [4].

2. Sensor operating principle

A new type of sensor has been developed. Temperature variations in a sample are transformed directly to an electrical signal by means of a thermistor. A cross sectional view of the sensor is shown in figure 1.

Visible radiation in the light guide passes through the sapphire disc and penetrates into the sample. The generated heat diffuses back through the sapphire to the thermistor.

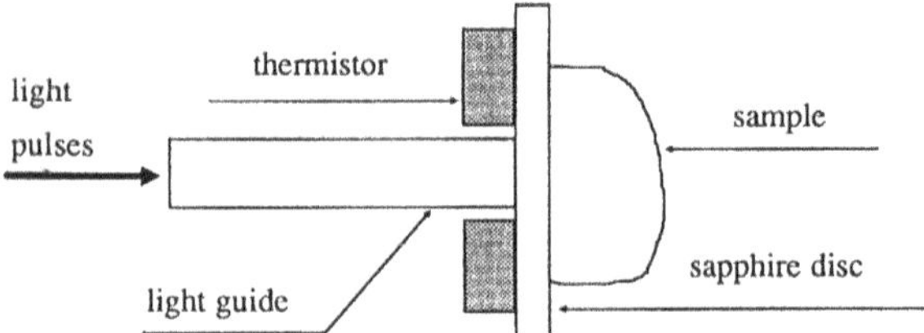

Figure 1. Cross sectional view of the sensor

Springer Series in Optical Sciences, Vol. 69 Photoacoustic and Photothermal Phenomena III
Editor: D. Bićanić

A unique feature of the new sensor compared to other techniques mentioned above is that the temperature variations are sensed outside the illuminated area. Thus the heat has to diffuse over a distance comparable to a diameter of the light beam. This is possible only if the illuminated region is small, the modulation frequency rather low and the thermal diffusion length in the disc material is large.The thermistor also has to be placed very close to the illuminated region.

Usually a 1 mm diameter optical fiber is used as a light guide and the disc is made of sapphire. Sapphire has a large thermal diffusion length as well as superior optical, mechanical and chemical properties.

3. Sensor features

The main advantage of the geometry used is that (for instance) a thermistor can be used. A thermistor is very useful for the measurement of both the average temperature as well as of small temperature variations. Furthermore it is not sensitive to vibrations which can be a large problem associated with the use of other techniques.

The geometry used also allows for screening of the thermistor from both scattered light and external electrical fields. The sapphire disc provides good mechanical, electrical and chemical isolation between the sample and the thermistor.

Another advantageous feature of the sensor is the fact that no closed chamber is needed. Changing sample is therefore fast and easy. Only a small sample volume is needed and the signal is independent of the size (for volumes large enough).

The signal is also independent of the small periodic thermal expansion of the sample.

For the analysis of samples containing sedimenting particles it is especially useful that the whole sensor can be turned around so that the light penetrates into the sample from different directions.

With a suitable preamplifier the signal strength is independent of the mean temperature of the analyzed volume.

If calibrated, the sensor can also measure the mean temperature of the analyzed volume.

The critical point of the sensor is its frequency response. The heat has to diffuse both through the sapphire (with a radial spreading) and into the thermistor. This is only possible at rather low frequencies. Furthermore both the sapphire and the thermistor must be thin and the later must be placed very close to the illuminated area.

The frequency response of the sensor was measured for a saturated signal (black ink). The frequency response of the measured signal depends both on the sensor and the fact that the temperature variations in the sample are smaller at higher frequencies.

The frequency response of the temperature variations exhibits mainly a 1/f dependence. The frequency response of the measured signal also showed a 1/f dependence up to about 70 Hz. The sensor thus follows the temperature in the sample up to about 70 Hz.

In most applications, modulation frequencies well below 70 Hz are used. This is due to the fact that the thermal diffusion length in the sample is usually short enough even at low frequencies.

4. Applications

Typical applications of the new sensor are measurements of concentrations, sedimentation studies and analysis of layered samples.

The sensor has been tested for measurements of concentrations of dyes in solutions. Just as with the previously designed open cell [4] it is easy to obtain an absolute determination of a concentration. This is done by using the amplitude and phase of the signal to calculate the concentration [5].

The sensor operates well in routine measurement. An example is for instance determination of hemoglobin in whole blood.

The sensor can also be used in the infrared or ultraviolet wavelength regions. By using infrared radiation of a suitable wavelength it is possible to measure water content of different samples.

The sensor is most suitable for the analysis of liquid samples. Solid samples can be studied as well provided there exists good thermal contact between the sample and the sapphire disc.

5. Summary

A rugged and reliable sensor has been constructed. Due to its practical design and low sensitivity to disturbances the sensor is easy to use.

The sensor is well suited for the analysis of liquid samples at low frequencies.

The measured signal is also easy to interpret. A linear relation between the recalculated optothermal signal and concentration of a substance can be obtained.

References

1 A.Rosencwaig and A.Gersho J. Appl. Phys. 47, 64 (1976)
2 P.-E.Nordal and S.O.Kanstad Appl.Phys.Lett. 11, 486 (1981)
3 K.Tanaka, Y.Ichimura and K.Sindoh J.Appl. Phys. 63, 1815 (1988)
4 P.Helander J. Photoacoustics 1, 103 (1981)
5 P.Helander J. Appl. Phys. 59, 3339 (1986)

Transferable Measurement System for Fast Non-Destructive Evaluation

J. Varis, J. Hartikainen, R. Lehtiniemi, J. Rantala, and M. Luukkala

Department of Physics, University of Helsinki,
Siltavuorenpenger 20D, SF-00170 Helsinki, Finland

Abstract. Our transferable system consists of a remote unit and a hand held IR measurement head that can be brought to the object under inspection. The object is heated with a laser beam focused on a line and the surface temperature of the sample is monitored with an IR line scanner. The compact size of the measurement head is achieved because of the use of an optical fibre, which allows the placement of the heating laser in the remote unit. The measurement head is moved over the inspected area and the results are displayed as pseudo-color images on the screen of a microcomputer. The equipment is designed for industrial maintenance applications.

1. Introduction

An equipment for thermal non-destructive testing (NDT) in industrial environment has to be fast and rugged. Most of the already existing thermal NDT techniques are too slow to be used in industrial applications like maintenance work. Likely candidates for industrial applications are the so called fast photothermal methods based on infrared cameras or scanners. Our method uses a simple IR line scanner with a laser for heating.

Generally, a photothermal measurement system using a laser for heating is not suitable for maintenance work. Although this kind of system can be made fast and in some cases even rugged enough [1-4], the final problem is usually the large size of the heating laser and its cooling requirements. In order to build a transferable measurement system for maintenance applications, the equipment has to be reasonably small and the operator has to be able to move around the object under inspection. This can be achieved by using an optical fibre. It is possible to construct a compact hand held measurement head and place the laser in a separate unit. The optical fibre can have considerable length without causing a substantial loss in the heating power.

2. Measurement system

A schematic diagram of a transferable measurement system is shown in Figure 1. The system consists of a remote unit and a hand held measurement unit that can be brought near the object to be tested. The sample is heated with a line focused laser beam. The laser beam is transferred to the measurement head through an optical fibre. All the necessary optics for heating the sample and collecting the infrared radiation from the sample surface are integrated in the measurement head. The

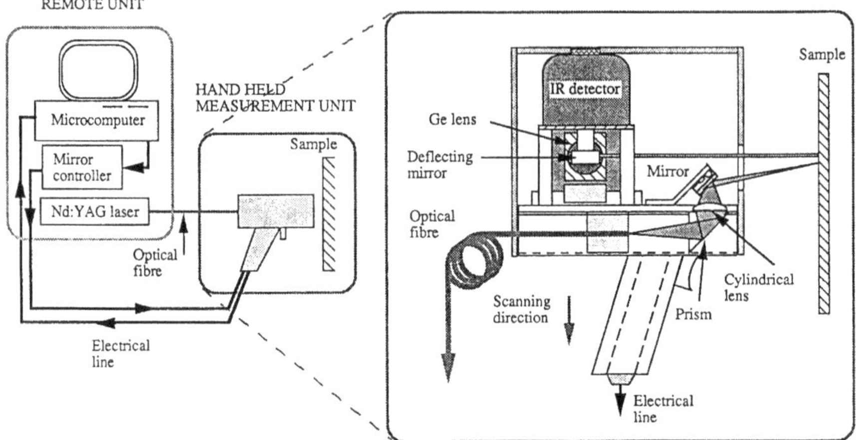

Figure 1. A diagram of the measurement system.

surface temperature of the inspected object is monitored with an IR line scanner consisting of a liquid nitrogen cooled HgCdTe detector, a germanium lens and a deflection mirror. The measured area is covered by combining a horizontal scan of the deflection mirror and a vertical scanning made by the operator.

In this kind of measurement system the spatial resolution depends on the width of the heating beam and the size of the infrared detector. For this reason a small sized detector (25 x 25 μm^2) was chosen. The signal from the HgCdTe detector is led to a microcomputer for processing and storage. The measurement data is displayed as a pseudo-color image. The microcomputer is equipped with a multi-function I/O board for data acquisition and controlling the position of the deflection mirror. The microcomputer is situated in the remote unit with the heating 35 W Nd:YAG laser and a mirror controller amplifier.

The measurement software has some basic image processing functions like median averaging and histogram equalization. Also different color maps including grey scale are available. The measurement data can be stored in a data format suitable for later analysis with commercial image processing programs. In addition to good graphics capabilities, considerable attention has been given to the user friendliness of the software.

3. Example of results

A composite plate consisting of several layers of carbon and glass fibres was used as a test sample. The sample is optically homogeneous both in the visible and infrared regions. However, the thermal conductivity of the carbon fibres is better than the glass fibers, thus an inner structure can be resolved in the thermal image (Figure 2 (a)). A schematic drawing of the sample is shown in Figure 2 (b). The wider appearance of the second carbon fiber layer results from the fact that the resolution decreases as the depth of the features increases. The dimensions of the measured area are 6.0 x 2.0 cm^2 and the measurement time was approximately 6s.

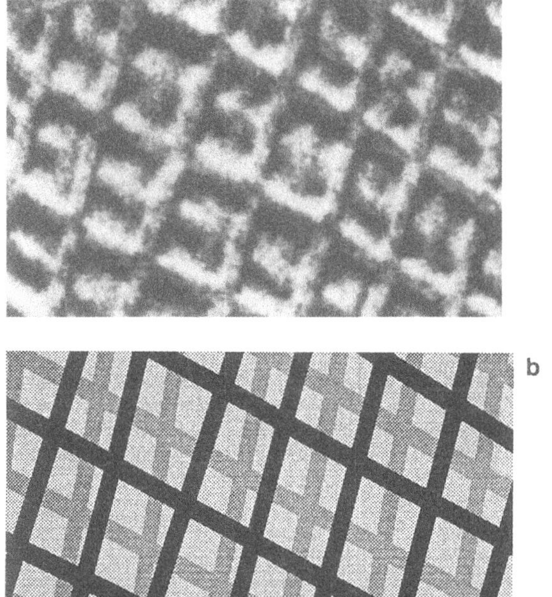

Figure 2. A thermal image of a carbon fibre composite plate (a) shown schematically in (b).

4. Conclusions

We have demonstrated a prototype of a fast transferable photothermal NDE system for industrial maintenance applications. Major goals have been transportability, user friendliness, good graphics capabilities and low cost of the equipment, all of which are important factors in industrial applications.

Acknowledgements

This work was supported by the Ministry of Trade and Industry.

References

1. J. Hartikainen, *Rev. Sci. Instr.* **60**, 670 (1989).
2. J. Hartikainen, *Rev. Sci. Instr.* **60**, 1334 (1989).
3. J. Hartikainen, *Appl. Phys. Lett.* **55**, 1188 (1989).
4. J. Hartikainen, R. Lehtiniemi and M. Luukkala, *Review of Progress in Quantitative NDE*, edited by D. O. Thompson and D. E. Chimenti (Plenum Press, New York, 1991), Vol. 10A, p. 539.

High-Sensitivity Cell for Pulsed Photoacoustic Spectroscopy in Gases and Liquids

I. Carrer, L. Fiorina, and E. Zanzottera

CISE Tecnologie Innovative, P.O. Box 12081, I-20134 Milano, Italy

Abstract Researchers in the field of photoacoustic spectroscopy have attenpted, for several years now, at improving the sensitivity of their photoacoustic apparatuses. Progress has been made in the use of lamps and CW lasers /1-5/, no particular arrangements, except when a special detector is used /6/, are, however, adopted when a pulsed laser source is utilized. We describe herein novel solutions /7/ which improve the sensitivity of pulsed photoacoustic detection to a large extent.

1. Theory

When a pulsed laser beam is directed into a cell containing an absorbing gas, an acoustic signal is generated which has the typical pattern of a compression peak followed by a rarefaction one /8/. The time delay T between the two peaks is only dependent on the radius R of the beam when dealing with short pulses, i.e., $T = R/v$ and the laser pulse duration τ is shorter than R/v (v = sound velocity). In the opposite case ($\tau > R/v$) T is equal to τ /9-11/.

The following considerations apply to both cases, provided that vT is small in comparison with the cell size and not larger than the microphone dimension; in this case we can utilize purely geometrical considerations to describe the propagation of the photoacoustic signal.

The acoustic wavefront generated by a cylindrical light beam is cylindrical too; if the cell has the same cylindrical shape where the beam is directed on its centre and the microphone is housed on the inner surface, a fraction of the signal (equal to the ratio between the microphone area and the cylindrical cell area) will be detected. Generally, small size (of the order of few millimeters) microphones are utilized, in order to guarantee the relatively high frequency response required; this implies that the signal is not fully collected. To improve the sensitivity it is necessary to solve how will it be possible to collect the signal generated in a large cell and to convey it onto a small microphone. The solution we have adopted is the following: the cylindrical wavefront must be turned into a spherical wavefront with its centre located on the cell walls. The efficiency of each solution can be evaluated, in a first approximation, by the ratio between the amount of the photoacoustic signal collected by the reflector surfaces to the total signal generated in the cell. Other effects on the signal efficiency are diffraction (when the size of the microphone is of the order of the photoacoustic wavelength, which is, however neglected in our approximation) and phase differences (due to laser beam misalignments).

Springer Series in Optical Sciences, Vol. 69 **Photoacoustic and Photothermal Phenomena III**
Editor: D. Bićanić

2. Experiments

Two kinds of cells (patent pending) have been designed and tested: one with a toroidal wall (Fig. 1), the other with two paraboloidal walls (Fig. 2). The toroidal cell is about cubic, the toroidal wall collects about 25% of the cylindrical wavefront and focuses it on the microphone housed in the opposite wall. The efficiency of the latter cell, which consists of two faced paraboloidal walls is higher, about 70%. One wall is a translation paraboloid with a linear focus coincident with the laser beam; it changes the cylindrical wavefront into a plane wavefront. The opposite reflector is a rotation paraboloid which focuses the plane wavefront on the microphone located in the middle of the former reflector. The experimental tests show remarkable improvement in the sensitivities of both cells, i.e., by one order of magnitude. Other advantages are good temporal discrimination of the signal against the window noise and the possibility of placing the microphone far from the beam without affecting the sensitivity.

Fig. 3 shows a typical signal generated in a parabolic cell. The dimensions of such a cell are: 20 mm translation paraboloid focal length, 100 mm rotation paraboloid focal length and 150 mm cell

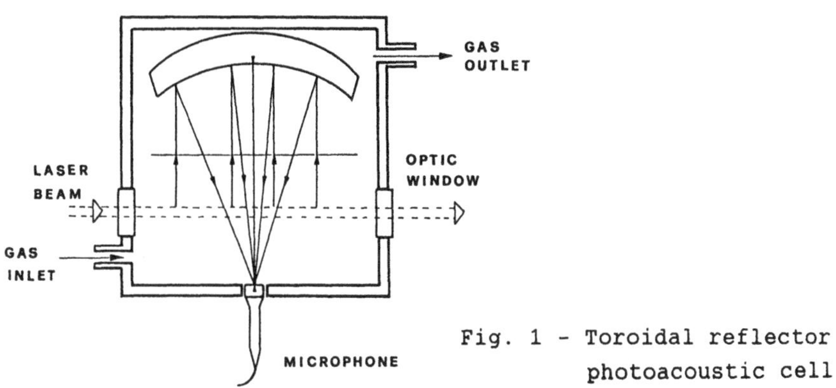

Fig. 1 - Toroidal reflector photoacoustic cell

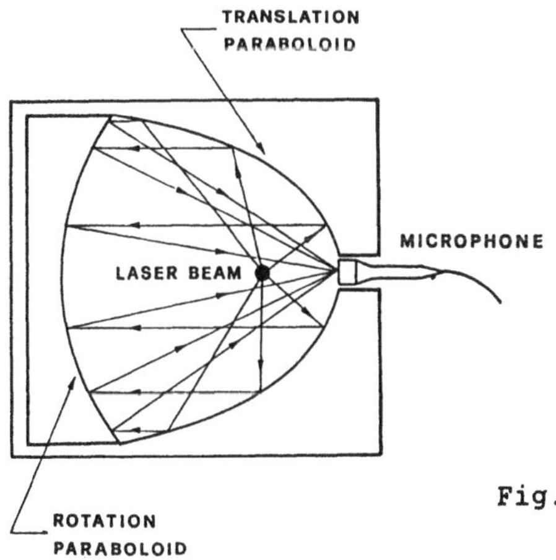

Fig. 2 - Parabolic reflector photoacoustic cell.

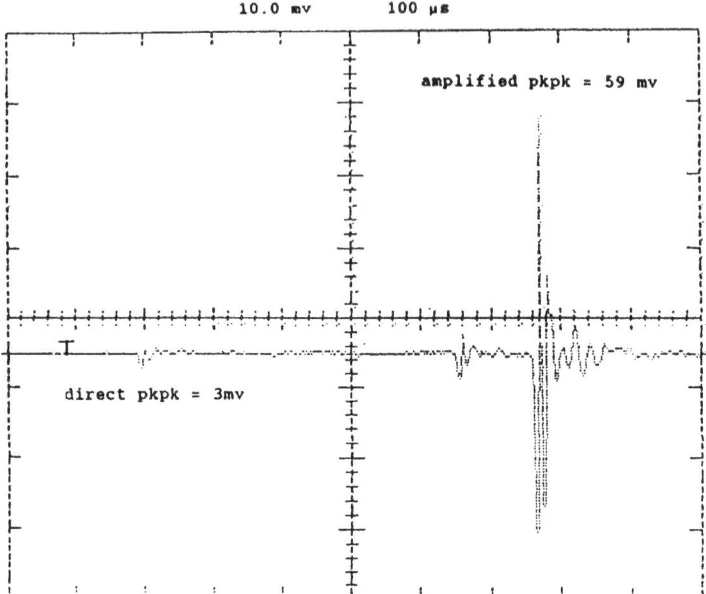

Fig. 3 - Typical signal generated in a parabolic cell with 20 mm translation paraboloid focal length, 100 mm rotation paraboloid focal length and 150 mm cell length. The ratio between the amplitudes of the enhanced signal and the direct one is about 20.

length. In this case the theoretical efficiency of the direct signal, which is the same we would collect if the cell were cylindrical, is $1.17 \cdot 10^{-3}$ and that of the enhanced signal is 0.67. The sensitivity improvement (that is, the root of the ratio between the enhanced signal and the direct one) is about 20. This value has been found also experimentally: in fig. 3, the first peak is the direct signal, the second is generated by the direct reflection of the rotation surface, and finally, the third is the enhanced one.

3. Literature References

1. A. Rosencwaig: Photoacoustics and Photoacoustic Spectroscopy, John Wiley & Sons, (1980).
2. P. Hess. (Ed.): Photoacoustic, Photothermal and Photochemical Processes in Gases, Springer Verlag, Berlin Heidelberg, (1989).
3. I.E. Jorgensen, S.A. Nexo, J. Christensen: UK Patent Application, Denmark, (1987).
4. O. Oehler, K. Mosbach, M. Seifert, H. Kunz, N. Kuster: US Patent N. 4,740,086, (1988).
5. M.B. Denton, S.B. Tilden: US Patent n. 4,688,942, Washibngton D.C., (1987).
6. J.M. Heritier, J.E. Foquet, A.E. Siegman: Appl. Opt., 21, 90, (1982).
7. I. Carrer: Optolaser, 2, 31, (1990).

8. C.K.N. Patel, A.C. Tam: Rev. of Mod. Phys., 53, 517, (1981).
9. K.A. Naugol'nykh: Sov. Phys. Acoust., 23, 98, (1987).
10. J.M. Heritier: Opt. Comm., 44, 267, (1983).
11. H.M. Lai, K. Young: J. Acoust. Soc. Am., 72, 2000, (1982).

Modification of a Conventional Fourier Transform Infrared Spectrometer for Enhanced Photoacoustic Detection

S.H. Gordon[1], *R.V. Greene*[1], *and C. Pareja*[2]

[1]National Center for Agricultural Utilization Research, USDA, ARS, 1815 N. University Street, Peoria, IL 61604, USA

[2]Laser Precision Analytical, A Division of Laser Precision Corporation, 17819 Gillette Avenue, Irvine, CA 92714, USA

Abstract. A commercially available Fourier transform infrared (FTIR) spectrometer has been modified to achieve a large improvement in the signal-to-noise (S/N) ratio and improved reproducibility of photoacoustic spectroscopy (PAS). Use of a high sensitivity photoacoustic detector increased the S/N ratio 10 times over the original detector. Another 8-fold improvement over this increase, resulting in an total 80-fold improvement, was accomplished by modifying the existing FTIR spectrometer for optimum performance in photoacoustics. Technical details of the modification are described. Photoacoustic measurements before and after modification are presented. Advantages of the instrument modification and preliminary results of its application to measuring microbial contamination in cereal grains are discussed.

1. Introduction

Over the past decade Fourier transform Infrared photoacoustic spectroscopy (FTIR-PAS) has been developed as a non-destructive technique for investigations of solids and other materials that are difficult to study by common spectrometric methods [1-3]. In a recent application the potential of FTIR-PAS for detection of fungal infections in cereal grains has been demonstrated [4]. In some of our previous work [5] FTIR-PAS suffered from a lack of sensitivity and reproducibility caused by poor S/N ratios and instrumental instability. Other workers have obtained better and more reliable photoacoustic spectra by using relatively expensive commercial FTIR spectrometers such as those few capable of step-scan operation [6]. In this report we describe how, in collaboration with the instrument manufacturer, an ordinary research grade FTIR spectrometer was modified to achieve photoacoustic S/N ratios and stability approaching those of much more expensive instruments.

2. Description of Instruments

The instrument used in our previous work was a Laser Precision Analytical (LPA), formerly Analect, model RFX-75 FTIR spectrometer. It was equipped with a water-cooled IR source (75 watts), a cube-mirror interferrometer with folded-path design, and a mechanical bearing for the moving mirror. The optical unit was controlled by a DCM 200 electronic module which gave 0.125 cm^{-1} resolution but limited the scan velocity to not better than 0.20 cm/sec optical path differential (OPD). The instrument was equipped with an external sample compartment dedicated for photoacoustic measurement. The photoacoustic detector was an LPA FX-400 sample cell with preamplifier. This instrument produced spectra of carbon black (100% lines) which, after 32 scans at 8 cm^{-1} resolution, contained as much as 6% noise (peak-to-peak) at 2100-2200 cm^{-1}.

Springer Series in Optical Sciences, Vol. 69 Photoacoustic and Photothermal Phenomena III
Editor: D. Bićanić

2.1 Instrument Modifications

The most inherent and unavoidable source of noise in any FTIR spectrometer is the detector [7]. This is particularly true with photoacoustic detectors. Therefore, in our first instrument change, the original detector was replaced by a more sensitive photoacoustic detector, the MTEC model 200 and preamplifier. The MTEC detector was selected because of its outstanding S/N characteristics. All further changes and comparisons were made with this detector as the reference.

Modifications of the existing FTIR spectrometer were inspired mainly by experimental observations that the photoacoustic signal strength is directly proportional to the intensity of the IR source. Figure 1 shows the increase in the PAS signals from cellulose as the IR source power is increased from 3.8 watts to 23.6 watts. These four spectra were acquired with the original FTIR instrument and PAS cell. The poor S/N ratio is evident, especially in the lower energy spectra. Consequently, the beneficial effect of increasing the IR source power in PAS was clearly indicated.

Therefore, the RFX-75 spectrometer was modified by 1) boosting the IR source power from 75 to 150 watts; 2) upgrading the electronic module to the DCM 250 version for better stability of operation of the FTIR; 3) rebuilding the interferometer to minimize acoustic noise from vibration in the mechanical bearing; 4) refurbishing the optics by adding a new beam splitter, a better laser, and new mirrors to maximize throughput of the IR beam; and 5) optimizing the operating conditions such as IR beam alignment, using Helium to purge the PAS cell, and adjusting parameters such as resolution, gain ranging and apodization for best performance in PAS.

The three most important factors affecting the photoacoustic S/N ratio were identified as: Infrared source power (IR beam intensity and stability); scan speed (minimum achievable optical velocity and stability); and frequency of operation (bandwidth and lowest Hertz number). Hence, the specific strategies followed in making the instrument modifications involved three steps:

In the first step, the power supply to the IR source was upgraded to a 150 watt model that operates at 1600^{0} Kelvin with a 24 volt, 6.25 amp element. In

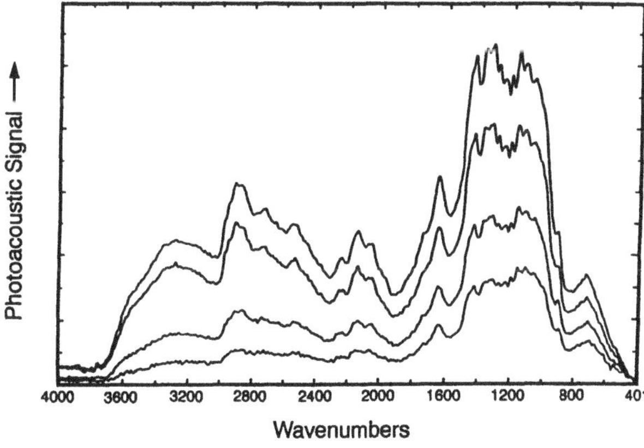

Figure 1. FTIR-PAS spectra of cellulose at four intensities of the infrared source power. From bottom to top, the spectra were acquired using 3.8, 7.3, 15.0 and 23.6 watts, respectively.

addition, we built and attached a variable power supply capable of 40 volts and 10 amps output. The power supply was tested at up to 28 volts and 7 amps without burning out the source element. However, the intensity of the glow from the source required a white light shield to be added to the Jacquinot.

In the second step, the servo motor controller or optical head controller (OHC) board was redesigned to extend the range of stable interferometer operation to slower optical velocities. Two redundant driver transistors were wired onto the board in place of the single original transistor. The driver transistors were balanced to supply more power and constant output voltage to the solenoid driving the moving element. This allowed operation at scan velocities as slow as 0.05 cm/sec. In this type of folded-path interferometer the mechanical velocity of the moving mirror as derived from the OPD is 1/4 of the optical velocity. Therefore, operation at 0.10 cm/sec OPD implies a mechanical velocity of 0.025 cm/sec. A servo controller (OHC) capable of responding to such low values and still possess the ability to provide quick settling during the turnarounds was required and designed. The corresponding operating frequencies achieved for the mid-infrared range were 40 Hz to 400 Hz.

Finally, in the third step, because the sensitivity of the microphone in the PAS cell varies inversely with the sampling frequency [7], the A/D controller on the optical head signal (OHS) board also needed to be redesigned. The antialiasing filter on the OHS board was optimized for the 40 Hz to 400 Hz range after stable servo motion was obtained and the laser channel delay was defined based on the filter characteristics.

3. Results and Discussion

Tests of the photoacoustic sensitivity and stability of the modified FTIR spectrometer were conducted over a period of three weeks during which data were collected using automated software routines to allow unattended operation. The use of Helium as a purge gas (8,9) increased the signal output by nearly three times but the stability on a scan-to-scan basis was not the equivalent of a dry air purge. The change in stability may be due to changes in the dynamic sensitivity of the microphone when a more buoyant medium is present. The phase stability at the lowest frequencies was compromised by the phase modulation present in the servo controller circuitry.

By convention the photoacoustic technique involved the use of a carbon black standard to provide the reference (background) spectrum, and the sample spectrum itself was ratioed against this reference. 100% lines were ratios of carbon black spectra to carbon black backgrounds. The dramatic improvement in the stability of the FTIR-PAS instrument is illustrated in Figure 2 and Figure 3. Plots of carbon black 100% lines were made at 1 hour intervals over 24 hours of continuous scanning cycles in which each spectrum was ratioed against an averaged background collected at the start. Hence, the waterfall plots show the variation in PAS signal strength over time. Figure 2 was produced on the RFX-75 FTIR spectrometer before modification, while Figure 3 was produced after modification. The fluctuations and drift in the PAS signal that plagued our earlier experiments were virtually absent with the modified and optimized FTIR-PAS instrument. This translates into improved reproducibility of PAS measurements.

In a typical example the noise level at 2100-2200 cm^{-1} in the 100% line of carbon black was reduced 8 times; from 0.56% before modification to 0.07% after modification of the RFX-75 FTIR spectrometer. With the original LPA FX-400 detector the noise level, as estimated from Figure 1, was about 10 times the

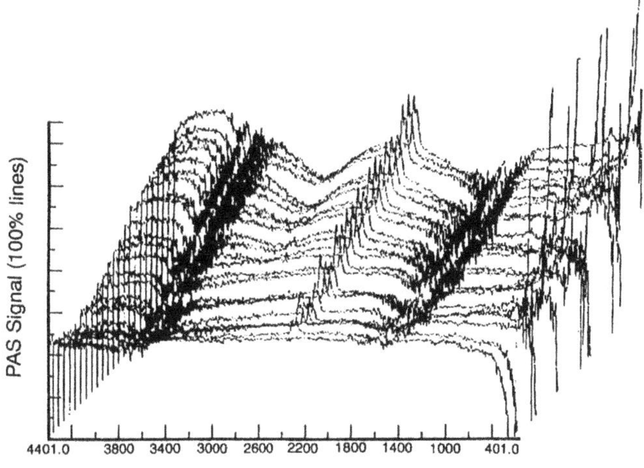

Figure 2. FTIR-PAS spectra obtained before instrument modification. Waterfall plots of 100% lines from carbon black spectra acquired at 1 hour intervals, from bottom (t = 0) to top, and ratioed against a carbon black background collected at the start.

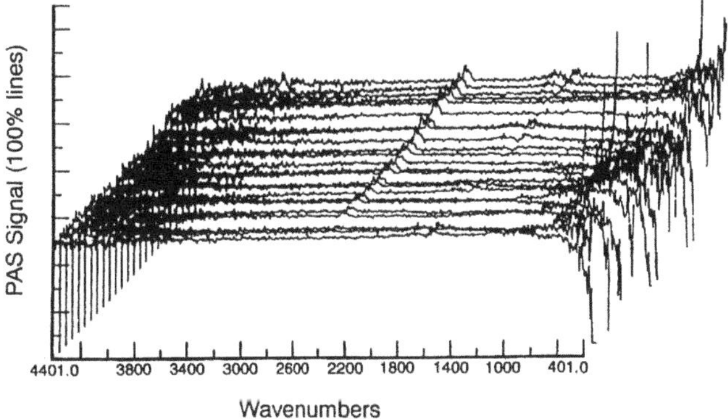

Figure 3. FTIR-PAS spectra obtained after instrument modification. Waterfall plots of 100% lines from carbon black spectra acquired at 1 hour intervals, from bottom (t = 0) to top, and ratioed against a carbon black background collected at the start.

noise level with the MTEC photoacoustic detector. Therefore, the S/N performance of the modified FTIR-PAS spectrometer was improved 80-fold over that of the original instrument.

With the present enhancements it is now feasible to obtain FTIR-PAS spectra having S/N ratios comparable to the sensitivity of diffuse reflectance or attenuated total reflectance spectroscopy [10]. Although PAS signal strength differences from background to sample can be quite large and can produce baseline shifts from use of the large gains assumed, PAS tests of polystyrene and polyethylene indicated that if the interferogram level is at least 25% of

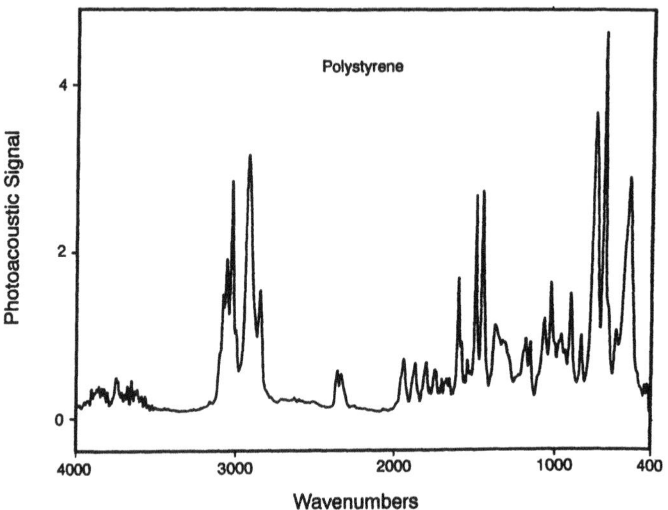

Figure 4. FTIR-PAS spectrum of polystyrene (50 um film) obtained after instrument modification. The spectrum was acquired at 0.10 cm/sec scan speed and 8 cm^{-1} resolution after a Helium purge.

the background level the errors created by quantitization limits of the A/D conversion can be avoided. Figure 4 shows a spectrum of a thin polystyrene film obtained with the modified FTIR-PAS spectrometer and illustrates a high spectral quality approaching that attainable with expensive step-scan FTIR spectrometers [6].

The levels of stability and S/N achieved on the relatively inexpensive commercial FTIR used in this work combine to increase the potential of FTIR-PAS for quantitative measurement of microbial growth on solid substrates. Thus, the prospect for detecting and measuring fungal contamination in whole cereal grains by FTIR-PAS is greatly improved.

4. Acknowledgments

The authors appreciate the expert and helpful assistance of Mr. John Baker and Mr. Jeff Teel in making this work successful. Support for the project was received from The Biotechnology Research and Development Corporation (Contract No. 1-5-019).

References

1. A. Rosencwaig, Annu. Rev. Biophys. Bioeng. 9, 31-54 (1984).
2. D.W. Vidrine, Appl. Spectrosc. 34, 314-319 (1990).
3. D.J. Gerson, J.S. Wong, and J.M. Casper, Amer. Lab. November, 63-71 (1984).
4. R.V. Greene, S.N. Freer, and S.H. Gordon, FEMS Microbiol. Lett. 52, 73-78 (1988).
5. S.H. Gordon, R.V. Greene, S.N. Freer, and C. James, Biotechnol. Appl. Biochem. 12, 1-10 (1990).

6. R. Rubinovitz, J. Seebode, and A. Simon, Brucker Report 1, 35-38 (1988).
7. P.R. Griffiths, and J.A. de Haseth, Fourier Transform Infrared Spectroscopy, (John Wiley & Sons, New York) p. 248 (1984).
8. K.Y. Wong, J. Appl. Phys. 49, 3033 (1978).
9. N.L. Rockley, M.K. Woodard, and M.G. Rockley, Appl. Spectrosc. 38, 329 (1984).
10. R.V. Greene, S.H. Gordon, M.A. Jackson, and G.A. Bennett, in preparation.

Why Phase in Phase Fourier Transform Infrared Photoacoustic Spectroscopy?

L. Bertrand, A. Cournoyer, A. Domingue, and H. Marchand

Department of Engineering Physics, Ecole Polytechnique, Montréal, Québec, Canada H3C 3A7

Abstract. By use of phase detection in Fourier transform infrared photoacoustic spectroscopy, we are able to obtain new information on materials difficult and often impossible to obtain otherwise. With our technique, a more precise and complete analysis is hereby possible. Applications of our approach to polyethylene, asbestos fibers and a thin layer of polymer illustrate well this capability.

1. Introduction

Fourier Transform Infrared Photoacoustic Spectroscopy (FTIR-PAS) is a well known technique and is widely used for the analysis of many materials [1]. Coupling the advantages of FTIR to the useful properties of photoacoustic or photothermal spectroscopy (PAS) makes possible the analysis, by infrared spectroscopy, of opaque or diffusing materials. Moreover, the high sensitivity of PAS allows the measurement of very low absorbances (weak absorptions or thin samples), and the nondestructive feature of PAS may be very important in applications concerning biochemical materials. Finally, depth profiling of the sample is possible by changing the thermal diffusion length via the variation of the acoustic frequency, i.e., the mirror speed.

But PAS presents some inconveniences such as signal saturation and band distortion due to sample dilation, optical diffusion or various thermal and optical heterogeneities. Even in the case of one material, depending on its physical nature (powdered or bulk), the relative intensities of the band spectra and the background signal between bands could vary, thus making quantitative PAS analysis difficult.

With FTIR-PAS, the depth profiling capability is in the range of a few μm, due to the limited frequency domain available with FTIR interferometers. Near surface analysis is thus inaccessible. Improvement of this advantage of PAS is made possible by use of the phase of the PA signal.

2. Phase Fourier transform photoacoustic spectroscopy

We have improved the performances of the FTIR-PAS technique by using the phase and the amplitude of the photoacoustic signal [2] or more precisely by using a complex Fourier transform to obtain the in-phase and in-quadrature components. First we correct the interferogram from the phase of the apparatus (using a thermally thick carbon black sample) and later take the complex Fourier transform. An iterative procedure is still necessary to correct the phase from any disturbance resulting from discrepancies between the apparatus phase obtained with the carbon black sample and the apparatus phase for the analysed sample.

The two components of the photoacoustic signal are now used to obtain the actual spectrum of the sample (more precisely the $\beta\mu$ product [3], β being the absorption coefficient and μ the thermal diffusion length) and to distinguish between absorption coming from the sample surface (less than a

Springer Series in Optical Sciences, Vol. 69 Photoacoustic and Photothermal Phenomena III
Editor: D. Bićanić

tenth of μ) and absorption coming from the remaining thermal diffusion length (or the remaining sample thickness if the sample is thermally thin).

With PHAS-FTIR-PAS we are able now to overcome all the photoacoustic artifacts cited previously as well as to improve greatly the depth profiling capability of PAS. Photoacoustic saturation now appears at higher values of absorbance resulting in better quantitative measurements. Sample dilation is also suppressed, hereby increasing the quality of the spectra (no background signal, improved resolution and more accurate band intensities).

Our PHAS-FTIR-PAS technique greatly improves the depth profiling of the sample absorption since we can now probe the thermal diffusion length itself in addition to modifying it conventionally. Moreover, PHAS-FTIR-PAS can be used with thermally thin samples and allows discrimination between surface absorption (a few Å) and bulk absorption (the remaining sample thickness).

3. Examples

The following examples illustrate the powerful capabilities of PHAS-FTIR-PAS.

Figure 1 shows the in-phase and in-quadrature components of the photoacoustic spectrum of polyethylene between 1000 and 4000 cm^{-1} and figure 2 shows the photoacoustic (amplitude) and $\beta\mu$ spectra, the latter being obtained from the data of Fig. 1. Clearly the $\beta\mu$ spectrum shows less photoacoustic saturation and a better resolution than those of the amplitude spectrum without the caracteristic photoacoustic background signal. Moreover, in Fig. 1 a large absorption band between 3000 and 3300 cm^{-1} appears only in the in-quadrature spectrum. This indicates that this absorption occurs on the sample surface only. This absorption is due to the water adsorbed on the sample surface corresponding, for this non-polar polymer, to a thickness of about a few hundred ångstroms.

Figure 3 is the photoacoustic and $\beta\mu$ spectra of asbestos, a very heterogeneous sample. Sample dilation is very important since interstitial gas will expand more than the solid. The photoacoustic spectrum demonstrates a large background signal and important photoacoustic saturation. Our $\beta\mu$ spectrum does not possess these photoacoustic artifacts. Figure 4 shows the spectra obtained from two asbestos samples, one being more compacted than the other. $\beta\mu$ spectra are quite the same, whereas the two photoacoustic

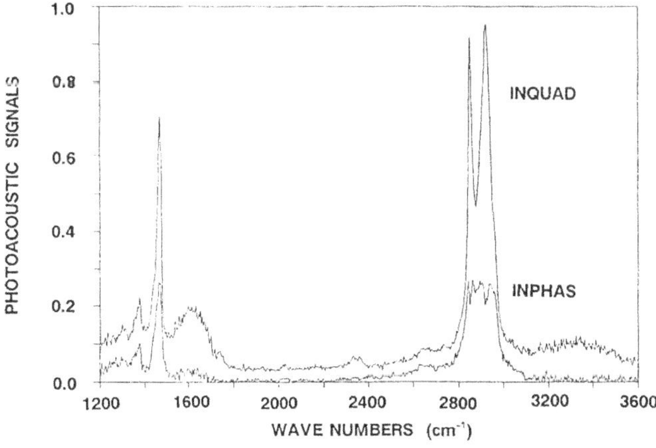

Figure 1: In-phase and in-quadrature photoacoustic spectra of polyethylene.

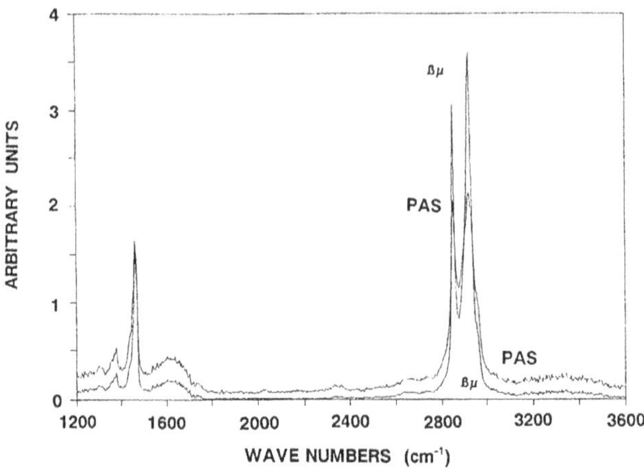

Figure 2: Amplitude and $\beta\mu$ spectra of polyethylene.

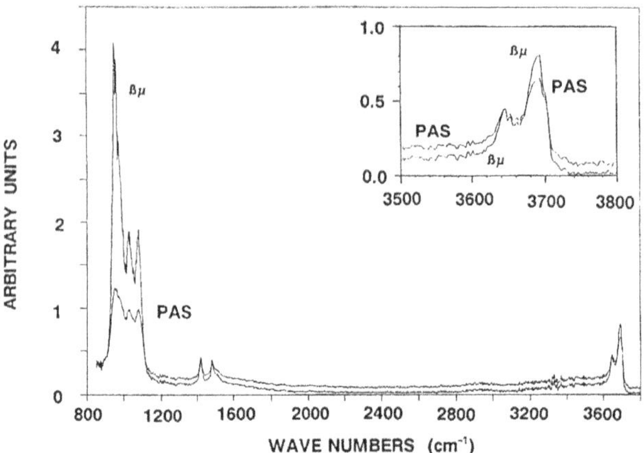

Figure 3: Amplitude and $\beta\mu$ spectra of asbestos.

spectra have different background signals and relative band intensities. This clearly demonstrates the effect of sample texture on the photoacoustic signal and also the necessity to have a method that corrects the photoacoustic signal for the photoacoustic artifacts.

The last example is a thermally thin sample: a thin layer (less than 1 μm) of plasma-polymerized hexamethyldisiloxane, a polar polymer, coated on an CaF_2 substrate. In this case, sorption results in a monolayer of water (1.5 Å thick) and a penetration of water into the bulk of the polymer (a volume signal). Studying the water band absorbance by PHAS-FTIR-PAS we are able to measure surface and volume water absorption. Figure 5 shows this absorption as a function of the relative humidity. Our results are in very good agreement with already published results about water sorption on polymers [4]. It is important to note that, in this case, the sample is thermally thin and that the surface signal arises from a thin layer only a few ångstroms thick.

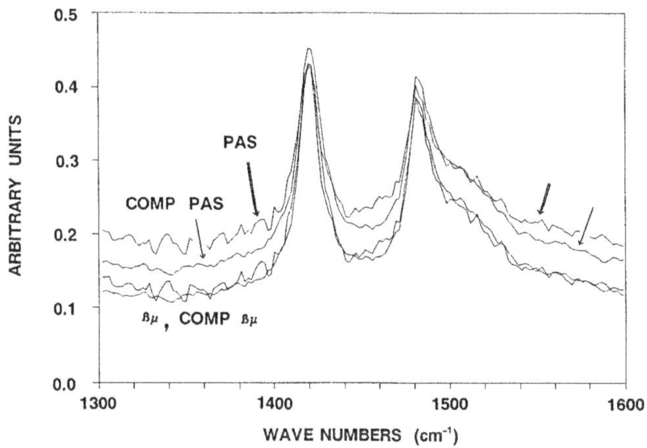

Figure 4: Amplitude and $\beta\mu$ spectra of two asbestos samples; COMP: compacted sample.

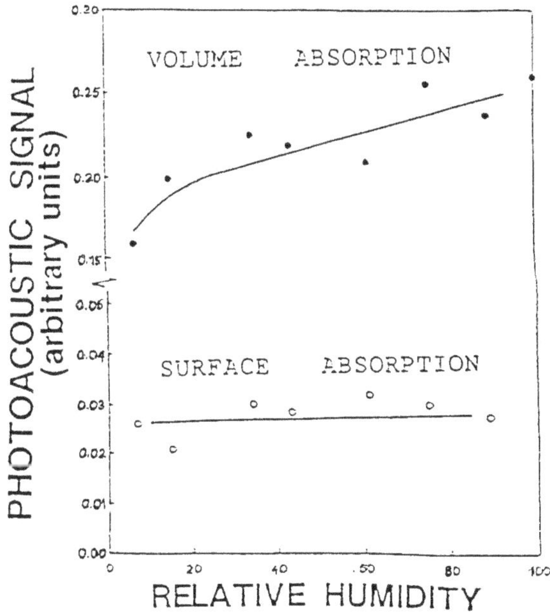

Figure 5: Surface and volume water absorption for a thin layer of PP-HMDSO on CaF_2.

4. Conclusion

We have thus demonstrated that, using a complex Fourier transform which gives the two components of the photoacoustic signal, we are able to overcome all the photoacoustic artifacts such as photoacoustic saturation, background signal and band distorsion due to the optical diffusion. Moreover, our PHAS-FTIR-PAS technique permits us to improve greatly the depth profiling capability of the photoacoustic method, a technique useful even with thermally thin samples.

References

1. McClelland, J.F., Luo, S., Jones, R.W., Seaverson, L.M., 7th International Topical Meeting on Photoacoustic and Photothermal Phenomena, August 26-30, Doorwerth (The Netherlands), (1991).
2. Bertrand, L., Applied Spectroscopy, 42, (1988), 134.
3. $\beta\mu$ is calculated from two complex equations. See Bordeleau, A., Rousset, G., Bertrand, L., Can. J. Phys., 64, (1986), 1093; or Rousset, B., Bertrand, L., Mongeau, B., Proc. Int. Conf. Fourier and Computerized IR Spectroscopy, Ottawa, Ont., (1985), paper 2.37.
4. Sacher, E. and Susko, J.R., J. Appl. Polym. Sci., 24, (1979), 1997.

Photoacoustic Investigation of the Evaporation Resistance of Monomolecular Lipid Layers

W. Micheler[1], *R. Osiander*[1], *P. Korpiun*[1], *J. Schmidt*[2], *and W. Knoll*[2]

[1]Physik-Department E13, Technische Universität München, James-Franck-Strasse, W-8046 Garching, Fed. Rep. of Germany

[2]Max-Planck-Institut für Polymerforschung, Jakob-Welder-Weg 11, W-6500 Mainz, Fed. Rep. of Germany

Periodic evaporation enhances the photoacoustic signal of liquids the more the higher the temperature /1/. Thin lipid layers and oilfilms, spread on liquid surfaces, act as evaporation resistances and attenuate this effect /2/. We have developed a model to describe the photoacoustic signal of liquids, covered by thin films which reduce the contribution of the vapor to the pressure variation /3/. But this model is only valid for homogeneous films whose thicknesses are in the order of micrometers in which the transport of mass is characterized by a diffusion coefficient. The concentration inside of such films can be related by the solubility to the vapor pressure at both boundaries of the film. If films like lipid monolayers have an inhomogeneous structure (Fig.1), a concentration gradient is not well defined. Therefore, it is more convenient to describe the reduction of a molecular flow by the permeability or its reciprocal, the diffusion resistance r and not by the diffusion coefficient. The molecular flux is with this assumption

$$j(t) = [c_1(t) - c_2(t)] / r \tag{1}$$

where $c_1(t)$ is the temporal variation of the equilibrium vapor concentration at the liquid surface and $c_2(t)$ the time dependent vapor concentration at the film to gas boundary (Fig. 1). Using the Clausius Clapeyron relation c_1 is related by $c_1(t) = (P_s L/R^2T^3)\,\vartheta(t)$ to the temperature variation $\vartheta(t)$ at the film boundary, where P_s is the vapor pressure of the liquid, L the heat of vaporization, R the gas constant and T the ambient temperature. The pressure variation in the gas of a photoacoustic cell for a thermally thick gas column then is given as

$$p(t) \propto \left[1 + \frac{\sqrt{\frac{D_g}{\alpha_g}}\,\frac{L\,P_s}{P\,RT}}{1 + r\sqrt{D_g f/2}\,(1+i)}\right]. \tag{2}$$

D_g is the diffusion coefficient for the vapor in the gas, α_g its thermal diffusivity, P the total pressure of vapor and air and f the modulation frequency. The temperature

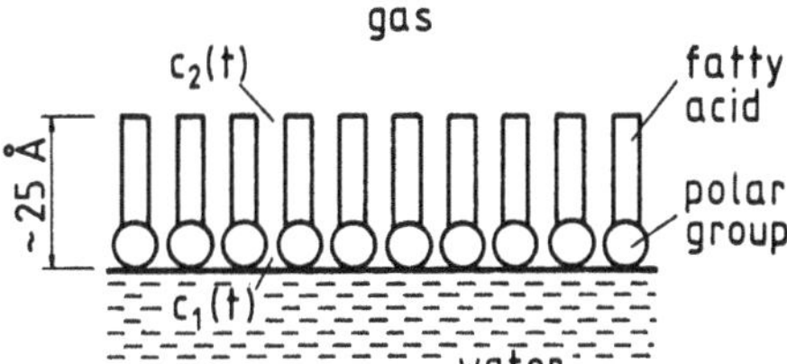

Fig. 1
Concentrations at the boundaries of the lipid film

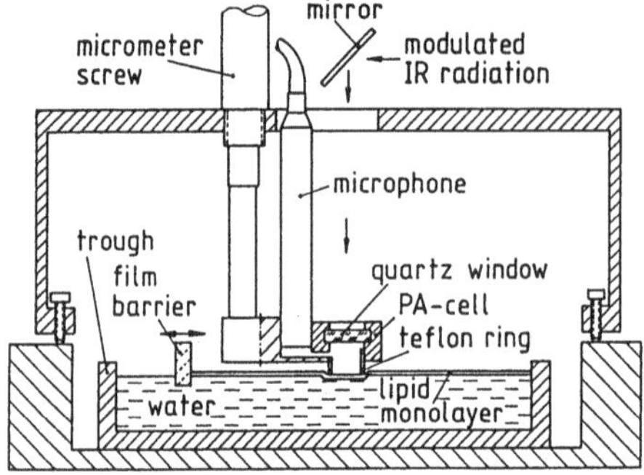

Fig. 2
Film balance with photoacoustic cell. Vertical position of PA cell is varied by a micrometer screen.

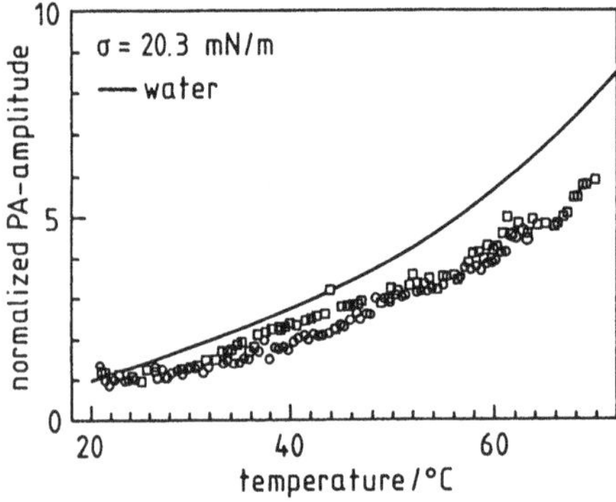

Fig. 3
Temperature dependence of the PA amplitude beasured on water (solid line) and on water covered with a monolayer of arachidic acid for a surface tension of 20.3 mN/m. Data of both measurements (o), (▫), are normalized to those at 20 ^{0}C.

dependence of the signal, eq. (2), at elevated temperatures is dominated by the temperature dependence of the vapor pressure. For large resistances r the second term will become smaller and the signal will increase less with temperature than for the uncovered liquid which can be described by eq. (2) for r = 0.

Lipids prepared in a proper way spread to a monomolecular film on a water surface. There is an equation of state for such systems which depends on the surface area, the number of molecules and the surface tension. These quantities can be varied in a wide range in a film balance. Such lipid films reduce the rate of evaporation of the water /4/.

For experiments we used an open photoacoustic cell which was mounted on a film balance (Fig. 2). This cell could be lowered by a micrometer screw in such a way that the lower part formed by a teflon ring touched the lipid film. The temperature of the water could be varied between 20 and 70 ^{0}C by pumping water from a thermostat throught a glass serpentine tube on the bottom of the trough.

We have measured the temperature dependence of the photoacoustic signal of water covered with monomolecular layers of arachid acid and DMPE in the temperature range between 20 and 70 ^{0}C for several surface tensions. The amplitude data measured with arachid acid are plotted in Fig. 3. The resistance is calculated by eq. (2) relating the photoacoustic signal to the signal measured on water in the same temperature range. The latter is given by eq. (2) for r = 0. We find that the diffusion resistance increases with rising temperature (Fig. 4). From the temperature dependence of the resistance (solid line in Fig. 4) we obtain an activation energy for diffusion of 30 kJ/mol. This value is consistent with the results of other authors /5/.

For DMPE, a phospholipid, the resistance below 50 ^{0}C is too small to be determined. Above this temperature the resistance increases for all surface tension to a constant

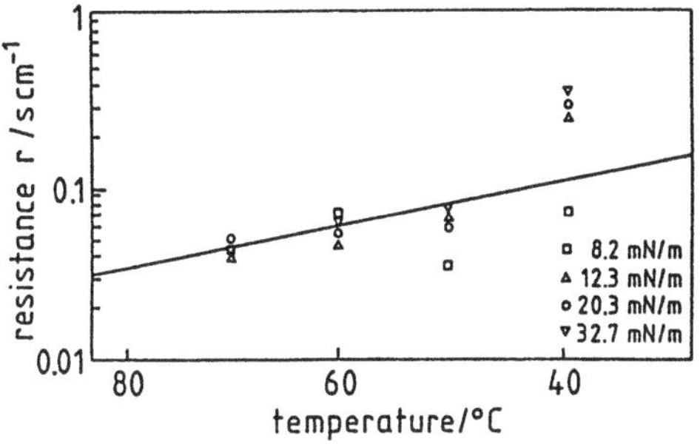

Fig. 4
Diffusion resistance and permeability of arachidic acid for water vapor for various surface tensions σ.

value of about 0.07 s/cm. A resistance increasing with temperature points to a negative activation energy. This is in good agreement with results of Vanderveen and Barnes /4/. For DMPE with a rather complicated structure defects in the monolayer diminish the resistance at low temperatures. At higher temperatures a structural conformation exists which is related to a reduction of free surface, i.e. a higher resistance.

In this work we show that it is possible to determine the permeability of monomolecular films for water vapor from photoacoustic experiments. The data which we obtained for the evaporation resistance, i.e. the reciprocal of the permeability, of various lipids are in good agreement with results of other authors.

/1/ P. Korpiun, R. Osiander, H. Schmitt, W. Micheler, Appl.Phys. A52, 223 (1991)

/2/ J. Langmuir, V. Schaefer, J.Franklin Institute, 235, 119 (1943)

/3/ R. Osiander, B. Helm, P. Korpiun in Photoacoustic and Photothermal Phenomena II, Springer Series in Optical Sciences; J.C. Murphy, J.W. Maclachlan–Sicer, L.C. Aamondt, and B.S.H. Royce eds., Springer Heidelberg, New York 1990, p.107

/4/ R. Vanderveen, G. Barnes, Thin Solid Films, 134, 227 (1985)

/5/ R. Archer, V. LaMer, J.Phys.Chem. 59, 200 (1955)

Photoacoustic Detection of Temperature Phenomena in Beeswax

H. Jalink[1], *D. Bićanić*[2], *W.J. van der Burg*[1], *R.A. van Zwol,*[1], *and J.W. Aartse*[1]

[1]Centre for Plant Breeding and Reproduction Research, CPRO-DLO, P.O. Box 16, 6700 AA Wageningen, The Netherlands

[2]Department of Agricultural Engineering and Physics, Agrotechnion, Wageningen Agricultural University, Bomenweg 4, 6703 HD Wageningen, The Netherlands

1. Introduction

The need for non-destructive monitoring of phase transitions in biological specimens is an important issue in agricultural practice for the status of the membrane stability of various plant materials [1]. This paper is concerned with flashlight photoacoustic detection of phase transitions in beeswax, a biological material that exhibits a pronounced phase transition (change in colour from transparent to milky near 32° C). The same technique was applied to observe phase transitions taking place in a coating of a familiar battery tester. These preliminary studies are initial steps towards development of the flashlight photoacoustic technique for investigations of more complex samples such as seeds and pollen of plants.

2. Experimental

The open photoacoustic cell used in this experiment basically relies upon a design originally proposed by da Silva *et al.* [2]. The cell body is a cylinder (diameter 20 mm, height 5 mm) that was manufactured from plexiglass material. The actual cell volume is a cylindrical cavity (diameter 2 mm, height 5 mm) open at one end. The sample under investigation was mounted (using silicon grease) atop the cell, closing the opening and providing simultaneously the airtight seal. A hole (diameter 1.3 mm) was drilled in the cylindrical side wall of the cell to allow insertion of the probe tip (25 mm) of a microphone (Brüel&Kjaer™ 4282). The signal detected by the microphone was fed into a Dual Channel Signal Analyzer (Brüel&Kjaer™ 2032) that digitizes the signal. The time signal was stored using a personal computer. The light produced by an electronic flashgun (Metz™ Mecablitz 28CT5) was directed onto the sample by means of a condenser lens (diameter 75

Springer Series in Optical Sciences, Vol. 69 Photoacoustic and Photothermal Phenomena III
Editor: D. Bićanić

mm, focal length 50 mm). An iris diafragm inserted just in front of the irradiated sample defined the size of the heated region (1 mm in diameter). The temperature is measured at a distance of 2 mm from the irradiated area by a thermocouple.

The sample was prepared from bulk beeswax material placed on a sheet of paper (approximately 110 μm thick) by exposing it to a heatgun. This procedure led eventually to a beeswax sample of uniform thickness (roughly 100 μm) and provided good thermal contact with the paper. In an actual experiment the front side of the paper (facing the incoming light) that carried the sample was blackened (commercially available copy machine toner) in order to enhance the absorption emitted by the source. The non-painted rear side of the paper was fixed onto the open photoacoustic cell by means of silicon grease. In this configuration the measurements were carried out in the transmission mode [3].

As stated above the study was extended with a battery tester (Duracell™). This latter undergoes a pronounced colour change (from black to yellow) when the increasing temperature reaches a value of 28° C. The battery tester is actually a sandwich consisting of several layers with a total thickness of 200 μm. The bottom layer is a thin metal serving as a resistive heater to increase the temperature of the tester. The other layers are made of plastic, paper and a yellow paint. A coating which covers the yellow paint changes from black to transparent in the vicinity of 28° C. During the experiment the coating was facing the incoming radiation. A thin (40 μm) piece of black carbon paper was attached on top of the coating by means of diffusion pump oil. This was done to assure that no change in the absorption of the light could take place, should the colour of the coating change.

3. Results and Discussion

The insert in figure 1 displays a time signal derived from the beeswax sample irradiated by a single flash. A fast drop followed by a much slower positive rise is clearly observed. Presumably such a rapid decrease can be ascribed to thermoelastic bending [4]. The observed behaviour may be explained by a temperature gradient across the sample which causes it to bend. As a consequence the volume of the photoacoustic cell increases and the corresponding pressure drops. In addition the

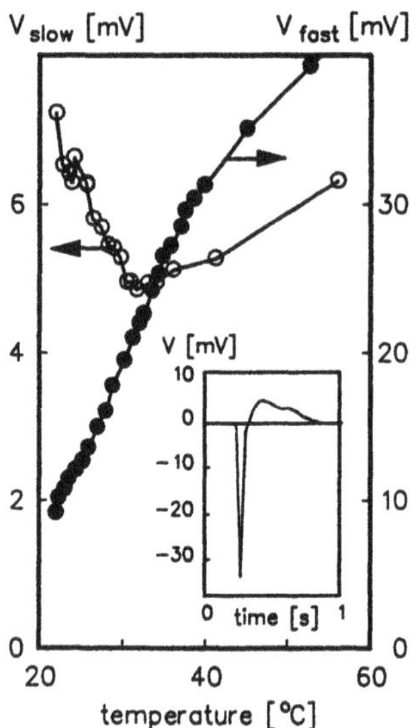

Figure 1. Temperature and time dependence of the PA signal observed in beeswax.

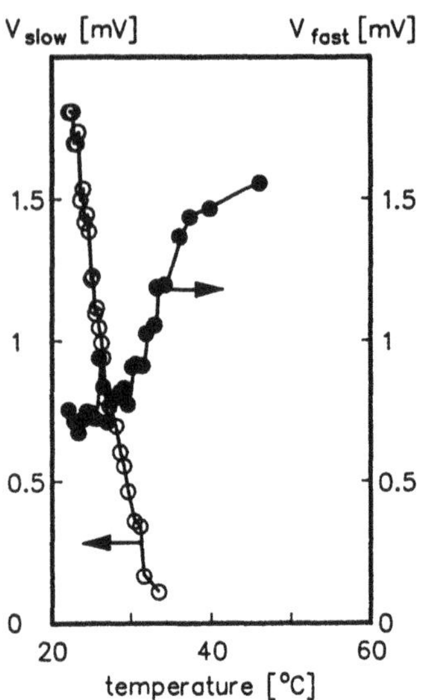

Figure 2. Temperature dependence of the PA signal observed in the battery tester.

linear coefficients of thermal expansion of the several layers may be different, enhancing or lowering the signal of the fast component. After a certain time interval the thermal wave reaches the back side of the strip and warms up the gas which causes a pressure rise. As the thermoelastic bending is completed within an interval being much shorter than the time that characterizes diffusion of heat through the sample, the negative signal is followed by a positive one. The maxima of the slow and fast signal components are temperature dependent and plotted versus the temperature in figure 1. As far as the slow component is concerned a discontinuity near 32° C is observed in the first derivative of the signal versus temperature. This is explained as an apparently abrupt change of the heat capacity of the sample, *i.e.* a phase transition [5]. Closer inspection of the curve recorded with the fast component also reveals the existence of a discontinuity (knee) but at a different temperature (close to 35° C). The origin of this discrepancy forms the content of study that is currently in progress.

Figure 2 shows the result obtained with the battery tester. The slow signal component rapidly decreases to zero. There is practically no heat diffusion close to 34°

C. Initially the fast component is constant but shows an increase as the temperature approaches 28° C most likely due to the phase change associated with the alteration in the coating's colour. Apparently all flash energy is used to induce a phase transition and an expansion of the sample's surface.

In conclusion, these preliminary results suggest the usability of the proposed technique for characterization of dynamical features of a phase transition.

References

[1] F.A. Hoekstra, J.M. Crowe, and L.M. Crowe, *Recent Advances in the Development and Germination of Seeds*, editor R.B. Taylorson, Plenum Press, New York, (1989).

[2] M.D. da Silva, I.N. Bandeira, and L.C.M. Miranda, J. Phys. E. Sci. Instrum. 20, 1476 (1987).

[3] N.F.Leite, N. Cella, H. Vargas, and L.C.M. Miranda, J. Appl. Phys. 61, 3025 (1987).

[4] G. Rousset, F. Lepoutre, and L. Bertrand, J. Appl. Phys. 54, 2283 (1983).

[5] K.L. Koster, Plant Physiol. 96, 302 (1991).

Three Dimensional Acoustical Modelling of Resonant Photoacoustic Cells

Z. Bozóki[1], *I. Gáspár*[2], *and A. Miklós*[1]

[1]Institute of Isotopes, Hungarian Academy of Sciences,
P.O. Box 77, H-1525 Budapest, Hungary

[2]Hungarian Dubbing and Video Studios,
P.O. Box 58, H-1021 Budapest, Hungary

Abstract. A true three-dimensional model was developed for determining the acoustical behavior of a complex photoacoustic (PA) system under realistic conditions. The behavior of the PA cell is modelled with an impedance matrix whose elements may be calculated via a serial expansion based on the eigenmodes of the PA resonator. A demonstration of the ability of this modelling is given.

1. Introduction

Numerous calculations have been made on the acoustical behaviour of different PA cells geometry like Helmholtz type [1,2], one dimensional [3] and cylindrical [4] resonators. A model for the acoustical properties of the acoustic filters connected to a PA cell [5] is also avialable. However a general model of three-dimensional cells is also needed, which regards the cell, the sample, the microphone, the windows and/or the acoustic filters adjacent to the cell openings and even the chopper as a noise source as one compact acoustic device. With such a general calculation the spatial structure of the acoustic field inside the cell can be taken into account, which is necessary, when optimising the microphone and filters position.

We developed a general computer program which model the response of a compact PA cells to both acoustical and photoacoustical excitations. The main features of the applied acoustical model is given in this paper.

2. Theory

The basic idea of the modelling is based on the so-called reciprocity theorem of acoustics [6]. In case of periodic excitation the sound pressure and particle velocity in an arbitrary acoustic resonator may be given in the following form:

$$p(r,t) = [p_0(r) + \sum_{n=1}^{\infty} A_n(\omega)p_n(r)]\exp(-i\omega t) \tag{1}$$

$$v(r,t) = [v_0(r) + \sum_{n=1}^{\infty} B_n(\omega)v_n(r)]\exp(-i\omega t) \tag{2}$$

where $p_n(r)$ and $v_n(r)$ are the pressure and velocity eigenfunctions of the n.-th eigenmode of the idealised resonator having perfectly hard walls (i.e. the normal velocity at the wall is zero) with no inhomogenities like microphone and filters

Springer Series in Optical Sciences, Vol. 69 Photoacoustic and Photothermal Phenomena III
Editor: D. Bićanić

(windows). The A_n and B_n coefficients may be determined by using the orthogonality of the eigenmodes. The real PA cell may be modelled as an acoustic N-port characterized by an NxN impedance matrix:

$$p_j = Z_{jk} U_k \tag{3}$$

where p_j is the sound pressure at the port j, and U_k is the normal velocity at port k times the surface of the k.-th port (volume velocity). Using Eq. 1. and 2. p_j and U_k can be determined. Then from Eq. 3. Z_{jk} can be calculated, using the characteristic impedance of every port. The noninfinite quality factors of a resonance in a real cell can be calculated [4] and included in the model phenomenologically.

After specifying the acoustical excitation, the Z_{jk} acoustic impedance matrix describes the acoustic behaviour of the PA cell completely. It includes all the informations about the resonant properties of the cell, about the positions of the ports, about the wall material and surface quality, etc. The calculation of the elements of the matrix can be made by computers.

3. Results

In a previous work [4] an open cylinderical PA cell (radius 50 mm and length 42mm) equipped with filters was described. The microphone and the filters were connected to the end plates of the cell (both were 25 mm far from the symmetry

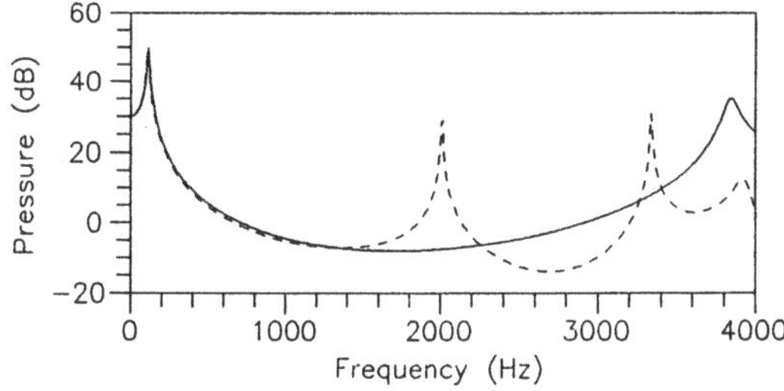

Fig. 1. Acoustical response of a cylindrical cell [4] to outside acoustical excitation, calculated with a one (solid line) and a three (dashed line) dimensional calculation.

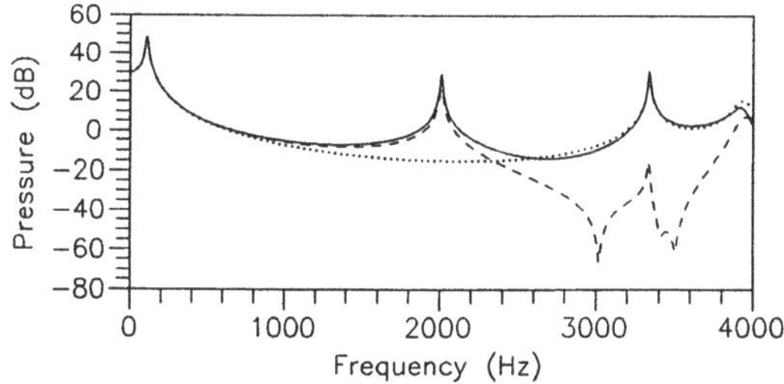

Fig. 2. Acoustical response of a cylindrical cell [4] for different microphone-filter position. The polar-angle on the endplate, between the microphone and the filters varies: 180° (original case) solid line, 135° dashed line, 90° dotted line.

axis). As the cell works on its first asymuthal resonance the three dimensional modelling is very important. In Fig. 1. we compared the one dimensional modelling of Ref. 4. with the result of three dimensional calculation. The influence of the microphone position on the acoustic transfer function is shown on Fig. 2. Experiments made in an anechoic room with acoustical excitation [4] verifies the three dimensional model.

4. Conclusion

The calculation presented in this work has its importance being a model which is able to describe realistic three-dimensional resonant PA cells. The model is valid as long as no transversal resonancies are excited in the acoustical filters. Preliminary experiments verify this model.

References

1. O.Nordhaus and J.Pelzl: Appl. Phys. **25**, 221-229 (1981)
2. Cheal H.So and H.C.Kim: J.Phys. **E22**, 626-630 (1989)
3. S.Bernegger and M.W.Sigrist: Infrared Phys. **30**, 375-429 (1990)
4. A.Miklós and A.Lörincz: Appl. Phys. **B48**, 213-218 (1989)
5. Gy.Z.Angeli et al.: Rev.Sci.Instr. **62**, 810 (1991)
6. P.M.Morse and K.U.Ingard: *Theoretical Acoustics*, (McGraw-Hill, New York 1968)

Photoacoustic Signal Generation in a Long Open Tube

A. Miklós[1], *Z. Bozóki*[1], *I. Gáspár*[2], *and Gy.Z. Angeli*[3]

[1]Institute of Isotopes, Hungarian Academy of Sciences, P.O. Box 77, H-1525 Budapest, Hungary
[2]Hungarian Dubbing and Video Studios, P.O. Box 58, H-1021 Budapest, Hungary
[3]Analtron Applied Research Company, P.O. Box 63, H-1311 Budapest, Hungary

Abstract. The well known theories of photoacoustic signal generation seem to be insufficient for explaining the results of some recent PA experiments. For investigating the PA signal generation, the simplest possible device, a long open tube was chosen as PA cell. The PA signal generated by the absorption of the light of a CO_2 laser in air was investigated as a function of the chopping frequency. The whole set-up was placed into an anechoic room in order to reduce the effect of the outside noises. The acoustic properties of the open tube were measured by using the external white noise source, then the PA signal and the background noise were measured by opening and closing the beam shutter of the laser. According to the present theory of PA signal generation only the odd longitudinal eigenmodes may be excited photoacoustically, because of the perfect symmetry of the tube. The results of the experiment clearly show, that the even modes will be excited, too, and the effectiveness of the excitation is quite the same for odd as for even modes. The experiment may be modelled by assuming simultaneous pressure and velocity generation.

1. Introduction

Although the photoacoustic methods were found to be very effective on several applications, the theoretical basis of photoacoustical signal generation is not properly founded yet. In the case of solid samples the absorbed light generates velocity at the sample surface [1], but for gases two existing theories contradict each other. Usually it is assumed, that the absorbed light energy is converted into heat via nonradiative processes. The periodic heating generates periodic pressure changes, that is sound in the medium [2]. This description corresponds to the statement of acoustics about sound generation by periodic heating [3]. Nevertheless an other theory also exists, which assumes particle velocity generation along the laser beam [4], and this theory proved to be useful for determining the acoustical behaviour of a resonant PA cell.

2. Experimental

In order to reduce the misleading factors of the PA experiments, the simplest possible photoacoustic cell, a long, open metal tube was used as a PA cell in our experiment. The ends of the tube were covered by a circular metal plate with a hole of 12 mm diameter bored on the middle. The light beam of a CO_2 laser could pass along the axis of the tube without hitting any metal surface, thus the only mechanism of PA signal generation is the periodic absorption of the light energy by the air. From acoustical point of view, the tube with the holes at its ends may be regarded as an open resonator with larger length than the physical one. Strictly speaking, the resonance frequencies of such a tube can be calculated from the following equation [5]:

$$\tan kL = -2\delta/(1 + \delta^2) \qquad \text{where} \qquad \delta = 0.57\pi R^3/LS \tag{1}$$

where $k=\omega/c$ is the wavenumber, R and L are the radius and the length of the tube, respectively, and S is the area of the hole at the ends. This formula gives very accurate values for the resonance frequencies of any type of open tubes, including organ pipes, of weak attenuation. In case of big attenuation the more general formula given in [5] must be used.

The acoustic resonances of the tube (L=49.5 cm, R=2.5 cm) were determined by an acoustic measurement. The tube was placed into the anechoic room of the Acoustic Research Laboratory of the Hungarian Academy of Sciences, and white noise was fed onto a loudspeaker placed near to the tube. The sound pressures outside and inside the tube were monitored by BK 4165 type condenser microphones. The inside microphone was located near to one end of the tube in order to avoid misleading peaks. The measured resonance frequencies were in very good agreement with the values calculated according to (1).

The photoacoustic signal generation was investigated then by chopping the laser beam by a rotating beam chopper whose chopping frequency was synchronized by a Wavetek crystal-controlled function generator. The point-by-point measurements were performed around the resonance frequencies of the tube. The background signal due to chopper noise and other instrumental noises was measured at every frequency step by closing the beam shutter of the CO_2 laser.

3. Discussion

The periodic heating of the air generates a temperature and pressure rise along the laser beam. This periodic pressure change radiates spherical sound wave from every point of the laser beam. The resultant wave will have cylindrical symmetry, because the axial waves cancel each other. This cancellation mechanism is not complete when inhomogenities like tube ends are present, but the resultant sound field may be determined by taking into account the reflections from the tube ends, too. Nevertheless, such distributed pressure source can generate only the odd modes of the tube.

The even modes of the tube can be excited only if distributed velocity sources are assumed. The pressure distribution inside the tube can be calculated by taking the reflections into account and by integrating the distributed source along the length of the tube. The result of such a calculation is shown on Fig.1.

It is very clear, that pure pressure generation can excite only odd modes, and pure velocity generation excites only even modes. The response of the tube is bigger for velocity excitation because the microphone is placed near to the end, where the velocity is big, while the pressure is small.

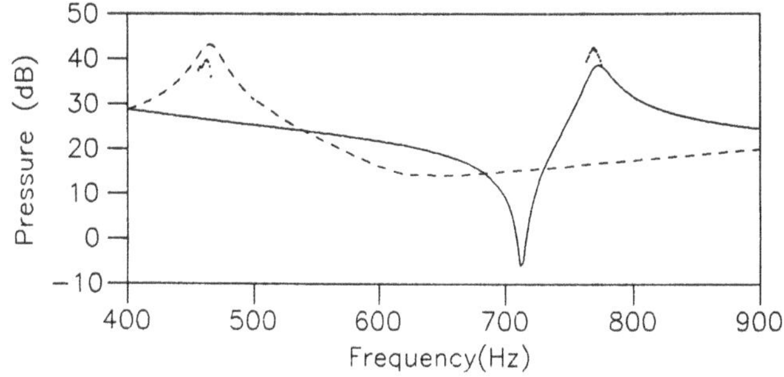

Fig.1.: Sound pressure in an open tube. Solid line: computed by pressure, broken line: by velocity generation; dots: experimental data.

The result of the measurement proved [6], that both the odd and even resonances can be excited in an open tube by axial laser beam. That is both mechanisms of signal generation exist. The strength of the sources is approximately the same for our case. However, further efforts would be needed for determining accurately the source strengths for different cases.

Such measurements could give the necessary data for establishing an appropriate theory of PA signal generation, which would be a very important step towards quantitative photoacoustics of gases.

References

[1] O.Nordhaus and J.Pelzl: Appl. Phys. 25, 221-229 (1981)

[2] A.Rosencwaig: *Photoacoustics and Photoacoustic Spectroscopy* (John Wiley and Sons New York, 198O) pp. 26-3O

[3] P.M.Morse and K.U.Ingard: *Theoretical Acoustics*, (McGraw-Hill, New York 1968)

[4] S.Bernegger and M.W.Sigrist: Infrared Phys. **30**, 375-429 (1990)

[5] J.Angster and A.Miklós: Journ. Acoust. Soc. Amer. To be published

[6] Z.Bozoki et al.: Proc.10th FASE Symp. B.Fured, 122-126 (1991)

Dependence of the Photoacoustic Cell Constant on the Sample Gas

Gy.Z. Angeli[1], *A.M. Sólyom*[1], *A. Miklós*[2], *and D. Bićanić*[3]

[1]Analtron Applied Research Company, P.O. Box 63, H-1311 Budapest, Hungary

[2]Institute of Isotopes, Hungarian Academy of Sciences, P.O. Box 77, H-1525 Budapest, Hungary

[3]Department of Agricultural Engineering and Physics, Agrotechnion, Wageningen Agricultural University, Bomenweg 4, 6703 HD Wageningen, The Netherlands

Abstract. The CO_2 laser and the windowless resonant cell equipped with the electronic resonance tracking system were combined into the photoacoustic spectrometer. In order to facilitate the comparison of various analytical purpose PA systems and other spectroscopic methods, we suggest the introduction of instrumental figure characterizing the PA instrument. The photoacoustic spectrometer was calibrated using a series of certified precision mixtures of ammonia, ethylene and carbon-dioxide in a high purity grade nitrogen. The absorbance as low as $2*10^{-6}$ could be measured. The instrumental figures were found different: 0.98 V/W for the carbon-dioxide, 1.68 V/W for ethylene and 2.4 V/W for the ammonia.

1. Introduction

The CO_2 laser and the windowless resonant cell (Q=300) were combined into the photoacoustic spectrometer for trace analysis of pollutant gases in the ambient air [1].

The acoustic amplification of the resonant chamber of high quality factor is a sensitive function of temperature and gas composition as these influence the speed of sound and hence the resonance frequency. In order to eliminate the signal degradation due to the uncontrolled drifting of the cell's resonance frequency, we constructed an electronic resonance tracking system. The system is capable of adjusting the light modulation frequency to the instantaneous resonance frequency of the chamber [1]. The measured frequency regulation of the resonance tracking system can be seen in Fig 1.

In order to facilitate the comparison of various analytical purpose PA systems, we suggest the introduction of a different quantity characterizing the PA instrument. This new parameter is consistent with the traditional terminology of analytical absorption spectroscopy enabling a direct comparison of PA method to other spectroscopic methods. The energy absorbed by the sample I_a (W) can be calculated easily for low absorption from Bouguer-Lambert-Beer's Law by means of transmitted intensity I_t (W) and absorbance A as:

$$I_a = I_t \, A \ln 10$$

Springer Series in Optical Sciences, Vol. 69 Photoacoustic and Photothermal Phenomena III
Editor: D. Bićanić

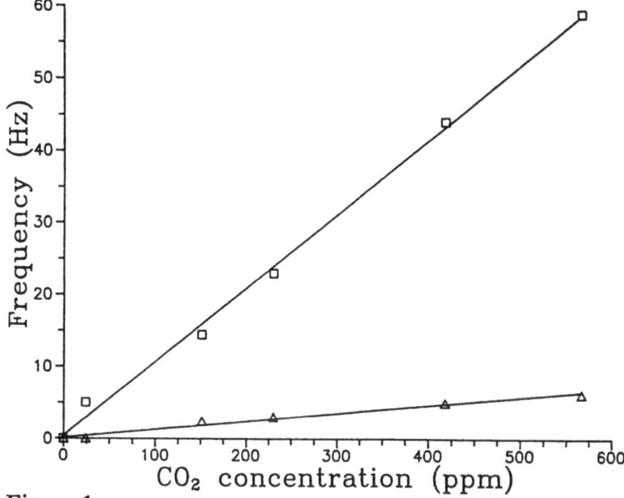

Figure 1
The difference between the cell's resonance frequency and the chopper frequency in function of gas composition (□) without resonance tracking and (△) with resonance tracking

We can assume that the photoacoustic signal S (V) is directly proportional to the absorbed radiation power I_a. The normalized signal $S_N = S/I_t$ (V/W) is thus directly related to the absorbance A with the proportionality factor called the instrumental figure F (V/W) since

$$S_N = F A$$

The relationship between the widely used "cell constant" C (Vcm/W) and F can be obtained by means of the definition of "cell constant" [2]:

$$F = C / d \ln 10$$

with d (cm) being the interaction pathlength.

2. Results and discussion

The photoacoustic spectrometer was calibrated using a series (concentration varying from few ppm to few hundred ppm) of certified precision mixtures of ammonia, ethylene and carbon-dioxide in a high purity grade nitrogen. The normalized photoacoustic signal was recorded; simultaneously the magnitude of absorbance A_c was computed at each concentration using the literature data of the absorption cross sections. The measured data could be linearly fitted with high correlation factors. Obviously the plot of measured absorbance versus concentration depend on the supposed value of the instrumental figure, because definitely the measured absorbance

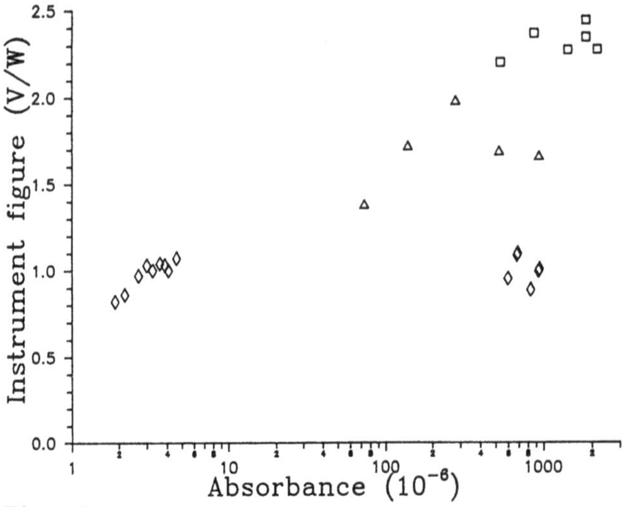

Figure 2
Dependence of instrumental figure F on absorbance of (◇) carbon-dioxide, (□) ammonia, and (△) ethylene.

Table 1: The achived detection limits

Sample	Laser line	Concentration (ppm)	Absorbance (10^{-6})	Extrapol.det.limit
CO_2	9R16	183-79800	2-890	4.6 ppm
C_2H_4	10P14	1.3-13.9	73-934	1.3 ppb
NH_3	9R30	4-15.7	538-2.15	0.4 ppb

is the ratio of S_N normalized photoacoustic signal and F instrumental figure. The appropriate instrumental figure could be determined by minimizing the difference between the fitted curve and the computed absorbance values. The instrumental figures were found different (with significance of 99.9 %): 0.98 V/W for the carbon-dioxide, 1.68 V/W for ethylene and 2.4 V/W for the ammonia (see the Figure 2).

Contrary to reported literature [2] this experiment provides the evidence that the instrumental figure (related to a more familiar cell constant) is not independent of the absorbing gas. The absorbance as low as $2*10^{-6}$ could be measured in our experiments (Table 1). In comparison, the absorbance of $5*10^{-4}$ was achieved by TDLS method, and the value of 10^{-5} was measured by DOAS system [3].

References

[1] Gy.Z. Angeli, Z. Bozóki, A. Miklós, A. Lőrincz, A. Thōny and M.W. Sigrist: Rev. Sci. Instrum. **62**, 810-813 (1991)
[2] P.L. Meyer, M.W. Sigrist: Rev.Sci. Instrum. **61**, 1779-1807 (1990)
[3] B.J. Finlayson Pitts, J.N. Pitts Jr.: *Atmospheric Chemistry* (Wiley, N.Y., 1986)

Characterisation of a Photochromic Triarylmethane Dye Sulphite and Its Application to the Visualisation of Water Flows

P. Douglas[1], *R.D. Enos*[1,*], *B. Azzopardi*[2,**], *and C.B. Hope*[2]

[1]Chemistry Department, University College Swansea, Swansea, SA2 8PP, UK
[2]Thermal Hydraulics Division, Harwell Laboratory, Oxfordshire, OX11 0RA, UK
Present addresses: *Albright and Wilson, Trinity Street, Oldbury, West Midlands, B69 4LN, UK
**Department of Chemical Engineering, Nottingham University, Nottingham, NG7 2RD, UK

Abstract. The application of photochromic dye tracing in flow visualisation studies relies on the use of a laser pulse to 'instantaneously' generate a dye in a localised volume of solution with subsequent monitoring of the position of the dye as a function of time. Until now, photochromic dye tracing has been limited to organic liquids [1,2] which has restricted use of the technique because of the toxic and flammable nature of these materials. Recent research has identified a useful water soluble photochromic for flow visualisation, [3,4], which relaxes many of the constraints within which the technique has previously been bound. In this report we summarise our finding on the chemical and photochemical properties of this photochromic system and illustrate its application to studies of single phase and two phase air-water flow regimes.

Results

Addition of sulphite to an aqueous solution of the triarylmethane dye Acid Magenta Monomethyl Magenta, ($AMMM^{2-}$), (a methylated derivative of Acid Magenta, C.I.Acid violet 19) leads to formation of the colourless sulphite complex, $AMMM.SO_3^{4-}$ [5].

Pulsed irradiation of a solution of the dye sulphite, with light of wavelength less than ≃ 340 nm, causes the transient production of the parent dye and sulphite ions. Once produced, the dye recombines with sulphite over a period of time, and this regenerates the colourless dye sulphite. The lifetime of the photoproduced parent dye is influenced by pH, sulphite concentration, temperature and solution ionic strength. Maximum useful lifetimes are of the order of several seconds. The parent dye $AMMM^{2-}$ is intensely coloured ($\varepsilon = 4.5 \times 10^4\ M^{-1}\ cm^{-1}$ at 545 nm), and the quantum yield of dye production for 308 nm excitation is ≃ 0.5-1.0. These properties result in very efficient colour production following UV excitation, and an 8 mJ pulse of 308 nm radiation from a XeCl excimer laser will generate an easily detected trace. However 355 nm radiation from a Nd/YAG laser is much less effective, and the action spectrum for dye production from $(AMMM.SO_3)^{4-}$ suggests that there is a non photochromic absorption band at about 350 nm.

The photochromic system identified above has been applied to studies of horizontal single and two phase flows in a conduit of circular cross-section with an internal diameter of 32 mm [6]. A XeCl eximer laser generating 8 mJ pulses at

Springer Series in Optical Sciences, Vol. 69 Photoacoustic and Photothermal Phenomena III
Editor: D. Bićanić

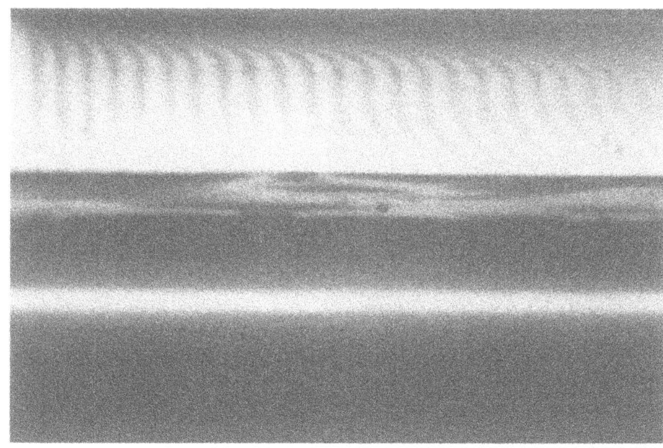

Figure 1. Characteristic s-shaped profile for a smooth stratified flow. The liquid is flowing from right to left, and the air-water interface can be clearly seen.

308 nm operating at 10-100 Hz has been employed as the excitation source. By focussing the laser into the aqueous solution containing the colourless dye sulphite complex 'lines' of colour can be generated virtually instantaneously and non-intrusively in the liquid. These lines then move with the liquid, revealing both qualitative and quantitative information relating to fluid motion. The motion of the dye traces was recorded using backlight illumination and a high speed video system, with a filming rate of up to 6000 frames/second. Refraction related distortion, caused by the curvature of the conduit walls, was minimised by manufacturing the conduit from a transparent polymer material with a refractive index very close to that of water [7].

The technique has been applied to horizontal single phase laminar flow, and has shown the characteristic parabolic velocity profile where the peak velocity occurs on the axis of the tube, and the velocity at the wall is zero, demonstrating the no-slip boundary condition. Further studies have been conducted with single-phase flows, where the exit to the conduit has been partially obstructed. The dye traces in this case show that the peak velocity has been displaced away from the axis, the asymmetry being caused by the obstruction. The technique has also been applied to two-phase air-water flows, stratified, slug and annular. In the stratified case, results were obtained when the interface was both smooth and wavy. Figures 1 and 2 show the dye trace profiles for both cases. In the smooth case an S-shaped profile is observed. This would be expected since a significant shear from the gas would accelerate the liquid at the interface. In the wavy case however it can be seen that the peak velocity in the liquid phase is unexpectedly found to be located below the interface. Using a dual camera arrangement this flow was filmed from both the front and below at the same time, and no secondary action was observed. The peaking of the velocity away from the interface is therefore thought to be a consequence of a partial recirculation beneath the waves. The computational fluid dynamics computer program, FLOWED, was used to simulate the motion of the liquid in a wave of this type, and the results support the idea of recirculation. More work is however needed to fully investigate the reasons for this unusual velocity profile.

Figure 2. Velocity peaking away from the interface in a wavy stratified flow.

References

1) Delhaye J. M., and Cognet C., (Eds.) *'Measurement Techniques in Gas-Liquid Two-Phase flow'*, Springer-Verlag, 1984.
2) Merzlirch W., (Ed.) *'Flow Visualisation II'*, Hemisphere Publishing Corp., 1980.
3) Douglas P., Accepted for publication in *Chemical Engineering and Technology*, 1991.
4) Douglas P. and Enos R. D., Submitted for publication in *Chemical Engineering and Technology*, 1991.
5) Bertleson R. C., in *'Photochromism, Techniques in Chemistry, Vol. 3'*, (Brown G. H., Ed.), Wiley-Interscience, p. 45, 1971.
6) a) Fogwell T. W. and Hope C. B., *UKAEA Report AERE R12652*, 1987.
   b) Hope C. B., *'The development of a water soluble photochromic dye tracing technique and its application to horizontal two phase flows'*, *Ph.D. Thesis*, University of London, 1991.
7) Fogwell T. W., *UKAEA Report AERE R12313*, 1986.

Photorefractive Lock-In Amplifier and Thermal Wave Microscope

A. Lörincz[1] *and O. Dóka*[2]

[1]Institute of Isotopes, Department of Photophysics, P.O. Box 77, H-1525 Budapest, Hungary

[2]Pannon University of Agriculture, Department of Physics, P.O. Box 90, H-9201 Mosonmagyaróvár, Hungary

Abstract. In this paper it is shown that optical lock-in detection based on two-beam coupling in photorefractive medium is alternative to the electronic lock-in scheme. Photorefractive thermal wave microscope is considered as possible application.

1. Two-beam coupling

The photorefractive effect is a result of the presence of imperfections, donors and acceptors having localized electronic states within the bandgap. Incident photons create free charge carriers which may be trapped by these levels. When two light beams enter the crystal interference results in spatially varying excitation. Then, the distribution of the trapped charge carriers varies with space too, resulting in space dependent charge density, thus space dependent electric field, and eventually a refractive index grating through the electrooptic effect [1].

The coupled equations of the electromagnetic field and grating amplitude are as follows [2]:

$$\frac{\partial E_1}{\partial z} = GE_2 - \frac{\alpha}{2}E_1 \tag{1}$$

$$\frac{\partial E_2}{\partial z} = -G^*E_1 - \frac{\alpha}{2}E_2 \tag{2}$$

$$\frac{\partial G}{\partial t} = \gamma[-G + \frac{\Gamma}{2}\frac{E_1E_2}{|E_1|^2 + |E_2|^2}] \tag{3}$$

where E_1 and E_2 are the envelope functions of the electric fields, G is the "scalar" grating field, γ is the decay constant, Γ is the coupling constant, and α is the absorption coefficient.

It has been shown, that photorefractive media are capable of low pass and high pass filtering, that is, photorefractive media may inhibit the transfer of fast changes of the image or, alternatively, it is possible to transfer only the fast changes of the image. These options are called monotony and novelty filters. In the following, other possibilities; the method of converting a small sinusoidal modulation to d.c. intensity similarly to electronic lock-in detection and a thermal wave microscope scheme, will be discussed.

2. Lock-in detection

In the electronic lock-in detection scheme the goal is to find the amplitude and the phase of a small modulated signal of known frequency on top of a large noisy background. It is achieved by multiplying the signal with a modulated reference signal of the known frequency.

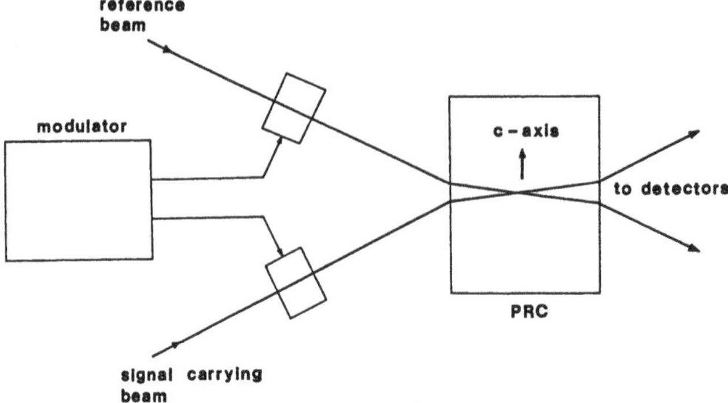

Fig. 1. Photorefractive lock-in amplifier: the signal carrying beam and the reference beam undergo identical chopping and enter the photorefractive crystal (PRC). The outputs undergo electrical differencing and low-pass filtering.

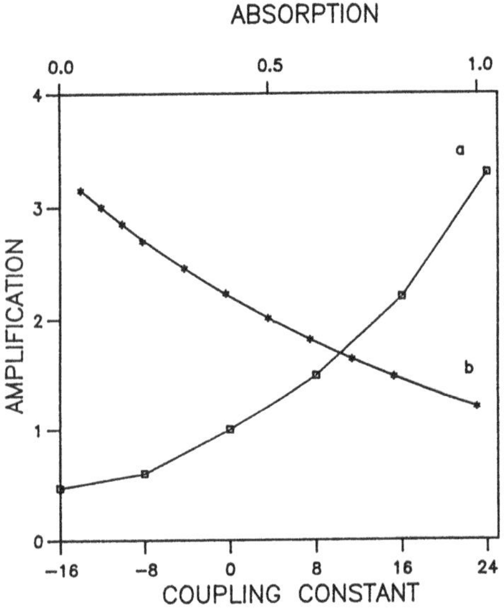

Fig. 2. Photorefractive lock-in amplification as a function of the imaginary coupling constant for zero absorption (Curve a), and as a function of absorption for coupling constant (0,24i) (Curve b).

The electronic multiplication produces sum and difference frequencies and in the multiplied frequency spectrum the signal shall develop a d.c. component, which is pulled out of the noise by low-pass filtering.

Consider now, that the time dependence of the grating formation of Eq. (3) has a nonlinear $E_1E_2^*$ term. This term is, in fact, the product of the envelope functions of the two electric fields. This term may then produce sum and difference frequencies and may create the desired d.c. output. There are then two conditions to be satisfied to have a lock-in detection scheme: (i) the small sinusoidal modulation of the signal carrying beam should be converted to d.c. signal, and (ii) the output of the device should be directly proportional to the amplitude of the original small modulation with no other contribution.

In the following a scheme is introduced which satisfies both conditions according to the numerical results. Assume now, that the amplitude of the signal carrying laser beam has originally the following form:

$$E_1^{(0)}=E_{1,0}(1+qe^{i\omega t}) \tag{4}$$

where ω is the frequency of the a.c. signal, $E_{1,0}$ is a constant, and $q=|q|\ \exp(i\Phi_q)$ describes the small sinusoidal signal on the top of a large back ground ($|q|<<1$). To produce the photorefractive energy transfer another beam is used. In the photorefractive scheme this beam is chopped at the known frequency and shall be called as the "reference" beam, in accord with the reference signal of the electronic scheme. It turns out, that for the case of photorefractive lock-in amplification conditions (i) and (ii) are fulfilled if the signal beam undergoes an identical chopping procedure at the known signal frequency. That is the scheme is different from the electronic one, it is sketched in Fig. 1. Lock-in amplification properties are shown in Fig. 2 as a function of the coupling constant for the case of zero absorption (Curve a), and as a function of the absorption coefficient for the case of coupling constant (0,24i) (Curve b). These results show that lock-in amplification (a.c. to d.c. conversion and amplification) may be achieved with the arrangement of Fig. 1 under moderate conditions such as an absorbing crystal and reasonable coupling constant. It may be worth to emphasize that the photorefractive lock-in scheme is not restricted to scanning mode but might be used for image carrying beams.

3. Thermal wave microscope scheme

The effects of thermal waves on optical measurements may be summarized as follows:

- there is a periodic change in the optical path length as a result of the thermoelastic deformation,
- there is an apparent periodic change in the optical path length, originating from the temperature dependence of the phase change of the reflected beam,
- the temperature dependence of the index of refraction gives rise to modulation in the reflected (transmitted) intensity.

As a result, thermal waves modulate both the amplitude and the phase of a reflected (or transmitted) probe beam. In order to subtract the information, lock-in detection should be used.

The investigated surface may have areas of different reflectivities and the lock-in scheme alone would lead to spurious results. This problem is solved by inlcuding an a.c. signal transmitting scheme.

The first stage of the TWM scheme is shown in the Fig. 3. The modulated pump beam causes periodic heating in the sample. The probe beam hits the sample, its reflected (or transmitted) component is modified by the non-uniform modulated reflectivity of the sample. The reflected part of the beam and an auxiliary beam of uniform intensity density are directed to the first photorefractive crystal. The crystal couples the beams and the a.c. content of the reflected beam is partly transferred to the auxiliary beam, but the d.c. amplitude of the auxiliary beam stays uniform. The auxiliary beam then enters the lock-in stage of Fig. 1. Then the outputs undergo differencing and electronic low pass filtering. The resulting signals will be proportional to the original modulation at every point if the interactions were uniform in the coupling processes. We would like to point to the strength of this scheme; it suggests to do the thermal wave imaging simultaneously for the whole image. That requires image digitization for differencing and electronic low pass filtering stages. Nevertheless, the a.c. to d.c. conversion, the crucial part of the procedure, happens in one step for the whole beam. Low-pass filtering schemes are also possible by photorefractive methods [2].

Resolution is limited by the thermal wave process and it is an advantage that the scheme allows high modulation frequencies. Present day photorefractive crystals allow a few micron

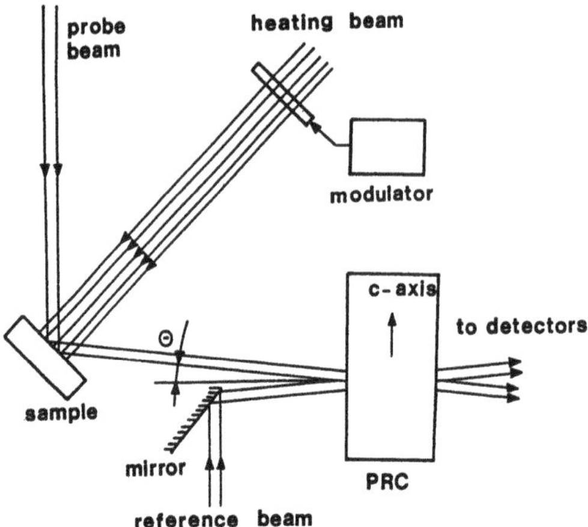

Fig. 3. First stage of a photorefractive thermal wave microscope: periodic heating results in a modulated beam. The modulation is transferred to an other beam for reflectivity correction.

resolution and barely limit the method. The sensitivity of the method is limited by the stability of the different components and that is an open question at the moment.

References

1. N. Kukhtarev, V. Markov, S. Odulov: Opt. Comm. **23**, 338-343 (1978)
2. D.Z. Anderson, J. Feinberg: IEEE J. Quantum Electron. **25**, 635-647 (1989)

Neural Networks in the Framework of Non-Destructive Evaluation from Photothermal Experiments

S. Knerr, D. Fournier, G. Dreyfus, and A.C. Boccara

Non-Destructive Evaluation Group, ESPCI,
10, rue Vauquelin, F-75005 Paris, France

Abstract. We present a first attempt at using neural networks as classifiers of subsurface defects evidenced by photothermal signals. The robustness of the classification performance with respect to noise and distortions is very encouraging.

Signature analysis is a popular tool for the interpretation of physical measurements. Because of (i) quantum and thermal noises, and of (ii) roughness and granularity of the samples, physical signatures exhibit, in general, low signal-to-noise ratios. Therefore, the interpretation of the signatures requires the use of sophisticated signal processing and data analysis methods. Among the latter, *neural network* techniques have proven to be fairly robust with respect to noise and extraneous distortions. In addition, they easily lend themselves to highly parallel, fast hardware implementations. The use of a neural network classifier might lead to an automated identification of the signature: in a first step, the classifier is trained on a collection of examples corresponding to known signals; once training is completed, the efficiency of the classifier is estimated using a set of curves which were not used for training. This is similar in spirit to other tasks which have been successfully performed by neural network classifiers, such as the automatic recognition of handwritten characters [1].

Despite a high level of noise and distortion, the signals used for diagnostics in NDE carry a specific information which appears through the shape of the recorded signals. For instance, when scanning the surface of a sample exhibiting a subsurface crack by means of *photothermal techniques*, the variation of the amplitude of the local temperature gives rise to characteristic curves [2]. Their shape depends on the orientation of the crack with respect to the surface (Figure 1a.). The signatures may exhibit a lot of noise; in addition, the position of the peaks is only approximately known (Figure 1b.). The problem to be solved by the neural network classifier is that of associating a given photothermal signal with one out of N orientations of slanted cracks. Here, we restricted the number of classes (associated orientations) N to four: 0, 30, 45 and 60 degrees. The amplitude of the signal at 80 sampling points is input to the network.

We trained a *Multilayer Perceptron* on a set of preclassified artificial examples [3]. In the training data, the signal-to-noise ratio with respect to the amplitude of the smaller peak of the signatures varied between 0.6 for class 4 and 3.2 for class 1.

The position of the peaks varied by about half the distance S_1S_2 (see Figure 1a.). Figure 2 shows the network architecture which features 80 inputs (corresponding to the 80 points of the signal), 10 hidden units and 4 output units, corresponding to the

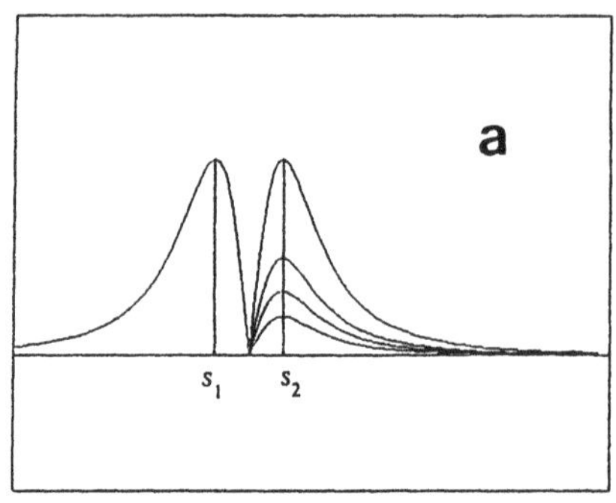

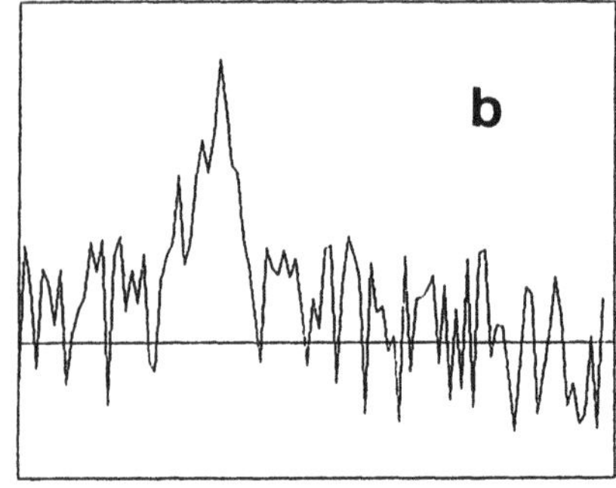

Figure 1a: Signatures of subsurface cracks with four different orientations with respect to the sample surface. 1b: Signature of a subsurface crack with a signal-to-noise ratio of 0.73.

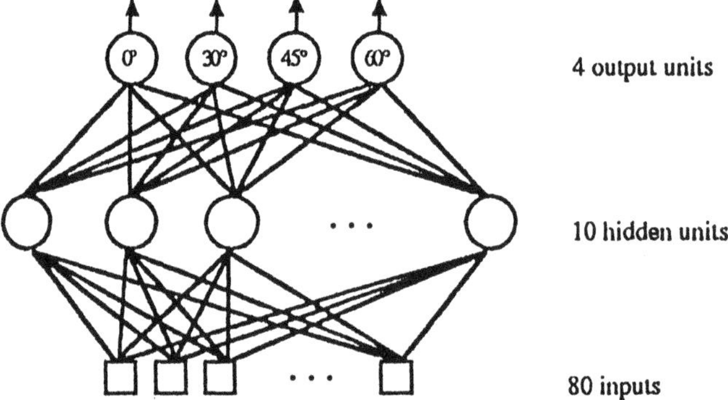

Figure 2: The neural network classifier. The four output units code for the four classes of orientations of the cracks.

four possible orientations of the crack. During training, five hundred examples of each class were repeatedly presented to the network. The weights of the network were adapted using the standard backpropagation algorithm [3].

It is important to train the classifier with rather noisy data, even when data without or with less noise is available. Because the training set must be representative of the data, a network trained on noiseless data performs well on samples with no or little noise, but performs poorly as soon as the data becomes noisy; on the contrary, when the network is trained with noisy data, it has a satisfactory performance on both, noisy or noiseless data. Figure 3 shows the classification rate of the neural network (measured on 500 test curves per class) as a function of the signal-to-noise ratio. Even for signal-to-noise ratios smaller than 1, one still gets approximately 80% correctly classified samples.

Since noiseless examples are available, we also applied a non-trainable *pattern matching* classifier to the problem: one noiseless example per class is used as a prototype signature for this class. The unknown signals are simply compared (matched) to the prototypes by computing a euclidean distance. The class of the

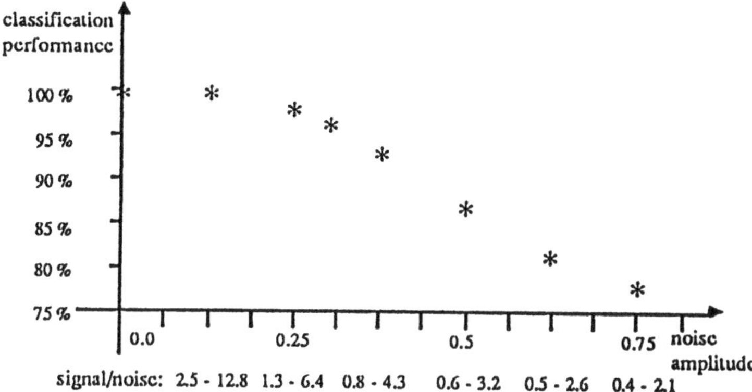

Figure 3: Classification performance of the neural network as a function of the noise amplitude. For each performance measure, the range of the signal/noise ratio with respect to the four classes is given.

prototype with the minimal distance is associated with the unknown signal. For small variations in the absolute position of the signatures with respect to the prototypes, this approach gives results comparable within 1 to 2 % to those obtained with the neural network. However, for larger variations in the absolute position of the signatures, this simple approach does not give satisfactory results: for variations of $\pm 2\ S_1S_2$ one obtained a 46 % rate of correctly classified samples, whereas the neural network trained on this data achieved a classification rate of 83 %.
However, a drawback of the neural network approach is the need of a relatively large number of training samples. Especially in the context of NDE, it is often difficult to provide such large data bases. We will therefore investigate the possibilities of using *a priori knowledge about the classification task*, either in order to perform appropriate preprocessing operations before classification, or in order to adapt the network architecture to the specific classification problem at hand. Both approaches are known to improve the data representation, thereby allowing for smaller training sets.

References

[1] S. Knerr, L. Personnaz, G. Dreyfus, Proceedings of IJCNN'91, Vol.1, I-91.
[2] F. Lepoutre, D. Fournier, A.C. Boccara, J. Appl. Phys. 57, 1009 (1985).
[3] D.E. Rumelhart, J.L. McClelland, Parallel Distributed Processing, Vol. 1, MIT Press, Cambridge Ma., 1986.

Pulsed Laser Machining by Training Artificial Neural Networks with the Help of the Thermal Diffusion Equation

T. Szakács and A. Lörincz

Institute of Isotopes, Department of Photophysics,
P.O. Box 77, H-1525 Budapest, Hungary

Abstract. Iron and gold evaporations produced by nanosecond laser pulses were modelled in two dimensions assuming rough one dimensional "surfaces". Computed data were used to train an artificial neural network for solving a material's processing problem; surface polishing. Training results of an extended Kohonen's model for the formation of topographically correct feature maps equipped with an error-correction scheme of Widrow-Hoff type learning rule are presented.

1. Introduction

Artificial neural networks (ANN) are in the focus of research interest as means for fast parallel processing. ANN's are simplified brain models trying to copy the brain's working principles. The structure of the brain makes pattern recognition and association feasible in time periods that can not be matched by present day computers, although the processing time of one unit is longer than a millisecond. This originates the motivation of concentrating of efforts in this field.

The application presented here assumes fast neural network processors that may be available in the near future. At present we are simulating the neural network instead of real parallel computing. The result is that the computer runs on our 25 MHz IBM PC 386 equipped with coprocessor take about 15 seconds for generating laser fields for different surfaces with the trained neural net. In these computations we are using only one dimensional "surfaces". In case of real two dimensional surfaces both training and laser field generation are expected to increase considerably. For comparison, the rate at excimer lasers are capable of firing is around 50 Hz. However, neural network processors may meet figure easily.

In this paper we present an approach that uses an extension of Kohonen's network architecture for the formation of topographically correct feature maps. It is based on the work of Ritter, Martinetz and Schulten [1] about learning visuomotor coordination. The analogy relies on topological equivalence of these problems. In this model, the ordering and the resolution of the topographic map evolve during learning. A layer of input weights that determines the distribution of neuronal units over the surface space is adjusted in a self-organizing fashion that may be considered as the optimization of non-uniform resolution.

For the adaptation of the output weights an error-correction scheme based on the Widrow-Hoff learning rule for adaptive linear elements was employed [2], just like in the case of visuomotor coordination problem.

2. Formulation of the problem

The problem to be considered here is laser machining. We assume a rough surface to be modified by our pulsed laser. We assume that the laser may provide short pulses of about nanosecond duration, and there is a way to modify the pulse duration, like using etalons or

Springer Series in Optical Sciences, Vol. 69 Photoacoustic and Photothermal Phenomena III
Editor: D. Bićanić

competing cavities, etc. As a rough approximation consider that sudden heat deposition to a structure of dimension n spreads to a distance of $(\kappa_d t_0)^{(3-n)/2}$ during time t_0, where κ_d is the thermal diffusivity of the substance. That is heat diffusion may be limited by decreasing pulse duration. This allows for us to determine a pulse duration for every substance in such a way that the temperature of a point at the surface at a given time is determined by the history of the heating up to that time and the temperature distribution of the immediate environment. Let us assume that the pulse duration is so short that the distance of lateral heat diffusion is negligible compared to the optical resolution during heating. The heat diffusion then becomes a problem of one degree of freedom, namely the height of the surface at a given point. Similarly for two dimensional surfaces, the limit of ultrashort pulses is a one degree of freedom problem. Longer pulses may involve more degrees of freedom and in general the longer the pulse the larger the environment is that takes part in the heat diffusion. There is an other point of view to be considered here and this is the role of stresses during evaporation. In case of picosecond heating it is well known that considerable stress may accumulate and may result in ablation for example. To avoid it one would like to keep the heating pulse long. The short the pulse the closer one may get to the optical resolution in forming structures on the surface on the other hand. This is a trade off, that seems to render the duration of the laser pulse to the nanosecond range and thus to an approximately two degree of freedom problem for one dimensional surfaces and to the three degree of freedom one for real two dimensional surfaces for best performance.

3. The neural net and its assumed environment

In these simulations we envision a laser system that detects surface roughness with the help of interferometric methods. The data are then plugged into the neural network that designs the laser field distribution. One may tailor the appropriate laser field distribution with a spatial light modulator to final form and fire the laser pulse. The result of the experiment is detected through the interferometric procedure and the neural network is adjusted accordingly. In this paper we consider the numerical simulations as simulations of the experiment, although in a real system it may serve as pretraining to shorten the training procedure if data are available about the material.

The two-dimensional neural net is built in the same fashion as given in [1] and the reader is kindly referred to that paper because of space limitations here. We may give the connections only. Each neuron is responsible for a small subset (its "receptive field") of the input space. The input space corresponds to two three dimensional vectors. The first of these three dimensional vectors represents the height of the surface point in question and its next neighbours. The other three dimensional vector represents the height of the aimed surface point in question and its next neighbours. This latter one was kept constant in the runs.

In case of topographically correct feature maps it is not only the winning neuron but its neighbours as well that are adjusted after each run. This neighbourhood is continuously decreasing as training goes on. It is accomplished by multiplying the correcting term of neuron s with a term that accounts for the distance from the winning neuron r. This function h_{rs} was taken to be Gaussian [1]:

$$h_{rs} = exp(-\|r-s\|^2/2\sigma^2(n)) \tag{1}$$

where n is the learning step number and σ has a power dependence:

$$\sigma(n) = \sigma_i(\sigma_f/\sigma_i)^{n/n_{max}} \tag{2}$$

with n_{max} being the maximum step number, 20000 in our case. $||...||$ is the norm in the input

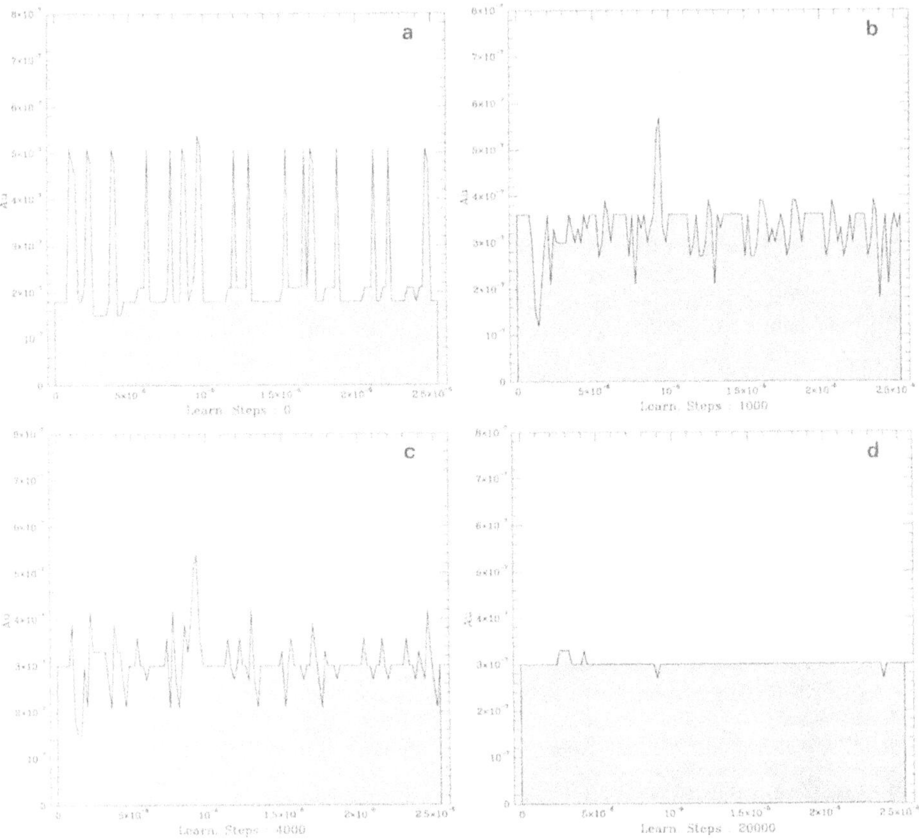

Fig. 1
Performance of the neural network after 0, 1000, 4000 and 20000 training steps. Units are given in meters. The initial surface (not shown) has a roughness that is similar to that of Fig. 1 (c), but that surface is at a higher position, it is 600 nm above the zero level the graph. The aimed depth was set to 300 nm above the zero level.

or output spaces. The parameters of the equations were set for both inputs and outputs according to [1]. It may be interesting to note, that using Cronecker's delta function instead of the Gaussian in Eq. 1, the system corresponds to a perceptron. Having the trained or partially trained neural nets the laser field distribution was computed for the whole surface point by point. Surfaces generated by the nets are shown in Fig. 1. Curves (a), (b), (c) and (d) correspond to training steps 0, 1000, 4000 and 20000 on aluminum, respectively. The initial surface (not shown) has a roughness that is similar to that of Fig. 1 (c), but that surface is at a higher position, it is 600 nm above the zero level the graph. The aimed depth is at 300 nm.

For computational simplicity we assumed a material in amorphous phase, that is the latent heat of melting was set to zero. In fact, this may be realistic situation for a wide range of boron doped irons. Taking melting into consideration would have lengthened not the neural net but the heat diffusion simulations substantially. Numerical data on specific heat, heat conductivity, heat of evaporation, evaporation temperature, density and absorption coefficient were taken from the literature and were approximated by the room temperature values [3].

References

1. H. Ritter, T. Martinetz and K. Schulten: Neural Networks, **2**, 159-168 (1988)
2. B. Widrow and M.E. Hoff: WESCON Conv. Record, **IV**, 96-104 (1960)
3. American Institute of Physics Handbook, edited by D.E. Gray (McGraw-Hill, New York, 1982)

Raster Thermowave Technique of Laser Spectroscopy

S.A. Vinokurov and *A.M. Skaleris*
Vavilov Optical Research Institute, 199034 St. Petersburg, Russia

Thermowave techniques attract more and more attention of investigators in the whole world. Such techniques have rather high sensitivity, are noncontacted and can be applied to the wide class of samples under investigation.

The attractiveness of these techniques impels to lock for such decisions which would allow to raise their sensitivity, their lack of receptivity to external influenses, convenience of realisation and exploitation.

We have proposed and realised /1/ a new thermowave technique using the image transfer of one linear raster on the other by coherent laser radiation to form the signal. The principle of action of the thermowave technique is represented in this paper; the versions of the technique realisation is considered as well.

The technique realisation scheme is shown in Figure 1. The probe beam 1 is resolved in diffraction spectrum by a linear raster 2. This spectrum is transformed in raster Fourier-image by a positive lens 3 in its focal plane. The second positive lens 5 recovers the real raster

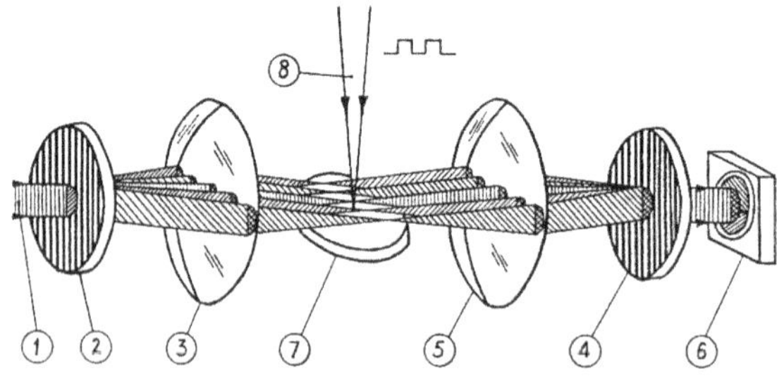

Figure 1. Raster thermowave technique scheme : 1 - probe beam; 2 , 4 - rasters; 3 , 5 - lenses; 7 - sample; 8 - heating beam.

Springer Series in Optical Sciences, Vol. 69 Photoacoustic and Photothermal Phenomena III
Editor: D. Bićanić

image. The second linear raster 4 is placed in the plane of recovered image so that its hatches are parallel to the hatches of the first raster. Radiation transmitted through the system falls on a photodetector 6. Modulated heating beam 8 is focused on a sample 7 to bring the region of periodical heating into coincidence with the Fourier-image of zero order. In so doing the periodic phase shift is brought in the zero order of spectrum distorting the raster image and resulting in periodic change of light flow falling on the photodetector.

The signal formation phase mechanism with little sensitivity to vibrations acts here. The scheme is tuned in such a manner that the first raster hatches image would fall at the intervals between the hatches of the second raster. Decreasing of constant illumination decreased the noise level by several times. The sensitivity as against that, realised by conventional deflection technique on the same conditions, increased more than tenfold. The surface absorption 10^{-5} corresponded to signal to noise relation S/N=1 at the following conditions: heating beam power 1W, modulation frequency 12Hz, transmission band 1Hz. We have used a black-and-white raster with a width of the hatches and intervals between them of 0.05mm and a lens with 60mm focus lengh. The signal was registered by a conventional photodiode when the raster hatches were strictly parallel. Under favourable disposition of the raster with regard to sensitivity the level of constant illumination undergoes a 12-times decreasing.

A small turning or using of a shutter for one of the rasters allowed to give rise to a spatial irregularity of a beam section and to apply a positioned two-elements photodetector with differencial signal recording. The constant illumination was trebled but this scheme allowed to decrease significantly the effect of the probe laser instability and of scattered radiation.

The scheme can function when the probe beam falls on the mirror surface of sample at an angle. In this case the second lens, raster and photodetector are situated along direction of the beam reflected by the surface.

The numerical Fourier analysis was used to analyse the scheme. Fourier-image of the first raster was multiplied by the phase shift function and then restored in the first raster image by the use of an inverse Fourier transform. The multiplication of this image by the transmission function of the second raster permitted the constant illumination of photodetector and the periodic signal to be found. As a result it was found that a sensitivity for our conditions is optimal on 4-mm shifting of one of the rasters from focal plane for one-element photodetector and on 8-mm shifting for the position photodetector. These values are independent of the lens focal distances and are defined by the wave lengh and raster spacing only.

1.S.A. Vinokurov, N.A. Pankratov, A.M. Skaleris, Pisma v Zhurnal Technicheskoi Fisiki (USSR) 16, 8 (1990).

Part VII

Phase Transitions, Heat and Mass Transfer

Photothermal Study of Heat Conduction, Diffusion, and Sorption in Thin Films and Porous Materials

P. Korpiun and R. Osiander

Physik-Department E13, Technische Universität München, James-Franck-Strasse, W-8046 Garching, Fed. Rep. of Germany

Abstract. Layers with thicknesses small compared to a diffusion length are considered as thermal resistances or permeabilities, respectively. The thermal resistance of grain boundaries in aluminium films is obtained from modulated optical reflectance measurements. The permeability of lipid monolayers for water vapor is determined from the attenuation of temperature induced concentration waves. Heat conduction in powders may be considered as anomalous diffusion. The frequency dependence of the photoacoustic signal of graphite powders can be explained by assuming a fractal structure of the sample. In a microporous material, e.g. charcoal, temperature waves generate concentration waves of the adsorbed gas. The effect of adsorption on the photoacoustic signal is discussed for carbon dioxide adsorbed on charcoal.

1. Introduction

To learn from photothermal experiments something on the transport of heat, charge carriers or molecules one has to solve the respective diffusion equations for appropriate boundary conditions. In this contribution we like to consider some special problems for which the theoretical description of the transport of heat and mass exhibit some pecularities.

In Sect.2 we introduce the concept of thermal resistance and permeability to describe the effect of submicron thick layers such as grain boundaries, lipid monolayers, and biological membranes on the diffusion of heat and molecules, respectively. We show in Sect.2.1 that it is possible to evaluate data for the thermal resistance of grain boundaries of polycrystalline metal films from measurements of the modulated optical reflectance. In Sect.2.2 we describe a method how to determine the permeability of lipid monolayers for water molecules. Values for the permeabilities of membranes for photosynthetically evolved oxygen in leaves may be obtained from photoacoustic measurements, as shown in Sect.2.3.

Porous samples give large photoacoustic signals, but their interpretation in terms of optical and thermal quantities is not as easy as for homogeneous samples. In Sect.3 we consider two aspects of the photoacoustic effect on porous materials. At first we describe in Sect.3.1 the photoacoustic signal of macroporous samples by considering the conduction of heat in an inhomogeneous structure. Therefore we use a heat conduction equation which includes a time dependent thermal diffusivity and a volume dependent heat capacity. Calculated data for a fractal structure of porous materials will be compared with experimental data measured on graphite powders. The photoacoustic signal of microporous materials like charcoal, silicagel, zeolite, etc. with pore sizes in the order of magnitude of nanometer can be governed by the temperature dependent adsorption of the cell gas. This phenomenon is briefely described in Sect.3.2.

2. Effect of Thin Layers on Thermal and Concentration Waves

Let us consider the temperature distribution in two regions labelled by m=1 and m=2 with different thermal diffusivities α_1 and α_2, Fig.1(a). We assume that at $z=-l_1$ there is a heat source extended in a plane parallel to the plane boundary z=0 between both regions. The heat supplied to region 1 at $z=-l_1$ varies sinusoidally with frequency $\omega/2\pi$.

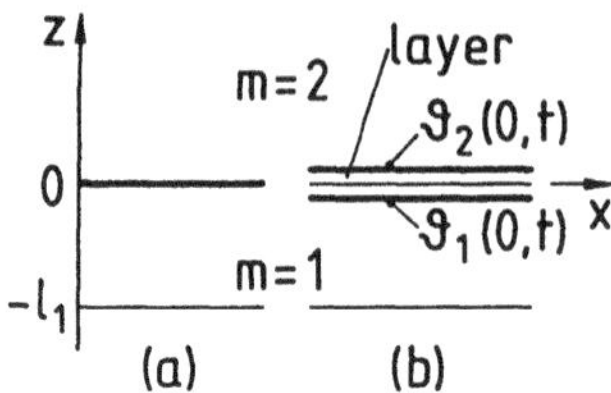

Fig.1 : Scheme of two media separated by (a) a boundary, (b) a thin layer.

If $\vartheta_m(z,t)$ is the temperature variation, we have to solve the heat conduction equation $\alpha_m(\partial^2\vartheta_m/\partial z^2) - \partial\vartheta_m/\partial t = 0$ for m=1 in the range $-l_1 \leq t \leq 0$ and for m=2 in the range $0 \leq z$. Solutions of this equation are plane thermal waves, and the temperature distributions in both regions are finally

$$\vartheta_m(z,t) = \Theta_m^+ e^{-i(k_m z-\omega t)} + \Theta_m^- e^{+i(k_m z+\omega t)}, \tag{1}$$

where $k_m=(1-i)/\mu_m$ is the thermal wave number and $\mu_m=\sqrt{2\alpha_m/\omega}$ the thermal diffusion length in region m. $\Theta_m^{\pm}$ are the temperature amplitudes of the thermal wave running in ±z–direction. The complex amplitudes Θ_1^+, Θ_1^- of region 1 are related to the amplitudes Θ_2^+, Θ_2^- in region 2 by boundary conditions at z=0 according to continuity of temperature:

$$\vartheta_1|_{z=0} = \vartheta_2|_{z=0} \rightarrow \Theta_1^+ + \Theta_1^- = \Theta_2^+ + \Theta_2^- \tag{2}$$

and continuity of heat flux:

$$-\lambda_1\frac{\partial\vartheta_1}{\partial z}\bigg|_{z=0} = -\lambda_2\frac{\partial\vartheta_2}{\partial z}\bigg|_{z=0} \rightarrow \lambda_1 k_1(\Theta_1^+ - \Theta_1^-) = \lambda_2 k_2(\Theta_2^+ - \Theta_2^-), \tag{3}$$

where λ_m is the thermal conductivity of region m.

Additional molecules may be generated photothermally by photochemical processes or may be released by heat induced evaporation or desorption. The transport of this molecules in the range labelled by m can be described by the diffusion equation $D_m(\partial^2\psi_m/\partial z^2) - \partial\psi_m/\partial t = 0$ for the concentration of the additional molecules. The concentration variation in the regions 1 and 2 with the diffusion coefficients D_1 and D_2 may be described by

$$\psi_m(z,t) = c_m^+ e^{-i(\xi_m z-\omega t)} + c_m^- e^{+i(\xi_m z+\omega t)}, \tag{4}$$

where $\xi_m=(1-i)/\kappa_m$ is the concentration wave number and $\kappa_m=\sqrt{2D_m/\omega}$ the concentration diffusion length. The amplitudes c_1^+, c_1^- of the concentration waves running upwards and downwards in region 1, respectively, are related to c_2^+ and c_2^- by the boundary condition

$$-D_1\frac{\partial\psi_1}{\partial z}\Big|_{z=0} = -D_2\frac{\partial\psi_2}{\partial z}\Big|_{z=0} \rightarrow D_1\xi_1(c_1^+ - c_1^-) = D_2\xi_2(c_2^+ - c_2^-) \tag{5}$$

for the flux, but the concentration is generally not continuous at a boundary. There may be a relation in the form of

$$\psi_1|_{z=0} = K_{12}\,\psi_2|_{z=0} \tag{6}$$

between the concentrations at both sides of the boundary. If, e.g., the medium (m=2) above the boundary z=0 is a gas, the medium in the region $z \leq 0$ (m=1) a liquid, the coefficient K_{12} is the solubility of the gas in the liquid /1/. If we consider the diffusion of a gas with concentration ψ_1 in a liquid labelled by m=1 for $z \leq 0$ crossing at $z = 0$ the boundary to another liquid labelled by m=2 for $z \geq 0$, K_{12} is the relative partition coefficient /2, 3/.

There are layers which influence the diffusion of heat and mass, respectively. Therefore, they can be studied by thermal or concentration waves. Such layers which were studied photothermally are grain boundaries in polycrystals which reduce the heat conduction in solids /4/, lipid monolayers spread on water which reduce the evaporation /5/, and cell membranes which control the transport of oxygen in leaves /6/.

The thicknesses of these layers are in the submicron regime, i.e., they are much thinner than one diffusion length μ_m or κ_m, respectively. Furthermore, the structure of these layers is generally very inhomogeneous, so that a diffusion coefficient can not be defined. Thus, it makes no sense to describe the effect of very thin layers on the propagation of diffusion waves by solving diffusion equations as described above.

If one knows the temperatures directly below and above the layer (Fig.1(b)), it is reasonable to assume that the heat flux j_q across the layer is proportional to the difference of the temperatures /7/:

$$j_q = \frac{1}{r}\left[\vartheta_1(0) - \vartheta_2(0)\right]. \tag{7}$$

The constant r is called thermal resistance. Similiar considerations lead to the molecular flux /1–3/

$$j_c = P\left[\psi_1(0) - \psi_2(0)\right], \tag{8}$$

where P is the permeability of the layer. Thus, the effect of a thin layer on the propagation of heat or concentration waves is taken into account by a second boundary condition for the flux at z=0, so that

$$-\lambda_1\frac{\partial\vartheta_1}{\partial z}\Big|_{z=0} = -\lambda_2\frac{\partial\vartheta_2}{\partial z}\Big|_{z=0} = \frac{1}{r}\left[\vartheta_1(0) - \vartheta_2(0)\right]. \tag{9}$$

If λ is the thermal conductivity of both regions m=1 and 2, the layer with thermal resistance r reduces the amplitude by

$$f = |\vartheta_2(0)| \,/\, |\vartheta_1(0)| = 1 \,/ \sqrt{1 + 2\lambda r/\mu(1 + \lambda r/\mu)} \tag{10}$$

and shifts the phase angle by

$$\Delta\varphi = -\arctan\{(\lambda r/\mu)/(1 + \lambda r/\mu)\}. \tag{11}$$

For very impermeable layers the phase angle may be shifted up to –90 degrees.

2.1 Thermal Resistance of Grain Boundaries

Grain boundaries in polycrystals reduce the conduction of heat appreciably. Their thickness is in the order of magnitude of nanometer. The Therma Wave group showed that a scan of the modulated optical reflectance (MR) is correlated to the grain size /8/. We performed similiar measurements on 800 nm thick aluminium (1% Si, 0.5% Cu) films sputter deposited on a thin layer of SiO_2 on a silicon crystal (Fig.2). If we move pump and probe beam with a constant offset of 1.5 μm along the surface of such a polycrystalline film, there are two different extreme positions. If both beams hit the sample within the area of one single grain (Fig.3, position 1), the amplitude has a maximum. If there is a grain boundary between both beams (Fig.3, position 2), the amplitude has a minimum.

We simulated such a scan using the thermal resistance of the grain boundary as a variable parameter /4/. The best fit of such simulations to the data measured on metal films with different grain sizes we obtained for a thermal resistance of (1.1±.3) 10^{-3} Kcm^2/W. As far as we know this is the first direct measurement of the thermal resistance of a grain boundary.

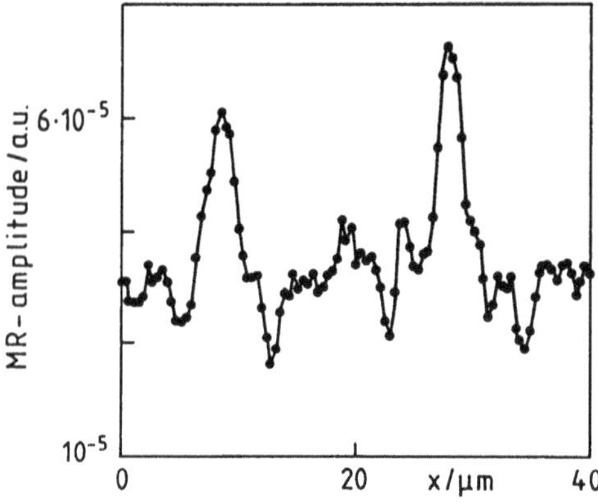

Fig.2 : MR amplitude measured on polycrystalline AlSiCu along a distance of 40 μm. Average grain size: 2.23 μm.

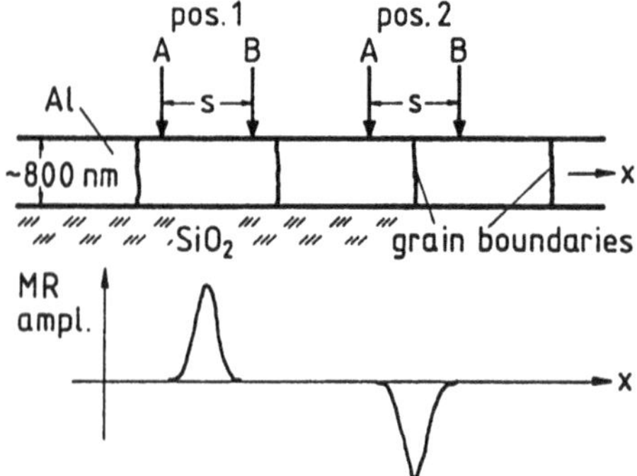

Fig.3 : Above: MR measurement (schematic); A: pump beam (Ar–laser); B: probe beam (HeNe–laser). Below: MR amplitude around position 1 and 2 (schematic)

2.2 Water Permeability of Lipid Monolayers

Lipids may form monolayers on a water surface. We measured the permeability of lipid monolayers for water vapour in an experimental arrangement described elsewhere /5/. At first, we measured the PA amplitude of water without the lipid film as a function of temperature (dotted line in Fig.4). With a monolayer of the phospholipid DMPE (L–α–dimyristoylphosphatidyl–ethanolamine) spread on water we obtain the data plotted in Fig.4. The permeability is calculated from the reduction of the amplitude by using a formalism derived earlier /9/. At temperatures <40°C the permeability is relatively large and can not be determined from the experimental data. The decrease of the permeability above 45°C points to a structural transition. The solid line fitted to the experimental data corresponds to a permeability of P=(14.3±2)cm/s. This result is consistent with that obtained by other authors /10/. Results of our measurements on arachidic acid are given elsewhere /5/.

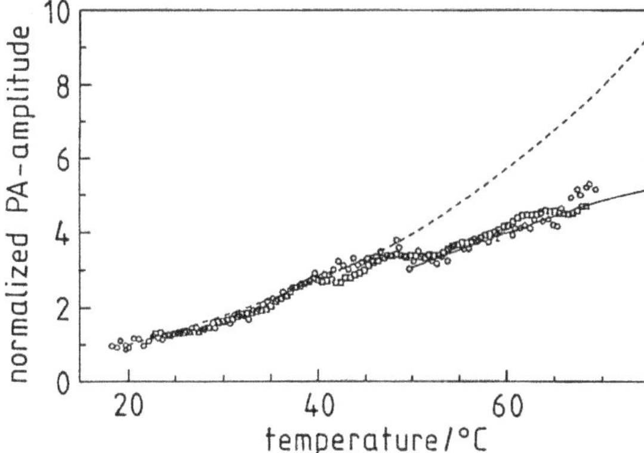

Fig.4: Temperature dependence of the PA amplitude measured on water (dotted line) and on water covered by a monolayer of DMPE for 32 mN/m surface tension. Data of both measurements (circles and squares) are normalized to those at 20°C.

2.3 Oxygen Diffusion Across Cell Membranes in Leaves

Malkin, Cahen and coworker /11, 12/ showed that oxygen evolved periodically by photosynthesis in the chlorophyll of leaves can be determined by a microphone as soon as the oxygen molecules reach the intercellular air space (Fig.5). We calculated the PA signal by considering the propagation of oxygen concentration waves in the several micron thick tissue of the meosphyll cell /6/. It consists of different layers: the chloroplast containing the chlorophyll, the cytoplasm, and the cell wall. There are also two membranes: the limiting membrane and the plasmalemma.

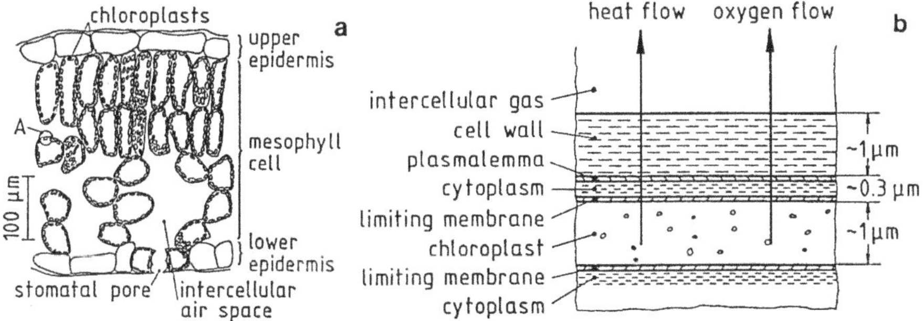

Fig.5: Leaf (lateral section) with simplified model of region A of mesophyll cell used to calculate oxygen diffusion.

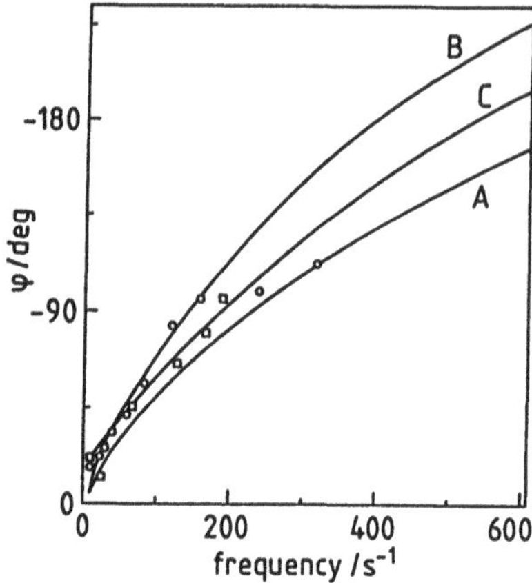

Fig.6: Phase angle difference φ between photobaric and photothermal part of the PA signal; circles: expt./12/, squares: expt./13/. Solid lines: calculated for limiting membrane permeability 1cm/s (A,B), 0.1 cm/s (C) and rate constant $2 \cdot 10^5$ (A, C), $2 \cdot 10^3$ (B).

In Fig.6 the difference φ between the phase angle of the oxygen (photobaric) and the photothermal component of the pressure variation in the intercellular gas is plotted as a function of modulation frequency. The symbols are experimental data /12, 13/. Our calculations show that φ mainly depends on the thickness of the chloroplast, the cytoplasm, the cell wall, and on the rate constant of the photochemical reaction /6/ (solid lines A and B in Fig.6). Permeabilities of limiting membranes and plasmalemmas are not well known. It comes out that the limiting membrane influences the signal more than the plasmalemma, as may be seen from the solid lines A and C in Fig.6. Line A was calculated for a permeability of 1cm/s of the limiting membrane, line C for 0.1cm/s. A smaller permeability shifts φ to higher negative values. If the geometry of the cell would be known, one could study the permeabilty of limiting membranes by photoacoustic measurements.

3. To Photothermal Phenomena on Porous Materials

Since the rediscovery of the photoacoustic spectroscopy in the seventies it has been commonly assumed that this experimental technique is a valuable tool for the investigation of porous samples /14,15/ The interest of most scientists was mainly directed to the optical spectroscopic aspect of the problem, namely to the scattering of light /16–18/. The influence of an interstitial gas in powdered optically strong absorbing samples was studied by Monchalin et al. /19/ and by McGovern et al. /20/. In the present work we consider two further aspects: (i) the conduction of heat in an inhomogeneous structure and (ii) the temperature dependent adsorption of gas in microporous materials.

3.1 Powdered Samples

We measured the PA signal as a function of modulation frequency for various metallic and graphite powders in a gas–microphone cell with air as transducer gas. The diameter of the illuminated area of the sample was much larger than the grain size and the thermal diffusion length. Experimental results obtained on graphite powders with grain sizes of 2, 30, and 250 μm are shown in Fig 7, where amplitude times frequency is plotted versus frequency. A straight line parallel to the frequency axis means: the amplitude decreases as $1/\omega$. There is a range between 300 Hz for the smallest and 1 kHz for the largest grains where the frequency dependence deviates from the $1/\omega$ behaviour. It can be explained by anomalous diffusion, i.e., the diffusion coefficient is time dependent. One can solve the problem analytically, if one assumes that a powder has a fractal structure /21,22/. In this case for a linear sample size l its density scales as $\rho \propto l^{3-d_f}$ where d_f is the fractal dimension and we obtain for the frequency

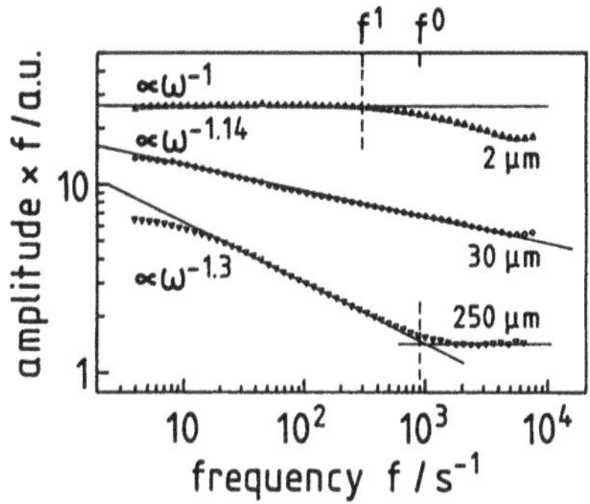

Fig.7 : PA amplitude × modulation frequency f versus f measured on graphite powders with grain sizes of 2, 30, and 250 μm ($\omega=2\pi f$).

dependence of the thermal diffusivity $\alpha \propto (i\omega)^{1-d_s/d_f}$, where d_s is the spectral dimension /22/. The amplitude of the temperature oscillation at the sample to gas boundary becomes $\Theta(\omega) \propto (i\omega)^{-\{2+d_w(2-d_f)\}/2}$. The amplitude p^0 of the pressure variation in the gas depends as $p^0 \propto (i\omega)^{-\{3+d_w(2-d_f)\}/2}$ on frequency /23/. For $d_f=2.8$ and $d_s=1.2$ we obtain $p^0 \propto (i\omega)^{-1.34}$ which agrees with the frequency behaviour measured for the largest grains (Fig.7). One also can show that the interstitial gas reduces the exponent observed experimentally. With the concept of fractals we also can explain the upper and lower limit for anomalous diffusion. At $f^0=1$ kHz the thermal diffusion length of graphite is equal to the average grain size of $l^0=250$ μm, i.e., $f^0=\alpha^0/\pi l^{0^2}$, where α^0 is the thermal diffusivity of a grain. It is obvious that the amplitude for higher frequencies is $\propto 1/\omega$. The transition to the $1/\omega$ behaviour for frequencies below $f^1=300$ Hz which we found for the small grains points to the existence of a correlation length ξ expected in a percolation system. Since the thermal diffusion length scales as $\mu \propto f^{-d_s/d_f}$, ξ can be estimated to $\xi/l^0=(f^0/f^1)^{d_s/d_f}=1.27$. It corresponds to a critical cluster size of 159 grains with 2 μm diameter.

3.2 Microporous Materials

Another type of porous substances are charcoals, silicagel, zeolites etc. They are characterized by large inner surfaces (~10^3 m^2/g) of micropores with diameters less than 10 nm. The amount of gas adsorbed depends on temperature. A temperature variation, therefore, generates not only a heat wave but also a concentration wave. This concentration wave comes from a layer with a thickness of about one concentration diffusion length $\kappa=\sqrt{2D_k/\omega}$ where D_k is the Knudsen diffusion coefficient. The total pressure variation in the cell gas is /24/

$$p(t) \propto P^0/\sqrt{\rho c}\left[1+\sqrt{\frac{D_k}{\alpha}}\frac{h\rho_s}{M}\left[\frac{\partial X}{\partial P}\right]_T\right], \qquad (12)$$

where P^0 is the pressure, ρ the density, c the isobaric specific heat, α the thermal diffusivity and M the molecular weight of the cell gas; ρ_s is the density of the sample, h the heat of adsorption , X the adsorbed amount per unit sample mass; $(\partial X/\partial P)_T$ is the slope of the adsorption isotherm.

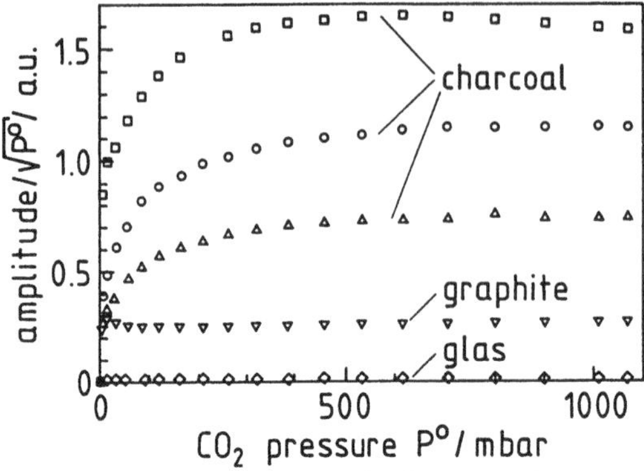

Fig.8 : Experimental data of PA amplitude over square root of carbon dioxide pressure P^0 versus P^0 for various samples.

We measured the photoacoustic signal for glass, graphite powder, and various charcoals in a carbon dioxide filled gas–microphone cell as a function of the gas pressure. In Fig.8 are plotted the amplitude data divided by the square root of the gas pressure versus the gas pressure. This reduced amplitude should be constant if no gas is adsorbed.(Fig.8),because the density ρ of the gas in (12) is proportional to P^0. We, indeed, find this behaviour for samples of glass and powdered graphite. Both materials do not adsorb appreciable amounts of gas and thus, the second term in the brackets of (12) can be neglected. For charcoal samples, however, the reduced amplitude increases with pressure: since charcoal adsorbs, the second term in (12) contributes to the signal. We can also argue vice versa: if the reduced amplitude increases with the gas pressure, the corresponding substance does adsorb appreciable amounts of cell gas. From such measurements one may obtain quantities like the slope $\partial X/\partial P$ which characterizes the adsorption /24/.

In conclusion, the signal detected by a microphone is very sensitive to the adsorption of gases. Whether adsorption does occur or does not is detected most effectively by varying the pressure of the adsorptive.

References

1. J. Crank, The Mathematics of Diffusion, Clarendon Press, Oxford, 1975
2. W.D. Stein, Transport and Diffusion across Cell Membranes, Academic Press, Orlando, 1986, p.39
3. P.S. Nobel, Introduction to Biophysical Plant Physiology, W.H. Freeman and Co., San Francisco, 1973
4. G. Meyer–Berg, R. Osiander, P. Korpiun, R. Kakoschke, H. Joswig, this issue
5. W. Micheler, R. Osiander, P. Korpiun, J. Schmidt, W. Knoll, this issue
6. P. Korpiun, and S. Malkin, this issue
7. H.S. Carslaw and J.C. Jaeger, Conduction of Heat in Solids, Oxford Univ.Press, London, 1959
8. Technical Report of Therma–Wave Inc. Freemont/Calif. (USA) 1989
9. R. Osiander, B. Helm, P. Korpiun, in Photoacoustic and Photothermal Phenomena II, eds. J. Murphy, J. Maclachlan–Spicer, L. Aamodt, B.S.H. Royce, Springer Ser. Opt.Sci. Springer, Heidelberg, 1990, p.107
10. R. Vanderveen, G. Barnes, Thin Solid Films, 134, 227 (1985)
11. G. Bults, B.A. Horwitz, S. Malkin, D. Cahen, Biochim.Biophys.Acta, 679, 452 (1982)
12. P. Poulet, D. Cahen, S. Malkin, Biochim.Biophys.Acta, 724, 433 (1983)
13. C. Geel, J.F.H. Snel (unpublished data) (1990)
14. A. Rosencwaig, A. Gersho, J.Appl.Phys. 47, 64 (1976)
15. E.M. Monahan, Jr., A.W. Nolle, J.Appl.Phys. 48, 3519 (1977)
16. P. Helander, J. Lundström, D. McQueen, J.Appl.Phys. 51, 3841 (1980)
17. Z.A. Yasa, W.B. Jackson, N.M. Amer, Appl.Opt. 21, 21 (1982)

18. R. Tilgner, Appl.Opt. 20, 3780 (1981)
19. J.F. Monchalin, L. Bertrand, G. Rousset, and F. Lepoutre, J.Appl.Phys. 56, 190 (1984)
20. S.J McGovern, B.S.H. Royce, J.B. Benzinger, J.Appl.Phys. 57, 1710 (1985)
21. A. Boccara and D. Fournier in: Photoacoustic and Photothermal Phenomena, P. Hess and J. Pelzl. eds., Springer Series in Optical Sciences, Springer, Heidelberg 1988, p.302
22. R. Osiander, R. Haberkern, P. Korpiun, and W. Schirmacher in Photoacoustic and Photothermal Phenomena II, eds. J. Murphy, J. Maclachlan–Spicer, L. Aamodt, B.S.H. Royce, Springer Ser. Opt.Sci. Springer, Heidelberg, 1990, p.309
23. R. Osiander, J. Lobermeier, P. Korpiun, this issue
24. P. Korpiun, W. Herrmann, R. Osiander, in Photoacoustic and Photothermal Phenomena II, eds. J. Murphy, J. Maclachlan–Spicer, L. Aamodt, B.S.H. Royce, Springer Ser. Opt.Sci. Springer, Heidelberg, 1990, p.313

Analysis of Photothermal Sensitivity in the Detection of Phase Transitions

A. Hadj-Sahraoui[1], *G. Louis*[2], *P. Peretti*[1,2], *and J. Billard*[3]

[1]Laboratoire de Physique et Biophysique des Milieux Mal Ordonnés, Université René Descartes (Paris V), 45 Rue des Saints Pères, F-75270 Paris Cedex 06, France

[2]DRP (UA CNRS 71), Université Paris VI, F-75252 Paris Cedex 05, France

[3]LPMC (UA CNRS 571), Collège de France, F-75232 Paris Cedex 05, France

Abstract. The photothermal signal sensitivity in detection of phase transitions is defined. The theoretical and experimental dependence of this sensitivity on frequency are investigated for two experimental configurations (front excitation and rear excitation). The effect of the nature of the probed medium on the sensitivity is analyzed.

1 - Definition of the photothermal signal sensitivities in the detection of phase transitions.

In previous papers [1, 2] we have shown that both amplitude and phase photothermal signal are sensitive to the first and second order phase transitions. In the following we present the way we estimate this sensitivity.

Calculation of the 3D probe beam deflection has been presented elsewhere [3]. The normal component $\Phi_n(x_0,z)$ of the probe beam deflection may be written in the form

$$\Phi_n(x_0,z,t) = a_n(x_0,z) \exp [-j(\omega t + \phi_n)]$$

where x_0 is the transverse probe beam offset, $f = \omega/2\pi$ is the modulation frequency and z the distance between the probe beam and the heated surface. This deflection is related to the sample heat capacity c_s and thermal conductivity K_s which depend on the average mean temperature T. The asymptotic behavior [4] of the heat capacity $c_s(T)$ near temperature transition T_c is :

$$c_s(T) = \frac{A}{\delta}\varepsilon^{-\delta} + B \quad \text{with } \varepsilon = \frac{T - T_c}{T_c} \text{ for } T > T_c$$

$$c_s(T) = \frac{A'}{\delta'}\varepsilon^{-\delta'} + B' \quad \text{with } \varepsilon' = \frac{T_c - T}{T_c} \text{ for } T < T_c$$

Values of A, B, A', B', d and d' are available for the smectic A-nematic and nematic-liquid phase transitions of 4-n-octyl-4'-cyanobiphenyl 8CB [5]. We have neglected the change in the thermal conductivity at the smectic A-nematic transition. In the case of the nematic-liquid transition the thermal conductivity [5] is $K_s = 0.17$ W / m/K for $T < T_c$ and $K_s = 0.134$ W/ m / K for $T > T_c$. In order to quantify the sensitivity of the amplitude a_n in the detection of phase transition, we introduce the relative variation S_a of a_n between two values of T below the transition temperature T_c, T_1 far away from T_c and T_2 close to T_c :

Springer Series in Optical Sciences, Vol. 69 Photoacoustic and Photothermal Phenomena III
Editor: D. Bićanić

$$S_a = [a_n(T_2) - a_n(T_1)] / a_n(T_1)$$

where $T_1 = 33.70$ °C, $T_2 = 33.76$°C, $T_c = 33.77$°C for the smectic A-nematic transition and $T_1 = 40.30$°C, $T_2 = 40.42$°C, $T_c = 40.43$°C for the nematic-liquid transition [4].

Note that, $S_a > 0$ corresponds with an increase of the signal amplitude during the phase transition while $S_a < 0$ corresponds with a decrease of the signal amplitude. The sensitivity of the signal phase $S\phi$ is defined as an angle measuring the variation of the phase shift near the temperature transition :

$$S_\phi = \phi(T_2) - \phi(T_1)$$

2. Results and discussion

We have studied the influences of the modulation frequency on S_a and S_ϕ within two experimental configurations : the front excitation and the rear excitation.

We have calculated the dependence of S_a and $S\phi$ versus the modulation frequency. The range of frequency has been chosen so that the sample thickness remains smaller than the thermal diffusion length for all frequencies. Calculations have been made for opaque samples which are thermally thin, using the standard values z = 60 µm and x_o = 0.5 mm. The probed medium is air with thermal diffusivity $\alpha = 2.15.\ 10^{-5}\ m^2/s$ and conductivity $K = 0.026\ W.m^{-1}K^{-1}$ and the backing is fluorite ($\alpha = 3.6.\ 10^{-6}\ m^2/s$; $K = 9.7\ Wm^{-1}.K^{-1}$). Figures 1a and 1b show the frequency dependence of S_a and $S\phi$ in the case of nematic-liquid transition in 8CB. It is seen that S_a can be either negative or positive and the photothermal signal may be insensitive ($S_a = 0$, $S_\phi = 0$) to the phase transitions for certain values of frequency. One can observe that the photothermal signal is more sensitive to the phase transitions in the rear configuration than in the front configuration. This is true for any value of frequency.

The effect of the nature of probed medium has also been investigated [2]. Calculations of S_a and Sf have been made for three probed media (air, CCl4, fluorite) in

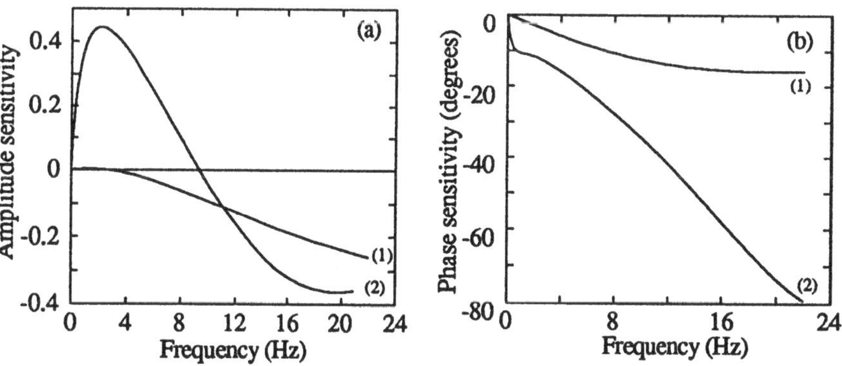

Fig. 1: Amplitude sensitivity (a) and phase sensitivity (b) dependence on frequency in the front excitation (1) and rear excitation (2).

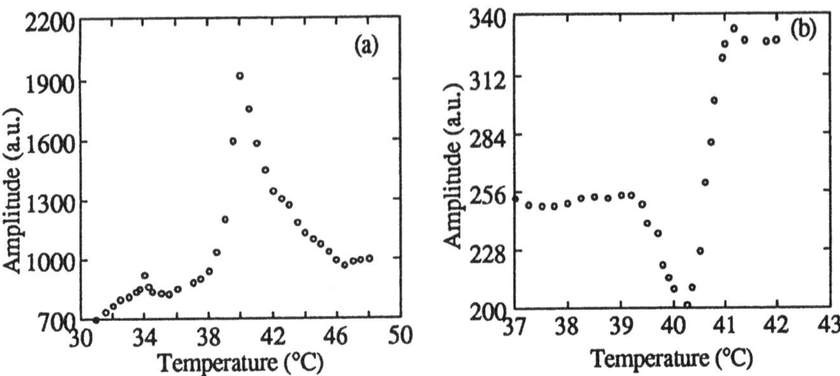

Fig. 2: Experimental plot of amplitude vs temperature. a) f=1Hz. b) f=20 Hz.

the rear pumping configuration with fluorite as backing. Using air instead of fluorite or CCl_4 as probed media results in high sensitivity in detection of phase transitions. Improved sensitivity is obtained when the probed medium has smaller effusivity than the sample.

The 8CB sample undergoes two phase transitions at 33.7°C and 40.5°C according to the following sequence: smectic A-nematic-isotropic. Figures 2a and 2b show the direct experimental plot of the photothermal signal amplitude as a function of temperature along the nematic-isotropic phase transition, at two different modulation frequencies. As expected theoretically the signal magnitude increases during the phase transition at a low frequency (f =1Hz), while it decreases at a higher frequency (f =20 Hz).

References

1. A. Hadj-Sahraoui, G. Louis, B. Mangeot, P. Peretti, and J. Billard, Liq. Crystals, 115, 579 (1989)
2. A. Hadj-Sahraoui, G. Louis, B. Mangeot, P. Peretti, and J. Billard, Phys. Rev. A, to be published (1991).
3. F. Lepoutre, B.K. Bein, L.J. Inglehart, Can. J. of Physics, 64, 1037, (1986)
4. J. Thoen, H. Marynissen, and W. Van Dael, Phys. Rev. A, 26, 2886, (1982)
5. U. Zammit, M. Marinelli, R Pizzoferrato, F. Scudieri and S. Martellucci, J. Phys. E : Sci. Instrum. 21, 935, (1988).

Laser Induced Photoacoustic Effect During Ferroelectric Phase Transitions

J.O. Tocho, F. Cussó, R. Ramirez, and J.A. Gonzalo*

Departamento di Física Aplicada, C-IV,
Universidad Autónoma de Madrid, E-28049 Madrid, Spain
*On leave from: Centro de Investigaciones Opticas and
Departamento de Física, La Plata, Argentina

Abstract. Measurements of the temperature dependent photoacoustic signals were performed in properly oriented ferroelectric crystals. Results have clearly shown the second order phase transition manifested by the pronounced thermal expansion anomalies. The method was applied to study triglycine sulfate (TGS) and the related compounds triglycine selenate (TGSe) and deuterated triglycine sulfate (DTGS).

1. Introduction

Detection of acoustic waves after absorption of modulated or pulsed optical radiation provides information of different thermoelastic properties of materials in a very attractive way.

In condensed matter, acoustic wavwes can be efficiently detected by piezoelectric transducers. The good acoustic impedance matching between detector and sample allows compact and simple experiments [1].

In the case of pulsed laser irradiation the amplitude of the first peak in the time-resolved photoacoustic signal (PAS) can be written as

$$PAS = K\ (\alpha_i / C_p\ \rho)\ f(v_i), \tag{1}$$

where K includes all the experimental parameters which do not depend on temperature, whereas the other factors (thermal expansion coefficient α_i, specific heat at constant pressure C_p, density ρ and the sound velocity in the material v_i) are temperature dependent. The subscript i has been explicited to indicate that for a non-isotropic material the sound velocity and the expansion coefficient depend on the particular direction of propagation chosen.

In this way the detection of phase transitions is possible by using the sensitivity of the PAS to follow the abrupt changes in the physical magnitudes associated with the phase transition.

2. Materials and method

TGS is a ferroelectric material with a second order phase transition at 49.3 ^{o}C. Single crystals of good optical quality can be easily grown from water solution.

The related compounds triglycine selenate (TGSe) and deuterated triglycine sulfate (DTGS) have also a good optical quality.

Properly oriented crystal slabs with typical dimensions 5x5x1 mm^3 were polished and coupled to the transducer. This set-up was introduced into a thermostatic enclosure were the temperature was stabilized better than 0.5 ^{o}C. Light pulses from a Nd:YAG laser with fundamental, second, third and fourth harmonic emissions were used. Pulse energy was attenuated to near 100 μJ and the light was focused on the sample by a cylindrical lens in order to produce acoustical waves with one-axis propagation. Fig.1 shows the experimental set-up.

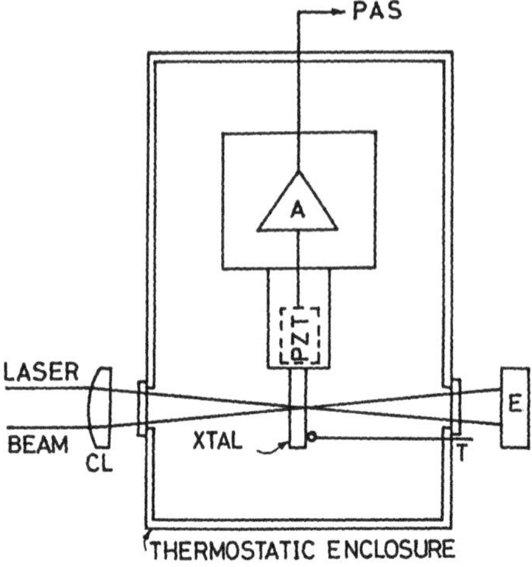

Fig. 1. Top view of the experimental set-up. CL, cylindrical lens; PZT, piezoelectric transducer; E, energy meter; T, temperature sensor.

3. Results

Fig. 2a shows several PAS at different temperatures for a TGS sample, where the experimental x direction corresponds to the crystallographic b-axis. Small IR absorptions found in TGS and TGSe samples allow to use 1.06 μm excitation.

Change in the polarity of the first peak of the PAS can be clearly seen when the phase transition occurs. The amplitude and sign of the PAS reflects mainly the thermal expansion coefficient which has an abrupt change during the transition.

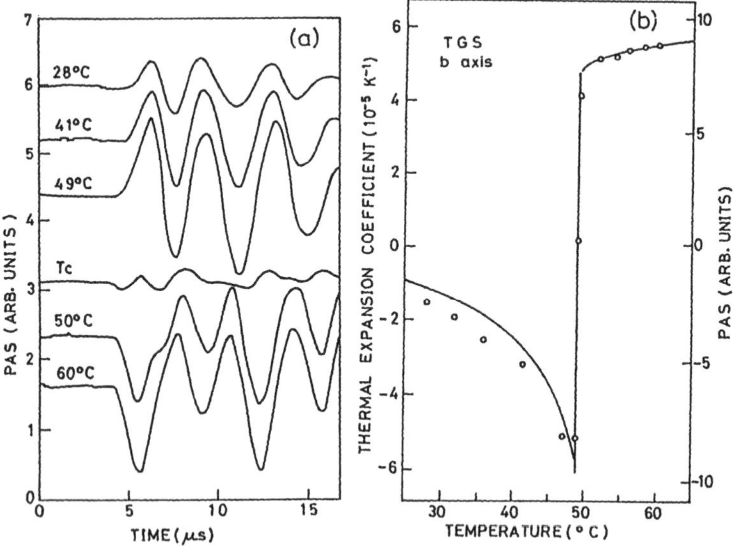

Fig. 2. a) Time resolved PAS of b-cut TGS at different temperatures. b) PAS amplitude vs. temperature compared with literature values of thermal expansion coefficient for b-axis of TGS. PAS is normalized to fit the thermal expansion coefficient at 60°C.

Fig.2b shows the behaviour of the PAS (circles) vs. temperature for a TGS sample. Literature values [2] of the relevant thermal expansion coefficient are also given (full line) for comparison purposes.

In the same way c-axis oriented TGSe and DTGS crystals were also studied. The polarity change observed in TGS at 49.3°C is also apparent in these materials, now at 20°C (TGSe) and 57°C (DTGS), in accordance with the reported critical temperatures in these materials.

In DTGS, grown from ultra pure materials, and where no IR absorption was detected, the fourth harmonic of Nd:YAG at 265 nm which reaches the UV absorption edge, was used as excitation.

4. Conclusions

The results have shown that the PAS reflects mainly the changes in the thermal expansion coefficients along the direction of the acoustic wave propagation. Although the specific heat and the sound velocity also change at the phase transition, in the materials studied in this work these changes are much smaller than those of the thermal expansion coefficients.

Sound velocities, with axis selection, have also been measured in the three crystals. Transparent materials allow to register the time delay of acoustic waves and the irradiation position with high accuracy, giving better resolution than in the case of non transparent materials.

It can be concluded that laser induced photoacoustic effect provides a sensitive method to study ferroelectric phase transitions.

Acknowledgements

One of us (J.O.T.) acknowledges financial support from C.I.C.Y.T. (Spain) under a sabbatical year program.

References

[1] S.E.Braslavsky and K.Heihoff, Handbook of organic photochemistry (CRC Press, New York 1989), Vol.1, p.327 and references therein.

[2] I.Shibuya and S.Hoshino, Japan J. Appl. Phys. 1, 249 (1962).

A Nonlinearity in the Photoacoustic Effect at a First Order Phase Transition

P. Korpiun and J. Weiser

Physik-Department, Technische Universität München,
James-Franck-Strasse, W-8046 Garching, Fed. Rep. of Germany

If the photoacoustic effect is measured as a function of temperature one observes in a limited range about a first or second order phase transition a distinct change in amplitude and phase angle of the PA singal /1–5/. Different models were proposed to relate this temperature behavior of the signal to the latent heat L, the heat capacity ρc and the thermal conductivity λ of the sample.

The model of oscillating boundary /2,3/ which is valid for first and second order transitions, relates the temperature behavior of the signal to the latent heat and to ρc and λ below and above the transition. It is valid for all values of the optical absorption coefficient.

The photoacoustic signal is proportional to the pressure variation p(t) in the transducer gas which again is proportional to the variation in enthalpy, Q(t), supplied to the gas. Q(t) is related to the heat flux at the sample to gas boundary which is proportional to the negative of the gradient in the temperature wave. The pressure variation, therefore, depends linearly on the temperature amplitude at the surface of the sample. If the heat in the sample is generated by the absorption of periodically modulated light, we can write

$$p(t) \propto Q(t) \propto \vartheta(t) = \theta' \sin \omega t + \theta'' \sin 2\omega t,$$

where the last expression means that we consider the first and the second harmonic of the temperature variation. If the temperature $\vartheta(t)$ is the linear response to the heat q(t) supplied by the absorbed light: $\theta'' = 0$. The temperature amplitude θ' generally is complex and contains all informations on amplitude and phase angle of the temperature variation in the sample. The amplitude θ', therefore, also depends on phase transitions in the sample /2,3/. In the following considerations we use the temperature amplitude of the temperature oscillation at the sample to gas boundary. This amplitude is denoted by θ and assumed to be real for simplicity (Fig. 1a).

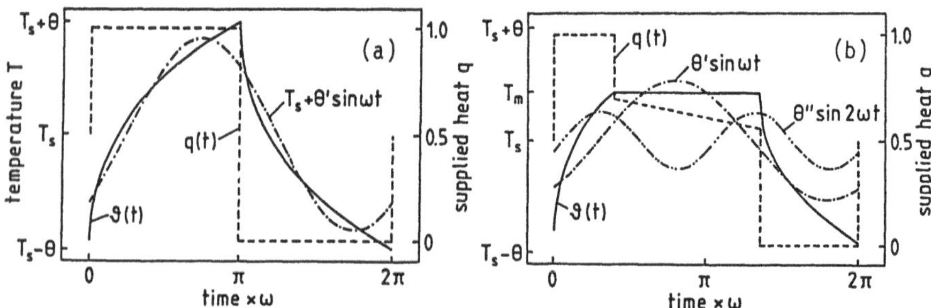

Fig. 1: Oscillation of heat q generated by light absorption and temperature ϑ plotted versus time (a) outside phase transition region with first harmonic, (b) in the transition region with first and second harmonic (T_m: melting temperature)

Springer Series in Optical Sciences, Vol. 69 Photoacoustic and Photothermal Phenomena III
Editor: D. Bićanić

At a first order phase transition it is the total temperature

$$T(t) = T_s + \vartheta(t)$$

at the sample to gas boundary which has to be considered. T_s is the mean temperature at the sample surface. This temperature is varied usually if phase transitions are investigated.

The model of oscillating boundary is based on the fact that the PAE always generates a stationary temperature gradient in the sample to which a temperature oscillation is superposed /2,3/, i.e. the temperature is the highest at the surface of the sample. This is correct for optically strong absorbing materials as well as for optically transparent samples. According to our model the temperature oscillation $\vartheta(t)$ is related linearly to q(t) for all temperatures T_s which are by the amount θ smaller or greater than the transition temperature T_m. That means, for mean temperatures $T_s \lesssim T_m - \theta$ and $T_s \gtrsim T_m + \theta$ one only measures the first harmonic of $\vartheta(t)$ if this is detected by lock–in technique. In the temperature interval $T_m - \theta \lesssim T_s \lesssim T_m + \theta$ the response of the surface temperature is not further linearly related to the heat q(t) supplied to the sample by absorption of intensity modulated light. This may be explained qualitatively as follows.

We consider the total temperature T(t) varying the mean temperature slowly from $T_m - \theta$ below the melting point to $T_m + \theta$ above the transition temperature. For all temperatures outside the intervall $\pm\, \theta$ around T_m the temperature at the sample to gas boundary varies periodically as shown in Fig. 1a. With lock–in technique one measures the first harmonic of the pressure variation in the PA cell gas which is proportional to $\theta' \sin \omega t$. There is no second harmonic because the temporal temperature variation is symmetric with respect to the time axis. But for temperatures T_s which differ less than θ from T_m the periodical temperature variation becomes asymmetric as it is shown in Fig. 1b for an example where $T_s < T_m + \theta$. This can be explained by the fact that the temperature rise stops at the melting temperature because latent heat is consumed for the melting process. The response of the surface temperature is not further linearely related to the heat q(t) supplied to the sample. Because the temperature oscillation now becomes asymmetric with respect to time axis, Fig. 1b, the response of $\vartheta(t)$ to q(t) is not further linear and the second harmonic $\theta'' \sin 2\omega t$ exists. It is now easy to understand that the second harmonic should occur in the entire temperature intervall $T_m - \theta \lesssim T_s \lesssim T_m + \theta$.

We measured the PAE as a function of temperature T_s on indium films of various thicknesses vapor deposited on a quartz substrate. Using lock–in technique we could indeed observe the second harmonic of the pressure variation in a narrow temperature region around the melting point. Amplitude and phase angle of the second harmonic measured on a 200 nm thick In film are plotted as a function of temperature in Fig. 2a and 2b, respectively.

With a simplified version of our one–dimensional model we have qualitatively calculated amplitude and phase angle of the second harmonic of the PA signal. The results are plotted in Fig. 3 as a function of the mean temperature T_s. The calculated temperature behaviour of amplitude and phase angle agree fairly well with the experimental results in Fig. 2a and 2b.

The main feature of the measured temperature behaviour of the phase angle is the asymmetry with respect to the melting temperature $T_m = 156.6\ ^oC$. A similar asymmetry is seen in the plot of the phase angle which we have calculated. The existence of a second harmonic in the transition region cannot be explained by a model used by other authors /4,5/.

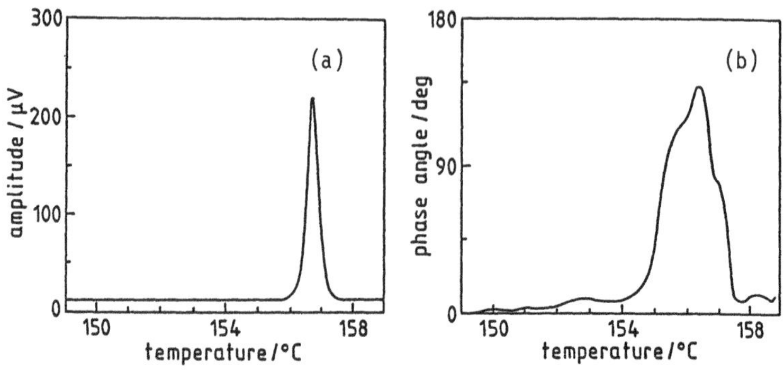

Fig. 2: Temperature variation of (a) photoacoustic amplitude and (b) phase angle of the second harmonic measured on a 200 μm thick indium layer around the melting point (T_m = 156.6 ^{0}C). Modulation frequency 28 Hz.

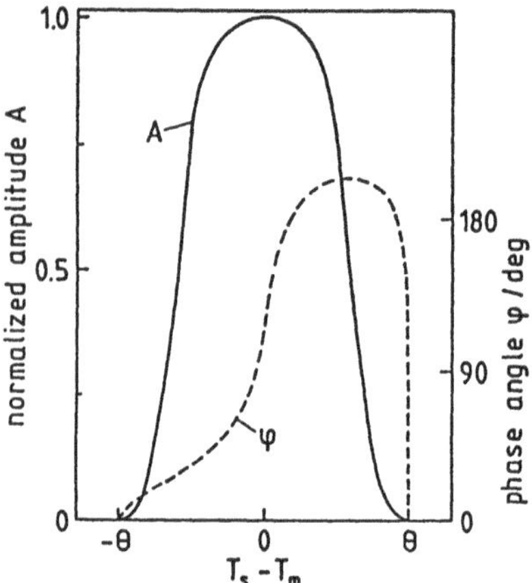

Fig. 3: Temperature variation of amplitude A and phase angle φ of the photoacoustic second harmonic calculated at a first order transition

/1/ R. Florian, J. Pelzl, M. Rosenberg, H. Vargas, R. Wernhardt, phys.stat.sol.(a), K35 (1978)

/2/ P. Korpiun, R. Tilgner, J.Appl.Phys. 51, 6115 (1980)

/3/ P. Korpiun, R. Tilgner, D. Schmidt, J.Phys.(Paris) 44, Suppl. C6–43 (1983)

/4/ U. Zammit, M. Marinelli, R. Pizzoterrator, F. Scudieri, S. Martelluci, Phys.Rev. A41, 1720 (1990)

/5/ C. Glorieux, F. Schoubs, J. Thoen, Mat.Scie.Eng. A122, 87 (1989)

The Anomalous Behaviour of the PPE Signal in the Critical Region of a Phase Transition – New Support for the General Theoretical Model

D. Dadarlat[1], *M. Chirtoc*[1], *R.M. Candea*[1], *A. Frandas*[1], *and D. Bićanić*[2]

[1]Institute of Isotopic and Molecular Technology, P.O. Box 700, Cluj-Napoca 5, R-3400 Romania

[2]Department of Agricultural Engineering and Physics, Agrotechnion, Wageningen Agricultural University, Bomenweg 4, 6703 HD Wageningen, The Netherlands

Abstract. The structural (tetragonal-orthorombic) phase transition of $La_{1.94}Sr_{0.06}CuO_4$ was detected using the PPE method. The experiment, performed in the vicinity of the singular point of the operation diagram, allows for the indirect determination of the material optical absorption coefficient.

1. Introduction

The theory of the reflection photopyroelectric (RPPE) method was developed by Chirtoc and Mihailescu [1] for a PPE cell consisting of six layers and taking into account the possible reflection of light at the sensor's electrode. The general expressions for the PPE signal amplitude and phase were derived, particularized and verified for special cases of experimental interest. However, some theoretical predictions of the PPE theory have received no experimental support yet, such as a local dip of the amplitude and a sharp phase variation at the (1b)-(1c) boundary on the operation diagram surface [1], where $a_mL_m=\beta L_m$ (see Fig.1 and Ref.2), with a_mL_m and βL_m the dimensionless thermal and optical thicknesses of the sample. On the other hand, a recent application of the PPE technique has been the detection of phase transitions. Ferroelectric [3,4], magnetic [5] and superconducting [6.7] phase transitions were investigated and in all cases the signal jump in the critical region was found to be opposite to that of the specific heat.

In this paper we report on an anomalous behaviour of the PPE voltage amplitude in that the jump has the same direction with that of the specific heat, in the critical region of the structural (tetragonal-orthorombic) phase transition of a $La_{2-x}Sr_xCuO_4$ (x=0.06) single crystal. Following the theoretical

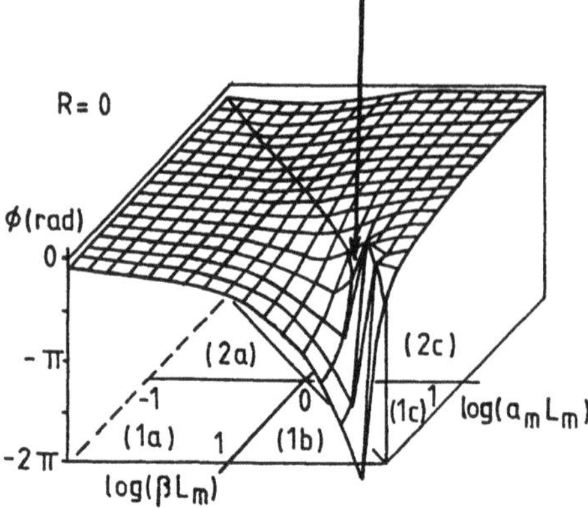

Fig.1 Example of an operation diagram surface for the PPE signal phase [1] for zero reflection (R=0) at the sensor's electrode. Note the sharp phase variation at the singular point indicated by the arrow defined by $a_m L_m = \beta L_m = 5.3$. The regions (1a)...(2c) correspond to various cases of thermally and optically thick/thin sample.

model, such an inversion is possible only under restrictive experimental conditions, associated with the sharp variation of the PPE amplitude and phase mentioned before.

2. Results

A 0.4 mm thick and 10 mm^2 area $La_{1.94}Sr_{0.06}CuO_4$ single crystal was placed on top of a 1 mm thick and 10 mm in diameter PZT pyroelectric sensor, in intimate thermal contact. The setup used for phase transition investigations was described in Ref.4 and the experimental results are presented in Fig.2. The signal

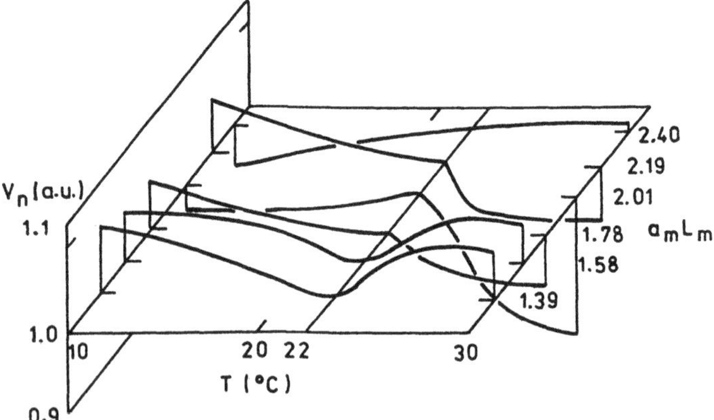

Fig.2 Normalized PPE signal amplitude V_n as a function of the sample temperature at various sample thermal thickness $a_m L_m$. The signal discontinuity due to the phase transition at 22°C changes the sign at $1.78 \le a_m L_m \le 2.40$.

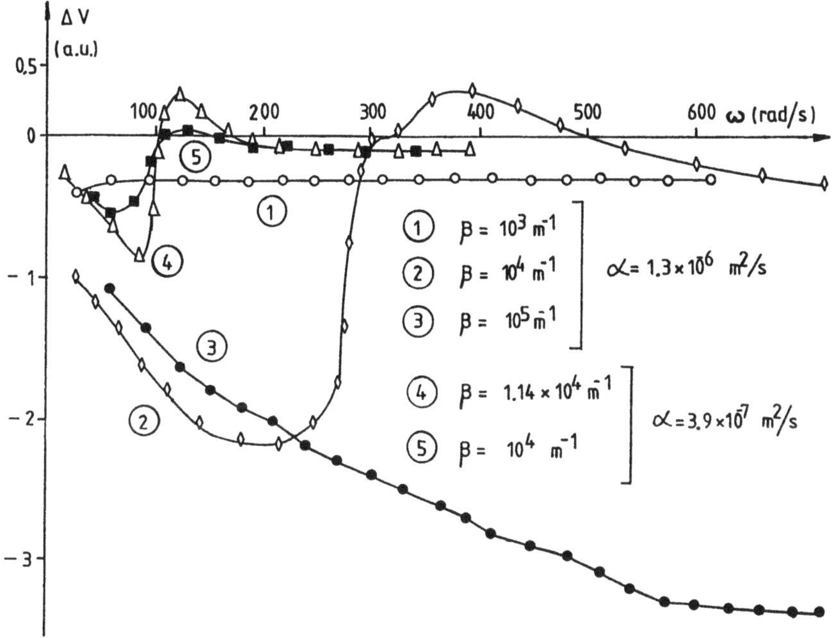

Fig.3 Evolution of the PPE signal jump ΔV at the phase transition temperature for a 1% increase in the specific heat, vs. chopping frequency, calculated with the model of Ref.[1]. The best fit for the measured inversion effect is curve 4. Curves 1-3 indicate how sensitive is this anomalous effect to parameter β.

anomaly occurs at 22°C due to the structural phase transition observed for this material also by calorimetric methods. What is particular is the positive jump of the PPE voltage amplitude for chopping frequencies ranging between 16Hz and 30Hz (ω between 100rad/s and 180rad/s) as contrasted to the usual, negative jump for higher or lower frequencies. In order to explain these data a fit was performed using the exact formula for the PPE signal [1] with the optical absorption coefficient β as a fit parameter. Fig.3 indicates that such an anomalous behaviour, which was not observed until now, is possible for this material in a limited chopping frequency range, for $\beta \approx 10^4 m^{-1}$. The discontinuity inversion effect is very sensitive to the value of β. The best fit was obtained with $\beta = 1.14 x 10^4 m^{-1}$, whereas the rough estimation based on the condition $a_m L_m = \beta L_m$ yields $\beta \approx 10^4 m^{-1}$.

3. Conclusions

The specific optical and thermal parameters of $La_{1.94}Sr_{0.06}CuO_4$ single crystal offered the possibility to perform a phase transition experiment in the proximity of a singularity of the PPE operation diagram. The sharp local variation of the PPE amplitude and phase can explain the anomaly inversion of the PPE voltage amplitude upon the sample phase transition. From the modulation frequencies corresponding to the appearance of the inversion, the sample average optical absorption coefficient for the wavelength range of the radiation source used in the experiment (0.4-4μm in our case), was obtained.

References

[1] M.Chirtoc and G.Mihailescu, Phys.Rev. **B40**, 9606 (1989).

[2] M.Chirtoc, D.Dadarlat and D.Bicanic, paper in this Conference.

[3] A.Mandelis, F.Care, K.K.Chan and L.C.M.Miranda, Appl.Phys. **A38** , 117 (1985).

[4] D.Dadarlat, M.Chirtoc and D.Bicanic, Appl.Phys. **A50**, 357 (1990).

Photopyroelectric Study of a Phase Transition in Liquid Crystals

F. Mercuri, M. Marinelli, U. Zammit, R. Pizzoferrato, and F. Scudieri

Dipartimento di Ingegneria Meccanica, 2° Università di Roma
"Tor Vergata", Via E. Carnevale, I-00173 Roma, Italy

Abstract A comparison between the gas microphone and photopyroelectric techniques for the study of liquid crystal phase transitions is reported. Both the techniques have been used to measure simultaneously the behaviour of the specific heat, thermal conductivity and thermal diffusivity over the S_A-N phase transition in 9CB liquid crystal. As expected from the theory, an increase in the sensitivity for the Photopyroelectric technique has been demonstrated.

Photothermal techniques in different configurations have been used to study the behaviour of thermal parameters at a liquid crystal phase transition /1,2/. In particular the gas microphone configuration (PA) has been used to monitor simultaneously the critical behaviour of the specific heat (c), the thermal conductivity (k) and the thermal diffusivity (α) over the 9CB liquid crystal S_A-N phase transition.

In the present paper a photopyroelectric (PE) configuration that allows simultaneous determination of c, k and α at the same transition of the same compound studied in Ref.1 is presented. A comparison of the PE results with data previously obtained in the gas microphone configuration shows an increase in sensitivity.

It has been shown that for an optically opaque and thermally thick sample, the signal amplitude and phase for the PE are given by:

$$A \propto \frac{1}{\sqrt{1+(\omega\tau_e)^2}} \frac{\exp\left(-\sqrt{\frac{2\omega}{\alpha_s}}L_s\right)}{(e_m+e_s)(e_p/e_s+1)} \qquad 1$$

$$\phi = -\tan^{-1}(\omega\tau_e) - \sqrt{\frac{2\omega}{\alpha_s}}L_s \qquad 2$$

where $\omega = 2\pi f$; f is the modulation frequency, τ_e is the electronic cut off frequency; L_s is the sample thickness; e_m, e_p, e_s are the front medium, pyroelectric and sample thermal effusivities.

Comparing the expression given in Eq.1 with the corresponding PA signal amplitude for a thermally and optically thick sample, it is clear that the exponential term acts as an amplification factor near a phase transition where α_s is supposed to vary. Moreover, since PE is a back detection technique all the sample thickness gives a contribution to the signal phase, while in PA this contribution comes only from the sample part contained in a thermal diffusion length. In other words the PE phase sensitivity is given by

$$\frac{d\phi}{d\alpha_s} = \frac{1}{2\alpha_s}\frac{L_s}{\mu_s}$$

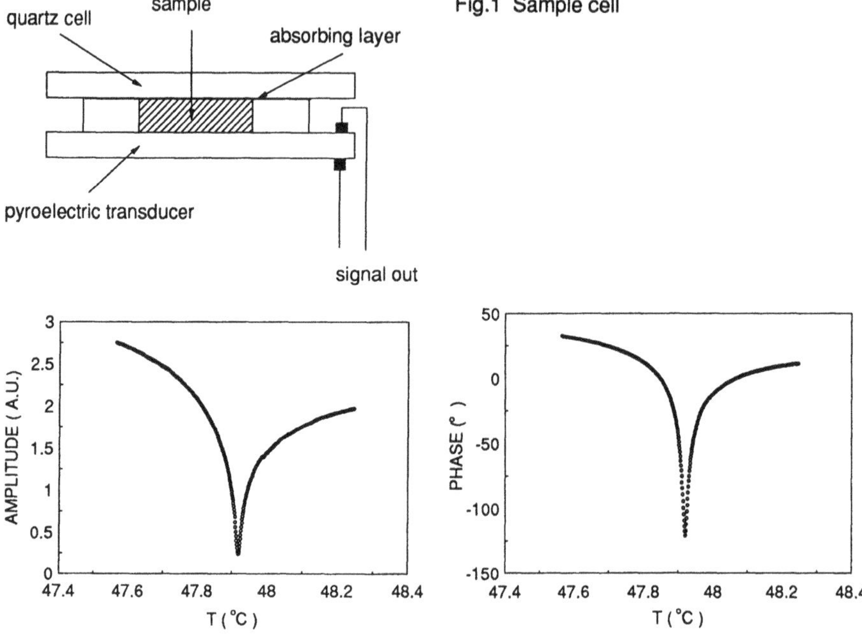

Fig.1 Sample cell

Fig.2 PE signal phase

Fig.3 PE signal phase

where μ_s is the sample thermal diffusion length, can be increased increasing L_s or decreasing μ_s. The ratio L_s/μ_s cannot be made too large, of course, because of the decrease in the PE signal amplitude.

The sample cell, which contains a $40\mu m$ thick 9CB liquid crystal sample is sketched in Fig.1. It was illuminated by 1mW acoustooptically modulated He-Ne operating at $\lambda = 633nm$. Since the sample is transparent to visible radiation, this was possible because of the absorbing layer. The layer consists of Ni-Cr and Au evaporated films, few hundreds of nanometers thick, thus being thermally thin. The transducer was a $280\mu m$ thick $LiTaO_3$ single crystal and the PE signal was analyzed by means of a two phase lock-in amplifier. Data have been taken every mK and the temperature rate change was less than 1 mK/min. The modulation frequency was 89 Hz. This corresponds to a thermal diffusion length in the nematic phase of about $16\mu m$ which is sufficiently small to fulfil the condition of a thermally thick sample. Scanning the temperature across the transition region it is known that the thermal diffusivity has a dip. This ensures that the model remains valid all over the investigated temperature range.

Fig.2 and 3 show the behaviour of amplitude and phase near the transition region. Both show a dip as in the case of the PA. In the last case we had a variation of the signal amplitude of about 70% and a decrease of the phase of about 5° going from temperature far away the transition temperature to T_c. In the present case we have a variation in the PA signal amplitude of 90% as expected because of the exponential term present in Eq.1. More strikingly we have a phase variation of about 150°.Those observations confirm the expected increase in sensitivity obtained with PE with respect to PA.

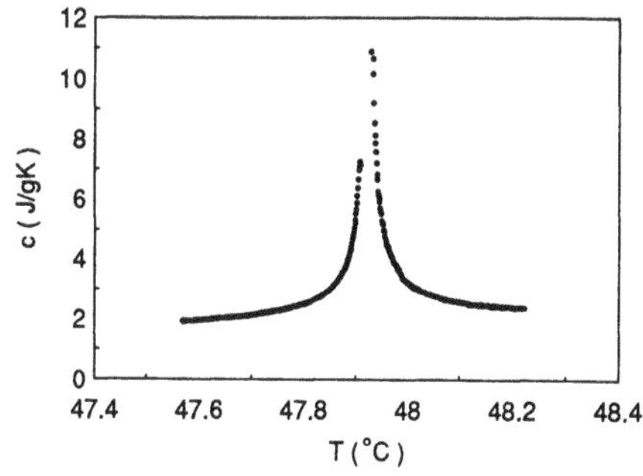

Fig.4 Sample specific heat

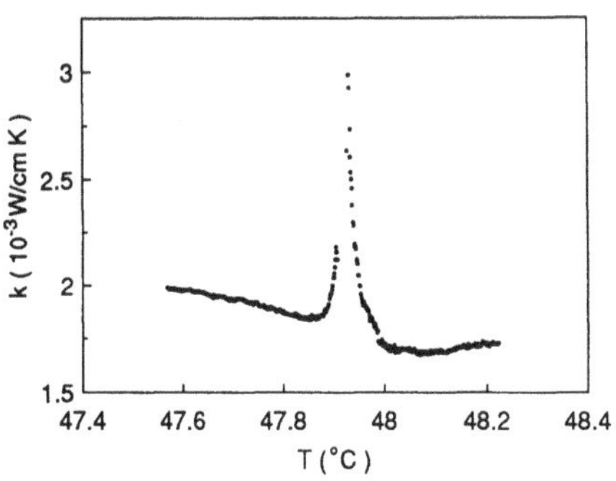

Fig.5 Sample thermal conductivity

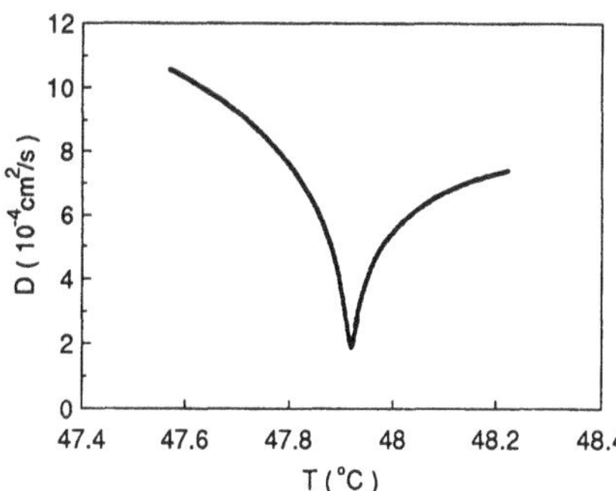

Fig.6 Sample thermal diffusivity

Fig.4, 5 and 6 show the calculated behaviour of c, k and α. Points within a temperature range of 7mk from T_c have been removed because they are affected by rounding off effects, due to residual temperature gradients in the sample. The obtained data are in good agreement with those found in the literature /1/.

REFERENCES

1. U. Zammit, M. Marinelli, R. Pizzoferrato, F. Scudieri, S. Martellucci, Phys. Rev., **A41**, 1153 (1990)
2. Massimo Marinelli, F. Murtas, M.G. Mecozzi, U. Zammit, R. Pizzoferrato, F. Scudieri, S. Martellucci, Marco Marinelli, Appl. Phys., **A51**, 387 (1991)

Photopyroelectric (PPE) Detection of the Antiferromagnetic 3D–2D Phase Transition in Single Crystal La_2CuO_4

D. Dadarlat[1], *R.M. Candea*[1], *A. Frandas*[1], *M. Chirtoc*[1], *and D. Bićanić*[2]

[1]Institute of Isotopic and Molecular Technology, P.O. Box 700, Cluj-Napoca 5, R-3400 Romania

[2]Department of Agricultural Engineering and Physics, Agrotechnion, Wageningen Agricultural University, Bomenweg 4, 6703 HD Wageningen, The Netherlands

Abstract. The 3D-2D antiferromagnetic phase transition in La_2CuO_4 single crystal was detected by the PPE technique. The critical discontinuity of the specific heat found by this method is about 1%.

1. Introduction

As a recent application, the photopyroelectric (PPE) method was used in order to detect phase transitions in ferroelectrics [1,2], ferromagnets [3], itinerant electron antiferromagnets [3] and high T_c superconductors [4]. The capability of the PPE technique for evincing both first and second order phase transitions is due to the dependence of the PPE signal amplitude on the sample specific heat. This dependence is hyperbolical or exponential, according to the sample optical and thermal thicknesses, and to the relative effusivity ratios of the PPE cell components [2,3]. In the thermally thick limit, the exponential dependence type allows for a convenient amplification effect of the signal jump. As a consequence, this case is suitable for investigating phase transitions associated with small specific heat critical anomalies. Such a situation is the 3D-2D antiferromagnetic phase transition in La_2CuO_4 single crystal where the small specific heat critical anomaly could not be evidenced before by classical calorimetric techniques [5].

2. Results

A 0.4mm thick and $8mm^2$ area single crystal La_2CuO_4 was placed on top of a 1mm thick and 10mm in diameter PZT pyroelectric sensor, in intimate thermal contact. The experimental setup used for phase transitions investigations was described in Refs.2,3. The single crystals La_2CuO_4 were

Springer Series in Optical Sciences, Vol. 69 Photoacoustic and Photothermal Phenomena III
Editor: D. Bićanić

grown in air by the top seeded solution growth method using CuO flux [6]. The samples were annealed in vacuum at 900°C for 30 min. The procedure yielded highly insulating samples with a minimal oxygen content.

Due to the optical and thermal parameters of the PPE cell components, for chopping frequencies f>5Hz, both the sensor and the sample are thermally thick. The sample is optically opaque and the following conditions for the thermal effusivity ratios are satisfied: $b_{air,m}$, $b_{air,p} \ll 1$; $b_{m,p} > 1$ (m=material, p=pyroelectric). As a consequence, the PPE voltage amplitude is given by [2,3]:

$$|V| \approx b_{p,m} \exp(-a_m L_m), \tag{1}$$

were $a_m L_m$ is the sample thermal thickness. In the critical region, the sample specific heat presents an anomaly $\Delta c_m / c_m$ accompanied by a jump in the PPE voltage amplitude:

$$\frac{\Delta |V|}{|V|} = - \frac{a_m L_m + 1}{2} \frac{\Delta c_m}{c_m} . \tag{2}$$

Eq.(2) shows that, for $a_m L_m > 1$, a given specific heat anomaly produces an enhanced signal jump.

The experimental results are presented in Fig.1. As one can see, the PPE voltage amplitude presents a jump at T_N=53°C which is enhanced with increasing chopping frequency, as predicted by eq.(2).

Table 1. Experimental signal jump and resulting specific heat critical discontinuity for various modulation frequencies.

f(Hz)	$\Delta\|V\|/\|V\|$(%) (experimental)	$\Delta c_m/c_m$(%) (from eq.(2))
5	1.7	1.0
8	2.0	1.0
15	2.4	1.0
20	3.5	1.3

By using the experimental values for $\Delta|V|/|V|$ at various frequencies, we have derived the relative critical anomaly of the specific heat as being of about 1% (Table 1).

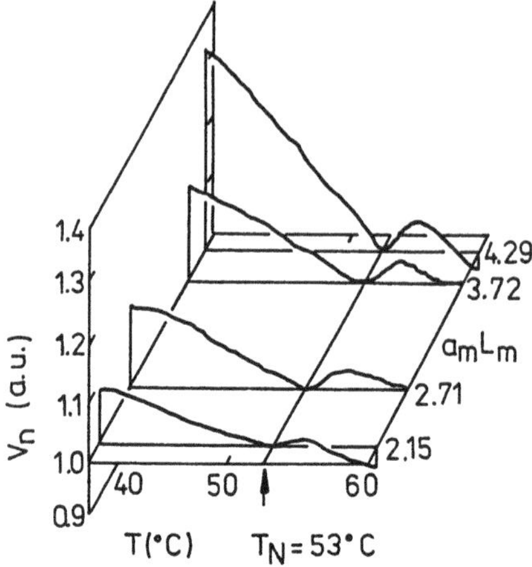

Fig. 1. Normalized PPE signal amplitude $V_n=|V|/|V(T_N)|$ vs. temperature, for various values of the thermal thickness a_mL_m, which is proportional to the square root of the chopping frequency.

3. Conclusions

The 3D-2D antiferromagnetic phase transition in La_2CuO_4 was detected by the PPE technique. This fact was possible due to the capability of the method to enhance the jump of the PPE signal in the critical region, compared to the one of the specific heat. In this experiment, the maximum value for the amplification factor $(a_mL_m+1)/2$ is 2.6, enough for evincing the transition. The discontinuity of the specific heat obtained by this method ($\Delta c_m/c_m \approx 1\%$) is larger than that predicted for single crystals [7], but much smaller than the one found for polycrystalline La_2CuO_4 samples [5].

References

[1] A.Mandelis, F.Care, K.K.Chan and L.C.M.Miranda, Appl.Phys, **A38**, 117 (1985).

[2] D.Dadarlat, M.Chirtoc and D.Bicanic, Appl.Phys. **A50**, 375 (1990).

[3] D.Dadarlat, M.Chirtoc, C.Neamtu and D.Bicanic, J.Phys.Chem. Solids **51**, 1369 (1990).

[4] M.Marinelli, F.Murtas, M.G.Mecozzi, U.Zammit, R.Pizzoferrato, F.Scudieri, S.Martellucci and M.Marinelli, Appl.Phys **A51**, 387 (1990).

[5] K.Sun, J.H.Cho, F.C.Chou, W.C.Lee, L.L.Miller, D.C.Johnston, Y.Hidaka and T.Murakami, Phys.Rev.B, in press.

[6] P.J.Picone, H.P.Jenssen, and D.R.Gabbe, J.Cryst.Growth. **85**, 576 (1987); **91**, 463 (1988).

[7] Jing Rong-Ying, Zhao Guo-Meng and Qi-Ze Ran, Solid State Commun. **67**, 415 (1988).

Acousto-Thermal Detection of the Magnetocaloric Effect

W. Otowski, C. Glorieux, R. Hofman, and J. Thoen

Laboratorium voor Akoestiek en Warmtegeleiding, Departement Natuurkunde, K.U. Leuven, Celestijnenlaan 200D, B-3001 Leuven, Belgium

Abstract We present a new photoacoustic–like detection technique which allows for the rapid detection of small periodic temperature variations induced by small periodic adiabatic magnetic field variations. Results are presented for gadolinium near the Curie point.

From the following Maxwell relation, derived from the first law of thermodynamics for reversible changes in magnetic systems, one can see that periodic adiabatic magnetic field variations induce periodic temperature variations [1] :

$$\left[\frac{\partial T}{\partial B}\right]_S = -\left[\frac{\partial M}{\partial S}\right]_B = -\frac{T}{C_B}\left[\frac{\partial M}{\partial T}\right]_B , \tag{1}$$

with T the temperature, B the magnetic field, M the magnetization, S the entropy and C_B the heat capacity at constant field.

Standard measuring techniques for this magnetocaloric effect use rather large steplike or pulsed fields and temperature detection with long time constant temperature sensors (e.g. thermistors) in contact with the sample [2]. Our new method allows for the rapid detection of small periodic temperature variations at the surface of the sample induced by small periodic field variations without a temperature sensor in direct contact with the sample. The sample is placed in a small gas (air) tight cell (Fig.1) and subjected to a periodically changing magnetic field of the following form:

$$B(t) = B_0 + B_1 \exp(i\omega t) \tag{2}$$

with B_0 a static field (from permanent magnets) and $B_1 (<< B_0)$ the amplitude of the harmonic field variations (from coils). The frequencies normally used are in the range

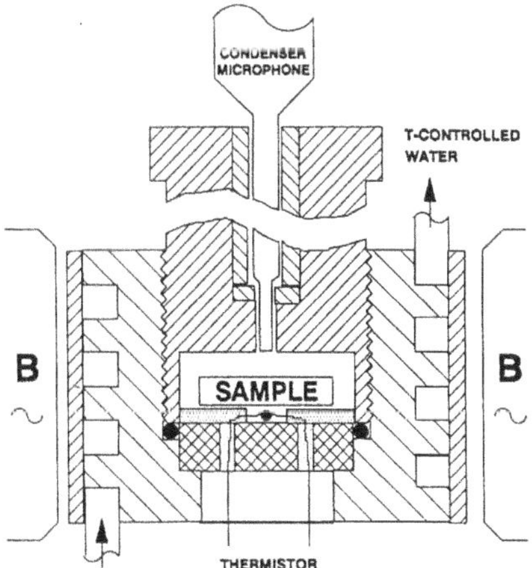

Fig.1 Schematic diagram of the acoustical cell.

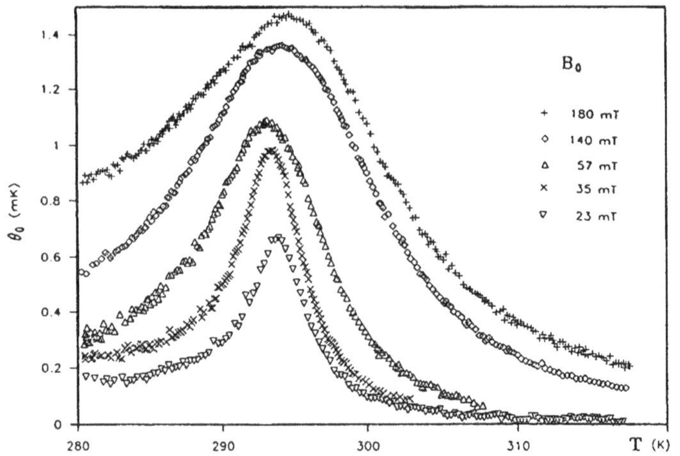

Fig.2 Temperature dependence of the amplitude θ_0 for different values of B_0

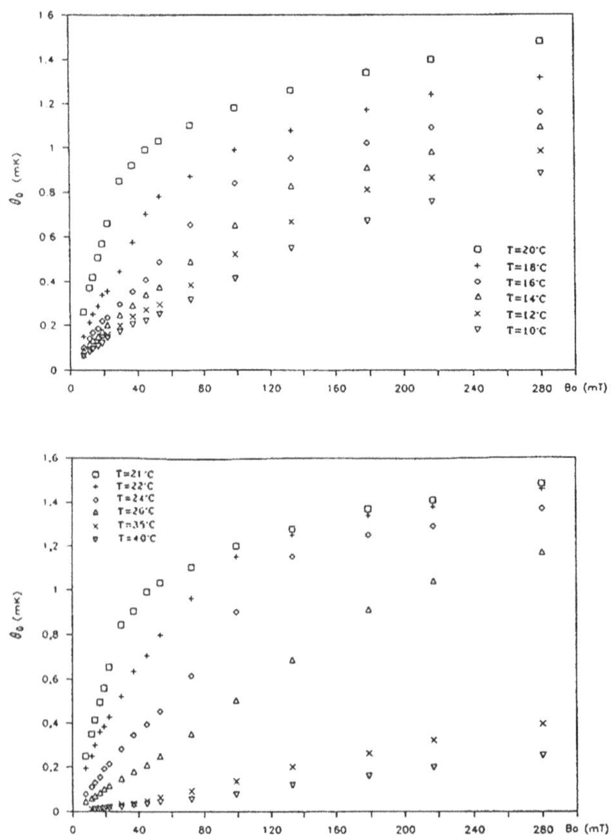

Fig.3 Field (B_0) dependence of the amplitude θ_0 for several isotherms above and below the Curie temperature (20.9° C) of gadolinium.

between a few Hz and a few tens of Hz. The temperature variations at the surface of the sample result in an exponentially decaying thermal wave in the surrounding gas which causes periodic pressure changes in the cell. These pressure variations can be detected with a sensitive microphone in the wall of the cell. The heat loss from the sample is very small over the time interval of the period of the field change, thus one can assume to work in a nearly adiabatic situation. The use of periodic field variations offers the advantages of lock–in detection resulting in an increased signal–to–noise ratio. The temperature is measured by means of a calibrated thermistor in contact with the cell wall. Temperature control of the sample is achieved by means of a temperature controlled water flow through a helical cavity in the outer wall of the sample holder.

Using the Rosencwaig–Gersho [3] theory for a homogeneously heated sample (with density ρ), subjected to the periodic power $P = M\,(dB/dt)$, results for the amplitude variation in :

$$\theta_0 = B_1 M(B_0)(\rho C_B)^{-1} \tag{3}$$

The amplitude R of the acoustical signal detected by the microphone is proportional to θ_0 and given by relation $R=f_{syst}\theta_0$. The system function f_{syst} depends on the microphone and cell characteristics and is ω and T dependent and can be derived from an electric heating calibration procedure.

We have carried out a series of measurements on a polycrystalline thin film of Gd (0.25x5x9mm) obtained from Johnson Matthey (England) with a purity better than 99.9%. Data have been obtained for θ_0 as a function of temperature for 16 values of the static magnetic field B_0. For the amplitude B_1 a value of 1.1 mT was chosen which was still about 7 times smaller than our lowest B_0 value (8 mT). In Fig.2 an overview of θ_0 as a function of temperature (near the Curie point at 20.9°C) is given for different choices of B_0. The temperature resolution in the present configuration is of the order of 5 μK.

From equation (3) it follows that θ_0 is proportional to the induced (by the static field B_0) magnetization M and inversely proportional to the product of the specific heat capacity C_B and the density ρ. M and C_B are both functions of T and B_0. In Fig.3 data above T_c as well as below T_c are in qualitative agreement with (3), in the sense that θ_0 increases as B_0 increases similarly to M(B) isotherms. However, below T_c the $\theta_0(B_0)$ isotherms also start from zero and not from a value corresponding with the spontaneous magnetization. We believe this to be a consequence of the polycrystalline nature of our sample, resulting in a zero average spontaneous magnetization. The critical behavior of the data can be described by power laws with the critical exponents related by scaling law relations. For the field dependence along the critical temperature one has :

$$M(T_c,B)=mB^{1/\delta}, \qquad C_B(T_c,B)=aB^{-\alpha/\beta\delta}+C_0 \tag{4}$$

with C_0, the non-magnetic background heat capacity, a and m critical amplitudes, and δ, β and α critical exponents [1]. Combining these equations with (3) gives for the field dependence of θ_0

$$\theta_0 = \rho^{-1}B_1 m B_0^{1/\delta}(aB_0^{-\alpha/\beta\delta}+C_0)^{-1} \tag{5}$$

In the limit $B_0\to0$ and neglecting C_0 gives $(\alpha+\beta)/\beta\delta$ as a critical exponent. From our data for $B_0>25$mT we obtain a value of 0.22 for the critical exponent $(\alpha+\beta)/\beta\delta$ which is in between the Ising (0.278) and the Heisenberg (0.142) value. The data below 25mT show substantial curvature. This is probably due to rounding off effects, possibly related also to demagnetization effects.

Since according to equation (3) θ_0 and M are proportional, one can combine the power law expressions

$$C_B(T,B=0)=A_\pm\left|\frac{T-T_c}{T_c}\right|^{-\alpha}+C_0, \qquad \chi_T(T,B=0)=\left(\frac{\partial M}{\partial B}\right)_T=\Gamma_\pm\left|\frac{T-T_c}{T_c}\right|^{-\gamma} \tag{6}$$

for the temperature dependence of the heat capacity C_B and the susceptibility χ_T (with critical exponent γ) into the following power law expression (assuming C_B and ρ to be field-independent) :

$$\left(\frac{\partial \theta_0}{\partial B_0}\right)_{T,B_0 \to 0} = B_1 \frac{\partial}{\partial B_0}\left(\frac{M}{\rho C_B}\right)_{T,B_0 \to 0} \cong \frac{B_1}{\rho C_B} \chi_T(T,B_0 \to 0) \tag{7}$$

For the limiting exponent ($\psi-\alpha$) (neglecting C_0) we obtain, for the reduced temperature range larger than 4×10^{-3}, a value of 1.15 which is closer to the Ising (1.13) than to the Heisenberg (1.503) value.

We also want to point out here that a straightforward application of equation (1), identifying dT with θ_0 and dB with B_1 results in a temperature and field dependence of θ_0 which is inconsistent with the experimental results. For the critical isotherm e.g. one can show [4] that $(\partial M/\partial T)_B$ has a power law behavior as a function of the magnetic field with an exponent $(\beta-1)/\beta\delta$. For θ_0 this would give an exponent $(\alpha+\beta-1)/\beta\delta$, which is negative, resulting in a divergence of θ_0 for B_0 going to zero. This is in contradiction with the θ_0 results in Fig.2 showing also θ_0 going to zero when B_0 is going to zero.

References

1. H.E. Stanley, "Phase Transitions and critical Phenomena", University Press (Oxford, 1971)
2. Ponomarev B.K., J. Magn. and Magn. Mat. 61, 129 (1986)
3. Rosencwaig A. and Gersho A., J. Appl. Phys. 47, 64 (1976)
4. Rogiers J., Private communication (1990)

Physical Adsorption Detected by the Mirage Effect

M.Z. Silva[1,*], *Ph. Forge*[1], *F. Lepoutre*[1,**], *and P. Korpiun*[2]

[1]Laboratoire d'Optique Physique, UPRA0005 CNRS, ESPCI, 10, rue Vauquelin, F-75005 Paris, France

[2]Physik-Department E13, Technische Universität München, James-Franck-Strasse, W-8046 Garching, Fed. Rep. of Germany

*On leave from: Dep. de Fisica, Universidade do Minho, 4700 Braga, Portugal

**Present address: ONERA, Division DES/SB, B.P. 72, F-92322 Châtillon Cedex, France

ABSTRACT. The mirage effect is shown to be a quantitative method to study physical adsorption. A theoretical model is developed to take into account the simultaneous thermal and concentration gradients which contribute to the mirage deflection. To test this model we used samples of reduced TiO_2 and glass and a mixture of air with an organic vapor condensable close to the normal temperature and pressure. From the signal amplitude and phase variations with the vapor pressure, the probe beam-surface distance and the square root of the modulation frequency it is possible to deduce the adsorption isotherm.

In the last few years several experiments /1-4/ have demonstrated that the photothermal methods are sensitive to adsorption on solid surfaces.

The periodic heating of a sample caused by absorption of a modulated light beam induces a periodic desorption of molecules from the solid surface when in presence of an adsorbable vapor. This creates a concentration gradient near the sample surface that deflects a probe beam skimming the surface (mirage effect).

In our experiments the solid sample is in a closed cell where we let in a mixture of air and a gas condensable at conditions close to the normal temperature and pressure. Let the mole concentrations of the two gases be C_A and C_B, respectively, and T the temperature of the experiment. A probe beam at a distance z from the sample surface is deflected by the gradient of the gas refractive index n produced by both the temperature gradient $\partial T/\partial z$ and the molecular gradients $\partial C_A/\partial z$ and $\partial C_B/\partial z$:

$$\Phi_n = \frac{1}{n}\frac{\partial n}{\partial T}\frac{\partial T}{\partial z} + \frac{1}{n}\frac{\partial n}{\partial C_A}\frac{\partial C_A}{\partial z} + \frac{1}{n}\frac{\partial n}{\partial C_B}\frac{\partial C_B}{\partial z} \tag{1}$$

Using the expression for the refractive index of a binary mixture /5/ and assuming a low modulation frequency is used so that the sound wavelength in the cell is much larger than the cell length, meaning that the total mole concentration $C = C_A + C_B$ is independent of the position z, we can write for the mirage deflection:

$$\Phi_n = -\frac{1}{n}\frac{RT_0}{P_0}\left\{ [C_A\,(n_A^\circ\text{-}1) + C_B\,(n_B^\circ\text{-}1)]\,\frac{1}{T}\frac{\partial T}{\partial z} - (n_B^\circ - n_A^\circ)\,\frac{\partial C_B}{\partial z} \right\} \tag{2}$$

where R is the ideal gas constant and the superscript ° denotes quantities evaluated at standard temperature (T_0) and pressure (P_0).

To calculate $\partial T/\partial z$ and $\partial C_B/\partial z$ we have assumed that the gases are perfect, no condensation occurs on the cell window and the Clapeyron law is valid. Then (2) can be written:

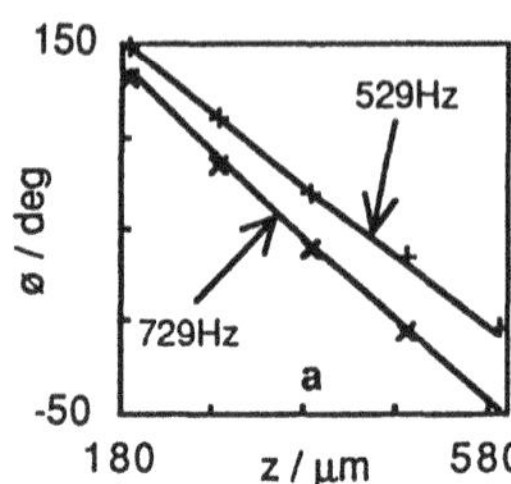

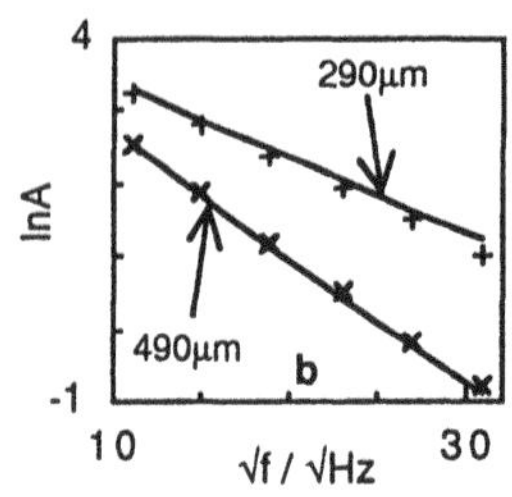

Fig. 1: For TiO_2 sample with air/acetone mixture (X = .9); + and x : experimental; — from eq.(3) with tabulated value of D /7/ and $Y_m = 5x10^{-5}$ mole/m^2: (a) ø(z) when x = .95; (b) lnA($\sqrt{f}$) when x = .16.

$$\Phi_n = -\frac{(1+j)A_gRT_o}{nTP_o}\left\{[C_A(n_A°-1)+C_B(n_B°-1)]\frac{e^{-(1+j)\frac{z}{\mu_g}}}{\mu_g} - C_B(n_B°-n_A°)G\,\frac{e^{-(1+j)\frac{z}{\mu_D}}}{\mu_D}\right\} \tag{3}$$

μ_g and μ_D are related to the thermal diffusivity α_g and the diffusion coefficient D by $\mu_g = \sqrt{\alpha_g/\pi f}$ and $\mu_D = \sqrt{D/\pi f}$, A_g is the solid surface temperature and G is a function of: L_{st} the isosteric heat of adsorption, $(\partial Y/\partial P_B)_T$ with Y the adsorbed amount of B per unit surface - P_B the partial pressure of B, X the mole fraction of B in the mixture, μ_D and T.

By measuring the amplitude A and phase ø of the mirage signal as function of the total gas pressure P we obtain experimental curves which can be interpreted with the theoretical equation (3) to recover the adsorption isotherm $Y(P_B)$. Additional information on the adsorption process can be obtained from the study of ln(A) and ø as functions of z and $\sqrt{f}$.

The experimental setup has been specially prepared in order to verify the hypothesis of the model: - to produce the adsorption preferentially on the sample surface, the sample holder is cooled a few degrees below the cell temperature which itself is maintained at room temperature; - to keep the gas and the adsorbed phase close to equilibrium the gas mixture is admitted very slowly in the cell.

In a first approach we take L_{st} = const. ≅ L_o (the vaporization heat) - valid in the multilayer adsorption domain - and we use for $Y = f(P_B)$ an equation derived from an adsorption model, the B.E.T. model /6/.

Using a sample of reduced TiO_2, with a vapor pressure P_B close to the saturation pressure P_{sat} ($x = P_B/P_{sat}$ = .95) the concentration term dominates the mirage effect; ø and ln(A) vary lineary with z and from their slope we get the interdiffusion coefficient of the two gases D at the given pressure (Fig. 1a). By fitting the experimental curves ø($\sqrt{f}$) and lnA($\sqrt{f}$) to the theoretical expression (3) we can determine the parameters of the adsorption equation $Y(P_B)$, in particular Y_m, the adsorbed amount in the first monolayer. For the TiO_2 sample, with the experimental results for x = .95, we obtained $Y_m = 5x10^{-5}$ mole/m^2. Introducing this value in eq.(3) we can reproduce the results for other experimental conditions, even when the thermal contribution dominates (Fig. 1b).

Using a less adsorbing sample (Shott glass RG830) we find that the thermal contribution dominates even close to the saturation pressure. We are sensitive to this small adsorption and found $Y_m = 2x10^{-7}$ mole/m^2 for this sample.

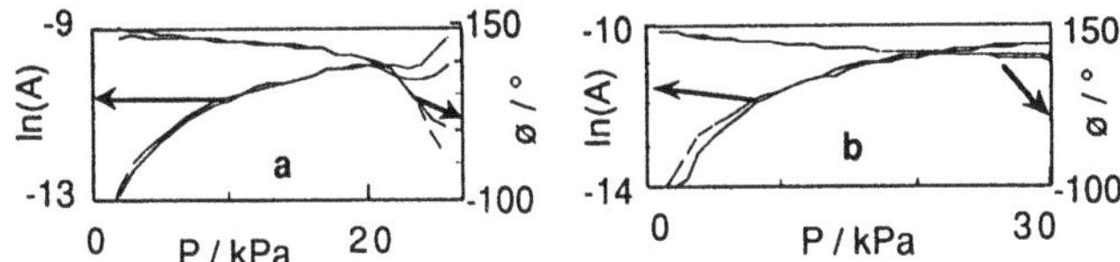

Fig. 2: lnA(P) and ø(P) for air/acetone mixture (X = .9);— experimental; - - eq.(3): (a) TiO_2 sample; (b) Shott glass RG830

Figure 2 presents the variation of lnA and ø with the total pressure P for the two samples used. For the TiO_2 sample there is an amplitude minimum accompained by the rapid phase shift when x approaches 1 indicating the transition between a dominant thermal contribution (low pressures) and a dominant concentration effect (high pressures).

The agreement between the experimental results and the theoretical expression (3) is not perfect in the case of the TiO_2 sample when $x \rightarrow 1$. There are several possible causes for this: i) The B.E.T. equation predicts too much adsorption when $x \rightarrow 1$ /6/; ii) A large adsorption on the cell walls can change the gas mixture composition by reducing the vapor mole fraction.

REFERENCES

/1/ T.Somasundaram, and P.Ganguly, *J.Phys.(Paris)*, 44, C6-239 (1983)
J.Srinivasan, R.Kumar, and K.S.Gandhi, *Appl.Phys.* B43, 35 (1987)

/2/ P.Korpiun, *Appl.Phys.Lett.* 44, 675 (1984)

/3/ F.Lepoutre, G.Rousset, V.Plichon, and N.Rollat, in *Photoacoustic and Photothermal Phenomena*, P.Hess, J.Pelzl eds.,(Springer, Berlin, 1988), p187

/4/ P.Korpiun, R.Osiander, H.Schmitt, and W.Micheler, to be published

/5/ M.Born and E.Wolf, *Principles of Optics*, (Pergamon, London, 1959), p.86

/6/ A.W.Adamson, *Physical Chemistry of Surfaces* (Wiley, N.Y., 1982), p.517

/7/ Landolt-Börnstein, *Zahlenwerte und Functionen aus Physik, Chemie, Astronomie, Geophysik, Technik* , (Springer, Berlin, 1961), Vol.2, part 5a

Part VIII

Surfaces, Interfaces, Thin Films, and Heterogeneous Media

Photothermal Detection at Surfaces and Interfaces: Developments in Non-Conventional Diagnostics

A. Mandelis

Photoacoustic and Photothermal Sciences Laboratory,
Department of Mechanical Engineering and Ontario Laser
and Lightwave Research Center, University of Toronto,
Toronto, Ontario, Canada M5S 1A4

Abstract. A review of certain photothermal-based non-conventional methodologies, pertaining to surface, interface and near-surface investigations is presented. These methodologies have been developed to satisfy increased requirements for diagnostic systems able to handle more sophisticated and complex materials and processes. The review primarily addresses long-term research and development toward non-conventional instrumentation and detection methodologies, centered on frequency-multiplexed photothermal impulse response generation and on photopyroelectric spectroscopy and detection. It also gives a brief introduction to new, promising non-conventional applications of well-known photothermal methods, highlighting a recent method for photoacoustic depth profiling of continuously inhomogeneous samples and the concept of rate-window transient spectrometry. Finally, it informs about the design and implementation of a new hydrogen gas sensor, the first photothermal solid-state device.

1. Introduction

As the field of Photoacoustic and Photothermal technologies advances toward maturity, more sophisticated, complex, and demanding applications must be addressed, which frequently require non-conventional approaches to signal generation and/or detection. A case in point is the ever increasing demand for experimentally simple, fast, reliable, non-destructive and non intrusive thermal wave analysis of surfaces, interfaces and near-surface material regions [1]. Constraints imposed by the geometry, degree of inhomogeneity, speed of response and possible optical damage threshold of condensed phase materials can be best satisfied by non-conventional experimental methodologies and/or new analytical tools for non-destructive characterization of surfaces and at surface-transducer interfaces. In this review I will survey i) non-conventional photothermal signal generation and detection methods introduced in our laboratory in order to measure thermal, thermodynamic, electronic and spectroscopic properties of surfaces and near-surfaces; ii) a few non-conventional applications of well-established photothermal techniques; and iii) new advances in photothermal solid-state surface sensitive device technology.

2. Non-Conventional Instrumentation Methodologies

2.1 Frequency-Multiplexed Photothermal Impulse Response Generation

A very powerful and useful photothermal technique, intermediate between time and frequency domains was introduced [2], capable of recovering the impulse response from materials non-destructively, without laser pulsing. This frequency-multiplexed, time-delay domain (FM-TDS) method depends on the generation of chirped (frequency-modulated, FM) intensity profile of the exciting laser beam intensity, coupled with cross-correlation and spectral photothermal analysis [2], to produce the transient impulse response of the sample while using a cw laser, which affords much higher optical damage thresholds than ordinary pulsed lasers. The FM-TDS method is based on a linear frequency sweep of the excitation function and, when applied to acoustical measurements has been shown to out-perform any other time selective technique with respect to noise rejection and nonlinearity suppression [3]. Comparisons between the TDS method and an m-series pseudo-random binary sequency (PRBS) modulation method [4] in terms of their spectral coherence functions and transfer functions as implemented in a photothermal beam deflection (Mirage effect) instrument, clearly demonstrated superior performance of the FM time delay spectrometer [5]. This excitation and detection methodology has been since utilized with a variety of photothermal schemes, including the photopyroelectric (PPE) technique [6] (see Sec. 2.2) and the laser thermoreflectance technique [7]. Very recently, the inherent high sensitivity of the FM-PPE technique was used as a back-surface photothermal diagnostic to measure small variations in the thermal diffusivity of high purity aluminum (99.99% pure) and correlate the results with the degree of recrystalization of the specimen [8]. For cold-rolled aluminum the thermal diffusivity was measured to be $\alpha = (3.10 \pm 0.09) \times 10^{-5} m^2/s$ for samples exhibiting the initial stages of grain growth. An important feature of the method is that precise thermal diffusivity measurements can be simply obtained from the temporal position of the peak of the current impulse response of the PPE signal, while the conventional frequency-domain (FD) PPE response requires good instrumental (and time consuming) frequency calibration in order to supply similar quantitative information from the slope of the $\ln | V_N |$ vs. $f^{1/2}$ curves, where V_N is the normalized PPE voltage amplitude and f is the modulation frequency [8]. Perhaps, the strongest impact of an interfacial diagnostic impulse response methodology such as FM-PPE spectrometry (i.e. measuring the spectral frequency-multiplexed response) is its recently proven ability to measure simultaneously the thermal diffusivity, α, thermal conductivity, K, and the specific heat, C, of a sample in contact with the detector. Unlike the related FD PPE method, which was also shown to be capable of simultaneous determination of these three thermal/thermodynamic parameters [9], no calibration is required for the measurements: absolute values for α, K, and C may be obtained without having to employ a reference sample. Results on the high-T_c superconductor $YBa_2Cu_3O_{7-x}$ were easily obtained [10] in the temperature range 50-300 K and the experimental PPE current impulse responses were fitted to a one-dimensional theoretical model of the photothermal Green's function [6]. The thermal diffusivity is determined from the peak delay time τ_p, and the thermal conductivity is obtained from the zero-crossing time τ_o, at which $I(\tau_o) = 0$. The specific heat may then be easily calculated from the relation $C = \frac{K}{\rho\alpha}$ with ρ the specimen density. The values obtained for $YBa_2Cu_3O_{7-x}$ compare favorably [10] with other reported measurements of α, K, and C, which previously required separate experiments. In a recent development, of the frequency-multiplexed photothermal impulse response generation technique, Power introduced amplitude and phase-modulation (AM-PM) wide-band PPE spectrometry [11], capable of isolating specific regions of the sample frequency response spectrum (i.e. the transfer function), with a corresponding high degree of spectral selectivity for analytical applications.

2.2 Photopyroelectric Spectroscopy and Detection

Among the photothermal detection schemes, a pyroelectric detector-based technique, photopyroelectric spectroscopy was developed in recent years [12,13], the popularity of which is steadily increasing as described in a recent review [14]. The advantages of this non-conventional detection technique are detector compatibility with most media - UHV, air, most liquids and solids -, high detection sensitivity when combined with compensation (zero background) techniques, open sample holder design and multiple degrees of freedom concerning sample geometry, very low cost (*ca.* \$0.12/detector active element) and highest possible time resolution, when the pyroelectric thin-film polymer polyvinylidene fluoride (PVDF) is used.

In terms of the use of PPE detection for surface kinetic studies, the technique has been applied spectroscopically to the detection of fractions of a monolayer (e.g., 5×10^{-3} of a monolayer for the NH_3-Ag system [15]. Furthermore, without employing compensation or signal averaging techniques, a sensitivity corresponding to a surface coverage of 2×10^{-3} of a monolayer of SF_6 on silver was achieved under UHV conditions [16]. Pyroelectric *in-situ* detection of the reduction of a film CuO surface oxide layer into metallic Cu by diluted HCl has further demonstrated the ability of the method to monitor interfacial chemistry and measure rate constants. [13]

The back-surface photothermal wave detection character of the PPE technique has allowed spatially resolved scanning imaging [17]. Thermal wave information was obtained with a PVDF film sensor unelectroded on one side facing a thin metal probe tip. A modulated photothermal source on the surface of the sample induced an oscillating temperature field in the pyroelectric material, which produced a displacement current proportional to the temperature change. The effects of this current on the charge density of the unelectroded surface were monitored as a synchronous change in the space capacitance between the PVDF film and the probe tip. The oscillating capacitive voltage was thus used to provide a means for measuring the *local* values of laser-induced diffractive [18] and/or interferometric [19] thermal wave fields in metallic samples. Quantitative comparisons with a diffraction-integral approach of the thermal wave field at the back surface of a sample [20] have been presented [19]. Coherent and anti-coherent thermal-wave interference and diffraction profiles at the back surface of a homogeneous sample, entirely analogous to Young's optical-wave experiment, have been produced [19], using two exciting laser beams generating thermal waves at the front surface in-phase (coherent) or out-of-phase (anti-coherent). The latter configuration was shown to be of optimal sensitivity to the presence of sub-surface defects, due to the self-compensating destructive interference of the two anti-coherent thermal wave sources in PVDF.

Using the aforementioned instrumentation, in a recent development we demonstrated PPE tomographic generation and reconstruction of a material cross-section with sub-surface defects using a ray-optic algorithm to map the thermal diffusivity function (see Figures in Ref. [21]). The results are encouraging for near-surface imaging and non-destructive evaluation, and suggestive of the importance of improved image resolution using the diffractive nature of thermal waves. A theoretical basis for PPE diffraction tomography has been further developed [22] in the limit of a perturbative first order Born (or Rytov) approximation for laterally translationally invariant solids and near-surface defects in solids. The ranges of validity of the pertinent two-dimensional spatial frequency/thermal wavenumber domain complex plane contours have been investigated in terms of the existence of inverse spatial Laplace transforms in the mean-square sense. The most important result in near-surface PPE diffraction tomography is the construction and proof of a Laplace Diffraction Theorem (LDT) relating spectrally (i.e. in spatial frequency domain) the two-dimensional Laplace transform of the Born field over the scanned input line and detection line coordinates (x_t, x_l), to the two-dimensional Laplace transform of the object scattering field $F(r_o)$ over the coordinates of the probe region R_o. The Laplace Diffraction Theorem links the transform of the (experimentally measurable) scattered thermal-wave field to the transform of the scatterer to be determined tomographically, modified by the spectral distribution of the incident photothermal aperture A_o, represented by the laser beam waist intensity profile at the material surface of incidence. The most severe shortcoming of the present theory impacting experimental situations is expected to be the Born (or Rytov) assumption of weakly scattering field, needed to validate the LDT. Extensive discussions of these approximations have been presented elsewhere [23].

Photopyroelectric spectroscopy (PPES) has been very successful in measuring very low absorbances in thin surface films using high-fluence laser sources [12], or with modest lamp fluences using the out-of-phase (quadrature) lock-in component at a fixed modulation frequency, by virtue of the fact that all direct transmission and background contributions affect the in-phase component only [24,25]. A combination of out-of-phase PPES (sample-PVDF coupled) and optical transmission spectroscopy (sample-PVDF uncoupled) yielded quantitative optical absorption coefficient spectra of amorphous Si thin films deposited on crystalline Si substrates [26] with instrumental sensitivity limit of $(\beta l)_{min} \approx 5 \times 10^{-3}$. This is very good in the light of $(\beta l)_{min} \approx 10^{-4}$ with the Mirage effect technique [27] by immersing the sample in CCl_4. The PPES advantage lies in the simplicity of the experimental set-up and the avoidance of a liquid interface, especially where delicate microelectronic and optoelectronic structures are concerned. Layers as thin as 500 Å were measurable with the possibility of improved $(\beta l)_{min}$ detectivity.

Finally, I wish to point out that non-radiative quantum yield, $\eta_{nr}(\lambda)$, spectra can be easily measured using PPES, an arduous task if PAS were to be used instead [28]. This can be done with thin surface films by combining high-modulation-frequency (substrate thermally thick limit) spectra with low-frequency (substrate thermally thin regime) spectra and deducing $\beta(\lambda_j)$ and $\eta_{nr}(\lambda_j)$ from the appropriate equations [24], in the sense of a system of two equations with two unknown parameters [29]. The same task can be accomplished using similar methods with thick samples (e.g. Ge crystals) [30], or from a single modulation frequency (thermally thin limit) [31] by use of the equations governing the in-phase and quadrature channels of the output normalized PPE signal [24]. The most interesting outcome of the $\eta_{nr}(\lambda)$ investigations in electronic solids has been the discovery of large spectral variations, indicating a wealth of underlying physics and calling attention to this quantity with regard to the usual simplifying assumptions about its value and its constancy.

3. Non-Conventional Applications of Conventional Methodologies

In this section I will identify two new promising photothermal methodologies implemented on near-surface well-established photothermal techniques.

3.1 Photoacoustic Depth Profiling of Continuously Inhomogeneous Condensed Phases

A theoretical development of the Hamilton-Jacobi formulation of thermal-wave physics [32] has recently led to an analytical, simple inverse problem solution of the PA frequency response of inhomogeneous condensed phases with arbitrary, continuously varying thermal diffusivity profiles. The method exhibits excellent profile reconstruction fidelity (See Figures in Ref. [33]). It has been used to obtain quantitative depth profiles of thermal diffusivity decreases extending 20-30 μm below the surface of the liquid crystal 8 CB, as a result of observed changes in the photoacoustic signal frequency response upon the application of a transverse magnetic field in the nematic phase at 37°C. These decreasing profiles are a measure of the extent of bulk reorientational effects due to the magnetic field, as well as the extent of the influence of the surface as a domain reorientation inhibitor. Ongoing work in this area includes investigations in the stress and plastic deformation profiles and mechanisms of steels, surface treated with shot-peening or laser hardening.

3.2 Rate-Window Transient Thermomodulation Spectrometry

This novel photothermal methodology can be implemented with various detection schemes and is suitable for the characterization of thermal and electronic transport parameters of wide classes of materials. In our original measurement set-up with a conventional photothermal reflectance (PTR) system (See Ref. [34]), a phase-sensitive, two-channel lock-in amplifier with fast (ms) time constant and an electronic filter with frequency tracking capability is referenced at a given frequency and tracks the temporal evolution of the Fourier components of the photothermal response of a sample during and after the end of an optical excitation pulse. Using a "rate-window" concept [35], the instrumental output produces a maximum when the reference frequency equals a characteristic system frequency, f_s, which may depend on thermal and/or electronic carrier transport near-surface processes, as well as on pump beam spot size. Recent applications with PPE detection have produced purely thermal-wave transport information and have demonstrated the utility of the lock-in as a frequency filtering instrument.

4. Advances in Photothermal Solid-State Device Technology

A recent development is a thin-film photopyroelectric sensor, a new type of solid-state device for the detection of minute concentrations of hydrogen gas under environmental, flow-through conditions [36]. The sensor was made of thin, commercially available PVDF film, sputter-coated with Pd. An infrared beam from a semiconductor laser diode was coupled to an optical fiber and irradiated the PVDF, thus generating a carrier wave ac photopyroelectric voltage. Sensitivity to exposures in a hydrogen partial pressure has been demonstrated down to concentrations as small as 40 ppm, in a flowing $H_2 + N_2$ mixture.

For a detailed photothermal electrostatic consideration of the Pd-pyroelectric junction H_2 sensor see Ref. [37]. Experimental evidence is in agreement with the fundamental features of the theory, which supports two possible mechanisms of ac-mode device operation: pyroelectric coefficient dependence on the hydrogenic dipole induced charge density at the Pd-insulator (polyvinylidene fluoride, PVDF) interface, and Pd work-function shift due to hydrogen absorption. The dominant operating mechanism is over thermostatistical vibration and libration in the pyroelectric matrix was successfully used to explain the temperature dependence of the photopyroelectric signal.

The implications of these observations will be discussed in terms of sensor design and sensitivity optimization for this first photothermal solid-state device.

References

[1]. Hess, P. (Ed.), *Photoacoustic, Photothermal and Photochemical Processes at Surfaces and in Thin Films, Topics in Current Physics*, Vol. **47**, Springer-Verlag, Berlin, (1989).

[2]. Mandelis, A., IEEE Trans. UFFC - **33**, 590 (1986).

[3]. Biering, H. and Pedersen, O.Z., Brüel and Kjaer Tech. Rev. **1**, 5 (1983).

[4]. Sugitani, Y., Uejima, A. and Kato, K., J. Photoacoust. **1**, 217 (1982).

[5]. Mandelis A., Borm L.M-L. and Tiessinga, J., Rev. Sci. Instrum. **57**, 630 (1986).

[6]. Power J.F. and Mandelis A., Rev. Sci. Instrum. **58**, 2018 (1987); **58**, 2024 (1987); and **58**, 2033 (1987).

[7]. Mandelis, A. and Power, J.F., Appl. Opt. **27**, 3397 (1988); and **27**, 3408 (1988).

[8]. Peralta, S.B., Ellis, S.C., Christofides, C., Mandelis, A., Sang, H. and Farahbakhsh, B, Res. Nondestr. Eval. **3**, 69 (1991)

[9]. Marinelli, M., Murtas, F., Mecozzi, M.G., Zammit, U., Pizzo-ferrato, R., Scudieri, F. Martellucci, S. and Marinelli, M., Appl. Phys. A **51**, 387 (1990).

[10]. Peralta, S.B., Chen, Z.H. and Mandelis, A., Appl. Phys. A **52**, 289 (1991)

[11]. Power, J.F., Rev. Sci. Instrum. **61**, 90 (1990); and **61**, 101 (1990).

[12]. Coufal, H., Appl. Phys. Lett. **44**, 59 (1984).

[13]. Mandelis, A., Chem. Phys. Lett. **108**, 388 (1984).

[14]. Coufal, H., and Mandelis, A., Ferroelectrics (In the press)

[15]. Coufal, H., Trüger, F., Chuang, T.J. and Tam, A.C., Surf. Sci. Lett. **145**, L504 (1984).

[16]. Coufal, H., Chuang, T.J. and Träger, F., IBM Res. Rep. RJ 4344 (1984).

[17]. Mieszkowski, M., Leung, K.F. and Mandelis, A., Rev. Sci. Instrum. **60**, 306 (1989).

[18]. Mieszkowski, M. and Mandelis, A., J.O.S.A. **A7**, 552 (1990).

[19]. Mandelis, A. and Leung, K.F., J.O.S.A. **A8**, 186 (1991).

[20]. Mandelis, A., J.O.S.A. **A6**, 298 (1989).

[21]. Munidasa, M. and Mandelis, A., this Conf. Digest.

[22]. Mandelis, A., J. Phys. A: Math. Gen. **24**, 2485 (1991)

[23]. Kak, A.C., and Slaney, M., *Principles of Computerized Tomographic Imaging*, (IEEE Press, New York, 1988).

[24]. Mandelis, A. and Zver, M.M., J. Appl. Phys. **57**, 4421 (1985).

[25]. Coufal, H., Appl. Phys. Lett. **45**, 516 (1984).

[26]. Christofides, C., Mandelis, A., Engel, A., Bisson, M. and Harling, G., Can. J. Phys. **69**, 317 (1991)

[27]. Tanaka, K. in *Photoacoustic and Thermal Wave Phenomena in Semiconductors*, A. Mandelis, Ed. (North-Holland, New York, 1987) p. 441.

[28]. M. Kitamura, T. Ogawa and T. Arai, J. Phys. Soc. Jpn. **52**, 2561 (1983).

[29]. Mandelis, A., Wagner, R.E., Ghandi, K., Baltman, R. and Dao, P., Phys. Rev. B15, **39**, 5254 (1989-I).

[30]. Christofides, C., Mandelis, A., Ghandi, K. and R.E. Wagner, Rev. Sci. Instrum. **61**, 2360 (1990).

[31]. Christofides, C., Engel, A. and A. Mandelis, Ferroelectrics **118**, 411 (1991)

[32]. Mandelis, A., Math., J., Phys. **26**, 2676 (1985).

[33]. Mandelis, A., Schoubs, E., Peralta, S.B. and Thoen, J. this Conf. Digest.

[34]. Chen, Z. and Mandelis, A., this Conf. Digest.

[35.] Lang, D.V., J. Appl. Phys. **45**, 3023 (1974).

[36]. Mandelis, A. and Christofides, C., J. Vac. Sci. Technol. A **8**, 3980 (1990).

[37]. Mandelis, A. and Christofides, C., this Conf. Digest.

Heat Diffusion at a Micronic Scale

F. Lepoutre[1,*], *S. Bouchoule*[1], *G. Bäckström*[2], *and D. Balageas*[1]

[1]ONERA, Thermophysics Division,
B.P. 72, F-92322 Chatillon Cedex, France
[2]Umea University, Physics Department, S-901 Umea, Sweden
*On leave from: ESPCI, CNRS ER05,
10, rue Vauquelin, F-75005 Paris Cedex 05, France

Abstract. This paper presents several thermal images at a micronic scale obtained using a photothermal microscope. The experimental set-up is based on the modulated laser induced photoreflectance. The spatial resolution is achieved, optically by focusing the pump and probe beams with a microscope, and thermally by working at modulation frequencies larger than 1MHz. The phase images are thermal maps that reveal the local variations of the thermal diffusivity and the thermal barriers.

1. Introduction

The Non Destructive Evaluation and the process control of new materials sometimes need experimental methods of investigation with spatial resolutions of the order of 1 μm to determine the microscopic origins of macroscopic defects. As a possible answer to this problem, we have built a photothermal microscope which can be considered as a thermal probe having a micronic resolution [1], and which is probably the unique method able to characterize the thermal interfaces that exist, for instance, at grain boundaries or between fibers and matrix in composites.

2. Principle of the photothermal microscope

The basic principle used is called the photoreflectance [2]: the optical reflection coefficient of a solid surface is modified by the increase of its temperature produced by the absorption of a light beam (pump). To increase the signal to noise ratio, the pump beam intensity is modulated and the reflectance variations are detected at this modulation frequency. The optical resolution is obtained by focusing the pump beam (down to 0.5μm) on the sample surface with an optical microscope. The modulation frequency must be chosen large enough not to reduce the thermal resolution by heat diffusion during the time of observation (i.e. a period of modulation). The detection of the modulated surface temperature is achieved by a second light beam (probe) focused by the microscope on the center of the heated region and reflected by the sample surface. The part of the reflected probe intensity that is modulated at the frequency of the pump is measured by a photo cell and demodulated by a lock-in amplifier [3].

A sketch of the experimental set-up is shown in figure 1. The pump beam comes from an Ar^+ laser (10mW to 2W) and is modulated by an opto-acoustic modulator (up to 10 MHz). The probe beam (He-Ne laser) reflected on the specimen is sent to a silicon cell through an interference filter and a Faraday isolator which prevents any return into the laser cavity.

Springer Series in Optical Sciences, Vol. 69 Photoacoustic and Photothermal Phenomena III
Editor: D. Bićanić

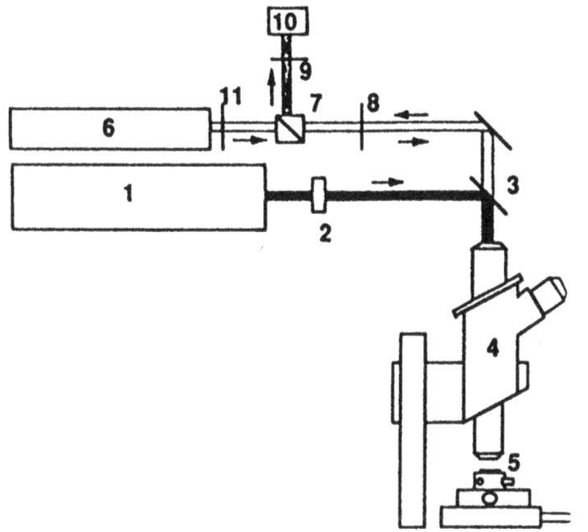

1: Pump (Ar+ laser)
2: Opto-acoustic modulator
3 Dichroic mirror
4: Microscope
5: Specimen
6: Probe (He-Ne laser)
7: Beam splitter
8: Polarizer
9: Interferential filter
10: Photocell
11: Faraday isolator

Figure 1 : Sketch of the experimental set up

The sample held on two translation stages is scanned under the microscope objective. The phase images alone are used because they are purely thermal images (when the absorption is located at the sample surface). On the contrary, the amplitude leads to rather complex images which depend, not only on the thermal local properties of the sample, but also on the local variations of the coefficients of optical reflection (at the probe wavelength) and absorption (at the pump wavelength).

3. Experimental results

The temperature distribution at the sample surface can be calculated very easily in the case of a bulk sample. The comparison of this theoretical result with the measurement of the actual experimental distribution allows to check the experimental set-up. Figure 2 (Co sample) proves that the probe and pump beams are parallel to the microscope axis The comparison with the calculation also gives the value of the pump beam radius r_0. The knowledge of r_0 is essential for the understanding of the thermal images as explained by figure 3. In this figure, the calculated phase of the surface temperature is plotted as a function of the modulation frequency for two values of the sample thermal diffusivity and for two pump beam radii. The first information brought by this calculation is that the phase images will be, at least, a representation of the local variations of the thermal diffusivity α. The second information is that the maximum of the contrast in phase is reached when r_0 is of the same order of magnitude as the thermal diffusion length $\mu=\sqrt{(\alpha/\pi f)}$. To obtain a contrasted thermal map with a 1 µm resolution, it is then necessary to use $r_0<1$ µm and $f \approx \alpha/\pi r_0^2$, α being a mean value of the local thermal diffusivity. For instance, in the case of figure 3, for a beam radius $r_0=1\mu m$, the phase variation is 20° for a thermal diffusivity ranging from $10^{-5} m^2 s^{-1}$ to $10^{-6} m^2 s^{-1}$. If the pump is unfocused (for instance $r_0=5\mu m$), the phase shift between these two positions is reduced to less than 5°.

Figures 4 and 5 show two examples of application of this instrument:

- The sample of figure 4 is a cross section of a bundle in a three directional Carbon-Carbon composite. The phase shift observed between the fibers and the

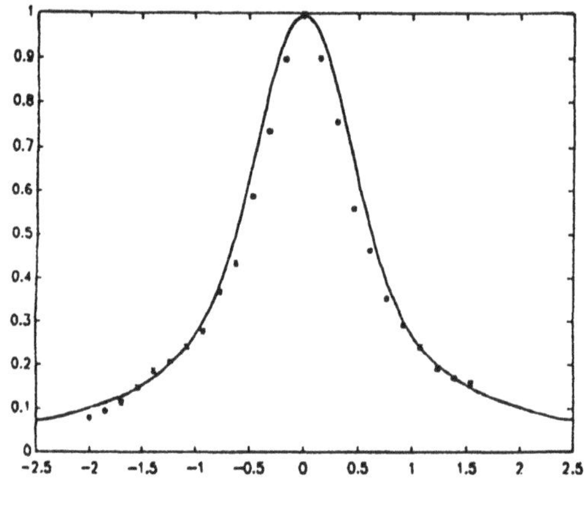

Figure 2 : Amplitude (u.a.) vs offset between probe and pump (μm)

---: calculated(r_o=0.5μm).
x: experiment

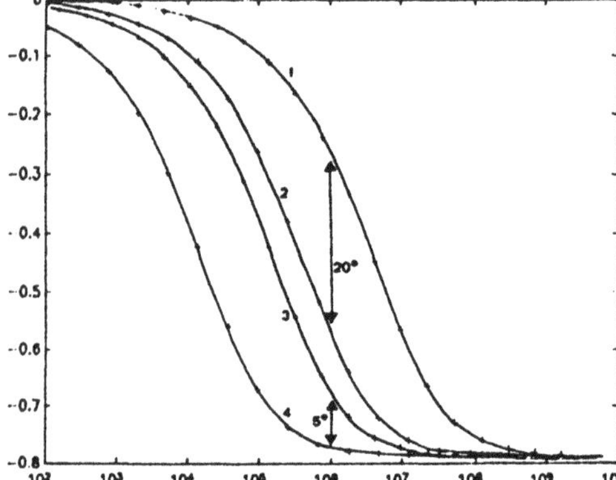

Figure3 : Calculated phase (rd) vs frequency (Hz).
Upper curves: r_o=1μm.
Lower curves: r_o=5μm
Curves 1,3 : a=$10^{-5}m^2s^{-1}$.
Curves 2,4: a=$10^{-6}m^2s^{-1}$

matrix is essentially produced by the difference of their diffusivities because the distances separating the interfaces are larger than the thermal diffusion lengths in our experimental conditions and the fibers are highly anisotropic (radial diffusivity ten times lower than the longitudinal one). This phase variation of about 20 degrees can be related to a thermal diffusivity difference of an order of magnitude, which is in good agreement with diffusivity measurements made on the same material by the flash method [4]. In this reference, for this material the matrix diffusivity was found equal to 2.7 $10^{-5}m^2s^{-1}$ and the axial diffusivity of the bundle (obviously smaller than the axial diffusivity of a fiber) was equal to 3.0 $10^{-4}m^2s^{-1}$. Let us also point out the effect of the unfocusing of the beams in the top and the bottom of the field (here due to a small curvature of the surface) which produces the decrease of the contrast as previously mentionned. The black spot in the right top corner is a small particle of dust. Its very strong contrast with respect to the specimen is the result of its non-contact with the surface . It proves that these images are also very sensitive to the presence of thermal barriers. This image is to be compared to those previously published [5] using the photoacoustic method by photodeformation with a YAG laser as a pump. The quality of the present image is better (enhanced resolution, lower noise) and the principle of the method itself must allow quantitative evaluations in an easier way.

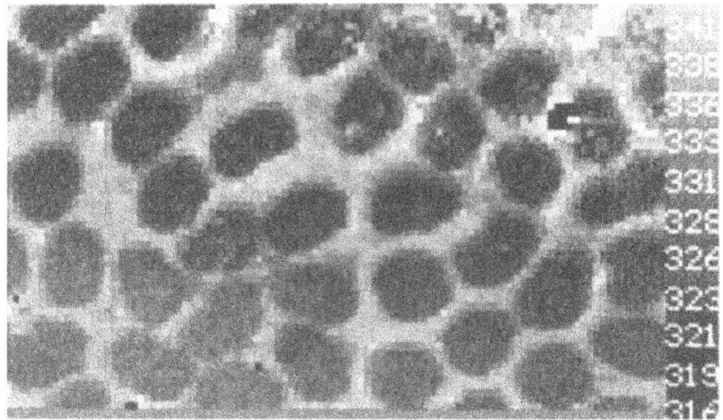

Figure 4 : Carbon fibers in a bundle of a 3-D C/C composite. Frequency: 2MHz. Power: 250mW. Scale: 50µm X 30µm. Step: 0.5µm. Scale : phase in degrees.

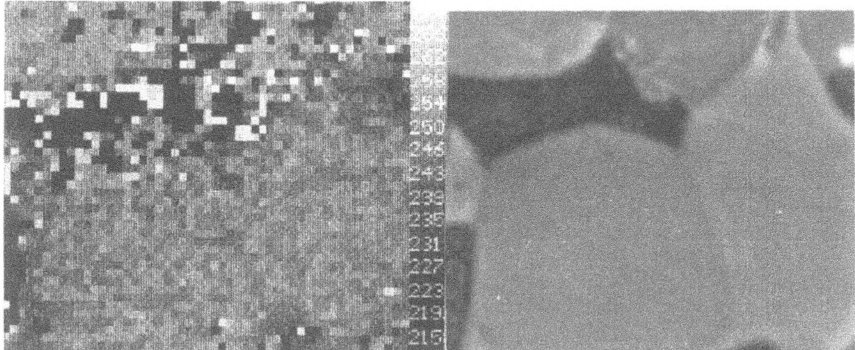

Thermal image (25x25µm) *Optical image (25x25µm)*

Figure 5: AlN ceramic. Thermal image (scale : phase in degrees): Frequency: 2MHz. Power: 100mW. The grains appear in gray. The thermal resistances appear in the dark areas. Optical image: The grains are clear and the secondary phasis is dark.

- Figure 5 is an image of grains in an AlN ceramic. This ceramic was prepared at high temperature (about 2000°C) so that a secondary phasis is present in a few grain boundaries. As in the previous example, these interfaces play an important role in the observed values of the phase. In this particular case, it is possible, with a rather simple model [6], to show that the phase shifts observed at the grain boundaries increase with the thermal resistance of the interfaces. For instance, the two grains in the bottom of the image are in very good contact (no phase shift between them) while, on the contrary, the grains of the lower left part and of the upper left part are separated by a thermal resistance (associated with the secondary phasis) which produces a large phase shift.

4. Conclusion

The images here presented, obtained by photoreflectance, reveal the two main possibilities of the method: diffusivity mapping and interface thermal characterization. Thus, these images are purely thermal, but let us recall that with very few changes in the experimental set-up, the photodeformation of the sample can be detected. This second method must be sensitive to the mechanical properties of interfaces and thus probably leads to a mechanical interface characterization which is of more practical interest. This needs a more complex modelization which is now in progress in our laboratory.

References

[1] L.J. Inglehart, A. Broniatowski, D. Fournier, A.C. Boccara, and F. Lepoutre, Appl. Phys. Lett. **56**,1749 (1990)

[2] M. Cardona, *Modulation Spectroscopy* p.131, Academic Press, New York (1969)

[3] A. Rosencwaig, J. Opsal, W.L. Smith, and D.L. Willenborg, Appl. Phys. Lett. **46**, 1019 (1985)

[4] R.M. Pujolà, and D.L. Balageas, High Temperatures - High Pressures **17**, 623 (1985).

[5]A. Déom, D. Boscher, and D. Balageas, in "Photoacoustic and Photothermal Phenomena II", Springer Series in Optical Sc. **62**,13 (1990)

[6] F.A. Mc Donald, G.C. Wetsel, and G.E. Damieson, Can. J. Phys. **64**,1265 (1986)

Photothermal Investigations of Realistic Interfaces

H.G. Walther[1], *K. Friedrich*[1], *U. Seidel*[1], *G. Busse*[2], *and W. Karpen*[2]

[1]Institut für Optik und Quantenelektronik, Friedrich-Schiller-Universität Jena, Max-Wien-Platz 1, O-6900 Jena, Fed. Rep. of Germany

[2]Institut für Kunststoffprüfung und Kunststoffkunde, Universität Stuttgart, W-7000 Stuttgart 80, Fed. Rep. of Germany

Abstract. The interpretation of photothermal measurements of layered systems demand the modelling of different kinds of interfaces. We found a method to receive the thermal wave reflectivity and the thermal contact resistance of the interface of a system layer/substrate. Furthermore, theoretical and experimental results of the investigation of thick interfaces ('smeared transition' between two materials) are given.

1. Introduction

The detection of hidden boundaries is a significant task for nondestructive evaluation (NDE). Photothermal methods can be used successfully to find boundaries between regions with different thermal properties. To achieve a qualitative interpretation of photothermal signals it is not enough to take into account an impedance discontinuity, one rather needs a mathematical model describing the boundary in a realistic way.

2. Thermal reflectivity and contact resistance

Deposited layers (e.g. galvanic or evaporated coatings, paints, laminates) may have a thermal contact resistance Ω' with respect to the substrate /1/ which is generally correlated with a reduced layer adhesion. The value Ω' is defined by the temperature difference between layer and substrate $T_l - T_s$ and the heat flow through the interface $\Phi_l = \Phi_s$ by $T_l - T_s = \Omega'\Phi_s$. This thermal contact resistance modifies the interference of thermal waves within the layer and generates a significant distortion of the signal phase measured at the sample surface. Fig.1 shows the frequency dependence of the normalized signal phase for a constant normalized contact resistance $\Omega = \Omega'\kappa_l/d$ (κ–thermal conductivity, d–thickness of the layer) for different thermal wave reflectivities $R = (e_l - e_s)/(e_l + e_s)$ with $e = \sqrt{\kappa C}$ (C–heat capacity per volume). Each of these curves belonging to various combinations of R and Ω has certain characteristic features, e.g. the height of the first extremum of phase ϕ_E and the abscissa ratio from the zero crossing X_0 to the position X_E of the phase extremum. In fig.2 curves of constant Φ_E and constant X_0/X_E–ratio are plotted in the R–Ω–plane. Measuring the values of Φ_E and X_0/X_E one can determine R and Ω.

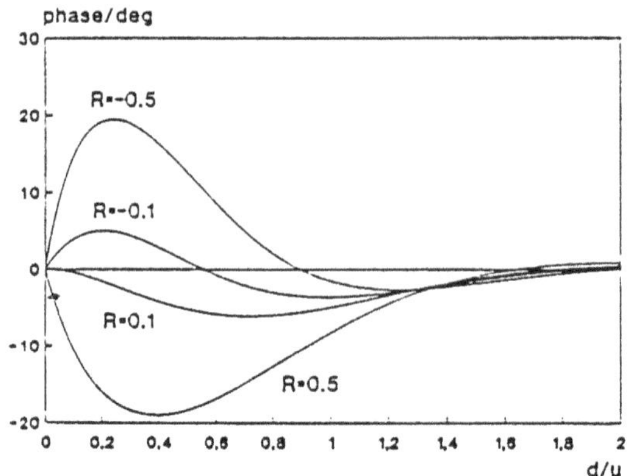

Fig.1: Phase angle difference vs. normalized thickness with the thermal reflectivity R as the parameter and $\Omega = 0.5$.

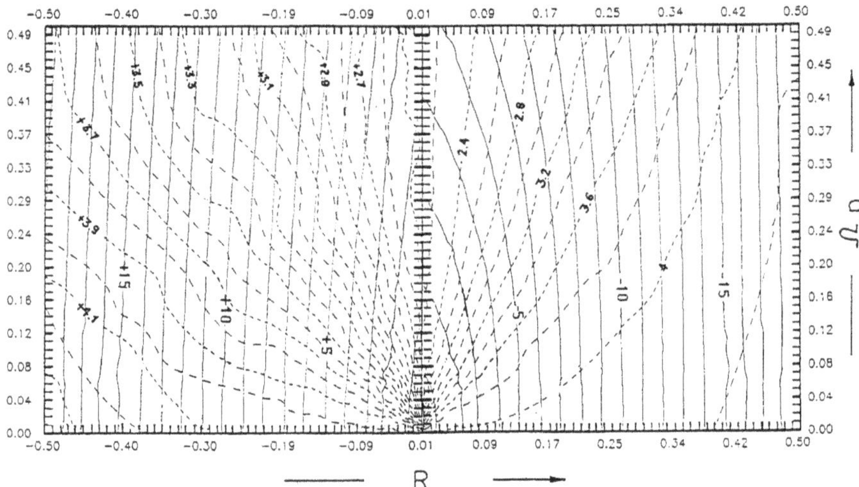

Fig.2: A topographical plot of curves with constant phase extremum Φ_E (continuous line) and constant X_0/X_E-ratio (dashed line). Measuring Φ_E and the X_0/X_E-ratio it is possible to determine the thermal reflectivity R and the normalized thermal contact resistance Ω.

3. Thick interface

Layers generated by interdiffusion or by phase transition may have a smooth curve describing how thermal wave impedance gradually changes. Examples for 'smeared transitions' can be doped semiconductors or surface hardened steel. In all these situations the depth dependence of thermal properties needs to be considered in the heat diffusion equation

$$\frac{\partial}{\partial z}\left(\kappa(z)\frac{\partial}{\partial z}T\right) - i\omega C(z)T(z) = 0. \tag{1}$$

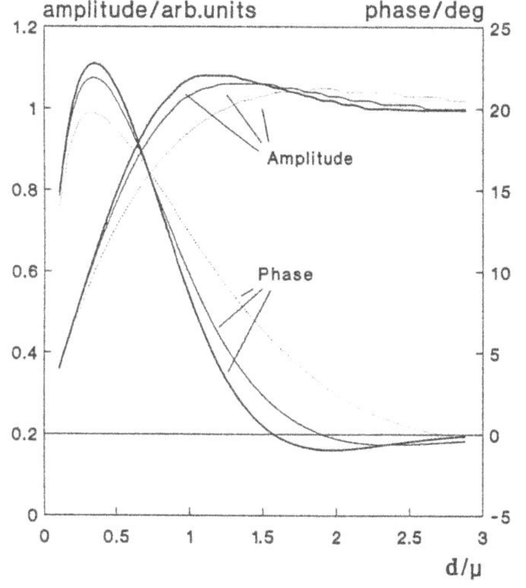

Fig.3: Calculated amplitude and phase curves for a steep transition (thick continuous line) and 'smeared transitions' (thin continuous and dashed lines) of the thermal conductivity κ.

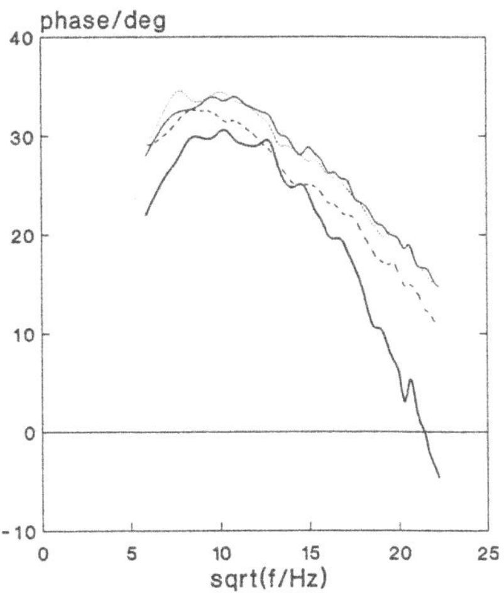

Fig.4: Measured normalized phase curves for paint on rough copper without (thick continuous line) and with (thin continuous and dashed lines) an interface layer.

For linear variation of either thermal conductivity κ

$$\kappa = \kappa_1 + \frac{\kappa_2 - \kappa_1}{z_2 - z_1}(z - z_1) \tag{2}$$

or thermal density per volume $C(z) = (\rho c)(z)$ we found analytical solutions allowing us to calculate the photothermal signals. Theoretical results are presented in fig.3 where changes in the amplitude and phase curves are observed which depend on the width of the thermal transition. Experimental observation of such changes should allow some conclusions about the nature of the transition. Fig.4 shows experimental results that were obtained on paint/copper systems with transition zones fabricated in various ways. The soft impedance changes are correlated with broad curves as is expected from fig.3.

Reference

/1/ Carslaw, Jaeger, Conduction of heat in solids, Oxford (1959), 1.9

Pulsed Photothermal Radiometry to Measure the Properties of Thin Layered Films

A.C. Tam[1], *W.P. Leung*[2], *and H. Sontag*[3]

[1]IBM Research Division, Almaden Research Center, 650 Harry Road, San Jose, CA 95120-6099, USA

[2]IBM Corporation, Storage System Products Division, 5600 Cottle Road, San Jose, CA 95193, USA

[3]Dornier GmbH, W-7990 Friedrichshafen 1, Fed. Rep. of Germany

Abstract. The technique of pulsed radiometry can be used to monitor thermal transients in a layered thin film sample, which can provide information on certain properties such as film thickness, adhesion or tendency for delamination, and composition. In a typical measurement, the surface of the material under investigation is heated by short pulsed radiation, typically from a flashlamp or a laser of moderate energy. This leads to an increase of the thermal emission from the sample, which is detected by a broadband infrared detector. From the decay of the infrared emission signal, information on the thermal properties and on the internal structures can be derived. A practical application is given here for the use of pulsed radiometry to monitor the layered structure and composition of printer ribbon samples.

Introduction

Thermal diffusivity is traditionally measured by contact methods, i.e., methods that require the attachment of electrical leads and/or thermometric probes onto the sample. Such methods suffer from the following disadvantages when used as an industrial quality control monitor: (1) it is slow, requiring attachment, detachment, and clean-up steps; (2) it may cause scratching or contamination and (3) it cannot be used for nonstationary products, like reeling ribbons, or moving parts on a conveying belt.

It is obvious that for industrial quality control applications, noncontact, fast and nondestructive testing methods will be much superior. "Noncontact" means no attachment, detachment or contamination. "Fast" means applicable to moving objects provided that during the time of measurement, the sample does not move significantly. "Nondestructive" means that the integrity of the products is not degraded by the test, and hence they can be tested 100% instead of on a sampling basis.

We describe here the practical application of the established flash radiometry technique [1-4] for noncontact, nondestructive, and fast monitoring of printer ribbon properties in a single-ended measurement scheme. Physical properties that can be monitored include the potential of subsurface delaminations, coat weight variations, thickness variations, changes in carbon loading and moisture absorption.

Experiment

The schematics of the structure of a resistive printer ribbon is indicated in Fig. 1. In such kind of layered film, many possible problems may occur, for example, inhomogeneous composition or abnormal aggregation of carbon particles, poor adhesion between the different layers and excessive variations of composition and thicknesses of the various layers. All these problems affect the thermal diffusivity or its depth dependence.

The simplest boundary condition in the monitoring scheme is obtained by using a short-pulsed radiation source like a flashlamp or a pulsed laser beam, that is strongly absorbed by the material being tested. "Strongly absorbed" means that the heating is restricted to the top layer, while "short-pulsed" means, that during the irradiation time, no appreciable heat diffusion occurs. Although these conditions are not absolutely necessary [2], they simplify the discussion and the mathematical treatment. For our experimental situation, as shown in Fig. 2, we use a pulsed N_2 laser giving laser pulses of 8 ns duration

Springer Series in Optical Sciences, Vol. 69 Photoacoustic and Photothermal Phenomena III
Editor: D. Bićanić

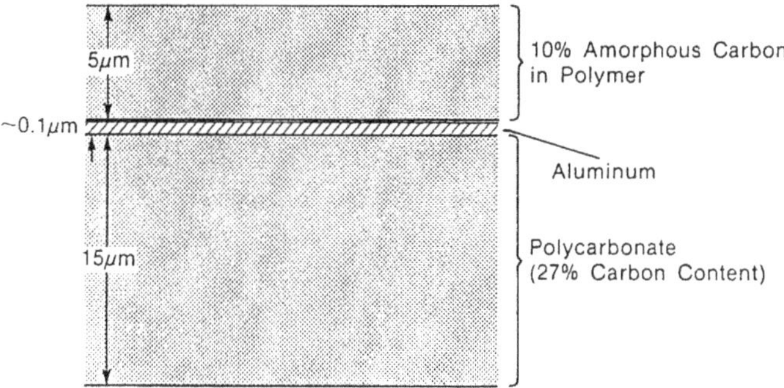

Figure 1. Schematic composition of a sample of resistive printer ribbon. The ink layer is at the top while the resistive or substrate layer is at the bottom. The very thin aluminum layer in the middle serves as a electrical ground plane.

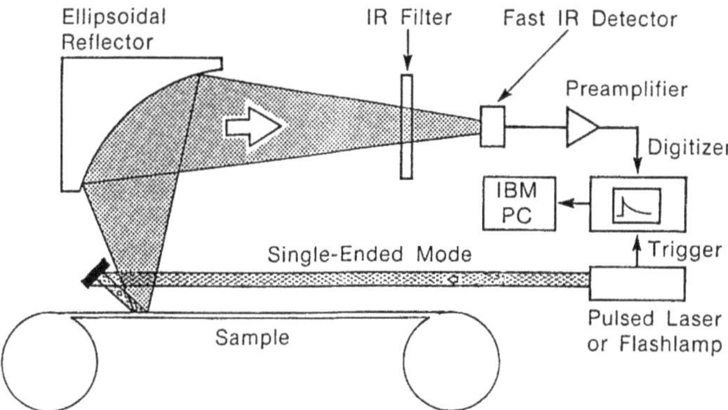

Figure 2. Experimental scheme to detect pulsed radiometry from a thin film on spools.

at a wavelength of 337 nm, which fulfills the above requirements. We note that no sample damage by the laser pulse (typically 0.1 mJ and a 4 mm spot size) can be observed in all the experiments described in this paper.

The initial temperature gradient drives heat into the sample, causing top surface cooling. At early times, the surface cooling is only affected by the material composition immediately underneath the top surface. The subsurface structure is not yet manifested, and the heat diffusion proceeds as if it was a homogeneous semi-infinite solid instead of a film. After a typical diffusion time $\tau = h^2/4D$ where h is the top surface layer thickness, the heat diffusion is affected by the layered structure if there is a change in the thermal diffusivity of the layers or a thermal resistance at the interface. Effectively, this means that the surface cooling is slowed down compared to the case of a semi-infinite solid. Properties such as subsurface contact resistance, delamination or thermal diffusivities of the individual layers will therefore influence the decay of the radiometry signal [5].

At long times, the heat deposited by the laser pulse is uniformly distributed through the thickness of the film, and further cooling can only proceed by radiation or conduction to air. Such cooling is typically very slow, and may take 0.1 s for these samples to approach room temperature. However, if the film is touching a substrate material, this cooling rate at long times can be much faster. This is shown in Fig. 3, where the long time decay for a free-standing ribbon and a ribbon wrapped on a spool are compared. In the following, we present results observed only for free-standing ribbon samples. All laser irradiation and IR detections are done from the ink layer side, unless otherwise specified.

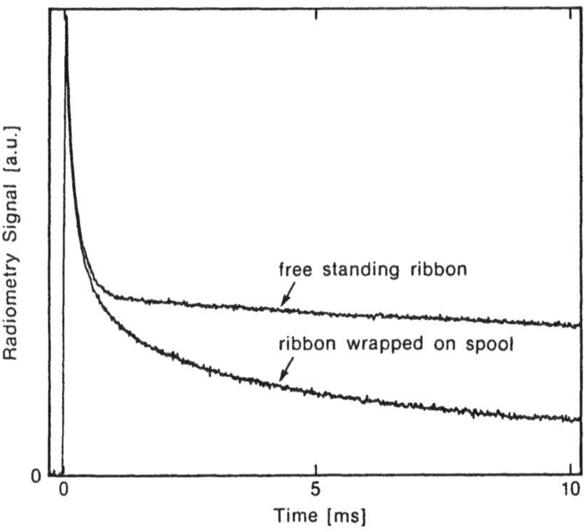

Figure 3. Radiometry signals for a free standing film and a film wrapped on a spool.

Detection of Poor Adhesion or Tendency for Delamination

As pointed out before, at early times the signal intensity is only determined by the top layer composition, mainly its optical absorption coefficient at the excitation wavelength and its infrared absorption and emissivity. Poor adhesion or even delamination is physically manifested as an increased thermal resistance at the interface. This in turn means a slower cooling of the top surface and a corresponding higher temperature for times which exceed the thermal diffusion time for the top layer. The signal is therefore expected to show a kink at the point where the heat diffusion reaches the interface and subsequently show a larger intensity for samples with poor adhesion. This is shown in Fig. 4 for two samples, one of which shows good bonding between the ink and substrate layers while the other tends to exhibit delamination.

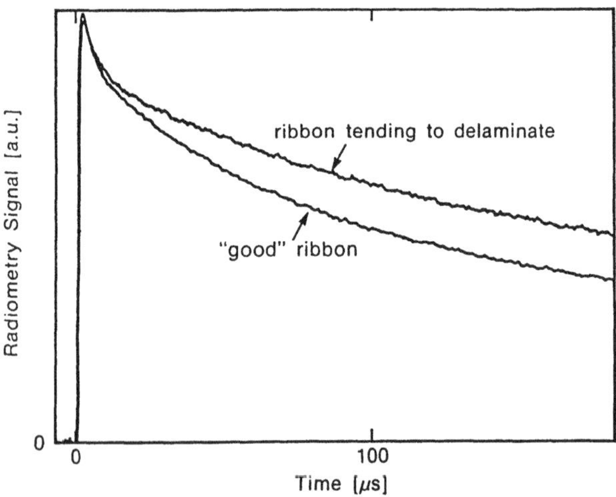

Figure 4. The effect of good thermal contact at the first interface is demonstrated for two samples which show good bonding and a tendency to delamination, respectively. The interface thermal resistance for the ribbon tending to delaminate is much higher, which consequently yields a higher surface temperature at later times.

Detection of Different Coat Weights

The "coat weight" of the ink layer as used here refers to its thickness. Variations of the ink layer coat weight occurs due to changes in the coating conditions of the ink on the aluminized substrate (which is of constant composition and thickness). Coat weight of the top ink layer is an important parameter of the ribbon. Variations in the top layer thickness can be easily detected, as the heat diffusion takes longer to reach the first interface for a thicker layer. This in turn yields a cooler surface temperature as compared to samples with a thinner top layer. As shown in Fig. 5, this shows up most strongly by looking at the difference between the radiometry signals from different samples. As expected, the difference in surface temperature is most pronounced after a delay time, when the heat diffusion has reached the first interface. Variations down to 50 nm in the top layer thickness can typically be detected.

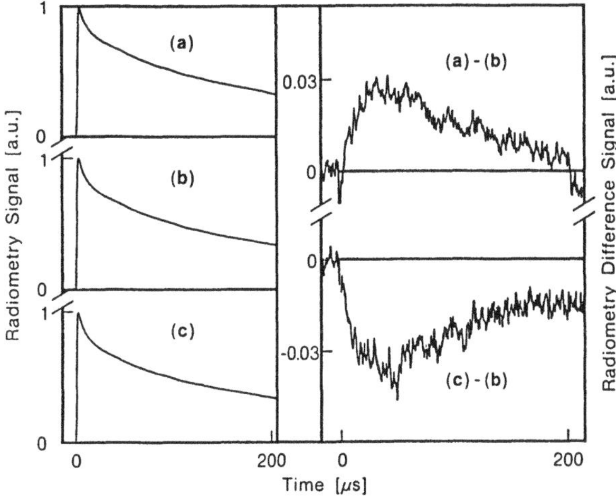

Figure 5. Differences in coat weight show up clearly by substraction of radiometry signals. (a), (b) and (c) give radiometry signals for ribbons with ink layer coat weights of 2.60, 2.75 and 2.95 lb/ream, respectively. On the right, the difference of curves (a) and (b), and the difference of curves (c) and (b) are shown. The temperature difference is largest for certain delay times when the heat has diffused to the interface.

Detection of Different Carbon Loading

The maximum surface temperature reached depends strongly on the optical absorption coefficient at the excitation wavelength. As the polymer is highly transparent to the excitation wavelength, the absorption coefficient is mainly determined by the carbon loading. The IR absorption coefficient on the other hand controls the intensity of the IR signal emitted from the surface. If the IR absorption length ℓ_{ir} is large compared to the absorption length ℓ_{exc} at the excitation wavelength, the radiometry signal would be independent of ℓ_{exc} for small temperature rises ΔT. In our case, however, $\ell_{ir} << \ell_{exc}$ as ℓ_{ir} is mainly determined by the absorption coefficient of the polymer. The radiometry signal therefore comes out of a surface layer which is much thinner than the optical penetration depth. This however means that the signal is a direct measure of the actual surface temperature and thus of the optical penetration depth, which is controlled by the carbon loading. Figure 6 shows signals taken at the front and back side of the ribbon, where the surface layer carbon content are 10% and 27%, respectively. Apart from the signal magnitude, which is strongly affected because of the arguments outlined above, the thermal diffusivity is also affected, as carbon has a higher thermal conductivity than the surrounding polymer. Overall, variations in carbon loading can easily be distinguished to better than 1%.

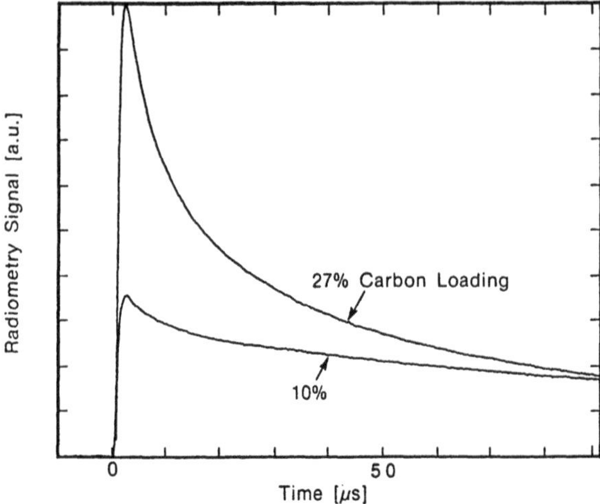

Figure 6. Radiometry signals observed from the ink side (10% carbon loading) and from the substrate side (27% carbon loading) with the same vertical scale. The difference in amplitude is mainly due to the shorter optical depth in the substrate side. The rapid change in layered structure in the ink side also contributes to the signal shape at early times. The signals also indicate that the difference in optical penentration depth and (to a smaller extent) the thermal diffusivity cause a faster decay for polymers with higher carbon content.

Detection of Sample Thickness Variations

At late times, the temperature distribution has been equilibrated across the ribbon, independent of different thermal interface resistances. The temperature reached is mainly determined by the heat capacity of the layers and their thicknesses. Figure 7 shows the measured signal intensity at late times for samples with various coat weights, which shows a linear coat weight dependence.

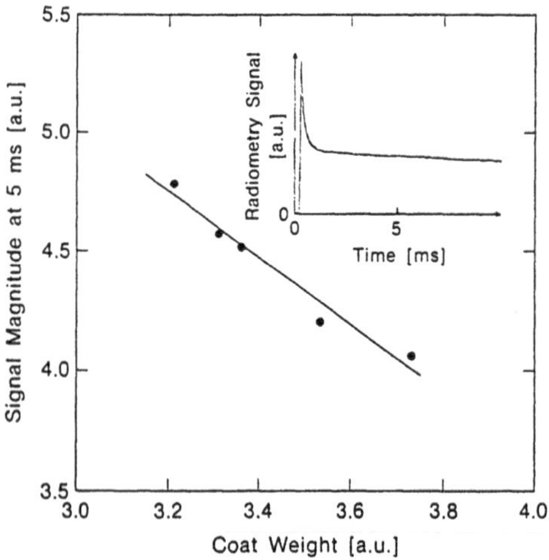

Figure 7. Absolute intensity of the radiometry signal at late times, after a temperature equilibrium across the ribbon has been reached. This intensity then is a measure of the ribbon total thickness. The inset shows one of the experimental curves, from which the data points shown are derived.

Effects of Ambient Conditions

Ambient conditions such as temperature or humidity can have a strong influence on ribbon quality. As an example, the humidity controls the water uptake by a polymer film, which changes its physical properties. Pulsed radiometry provides a noncontact measurement technique to detect these changes. This can be seen from Fig. 8. The experimental data suggest that the thermal diffusivity is hardly affected, but the heat capacity is linearly dependent on the ambient humidity conditions. These measurements can be extended to measure in a noncontact manner how fast water uptake or release do occur.

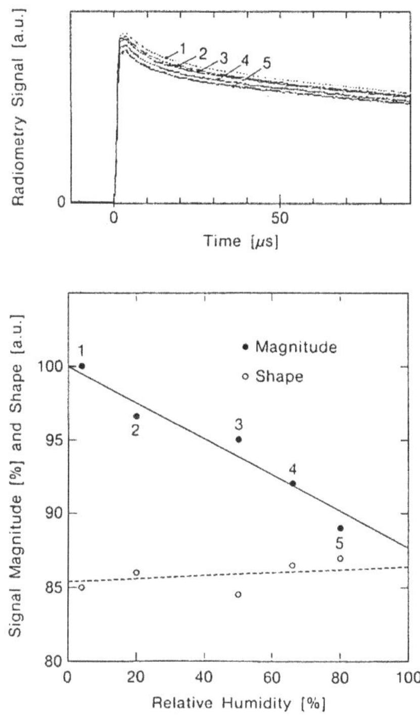

Figure 8. Magnitude and shape of the pulsed radiometry signals as a function of relative humidity of the ambient air. In the top figure the experimental curve taken at five different humidity conditions are shown. Below, the magnitude and a shape parameter are plotted separately. The shape is a measure of the ribbon thermal diffusivity, while the magnitude is mainly controlled by the heat capacity.

Summary

The pulsed radiometry technique is of practical use to characterize layered structures such as ribbons as described in this paper. Thermal properties, which are monitored by this method, are intimately connected with many important functional parameters of the printer ribbon. Because of its noncontact and nondestructive nature, it extends the broad range of optical techniques, that are applicable to optically transparent samples, to opaque materials. Because it is also a fast measurement technique, many testing procedures can be performed in real-time. Additional advantages of this technique include the simplicity of the set-up and its insensitivity to the optical properties of the sample under investigation. If stationary objects are encountered, a variation of this technique employing modulated cw instead of a pulsed excitation can also be employed. This report utilizes some examples of samples of resistive printer ribbons to suggest the broad range of thin-film materials characterizations possible by pulsed radiometry monitoring.

References

1. A. C. Tam and B. Sullivan, *Appl. Phys. Lett.* **43**, 333 (1983).
2. W. P. Leung and A. C. Tam, *J. Appl. Phys.* **56**, 153 (1984).
3. R. E. Imkof, D. J. S. Birch, F. R. Thornley, J. R. Gilchrist and T. A. Strivens, *J. Phys E* **17**, 521 (1984).
4. P. Cielo, *J. Appl. Phys.* **56**, 230 (1984).
5. W. P. Leung and A. C. Tam, *J. Appl. Phys.* **63**, 4505 (1988).

Thermal "Superdiffusion" in the Vicinity of a Rough Surface

A.C. Boccara and D. Fournier

Laboratoire d'Optique Physique, UPR A0005 du CNRS, ESPCI, 10, rue Vauquelin, F-75005 Paris, France

ABSTRACT. We describe an experimental situation in which the probability of observation of *atypical* events can be large enough to dominate the heat diffusion processes. So the time (or frequency) evolution of the temperature field (when probed along a line) will exhibit anomalous behaviour compared to usual situations.

Diffusion processes (mass, heat...) are usually described by random walks in homogeneous spaces or inhomogeneous ones. In the first case, the probability for the random walker to be at a distance from its starting position is a gaussian law whose width is proportional to $t^{-0.5}$ (Brownian motion-like). In the second case, the walks are hindered by the randomness of the medium and the dependence scales as t^{-d_W} with $d_W < 0.5$. Fractal structures such as percolation networks have been found of great help in understanding such anomalous behaviour for heat diffusion experiments /1/.

A diffusion process faster than the brownian motion has been observed recently by Ott et al. /2/, in studying mass diffusion of polymerlike breakable miscelles. They have named this process "superdiffusion".

We have previously demonstrated that geometrical considerations (e.g. fractal versus Euclidean heat sources) can be used to explain the "slowering" of the diffusion processes on a fractal surface for instance. A plane source temperature exhibits a $t^{-0.5}$ dependence after a short pulsed excitation, whereas a fractal sources ($2 < d < 3$, d being the fractal dimension) will exhibit a $t^{-\alpha}$ ($\alpha = (3 - d) / 2) < 0.5$) dependence. The power dependence exponent α is smaller than 0.5 because the density of heat sources *decreases* as we go farther and farther from the point of observation (the density is not constant for a fractal but decreases when the dimensions are increased).

Let us imagine that we probe the temperature field not *of the surface itself* as we do by using e.g. IR detection but *on the top of it* as we do usually by using the mirage detection or an interferometer probe (Fig. 1). If the probability density to find the points of the surface at a distance r from the top goes as r^{β} ($\beta>0$), at least in a certain scale range, we will give a *larger importance to the most unprobable* points of the distribution, because they are much closer to the probe beam. One model for such behaviour is the so-called "Levy flight" /3/ which is the generalization of the Brownian motion.

Usually when a random surface is under investigation, the probe beam is considered to be at an average distance x_0 from the sample surface, and we thus

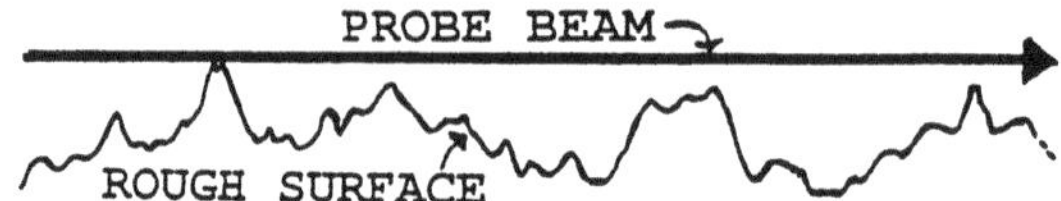

Fig. 1 - Probe beam skimming on rough surface : The most improbable points give the most active contribution to the temperature.

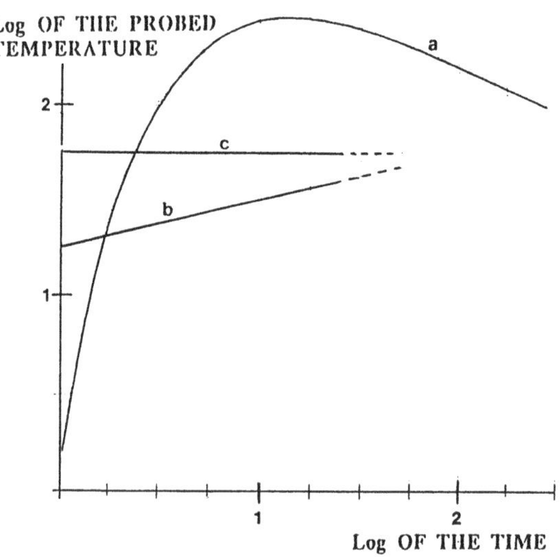

Fig. 2 - Dependence, at short time, of the probed signal close to a surface.Flat surface : a) $x_o \neq 0$
Rough surface probed as in Fig. 1 : b) ß = 1.5 ; c) ß = 1 (see text).

expect the well known variation exp $(-x_0^2/4Dt) / t^{1/2}$, D being the thermal diffusivity. With the above hypothesis (probe beam on the top of the rough surface and a r^β density probability law) one finds a power law. More precisely the average temperature at r = 0 (probe beam position) will go as $t^{-(1-\beta)/2}$ (Fig. 2). This function which is fairly different from the usual diffusion law is monotonically *decreasing* or *increasing* or *stationary* (which is rather unusual !) and is represented by a line in log-log scale, as long as the heat is diffusing from the rough random surface. When the heat reaching the probe beam starts to be emitted from the bulk, ($t >> x_0^2 / D$), we go back to the usual behaviour.

Let us underline that such situations can be also found with a classical euclidean source, e.g. the temperature field at the top of a cone will give a time independant signal when using a linear probe beam. This result is consistant with the above model in which $\beta = 1$ (linear variation of the number of heat sources with the distance).

We hope that these new considerations will help in understanding unusual temperature variations with time (or frequency) when using mirage or interferometric detection close to a rough sample surface.

References

/1/ D. Fournier and A.C. Boccara, Physica, 157 587 (1989).
/2/ A. Ott, J.P. Bouchaud, D. Langevin and W. Urbach, Phys. Rev. Lett. 65 2201 (1990).
/3/ B. Mandelbrot, "The fractal geometry of nature", Freeman, New-York (1983).

Thermal Diffusivities of Thin Gold Films

J. Jauregui[1], *Z.L. Wu*[1,2], *D. Schäfer*[3], *and E. Matthias*[1]

[1]Fachbereich Physik, Freie Universität Berlin, Arnimallee 14, W-1000 Berlin 33, Fed. Rep. of Germany

[2]Shanghai Institute of Optics and Fine Mechanics, Academia Sinica, P.O. Box 800-211, Shanghai, P.R. of China

[3]Zentralinstitut für Optik und Spektroskopie, Rudower Chaussee 6, O-1196 Berlin-Adlershof, Fed. Rep. of Germany

Thermal diffusivities of gold films on BK7 glass substrates ranging in thickness from 20-2500 nm have been investigated by time-resolved photodisplacement detection of laser-induced transient thermal gratings (TTG's) [1]. This technique, described in another contribution to this conference [2], is a promising tool for studying thermal diffusivities of thin absorbing films. The reason is that it measures predominantly the lateral heat diffusion. The application of this technique is only limited to film ticknesses that guarantee sufficient optical absortpion.

Using two beams of a pulsed dye laser (543 nm, 4 ns FWHM), TTG's were excited on gold films ranging in thickness from 20 nm to 2.5 µm. The gratings were examined by a HeNe probe laser beam of about 50 µm focal width, and their periodicity and decay time was observed. A typical grating pattern and its decay is shown in Fig. 1 for a 100 nm thick film. Solid lines are fits to the data yielding a grating period $\Lambda = 90$ µm and an effective decay time $\tau_{eff} = 11.7 \pm 0.4$ µs.

In Ref. [2] it has been shown that for bulk material an approximate 3-dimensional solution of the heat conduction equation can be found to describe the experimental data. The same model can be applied to thin gold films. It predicts that the effective decay rate τ_{eff}^{-1} of the TTG is composed of the decoupled decay rates for the three directions [2]

$$\tau_{eff}^{-1} = \frac{1}{2}\tau_x^{-1} + \tau_y^{-1} + \tau_z^{-1} \ .$$

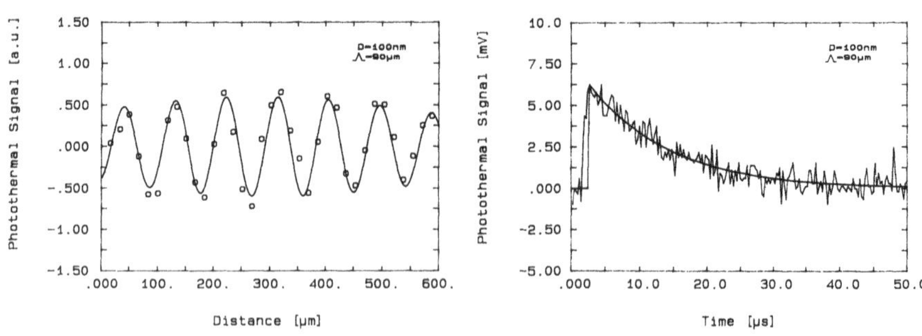

Fig. 1: Transient thermal grating pattern (left) and its time decay (right) for a 100 nm gold film on BK7 glass. Solid lines represent fits to the data.

Springer Series in Optical Sciences, Vol. 69 Photoacoustic and Photothermal Phenomena III
Editor: D. Bićanić

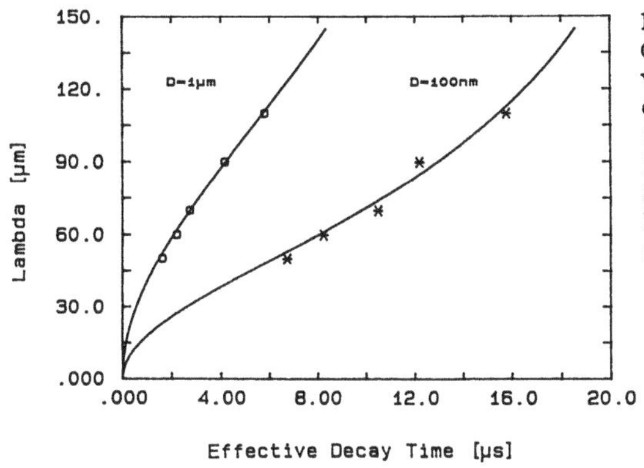

Fig. 2: Grating period versus effective decay time for gold films of 1 μm and 100 nm thickness. Solid lines are best fits to the data.

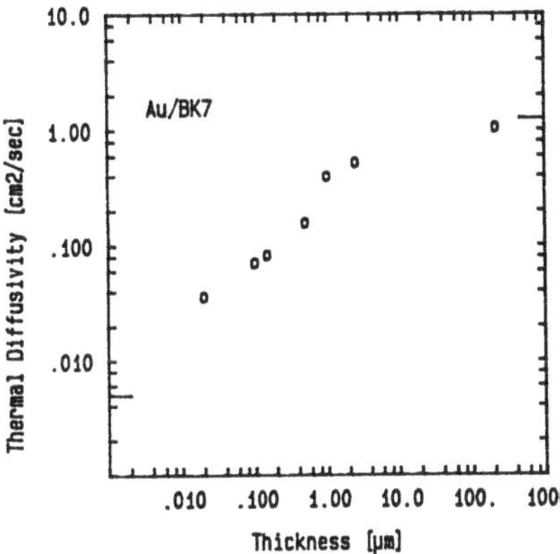

Fig. 3: Thickness dependence of thermal diffusivities of gold films on BK7 glass.

Here, the grating periodicity is along the y-axis and its wave vector $q = 2\pi/\Lambda$ is connected to τ_y^{-1} by the relation $\tau_y^{-1} = q^2\kappa$, where κ is the thermal diffusivity. The decay rate along the x-axis is given by $\tau_x^{-1} = 4\kappa/a^2$, where a denotes the grating width in the x-direction. τ_z^{-1} is the vertical heat diffusion rate. In Fig. 2 the predictions of this model (solid lines) are fitted to the experimental data for films of 100 nm and 1 μm thickness. The fits lead to values for the thermal diffusivities which are displayed in Fig. 3. The different shape of the curves in Fig. 2 is due to different contributions of the vertical heat flow for the two films. The calculated values are τ_z(100 nm) = 25.4 μs and τ_z(1 μm) = 20 μs, respectively.

The diffusivities for various film ticknesses are illustrated in Fig. 3. The literature values of bulk gold [3] and BK7 glass [4] are also indicated. In general the data show the typical decrease of thermal diffusivities with decreasing film thickness [5]. Such decrease is expected since both geometry and deposition-induced disorder of the films will impede the heat flow. However, in contrast to earlier work on thin films of Cu [5,6], Ag and Al [7], where deviations from bulk values occured only for thicknesses smaller than 1 μm, we find strong deviations for 2.5 μm (about 50 %) and 1 μm (about

70 %), and even the bulk value is not well reproduced by a measurement on a 210 μm gold plate (c.f. Fig. 3). Our technique is mostly sensitive to thermal transport parallel to the surface, therefore we suggest that the results indicate a strong anisotropy of the thermal diffusivity. This may be expected for films deposited on cold substrates, as in our case, which often show a columnar structure. The reduced value for the gold plate is blamed on the unpolished surface.

In conclusion, we have demonstrated that TTG's detected by the displacement technique can be utilized for determining thermal diffusivities of metal films as thin as the optical skin depth, independent of any knowledge of the thermal properties of the supporting substrate. Further the technique is capable of providing information about the quality of thin films.

This work was supported by the Deutsche Forschungsgemeinschaft, Sfb 337. Z.L.W. acknowledges a fellowship from the AvH-Foundation, and J.J. one from the Deutsche Akademische Austauschdienst (DAAD).

References

1. H. Eichler et al.: J. Appl.Phys, 44, 5383 (1973)
2. J. Jauregui and E. Matthias, paper at this conference
3. CRC Handbook of Chemistry and Physics, 66th Edition, editors: R.C. Weast et al., D-185
4. E. Welsch et al.: Can. J. Phys. 66, 638 (1988)
5. P. Nath and K.L Chopra: Thin Solid Films 20, 53 (1974)
6. F. Kelemen: Thin Solid Films 36, 199 (1976)
7. B.T. Boikov et al.: Thin Solid Films 17, 157 (1973)

Photoacoustic Investigation of Thin Solid Films in Contact with Liquids

C. Glorieux, E. Mens, and J. Thoen

Laboratorium voor Akoestiek en Warmtegeleiding, Departement Natuurkunde, K.U. Leuven, Celestijnenlaan 200D, B-3001 Leuven, Belgium

Abstract The photoacoustic (PA) technique can be very useful in non–destructive testing of defects in materials. It offers also a possibility for performing a depth profiling analysis of layered materials. At high modulation frequencies, the signal only depends on the thermal diffusivity of the top layer. In the low frequency range the signal contains information about the effusivity ratios of the subsequent layers. In this work, the frequency dependence of the PA signal is used to determine the thermal properties of two–layer structures composed of thin solid films in contact with liquids.

Photoacoustic measurements have been used in the past to determine the thermal diffusivity α of various materials. Both time domain and frequency domain experiments were performed with piezoelectric [1], mirage [2] and microphone [3] detection. However, thermal transport through a boundary between two materials depends on the thermal effusivity ratio $b=e_2/e_1$ between the layers, rather than on the thermal diffusivities. As a consequence, the photoacoustic signal is influenced by the thermal diffusivity of the surface layer, and by the thermal effusivity ratio b of the surface and background layer. The influence of the backing material effusivity on the signal becomes less important if the frequency increases, since then the thermal diffusion length μ, which is inversally proportional to $\sqrt{f}$, becomes smaller than the thickness l of the surface layer : the signal evolves to a thermally thick regime. In principle, careful analysis and fitting of the photoacoustic signal frequency dependence between a thermally thin and thick regime, can lead to the determination of the backing and sample effusivity and the sample diffusivity.

We have carried out photoacoustic experiments on two layer samples consisting of a black thin film, in contact with a liquid. The film was periodically heated by a chopped 60 mW Argon laser beam, the resulting pressure change in the photoacoustic cell was lock–in detected by an electret microphone. An automatic setup collected the amplitude and the phase of the signal at different modulation frequencies (5 Hz to 600 Hz). In this way the photoacoustic signal changed from a thermally thin to a thermally thick regime. The signal was photoacoustically saturated (optical absorption depth $1/\beta << \mu$). In that case, together with the approximation that $g<<1$ (g is the ratio between the gas effusivity e_g and the surface layer effusivity e_s), the Rosencwaig expression [4] for the pressure signal $P = q \exp[i(2\pi ft-\varphi)]$ can be simplified to :

$$q=f_{sys}\frac{\mu_g \sqrt{(1-Y^2)^2 + 4Y^2 \sin^2(B\sqrt{f})}}{e_s\sqrt{2}\pi f \ (1 + Y^2 + 2Y\cos(B\sqrt{f}))} \tag{1}$$

$$tg(\varphi+45^\circ-\varphi_{sys})= 2A\exp(-B\sqrt{f})\sin(B\sqrt{f})\,(1-A^2\exp(-B\sqrt{f}))^{-1} \tag{2}$$

with μ_g the thermal diffusion length in the gas, $A = (b-1)/(b+1)$ and $B = 2l\sqrt{\pi}/\alpha_s$ The amplitude factor f_{sys} and phase shift φ_{sys} depend on the sample reflectivity, and on the cell and microphone response, and are difficult to determine. Therefore, simultaneous measurements were carried out on two samples consisting of the same surface layer and two different background liquids (1 and 2), in good thermal contact with the film. Since the system response is the same for the two cases, the amplitude ratio $R=q_1/q_2$ and the phase difference $\Delta\varphi = \varphi_1-\varphi_2$ of the two signals is then independent of f_{sys} and φ_{sys}. It is clear in equations (1) and (2) that R and $\Delta\varphi$ only depend on the thermal diffusivity of the film, and on the thermal effusivities of the film and the two background liquids.

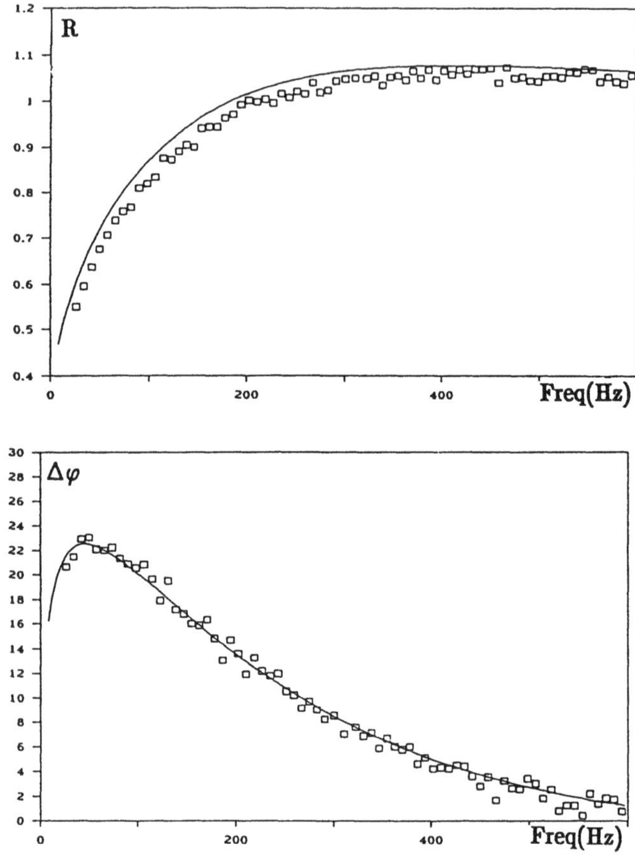

FIG.1 Experimental R(f) and Δφ(f) experimental data and the best fitting simulation curve for 15μm black polycarbonate film on the background liquids water and hexadecane

Fig. 1 shows the experimental data for the frequency dependence of R and $\Delta\varphi$ for a thin black carbon filled polycarbonate film with unknown thermal properties. The background liquids were water (e_1 = 1584 J/m²K√s) and hexadecane (e_2 = 507 J/m²K√s). The sample diffusivity can be determined by visually comparing or fitting different simulations to the experimental data. The parameters A and B in $\Delta\varphi$(f) were determined to give the best fit to the experimental data. The corresponding value for α_s was found to be $2.1\times10^{-7}m^2/s$.

A more simple, but less accurate way of determining α_s is the following : since for typical thermal properties of our sample the parameter A is much smaller than 1, the denominator in (2) can neglected away so that the maximum of $\Delta\varphi$(f) satisfies the simple relation :

$$f_{max} = \frac{\alpha_s}{64} \frac{\pi}{l^2}$$

With in the sample thickness l = 15 μm and the value for $f_{max} \cong 55$ Hz, the corresponding value for α_s is $2.5x10^{-7}m^2/s$, which is in reasonable agreement with the previously found value. It is evident, that in the case of a known surface layer diffusivity, one can in principle determine the film thickness in an analogue manner.

The frequency dependence of the signal also enables us to determine the thermal effusivity e_1 of one of the background liquids, if α_s and e_2 of the other liquid are known.

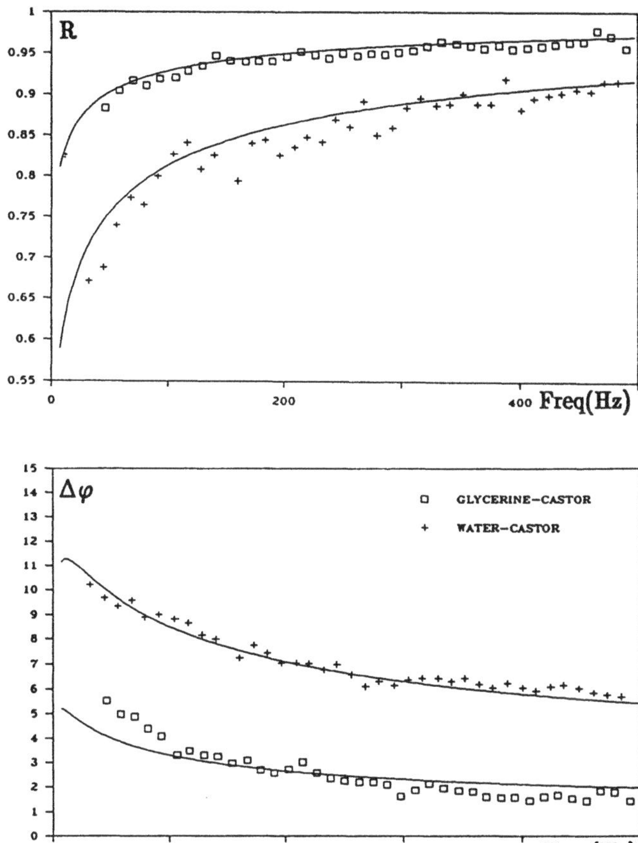

FIG.2 Experimental R(f) and Δφ(f) data for 75μm Si–film on combinations of different background liquids

In fig.2 results are shown for an optically very thick 75 μm silicon film on top of the combination water–castor oil($e = 603\ J/m^2K\sqrt{s}$) and glycerine($e = 938\ J/m^2K\sqrt{s}$)–castor oil. The corresponding simulations (solid curves) are in excellent agreement with the measured data.

References

1. M.Liezers and R.M.Miller, *Laseroptoacoustic spectroscopy* (Springer Series : Optical Sciences 37, 1986)
2. G.Rousset et F.Lepoutre, Revue Phys.Appl.17, 201 (1982)
3. U.Zammit, M.Marinelli, F.Scudieri, and S.Martellucci Appl.Phys.Lett. 50(13), 830 (1987)
4. A.Rosencwaig and A.Gersho, J.Appl.Phys. 47, 64 (1976)

In Situ Measurement of Thin Film Growth by Photothermal Induced Frequency Conversion

H.D. Geiler, N. Winkler, and D. Schiller

Institut für Festkörperphysik, Friedrich-Schiller-Universität Jena, Helmholtzweg 5, O-6900 Jena, Fed. Rep. of Germany

Abstract: A single laser beam double modulation method of photothermal modulated reflectivity is used to monitore the growth of thin metallic layers on glass during the deposition. The speciality of the method allows the autocalibration of the thickness scale.

1.Introduction

Nondestructive and contactless techniques for the measurement of material parameters are important for the tasks of control and monitoring processes [1]. The in-situ measurement of layer thickness is possible by the detection of the amplitude and phase of a thermal wave in the layer [2]. A new single laser beam technique of measuring the photothermal modulated refractive index [3] is used to monitor thin film growth. In competition with purely optical methods [4] this photothermal measurement has some advantages in the case of layers with high optical densities, high absorptive layers and layer systems with low differences of the refractive indices.

2.Experimental Method

The beam of an Ar^+-laser is subjected to intensity modulation with two frequencies Ω_1 = 1.04308 MHz, Ω_2 = 1.0 MHz and the corresponding degrees of modulation δ_1 and δ_2, respectively. The energy deposition due to the divergence of the Poynting vector of the electromagnetic field in the layer [5] causes the thermal wave, which modulates the complex refractive index with the two frequencies Ω_1 and Ω_2. Then the intermodulation with the same electromagnetic field in the layer creates a series of combinations of both frequencies. Particularly we are interested in the difference frequency Ω_1-Ω_2=Ω_-, which may be detected both in the foregoing and backgoing waves. The intensity I_- representing the part with Ω_-=43.08 kHz can be expressed following [3] by

$$I_-=I_o^2\delta_1\delta_2M_-\cos\left(\frac{\varphi_1+\varphi_2}{2}\right)\cos\left(\Omega_-t+\frac{\varphi_1-\varphi_2}{2}\right) \qquad (1)$$

with the phases of the photothermal response φ_1 and φ_2 corresponding to the two frequencies and the amplitude of the photothermal response M_-. Introducing the conversion coefficient K_-

$$K_- = \frac{I_-}{\delta_1\delta_2 I_o^2 R_o D_s}$$

and assuming negligible photothermal dispersion ($\varphi_1=\varphi_2=\varphi$) one gets $K_- = m_-\cos\varphi$ with the photothermal response factor m- independent on the laser intensity I_o, the optical reflectivity R_o and nomalized to the laser energy deposited in the layer under consideration D_sI_o. Note that the phase of the response is directly involved in the conversion coefficient by the cosine. The experimental arrangement to determine K_- is

Springer Series in Optical Sciences, Vol. 69 Photoacoustic and Photothermal Phenomena III
Editor: D. Bićanić

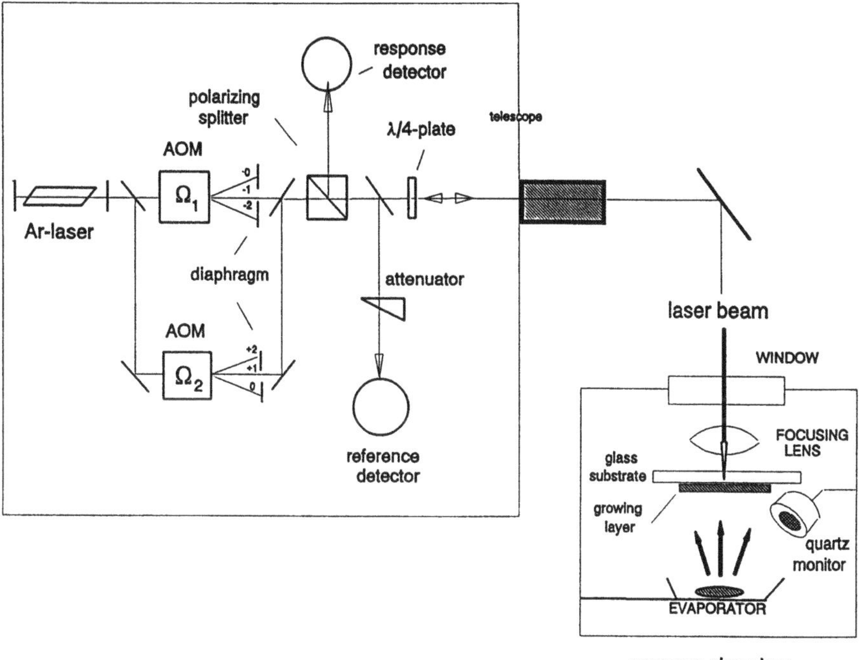

Fig.1 Experimental setup of the in situ measurement during evaporation

shown in fig.1 [3]. This equipment is applied to measure the photothermal response from the backside of a film growing on a transparent substrate during the vapour phase deposition from the front side (fig.1) in the vaccum chamber. The relative accuracy of the method is about 5 %.

3. Results and Interpretation

The growth of Au, Cu and Ge films on glass substrates were investigated. Tab.1 presents the collected optical and thermal parameter of the substances. The typical dependence of the conversion coefficient on the thickness of the growing film is shown in fig.2 for the deposition of Au on a glass substrate. Taking into account the thermal diffusion length l_T (see tab.1) the curve is well understood in the range $1/\alpha<d<l_T$. The monotonous increase of K_T for $d>0.1\times l_T$ is mainly due to the phase of the thermal response at the interface glass/layer. If the phase passes $-\pi/2$ the signal vanishes. This phase behaviour is due to the nonexponential energy deposition in the metal layer at thicknesses less than $1/\alpha$ (α- absorption coefficient). To study this effect and clarify the possibility of measuring the layer thickness $d < 0.1\times l_T$ detailed results of the in-situ measurement are shown in fig.3. Using the theory of frequency conversion [6] basing on the matrix formalism of electromagnetic wave amplitudes in layer systems [5] it is deduced that under the conditions $1/\alpha_2<<l_{T2}$ and $1/\alpha_1>>l_{T1}$ in the layer and the superstrate, respectively, the conversion coefficient is proportional the cosine of the phase of the thermal response at the interface, despite of the complex optical problem. The calculated phase dependence on the thickness and the $\pi/2$-passage are shown in fig. 4. The arrows inicate the measured zeros of K_-, which coincide with the calculated ones within the accuracy of the thickness determination by talystep measurement (5%).

Tab.1 Material parameter of the involved substances

substances	Au	Ge	Cu	glass
refraction index: n	1.3	4.1	1.1	1.463
absorption index: k	1.86	2.0	2.5	$<10^{-6}$
thermal capacity: ρc ($Jcm^{-3}K^{-1}$)	2.509	2.3	3.36	1.978
thermal conductivity: κ ($Wcm^{-1}K^{-1}$)	2.89	0.61	3.86	10^{-2}
thermal diffusion length: l_T (μm)	6.05	2.8	6.02	0.38
film thickness at the zero passage: $d_{\pi/2}$ (nm)	40	50	20	

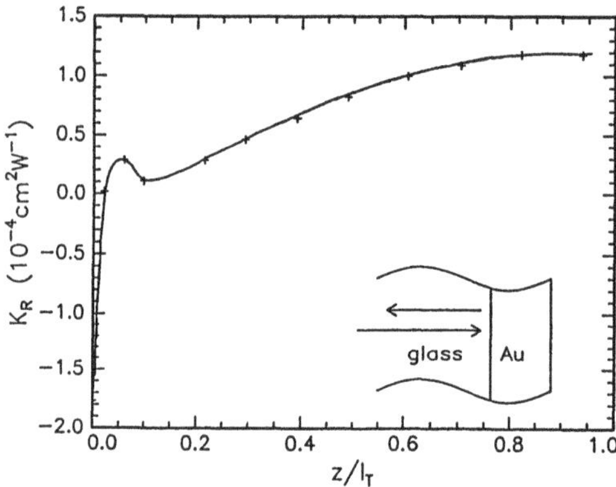

Fig.2 Typical dependence of the conversion factor K_- on the layer thickness of Au normalized to the thermal diffusion length l_T in the layer

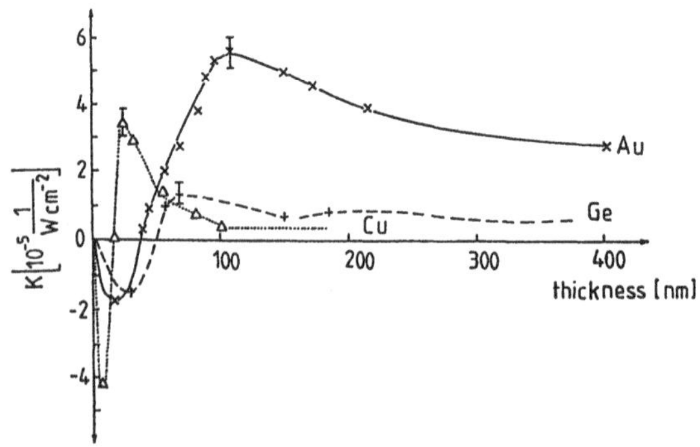

Fig.3 Conversion factor versus layer thickness in the region of $d<0.1\times l_T$ representing the zero passage of the signal

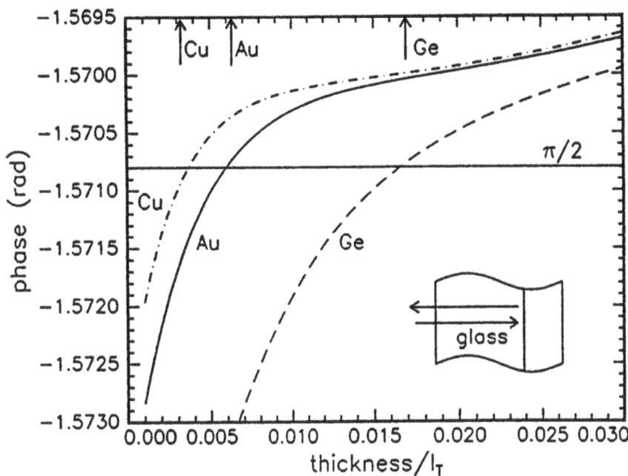

Fig.4 Calculated phase dependence of the termal response at the interface glass/layer (The arrows indicate the measured π/2-passage taken from fig.3)

Because the zero passage depends only on α and the thermal diffusion length l_T in the layer it can be used for the calibration of the thickness scala independent on the deposition rate.

4. Conclusions

The method allows in-situ measurements of the thickness of growing opaque films in the range $l_T/1000 < d < l_T$ with an accuracy of 5 %. The absolute range of measurement can be adjusted by changing the modulation frequency in the interval $100kHz < \Omega < 10MHz$.

References

[1] H.Ryssel, P.Eichinger, C.Schneider, L.Pfitzner in "Semiconductor Production",Fachtagung Productronica 1989, VDE-Verlag, Berlin 1989,p.61

[2] P.M.Patel, D.P.Almond, H.Reiter; Appl.Phys. B43(1987)9

[3] M.Wagner, H.D.Geiler, N.Winkler; Appl.Surf.Sci.50(1990)373

[4] H.A.Crostack, W.Jahnel, E.H.Meyer, K.J.Pohl; Thin Solid Films 181(1989)195

[5] H.D.Geiler, D.Stock, K.Hehl; phys.stat.sol.(a)78 (1983)193

[6] H.D.Geiler, D.Wolff; phys.stat.sol., in preparation

Thermal Characterization of Thin Layers

M. Beyfuss[1], *H. Reichl*[2], *and J. Baumann*[1]

[1]Siemens AG, Otto-Hahn-Ring 6, W-8000 München 83, Fed. Rep. of Germany
[2]Technische Universität Berlin,
Einsteinufer 17, W-1000 Berlin 10, Fed. Rep. of Germany

Abstract. We analyze the conditions for an accurate determination of the thermal properties of layers by photothermal radiometry. This discussion is based on a 3-dimensional theory of thermal wave propagation in layered samples. Therefrom a simple method of measurement and evaluation is deduced and applied to the thermal characterization of thin microelectronic layers.

1. Introduction

The thermal properties (conductivity k and volume-heat-capacity C), and especially those of thin layers, are of growing interest in microelectronics and microsystems, because the heat removal from highly integrated devices becomes more and more difficult. As shown by several authors /1-6/, thermal waves are very sensitive to a change in the material's thermal data.

2. Experimental set-up

In the photothermal experiments we present here, a chopped 3W Ar^+-laser is used to illuminate the sample's surface. The thermal wave temperature response (amplitude A and absolute phase ϕ vs. modulation frequency f) is measured at the front side of the sample focusing the emitted infrared radiation into a lN_2-cooled MCT infrared-detector. Then, the signal is fed into a lock-in-amplifier. With this set-up we obtain a very good accuracy with respect to amplitude (±20μK) and phase (±0.1 deg).

3. Evaluation of the thermal properties

In order to evaluate the thermal properties of thin layers we extend the 3-dimensional model of thermal wave diffusion /7/ to multi-layered samples. According to this model we calculate the phase change $\partial\phi$ induced by a small variation ∂p of a thermal parameter p at the value p_0. Here, p represents k and C or the effusivity $e=(k\cdot C)^{\frac{1}{2}}$ and the diffusivity $\alpha=k/C$ of the interesting layer, respectively. The relative sensitivity ϵ_p for the determination of p is, thus, given by $\epsilon_p=p_0\cdot\partial\phi/\partial p$.

From the frequency behaviour of ϵ_p, computed for a layer on a thermally thick substrate, we obtain, that for thermal wave reflection coefficients $|R_T|>0.9$ and plane waves both thermal parameters of the layer can be determined very sensitively. At low frequencies, where $\mu=(\alpha/(\pi\cdot f))^{\frac{1}{2}}\gg d$ (d: thickness, μ: diffusion length) holds, k can be measured selectively, because the sensitivity for the complementary parameter C vanishes. Within $0.8\le d/\mu\le 1.7$ ϵ_e tends to zero and α can be obtained selectively.

Springer Series in Optical Sciences, Vol. 69 Photoacoustic and Photothermal Phenomena III
Editor: D. Bićanić

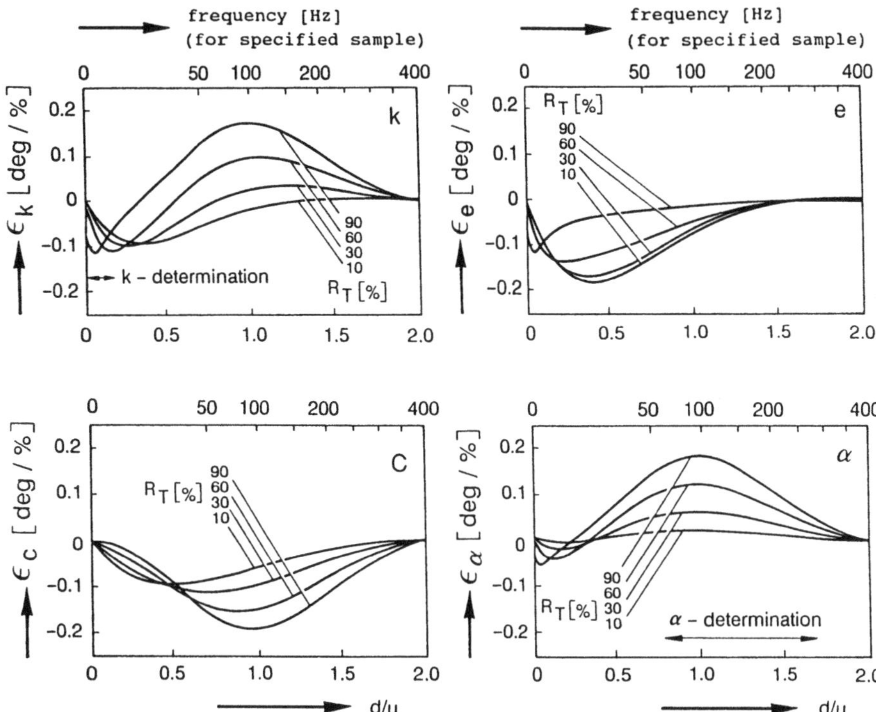

Fig. 1: Relative sensitivity ϵ for k, C, e, and α vs. thermal thickness. We assume $d=20\mu m$, $k=0.35 WK^{-1}m^{-1}$, and $C=2.8 JK^{-1}cm^{-3}$.

Thus, measuring the front side phase within $0.05 \leq d/\mu \leq 1.7$ permits a complete definite thermal characterization of the layer. It is important to mention, that for increasing lateral thermal diffusion the sensitivity decreases strongly. Even if the 1/e laser beam diameter is as large as 5 times the diffusion length of the substrate, ϵ_k is less than half the value for the plane wave case. This can be understood by the reason, that for lateral diffusing thermal waves no multiple interferences, which are essential for the sensitivity in thermal wave analysis, can occur within the small surface area seen by the detector.

4. Results

In the frequency range described above we investigated several thin microelectronic materials. The results are listed in tab. 1. A typical phase response is shown in fig. 2. In order to determine k and C the calculated curves were matched to the experimental data by a two-parameter least-square-fit. For an aluminum alloy and a polymer we compared the data obtained photothermally to values yield conventionally (stationary heat flux technique for k, and differential scanning calorimetry for C). The results exhibit an excellent agreement. Hence, the method presented here is not only accurate with respect to relative, but also to absolute values of thermal properties.

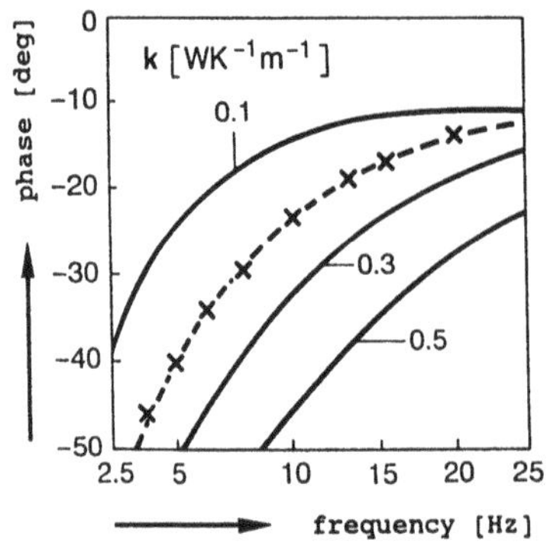

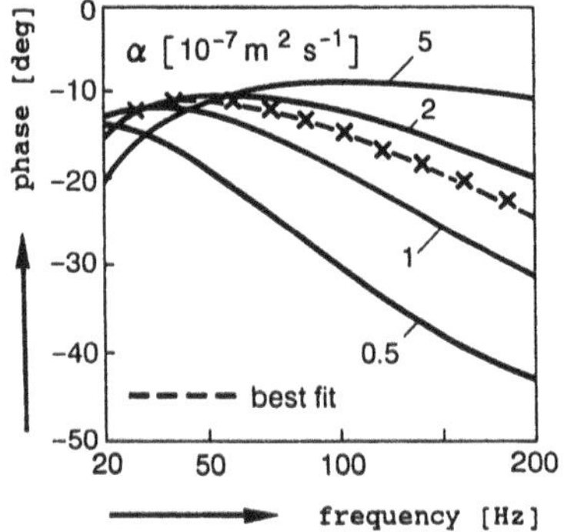

Fig. 2: Experimental front side phase data and computed curves of a 12 μm PB 96 passivation (coated by a 1 μm TiN_x absorber) on a silicon wafer. The best fit data are shown in tab. 1.

material term	d [μm]	k [$WK^{-1}m^{-1}$]	C [$JK^{-1}cm^{-3}$]	application
polybenz-oxyle 96	12	0.20±0.01	1.4±0.1	passivation
diamondlike carbon film	2	1.50±0.05	1.9±0.1	passivation
epoxy	100	0.65±0.02	1.5±0.1	DRAM device encapsulation
glaze	12	1.22±0.04	2.1±0.1	thermotransfer printhead
AlN	1000	85±5	2.8±0.2	insulator with large k
adhesive	48	1.30±0.04	1.8±0.1	device assembling

Tab. 1: k- and C-data of microelectronic materials determined as described before. The inaccuracy (±3% for k, ±7% for C) is computed assuming noise influence, systematic errors and errors due to the finite accuracy of other data of layers involved.

5. Conclusion

It has been shown, that both the thermal conductivity and the volume-heat-capacity of thin layers can be determined very accurately from the photothermal front side phase measured within a matched frequency range. The method described here permits to select microelectronic materials systematically and to get precise input data for microelectronic device modelling.

References

1. D. R. Green: J. Appl. Phys. **37**, 3095 (1966)
2. M. J. Adams, G. F. Kirkbright: Analyst **102**, 281 (1977)
3. P. Korpiun, B. Mertè, G. Fritsch, R. Tilgner, E. Lüscher: Coll. & Pol. Sci. **261**, 312 (1983)
4. C. A. Bennett, R. R. Patty: Appl. Opt. **21**, 49 (1982)
5. R. T. Swimm: Appl. Phys. Lett. **42**, 955 (1983)
6. A. Lachaine: P. Poulet: Appl. Phys. Lett. **45**, 953 (1984)
7. F. A. McDonald: J. Appl. Phys. **52**, 381 (1981)

Photodeflection Method Applied to Loss Characterization of Channel Waveguides

M. Bertolotti, L. Fabbri, R. Li Voti, and C. Sibilia

Dipartimento di Energetica, Università di Roma I,
Via Scarpa 16, I-00161 Roma, Italy and
GNEQP of CNR, Italy

A way is suggested to obtain an accurate evaluation of optical losses of channel waveguides by using a photodeflection method in its transverse configuration.Theory and numerical results are presented.

1-Introduction

Among the problems concerning the determination of channel waveguide properties, the exact determination of absorption losses is perhaps the main one. Several methods have been applied but most of them exhibit several limitations.
Recently [1] a photodeflection method in its collinear configuration has been applied,but no clear theoretical discussion has been reported.In this paper the possibility to use a photothermal deflection technique in a transverse configuration is considered from a theoretical point of view.In particular, numerical results are presented for the case of strong and weak losses.It is shown that the modulus of the deflection signal exhibits a minimum value as a function of the propagation axis, which is related to the optical losses of the guide.

2-Theory

We consider a modified version of the method usually named Photothermal Deflection in its transverse configuration. The pump beam is guided in the waveguide and modulated by means of a mechanical chopper;the probe beam travels in the air near to the surface of the guide (fig.1). The amplitude and phase of the deflected signal give information about losses in the propagation.The main problem consists in the analysis of the thermal field in the film-substrate system when a laser beam is propagating inside the channel-guide.Heat transfer equations with the appropriate boundary conditions and the heat source espression have to be solved. In particular, the heat source is due to light absorption in the channel during the modulated pump laser propagation and the energy absorbed by the film-substrate system is distributed along the channel direction (y-axis).

Being the thermal properties of the different parts of the waveguide substantially identical,the system results equivalent to the one in which there is only a substrate and a linear heat source along the y axis.Moreover if the system is assumed as a semi-infinite slab along the y direction, which is reasonable if the waveguide dimension is far greater than the thermal diffusion length ($l_t = \sqrt{2D/\omega}$),the following steady-state Fourier equations hold true:

$$\nabla^2 T_i - \frac{j\omega}{D_i} = - \frac{W_i}{\lambda_i} \qquad (1)$$

i=0 (air) ,1(medium),

being T the temperature rise ,ω the angular frequency of heating laser, D the thermal diffusivity and W the power per unit volume dissipated in the guide:

$$W_0 = 0 \qquad \text{(in air)} \qquad (2)$$
$$W = \alpha * I_o(x,z) * e^{-\alpha y} \qquad (y > 0) \qquad (3)$$

being I_o the input laser intensity and α the absorption coefficient of the waveguide.Continuity of temperature and heat flux as well, are assumed as boundary conditions. The thermal field distribution in air has been obtained from solution of eq.s (1) derived by means of Fourier transforms. In this way the expressions of the two perpendicular components of the deflection angle of probe beam skimming over the waveguide surface have been obtained:

$$\begin{cases} \Phi_y = \dfrac{\alpha * P}{n\pi\lambda_1} * \dfrac{dn}{dT} * \displaystyle\int_o^\infty \frac{\delta * [\delta cos(\delta y) - \alpha sin(\delta y)]}{\beta_1 * (\alpha^2 + \delta^2)} e^{-\beta_o z} d\delta \\ \Phi_z = -\dfrac{\alpha * P}{n\pi\lambda_1} * \dfrac{dn}{dT} * \displaystyle\int_o^\infty \frac{\beta_o [\delta sin(\delta y) + \alpha sin(\delta y)]}{\beta_1 * (\alpha^2 + \delta^2)} e^{-\beta_o z} d\delta \end{cases} \qquad (4)$$

Being $\beta_i^2 = \delta_x^2 + \delta_y^2 + \frac{j\omega}{D_i}$, δ is a transformation variable,P the input power and n the refractive index of the air.

Both expressions provide an interesting result if we study the behaviour of the deflection signal as a function of the vertical offset between probe and surface of the guide and of the distance y of the probe along the channel.

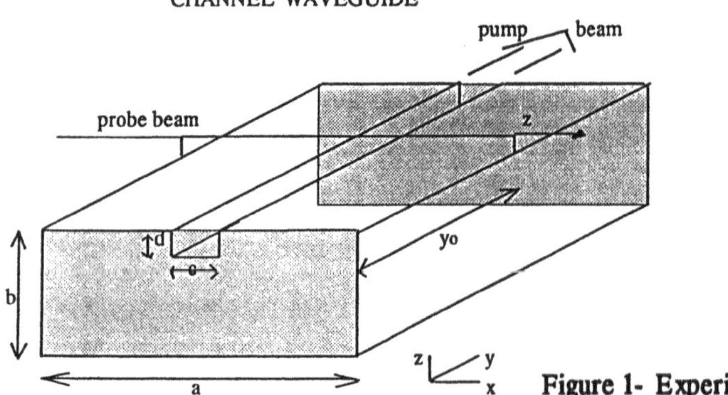

Figure 1- Experimental disposition

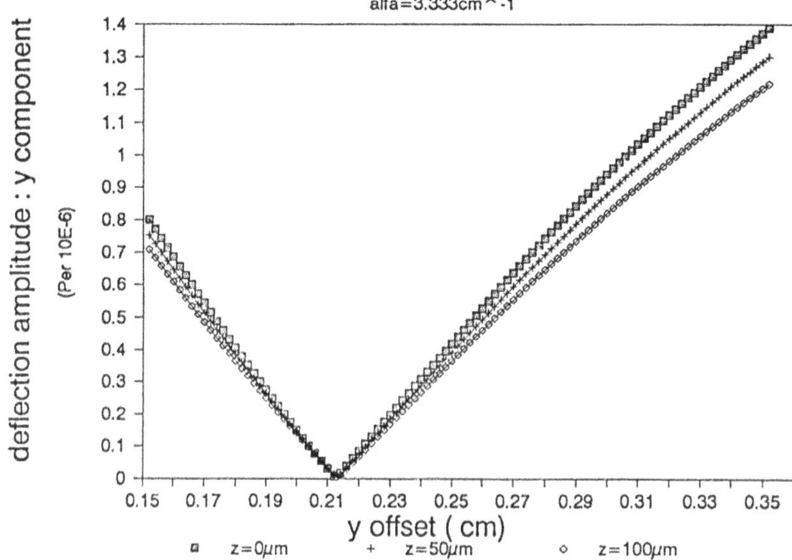

Figure 2 - Photodeflection signal vs y offset

The signal exbihits a minimum which is directly connected to the optical losses of the structure :

$$y = cost/\alpha \qquad (5)$$

where cost is a constant depending on the thermal diffusivity of the material. An example is shown in Fig.2 for y component of photodeflection signal for different vertical offset z .

Work performed with partial support under MPI 40% grant

REFERENCES

[1] R.K.Hickernell, D.R.Larson , R.J.Phelan , J.R., and L.Larson - Appl.Optics,27,2636(1988).

[2] M.Bertolotti et al. J.Phys.D21,S14,(1988)

High Frequency Photothermal Reflectivity and Displacement Measurements on Thin Film Samples

M. Reichling[1], *Z.L. Wu*[1,2], *E. Welsch*[1,3], *D. Schäfer*[4], *and E. Matthias*[1]

[1]Institut für Atom- und Festkörperphysik, Freie Universität Berlin, Arnimallee 14, W-1000 Berlin 33, Fed. Rep. of Germany

[2]Shanghai Institute of Optics and Fine Mechanics, Academia Sinica, P.O. Box 800-211, Shanghai, P.R. of China

[3]Institut für Optik und Quantenelektronik, Friedrich-Schiller-Universität Jena, Max-Wien-Platz 1, O-6900 Jena, Fed. Rep. of Germany

[4]Zentralinstitut für Optik und Spektroskopie, Rudower Chaussee 6, O-1199 Berlin, Fed. Rep. of Germany

Abstract. A new method for the determination of the thermal conductivity of thin film-on-substrate systems using the photothermal reflectance or displacement technique is presented. The method is based on a measurement of the frequency dependence of amplitude and phase of the photothermal signal. Characteristic features appear when the thermal length is equal to the film thickness. Quantitative information is extracted by fitting theoretical model curves to the frequency response data. Results of conductivity measurements on 1 μm films of optical coating materials are shown.

1. Introduction

Several photothermal methods have been utilized for thermal diffusivity measurements on thin film samples /1/. The most successfull measurements by Roger et al. and Skumanich et al. include a mirage thermal wave detection and an analysis of the frequency dependent phase behavior /1,2/. Both authors suggest to measure the mirage signals for front and backside illumination of the film. The thin film diffusivity is derived from a comparison of the phase difference between both measurements with a theoretical model calculation. However, in all of these experiments frequencies below 100 kHz were used and the thermal diffusion length was always larger than the film thickness. In that case the photothermal signal is dominated by the substrate thermal properties and the thin film causes only a relatively small correction what implies severe restrictions on the minimum film thickness that can be investigated and the overall accuracy of the measurement.

Provided that film and substrate diffusivities differ considerably from each other there will be a much more pronounced effect in the frequency response curve of the thin film system when the thermal diffusion length is scanned from values below to values above the film thickness. Model calculations /3/ and experimental results shown below provide evidence for the appearance of typical features in the phase curves that differ significantly from any bulk behavior. The exact curve form, however, is not only determined by the system thermal properties but also influenced by other parameters like the optical absorption, film thickness, and pump beam diameter. For a quantitative analysis it is, therefore, necessary to extract the thermal information from a fit of theoretical curves to experimental data. To avoid instrumental effects leading to a distortion of the phase curves most results are based on thin film measurements normalized to bulk response data.

2. Experimental

CW-photothermal reflectivity and displacement measurements are well established methods for bulk and thin film material characterization in the frequency range from 10Hz to 100kHz /4,5/. For the extension of these methods to thin films it is generally

Springer Series in Optical Sciences, Vol. 69 Photoacoustic and Photothermal Phenomena III
Editor: D. Bićanić

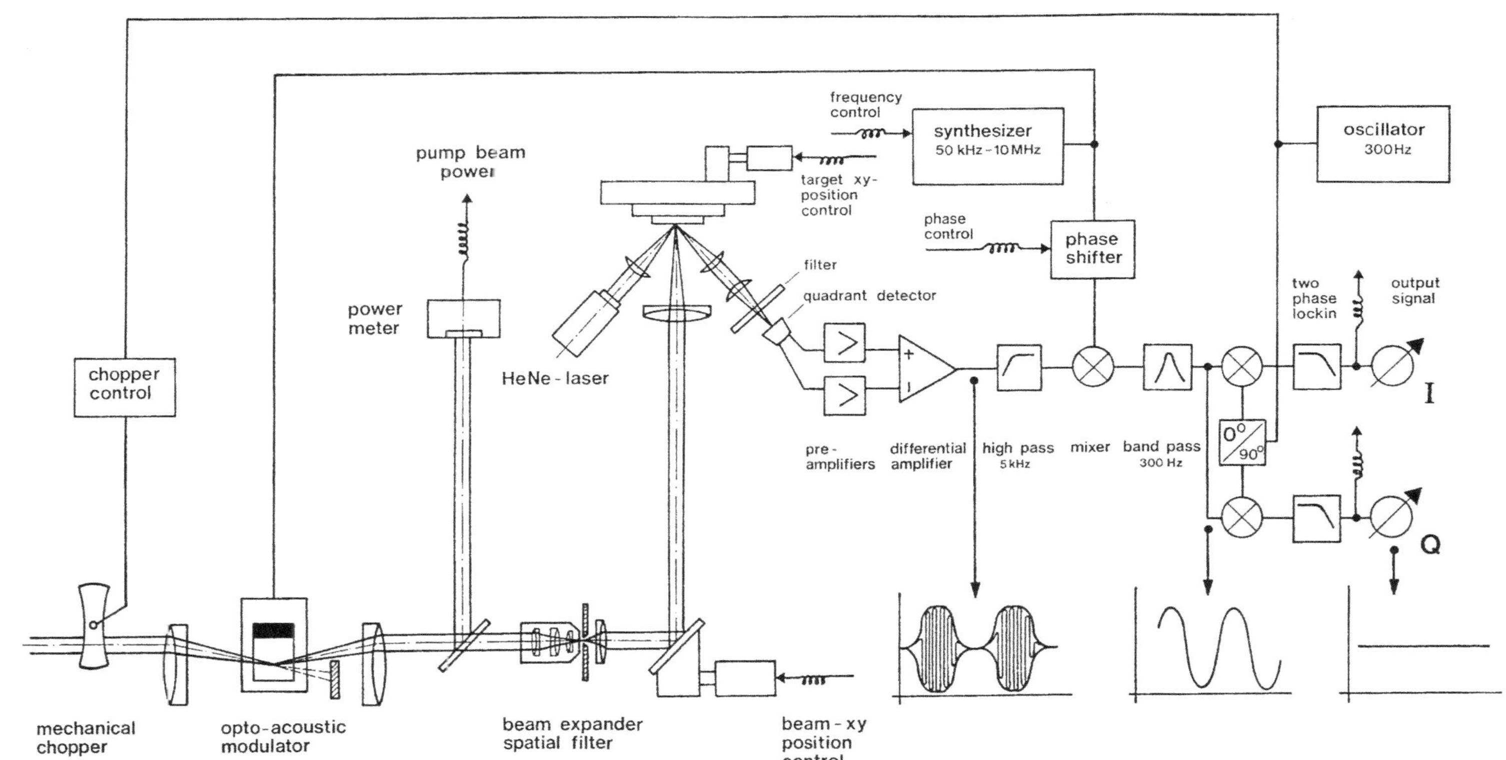

Fig. 1 Experimental setup for high frequency reflectivity and displacement measurements.

necessary to use higher frequencies. Opsal et al. /6/ and Mandelis et al. /7/ reported about measurements using MHz frequencies, however, there have been no systematic studies on the amplitude and phase response in that frequency regime. We developed instrumentation for photothermal reflectivity and displacement measurements covering the frequency range from 100Hz to 10MHz. For frequencies above 50kHz we use a double modulation technique applying a HF-Lock In-mixer in addition to the conventional Lock In-amplifier (Fig. 1). This does not only greatly enhance the signal/noise-ratio but also avoids spurious signals due to stray fields in the high frequency region. The usable dynamic range of the system is 140 dB.

3. Results

The method described above was used to determine thermal conductivities of optical thin film materials. Up to now there is only limited understanding of the thermal properties of these films, however, it is a well accepted fact that the thermal conductivity of many of these materials in the form of thin films differ considerably from bulk values while density and heat capacity are much less influenced by the structural compositon /8/. Conductivity measurements based on our technique yield much more precise values for the conductivity of optical thin films than previously available.

A typical result for a photothermal displacement measurement on a ZrO_2 film on a glass substrate is displayed in Fig. 2. It can be shown that the maximum at 10 kHz in the phase curve corresponds to a thermal length that is equal to the film thickness /3/, however, a complete quantitative description of the surface displacement is not yet available.

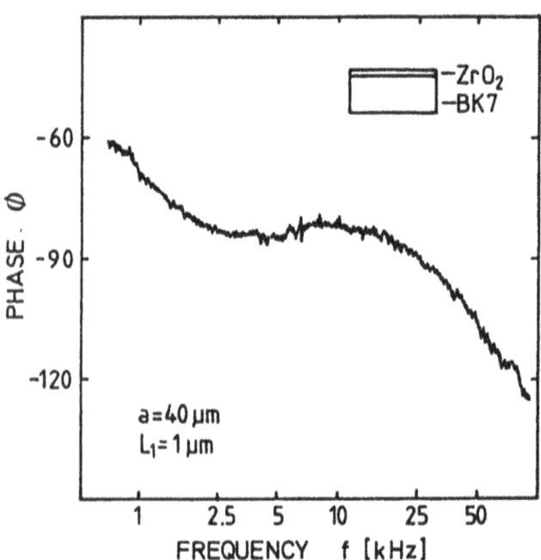

Fig. 2
Frequency dependent phase of the displacement signal for a 1 μm ZrO_2 film on a BK7 substrate.
Film thickness: $L_1 = 1\ \mu m$
Pump beam radius: $a = 40\ \mu m$.
The maximum at 10 kHz corresponds to that frequency where the diffusion length L_{th} in the film is equal to the film thickness L_1.

Therefore, at the present a higher degree of accuracy can be obtained by thermoreflectance measurements as shown in Fig. 3. For these measurements the optical film is overcoated with an additional thermally thin gold layer that acts as a surface absorber for the Ar^+ pump laser light and provides excellent reflectivity and thermoreflectance signals for the HeNe-probe laser wavelength.

To suppress any frequency dependent phase shifts introduced by the measurement system results are normalized against the response from a simple Au film on BK7.

Measurements were performed on 1μm films of SiO_2, Al_2O_3, ZrO_2 and Ta_2O_5 on BK7 glass substrates. We found a significant reduction of the film thermal conductivity compared to the bulk value for SiO_2:(1.18-▹0.25)$Wm^{-1}K^{-1}$, ZrO_2:(1.25-▹0.05)$Wm^{-1}K^{-1}$ and Ta_2O_5:(0.33-▹0.20)$Wm^{-1}K^{-1}$ and an increase for Al_2O_3:(27-▹150)$Wm^{-1}K^{-1}$.

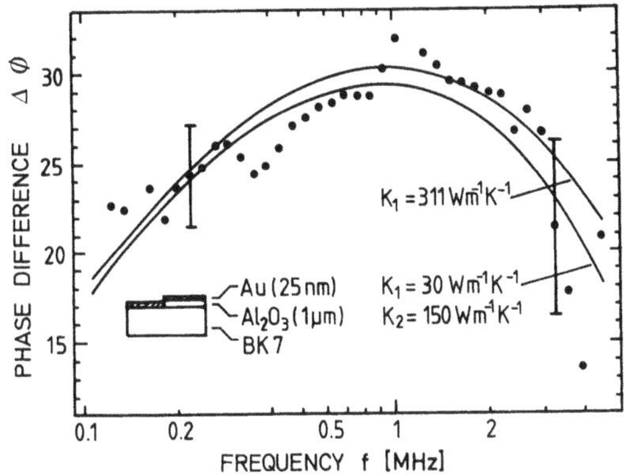

Fig. 3
Normalized phase of the thermoreflectance signal for Al_2O_3/BK7.
Solid lines represent best fits of the theory to the experimental points. The two curves for different overcoat conductivities K_1 demonstrate that the result is not sensitive to changes in K_1.
The fit yields a Al_2O_3 thin film conductivity of $K_2 = 150 Wm^{-1}K^{-1}$.

This work is supported by the Deutsche Forschungsgemeinschaft, Sfb 337.
E. Welsch and Z.L. Wu acknowledge fellowships from the A. v. Humboldt foundation.
The preparation of coating samples by Z.X. Fan is greatfully acknowledged.

References

1. J.P. Roger, F. Lepoutre, D. Fournier, A.C. Boccara, Thin Solid Films **155**, 165 (1987) and references therein.
2. A Skumanich, H. Dersch, M. Fathallah, N.M. Amer, Appl.Phys. **A43**, 297 (1987)
3. M. Reichling, H. Grönbeck, M. Schneider, paper presented at this conference.
4. M.A. Olmstead, N.M. Amer, S. Kohn, D. Fournier, A.C. Boccara, Appl.Phys. **A32**, 141 (1983)
5. E. Welsch, H.G. Walther, K. Friedrich, P. Eckhardt, J. Appl.Phys. **67**(10), 6575 (1990)
6. J. Opsal, A. Rosencwaig, D.L. Willenborg, Appl. Opt. **22**(20), 3169 (1983)
7. A. Mandelis, A. Williams, E.K.M. Sin, J. Appl.Phys. **63**(1), 92 (1988)
8. A.H. Guenther, J.K. McIver, Thin Solid Films **163**, 203 (1988)

Photoacoustic Measurement of the Energy Absorption of a Pulsed Nd:YAG Laser at Cu and Al Surfaces

Y. Ito, S. Nakamura, and F. Hirakawa

Department of Mechanical Engineering, Nagaoka University of Technology, 1603-1, Kamitomioka, Nagaoka, Niigata 940-21, Japan

Abstract. Energy absorption of pulsed Nd:YAG laser radiation at surfaces of Cu and Al were measured by a photoacoustic method as a function of laser energy fluence. Integrated photoacoustic signal intensity linearly correlated to irradiating laser pulse energy at low fluences. When the fluence increased, we found a threshold above which the signal intensity increased with the fluence. It reached the maximum at a certain fluence and then decreased with further increase in the fluence.

1. Introduction

Recently, intense and short pulse lasers have attracted much interest because of their capability in machining. Most lasers used in machining thus far are continuous or quasi-continuous infrared lasers. Even though these lasers can deposit an exceptionally large amount of energy into a limited volume, they cause thermally affected regions near the irradiated zone. It is believed that applying a short intense pulse laser would reduce this affected zone.

The photoacoustic method has been applied to determine damage thresholds of laser irradiated materials[1]. In most of these studies, however, a thermal diffraction or gas-microphone detection was used. These methods become inappropriate for high power laser irradiation because laser induced phenomena such as melt or plasma formation will interfere with the measurements. In the laser machining, however, the melting and/or the removal of materials are essential processes. It is very important to study the interaction between intense laser radiation and matters under these processes in optimizing the laser irradiating conditions to get a desired results. In this study, energy absorption of pulsed Nd:YAG laser radiation at the surfaces of Cu and Al were measured by photoacoustic method using a piezoelectric transducer.

2. Experimental

Commercially available copper (JIS C1100) and aluminum (JIS A5056) were cut into disks measuring 8 mm in diameter and 1 mm thick and were mechanically polished. The disk was mounted on a surface of aluminum plate with silicon vacuum grease. On the other side of the plate, a disk shape piezoelectric transducer (Murata Co. Piezotite 7D-16-1400) was attached with conductive silver paste. This assembly was put in a brass made cylinder with a shroud which allowed an inert gas to flow onto the sample during laser irradiation.

Springer Series in Optical Sciences, Vol. 69 Photoacoustic and Photothermal Phenomena III
Editor: D. Bićanić

A Q-switched Nd:YAG laser which generated pulses of 8 ns duration with the energy up to 900 mJ was used in its fundamental wave length. The energy fluence of the pulse was changed using lenses with focal lengths of 150 and 500 mm. The effect of the pulse energy was examined at 200, 400 and 800 mJ/pulse. A photoacoustic signal was recorded by a digital oscilloscope in pre-trigger mode and then transferred to a personal computer for further processing. The measurements were carried out by one-on-one method. In some cases, the luminescence spectra of laser induced plume were recorded by a multichannel photodetector (SMA).

3. Results and Discussion

The correlation between the photoacoustic signal and absorbed laser energy was checked by irradiating a graphite coated aluminum disk with unfocused laser pulses of different energy. It was found that a signal intensity integrated from 0 to 20 μs linearly correlated to the laser pulse energy.

The time integrated photoacoustic signal intensity (I) increased according to the laser pulse energy. For a certain pulse energy, the I was a function of the energy fluence (F) and had a threshold. When the F was lower than the threshold, the I was constant. At the F higher than the threshold, it increased to reach the maximum I_{max} and then began to decrease with further increase in the F. If we represent the relative changes of I at different pulse energies as the normalized intensity I/I_{max}, the difference due to pulse energy is ceased out and a universal curve is obtained. I/I_{max} as a function of the F is shown in Fig.1 for Cu.

Since the I corresponds to the amount of absorbed laser energy, the I/I_{max} represents the relative change of the absorbed energy. The energy absorption at the metal surface is separated into three F regions as represented in Fig. 1. In region 1, the absorption process is considered as the normal photon absorption by the metal. The absorbed energy is constant because the laser pulse energy is kept constant and the absorption coefficient remains unchanged. In region 2, the metal surface temperature increases high enough to melt and the photo-absorption coefficient greatly increases

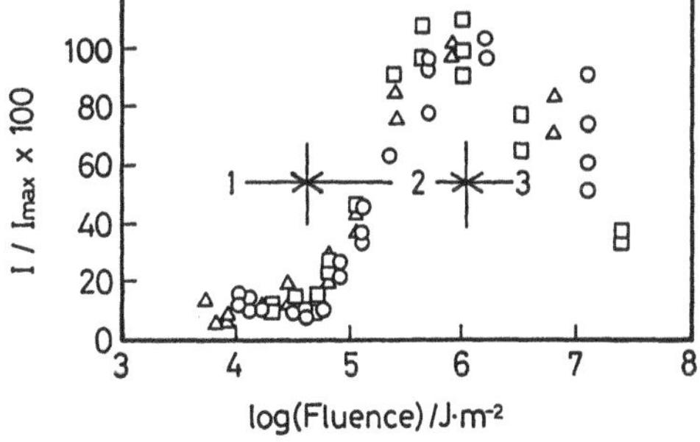

Figure 1 Normalized PA signal intensity I/I_{max} as a function of fluence for Cu. Pulse energy: Δ, 200 mJ; O, 400 mJ; □, 800 mJ. Three fluence regions are indicated.

during one laser pulse. The damage left on the irradiated surface becomes evident above the threshold. Further increase in the F in region 3 causes some processes which inhibit the energy absorption, such as plasma formation, and the energy absorption rate at the surface decreases. The emission spectra of laser induced plume support this inspection. The emission lines of the neutral atom dominated in region 2 while those of the ion became evident in region 3.

Table 1 Laser Fluences of Several Thresholds and I_{max} for Cu and Al (10^4 Jm^{-2})

	Phenomenon	Copper	Aluminum
This work	PA threshold	5.5	2.9
	I_{max}	~102	~51
Jee *et al.* [2]	Slip line	8.8	0.60
	Ripple	10.5	2.30
	Flat melting	11.3	2.99
	Boiling	12.7	3.32

Similar results were obtained for Al. But the threshold energy density and the F at which the I_{max} occurred are lower than those for Cu. They seem to be unique to the material. These data are listed in table 1 together with the corresponding results reported previously[2].

The photoacoustic method described here provides important information about the interaction between intense laser pulses and materials. The energy absorbed by the material increases with laser fluence in intermediate fluence region. In the high fluence region, however, it decreases with the fluence. There is an optimum laser fluence in the efficient energy transfer from laser radiation to matters.

References

[1] E.Matthias, R.W.Dreyfus, in 'Photoacoustic, Photothermal and Photochemical Processes at Surfaces and in Thin Films', ed. P.Hess, Springer-Verlag, (1989), pp. 89-128 and references therein; A.C.Tam, Rev. Mod. Phys., **58**, 381 (1986)

[2] Y.Jee, M.F.Becker, and R.M.Walser, J.Opt.Soc.Am., **B5**, 648 (1988)

Directivity Pattern of OA Signals Accompanying Laser Induced Damage of a Thin Cr Film

D. Maksimović[1], *M. Terzić*[2], *and J. Možina*[1]

[1]University of Ljubljana, Faculty of Mechanical Engineering, P.O. Box 394, 61000 Ljubljana, Slovenia

[2]University of Novi Sad, Institute of Physics, P.O. Box 224, 21000 Novi Sad, Yugoslavia

Abstract. Acoustic waves generated by a Q-switched Nd:YAG pulsed laser on a thin Cr film (80 nm) surface were detected in the free air surrounding the sample. The laser radiation caused visible degradation of the film. Deflection of a He-Ne laser probe beam [1,2] at several horizontal and vertical positions was used as a detection method. Dependence of the signal shape and amplitude on horizontal distance from the source was observed as well as amplitude dependence on source diameter. The amplitude directivity pattern also shows observable diminution of the first maxima amplitude at zenithal position.

We have studied the process of ablation and possible evaporation on the surface of a thin Cr film (80 nm) by optical probing of laser induced acoustic wave in free air space [2,3].

The experimental setup is shown in Fig. 1. Q-switched Nd:YAG laser radiation of 1.067 μm wavelength with 18 mJ of energy and duration of 16 ns (FWHM) was incident normal to the Cr film surface. The radiation used caused visible degradation of the film surface. The waves were detected by probe beam deflection with the knife edge method at various distances z from the source as well as at various horizontal positions x and y (Fig. 1), recorded on HP 54502A oscilloscope and stored in PC IBM computer.

As shown in Fig. 2, the shape of the optoacoustic signal varies with horizontal distance from the source leading to the occurrence of a second positive peak close to primary wave. This can be explained as an interference of acoustic waves due to the finite dimension of the source.

Also interesting for the study of laser induced ablation is the directivity pattern of optoacoustic wave amplitude which is presented in Fig. 3.

To investigate the influence of the source size on these characteristics we conducted experiments with a Q-switched Nd:YAG pulsed laser (FWHM=10 ns), having effectively 24 mJ of energy incident on the sample surface. Spot diameters from 0.5 mm to 3.5 mm were obtained on the sample surface by moving the focusing lens. In all cases the incident beam power density exceeded 10^{11} W/m^2 and visible degradation of the Cr film took place. Fig. 4 shows the angle dependence of peak to peak signal amplitude for three different source dimensions. By increasing the source

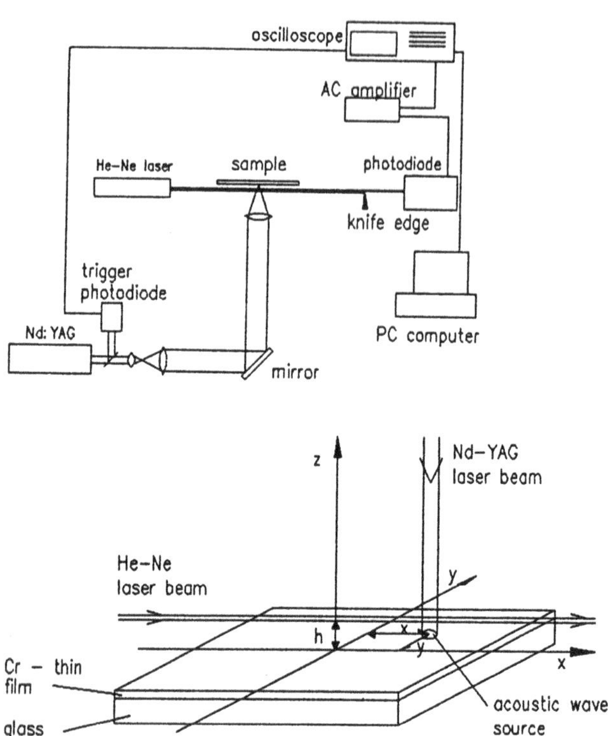

Fig. 1. Experimental setup and geometry

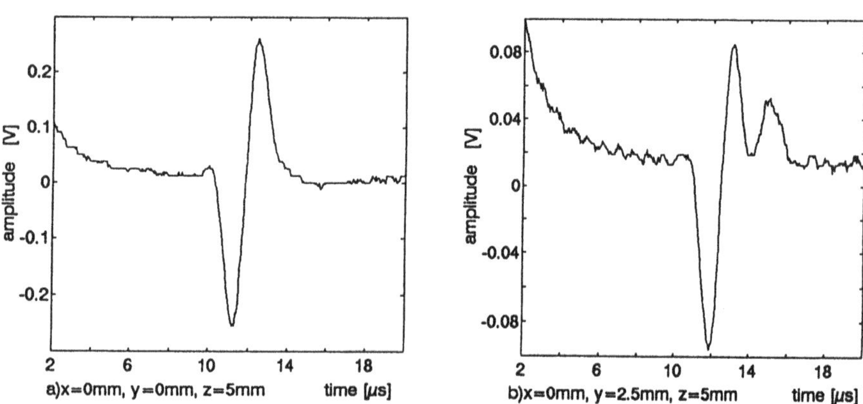

Fig. 2. Optoacoustic signals at different horizontal positions

diameter the directivity pattern becomes more centered around the direction normal to the sample surface.

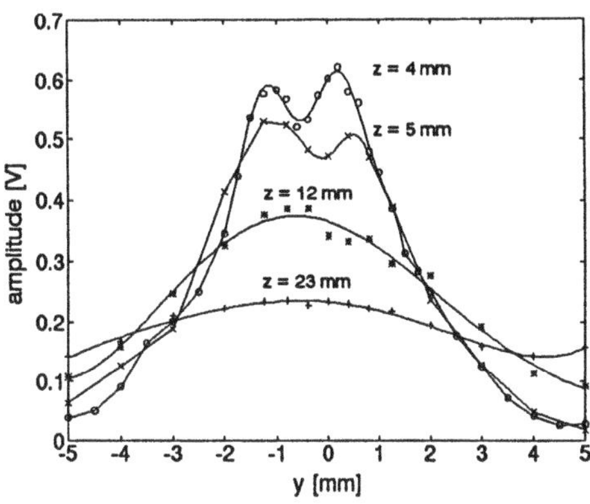

Fig. 3. Probe beam amplitude deflection at z=4,5,12,23 mm

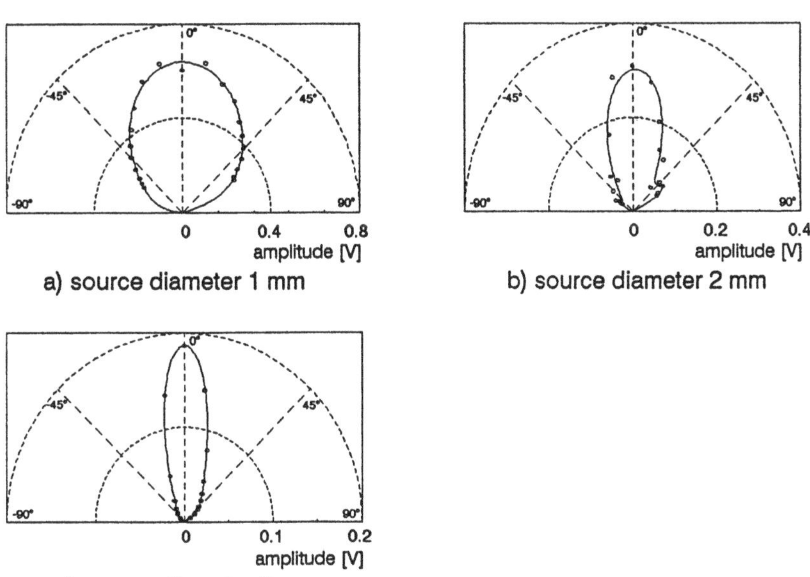

Fig. 4. Directivity pattern for source sizes a) 1 mm, b) 2 mm, c) 3 mm

REFERENCES

1. D. Bauerle ed. Laser Processing and Diagnostics, Springer Ser. Chem. Phys., Vol. **39** (Springer, Berlin, Heidelberg 1984)
2. A. C. Bocarra, D. Fournier, J. Badoz, Appl. Phys. Lett., **36**, 130 (1980)
3. L. C. Aamodt, J. C. Murphy, J. Appl. Phys., **54**, 581 (1983)
4. R. K. Aldhair, P. E. Dyer, Z. Zhu, Appl. Phys. **B 51**, 81 (1990)

Investigation of Surface Laser-Ablation Plumes Using High-Sensitivity Cavity-Enhanced Detection

E.C. Benck and H.A. Schuessler

Department of Physics, Texas A&M University,
College Station, TX 77843-4242, USA

Abstract. Cavity-enhanced detection is used to monitor minute vapor plumes produced by focusing a pulsed laser beam onto a surface placed inside a resonant optical cavity.

1. Introduction

Cavity-enhanced detection has been used as a sensitive probe of laser-ablation plumes [1,2]. The tiny plume ejected from a solid or liquid surface by a focused pulsed laser beam is monitored inside an optical cavity by a fixed frequency cw probe laser beam. Since the probe beam is resonant with the optical cavity, the weak absorption, scattering and optical pathlength changes in the plume are amplified by the quality factor of the high-finesse optical cavity.

2. Experimental Apparatus

The vaporization is produced by a tunable pulsed dye laser called the pump beam which is focused to a 20 μm diameter spot on a sample located inside a spherical etalon. The pulsed dye laser produces pulse energies in the range of 0.35 to 35 μJ around λ=640 nm and with a pulse length of 10 ns. The material removed from the surface is monitored by a weak HeNe laser probe beam. The mode-matched HeNe beam is split into two beams and both are injected into the etalon. One beam is used as the probe beam and grazes the sample surface above the focal spot of the pump laser beam. The other beam is used to stabilize the etalon by locking it to the side of a transmission fringe and, thereby, to the HeNe-laser frequency. The sample is advanced by a precision translation stage between each laser pulse of the pump beam, so that each time a new portion of the surface is exposed.

3. Results

Two types of signals were observed: the thermal and the acoustic ones.

The time evolution of the thermal signal occurs on the msec time scale. The amplitudes and temporal structures of thermal signals of more than 25 different materials were previously reported [1,2]. Fig. 1 shows the two most common shapes. The single peaked structure occurred for all materials at low laser fluences and at higher fluences for the more refractory metals. The double peaked structure was observed only in samples with melting points below 400 °C at the higher laser fluences. It is probably caused by the convective plume of the heated air/ablation product mixture.

Springer Series in Optical Sciences, Vol. 69 Photoacoustic and Photothermal Phenomena III
Editor: D. Bićanić

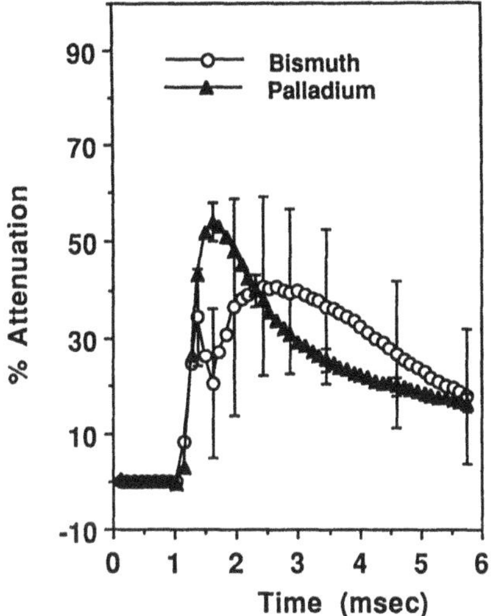

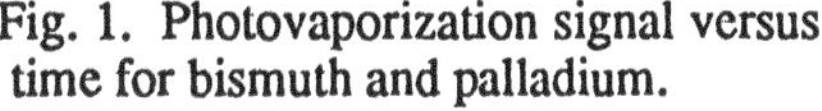
Fig. 1. Photovaporization signal versus time for bismuth and palladium.

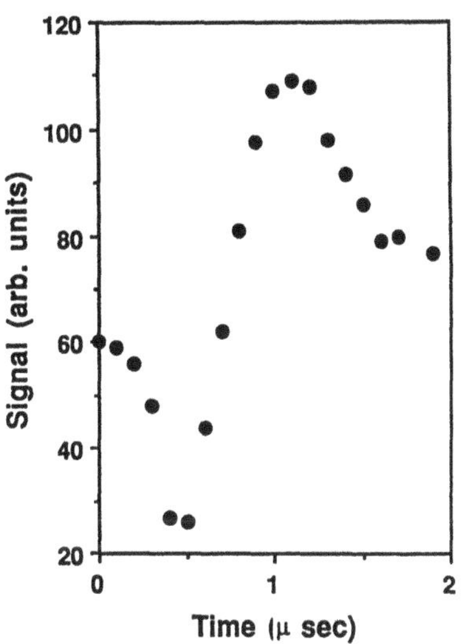

Fig. 2. Photoacoustic signal versus time for aluminum.

Although these double peaked signals varied considerably on a shot-to-shot basis, the average shows consistently the depicted behavior.

The maximum velocity of the convective heated air/ablation product mixtures leaving the surface was investigated by measuring the time delay between the dye laser pulse and the start of the thermal signal as a function of the height of the probe beam above the surface. For aluminum and lead, signals could be obtained up to heights of 1.5 and 6.5 mm, respectively. The corresponding maximum velocities of the leading edge were 2 and 9 m/s, respectively. These results display the same behavior as obtained in photothermal deflection measurements [3].

Fig. 2 displays the acoustic signal. It occurs on the μsec scale and displays sign reversals.

Examining the surfaces with a microscope showed surface damage in all samples. The width of the damage spots produced on low-melting-point targets, such as lead (T_m =328 °C), is three times as large as the focal spot of the laser. In the other extreme, the cratering produced in refractory metals, such as tungsten (T_m=3410 °C), was much smaller than the focal spot and had a diameter of only a few μm. For thin films, such as aluminum on glass, the film could be completely removed, leaving a clean substrate.

More quantitative measurements were made of the surface modifications produced by a single laser pulse on aluminum using stereoscopic scanning electron microscope images. In this case the apparent surface damage consisted of several shallow craters surrounded by tall narrow rims and is due to the transverse mode structure of the dye laser beam. Fig. 3a is a photograph of one of these craters. Approximately 25 μm^3, corresponding to 13 pg, of aluminum was removed from the surface. About twice as much material was pushed aside forming the crater rims. This behavior is expected from the theoretical prediction that the majority of material removed by a low energy laser pulse is due to lateral expulsion of liquid metal [4].

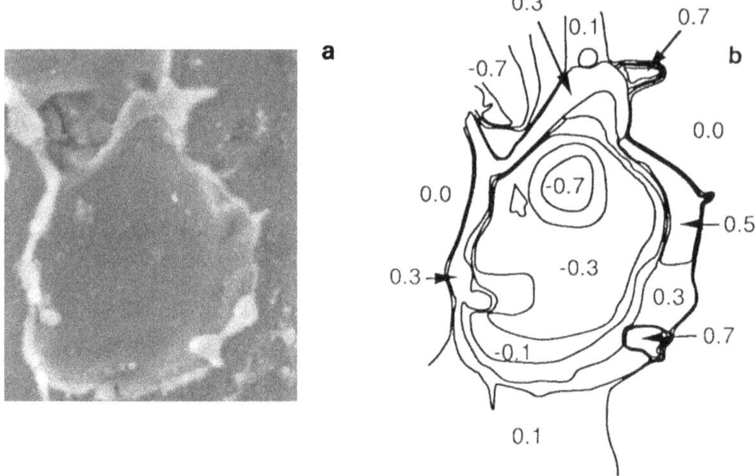

Fig. 3. Laser ablation crater on aluminum representing approximately one fifth of the cratering produced by a single laser pulse. a) Scanning electron microscope image. b) Topographic map of the same crater based on stereoscopic scanning electron microscope images. The numbers represent the height of the indicated region in μm.

This work was supported by the Texas Advanced Research Program, NSF, and the Center for Energy and Mineral Resources at Texas A&M University.

References

[1] E. C. Benck, Z. Rong, S. H. Chen, Z. C. Tang, and H. A. Schuessler, Appl. Phys. Lett. **58**, 1476 (1991).

[2] E. C. Benck, Z. Rong, S. H. Chen, and H. A. Schuessler, Appl. Surface Science **48**, 257 (1991).

[3] J. A. Sell, D. M. Heffelfinger, P. L. G. Ventzek, and R. M. Gilgenbach, J. Appl. Phys. **69**, 1330 (1991).

[4] C. L. Chan and J. Mazumder, J. Appl. Phys. **62**, 4579 (1987).

The Thermal Resistance of Grain Boundaries Determined by Modulated Optical Reflectance

G. Meyer-Berg[1], *R. Osiander*[1,*], *P. Korpiun*[1], *P. Kakoschke*[2], *and H. Joswig*[2]

[1]Physik-Department E13, Technische Universität München, James-Franck-Strasse, W-8046 Garching, Fed. Rep. of Germany

[2]Siemens AG, HL PE 21, Otto-Hahn-Ring 6, W-8000 München 83, Fed. Rep. of Germany

*Present address: Johns Hopkins University, Applied Physics Laboratory, Laurel, MD 20723, USA

The present work is based on the experimental technique of modulated optical reflectance (MR) developed by Rosencwaig et al. /1/ and its application to the characterization of sputter deposited metal films /2/.

Thin films of aluminium (1 % Si, 0.5 % Cu) of 800 nm thickness sputter deposited onto a 100 nm thick SiO_2–Si layer on a silicon substrate are polycrystalline. The average grain size depends on the temperature and the time of the thermal treatment. The average grain size of the samples we investigated varied between 0.15 and 2.54 μm.

We measured the modulated optical reflectance of such films by scanning an intensity modulated beam of an Ar laser, the pump beam (focus diameter 1.6 μm) and the beam of a HeNe laser (focus diameter 2.6 μm) with fixed beam separation of 1.5 μm along a distance of 40 μm crossing various grains (Fig. 1). The modulation frequency was 1 MHz. Each scan consisted of 100 measurements separated by 0.4 μm. The variation of the intensity in the scan about a mean value can be considered as a standard deviation as proposed by the Therma Wave Group /2/. This group found that the standard deviation increases with grain size. The results of our measurements on grains with sizes between 0.15 and 2.54 μm plotted in Fig. 2 confirm this observation.

It is obvious that the magnitude of the oscillation of the MR signal in a scan like that shown in Fig. 1 may be related to the reduced heat flow across a grain boundary. To interprete the experimental results we take into account that the thickness of the AlSiCu film is much smaller than the thermal diffusion length. In this case we can assume the temperature to be constant in the film along the z–direction. The thermal conductivity of the insulating SiO_2 layer is very small, so that heat flow

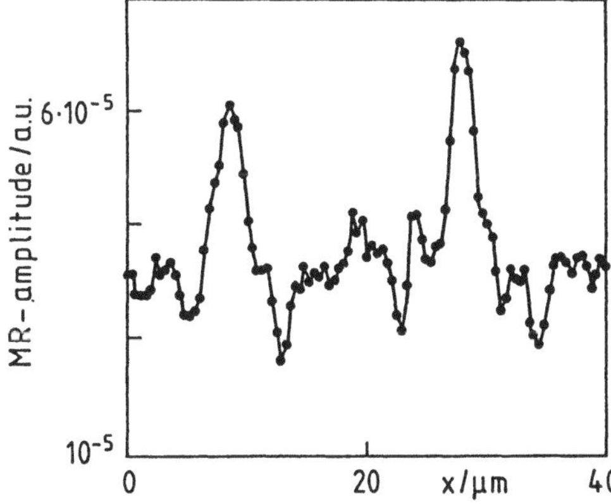

Fig. 1
MR amplitude measured on polycrystalline AlSiCu along a distance of 40 μm. Average grain size is 2.23 μm.

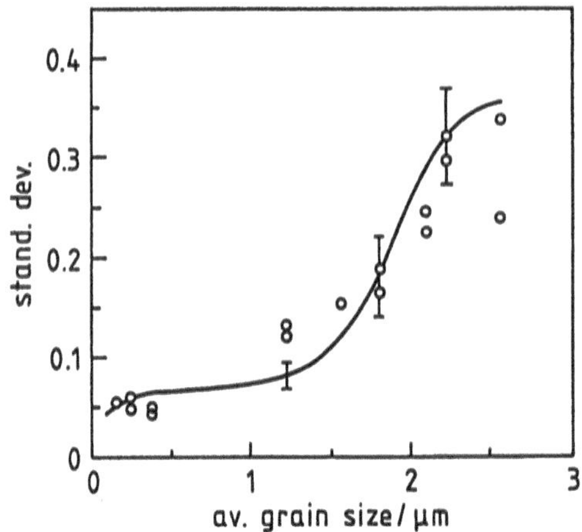

Fig. 2
Standard deviation of MR amplitude referred to the average value versus grain size.
o: experimental data; solid line with error bars: calculated for thermal resistance
$r = (1.1\pm0.3)\cdot10^{-2}\text{KcmW}^{-1}$

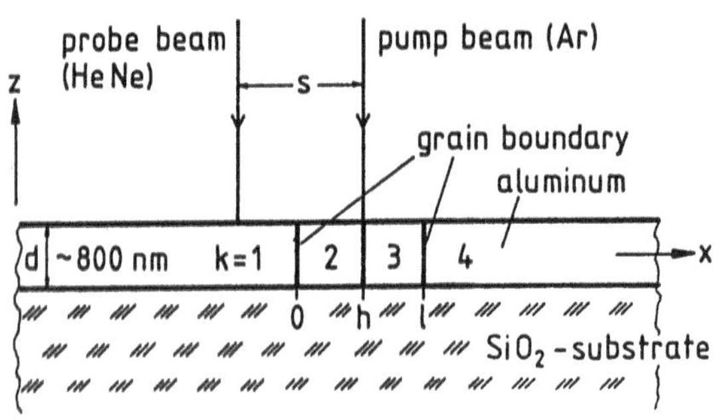

Fig. 3
Model for a single grain

into the substrate can be neglected. Thus, we only consider heat conduction along the x–axis. Because the thickness of the grain boundaries is very small compared to the thermal diffusion length in AlSiCu, the effect of the boundaries on the conduction of heat can be treated as a thermal resistance. In a first step, we approximate the grain by a strip grain with a width of l. The grain boundaries at x = 0 and x = l (Fig. 3) act as thermal resistances of magnitude r. The distance between both beams is s = constant.

Further we assume the laser beams to be infinitesimal small in x–direction but extended in y–direction. Thermal waves will run from the pump beam position h to both directions. For a grain labelled by k it may be approximately described by plane waves

$$T(x,t) = \left[A_k\, e^{(1+i)\sqrt{\omega/2\alpha}\,x} + B_k{}^{-(1+i)\sqrt{\omega/2\alpha}\,x}\right] e^{i\omega t}, \tag{1}$$

where α is the thermal diffusivity of the metal film and $\omega/2\pi = f$ the modulation frequency. The amplitudes A_k and B_k can be calculated using the boundary conditions for temperature and thermal flow $J = -\lambda\, dT/dx$ at the grain boundaries and the plane $x = h$ where the film is heated by the laser beam. Continuity of heat flow and temperature yields $J_1(0) = J_2(0)$, $J_3(l) = J_4(l)$ and $T_2(h) = T_3(h)$.

From the discontinuities, caused by the thermal resistances at x = 0, and x = l and the heat source at x = h we get

$$J(0)=\left[T_1(0)-T_2(0)\right]/r,\ J(l)=\left[T_3(l)-T_4(l)\right]/r \text{ and } J_3(h)-J_4(h)=Q_0. \qquad (2)$$

Q_0 is the heat supplied by the pump beam. T(x) can be calculated at the position x of the probe beam for a pump beam at the position h. It is obvious that the signal is maximal if the pump and the probe beam hit the same grain. Its amplitude increases with decreasing grain size. The signal becomes minimal, if the beams are seperated by a grain boundary. Thus, the difference in the magnitude between the maximum and the minimum is a function of the resistance between the grains and of the grain size itself. Because the grain size is in the order of magnitude of microns, it is difficult to measure on isolated grains. In order to determine the thermal resistance of grain boundaries from the experimental scans we extend our simple model to an assembly of several grains. Therefore, we express the temperature distribution, equ. (1), for an array of many grains by $T(x,s) = T_0(x \bmod l - s,\ x \bmod l)$, where l is equal to the average grain size.

In order to compare the caluclations to experiments with finite laser beam diameters we average the temperature variations at different probe and pump beam positions with the Gaussian weight function given by the beam profiles. The temperature amplitude A(x) now is proportional to the signal of the modulated reflectivity, and with the calculations done above we can simulate line scans over an assembly of grains with a constant average size l. We obtain the average and standard deviation of amplitude and phase angle for one scan as a function of grain size and thermal resistance. For s = 1.5 μm we find the best agreement between the standard deviation of the simulated scans and the measured scans for a thermal resistance of r = $(1.1 \pm 0.3)\ 10^{-3}$ Kcm^2/W as it is shown in Fig. 2. The error bars correspond to $\Delta r = \pm 3\cdot 10^{-4}\ Kcm^2/W$.

We may conclude that our model describes not only the standard deviation, but also the average of both, phase angle and amplitude. Though the assumptions made for the model are rough, the agreement between experiment and calculation is quite well. Furthermore the absolute value for the resistance seems to be quite reasonable. It is in the order of magnitude of the thermal resistance which Swartz and Pohl /3/ found for amorphous SiO2 in a layered system RH:Fe/SiO_2/Al_2O_3. We can estimate the thickness ε of the grain boundary to be about 25 to 50 Al atoms, if we use $r = \varepsilon/\lambda$ together with $\lambda = 10$ mW/cmK, a reasonable value for amorphous materials.

/1/ A. Rosencwaig, J. Opsal, W.L. Smith, D.L. Willenborg, Appl.Phys.Lett. 46, 1913, 1985

/2/ Technical report of Therma Wave Inc., Freemont / Calif. (USA), 1989

/3/ E.T. Swartz and R.O. Pohl, Appl.Phys.Lett. 51, 2200, 1987

Determination of the Thickness, Density and Elastic Properties of Thin Films with Laser Generated Surface Acoustic Waves

A. Neubrand, A. Mayer, and P. Hess

Institute of Physical Chemistry, University of Heidelberg, Im Neuenheimer Feld 253, W-6900 Heidelberg, Fed. Rep. of Germany

Abstract. A noncontact technique for the determination of the dispersion of surface acoustic waves (SAWs) is presented. It is demonstrated that the thickness, elastic properties and densities of films deposited on the surface can be derived from dispersion data. Results are given for very thin gold films and diamond-like-carbon films.

1. Introduction

Surface Acoustic Waves (SAWs) interact with films due to the fact that their energy is concentrated at the surface of a solid. Hence, surface acoustic waves can be used to determine the acoustic properties of a surface layer. Broadband SAW pulses with bandwidths exceeding 200MHz can be excited conveniently by absorption of laser radiation in a region close to the surface and their detection is facilitated by the development of new piezoelectric foil transducers [1]. However, in many situations it is desirable to study the acoustic properties of a surface by a noncontact technique such as the combination of photoacoustic generation and interferometric detection of SAW pulses [2,3]. In the present paper it is demonstrated that the SAW pulse shapes observed in such an experiment can be employed to determine the thickness, density or elastic constants of a layer deposited on the surface.

2. Experimental

SAW transients were excited by XeCl excimer laser pulses of 20ns pulse duration (FWHM). For this purpose, the 308nm radiation from this laser was passed through a variable slit. The image of this slit was projected onto the sample's surface by a cylindrical lens. In this way a rectangular area of 10-20μm width and 20mm length was heated by laser pulses of 20-400μJ energy. The SAW pulse generated by the excimer laser was detected after propagating a distance between 5 and 25mm using the sample as one of the mirrors of a Michelson interferometer. The interferometer could detect a displacement of 0.2Å in a 100MHz bandwidth using a HeNe laser as light source and a PIN photodiode with GaAs preamplifier as detector. A phase-sensitive scheme for the stabilization of the interferometer was used which permitted averaging of many single pulse experiments by a digital storage oscilloscope. The detection limit of the interferometer was thus sufficient to record SAW pulses of typically 1Å amplitude.

Fourier analysis of SAW pulse shapes at different distances from excitation line allowed the dependence of SAW velocity and attenuation on acoustic frequency to be computed. The elastic constants and densities of the films were determined by fitting the observed SAW dispersion curves with theoretical models [3,4].

Springer Series in Optical Sciences, Vol. 69 Photoacoustic and Photothermal Phenomena III
Editor: D. Bićanić

The gold films under study were sputtered on high quality fused silica substrates (Suprasil) of 3mm thickness. Diamond-like-carbon (DLC) films of 0.5-3.5μm thickness were prepared by electron beam evaporation of graphite in an ethylene atmosphere (10^{-4}mbar). During the deposition process amorphisation of the films was caused by argon and neon ions of 6keV energy, a method also known as ion beam assisted deposition (IBAD).

3. Results and Discussion

Due to the low sound velocity and high density of gold, even thin layers of this material cause a strong decrease of the SAW phase velocity with frequency, which was measured with a precision of a few m/s (Fig.1). For thick gold films (>0.3μm) the observed dependence of SAW velocity on frequency was nonlinear (Fig.1). Even for the thickest gold film, the dispersion curve could be fitted well with film densities and sound velocities of the bulk material known from the literature. However, the best fit was obtained for a film thickness of 957nm instead of the thickness of 982nm determined by a profilometer (Table 1). For a product of film thickness and acoustic wave number of less than 0.05 (i.e. a thickness of less than 0.3μm), the dependence of SAW phase velocity on frequency is almost linear, and thus only one of the parameters influencing the slope can be determined. Therefore, the dispersion curve was fitted for the thinner films with thickness as the only free parameter. The dispersion curves could be fitted well with film thicknesses slightly lower than those determined by the profilometer. The discrepancy was especially pronounced for the two thinner films (19nm and 26nm) and it is believed that this is at least partly due to gold island formation on the substrate surface, which occurs in these thin films. As the profilometer used a stylus of 14μm radius of curvature to scan the film surface, it sensed only the top of the islands leading to an apparently higher film thickness. This is further supported by film thicknesses determined by a quartz microbalance, which are much closer to the values determined by the SAW technique. It remains unclear whether the differences between the SAW and microbalance measurements are due to elastic anomalies of these extremely thin films [5] or whether models assuming a uniform film thickness lead to inaccurate results.

DLC films cause an increase of SAW phase velocity with frequency (Fig.1). Their elastic properties and densities were unknown, and thus, they were determined by fitting the dispersion curves with film density ρ, longitudinal and transverse sound velocities of the film v_l and v_t as free parameters. A fixed film thickness determined by a profilometer was used. The results for different ion species and ratios of bombarding ions to deposited atoms I/A are summarized in Table 2.

Table 1 Thicknesses of various gold films determined by acoustic and profilometric methods

	Film 1 h [nm]	Film 2 h [nm]	Film 3 h [nm]	Film 4 h [nm]
SAW	957 ± 2	65 ± 1	22 ± 1	12.5± 1
Microbalance	933 ± 5	60 ± 0.3	20 ± 0.1	10.0± 0.1
Profilometer	982 ± 20	65 ± 5	26 ± 3	19.0± 2

Table 2 Properties of DLC films

	Film 1	Film 2	Film 3	Film 4
Ion source	6keV Ar	6keV Ar	6keV Ne	6keV Ne
I/A	0.022	0.043	0.047	0.040
h [nm]	580 ± 40	958 ±140	928 ± 40	3430±120
ρ [kg/m^3]	1890±150	2010±100	2080±100	2250± 50
vt [m/s]	5600±1500	5700±1000	5700±1000	6700±1000

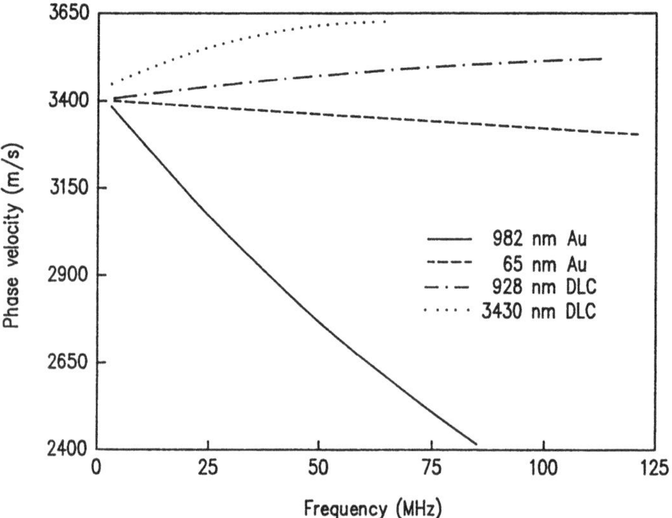

Fig.1 Measured dependence of SAW phase velocity on SAW frequency for gold and DLC films on a fused silica substrate. Continuous line: 982nm gold film; dashed line: 65nm gold film; dash-dotted line: 928nm DLC film; dotted line: 3430nm DLC film. The theoretical fits are in excellent agreement with the measured dispersion curves.

4. References

1. A.Neubrand and P.Hess: Mater.Sci.Eng. **A122**, 33 (1989) and A.Neubrand, P.Hess, H.Coufal, and R.K.Grygier: These proceedings
2. A.Neubrand, L.Konstantinov, and P.Hess: *Physical Acoustics,* ed. by O.Leroy, M.A.Breazeale (Plenum, New York, 1991)
3. A.Neubrand and P.Hess: accepted by J. Appl. Phys.**71**, (1992)
4. H.F.Tiersten: J.Appl.Phys. **40**, 770 (1969)
5. D.Wolf: Surf.Sci. **225**, 117 (1990)

Macro-, Meso- and Microscopic Heat Diffusion in Ceramics

F. Lepoutre, D. Fournier, and A.C. Boccara*

Laboratoire d'Optique Physique, ESPCI,
10, rue Vauquelin, F-75005 Paris, France
*Present address: ONERA, Thermophysics Division,
B.P. 72, F-92322 Châtillon Cedex, France

Abstract: Using three kinds of photothermal experiments, mirage, microscope and time resolved radiometry, we have been able to study heat diffusions at grain, macroscopic and intermediate scales (called mesoscopic). A theoretical analysis able to explain the transition between micro to macroscopic scales is also presented.

The ceramics such as Al_2O_3, SiC and AlN are used in microelectronics for their qualities of electrical insulation and good thermal conduction. But, even if the grains are very good thermal conductors, the macroscopic thermal properties of the polycristalline ceramics are quite sensitive to the physical processes used for their preparations. We propose to follow heat diffusion in these materials from microscopic to macroscopic scale in order to relate the microstructural defects to the macroscopic heat diffusion.

To achieve this local characterisation, we have used a photothermal microscope that can be considered as a thermal probe at micronic scale. Its principle lies on a modulated photothermal effect (photoreflectance) with the particularity of pump and probe beams highly focused by a microscope (typically $1\mu m^2$). As it is always the case in modulated photothermal effects, the phase of the signal only depends on the local thermal properties of the specimen. More precisely, the phase images can be related to the thermal barriers present in the material.

Figure 1 is an example of a thermal map which has been obtained with a photothermal scanning microscope. Each point ($0.25\ \mu m^2$) of this image represents

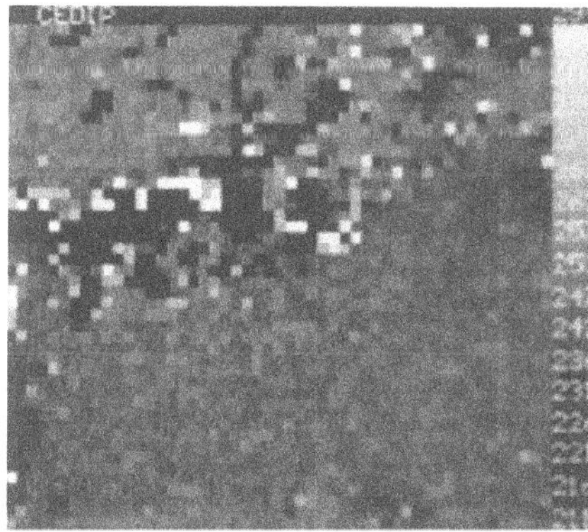

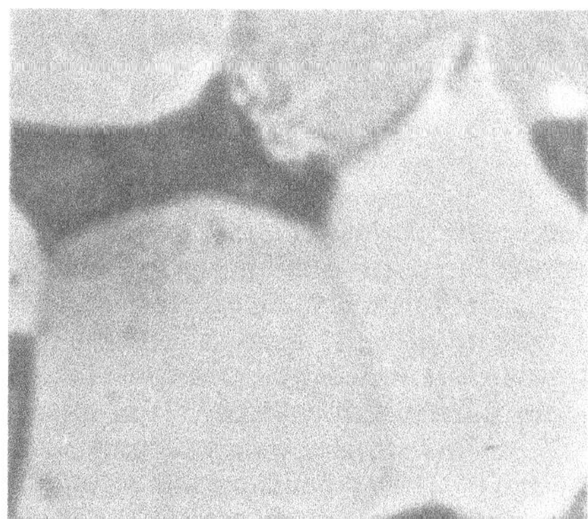

Figure1: Photothermal scanning microscope and optical images of grains of AlN (obtained at Div. Thermophysique of O.N.E.R.A.). The grain boundaries can contain secondary (YaG-like) phasis that appear in dark in the optical image. The scale of grays in the photothermal image is given in degrees. Scale 25µmx25µm.

the phase of the photothermal signal recorded at 2MHz with an AlN sample prepared at high temperature (2000°C). The grains on the lower part of the figure are separated from the ones of the upper part by an important thermal barrier ($R_T \approx 10^{-5}$SI) that can be associated with a secondary phasis located at the grain boundary as revealed by an X-ray microanalysis /2/. Between the grains of the lower part of the figure, the grain boundary without secondary phasis is not revealed in the photothermal image which proves that this grain boundary is not a thermal resistance.

In Table I, we summarize the mean values of the thermal resistances that we measured in three different samples of AlN. It is interesting to note that these values are three or four orders of magnitude smaller than the ones measured at macroscopic scales, since the reference value in the microscope is the thermal resistance of a grain.

TABLE I

Sample	$\overline{l_s}$	$\overline{l_s} / k_s'$	$\overline{R_T}$	k_s	$k_{s\,m}$
High purity T=1900°C	3μm	10^{-8}	10^{-9}	270	220
Impurities Y_2O_3 T=1800°C	8μm	$2\ 10^{-8}$	10^{-8}	220	140
Impurities Y_2O_3 T=1970°C	10μm	$3\ 10^{-8}$	10^{-7}	75	80

The second step of the study is to try to recover the macroscopic thermal properties from those microscopic results. The simplest way to achieve this relation consists to operate an average of the experimental microscopic results on a rather large number of grains randomly distributed in the specimen. For instance, if we deal with a ceramic in which the thermal barriers are located at the grain boundaries, we can relate the macroscopic conductivity k_s of the ceramic to the mean value of the thermal resistances of the boundaries, R_T, by: $l_s/k_s=l_s/k_s+R_T$ where k_s is the thermal conductivity of a grain and l_s the mean value of the grain size.

The values obtained by this procedure (k_s in Table I) were then compared with macroscopic measurements ($k_{s\,m}$ in Table I) of the thermal diffusivities determined using the mirage effect (see paper PX/4 in these proceedings). The agreement between k_s and k_{sm} can be considered as rather good accounting for the simplicity of the model of averaging used.

Nevertheless to try to improve the relation between the microscopic barriers and the macroscopic thermal diffusivities, we have inspected the specimens at a mesoscopic scale (typically millimetric scale). A Scanning Electron Microscope of retro-diffused electrons image of one of the specimens revealed that the secondary phasis is randomly distributed around the grain boundaries and looks like a percolation network. Since it has been demonstrated that in such a case, heat diffusion is strongly affected by the fractal dimension of the medium /3/, we have performed pulsed infrared thermography /3/ of these AlN ceramics.

The samples were uniformly illuminated by a YAG laser pulse (duration 10ns) and their infra-red emission was recorded as a function of time. In order to absorb the laser pump beam at the sample surface, the specimens were coated with a very thin layer of carbon (680A). Figure 2 gives the temperature vs time obtained with two different samples, (mean grain sizes 20μm, fig.2a, and 8μm, fig.2b, respectively) which can be considered as percolation networks. The slope difference observed between the two samples during the experimental time window cannot be explained by

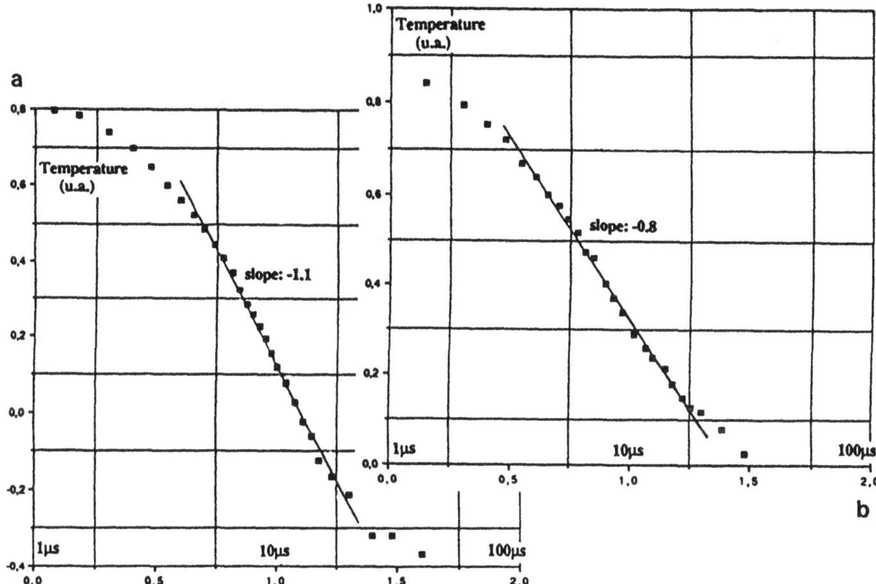

Figure 2a: Time dependence of an AlN sample (grain size 20µm) with a fractal distribution of thermal barriers. The slope is found equal to -1.1.

Figure 2b: Time dependence of an AlN sample (grain size 8µm) with a fractal distribution of thermal barriers. The slope is equal to -0.8.

the coating presence. On the contrary, if ones assumes that, at the time scale of the experiments, heat diffuses in percolation networks produced by the thermal barriers related to the secondary phasis, a slope variation is produced by the difference between the grain sizes of the two samples (fig.3). It must be noticed in figure 3 that the fractal diffusion can be seen only in the sample of figure 2a (large grain size): the grain size of the sample of figure 2b is too small and the slope remains the same as the one which would be obtained with a non fractal ceramic covered by a carboncoating (slope equal to -1).

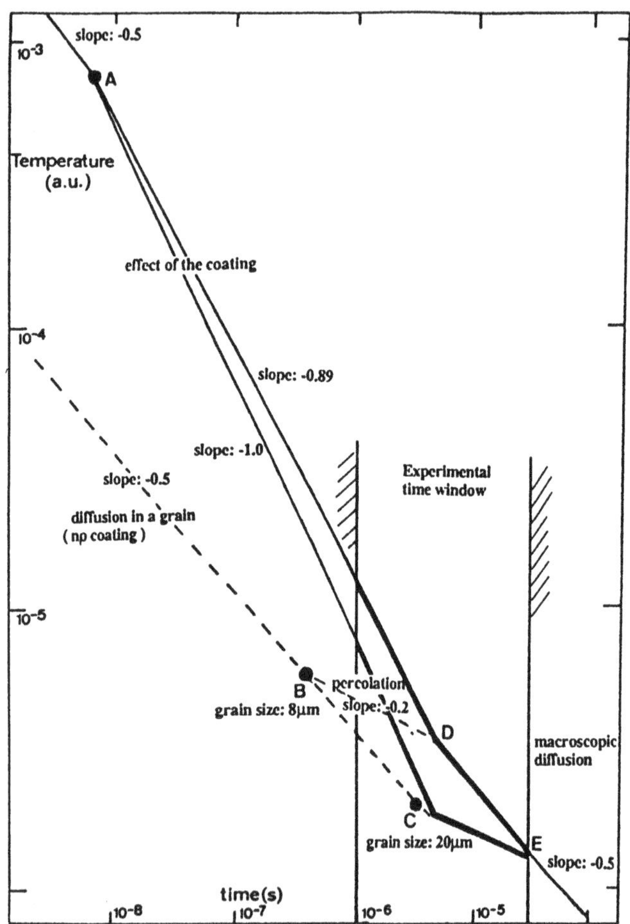

Figure 3: Interpretation of the slopes of the temperature vs time dependences of fig.2. Dashed lines: temperature decrease of the fractal AlN ceramics without carbon coating. Thin continuous lines: temperature decrease of the fractal AlN ceramics coated by 640A of Carbon. Thick continuous lines: part of the calculation in the experimental window of time. Upper curves correspond to the small grain size sample (fig.2b). Lower curves to large grain size sample (fig.2a). At times shorter than point A, heat diffuses in the Carbon coating (for coated AlN). At times shorter than B, heat diffuses in grains smaller than 8µm. At times shorter than C, heat diffuses in grains smaller than 20µm. Between B and D and between C and E, the diffusion is fractal in the percolation network of the thermal barriers. After D or E, the medium can be considered as homogeneous.

References:

/1/ L.J. Inglehart, A. Broniatowski, D. Fournier, A.C. Boccarra and F. Lepoutre, Appl.Phys. Lett. 56, 1749 (1990).

/2/ J.P. Sachet, F. Lepoutre, J.Y. Laval and A.C. Boccara J. Physique 51, C1, 921 (1990)

/3/ D. Fournier and A.C. Boccarra Physica C 157, 587 (1989).

Non-Destructive Evaluation of Microdefects in Ceramic Materials by Means of Photoacoustic Spectroscopy

K. Iwasaki

New Materials Research Laboratories, Nisshin Steel Co., Ltd., 7-1 Koya-shinmachi, Ichikawa, Chiba 272, Japan

Abstract. An automatic laser PAS system for the non-destructive evaluation (NDE) of micro-defects in solids has been developed. The promising feature of this system is demonstrated by showing some examples of observations of alumina particles dispersed in an aluminum nitride matrix. PAS can detect the particles by utilizing large difference in thermal conductivity between the matrix and the particles, while it is very difficult to detected them by ultrasonics because of their small difference in elastic properties. It is also shown that three-dimensional observations can be made by combining XY-scanning with the variation of modulation frequencies.

1. Introduction

Since the possibility of an application of photoacoustic spectroscopy (PAS) to the non-destructive evaluation (NDE) of solids was first suggested by Rosencwaig [1,2], there have been several trials to apply PAS to NDE of ceramic materials. In NDE by PAS the acoustic or the elastic waves which arise from the periodic illumination of a specimen with a chopped laser beam are utilized. As PAS reflects optical, thermal and elastic properties at the same time, it is possible to detect such defects that give any change to at least one of these properties. This should be contrasted to the cases of popularly used ultrasonic and X-ray methods, where only elastic properties and absorption coefficients are reflected, respectively.

Thermal diffusion length δ given by the equation,

$$\delta = \sqrt{\kappa / \pi \rho c f} = \sqrt{\alpha / \pi f}, \tag{1}$$

plays an important role in PAS, where κ, ρ, c, f and α are thermal conductivity, density, specific heat, modulation frequency and thermal diffusivity, respectively. The resolving power is determined by δ and the beam spot size, while the penetration (detection) depth for an opaque specimen is roughly equal to δ. It is to be noted here that δ is proportional to $1/\sqrt{f}$, which means that information in Z-direction (direction of the laser beam) is also obtainable by changing the modulation frequency. It is then possible to carry out three-dimensional observations of defects by combining this effect with XY-scanning of illuminating positions.

PAS signals consist of two parts, amplitude and phase angle. The former is known to have higher resolving power but shorter detection depth than the latter by a factor of about 2. Then the defects located in deeper positions from the surface can be detected more easily by the latter than by the former, though the sharpness of the obtained image is lower.

2. Experimental Procedures and Specimens

As the details of the measuring system have been given in the previous paper [3], they are not repeated here. PAS measurements are carried out in two scanning modes, X- and XY-scanning, with a point-focussed laser beam. The number of measuring points is 400 for X-scanning and 6400 (80 x 80) for

Springer Series in Optical Sciences, Vol. 69 Photoacoustic and Photothermal Phenomena III
Editor: D. Bićanić

XY-scanning. In the latter case the obtained signals are image-processed and the results are displayed in 25600 (160 x 160) pixels.

The specimen is a powder-sintered disk of 11.1mm in diameter and 1.8mm in thickness. Starting material is the mixture of 57.7% AlN and 32.3% Al ultrafine powders (mean grain size is about 100nm) and 10% alumina particles of about 100μm in diameter. The powder mixture is first mechanically pressed and subsequently reaction-sintered in nitrogen atmosphere, which results in a uniform dispersion of alumina particles in AlN matrix. This was confirmed by EPMA (electron-probe micro-analyzer) observations, where the oxygen signal was monitored to check the distribution of alumina particles.

3. Experimental Results and Discussion

The amplitude and the phase angle image of an alumina particle taken by XY-scanning are shown on the left and the right sides of Fig. 1, respectively, where the observed area is 400μm x 400μm. All the XY-maps shown in this paper are constructed by first dividing the obtained PAS signals into 8 equidistant levels between the maximum and the minimum values and then assigning a different tone to each level. Black and white colors correspond to the maximum and the minimum sides, respectively. The units of the maximum and the minimum values indicated in the figure are nV for amplitude and degree for phase angle.

Figure 2 shows the results of X-scanning along a line which passes through the center of the particle shown in Fig. 1. The particle is revealed as a peak both in amplitude and in phase angle. The half width of the amplitude peak is about 80μm and that of the phase angle peak is about 170μm. This agrees very well with the theoretical prediction that the ratio of the latter to the former should be about 2. It is then reasonable to assume that the value of δ for the phase image is equal to twice of that for the amplitude image calculated by Eq. (1).

A series of XY-maps of several alumina particles taken at different modulation frequencies f are shown in Fig. 3, where the observed area is 800μm x 800μm. The values of δ given in Fig. 3 are estimated by assuming α=0.2cm^2/sec. As the value of α depends largely on specimen conditions, the absolute values of δ quoted here should not be taken seriously.

As the resolving power is low (δ is large) at f=250Hz, particles cannot be observed in Fig. 3A. When f is increased up to 638Hz, several particles begin to appear in the amplitude image (Fig. 3B left) but only a few of them are observed vaguely in the phase image (Fig. 3B right). At f=1770Hz all the particles are observed very clearly both in the amplitude and in the phase images (Fig. 3C). It should be noted here, however, that they appear larger in the phase image than in the amplitude image. The phase

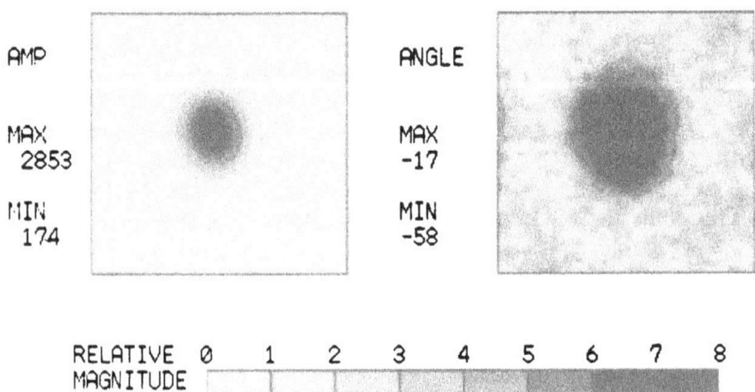

Fig. 1 Observation of an alumina particle by XY-scanning.

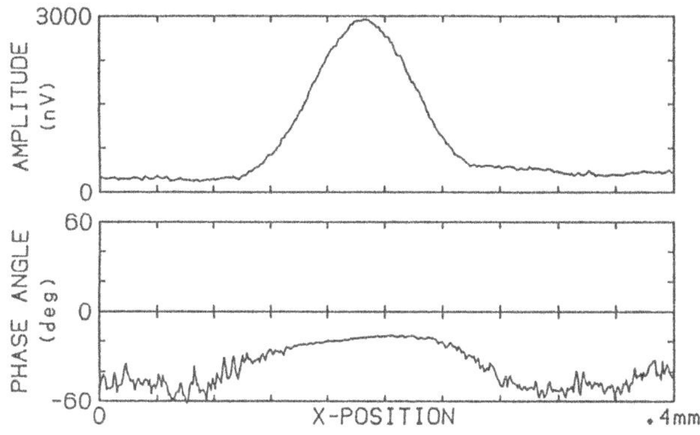

Fig. 2 X-scanning along the center of the particle of Fig. 1.

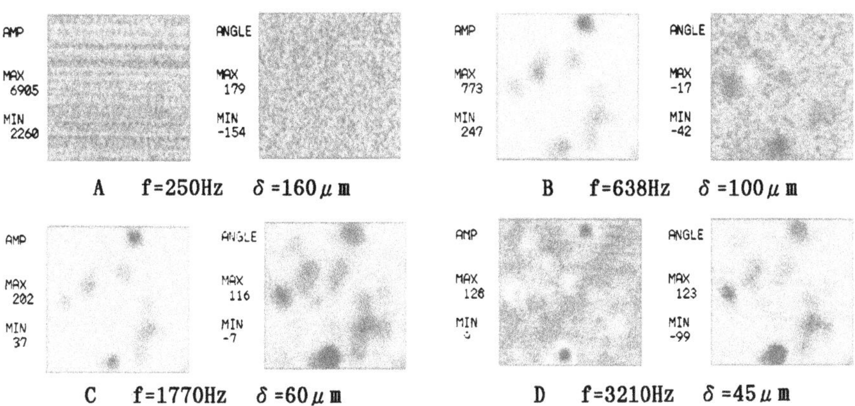

Fig. 3 Observation of alumina particles by XY-scanning at different modulation frequencies.

Table 1 Fundamental physical properties.

Properties	AlN	Alumina
Thermal conductivity (W/cmK)	0.62	0.17
Specific heat (J/gK)	0.81	0.99
Density (g/cm^3)	3.26	3.75
Thermal diffusivity (cm^2/s)	0.23	0.045
Young's modulus (GPa)	300	340

image becomes sharper and clearer at 3210Hz (Fig. 3D right), while the amplitude image of some particles begins to disappear (Fig. 3D left).

The behavior of both the amplitude and the phase images observed above can be explained by considering the dependence of δ on f (Eq. (1)) and the difference in δ between amplitude and phase angle. The distance from the top of the particle to the specimen surface is roughly equal to δ which corresponds to the frequency of image disappearance. For example, the particles which began to disappear in Fig. 3D are estimated to be located several tens of μm below the surface.

The ultrasonic methods have not revealed any image of the particles yet. This may be due to too small difference in elastic properties between the AlN matrix and the alumina particles as listed in Table 1. In contrast to this it is considered to be due to large difference in thermal conductivity (Table 1) that the particles can be detected clearly by PAS. If the images observed by PAS were pores, they could also be detected by ultrasonics. Then the images observed here by PAS are concluded to be the alumina particles dispersed in the AlN matrix.

As shown above, PAS can detect wider varieties of defects than other methods such as ultrasonics. This is due to the fact that PAS can reflect optical, thermal and elastic properties at the same time. With a little further improvements, especially in its measuring speed, PAS is expected to become a very promising tool for NDE.

References

[1] A. Rosencwaig and A. Gersho, J. Appl. Phys. **47**, 64(1976).

[2] A. Rosencwaig, Rev. Sci. Instrum. **48**, 1133(1977).

[3] K. Iwasaki, "Ceramics Today - Tomorrow's Ceramics", ed. by P. Vincenzini, Elsevier Sci. Pub. (1991) p. 1847.

Anomalous Thermal Waves in Powdered Samples

R. Osiander, J. Lobermeier, and P. Korpiun*

Physik-Department E13, Technische Universität München,
James-Franck-Strasse, W-8046 Garching, Fed. Rep. of Germany
*Present address: Applied Physics Laboratory, Johns Hopkins University,
Johns Hopkins Road, Laurel, MD 20723, USA

A.G. Bell (1881) mentioned that " the loudest signals are produced from samples in a loose, powdered, spongy condition" /1/. Two effects enhance the photoacoustic signal of powdered samples: the thermal expansion of the interstitial gas /2,3/ and the poor thermal conductivity of such samples. It has been shown earlier, that thermal diffusion in assemblies of spheres can be treated similar to diffusion in a percolation system /4,5/. In disordered systems the diffusion coefficient becomes time or frequency dependent /6/. We have deduced a heat conduction equation for disordered materials and we apply it to thermal wave measurements on powdered samples.

For a fractal sample, embedded in three-dimensional Euclidean space, the density ρ scales as $\rho \sim l^{(3-d_f)}$ with linear sample size l and fractal dimension d_f. Then the specific heat density $c\rho$ is a function of the temperature distribution $T(\vec{x})$ itself, and the heat contents Q can be written as

$$Q(\vec{x},t) = \int_{-\infty}^{\infty} c \cdot G(\vec{x}-\vec{x}')T(\vec{x}')d^3x' \quad (1)$$

where c is the specific heat and $G(\vec{x})$ is the density correlation function of the sample.

Temperature diffusion in a fractal sample can be treated analogous to diffusion in a fractal /4,5/. Corresponding to the diffusion flow, we get a temperature diffusion flow J_T,

$$J_T(\vec{x},t) = \int_0^t \tilde{\alpha}(t-\tau)\nabla T(\vec{x},\tau)d\tau \quad (2)$$

with the time dependent thermal diffusivity $\tilde{\alpha}$ /5/, which is the analog of the velocity autocorrelation function $<v(0)v(t)> \sim t^{(d_s/d_f-2)}$ in the diffusion case. Here d_s is the spectral dimension. To convert it into the heat flow J_Q, J_T has to be folded with $G(\vec{x})$ analogous to the heat content. If we use the equation of continuity,

$$\mathrm{div}\, J_Q + \dot{Q} = Q_S \quad (3)$$

together with (1) and (2) and a periodic heat source $Q_s = Q_0\delta(x)e^{i\omega t}$ with circular frequency ω at x=0, we get the heat conduction equation for the temperature T(x,t). For the one dimensional case its Fourier transform is /8/:

$$G(k)[\alpha(s)k^2 + s]T(k,s) = Q_0. \quad (4)$$

Here $\alpha(s)$ is the Laplace transform of the time dependent thermal diffusivity and G(k) is the Fourier transform of the density correlation function. The temperature distribution T(x,s) becomes

$$T(x,s) = \frac{e^{x\sqrt{s/\alpha(s)}}}{G(i\sqrt{s/\alpha(s)})\sqrt{s\alpha(s)}}. \quad (5)$$

Springer Series in Optical Sciences, Vol. 69 Photoacoustic and Photothermal Phenomena III
Editor: D. Bićanić

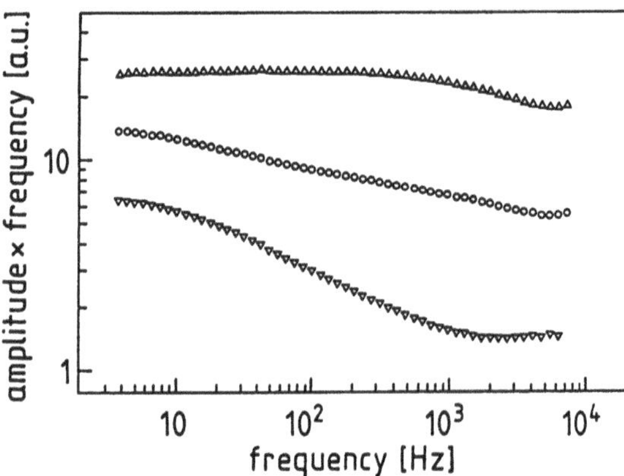

Figure 1: Amplitude of the photoacoustic signal for graphite powder with 2 µm (Δ), 30 µm (o) and 250 µm (∇) grain size.

This is a thermal wave with a wave number $k=\sqrt{s/\alpha(s)}$, which shows an anomalous frequency dependence. Having only replaced the material constants $c\rho$ and α by frequency dependent functions, the description to this point is not restricted to fractals. However, for the special case of a fractal, the frequency dependence of these quantities can be explicitly written in form of simple power laws /8/. Therefore, for $\alpha(s)\sim s^{(1-d_f/d_s)}$ and $G(k)\sim(ik)^{(3-d_f)}$ the temperature at x=0 is

$$T(0) \sim (i\omega)^{-[2+d_s/d_f)(2-d_f)]/2}. \tag{6}$$

For a percolation system well above the critical concentration, the dimensionalities are $d_f=3$ and $d_s=1.2$ (if the diffusion is considered in all clusters) /6/. Such a system is self similar for lengths greater than the diameter d of the particles and below the so called correlation length ξ /6/, which decreases with increasing particle or bond density. When the thermal diffusion length is in the selfsimilar regime we get $T(\omega)\sim(i\omega)^{0.8}$, whereas for an Euclidean system it is $T(\omega)\sim(i\omega)^{0.5}$.

To calculate the photoacoustic signal δP from the temperature distribution (6) in the sample, we use the models of Monchalin /2/ and McGovern et al. /3/, which take into account the thermal expansion of the interstitial gas. We get

$$\delta P \sim (i\omega)^{-[3+d_s/d_f)(2-d_f)]/2}\left[1+\frac{\varepsilon}{\gamma}\sqrt{\frac{\alpha_p(\omega)}{\alpha_g}}\right], \tag{7}$$

where α_p is the frequency dependent thermal diffusivity of the sample, α_g the thermal diffusivity of the gas, γ is the adiabatic exponent and ϵ is the porosity. For powders, $\epsilon=0.4$ /7/. Only if the thermal diffusivity itself is frequency dependent, the porosity would affect the frequency dependence of the signal.

We have performed photoacoustic measurements on metallic and graphite powders of various grain sizes in different gases. As a representative result, the frequency dependence of the amplitude of the photoacoustic signal of graphite powders of different grain sizes in air is shown in fig. 1.

For samples consisting of poorly conducting spheres /5/ or large grains the effect of the interstitial gas expansion is small. We find that for frequencies below a certain frequency f_0 the amplitude is proportional to $f^{-1.3}$, as calculated from equation (7) for a percolation system. For frequencies below $f_0=\alpha_s/\pi d^2$ the thermal diffusion length is larger as the grain size d and the thermal wave covers the contacts between the spheres. For higher frequencies it is proportional to f^{-1}.

For medium grain sizes, the frequency dependence of the amplitude is proportional to $f^{1.15...1.20}$ for different samples and gases over the entire measured frequency regime, as is seen for the 30 μm-graphite powder in figure 1.

For small grains we see a transition in the magnitude from f^{-1} at low frequencies to $f^{-1.14}$ for frequencies higher than a certain frequency f_ξ. This can be identified with the frequency for which the thermal diffusion length is equal to the correlation length ξ. For graphite powder it turns out that ξ is about 7 times the grain size.

From a percolation model for heat conduction in powdered samples we would expect only a shift of the transition frequencies for different grain sizes, but not a change in the power in the frequency dependence of the signal. Even if we take into account the interstitial gas expansion, the power in the frequency dependence should be about 1.2 to 1.3, larger than in the experimental results for small grains. It seems as if only the large grains behave as a percolating system.

The percolation system is a special case of a disordered sample with only a two status contact between the grains. A frequency dependent diffusion coefficient arises in all systems with a distribution of the transition coefficients over a few decades. Effective medium calculations /9/ show a power like behavior of the diffusion coefficient in the dispersive regime. Formally this power can be related to the spectral dimension d_s. It should increase if the density and therefore the width of the distribution of transition coefficients increases.

For our thermal system, the transition coefficients are the thermal contacts between the spheres. For large grains, there are only a few direct contacts, and the system appears to be percolative. For small grains, there are also contacts between the spheres across the pores, as now the thermal resistance of the small gas filled pores becomes smaller and the distribution of thermal resistances increases. As the spectral dimension increases, the power for the frequency dependence of the magnitude comes closer to -1, as observed for the powders with the small grains.

The frequency range of a photoacoustic experiment, less than three decades, is to small to claim fractal or percolative behavior. But the interpretation of the experimental results with our model can give some indication to the significance of those quantities for the heat transport, that characterize the state of disorder.

We grateful acknowledge the helpful discussions with M. Wagener and W. Schirmacher.

/1/ A.G. Bell, phil. Mag. 11, 510 (1881)
/2/ J. Monchalin, L. Bertrand, G. Rousset, and F. Lepoutre, J. Appl. Phys. 56, 190 (1984)
/3/ S. McGovern, B. Royce, and J. Benzinger, J. Appl. Phys.57, 1710 (1985)
/4/ A. Boccara and D. Fournier in: Photoacoustic and Photothermal Phenomena II, P. Hess and J. Pelzl eds., Springer Series in Optical Sciences, Springer 1988, P.309
/5/ R. Osiander, R. Haberkern, P. Korpiun and W. Schirmacher in: Photoacoustic and Photothermal Phenomena II, J.C. Murphy, J.W. Maclachan Spicer, L.C. Aamodt and B.S.H. Royce eds., Springer Series in Optical Sciences, Springer 1990, P.302
/6/ D. Stauffer, Introduction to Percolation,Theory, Taylor and Francis, London 1985
/7/ J. Dod and M. Leitzelement in: Physics in finely divided matter, A. Boccara and M. Daoud eds., Springer 1985, p.56
/8/ R. Osiander, PhD-Thesis, Technische Universität München 1991
/9/ W. Schirmacher, Ber. Bunsenges. Phys. Chem, 1991

Index of Contributors

Aamodt, L.C. 245, 516
Aartse, J.W. 586
Abdel-Mottaleb, M.S.A. 283
Ache, H.J. 41, 46
Adelhelm, K. 41
Agostini, A. 393, 411
Allen, J.E. 276
Almond, D.P. 235, 238, 241, 522
Amato, G. 387, 408
Ando, M. 135, 138
Angeli, Gy.Z. 20, 88, 593, 596
Ardeleanu, M. 81
Asselt, K., van 20
Azzopardi, B. 600

Bäckström, G. 664
Bain, C.D. 158
Bakker, R.A. 257
Balageas, D. 504, 537, 664
Balconi, L. 38
Baldwin, D.P. 3
Balk, L.J. 384
Baumann, J. 692
Bein, B.K. 497, 504
Benck, E.C. 482, 708
Benedetto, G. 387, 408
Berrehar, J. 202
Bertolotti, M. 217, 695
Bertrand, L. 81, 151, 578
Beyfuss, M. 692
Bićanić, D. 20, 85, 88, 103, 586, 596, 637, 644
Bijnen, F.G.C. 34
Billard, J. 628
Bindra, R.M.S. 95, 209
Birch, D.J.S. 95, 209, 213
Bleiß, R. 477
Blom, C.W.P.M. 56
Boarino, L. 387, 408
Boccara, A.C. 228, 396, 474, 607, 679, 717
Bohnacker, A. 248
Bohnert, B. 46
Bouchoule, S. 664

Bozóki, Z. 590, 593
Brand, B. 28
Braslavsky, S.E. 73, 77
Brugman, T. 34
Büchner, B. 378, 403, 406
Buijs, J. 168
Bulakh, G.I. 354, 360
Burbelo, R.M. 354, 360
Burg, W.J., van der 586
Burt, J.A. 276
Busse, G. 248, 669
Buurmeijer,, W.F. 65

Cahen, D. 403
Caldwell, J.J. 276
Candea, R.M. 637, 644
Carrer, I. 568
Cella, N. 403
Chambron, J. 141, 144
Chang, G.Y. 363
Chao, J.L. 161, 492
Charbonnier, F. 161
Charland, M. 69
Charlebois, D. 69
Chen, Z. 556
Cheng, J.C. 366, 423
Chirtoc, I. 85
Chirtoc, M. 20, 85, 103, 637, 644
Christofides, C. 6, 417
Coufal, H. 317, 331
Cournoyer, A. 578
Cowan, N.D. 209
Crowther, D.J. 544
Cussó, F. 631

Dadarlat, D. 85, 103, 637, 644
Davies, P.B. 158
Déom, A.A. 504
Déom, A. 537
Diaci, J. 547
Diebold, G.J. 263
Dienstbier, M. 350
Dittmar, R.M. 492
Domingue, A. 151, 578
Dóka, O. 168, 603

Douglas, P. 600
Dreyfus, G. 607
Dunina, T.A. 294

Eccleston, G.M. 95
Eckardt, P. 196
Egée, M. 205
Egée, P. 205
Egerev, S.V. 50, 294
Eickmeier, A. 131
Eilers, H. 320
Enguehard, F. 537
Enos, R.D. 600
Eremenko, V.G. 420
Eyletters, M. 62

Fabbri, L. 695
Faubel, W. 41
Faubel, W. 46
Favier, J.P. 168
Favro, L.D. 519, 544
Fernández, J. 199
Ferro-Luzzi, M. 28
Fiedler, M. 31, 270
Figari, A. 53, 224
Fiorina, L. 568
Forge, Ph. 541, 651
Forget, B.C. 375, 414
Fournier, D. 339, 375, 396, 414, 607, 679, 717
Frandas, A. 637, 644
Franko, M. 273
Friedrich, K. 248, 429, 669

Gärtner, W. 77, 73
Gáspár, I. 590, 593
Gang, J. 474
Gasparrini, F. 393, 411
Geiler, H.D. 477, 688
Geisau, O., von 221
Gerkema, E. 20
Giese, K. 92
Givens, R.B. 363
Glazov, A.L. 525, 528
Glorieux, C. 647, 685
Gnappi, G. 217
Göbel, C. 509

Gölz, Ch. 31
Göring, R. 429
Gonzalo, J.A. 631
Gordon, S.H. 572
Greene, R.V. 572
Grigoryev, V.V. 390
Grönbeck, H. 436
Grygier, R.K. 317, 331
Gu, J.H. 504
Gulyaev, A.L. 354
Gusev, V.E. 297, 306, 323

Hadj-Sahraoui, A. 628
Hakem, K. 205
Halaška, E. 350
Halthore, R.N. 276
Hammerich, M. 9
Harendt, N. 92
Harren, F.J.M. 12, 56
Harren, F.J.M. 34
Hartikainen, J. 512, 565
Haupt, K. 429
Helander, P. 20, 562
Henningsen, J. 9
Hess, P. 270, 300, 317, 331, 714
Heuret, M. 205
Hirakawa, F. 702
Hoffmann, M. 221
Hofman, R. 647
Holzwarth, A.R. 77
Hope, C.B. 600
Hoshimiya, T. 553
Hrovatin, R. 313
Hüttner, R. 504
Hunger, E. 436
Hyoguchi, T. 303

Iida, A. 135, 138
Imhof, R.E. 95, 209, 213
Ito, Y. 702
Ivanov, D. 228, 399
Iwasaki, K. 721
Izumi, T. 357

Jaarinen, J. 189, 193
Jalink, H. 20, 586
Jauregui, J. 436, 531, 534, 682
Jesus, M.E. Pita, de 213
Jin, H.J. 519
Jones, R.W. 3, 113, 125
Joswig, H. 711
Jung, H. 270

Kakoschke, P. 711
Kaplanová, M. 128, 180
Karabutov, A.A. 309, 334
Karpen, W. 248, 669
Katuščáková, G. 180
Kawai, J. 357
Kawashima, K. 303
Kerns, W.D. 516
Khan, M.I. 263
Kieseling, F. 320
Knerr, S. 607
Knoll, W. 583
Kojima, M. 553
Kölmel, K. 92
Kolomenskii, A.A. 326
Korpiun, P. 59, 378, 583, 619, 634, 651, 711, 725
Kowalski, P. 477
Kucherov, I.Ya. 354, 360
Kudinov, I.A. 334
Kukreja, L.M. 300
Kumar, A.V.R. 489
Kuo, P.K. 175, 189, 251, 519, 544

Lannoye, R. 62
Lapersonne-Meyer, C. 202
Lau, S.K. 522, 235, 238, 241
Leblanc, R.M. 69
Lehtiniemi, R. 512, 565
Leite, N.F. 183, 559
Lepoutre, F. 541, 651, 664, 717
Leung, W.P. 672
Leyk, A. 384
Li, G. 384
Li, H.S. 366, 369
Liakhou, G. 217
Liu, M. 485
Lobermeier, J. 725
Locke, J. 209
Lörincz, A. 257, 603, 610
Louis, G. 202, 628
Lu, Y.S. 366, 369
Lu, Y.S. 423
Lubbers, M. 20, 88
Luo, S. 113
Luukkala, M. 512, 565

Makarova, L.N. 306
Maksimović, D. 705
Malkin, S. 59
Man, N. 228
Mandelis, A. 6, 186, 372, 417, 550, 556, 657
Manning, C.J. 161
Mansanares, A.M. 183
Marchand, H. 578
Marinelli, M. 393, 411, 641
Martellucci, S. 393
Martinelli, M. 38
Martinu, L. 151
Masujima, T. 135, 138
Matrosov, M.P. 309
Matsumori, T. 357
Matthias, E. 436, 509, 531, 534, 682, 698
Mayer, A. 714
Maznev, A.A. 326
McClelland, J.F. 3, 113, 125
Mérienne, E. 205
Mekhtiev, M.M. 390, 396
Mens, E. 685
Mercuri, F. 393, 641
Meyer-Berg, G. 378, 711
Micheler, W. 583
Miklós, A. 20, 88, 168, 257, 280, 590, 593, 596
Miranda, L.C.M. 183, 559
Mityurich, G.S. 165
Montenero, A. 217
Morisset, R. 81
Možina, J. 313, 547, 705
Mukhin, D.O. 381, 396
Munidasa, M. 550
Muratikov, K.L. 525, 528
Murphy, J.C. 363, 516

Nakamura, O. 183
Nakamura, S. 702
Nampoori, V.P.N. 489
Narayanan, K. 147
Negm, S. 283
Netzelmann, U. 501
Neubrand, A. 317, 331, 714

Oehler, O. 290
Ohta, M. 553
Olafsson, A. 9
Ong, T.H. 158
Orth, Th. 221
Osiander, R. 378, 583, 619, 711, 725
Otowski, W. 647
Ouzafe, M. 144, 141

Padmaja, G. 489
Palmer, A. 492
Palmer, R.A. 161
Pareja, C. 572
Pashin, A.E. 50
Pastore, H.O. 183
Patel, P.M. 235, 238, 241, 522
Pelzl, J. 221, 396, 497, 504
Peralta, S.B. 186
Peretti, P. 202, 628
Petrovsky, A.N. 381, 390, 396
Petzoldt, S. 436
Pietsch, H. 436
Pizzoferrato, R. 393, 411

Pizzoferrato, R. 641
Plamann, K. 92
Platonenko, V.T. 334
Platt, U. 31
Podymova, N.B. 309
Polm, M.W. 65
Poulet, P. 141, 144
Puchenkov, O.V. 50

Quelin, X. 202
Quist, J. 20

Radhakrishnan, P. 489
Raillard, D. 290
Ramirez, R. 631
Rantala, J. 189, 193, 565
Raveh, A. 151
Reichl, H. 692
Reichling, M. 436, 509, 698
Reuss, J. 12, 34, 56
Richter, W. 196
Roger, J.P. 228, 474
Rohr, M. 73, 77
Rooth, R.A. 16

Sánchez-Lavega, A. 199
Sabbioneda, L. 53
Salazar, A. 199
Salnick, A.O. 381, 390, 396
Sanger, D.G. 209
Sauren, H. 20
Schäfer, D. 509, 682, 698
Schiller, D. 688
Schmidt, J. 583
Schock, H.W. 406
Schollmeyer, E. 131
Schoubs, E. 186
Schuessler, H.A. 482, 708
Schumacher, M. 290
Schweitzer, G. 77
Scudieri, F. 393, 641
Seaverson, L.M. 113
Seidel, U. 248, 429, 669
Sennhenn, B. 92
Seo, H.J. 231
Serdyukov, A.N. 165
Shen, Q. 366
Shiwaku, H. 135, 138
Sibilia, C. 217, 695
Sigon, F. 38
Sigrist, M.W. 28, 287
Silva, M.Z. 541, 651
Skagerberg, B. 193
Skaleris, A.M. 614
Skvortsov, A.T. 294
Sladky, P. 350
Snel, J.F.H. 65
Sogoyan, M.A. 334
Somekh, M.G. 485
Somlói, J. 257
Sontag, H. 672
Sólyom, A. 20, 88, 596
Spagnolo, R. 387, 408
Spicer, J.W.M. 363, 516
Stehlé, J.L. 474
Strauss, E. 320
Strivens, T.A. 213
Suddendorf,, M.B. 485
Sun, T. 263
Surnev, S. 399
Szakács, T. 610
Szurkowski, J. 155

Talaat, H. 283
Tam, A.C. 447, 672
Tang, J. 482
Telenkov, S.A. 297
Terzić, M. 705
Thakur, S.N. 147
Thier, R.H. 56
Thoen, J. 186, 647, 685
Thöny, A. 28
Thomas, R.L. 175, 251, 519, 544
Tocho, J.O. 631
Toyoda, T. 135, 138
Tran, C.D. 273, 463

Uchida, M. 357
Uehara, F. 357
Utrobin, Yu.B. 354

Vallabhan, C.P.G. 489
Vargas, H. 20, 183
Varis, J. 512, 565
Veeranjaneyulu, K. 69
Vegetti, G. ·38
Verhage, A.J.L. 16
Vichi, E.J.S. 183
Vinokurov, S.A. 614
Voesenek, L.A.C.J. 56
Volchanski, O.V. 360
Voti, R. Li 217, 695
Vredenberg, W.J. 65
Vries, H.S.M., de 12

Wagner, M. 477
Wagner, R.E. 372
Walther, H.G. 196, 248, 429, 497, 669
Wang, Y.X. 519
Wang, Z.Q. 423
Webb, J.F. 213
Wei, L. 175, 251
Weiser, J. 634
Weiss, P. 287
Welsch, E. 436, 509, 698
Wertheimer, M.R. 151
Wieland, J. 290
Willson, P.H. 209, 213
Winkler, A. 270
Winkler, N. 688
Woelker, M. 497
Wouters, L.W. 16
Wright, O.B. 303
Wu, Z.L. 436, 509, 531, 682, 698

Yang, K. 276
Yen, W.M. 320
Yoshinaga, H. 357
Yun, S.I. 231

Zammit, U. 393, 411, 641
Zanella, A. 53
Zanzottera, E. 568
Zelyony, V.P. 165
Zhang, S.Y. 366, 369, 423
Zhdanov, B.V. 297
Zozulya, O.M. 50
Zuev, V.V. 381, 390, 396
Zwol, R.A., van 586

Springer Series in Optical Sciences

1 **Solid-State Laser Engineering**
2nd Edition By W. Koechner

2 **Table of Laser Lines in Gases and Vapors**
3rd Edition
By R. Beck, W. Englisch, and K. Gurs

3 **Tunable Lasers and Applications**
Editors: A. Mooradian, T. Jaeger, and P. Stokseth

4 **Nonlinear Laser Spectroscopy** 2nd Edition
By V. S. Letokhov and V. P. Chebotayev

5 **Optics and Lasers** Including Fibers and Optical Waveguides 3rd Edition
By M. Young

6 **Photoelectron Statistics**
With Applications to Spectroscopy and Optical Communication By B. Saleh

7 **Laser Spectroscopy III**
Editors: J. L. Hall and J. L. Carlsten

8 **Frontiers in Visual Science**
Editors: S. J. Cool and E. J. Smith III

9 **High-Power Lasers and Applications**
2nd Printing
Editors: K.-L. Kompa and H. Walther

10 **Detection of Optical and Infrared Radiation**
2nd Printing By R. H. Kingston

11 **Matrix Theory of Photoelasticity**
By P. S. Theocaris and E. E. Gdoutos

12 **The Monte Carlo Method in Atmospheric Optics**
By G. I. Marchuk, G. A. Mikhailov, M. A. Nazaraliev, R. A. Darbinian, B. A. Kargin, and B. S. Elepov

13 **Physiological Optics**
By Y. Le Grand and S. G. El Hage

14 **Laser Crystals** Physics and Properties
By A. A. Kaminskii

15 **X-Ray Spectroscopy** By B. K. Agarwal

16 **Holographic Interferometry**
From the Scope of Deformation Analysis of Opaque Bodies
By W. Schumann and M. Dubas

17 **Nonlinear Optics of Free Atoms and Molecules**
By D. C. Hanna, M. A. Yuratich, and D. Cotter

18 **Holography in Medicine and Biology**
Editor: G. von Bally

19 **Color Theory and Its Application in Art and Design** 2nd Edition By G. A. Agoston

20 **Interferometry by Holography**
By Yu. I. Ostrovsky, M. M. Butusov, and G. V. Ostrovskaya

21 **Laser Spectroscopy IV**
Editors: H. Walther and K. W. Rothe

22 **Lasers in Photomedicine and Photobiology**
Editors: R. Pratesi and C. A. Sacchi

23 **Vertebrate Photoreceptor Optics**
Editors: J. M. Enoch and F. L. Tobey, Jr.

24 **Optical Fiber Systems and Their Components**
An Introduction By A. B. Sharma, S. J. Halme, and M. M. Butusov

25 **High Peak Power Nd: Glass Laser Systems**
By D. C. Brown

26 **Lasers and Applications**
Editors: W. O. N. Guimaraes, C. T. Lin, and A. Mooradian

27 **Color Measurement**
Theme and Variations
2nd Edition By D. L. MacAdam

28 **Modular Opticsl Design**
By O. N. Stavroudis

29 **Inverse Problems of Lidar Sensing of the Atmosphere** By V. E. Zuev and I. E. Naats

30 **Laser Spectroscopy V**
Editors: A. R. W. McKellar, T. Oka, and B. P. Stoicheff

31 **Optics in Biomedical Sciences**
Editors: G. von Bally and P. Greguss

32 **Fiber-Optic Rotation Sensors**
and Related Technologies
Editors: S. Ezekiel and H. J. Arditty

33 **Integrated Optics: Theory and Technology**
2nd Edition By R. G. Hunsperger
2nd Printing

34 **The High-Power Iodine Laser**
By G. Brederlow, E. Fill, and K. J. Witte

35 **Engineering Optics**
2nd Edition By K. Iizuka

36 **Transmission Electron Microscopy**
Physics of Image Formation and Microanalysis
2nd Edition By L. Reimer

37 **Opto-Acoustic Molecular Spectroscopy**
By V. S. Letokhov and V. P. Zharov

38 **Photon Correlation Techniques**
Editor: E. O. Schulz-DuBois

39 **Optical and Laser Remote Sensing**
Editors: D. K. Killinger and A. Mooradian

40 **Laser Spectroscopy VI**
Editors: H. P. Weber and W. Lüthy

41 **Advances in Diagnostic Visual Optics**
Editors: G. M. Breinin and I. M. Siegel

www.ingramcontent.com/pod-product-compliance
Ingram Content Group UK Ltd.
Pitfield, Milton Keynes, MK11 3LW, UK
UKHW021332070726
13610UKWH00011B/46

* 9 7 8 3 6 6 2 1 3 8 7 5 5 *